# Conference Proceedings of the Society for Experimental Mechanics Series

*Series Editor*
Tom Proulx
Society for Experimental Mechanics, Inc.,
Bethel, CT, USA

For further volumes:
http://www.springer.com/series/8922

Vijay Chalivendra • Bo Song • Daniel Casem
Editors

# Dynamic Behavior of Materials, Volume 1

## Proceedings of the 2012 Annual Conference on Experimental and Applied Mechanics

Springer

*Editors*
Vijay Chalivendra
University of Massachusetts
Dartmouth, North Dartmouth MA, USA

Bo Song
Sandia National Laboratories
Department of Mechanics of Materials
Livermore, CA, USA

Daniel Casem
U.S. Army Research Laboratory
Aberdeen Proving Ground, MD, USA

ISSN 2191-5644 ISSN 2191-5652 (electronic)
ISBN 978-1-4614-4237-0 ISBN 978-1-4614-4238-7 (eBook)
DOI 10.1007/978-1-4614-4238-7
Springer New York Heidelberg Dordrecht London

Library of Congress Control Number: 2012948881



Printed on acid-free paper

Springer is part of Springer Science+Business Media (www.springer.com)

# Preface

*Dynamic Behavior of Materials, Volume 1: Proceedings of the 2012 Annual Conference on Experimental and Applied Mechanics* represents one of seven volumes of technical papers presented at the Society for Experimental Mechanics SEM 12th International Congress & Exposition on Experimental and Applied Mechanics, held at Costa Mesa, California, June 11–14, 2012. The full set of proceedings also includes volumes on Challenges in Mechanics of Time -Dependent Materials and Processes in Conventional and Multifunctional Materials, Imaging Methods for Novel Materials and Challenging Applications, Experimental and Applied Mechanics, Mechanics of Biological Systems and Materials MEMS and Nanotechnology, and Composite Materials and Joining Technologies for Composites.

Each collection presents early findings from experimental and computational investigations on an important area within Experimental Mechanics. The Dynamic Behavior of Materials conference track was organized by Vijay Chalivendra, University of Massachusetts Dartmouth; Bo Song, Sandia National Laboratories; Daniel Casem, US Army Research Laboratory; and sponsored by the SEM Dynamic Behavior of Materials Technical Division.

This volume represents an ever growing area of broad interest to the SEM community, as evidenced by the increased number of papers and attendance in recent years. This track was initiated in 2005 and reflects our efforts to bring together researchers interested in the dynamic behavior of materials and structures, and provide a forum to facilitate technical interaction and exchange. The papers within this volume are organized to cover the wide range of experimental research being conducted in this area by scientists around the world. This year we have also included two sessions discussing multiscale modeling. The following general technical research areas are included:

Composites
Dynamic Fracture and Failure
Dynamic Materials Response
Novel Testing Techniques
Low Impedance Materials
Dynamic Behavior of Geo-materials
Dynamic Behavior of Brittle Materials
Shock and Blast Loading of Materials
Multiscale Modeling of Dynamic Behavior of Materials
Optical Methods for High Strain Rate Testing
Materials Response
Novel Testing Techniques

The contributed papers span numerous technical divisions within SEM. It is our hope that these topics will be of interest to the dynamic behavior of materials community as well as the traditional mechanics of materials community.

The track organizers thank the authors, presenters, organizers, and session chairs for their participation and contribution to this track. We are grateful to the SEM Technical Division chairs who cosponsored and organized sessions in this track (e.g., Composite, Hybrid and Multifunctional Materials, and the Optical Methods Technical Divisions). The SEM support staff is also acknowledged for their devoted efforts in accommodating the large number of submissions this year.

The opinions expressed herein are those of the individual authors and not necessarily of theSociety for Experimental Mechanics, Inc.

North Dartmouth, MA, USA
Livermore, CA, USA
Aberdeen Proving Ground, MD, USA

Vijay Chalivendra
Bo Song
Daniel Casem

# Contents

# Chapter 1
# Rate Dependent Response of Crosslinked Epoxy Networks

**Daniel Knorr, Andrew Schoch, Joseph L. Lenhart, and Daniel Fischer**

**Abstract** Epoxy resins are utilized extensively in coatings, paints, adhesives, electronics, and composites due to a combination of processability, stability, and property tunability. Over the past several decades, substantial research has focused on manipulating the mechanical properties of epoxy resins via control of the polymer structure and chemistry and the incorporation of various additives. While the quasi-static behavior of these systems has been thoroughly investigated, minimal research has investigated high rate mechanical response. This presentation will highlight the rate dependent behavior of model epoxy resins as a function of the monomer chemistry, chain structure, and resin morphology. The mechanisms for improving high rate energy dissipation in bulk epoxy resins are being translated to surface modification approaches that can be exploited to tune the mechanical behavior of epoxy-substrate interfacial regions. The interfacial chemistry and structure is being characterized utilizing the combination of synchrotron based X-ray Photoelectron Spectroscopy and Near Edge X-ray Absorption Fine Structure, and correlated with characterization of rate dependent adhesive performance.

**Keywords** Epoxy • Interface • Sizing • NEXAFS

## 1.1 Introduction

Epoxy resins are utilized extensively in coatings, paints, adhesives, electronics, and composites due to a combination of processability, stability, and property tunability. Over the past several decades, substantial research has focused on manipulating the mechanical properties of epoxy resins via control of the polymer structure and chemistry and the incorporation of various additives. While the quasi-static behavior of these systems has been thoroughly investigated, minimal research has investigated high rate mechanical response. This presentation will highlight the rate dependent behavior of model epoxy resins as a function of the monomer chemistry, chain structure, and resin morphology. The mechanisms for improving high rate energy dissipation in bulk epoxy resins are being translated to surface modification approaches that can be exploited to tune the mechanical behavior of epoxy-substrate interfacial regions. The interfacial chemistry and structure is being characterized utilizing the combination of synchrotron based X-ray Photoelectron Spectroscopy and Near Edge X-ray Absorption Fine Structure, and correlated with characterization of rate dependent adhesive performance.

Near edge X-ray absorption fine structure (NEXAFS) is a synchrotron based interfacial analysis technique, useful for characterizing the presence and bonding states of organic molecules at surfaces. The particular beamline exploited for this work has an energy range of 200–1,000 eV and is sensitive to the *K*-edges from carbon, nitrogen, oxygen and fluorine. NEXAFS is particularly sensitive to the $\pi^*$ transitions from the carbon-carbon double bond, which are present in most epoxy formulations.

D. Knorr • A. Schoch • J.L. Lenhart (✉)
US Army Research Laboratory, Aberdeen Proving Ground, Gaithersburg, MD 21005, USA
e-mail: joseph.l.lenhart.civ@mail.com

D. Fischer
National Institute of Standards and Technology, Gaithersburg, MD 20899, USA

V. Chalivendra et al. (eds.), *Dynamic Behavior of Materials, Volume 1: Proceedings of the 2012 Annual Conference on Experimental and Applied Mechanics*, Conference Proceedings of the Society for Experimental Mechanics Series, DOI 10.1007/978-1-4614-4238-7_1, 

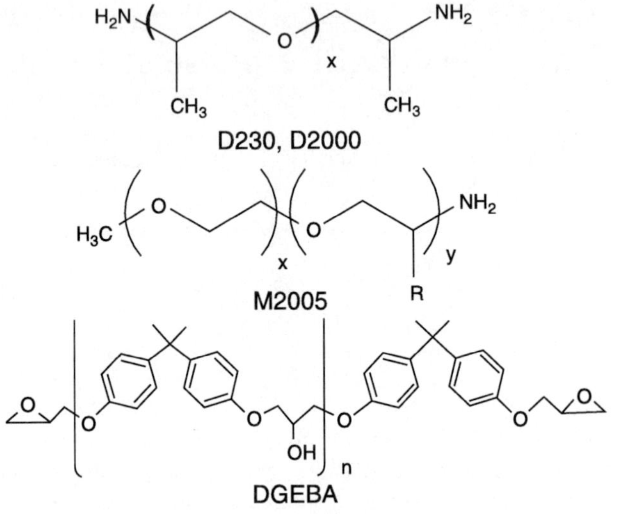


**Fig. 1.1** Molecular structures for the polyetheramines and DGEBA used in this study

The approach taken in this study is to carefully control the interface by varying the amount of amine available at the surface to react with the epoxide groups in poly(bisphenol A-co-epichlorohydrin) (DGEBA). DGEBA is well suited for these studies not only because it is a common component of epoxy resin formulations, but also because it contains ample amounts of carbon-carbon double bonds and therefore is easily detected with NEXAFS.

Controlling the density, flexibility, and organization of the epoxy-substrate interfacial bonding, provides a mechanism to tune the interfacial properties. Our method of controlling this involves using mixed layers of short and long polyetheramines which can react with the (3-glycidoxypropyl) trimethoxysilane (GPS) surface. Specific polyetheramines can be chosen to create reactive or passivated surfaces, as well as to control the flexibility and effective crosslink density in the interfacial region.

## 1.2 Experimental

In preparation of the layered samples used in this study, silicon wafers (<100> orientation, EPI, SSP, Rogue Valley Microdevices, Medford, OR) after being rinsed with ultrapure water (Barnstead NANOpure Infinity) were sonicated in ethanol and toluene using a 170W, NEY 300 Ultrasonik bath sonicator and subsiquently exposed to UV-ozone in a UVO-cleaner (Jelight model 42) for 30 min. The activated silicon wafers were then immersed in 1 wt.% GPS (Gelest, Morrisville, PA) solution in toluene for 24 h and cleaned thoroughly thereafter with toluene and ethanol in the bath sonicator and dried with nitrogen. The GPS/Si surfaces were then reacted at 60°C for 24 h with mixtures of three amine molecules, poly (propylene glycol) bis(2-aminopropyl ether) in two different molecular weights 230 g/mol (D230) and 2,000 g/mol (D2000) and a mono-aminated polyetheramine which is a copolymer of propylene glycol and polyethylene glycol in a 29/6 ratio which has a molecular weight of 2,000 (M2005). Huntsman, The Woodlands, TX (Fig. 1.1).

A schematic of the layered structure with mixtures of D230 and D2000 is shown in Fig. 1.2. The pure D230 (A) and D2000 (C) samples are depicted as well as a blend of the two (B) where they have been covalently bound to the underlying GPS layer. The Amine/GPS/Si surfaces were then reacted at 60°C for 24 h with DGEBA (glycidyl end-capped, Mn ~348, Sigma-Aldrich, Bellefonte, PA). Following adsorption of either the polyetheramine or DGEBA, physisorbed material was removed by sonication and repeated washings in both ethanol and toluene.

NEXAFS characterization of these samples was conducted at the U7A NIST beamline of the National Synchrotron Light Source at Brookhaven National Laboratory because of its high intensity soft X-rays. Partial electron yield (PEY) spectra were collected at room temperature and $10^{-7}$–$10^{-9}$ Torr from 200–620 eV with a grid bias of 150 eV. The spectra were normalized by the incident X-ray intensity ($I_0$) as measured by the total electron yield from a gold coated grid. The calibration was conducted at 285.1 eV using the carbon *K*-edge from graphite. Each measurement was collected at the magic angle to eliminate polarization dependence. The pre carbon *K*-edge baselines were subtracted and the values between 270 and 280 were averaged and subtracted from the spectra to normalize the pre-edge region to zero. Post edge jump normalization was conducted by averaging the points between 312 and 317 eV, and dividing the entire spectrum by this average [1].

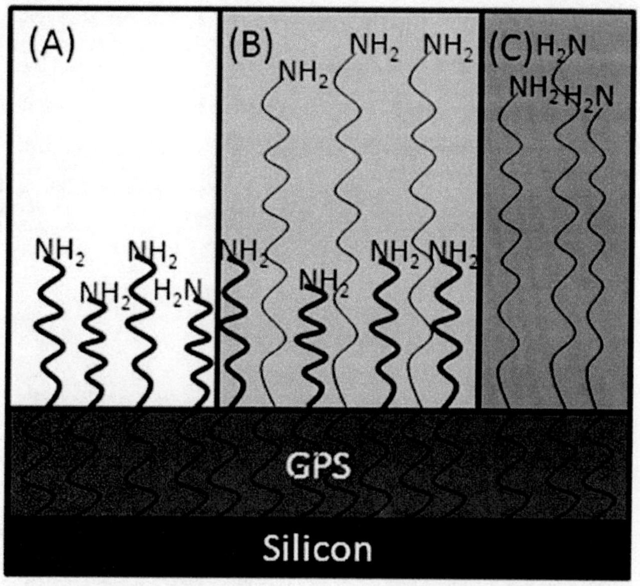


**Fig. 1.2** A schematic depiction of the various layered structures produced and examined in this study including (**a**) short diamines, (**b**) mixed short and long diamines, and (**c**) long diamines on GPS coated silicon wafers

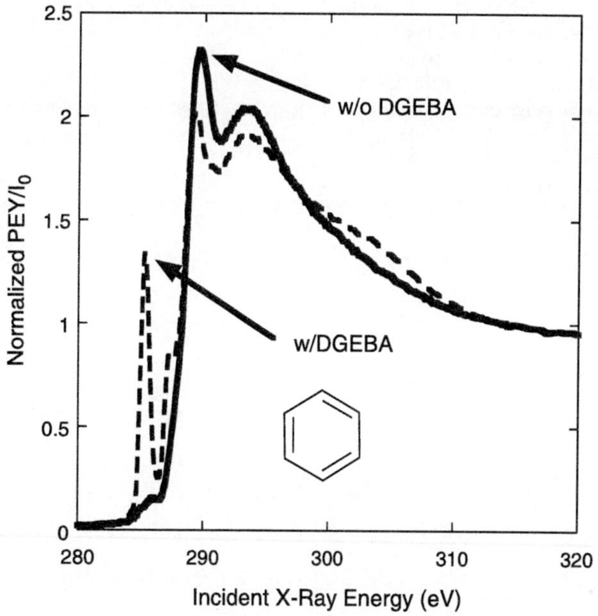


**Fig. 1.3** NEXAFS spectra for the carbon *K*-edge of the layered structure D230/GPS/Si which has been reacted with and without DGEBA

## 1.3 Results

NEXAFS sensitivity to the $\pi^*$ transition in the carbon-carbon double bond is evident from the carbon *K*-edge spectra in Fig. 1.3 where samples of D230/GPS/Si w/and w/o DGEBA are shown. A striking difference between the two curves comes at the peak near 285.3 eV, which is the C1s $\rightarrow \pi^*(C{=}C)$ peak from the phenyl rings on the adsorbed DGEBA. Neither GPS or D230 have carbon-carbon double bonds resulting in a negligible peak in this region, before the DGEBA is adsorbed. Therefore, the C1s $\rightarrow \pi^*(C{=}C)$ peak area is related to the amount of DGEBA adsorbed to the surface, and can be exploited in future experiments to quantify the effective reactivity of an engineered interfacial region.

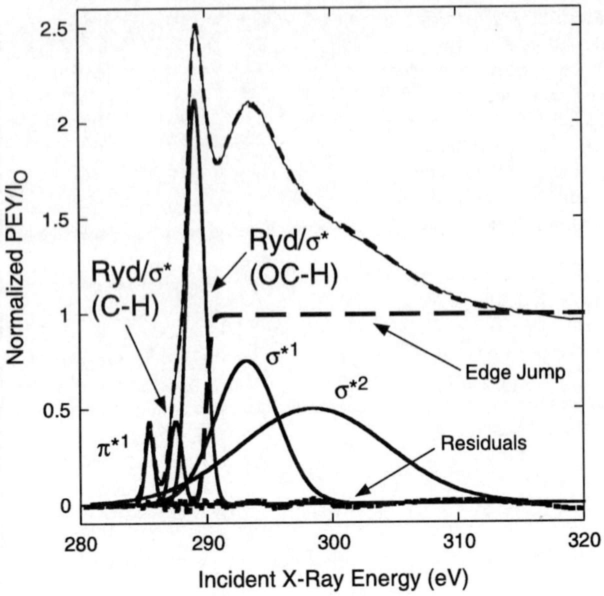


**Fig. 1.4** A fitted NEXAFS spectra for the carbon *K*-edge of the layered structure D2000/GPS/Si which has also been reacted with DGEBA. The peak assignments are based on work from a number of papers [1, 3–5]

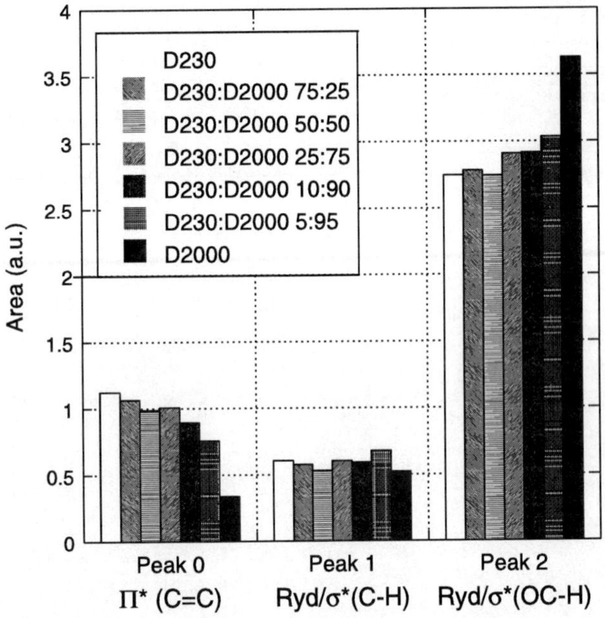


**Fig. 1.5** NEXAFS peak areas near the carbon *K*-edge derived from fits with IGOR for a range of concentration ratios from pure D230 to pure D2000 which have been reacted with DGEBA

After pre- and post-edge jump normalization, the NEXAFS spectrum is fit with a series of Gaussian peaks. An example of this analysis is shown in Fig. 1.4, which was taken from a sample of D2000/GPS/Si with adsorbed DGEBA on the surface. The individual peaks are shown as solid lines. The transitions at 285.4, 287.6, 289.2, 293.2, and 298.5 eV represent the C1s → π*(C=C), a combination of the Rydberg/C1s → σ*(C-H), Ryd/C1s → σ*(OC-H), C1s → σ*(C=C), and C1s → σ*(C-C) respectively. These peaks will be abbreviated as $\pi^{*1}$, Ryd/σ*(C-H), Ryd/σ*(OC-H), σ* [1], and σ* [2]. The carbon *K*-edge jump is also depicted in Fig. 1.4 as a long dashed curve. The sum of these five peaks and the edge jump is plotted as medium dashes over the actual intensity which is a hairline. The sum fits the data well with the residuals plotted in a small dashed curve near zero. Similar quality fits were obtained for all the data.

Areas from the first three transitions ($\pi^{*1}$(C=C), Ryd/σ*(C-H), Ryd/σ*(OC-H)) are plotted in Fig. 1.5 for mixtures of D230 and D2000. A clear trend can be seen in both the $\pi^{*1}$(C=C) and Ryd/σ*(OC-H) peak areas. The $\pi^{*1}$(C=C) transition shows a decrease in area with increasing concentration of D2000, illustrating a decrease in DGEBA adsorption with increasing D2000 content. The opposite trend is seen for the Ryd/σ*(OC-H) transition where an increase in peak area is seen with increasing D2000 content in the mixed amine addition. This is explained by the additional contribution from the C-H attached to the electron withdrawing oxygen atom with longer poly(propylene glycol) spacers in the D2000 molecules. We were able to observe similar trends when M2005 was used in place of D2000. Both of these molecules have the ability to change the concentration of amine at the interface and may then impact the interfacial mechanical properties.

## 1.4 Conclusions

The results of this research illustrate that attaching a mixture of long and short polyetheramines to a substrate can be utilized to control the adsorption of an epoxide functionalized monomer, DGEBA, to that interface. This technique can potentially be exploited to tune the mechanical, thermal, physical, and energy dissipative properties of the interfacial region in epoxy based composites. Future research will focus on linking the interfacial chemistry and structure to composite failure properties as a function of strain rate.

NEXAFS proved useful for analyzing the surface adsorption due to the sensitivity of the technique to π* (C=C) transitions typically present in epoxy formulations. Through careful preparation subtle variations in the concentration of certain di-functional polyetheramines could be detected. The technique is expected to be useful for analyzing adhesive fracture surfaces and providing insight into the location and mechanism of bonding failure.

**Acknowledgements** This work was performed at the US Army Research Labs. A.B. Schoch and D. Knorr were funded by ARL through a contract with the Oak Ridge Institute for Science and Education (ORISE). Certain commercial equipment and materials are identified in this paper in order to specify adequately the experimental procedure. In no case does such identification imply recommendations by the Army Research Laboratory or National Institute of Standards and Technology nor does it imply that the material or equipment identified is necessarily the best available for this purpose.

The authors would also like to thank Cherno Jaye for his assistance with the NEXAFS instrumentation on the NSLS at Brookhaven National Laboratory. Uncertainty in the NEXAFS spectra for these samples is estimated to be ±5% relative to the peak intensity or area value. Multiple measurements on the same sample exhibited similar uncertainty.

## References

1. Klein RJ, Fischer DA, Lenhart JL (2008) Systematic oxidation of polystyrene with ultra-violet ozone: Characterization by Near Edge X-ray Absorption Fine Structure and Contact Angle. Langmuir 24:8187–8197
2. Heise A et al (1997) Grafting of polypeptides on solid substrates by initiation of N–carboxyanhydride polymerization by amino terminated self assembled monolayers. Langmuir 13:723–728
3. Ade H, Urquhart SG (2000) Chemical applications of synchrotron radiation. World Scientific Publishing Co. Ltd, Singapore
4. Dhez O, Ade H, Urquhart SG (2003) Structural and compositional mapping of a phase-separated Langmuir–Blodgett monolayer by X-ray photoelectron emission microscopy. J Electron Spectrosc Relat Phenomena 128:85–96
5. Coffey T et al (2002) Characterization of the effects of soft X-ray irradiation on polymers. J Electron Spectrosc Relat Phenomena 122:65–78

# Chapter 2
# Fracture Response of Cross-Linked Epoxy Resins as a Function of Loading Rate

**Stephen Whittie, Paul Moy, C. Allan Gunnarsson, Daniel Knorr, Tusit Weerasooriya, and Joseph Lenhart**

**Abstract** The failure behavior of cross-linked polymer epoxies were investigated under Mode I fracture at low and high loading rates. By varying the monomer choices, the properties of the epoxies can be tailored to achieve a greater resistance to cracking and high impact toughness. For these experiments, a unique four-point bending specimen was developed. High rate experiments were conducted on a modified Split Hopkinson Pressure Bar (SHPB) with pulse-shaping. High speed digital imaging was used to determine the failure initiation, and allow for use of digital image correlation (DIC) to optically measure the crack opening displacement (COD) and crack propagation velocity. The experimental results are used to obtain the energy required to initiate fracture at various loading rates. The critical energy required to initiate fracture for this epoxy shows limited rate dependence from quasi-static to dynamic loading rates, but becomes rate sensitive when loaded at dynamic rates. The experimental procedures and results are discussed in this paper.

**Keywords** Epoxy • Fracture • Loading rate

## 2.1 Introduction

Polymer epoxies are frequently used in many applications to structurally bond components of similar or dissimilar materials. Some of the many uses can be found in the form of fiber reinforced polymer (FRP) for the aerospace and automotive industries, sporting goods, and marine crafts. When exposed to sudden impacts, these structural components and bonds are subjected to high loading rates. Typically, the failure of the resin matrix in FRPs is the dominating factor leading to interply delamination. Therefore, it is important to understand how these epoxies behave and fail at all loading rates, including high rates, especially as these high rates of loading are experienced during blast, ballistic impact, bird impacts, etc.

The fracture response of polymer epoxies has been investigated by many researchers. At quasi-static loading rates, studies have been performed to understand crack initiation, the effects of molecular weight between cross-links, and the fracture behavior of epoxy systems using various tougheners while using three or four point bend specimen geometries based on various ASTM standards [1–6].

Experimental techniques for dynamic fracture behavior have been developed and can be summarized into three categories: high rate bending, high rate tension, and dynamic wedging. Many different techniques exist; some of these have been summarized previously by Jiang et al. [7]. Previous studies have shown that there are two important criteria for valid and useful high rate fracture experiments: a state of dynamic equilibrium of the specimen, and a constant loading rate of the specimen during the experiment. Both of these criteria can be obtained by shaping the input loading pulse [8–14]. Additionally, low impedance materials, such as pre-cracked beams and polymer materials, frequently cause traditional strain gauge measurement techniques to be unusable. Other researchers, such as Chen et al. [9] have solved this problem by embedding quartz based loading devices in the loading train, which allows for direct measurement of dynamic equilibrium and loading rate. Weerasooriya et al. developed a high loading rate four-point bend fracture toughness experimental method, and used it to measure the dynamic fracture toughness of SiCN ceramic [10] and PMMA [11]. Gunnarsson et al. [12] further

S. Whittie (✉) • P. Moy • C.A. Gunnarsson • D. Knorr • T. Weerasooriya • J. Lenhart
Army Research Laboratory, 21005 Aberdeen Proving Ground, MD, USA
e-mail: stephen.d.whittie.ctr@mail.mil

V. Chalivendra et al. (eds.), *Dynamic Behavior of Materials, Volume 1: Proceedings of the 2012 Annual Conference on Experimental and Applied Mechanics*, Conference Proceedings of the Society for Experimental Mechanics Series, DOI 10.1007/978-1-4614-4238-7_2, 

developed this technique to measure the fracture toughness of human cortical bone, integrating ultra-high speed imaging and crack tip strain analysis.

Syn et al. [13] studied the surface morphology effect on the fracture behavior along dissimilar material interfaces (in this case, aluminum and epoxy) using a wing-like specimen. Weerasooriya et al. [14] used a similar method to study the dynamic failure behavior of adhesive bond interfaces. Both Weerasooriya et al. and Syn et al. showed an increase in failure load and energy rate for fracture initiation as loading rate increased from low to high rates.

In the present work, a unique fracture specimen was developed similar to the specimens used in [13, 14]. The specimen loading surfaces were designed to be curved instead of flat. This stiffened the loading noses, reducing the potential of specimen bending at loading arms during the experiments when compared to the original specimen design. This was necessary as the material studied here is much less stiff than the previous studies; it is important that the specimens do not bend and all energy absorbed is used to initiate failure and crack propagation. An additional benefit is the ability to use flat-faced loading fixtures instead of point loading fixtures. This allows for use of the bar ends to load the specimen, instead of special fixtures. This reduces experimental complexity, and improves wave transmission by removing changes in impedance in the loading fixture.

Using this modified geometry, in this study, Mode I fracture experiments were conducted on pre-cracked epoxy (DGEBA-PACM) specimens. Dynamic experiments were performed using an aluminum SHPB with high sensitivity semi-conductor strain gages and embedded quartz transducers, used to accurately measure the bar strain histories and stresses at the loading interfaces to verify dynamic equilibrium. These experiments were conducted to determine the loading rate dependence of fracture properties including critical failure load, failure energy and energy rate, and crack opening displacement (COD). Understanding the fracture behavior of epoxy at low to high loading rates will provide a quantitative fracture-resistance method for evaluation of epoxy systems as a function of loading rate.

## 2.2 Material

The epoxy used in this study was Di-Glycidyl of Bisphenol A cross-linked with Bis-(p-aminocyclohexyl)- Methane (DGEBA-PACM). The polymers DGEBA (resin) and PACM (curing agent) were acquired from Momentive and Air Products, respectively. PACM is an unmodified cycloaliphatic amine commonly used in filament winding and wet layup laminating for fabrication of fiber-reinforced polymers where high fracture toughness and better corrosive properties are necessary. The geometry of the unique fracture specimen is shown in Fig. 2.1 with dimensions in millimeters.

Because of the complexity of the fracture geometry, an aluminum mold to direct cast the epoxy into the fracture geometry were not feasible. This is due to the difficultly in extracting the epoxy from an aluminum mold without damaging the sample even if mold release is applied. Instead, a flexible silicon mold was created. To fabricate the mold, fracture specimens with the geometry as shown in Fig. 2.1 were machined from 25.4 mm thick aluminum and cast in silicone. Once the silicon mold was cured and since the silicone is flexible enough, the replicate aluminum specimens were removed from the mold. An example of the silicone mold is shown in Fig. 2.2. The epoxy mixture could then be poured into the silicon mold and cured.

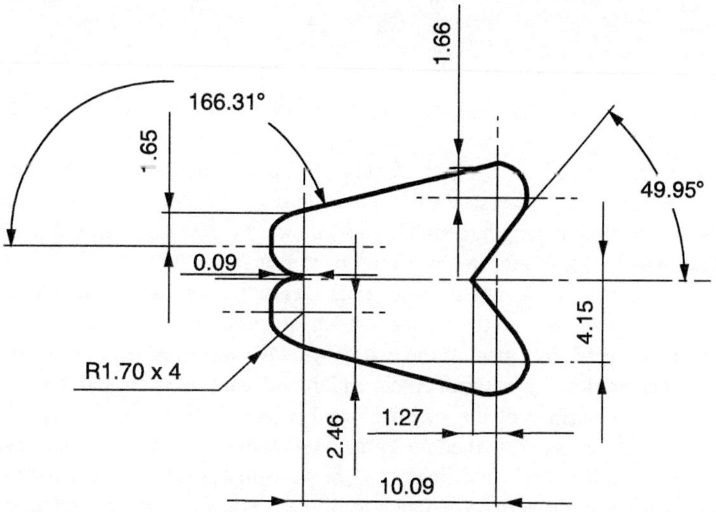


**Fig. 2.1** Fracture specimen geometry

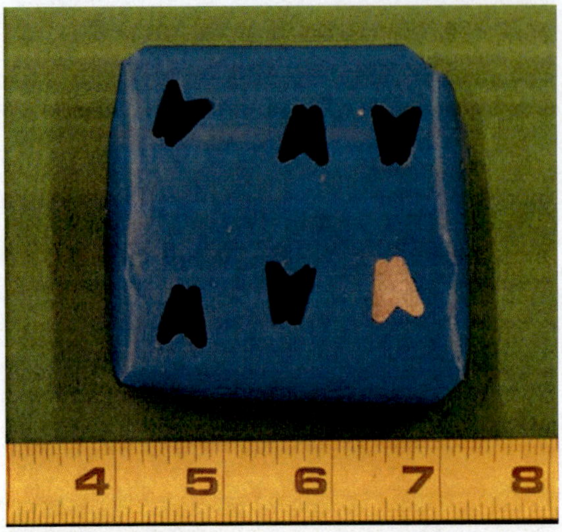


**Fig. 2.2** Silicon mold used to mold the epoxy

**Fig. 2.3** Custom notching fixture

Once the epoxy specimens were finished curing, they were pressed out of the mold and cut to a thickness of 20 mm using a precision cut-off saw. This process was necessary to remove the uneven material present near the mold surfaces after curing. With the specimens at the proper length, a notch was cut with a 300 μm width diamond saw blade. The study presented in this paper did not investigate other pre-crack profile other than the notch created from the saw blade. A custom fixture (Fig. 2.3) for specimen notching was designed to move parallel to the saw blade and hold each specimen at a constant height so a constant depth notch can be cut into the center of each specimen with a depth of 1.1 mm as shown in Fig. 2.4.

To accurately measure the relative displacement of the two wings, DIC was used. DIC is an optical method that is used to measure full-field displacement and deformation on an object under load using sequential images. To summarize the process, a speckled pattern is applied on the surface of the specimen and sequential images are captured using a digital camera throughout an experiment. The digital images are then post processed to compute the deformation and displacement field by tracking the movement of the speckles. Once the images were post-processed, by extracting the relative displacement between two selected points perpendicular across the crack (a digital extensometer), the COD was obtained as a function of time during the experiment.

Fig. 2.4 View of the notch and speckle pattern

Fig. 2.5 Experimental setup of mode I epoxy fracture specimen on an Instron

## 2.3 Low Rate Experiments

Quasi-static and intermediate loading rate experiments were conducted using an Instron 1331 servo-hydraulic test frame. The Instron test frame was controlled in displacement mode at $0.05*10^{-6}$ m/s (quasi-static) and 640*10–6 m/s (intermediate) rates, which corresponded to loading rates of 0.01kN/s and 7.07kN/s respectively. Silicon oil was used as lubrication on the four loading noses of the specimen to minimize friction during testing. Hardened steel compression platens with a swivel were used to compensate for minor misalignment in the loading train. The low rate test setup is shown in Fig. 2.5 below.

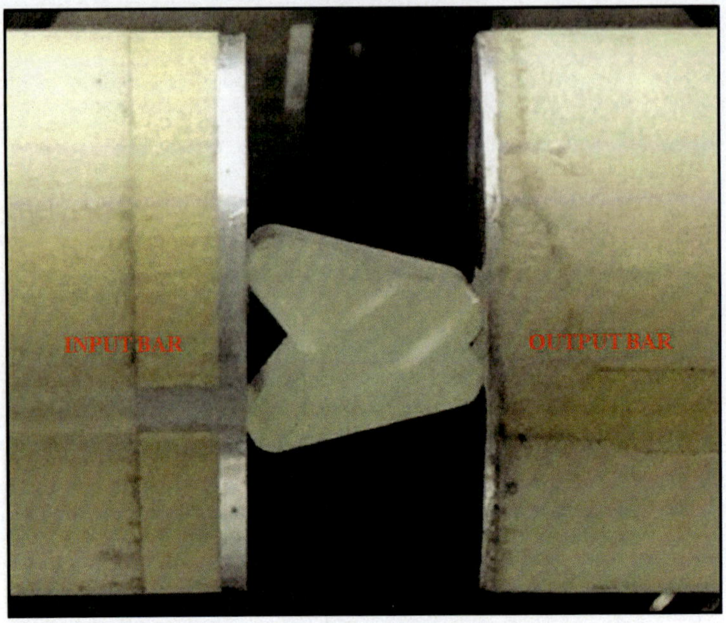


**Fig. 2.6** Mode I epoxy fracture specimen between the input and output bars in SHPB

A high-speed digital camera was used to obtain images during the experiment. The camera frame rate was at 60 k fps with a resolution of 128 by 128 pixels. The camera was triggered by a TTL pulse from the break detector system feature of the Instron test frame. This ensures that high speed digital images can be obtained to observe the crack initiation and propagation. The digital images were then post processed to determine the COD.

## 2.4 High Rate Fracture Experiments

Dynamic epoxy fracture experiments were conducted on a modified SHPB. The working principle of this experimental setup is well documented [4–6, 11]. The bars used for the incident and transmission bar in the test setup were made from Al 7075 with a diameter of 31.8 mm with a length of 3.66 m. A gas gun propelled Al 7075 striker with a length of 0.61 m was used to generate a stress pulse in the incident bar. Pulse shaping was used to shape the input pulse to achieve the necessary dynamic equilibrium, and a near constant loading rate. Semi-conductor strain gages were mounted at the midpoints of the input and output bars to record the bar stress histories. Semi-conductor gages were used for their high sensitivity, allowing for measurement of the low amplitude pulse in the transmission bar and the slight difference between incident and reflected pulse in the incident bar. Semi-conductor strain gages are much more sensitive to ambient temperature and humidity than traditional strain gages; therefore a calibration for the semi-conductor strain gages was conducted before each test series, using a known compressive strain pulse measured using standard resistive strain gages. The dynamic experiments had a relative bar velocity of 1.73 m/s, which corresponded to a loading rate of 72.54 kN/s. To capture the onset of crack initiation and propagation, an ultra-high-speed camera was used to record at a frame rate of one million fps with a resolution of 312 by 260 pixels. The specimen was oriented in the SHPB so that the pre-crack side was at the incident bar interface, as shown in Fig. 2.6.

## 2.5 Results and Discussion

### *2.5.1 Crack Propagation Velocity*

DIC was used to calculate the crack opening displacement at various locations along the specimen using digital extensometers. These measurements are basically like bonded crack detection gages. These extensometers and their locations are shown in Fig. 2.7a. The crack opening displacement at each of these locations was extracted and plotted together (Fig. 2.7b). By inspecting the images for crack tip location, and comparing this to the corresponding extensometer's

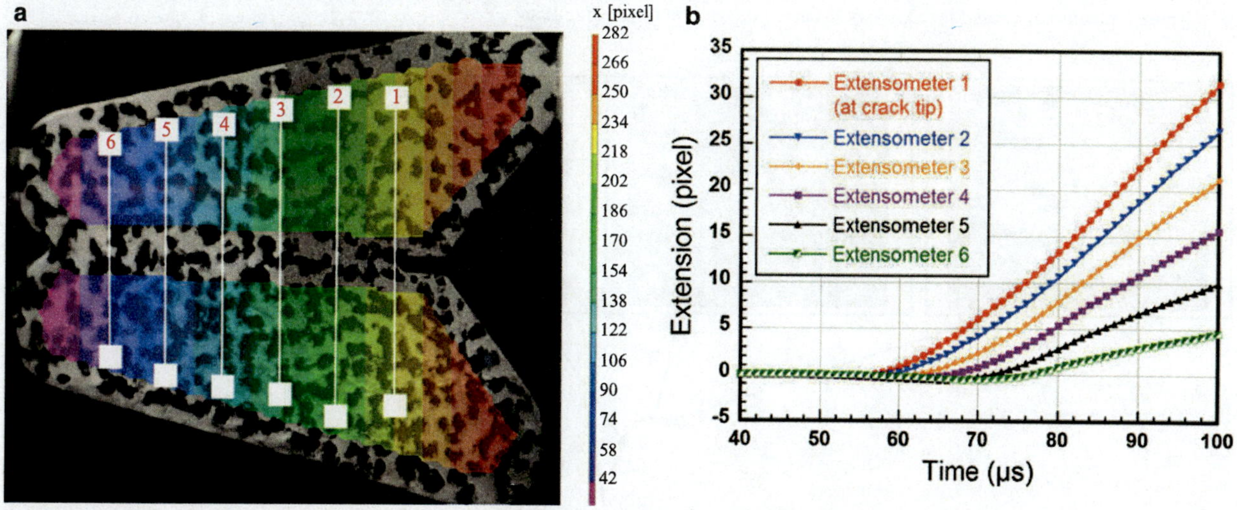


**Fig. 2.7** (**a**) Location of digital extensometers across specimen (**b**) COD of extensometers versus time plotting crack location as a function of crack arrival time, it is shown that the crack traveled at a linear rate (velocity) along the specimen (Fig. 2.8a–c). These figures show the crack location as a function of time for (Fig. 2.8a) high loading rate experiment (341 m/s), (Fig. 2.8b) intermediate loading rate experiment (186 m/s), and (Fig. 2.8c) quasi-static loading rate (192 m/s). The slopes of the linear fitted lines of the points indicate that the crack propagation is clearly higher than the quasi and intermediate rates and difference between low and intermediate loading rates is marginal. Figure 2.8d shows the crack propagation velocity as a function of loading rate. Similar to the other properties investigated here, there is little rate dependency at low to intermediate rates, and a large increase as the loading rate increases to dynamic

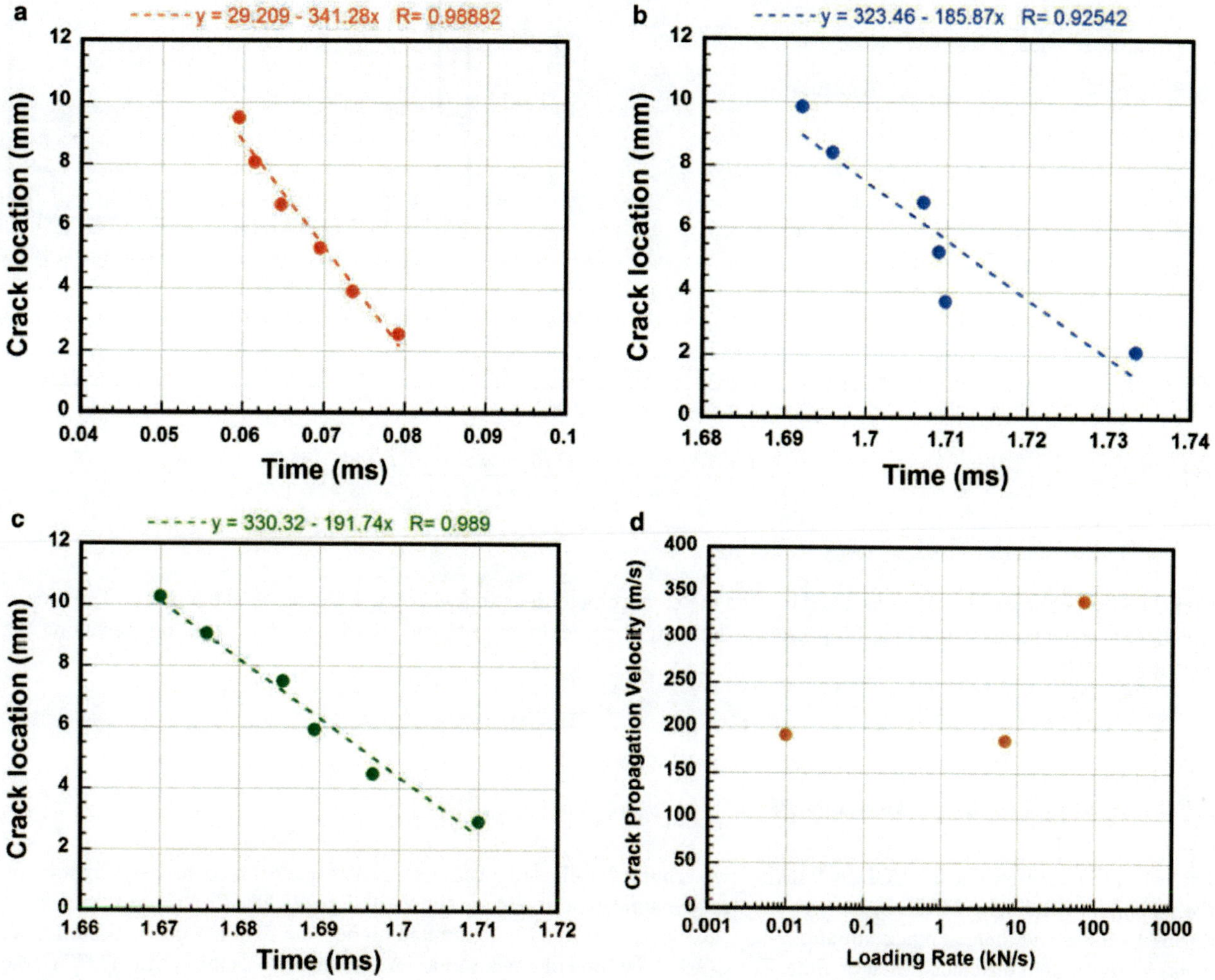


**Fig. 2.8** Crack location versus time for loading rates: (**a**) High, (**b**) Intermediate, (**c**) Quasi-static; (**d**) Crack propagation velocity of DGEBA-PACM as a function of loading rate

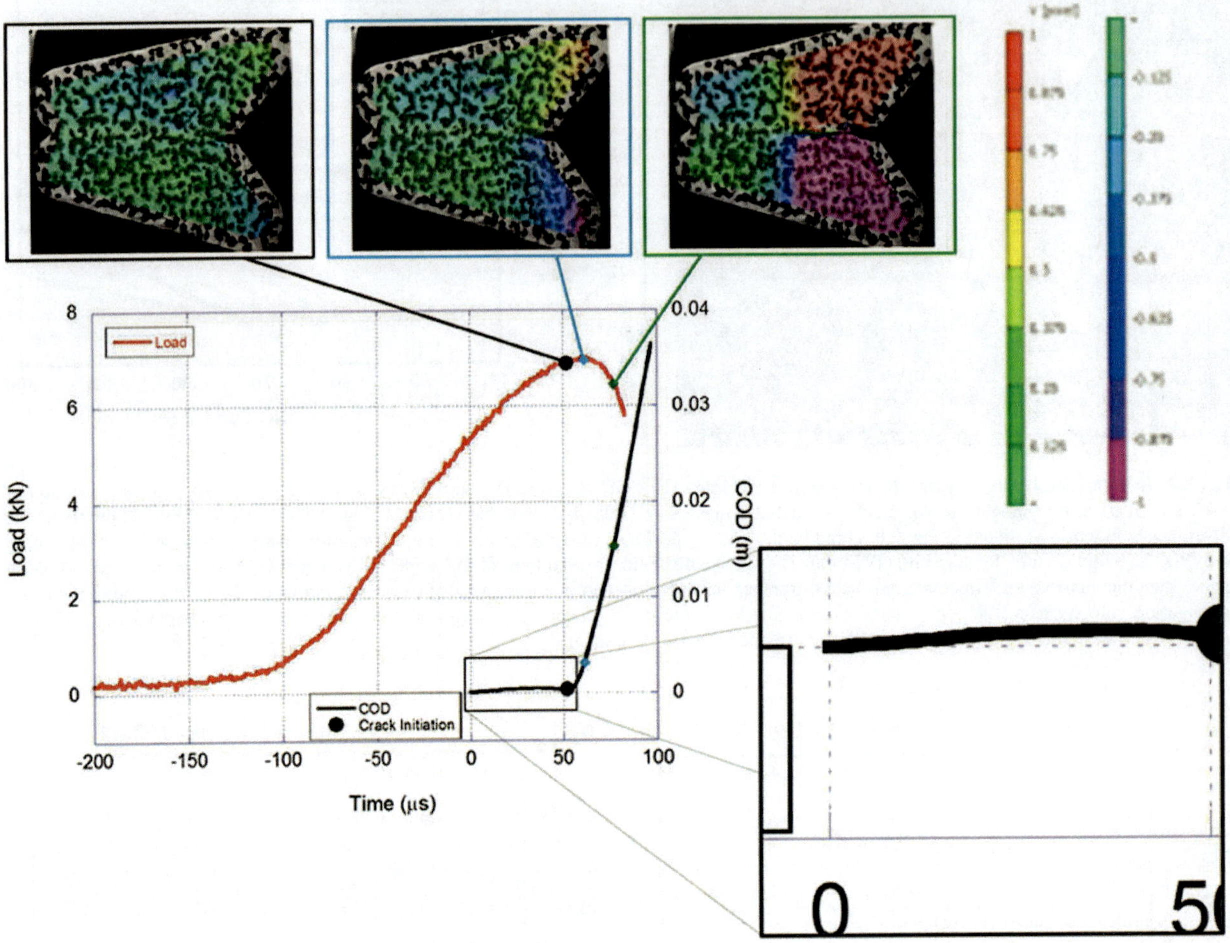


**Fig. 2.9** Load and COD history for a high rate experiment

**Table 2.1** Summary of results for the mode I fracture experiments of DGEBA-PACM epoxy

| Loading rate (kN/s) | Crack initiation load (kN) | Critical COD (mm) | Critical COV (m/s) | Failure energy (J) | Failure energy rate (J/m$^2$) |
|---|---|---|---|---|---|
| 0.01 | 4.11 ± 0.2 | 0.11 ± 0.1 | 86.95 ± 21.7 | 32.92 ± 10.9 | 213.79 ± 68.6 |
| 7.07 | 4.24 ± 0.4 | 0.23 ± 0.1 | 131.53 ± 18.8 | 74.14 ± 21.7 | 464.11 ± 130.1 |
| 72.54 | 8.21 ± 0.9 | 2.42 ± 1.4 | 400.86 ± 86.5 | 241.62 ± 45.8 | 1,539.64 ± 301.1 |

COD, it was determined that the crack had crossed the extensometer location when the COD reached 0.5 pixels. The time at which each extensometer reached 0.5 pixel was extracted, and compared to the location of the extensometer along the specimen length.

### *2.5.2 Crack Opening Displacement*

DIC was used to determine the COD by tracking two arbitrarily selected locations on each side of the outer load supports on the specimen. Loading and COD history for a typical dynamic experiment are shown in Fig. 2.9. The COD increased slightly during loading, and then, at crack initiation, the COD increased linearly at a much higher rate than before crack initiation. This clearly located the crack initiation time. The COD at the time of crack initiation is referred to as the "critical" COD, the amount of crack opening required for initiation of crack propagation. The DIC images illustrate the displacement field,

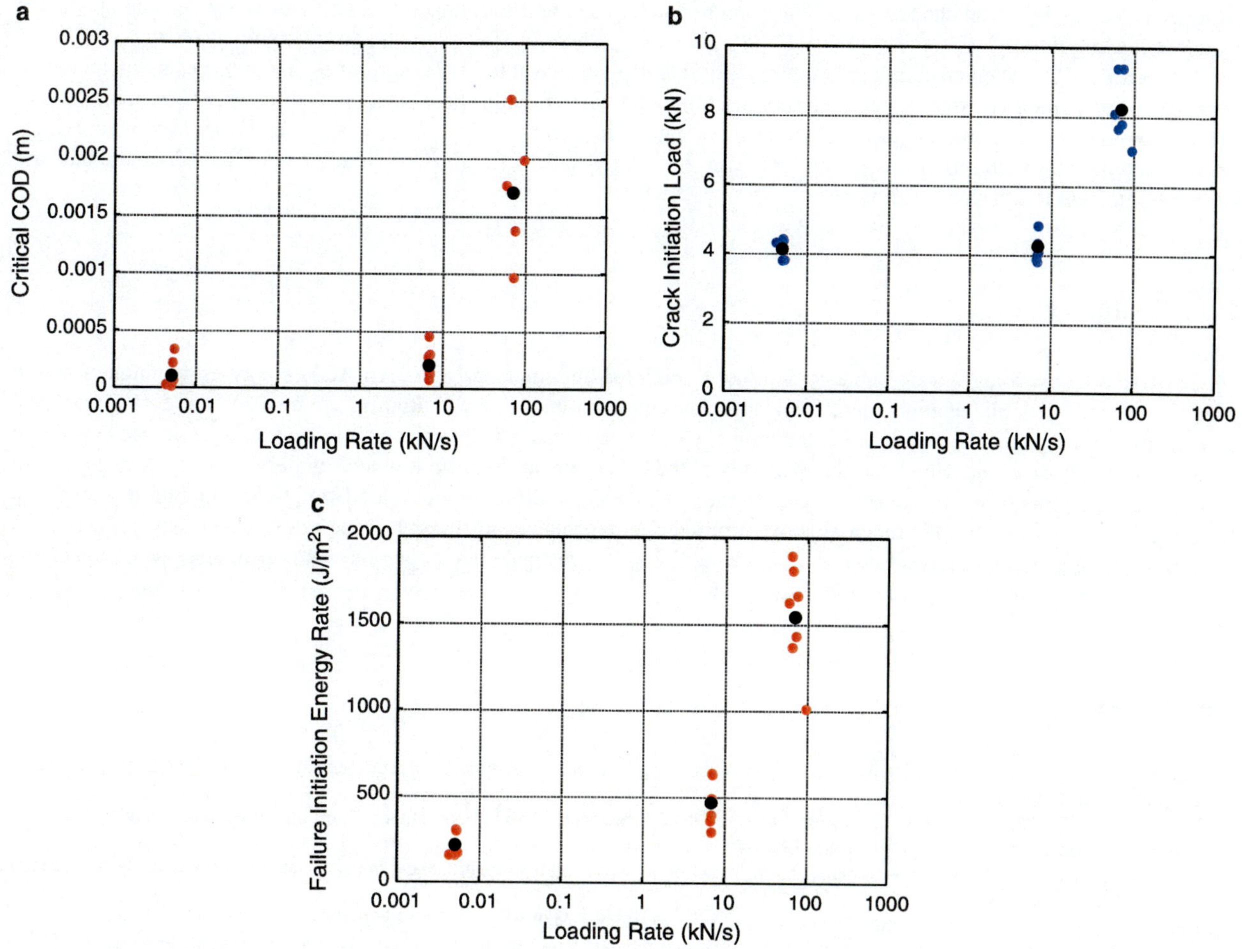


**Fig. 2.10** Relationship between the loading rate and (**a**) Critical COD, (**b**) Crack initiation load, and (**c**) Failure initiation energy rate

perpendicular to the axis of the pre-crack, of the wings as the crack propagates through the specimen as a function of time. DIC was unable to measure the crack propagation velocity; the crack tip moved too fast for reliable calculation, with more noise.

Table 2.1 shows a summary of the results of the Mode I fracture epoxy experiments at the various loading rates. Figure 2.10 shows these results graphically as a function of loading rate for (a) critical COD, (b) crack initiation load, and (c) failure energy rate. The crack initiation load increased slightly from 4.11 to 4.24 kN at loading rates of 0.01 and 7.07 kN/s, respectively. However, when the loading rate increased to dynamic rates (72.5 kN/s), the failure initiation load increased to about double the lower rates of 8.21 kN.

The failure energy, which is the energy required to initiate crack propagation, was calculated using Eq. 2.1 by integrating load history as a function of displacement to the beginning of crack growth.

$$FE = \int_{t_0}^{t_{critical}} P(x)dx \tag{2.1}$$

The failure energy rate is determined by the failure energy per unit of crack area. Failure energy and failure energy rate both increased with the loading rate and show results similar to critical COV with dynamic loading condition being three times higher than the intermediate loading rate of 7.07 kN/s.

Figure 2.10 shows the results as a function of the loading rate, for the values of (a) critical COD, (b) crack initiation load, and (d) failure initiation energy rate. The loading rate axis is in a log scale. In each plot, the black dots represent the average value for each loading rate. These results show the measured critical fracture parameters for DGEBA-PACM epoxy increase significantly at higher loading rates, compared to lower and intermediate loading rates. In addition, results show the critical

COD increases slightly from the low to the intermediate loading rate whereas, the critical COD increases considerably at the dynamic loading rate.

The DGEBA-PACM cured epoxy showed minimal rate dependence at lower loading rates, but becomes much more rate sensitive when loaded at dynamic rates. Crack initiation load was virtually the same across three orders of magnitude of loading rate; however when it increased to the dynamic rate, it increased by double. The same trends were shown in critical COD, and failure initiation energy rate: small increase from quasi-static to intermediate loading rates, then dramatic increase from intermediate to dynamic loading rates.

## 2.6 Summary

Mode I fracture experiments were conducted at low to high loading rates on DGEBA-PACM epoxy using unique fracture specimen geometry. Utilizing high-speed imaging, the accurate timing of crack initiation can be determined. DIC was used to measure the loading rate dependence of the critical COD. The critical COD for crack initiation show an increase with the increase of the loading rate. However, the increases of the CODs are subtle from low to the intermediate. The results from dynamic experiments show significant increase in the critical COD, failure initiation load, and failure initiation energy rate for crack growth initiation. This material shows limited rate dependence at lower loading rates, but becomes much more rate sensitive when loaded at dynamic rates. Future studies will investigate other epoxy systems with dispersed tougheners, changing the cross-link density, and molecular weight.

## References


1. Cho K, Lee D, Park CE, Huh W (1996) Effect of molecular weight between crosslinks on fracture behaviour of diallylterephthalate resins. Polymer 37(5):X13–817
2. Kang BU, Jho JY, Kim J, Lee S, Park M, Lim S (2001) Effect of molecular weight between crosslinks on the fracture behavior of rubber-toughened epoxy adhesives. J Appl Polym Sci 79:38–48
3. Hussain M, Nakahira A, Nishijima S, Niihara K (1996) Fracture behavior and fracture toughness of particulate filled epoxy composites. Mater Lett 27:21–25
4. Mostovoy S, Ripling EJ (1966) Fracture toughness of an epoxy system. J Appl Polym Sci 10:1351–1371
5. Wakako A, Kentaro N, Tadaharu A, Akihiko Y (2005) Fracture toughness for mixed mode I/II of epoxy resin. Acta Mater 53(3):869–875
6. Cho K, Huh W (1996) Effect of molecular weight between crosslinks on fracture behaviour of diallylterephthalate resins. Polymer 37(5):813–817
7. Jiang F, Vecchio K (2009) Hopkinson bar loaded fracture experimental technique: a critical review of dynamic fracture toughness tests. Appl Mech Rev Trans ASME 62:1–39
8. Wu XJ, Gorham DA (1997) Stress equilibrium in the split Hopkinson Pressure Bar test. J Phys IV 7(C3):91–96
9. Chen W, Lu F, Zhou B (2000) A quartz crystal embedded split Hopkinson bar for soft materials. Exp Mech 40(1):1–6
10. Weerasooriya T, Moy P, Casem D, Cheng M, Chen W (2006) A four point bending load technique for determination of dynamic fracture toughness for ceramics. J Am Ceram Soc 89(3):990–995
11. Weerasooriya T, Moy P, Casem D, Cheng M, Chen W (2006) Fracture toughness for PMMA as a function of loading rate. In: Proceedings of the 2006 Society of Experimental Mechanics Annual Conference
12. Gunnarsson CA, Sanborn B, Foster M, Moy P, Weerasooriya T (2012) Initiation fracture toughness for human cortical bone as a function of loading rate. In: Proceedings of the 2012 Society of Experimental Mechanics Annual Conference
13. Syn C, Chen W (2008) Surface morphology effects on high-rate fracture of an aluminum epoxy interface. J Compos Mater 42:1639–1658
14. Weerasooriya T, Gunnarsson CA, Jensen R, Chen W (2011) Strength and failure energy for adhesive interfaces as a function of loading rate. In: Proceedings of the 2011 Society of Experimental Mechanics Annual Conference
15. Davies EDH, Hunter SC (1963) The dynamic compression testing of solids by the method of the split Hopkinson pressure bar. J Mech Phys Solids 11(3):155–179

# Chapter 3
# Mechanical Response of Polycarbonate Nano-composites as a Function of Strain Rate and Weight Percent of Nano-silicate Clay Loading

**Paul Moy, Tusit Weerasooriya, and Wayne Chen**

**Abstract** In recent years, polymer-based nanocomposites have gradually found its way into mainstream applications such as automotive [1, 2] and food packaging [3]. Studies have shown that the inclusion of nano-particles in nanocomposites increases the tensile strength and modulus when compared to its parent base material [4–7]. Yet, the effect of nano-particles on the mechanical response and failure behavior of nano-composites at high strain rates is still not completely understood. This paper presents the results of a systematic investigation to understand the mechanical response for polycarbonate (PC) nanocomposites under uniaxial compression as a function of the strain rates and weight percent of the nano-silicate clay loading.

**Keywords** Nano-composites • High strain rate • Polycarbonate • Weight percent • Yield strength

The PC nanocomposites were processed by compounding Makrolon 2847 polycarbonate resin (Bayer Corp) and alkyl ammonium modified montmorillonite clays, Choisite® 25A. The weight percents used in this study are 1.5%, 2.5%, and 5% as well as a control that contains no clay. Typically, complete exfoliation of the nano silicates is desired for improved mechanical properties. In this case, however, the nano silicates were intercalated. Figure 3.1a, b shows TEM images of PC nanocomposite with 1.5%, 2.5%, and 5% by weight of nano-silicates. By observation, the lamellae groups of silicates can be seen dispersed throughout the monomer. Specimens were fabricated into right cylinder geometry with a 6 mm diameter and 3 mm gage length, yielding a 2:1 ratio of the diameter to length.

High rate experiments (greater than 1/s strain rates) were conducted using a modified split-Hopkinson Pressure bar (SHPB) with pulse-shaping technique to impose the compressive loading of the specimen at near constant strain rate under dynamic stress equilibrium [8–10]. The high strain rates were in the order of about $10^3$/s. In addition to the high rate experiments, lower and intermediate strain rate experiments were conducted at $10^{-3}$ and 1/s strain rates.

Figure 3.2a through d provide a summary of some of the results from the uniaxial compression experiments. Figure 3.2a shows the stress–strain behavior at 1/s strain rate for each of the PC nanocomposites materials with different amounts of nano-clay loadings. The trend reveals that the flow stresses increases with the increase of the weight percent of nano-clays with the exception of the compression results at 1.5% and 2.5%. It is possible that the 1% increase of the nanoclays were insignificant. Comparable to the results from the study by Moy et al. [9] on monolithic PC, the PC nano-composites exhibit intrinsic thermal softening after the yield and then followed by strain hardening effects.

The stress–strain response of 1.5% of nano-clays material at low to high strain rates is shown in Fig. 3.2b. Here the flow stresses increases with the increase in the strain rates. Figure 3.2c shows a set of high rate experiments for the 1.5% PC nanocomposites conducted from about 960 to 5,000/s strain rates. From these results, there is no discerning flow stresses and

P. Moy (✉) • T. Weerasooriya
US Army Research Laboratory, Aberdeen Proving Ground, Aberdeen, MD 21005, USA
e-mail: paul.moy.civ@mail.mil

W. Chen
Purdue University, School of Materials Engineering, West Lafayette, IN 47907, USA

V. Chalivendra et al. (eds.), *Dynamic Behavior of Materials, Volume 1: Proceedings of the 2012 Annual Conference on Experimental and Applied Mechanics*, Conference Proceedings of the Society for Experimental Mechanics Series, DOI 10.1007/978-1-4614-4238-7_3, © The Society for Experimental Mechanics, Inc. 2013

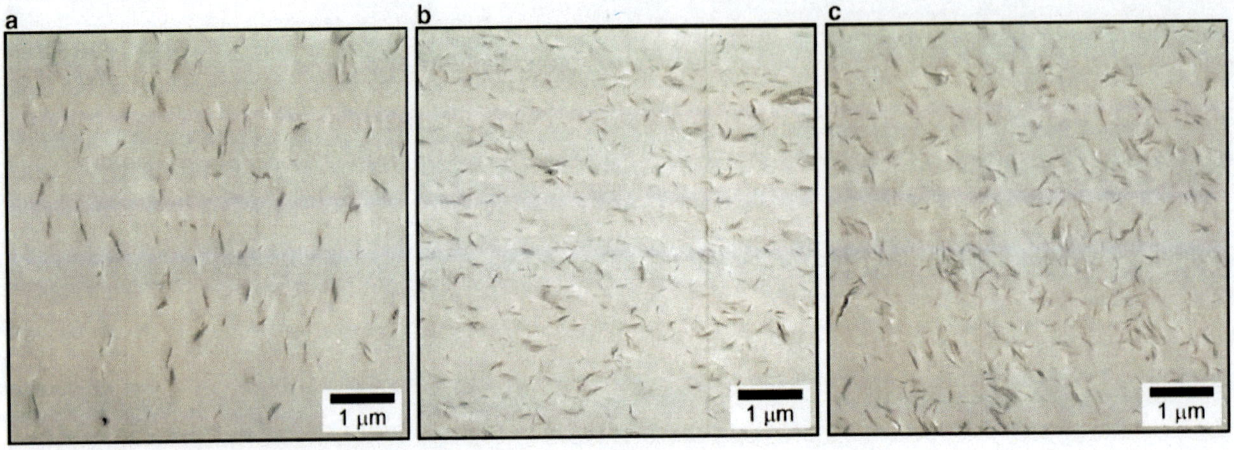


**Fig. 3.1** TEM micrographs of PC nanocomposites at 1.5 % (**a**), 2.5 % (**b**), and 5 % (**c**) by weight of nanoclay silicates

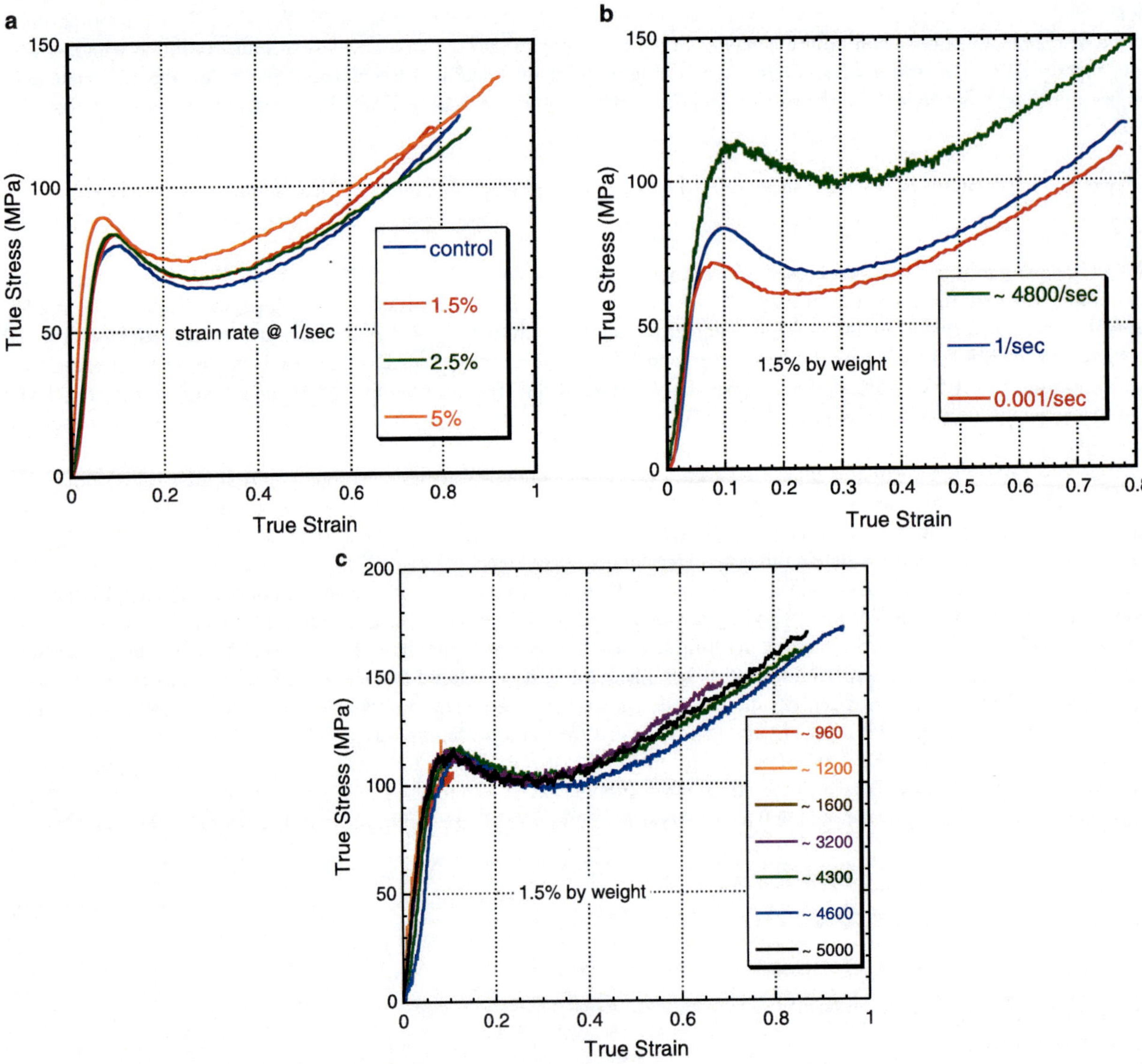


**Fig. 3.2** Experimental results of uniaxial compression: (**a**) Stress–strain behavior at 1/s for PC nanocomposites, (**b**) Stress–strain response of 1.5 % by weight of nano-silicate at low to high strain rates, (**c**) High strain rate experiments of 1.5 % by weight of nano-silicate

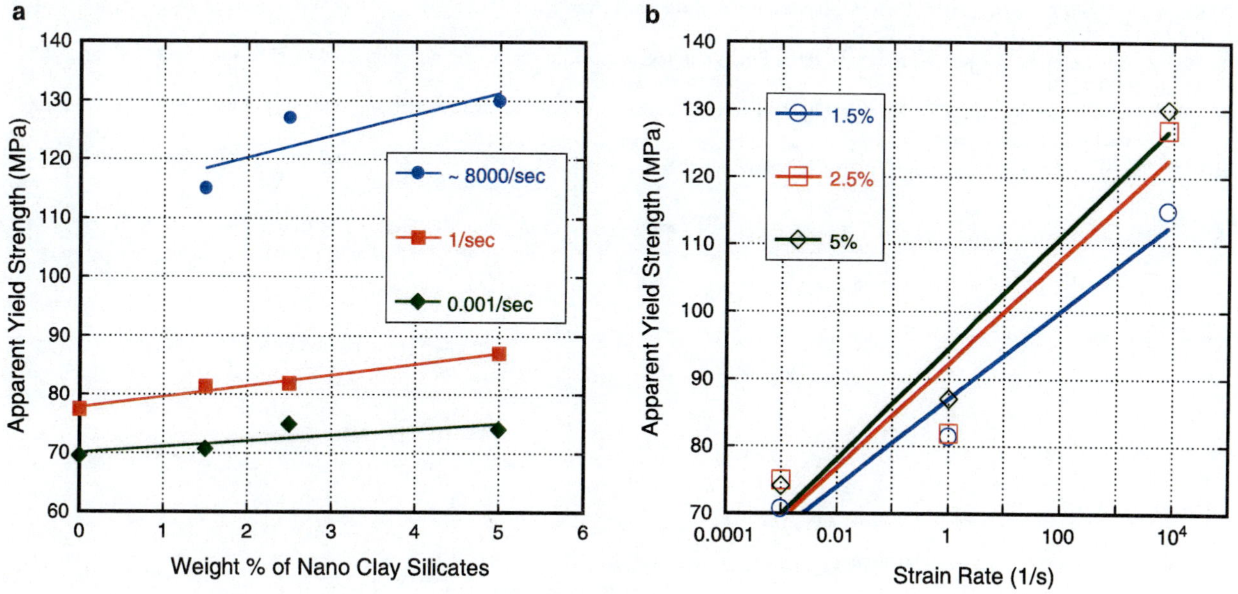


**Fig. 3.3** Apparent yield strengths as a function of the (**a**) Weight % of nano-silicates and (**b**) Strain rate

yield strengths due to the rate effects with the nano-composites. There is a possibility that the adiabatic conditions were dominating factors and thus the stress–strain results have nearly the same behavior. This is also evident for all PC nanocomposites at 2.5% and 5%. This data is not presented in this paper.

The result for the apparent yield stresses is summarized in Fig. 3.3a as a function of both strain rate and the weight percent of clay loading. Since there was no distinction between the apparent yield strengths from all high strain rate experiments, the results from ~8,000/s is presented in the graph. At the present, there is no high rate compression experiments conducted for this particular control (0%) baseline PC. The slope of the fitted linear line would indicate that the apparent yield strengths increase with the increase of the weight % of nanoclays. Nonetheless, there is additional study needed to understand how the nanoclays actually contribute to these effects.

The apparent yield strength as a function of the strain rate for each nanocomposite is presented in Fig. 3.3b. This graph provides a general summary of the stress–strain results from all experiments conducted at different strain rates and content of the weight percent of the nano material. As discussed earlier, the yield strength increases with the increase in strain rate, as well as increase with the weight percent of the nanoclay silicates.

Furthermore, no significant specimen fracture was evident even at strain rates as high as about 8,000/s and up to the applied maximum strains of about 0.8. Future efforts will investigate the 3.5% PC nanocomposites and in addition study the morphology of the tested specimen to determine if there is any effect from adiabatic heating at higher strain rates.

## References


1. Okada A, Usuki A (2007) Twenty-year review of polymer-clay nanocomposites at Toyota Central R&D Labs., Inc. SAE Technical Paper 2007-01-1017, doi:10.4271/2007-01-1017
2. The Future of Automotive Plastics (2003) PR Newswire. URL: http://www.scprod.com/gm.html. Accessed 30 April 2003
3. Butschli (2004) Nanotechnology packs promise. Retrieved on 13 Feb 2005 from Packaging World at http://www.packworld.com/cds_print.html?rec_id=17883
4. Hsieh AJ, Moy P, Beyer FL, Madison P, Napadensky E, Ren J, Krishnamoorti R (2004) Mechanical response and rheological properties of polycarbonate layered-silicate nanocomposites. Polym Eng Sci 44:825–837
5. Okamoto M (2006) Recent advances in polymer/layered silicate nanocomposites: an overview from science to technology. Mater Sci Technol 22(7):756–779
6. Kojima Y, Usuki A, Kawasumi M, Okada A, Fukushima Y, Kurauchi T, Kamigaito O (1993) Mechanical Properties of Nano 6-Clay Hybrid. J Mater Res 8:1185

7. Kojima Y, Usuki A, Kawasumi M, Okada A, Fukushima Y, Kurauchi T, Kamigaito O (1993) Sorption of Water in Nylon 6-clay Hybrid. J Appl Poly Sci 49:1259
8. Moy P, Weerasooriya T, Chen W, Hsieh A (2003) Strain rate response of polycarbonate under uniaxial compression. In: Proceedings 2003 ASME IMECE. Bethel, CT
9. Moy P, Weerasooriya T, Chen W, Hsieh A (2003) Strain rate response of polycarbonate under uniaxial compression. In: Proceedings 2003 SEM Annual Conference. Bethel, CT
10. Chen W, Lu F, Cheng M (2002) Tension and compression tests of two polymers under quasi-static and dynamic loading. Polym Test 21:113–121

# Chapter 4
# High Rate Characterization of Polymeric Closed-Cell Foams: Challenges Related to Size Effects

**S. Ouellet, D.S. Cronin, J. Moulton, and O.E. Petel**

**Abstract** Polymeric foams including expanded polystyrene and low-density polyethylene have been used extensively in the design of military protective systems to help mitigate threats that can range from low velocity impacts to explosive events. Polymeric foams are significantly rate dependent and have very low wave speeds, which can complicate their response in specific conditions. In the present study, two polymeric foams were characterized in compression at quasi-static and high strain rates. Rates from 1 $s^{-1}$ were obtained with a standard hydraulic test machine. Acrylic Hopkinson bars were used to generate compression rates on the order of $10^3$ $s^{-1}$. The two closed-cell polymeric foams investigated in this study were of similar density but with a significantly different macro-structure. Low and high strain rate testing on a relatively consistent cell-size material (low density polyethylene) demonstrated expected trends and results, while the effect of strain rate was masked for a material with high structural variability (expanded polystyrene).

**Keywords** Hopkinson bar • Polymeric compressive hopkinson bar • High strain rate • Expanded polystyrene foam

## 4.1 Introduction and Background

Polymeric foams are well known for their ability to sustain necessary deformation under low near-constant stress, making them very desirable for many protective applications such as fragile part packaging, sports equipment, transportation crashworthiness and different kinds of headwear. This unique characteristic provides means to absorb or manage a significant amount of energy within a given volume and at minimal weight, without transferring large stresses to surrounding structures. The energy absorption/management capability per unit volume of foam is usually estimated using the area under the stress–strain curve of the material, using the maximum allowable stress as the upper bound [1]. Polymeric foams are highly rate dependent materials, thus their energy absorption capacity is similarly dependent on the loading rate. More specifically, the elastic modulus, the collapse stress, and the slope of the collapse plateau increase with increasing compressive strain rate, while densification occurs at a lower level of strain [1–5]. It is therefore essential to characterize the foam response under the loading rate of interest to a particular application, as a miscalculation of the absorption capacity may lead to high transmitted stress as the foam reaches densification. Proper dynamic characterization of foams, including the range of peak strain rates related to automotive impact and military applications ($10^2$–$10^3$/s), is essential for evaluating the potential stress levels that could be transferred through a foam in a protective application as well as to build representative constitutive models for Finite Element Method codes [5].

Determining the dynamic response of a foam over a wide range of strain rates requires the use of multiple experimental techniques. While it presents a few challenges, characterization techniques for lower strain rates up to approximately $10^2$ $s^{-1}$ are relatively straightforward, generally relying on high inertia devices, such as a drop tower or a pendulum, accelerated to

S. Ouellet (✉)
Defence Research and Development Canada-Valcartier, 2459 Pie-XI Blvd North, Quebec, Canada
e-mail: simon.ouellet@drdc-rddc.gc.ca

D.S. Cronin • J. Moulton
Department of Metallurgical & Materials Engineering, University of Waterloo, 200 University Avenue West, Waterloo, Canada

O.E. Petel
Department of Mechanical Engineering, McGill University, 845 Sherbrooke Street West, Montréal, Canada

V. Chalivendra et al. (eds.), *Dynamic Behavior of Materials, Volume 1: Proceedings of the 2012 Annual Conference on Experimental and Applied Mechanics*, Conference Proceedings of the Society for Experimental Mechanics Series,
DOI 10.1007/978-1-4614-4238-7_4, 

low impact velocity and instrumented with a load cell and optical displacement measuring device. To obtain the higher rate compressive response of soft materials (~$10^3$ $s^{-1}$), Split Hopkinson bars, or Kolsky bars, are the most widely used experimental method [6]. Originally developed by Kolsky [7] to characterize metallic materials, the method simply involves compressing a material sample between two long elastic bars by sending a compressive pulse through the first (incident) bar, transmitting it through the specimen and into the second (transmitted) bar. By measuring the strain history of the incident and transmitted bars at specific locations and by assuming elastic wave propagation principles, the stress and strain history at the sample interfaces can be derived.

Unfortunately, the classical Kolsky bar technique includes some assumptions that may be challenging to meet when testing very soft low-density polymeric materials.

Inaccurate measurement of transmitted stress is the first challenge encountered when testing very soft materials with classic metallic bars. Because of the large impedance mismatch between the bar material and the soft test sample, compressive pulses are mainly reflected in the incident bar and the transmitted bar signal is so low that measurement away from the sample-bar interface becomes difficult due to lack of gage sensitivity and high signal/noise ratio [5–8]. Stress waves are also known to be dispersive in polymers and foams, and dispersion could be significant enough over a sample length to further limit the transmitted stress [6]. In order to solve this issue, two different strategies have been used. The first one, developed by Chen et al. [8, 9], consists in performing stress measurement directly at the sample-transmitted bar interface using embedded high-sensitivity quartz gauges in conjunction with a hollow aluminum transmitted bar to enhanced transmitted strain signal. Although this method introduces additional interface between the specimen and bars, the main advantage is that data reduction is still based on elastic wave propagation in the bars. A second common method is to use polymeric bars to address impedance mismatch between the sample and bars. However, due to the visco-elastic nature of the bar material, the propagating loading pulse attenuates and disperses as it travels along the length of the bar. Data reduction can no longer rely on elastic wave propagation principles to relate measured strains on the bar to the strain and stress in the specimen. However, an experimental method has been developed by Bacon [10] to determine the propagation coefficient of individual visco-elastic bars. The coefficient is a function of angular frequency, wave attenuation and wave phase velocity. Therefore, by performing free-end tests on individual bars and relying on one dimensional visco-elastic analysis in the Fourier domain, a scheme for predicting strain signal at any point on a bar from a single strain measurement done elsewhere is derived. The scheme can be further validated by performing strain measurements at several points on a bar and by doing bar-to-bar test with no test specimen. This approach has been used successfully with polymeric Hopkinson bars by other researchers [2, 5] and is the method of choice in the current study. Based on the implementation work of Salisbury [11], the polymeric Hopkinson bars have been used successfully in the past to obtain high rate compression data on polyurethane rubber [12], polymeric foams [4], bovine muscle tissue [13] and ballistic gelatine [14].

The second challenge encountered with the high rate testing of soft materials is achieving equilibrium within the test sample [15]. This condition involves a state of constant axial stress within the specimen at any given time, determined through front and back face stress measurements. It is a critical assumption behind the data reduction that is necessary to obtain valid sample stress. Uniformity of load and deformation is achieved through reverberation of the initial stress wave between the front and back face of the sample. This is often described as "sample ring-up" and illustrates the achievement of stress equilibrium. For hard specimens it is suggested that 3–4 transits of the stress waves inside the specimen are necessary to achieve equilibrium [6], however, it may take significantly more for soft dispersive materials. For this reason, early time measurements are often considered to be invalid, because they are made during a non-equilibrated state [5]. It has been found that the number of wave transits necessary to reach equilibrium also depends on the loading rate and the impedance mismatch between the specimen and the bars [15]. In general, the rise time of the bar signal, related to the bar stiffness and the sample length and wave speed determine if equilibrium can be achieved for a given test. To increase the rise time of the incident wave pulse shapers, consisting of a thin compressible copper disk have been used ahead of the incident bar. For polymeric bars, the use of pulse shapers is typically not necessary since the rise time of the signal is much longer (70–100 μs) compared to steel bars (approximately 15 μs) [13]. Of course, the time associated with a specific number of wave transits depends on wave speed and sample thickness. Since the first parameter cannot be controlled, it is advantageous to minimize sample thickness in order to obtain valid data for most of the duration of the test. For example, while characterizing a soft rubber material, Song et al. [15] had to use 1.52 mm thick specimens to achieve rapid equilibrium. However, the sample length is limited due to frictional effects and radial inertia. Further limitations on sample length are encountered for polymeric foam, since the macro-structure may include cell sizes on the order of 1–3 mm, hence rendering nearly impossible the testing of a representative volume of material. It is easy to understand that the main challenge in testing such material is the selection of a compromise between valid data and representative volume.

The goal of this study was to investigate specimen size effects for high rate compression testing using polymeric Hopkinson bars, for two polymeric foams of equivalent density but with different strength and macro-structure. The foams considered were expanded Polystyrene (EPS 4.4) and low density polyethylene (LDPE 70).

## 4.2 Material and Method

Two polymeric foams of equivalent density, but significantly different macro-structure were investigated in this study. The first foam was a low-density polyethylene (LD70, nominally 70 kg/m$^3$) with a simple closed-cell structure. The second foam was an expanded polystyrene foam (EPS4.4, nominally 70 kg/m$^3$), which is also a closed-cell foam, but manufactured from expanded polypropylene beads, creating a larger dual macro-structure comprised of grains and inner-grain cells. LD70 and EPS4.4 were received in stock sheets with nominal thicknesses of 19 mm and 21 mm respectively. For both foam materials, scanning electron microscopy was used to observe the macro-structure and qualitatively determine the level of regularity (Fig. 4.1).

Quasi-static compression tests were performed using an MTS hydraulic test machine instrumented with a 2.2 kN load cell and a LVDT displacement transducer. Four different sample sizes were tested, from 13 mm diameter circular samples to 75 mm large square samples, at a strain rate of 1 s$^{-1}$. Circular samples were cut using a hollow cutting drill and larger square samples were cut using a knife and jig. All samples were cut from random locations in the supplied sheet, away from the edges where foam skin was present. Materials were tested at their nominal sheet thickness. Aside from the prescribed thickness, tests on large samples were performed according to the ASTM D3575 standard. Strain levels exceeding 85% were achieved in all samples. The largest sample cross-section was chosen so that it would completely eliminate any macro-structural effects. Each sample was measured and weighed prior to testing. The data was then analysed to assess the level of variability caused by the use of small samples compared to the large sample size typically used in quasi-static characterization tests.

A schematic of the polymeric Hopkinson bars used in the high rate material characterization in the present study is shown in Fig. 4.2. The acrylic incident and transmitted bars were 2.44 m long with a diameter of 25.4 mm. The striker bar was propelled by a gas gun system, with the gas tank pressure adjusted to obtain the desired impact velocity on the incident bar, which in turn dictated the maximum obtained strain rate. The maximum attainable strain can be increased through the use of a longer striker. A shorter striker of 0.459 m was used to identify the time necessary to reach equilibrium. More specifically, a shorter compressive pulse ensured that there was no overlapping of the incident and reflected strain signal on the incident bar, which allowed the comparison of the force signal at both interfaces, as obtained using the classical 1-wave and 2-wave data reduction scheme. Full compression of the tested specimen required a striker length of 0.71 m, which created incident and reflected signal overlaps. Thousand Ohm semiconductor strain gauges, with a gauge factor of 100 were located at the mid-point of the bars. The strain gauge voltage signal was recorded by an Integra model 40 oscilloscope with a sampling rate of 1 MHz. Data was processed according to the method described by Salisbury [11, 14] to obtain sample strain, strain rate, and stresses. Aside from studying equilibrium in the sample, the 1-wave method using only the transmitted signal was used to obtain the stress history in the sample. A sample diameter of 11–12 mm allowed for relative ease in manufacturing consistent samples that would be large enough to transmit sufficient signal to characterize the material response, but not sufficiently large that the Hopkinson bar would be unable to densify the specimen. In order to ensure equilibrium and not compromise the validity of the data over a large range of strain, 4–5 mm thick specimens were used. This choice of sample thickness is at the core of the current study since thin samples may not result in a representative volume for the foam materials tested. Bar to specimen interfaces were lubricated to avoid friction and undesirable radial inertial effects. However, given the considerably low Poisson's ratio associated with low-density foam materials, these effects were not significant.

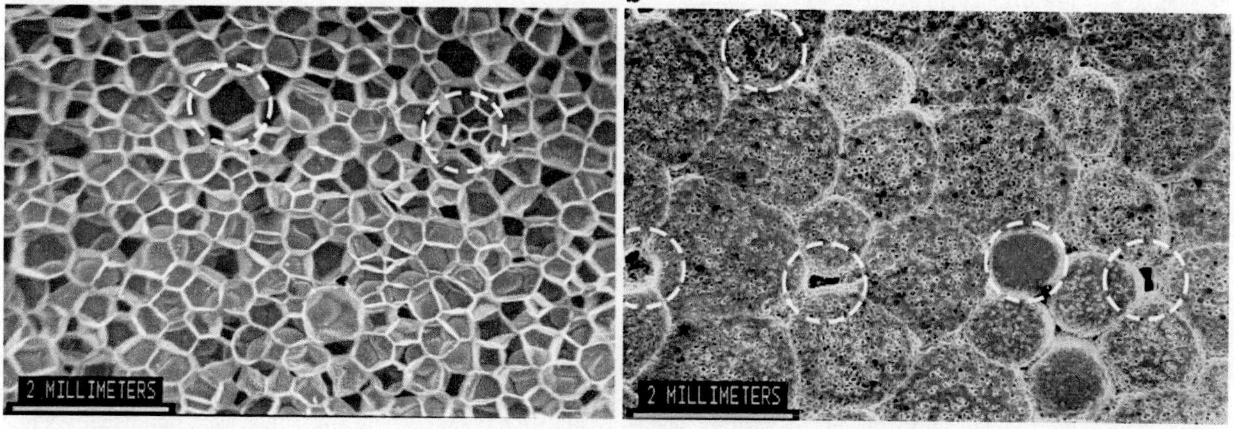


**Fig. 4.1** SEM images from (**a**) LD70 and (**b**) EPS4.4. Circles identify regions on visible variation in local density

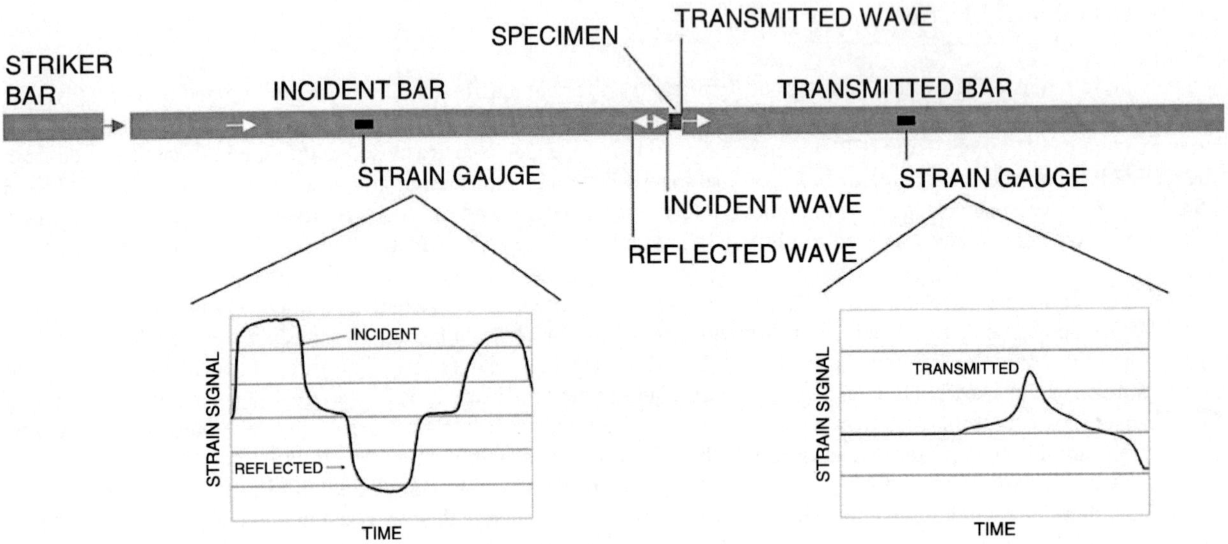


**Fig. 4.2** Schematic of the polymeric Hopkinson bars used in the present study

## 4.3 Sample Density Variation

From the SEM images of the foams, it was observed that LD70 had a much more regular macro-structure than EPS4.4, with less variability in local density and cell sizes (Fig. 4.1). Red circles on each image highlight the macro-structural "extremes". For the LD70, the highlighted regions point areas with cells that are both smaller and larger than the estimated average. For the EPS4.4, the voids present at grain interfaces, which would decrease the local density and potentially affect the response of a specimen, as well as important differences in inner grain density, with some grains that are noticeably denser than others. These grains, resulting from single expanded polystyrene beads, have an average diameter on the order of 1 mm. The variation in grain size and grain density is important, with grains observed at more than 2 mm and other less than 0.5 mm. In comparison, average cell size for LD70 is estimated around 0.4 mm.

Figure 4.3 shows the variation in density in randomly cut samples for each sample size used during the quasi-static compression testing. Complete material sheet density was included for comparison along with the manufacturer's specified density. It is again noticed that a larger variation exist in the EPS4.4 samples, which was expected from the nature of the respective manufacturing processes and from the SEM observations. While the LD70 density converged to the sheet density for the larger samples, there was still significant scatter in the EPS4.4 density for an equivalent size (70 $\times$ 70 mm).

There are likely two factors affecting the final sample density: variability in cell sizes and grain density within the foams, and the effect of cutting samples with large cell/sample diameter aspect ratio. When cutting a circular sample, cells may sometimes be completely removed from the cut face, giving a more irregular surface. Consequently, the density of the sample was lower for small samples, as seen for the EPS4.4. Also, voids present at grain boundaries are likely to have a more significant effect on small samples, as they represent a larger fraction of the total volume. It is important to note that there was no attempt to down select foam samples so that their density was the closest to the full sheet or quoted density from the manufacturer. One of the objectives of this study was to identify the effect of variability in foam structure on the measured response of the material. Prior to testing, it was however unclear how much of an effect that would have on the sample response in compression.

## 4.4 Quasi-static Compression Tests

For both foams, the family of stress–strain curves obtained for all sample sizes at a strain rate of 1 $s^{-1}$ is shown in Fig. 4.4. All curves followed the expected behaviour, described by a linear elastic compression phase below approximately 5% strain, a cell collapse regime up to approximately 70% strain, and a densification regime where stress increased significantly with further compression [1].

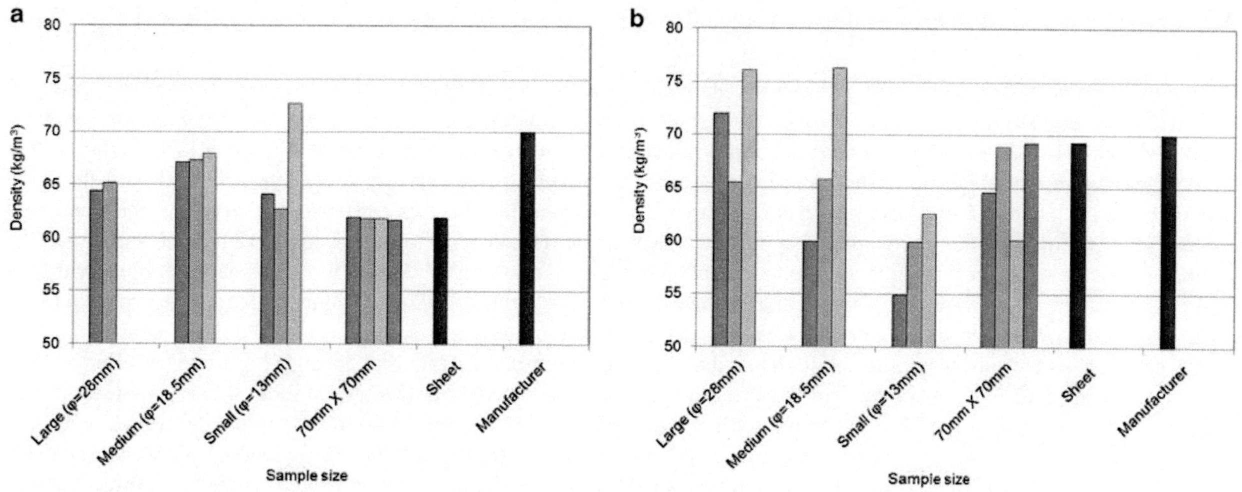


**Fig. 4.3** Density variation in different sample sizes for (**a**) LD70 and (**b**) EPS4.4

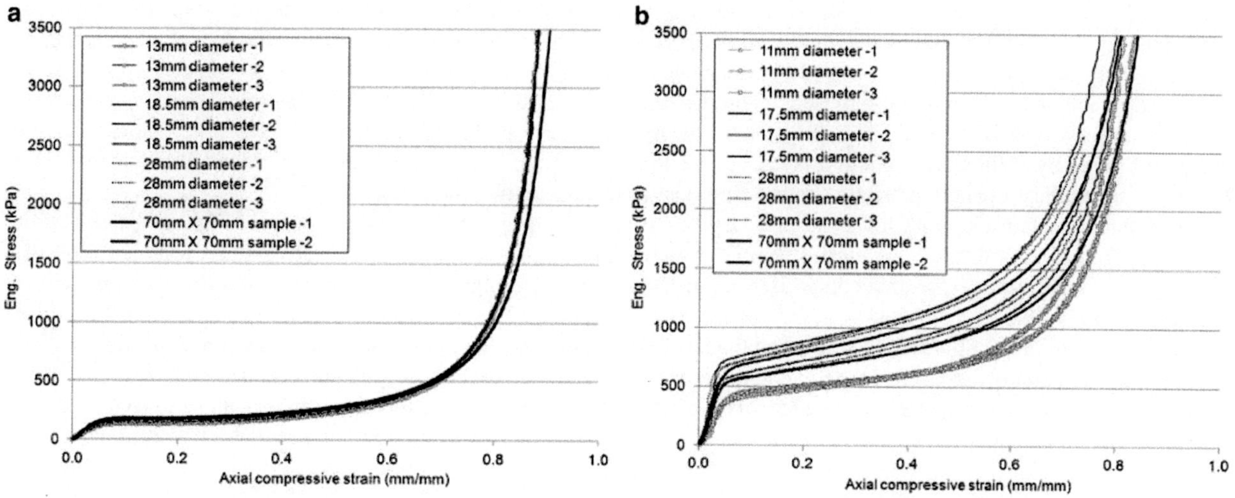


**Fig. 4.4** Stress–strain curves for different size sample at a strain rate of 1 $s^{-1}$ for (**a**) LD70 foam and (**b**) EPS4.4 foam

The LD70 results show limited scatter, with all circular sample responses converging towards the larger square samples. Densification occurred at lower strain for denser samples, which is a behaviour normally seen when increasing foam nominal density [1]. It was interesting to note that the smaller circular sample demonstrated a slightly lower collapse stress, even when a particular sample density was close to that of a the larger samples (small #2 vs. medium #1). This suggests that variation in the response was not solely due to variations in density. One hypothesis for this behaviour is that the proportion of broken cells could be larger for smaller test samples. While the number of broken cells only increases with the sample perimeter (linearly related to diameter), the number of total cells increases with the sample surface area (increases as diameter squared).

EPS4.4 exhibited notable scatter in the stress–strain curves with samples of higher density generally resulting in higher collapse stress and lower strain to densification. However, variation within a similar sample size, or between two samples of similar densities, was quite noticeable. This again suggested that scatter was not only a consequence of variability in sample density, but also affected by cell distribution and uniformity within samples, as well as cell integrity at the cut surface. Small sample sizes resulted in significantly lower collapse stress. There is a factor of 2 between the highest registered stress at 50% strain and the lowest one, which is significant knowing that rate effects from quasi-static to high rate compression may also be of that order [4].

## 4.5 High Rate Compression Tests

For the two very soft polymeric foams being investigated, the time needed to reach relative stress equilibrium was assessed by performing short striker tests. Given the bar configuration, it was impossible to generate a compressive pulse long enough to achieve sample deformation up to densification without causing incident and reflected strain signals to overlap at the incident bar strain gauge location. Short striker tests were performed to replicate the strain rate achieved with the long striker, but only induced a maximum strain of less than 50%. Given that the loading pulse rise time and the specimen configuration were the same, it was judged appropriate to evaluate the existence of stress equilibrium in this manner. To identify the range of strain for which valid data is recorded, the strain rate history of the test was analysed along with the strain history. Figure 4.5a shows an example of a test on a 4 mm thick sample of EPS4.4. Polymeric foams are expected to be highly rate sensitive and as such, only data obtained while the strain rate is relatively constant should be considered [6]. However, there is no established criterion to define the acceptable margin for strain rate during a high-rate test. Hence, this margin was defined as ±5%, which was sufficient for the purpose of this study. For this test on EPS4.4, a constant strain rate of 1,350 ± 5% was attained at 22.2 ms, and lasted until 24.9 ms. At 22.2 ms, the strain in the sample has already reached 0.13, which was significantly advanced in the compression process, and far beyond the elastic loading phase. Figure 4.5b shows the calculated force signals on both sides of the sample, using the 2-wave scheme for the "front face" force and the 1-wave scheme for the "back face" force, along with the identified range for constant strain. Although there was still a measurable difference between the two forces when constant strain was achieved, it is believed that the specimen was in relative equilibrium. It should be noted that both the incident and reflected waves typically contain inherent oscillations which may give rise to uncertainty in the calculation of the front force. These oscillations are subject to dispersion as a function of their frequency content and this can lead to asynchronization of the incident and reflected waves [6], which can cause inaccuracy when the waves are combined in the data analysis. This is emphasized by the fact that even though polymeric bars are used to reduce the impedance mismatch between the sample and bar material, the materials tested in the present study are very soft and still represent a significant mismatch (ratio ≈ $10^2$). Also, it was found that the front force calculation was very sensitive to the determination of the reflected wave starting point, which can be somewhat challenging to identify precisely. With that in mind, it is not believed that the remaining difference between the two force signals past 22 ms is due to a lack of equilibrium, but rather a product of the calculation method.

The dynamic response of both foams around 1,500 $s^{-1}$is shown in Fig. 4.6, where the processed high rate stress–strain curves are compared with the large square sample response in quasi-static compression. Quasi-static tests are typically carried out on large samples, so this comparison is made specifically to highlight how sample size effects can mask rate effects in compressible foams.

It was observed that for the LD70, which had the smallest variation due to sample size, rate effects are still discernable. Although some scatter in the results existed around 1,500 $s^{-1}$, the increase in collapse stress and the decrease in densification strain were noticeable and followed expected trends [1]. The scatter in strain to densification might again be caused by global or local density variations. The measured strain rate effect for LD70 was significant with the stress at 50% strain increasing

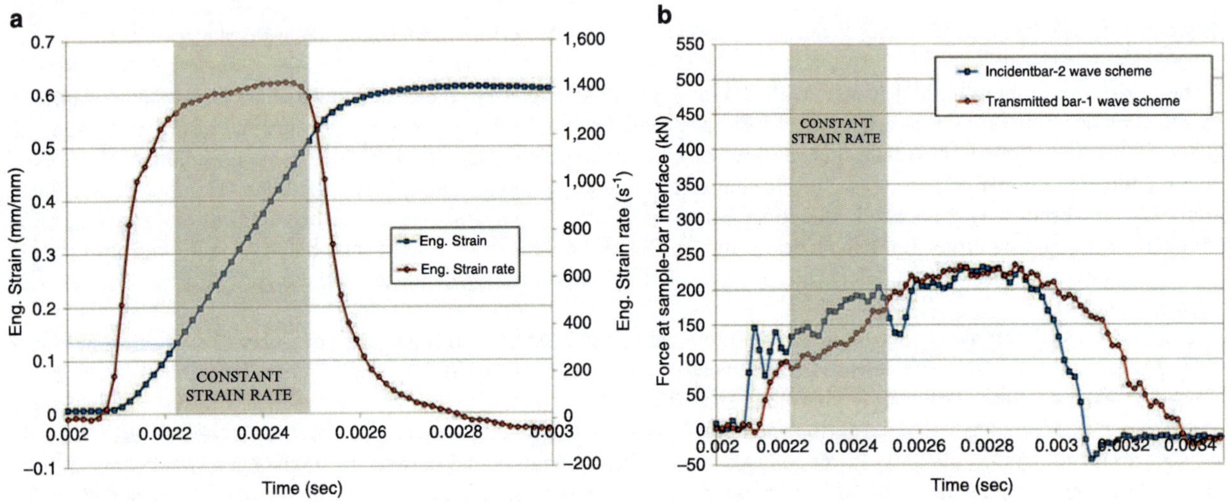


**Fig. 4.5** Validity check during short striker test (**a**) Strain at constant strain rate and (**b**) Force equilibrium in the sample

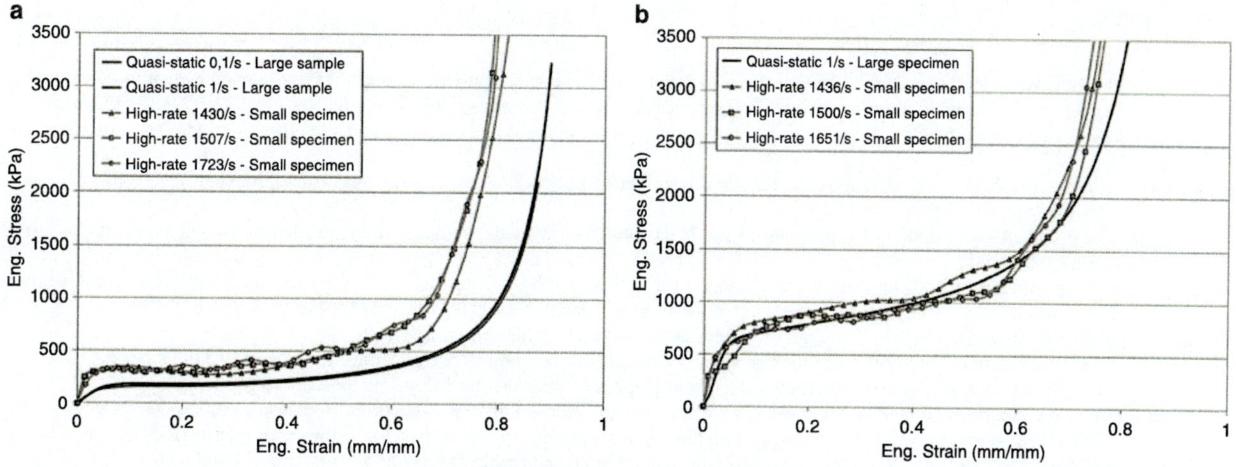


**Fig. 4.6** Obtained high rate stress–strain curves from small size samples of (**a**) LD70 and (**b**) EPS4.4

by a factor of 2. Typically, it is assumed that the rate dependence of polymeric foams scales linearly with the logarithm of strain rate, but it has been observed to become non-linear above $10^3$ $s^{-1}$ [4, 5]. An additional curve obtained with a large sample at a strain rate of 0.1 $s^{-1}$ is plotted in Fig. 4.6a to show that an order of magnitude difference in quasi-static strain rates caused a noticeable, albeit minor strength difference. This difference, even multiplied by 3, would not add up to the rate effects observed between 1 $s^{-1}$and 1,500 $s^{-1}$, confirming the existence of non-linearity.

Sample size effects for EPS4.4 are so significant that they completely mask the rate dependent behaviour of the material. The high rate compression response of the small samples overlaps the quasi-static compression response of the large foam samples. For this foam, the sample size requirement to produce valid high strain rate data leads to the use of an unrepresentative volume of material, masking the true material behaviour under high rates of loading. While sample diameter could be further increased to obtain a larger volume, sample thickness will still be limited by the necessity to achieve stress equilibrium during a Hopkinson bar test. Due to the EPP4.4 grain sizes, which are on the order of a millimetre, and the presence of voids and grain density variation, it is very challenging to obtain representative material data at high strain rates.

## 4.6 Conclusions

Two polymeric foams, LD70 and EPS4.4, of equivalent nominal density but different macro-structure were selected to study the effect of sample size during high rate characterization tests using a polymeric compressive Hopkinson bar. Samples of various sizes were randomly cut from sheets of both foams and the variability of their respective macro-structures was assessed using scanning electron microscopy and density measurements. Quasi-static compression tests were then performed and variability in the stress–strain response of both foams was analysed with regard to sample size and macro-structure. High rate tests were finally performed following sample size constraints imposed by the test method and it was shown that while rate effects could be observed for the LD70, the variability in the small samples masked rate effects for the EPS4.4. Due to the large grain sizes, inter-grain voids and inner grain density variations, the required sample size resulted in a non-representative volume.

The current study used expanded polystyrene foam to highlight the potential importance of sample size effects in high rate testing. However, the observations are valid for any type of soft polymeric foams whose macro-structure may show considerable variation. Expanded polystyrene foams and expanded polypropylene foams are widely used in impact mitigation systems and the knowledge of their high rate behaviour is critical to achieve an optimal design of protective components.

**Ackowledgments** The authors would like to thank Jacques Blais of Defence Research and Development Canada – Valcartier for producing the SEM images and some of the quasi-static compression data presented in this paper. The authors would also like to thank Ed Fournier from Biokinetics and Associates for his role in the organisation of the material testing.

## References


1. Gibson LJ, Ashby MA (1997) Cellular solids: structure and properties, 2nd edn. Cambridge University Press, United Kingdom
2. Zhao H, Gary G (2002) Behavior characterization of polymeric foams over a large range of strain rates. Int J Veh Des 20(1/2):135–145
3. Song B, Chen W, Dou S, Winfree NA, Kang JH (2005) Strain-rate effects on elastic and early cell-collapse response of a polystyrene foam. Int J Impact Eng 31:509–521
4. Ouellet S, Cronin DS, Worswick M (2006) Compressive response of polymeric foams under quasi-static, medium and high strain rate conditions. Polym Test 25:731
5. Bioux R, Viot P, Lataillade J-L (2009) Polypropylene foam behaviour under dynamic loading: strain rate, density and microstructure effects. Int J Impact Eng 36:329–342
6. Gray GT III, Blumenthal W (2000) Split Hopkinson pressure bar testing of soft materials. In: ASM International. Handbook Committee (ed) ASM handbook: mechanical testing and evaluation, vol 8. ASM International, Materials Park, pp 488–496
7. Kolsky H (1949) An investigation of the mechanical properties of materials at high rates of loading. Proc Phys Soc B 62:676–700
8. Chen W, Zhang B, Forrestal MJ (1999) Split Hopkinson bar techniques for low impedance materials. Exp Mech 39:81–85
9. Chen W, Lu F, Zhou B (2000) A quartz-crystal embedded split Hopkinson pressure bar for soft materials. Exp Mech 40:1–6
10. Bacon C (1998) An experimental method for considering dispersion and attenuation in a viscoelastic Hopkinson bar. Exp Mech 38:242–249
11. Salisbury CP (2001) Spectral analysis of wave propagation through a polymeric Hopkinson bar. M.Sc. thesis, University of Waterloo, Canada
12. Doman DA, Cronin DS, Salisbury CP (2006) Characterization of polyurethane rubber at high deformation rates. Exp Mech 46:367–376
13. Van Sligtenhorst C, Cronin DS, Brodland GW (2006) High strain rate compressive properties of bovine muscle tissue found using a split Hopkinson bar apparatus. J Biomech 39:1852–1858
14. Salisbury C, Cronin DS (2009) Mechanical properties of ballistic gelatin at high deformation rates. Exp Mech 49(6):829–840
15. Song B, Chen W (2004) Dynamic stress equilibration in split Hopkinson pressure bar tests on soft materials. Exp Mech 44(3):300–312

# Chapter 5
# Considerations for Numerical Modeling of Dynamic Behavior of Soft Materials

**John C. Fitzpatrick and Michael J. Scheidler**

**Abstract** Numerical simulations of split-Hopkinson pressure bar experiments on soft, nearly-incompressible specimens have been used to validate an "inertial correction" theory for specimens with a bulk-to- shear modulus ratio ($\kappa/\mu$) on the order of $10^4$. Applying the "inertial correction" theory to specimens with $\kappa/\mu$ on the order of $10^6$, we find that changes in stress computations in the simulation upon refining the mesh may be on the same order as the axial deviatoric stress necessary to separate a rate-dependent effect from the inertial material response. Simulations executed with a series of refined meshes were used to calculate nominal stress estimates at the bar-specimen interfaces and determine the most effective meshing and stress estimation techniques. Our results suggest that average cross-sectional average stresses computed from specimen mesh data at the bar-specimen interfaces are less likely to have mesh dependencies due to contact interactions in simulations. Additionally, it is suggested that radial element lengths in the bar mesh should be scaled to approximately 11% of the specimen length, and the element lengths in the specimen mesh scaled to less than 8% of the specimen length; this serves to increase the resolution of stress computations at an element centroid.

## 5.1 Introduction

Modeling and simulation efforts toward understanding warfighter injuries that result from high accelerative loadings on soft biological tissues require characterization and modeling of the mechanical behavior of these tissues at high strain rates. The split-Hopkinson pressure bar (SHPB), or Kolsky compression bar, experiment is commonly used to characterize the strain rate sensitivity of inelastic materials in a state of compressive uniaxial stress for moderate to large strains. The standard specimen shape for SHPB tests is a solid disk, whose length is relatively small compared to its diameter in order to reduce axial inertia effects.

For soft materials, the presence of relatively large radial and hoop stresses due to the high rates of lateral deformation invalidates the conventional analysis of experimental data, which is based on the assumptions of uniform uniaxial stress and uniform biaxial strain throughout the specimen. These inertial effects are more pronounced for nearly-incompressible materials, where the bulk modulus of a specimen, $\kappa$, is several orders of magnitude larger than its shear modulus, $\mu$. Scheidler et al. [1] derived an inertial estimate for the radial and hoop stresses that occur but cannot be measured in a SHPB test; see also Warren and Forrestal [2]. This estimate, together with the measured axial stress, may be used to estimate the deviatoric stress state; consequently, inertial effects can be separated from material strain rate effects. Since the deviatoric stress cannot be measured directly in a SHPB test, numerical simulations provide the best means for determining the validity of these estimates. Two simulations were reported in [1] that validated the approximate "inertial correction" theory for a particular nearly-incompressible material. The specimen material model in this case was calibrated to ballistic gelatin ($\kappa/\mu \approx 10^4$, $\mu = 80$ kPa); for these simulations, the magnitude of the axial deviatoric stress is approximately 30% of the axial stress.

However, for softer and even more nearly-incompressible materials, such as brain tissue ($\kappa/\mu \approx 10^6$, $\mu = 2$ kPa), we observe an approximately-hydrostatic stress state in the specimen for nearly-identical loading conditions, wherein the magnitude of axial deviatoric stress is less than 2% of the axial stress. Refining the mesh to confirm convergence of stresses may change the computed axial stress values by roughly 1–2%, which is of significant concern since mesh-dependent artifacts may "wash out" the axial deviatoric stress estimate for soft materials. As we develop and validate methods to extract

J.C. Fitzpatrick (✉) • M.J. Scheidler
U.S. Army Research Laboratory, Weapons and Materials Research Directorate, Aberdeen Proving Ground, MD 21005-5066, USA

V. Chalivendra et al. (eds.), *Dynamic Behavior of Materials, Volume 1: Proceedings of the 2012 Annual Conference on Experimental and Applied Mechanics*, Conference Proceedings of the Society for Experimental Mechanics Series, DOI 10.1007/978-1-4614-4238-7_5, 

useful data from SHPB tests, it is critical to minimize mesh-dependent effects in numerical simulations of split-Hopkinson Pressure Bar experiments. Here we examine mesh sensitivity of various estimates for the axial stress at the specimen-bar interfaces.

## 5.2 Numerical Simulation

In a SHPB test, a striker bar impacts a pulse-shaping material at a velocity $v_z^{\text{striker}}$ (see Fig. 5.1). As the pulse shaper deforms, a compressive wave is generated in the incident bar that propagates toward the specimen at the specimen-incident bar (S-IB) interface. The incident and transmission bars are selected to be linear elastic materials with a relative stiffness greater than the specimen material. The deformation of the linear elastic bars is detected by the strain gauges, and forces at the specimen-bar interfaces can be inferred from the strain gauge data. Since an impedance mismatch exists between the incident bar and the specimen, part of the compressive wave will propagate into the specimen, and the reminder will reflect as a tensile wave at the S-IB interface. As the motion of the incident bar deforms the specimen, a compressive wave travels through the specimen and into the transmission bar at the specimen-transmission bar (S-TB) interface. While we limit explanation of the SHPB technique to details relevant to the numerical simulation, the reader is referred to articles by Gray [3], Gama et al. [4], and Ramesh [5], and the recent book by Chen and Song [6] for more additional information.

### 5.2.1 *Simulation Geometry and Material Parameters*

A simulation with a simplified SHPB geometry (omitting the striker bar and pulse shaper) was executed in Sierra SolidMechanics (Presto 4.22; Sandia National Laboratory), an explicit finite element code. The incident and transmission bars are modeled as linear elastic, isotropic solids, and the specimen material is modeled as a nonlinear elastic, isotropic solid – a compressible version of the Mooney-Rivlin model. The bulk modulus of water (2.3 GPa) represents the hydrostatic response of a hydrated brain tissue specimen, and has been used in previous simulations to model ballistic gelatin. To reduce computational time, a quarter-geometry was used with appropriate boundary conditions imposed to ensure axisymmetry. The specimen shear modulus (2 kPa) is consistent with quasi-static bovine cerebral data [7]. The material properties used in the simulation are listed in Table 5.1.

### 5.2.2 *Conventional SHPB Loading Pulse*

The rate-dependence of a material is typically calibrated in a constitutive model using test data for various constant strain rates, so the velocity pulse in a SHPB test is shaped such that the specimen will deform at a constant rate. Though the striker and pulse shaper are omitted from the simulations, a shaped pulse is imposed as a velocity boundary condition at the striker bar end of the incident bar. The nominal axial strain rate, $\dot{e}_z$, is related to the specimen length, $L_S$, and respective velocities at the specimen-incident bar (S-IB) and specimen-transmission bar (S-TB) interfaces, $v_z^{[\text{S-IB}]}$ and $v_z^{[\text{S-TB}]}$, by

$$\dot{e}_z = \frac{v_z^{[\text{S-IB}]} - v_z^{[\text{S-TB}]}}{L_S} \approx \frac{v_z^{[\text{S-IB}]}}{L_S}. \tag{5.1}$$

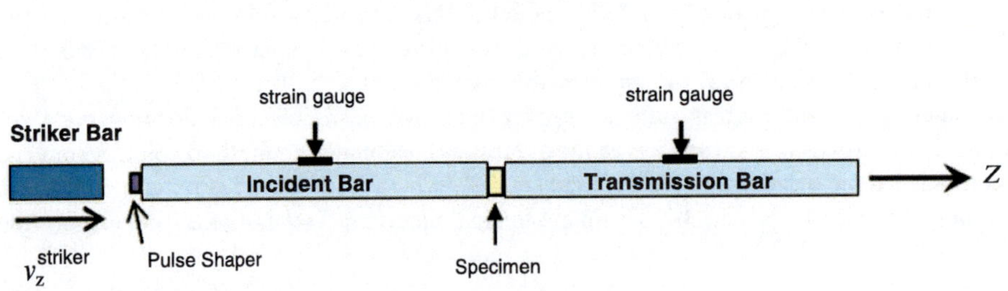


**Fig. 5.1** Geometry of specimen, incident and transmission bars in SHPB test

**Table 5.1** Material properties used in presto simulation

| | $\rho$ | $E$ | $\kappa$ | $\mu$ | $C_{01}$ | $C_{10}$ |
|---|---|---|---|---|---|---|
| Bar | 2,700 kg/m$^3$ | 68.0 GPa | 55.8 GPa | 25.7 GPa | – | – |
| Specimen | 1,000 kg/m$^3$ | – | 2.3 GPa | 2 kPa | 300 | 700 |

The approximation on the right follows from the fact that for soft specimens, the velocity of the transmission bar at the S-TB interface is several orders of magnitude lower than the velocity of the incident bar at the S-IB interface.

In a conventional SHPB test, the nominal axial strain rate of the specimen is ramped smoothly (by means of a pulse shaper) over a time interval [0, $t_1$] to a value $\dot{e}_z^*$, held at that value over the time interval [$t_1$, $t_2$], and then unloaded to zero over the time interval [$t_2$,$t_3$]. This strain-rate history can be expressed as a piecewise function, using a Hermitian smoothing spline for the loading and unloading,

$$\phi(\bar{t}) = 10\bar{t}^3 - 15\bar{t}^4 + 6\bar{t}^5, \quad 0 \le \bar{t} \le 1. \tag{5.2}$$

The Hermitian function $\phi$ has properties such that its initial value is zero ($\phi(0) = 0$), its first and second derivatives at its initial and final values are zero ($\dot{\phi}(0) = \dot{\phi}(1) = \ddot{\phi}(0) = \ddot{\phi}(1) = 0$), and its final value is 1, ($\phi(1) = 1$). Additionally, the integral of $\phi$, $\int_0^1 \phi d\bar{t}$, is evaluated to be 1/2. The strain rate history described above can be expressed with time parameters in a non-dimensional form,

$$\dot{e}_z(t) = \begin{cases} \dot{e}_z^* \phi\left(\frac{t}{t_1}\right) & 0 \le t \le t_1 \\ \dot{e}_z^* & t_1 < t < t_2 \\ \dot{e}_z^*\left[1 - \phi\left(\frac{t-t_2}{t_3-t_2}\right)\right] & t_2 \le t \le t_3 \end{cases} \tag{5.3}$$

Eq. (5.3) can be integrated to obtain the strains at time $t_1$, $t_2$, and $t_3$,

$$\begin{aligned} e_z|_1 &= \int_0^{t_1} \dot{e}_z(t)dt = \frac{1}{2}t_1\dot{e}_z^* \\ e_z|_2 - e_z|_1 &= \int_{t_1}^{t_2} \dot{e}_z(t)dt = \frac{1}{2}(t_2 - t_1)\dot{e}_z^* \\ e_z|_3 - e_z|_2 &= \int_{t_2}^{t_3} \dot{e}_z(t)dt = \frac{1}{2}(t_3 - t_2)\dot{e}_z^*. \end{aligned} \tag{5.4}$$

The times corresponding to these strain values can be obtained from Eq. (5.4),

$$t_1 = \frac{2e_z|_1}{\dot{e}_z^*}, \quad t_2 = t_1 + \frac{2(e_z|_2 - e_z|_1)}{\dot{e}_z^*}, \quad t_3 = t_2 + \frac{2(e_z|_3 - e_z|_2)}{\dot{e}_z^*}. \tag{5.5}$$

Since the impedance of the specimen is much lower than the bar, the velocity of the incident bar at the S-IB interface will be approximately that of a wave reflecting from a free surface, so that $_z^{[S-IB]} \approx 2\upsilon_z^{\text{striker}}$. Then it follows from Eq. (5.1) that

$$\dot{e}_z \approx 2\frac{\upsilon_z^{\text{striker}}}{L_S}. \tag{5.6}$$

Substituting Eq. (5.3) into Eq. (5.6), the striker velocity can be expressed as

$$\upsilon_z^{\text{striker}}(t) = \begin{cases} \frac{1}{2}\dot{e}_z^* L_S \phi\left(\frac{t}{t_1}\right) & 0 \le t \le t_1 \\ \frac{1}{2}\dot{e}_z^* L_S & t_1 < t < t_2 \\ \frac{1}{2}\dot{e}_z^* L_S\left[1 - \phi\left(\frac{t - t_2}{t_3 - t_2}\right)\right] & t_2 \le t \le t_3 \end{cases} \tag{5.7}$$

The velocity boundary condition corresponding to the "conventional" loading pulse is shown in Fig. 5.2. The striker velocity for the simulations are configured for a target strain rate $\dot{e}_z^* = 2,500/\text{s}$ and strains $e_z|_1 = 0.20$, $e_z|_2 = 0.46$, and

**Fig. 5.2** Striker velocity boundary condition for conventional loading pulse

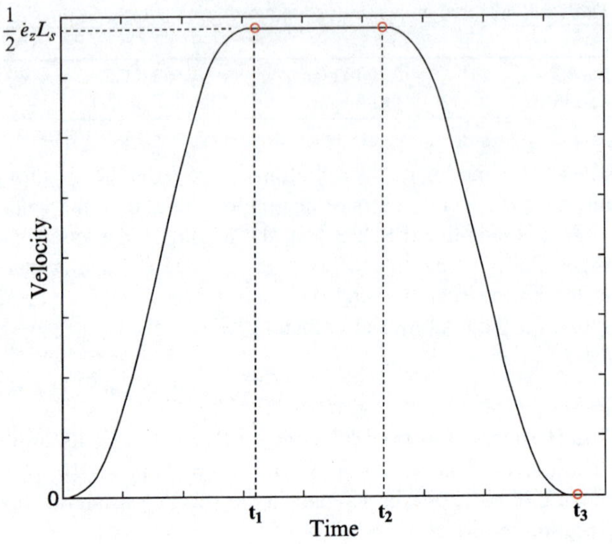


$e_z|_3 = 0.66$, respectively. Time values corresponding to the strains $e_z|_1, e_z|_2$, and $e_z|_3$ are $t_1 = 160\mu s$, $t_2 = 368\ \mu s$, and $t_3 = 528\ \mu s$, respectively. Note that $t = 0$ corresponds to the time at which the loading wave arrives at the S-IB interface.

## 5.3 Mesh Generation

The specimen, incident and transmission bars in the SHPB simulation were generated in CUBIT (Sandia National Laboratory) with 8-noded hexahedral elements. The bar and specimen radii are similar[1] to those in Moy et al. [8]; $R_{\text{IB}} = R_{\text{TB}} = 9.50$ mm, $R_S = 5.48$ mm. The low length-to-diameter ratio (0.114) of the specimen facilitates attainment of dynamic equilibrium by minimizing axial inertia effects.

The axial length of mesh elements, $l_z$, was set at the same value throughout the bar and specimen meshes; this value is determined by the number of axial elements, $n_z$, along the length of the specimen, $L_S$,

$$l_z = \frac{L_S}{n_z}; \quad n_z = 5 + 4(k-1), k = \{1, 2, 3\}. \tag{5.8}$$

The degree of mesh coarseness $k$ is represented by integers 1 (coarse), 2 (fine), and 3 (x-fine). The number of radial elements, $n_r$, in the bar and specimen meshes corresponds to the number of full axial element lengths that may fit within the radii of the bars and specimen, respectively. Mesh element lengths and the corresponding number of elements in the axial and radial directions for the bar and specimen meshes are listed in Table 5.2.

The length of the incident bar is set to 1,800 mm in the "coarse" mesh, but is reduced to a shorter length in the "fine" and "x-fine" meshes due to the computational expense of generating a "fine" mesh. The minimum length of the incident bar necessary to deliver a full loading pulse can be determined by the time duration of the loading pulse ramp and hold period, $t_2$, and the wavespeed of the incident bar (based on its elastic modulus, $E$, and density, $\rho$).

$$L_{\text{IB}} = \left(\frac{t_2}{2}\right)\sqrt{\frac{E}{\rho}}, \tag{5.9}$$

[1] The specimen radius was decreased so the deformed lateral surface of the specimen would not extrude beyond the bar radii at axial nominal strains $e_z$ of $\frac{2}{3}$.

**Table 5.2** Mesh element lengths and number of elements in specimen and bar geometries

| | | Coarse | Fine | Fine/X-fine |
|---|---|---|---|---|
| $l_z$ ($n_z$) | S | 0.250 mm (5) | 0.1389 mm (9) | 0.0962 mm (13) |
| | IB | 0.250 mm (7,200) | 0.1397 mm (4,882) | 0.1397 mm (4,882) |
| | TB | 0.250 mm (760) | 0.1397 mm (1,368) | 0.1397 mm (1,368) |
| $l_r$ ($n_r$) | S | 0.261 mm (21) | 0.1405 mm (39) | 0.0978 mm (56) |
| | IB | 0.250 mm (38) | 0.1397 mm (68) | 0.1397 mm (68) |
| | TB | 0.250 mm (38) | 0.1397 mm (68) | 0.1397 mm (68) |

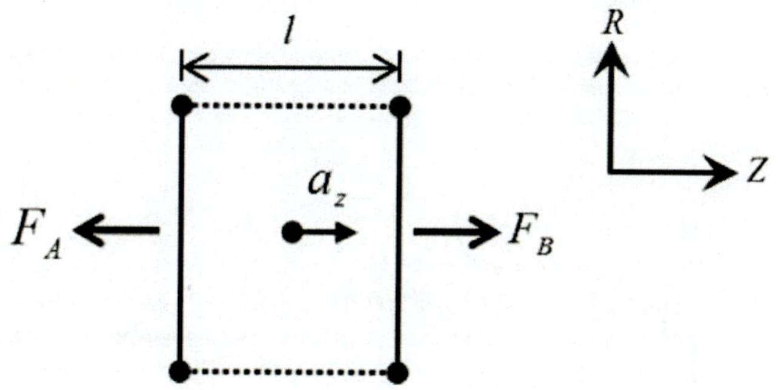


**Fig. 5.3** A positive sign convention is used to relate the accelerations of the element block to the forces on its faces

In the interest of reducing the computational expense of generating meshes, the ramp time duration $t_1$ was used in place of $t_2$ in Eq. (5.9). While a bar length based on $t_1$ may not allow for full loading of the specimen, we choose to focus on results over this interval because the most significant inertial effects occur at the beginning of the loading pulse. To avoid reflected waves in the transmission bar that may also unload the specimen, a "non-reflecting" boundary was imposed at an axial distance of ten bar diameters from the S-TB interface ($L_{TB} = 20R_{TB}$).

### 5.3.1 *Postprocessing: Mesh-Inertia Correction*

The total axial forces and average nominal stresses at the S-IB and S-TB interfaces are measured in a SHPB test [8] and are used to determine the mean axial stress in the specimen [1]. For the 8-noded hexagonal elements in the numerical simulations, stress values are computed at – and valid at – the integration point of a mesh element, which is located at the mesh element's centroid. The faces of the bar and specimen at the S-IB and S-TB interfaces, however, occur at nodal locations, where kinematic quantities such as displacement, velocity, and acceleration are computed. Newton's law implies that for high accelerations, there must be a difference in forces (and therefore stresses) across the half-element axial distance between the node of the element face and the position of the element centroid.

The sign convention used for forces and stresses in the following analysis corresponds to the positive-tensile convention used in Presto, as illustrated in Fig. 5.3. The acceleration $a_z$ on a volume with density $\rho$, bounded by length $l$ and cross-sectional area $A$ can be related to the forces $F_z^m$ and $F_z^{[m+1]}$ on its faces by

$$F_z|_B - F_z|_A = (\rho l A)a_z. \tag{5.10}$$

Since the axial forces $F_z|_A$ and $F_z|_B$ are positive in tension, whereas the forces in a SHPB test will typically be compressive, $-F_z|_A$ and $-F_z|_B$ should generally be positive, so we rewrite Eq. (5.10) in the equivalent form

$$-F_z|_B = -F_z|_A - (\rho l A)a_z. \tag{5.11}$$

The total uncorrected axial force estimate $F_z^{\text{uncorr}}$ is calculated for an axial cross-section ($\hat{\text{R}}$-$\hat{\Theta}$ plane) from the uncorrected axial stress values that are computed at the mesh element's centroid (see Fig. 5.4a). Then, $F_z^{\text{uncorr}}$ is corrected for forces due

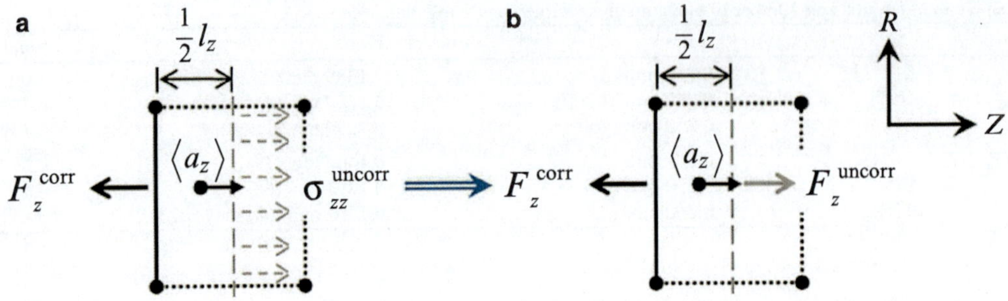


**Fig. 5.4** (a) The uncorrected axial stress $\sigma_{zz}^{\text{uncorr}}$ at the element centroid is multiplied by the element area $A|_{[n]}$ between an interval of the radial nodes $[n, n+1]$ to produce a force, $F_z^{\text{uncorr}}$, that is summed over the radial elements (b)

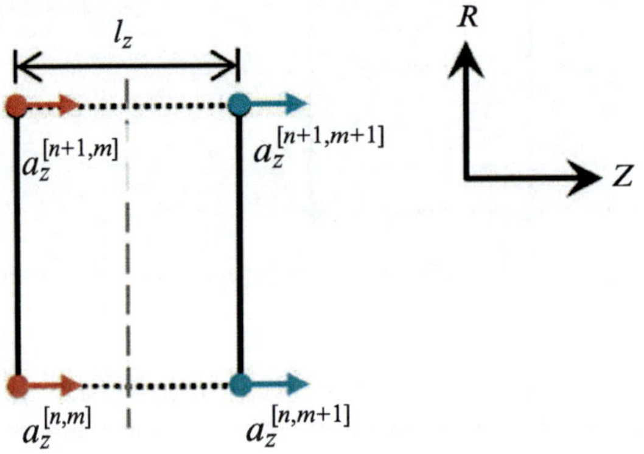


**Fig. 5.5** Average acceleration $\langle a_z \rangle$ computed from axial nodes at the bar-specimen interface $[m]$ and away from the bar-specimen interface $[m+1]$, and the adjacent radial $[n, n+1]$ nodes

to inertia that results from axial acceleration, $\langle a_z \rangle$, between the axial location of the element centroid and the nodal location corresponding to the face of the mesh block (Fig. 5.4b).

The total force from the contributions of uncorrected axial true stress, $\sigma_{zz}^{\text{uncorr}}$, are summed over $n_r$ radial elements in the deformed configuration using the midpoint rule,

$$-F_z^{\text{uncorr}} = \sum_{n=1}^{n} \left( -\sigma_{zz}^{\text{uncorr}}|_{[n]} \right) A_{[n]}, \tag{5.12}$$

where $A_{[n]}$ is the deformed area of each element between the radial nodes $[n, n+1]$, calculated from the radial position in the reference configuration, $R$, and its radial displacement $u_r$,

$$A_{[n]} = \pi \left[ \left( R_{[n+1]} + u_r|_{[n+1]} \right)^2 - \left( R_{[n]} + u_r|_{[n]} \right)^2 \right]. \tag{5.13}$$

Inertial force corrections relate $F_z^{\text{uncorr}}$ to the corrected average axial force, $F_z^{\text{corr}}$, as shown in Fig. 5.4. Assuming a constant mesh density $\rho$, the rightmost term in Eq. (5.11), $\rho lAa_z$, is calculated from the *nodal* quantities of displacement and acceleration determined from adjacent radial $[n, n+1]$ and axial $[m, m+1]$ positions (Fig. 5.5). Note that in Fig. 5.5 the nodes positioned at the axial coordinate $[m]$ represent the bar-specimen interface; the nodes positioned at the axial coordinate $[m+1]$ adjacent nodes within the respective geometry.

To account for the fact that the average acceleration used for the mesh-inertia correction, $\langle a_z \rangle$, is located halfway between the location of the element centroid and the element face (compare Fig. 5.4b with Fig. 5.5), a weight is applied to scale the average accelerations at $m$ and $(m+1)$ nodes,

$$\langle a_z \rangle = \frac{1}{4} \left[ 3 \left( \frac{a_z|_{[n,m]} + a_z|_{[n+1,m]}}{2} \right) + \left( \frac{a_z|_{[n,m+1]} + a_z|_{[n+1,m+1]}}{2} \right) \right]. \tag{5.14}$$

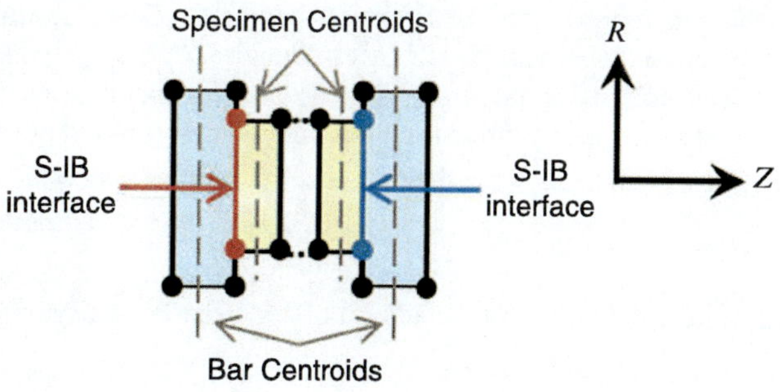


**Fig. 5.6** The mesh element centroid locations for bar and specimen meshes, at S-IB and S-TB interfaces

The average of acceleration values at adjacent radial nodes $[n,n + 1]$ are used in the form of the trapezoidal rule below.

Large axial and radial displacements are typical for a specimen mesh modeled as a soft, nearly-incompressible solid. The deformed axial element length, $l$, is calculated in terms of the axial reference position, $Z$, and displacement in the deformed configuration, $u_z$, for adjacent radial nodes $[n,n + 1]$,

$$l = \left(Z_{[n+1]} + u_z|_{[n+1]}\right) - \left(Z_{[n]} + u_z|_{[n]}\right). \tag{5.15}$$

We substitute $A_{[n]}$ from Eq. (5.13) for $A$, $\langle a_z \rangle$ from Eq. (5.14) for $a_z$, and $l$ from Eq. (5.15) into the rightmost term of Eq. (5.11) to calculate the total force due to accelerations within the half mesh-element volume between the centroid and S-IB face,

$$F_z^{\text{inertia}} = \sum_{n=1}^{n+1} \rho \left(\frac{1}{2} l\right) A_{[n]} \langle a_z \rangle. \tag{5.16}$$

The total axial force of the cross-section is corrected by substituting the uncorrected axial force $-F_z^{\text{uncorr}}$ and the inertia force $F_z^{\text{inertia}}$ into Eq. (5.11) as $F_z|_A$ and $F_z|_B$; the relative positions of the element centroids and bar-specimen interfaces (Fig. 5.6) determine how $-F_z^{\text{uncorr}}$ and $F_z^{\text{inertia}}$ are substituted. The corrected forces from bar mesh data follow as

$$\begin{aligned} -F_z^{\text{corr}}|_{[\text{S-IB}]} &= -F_z^{\text{uncorr}}|_{[\text{S-IB}]} - F_z^{\text{inertia}}|_{[\text{S-IB}]} \\ -F_z^{\text{corr}}|_{[\text{S-TB}]} &= -F_z^{\text{uncorr}}|_{[\text{S-TB}]} + F_z^{\text{inertia}}|_{[\text{S-TB}]}, \end{aligned} \tag{5.17}$$

and similarly, specimen mesh data has the form

$$\begin{aligned} -F_z^{\text{corr}}|_{[\text{S-IB}]}^{\text{spatial}} &= -F_z^{\text{uncorr}}|_{[\text{S-IB}]}^{\text{spatial}} - F_z^{\text{inertia}}|_{[\text{S-IB}]} \\ -F_z^{\text{corr}}|_{[\text{S-TB}]}^{\text{spatial}} &= -F_z^{\text{uncorr}}|_{[\text{S-TB}]}^{\text{spatial}} + F_z^{\text{inertia}}|_{[\text{S-TB}]}. \end{aligned} \tag{5.18}$$

The **spatial** axial force approximation obtained from specimen mesh data (i.e. axial true stress summed over a deformed axial cross sectional area) is distinguished from a referential average cross-sectional approximation (i.e. axial true stress summed over an axial cross-sectional area in the reference configuration).

### 5.3.2 *Postprocessing: Average Nominal Axial Stress*

The cross-sectional average nominal axial stress *at the specimen faces*, computed from the cross-sectional forces is

$$-\langle \Sigma_{zZ}^{\text{corr}} \rangle = \frac{-F_z^{\text{corr}}}{\pi (R_0)^2} \tag{5.19}$$

where $R_0$ represents the undeformed specimen radius. Substituting $F_z^{\mathrm{corr}}|^{\mathrm{spatial}}$ for $F_z^{\mathrm{corr}}$ in Eq. (5.19) yields $\langle\Sigma_{zZ}^{\mathrm{corr}}\rangle^{\mathrm{Spatial}}$ for the respective [S-IB] and [S-TB] faces; likewise, $F_z^{\mathrm{corr}}|_{[\mathrm{S-IB}]}$ yields $\langle\Sigma_{zZ}^{\mathrm{corr}}\rangle_{[\mathrm{S-IB}]}$ and $F_z^{\mathrm{corr}}|_{[\mathrm{S-TB}]}$ yields $\langle\Sigma_{zZ}^{\mathrm{corr}}\rangle_{[\mathrm{S-TB}]}$.

The nominal axial stress may also be computed using the first Piola-Kirchoff stress tensor, $\Sigma$, calculated from the Jacobian $J$, Cauchy stress tensor $\sigma$, and the deformation gradient $\mathbf{F}$,

$$\Sigma = J\sigma\mathbf{F}^{-\mathrm{T}} = \sigma\left\{(\det\mathbf{F})\left(\mathbf{F}^{-1}\right)^{\mathrm{T}}\right\} = \sigma(\mathrm{adj}\mathbf{F})^{\mathrm{T}}, \tag{5.20}$$

using the definition for the adjoint of a matrix. For axisymmetric deformations, $\mathbf{F}$ has the form

$$[\mathbf{F}] = \begin{bmatrix} \frac{\partial r}{\partial R} & 0 & \frac{\partial z}{\partial R} \\ 0 & \frac{r}{R} & 0 \\ \frac{\partial r}{\partial Z} & 0 & \frac{\partial z}{\partial Z} \end{bmatrix}, \tag{5.21}$$

$J$ is given by

$$J = \frac{r}{R}\left(\frac{\partial r}{\partial R}\frac{\partial z}{\partial Z} - \frac{\partial z}{\partial R}\frac{\partial r}{\partial Z}\right), \tag{5.22}$$

and

$$(\mathrm{adj}[\mathbf{F}])^{\mathrm{T}} = \begin{bmatrix} \frac{r}{R}\frac{\partial z}{\partial Z} & 0 & -\frac{r}{R}\frac{\partial z}{\partial R} \\ 0 & \frac{\partial r}{\partial R}\frac{\partial z}{\partial Z} - \frac{\partial z}{\partial R}\frac{\partial r}{\partial Z} & 0 \\ -\frac{r}{R}\frac{\partial r}{\partial Z} & 0 & \frac{r}{R}\frac{\partial r}{\partial R} \end{bmatrix}. \tag{5.23}$$

Eq. (5.20) can be combined with Eq. (5.23),

$$\begin{bmatrix} \Sigma_{rR} & 0 & \Sigma_{rZ} \\ 0 & \Sigma_{\Theta\Theta} & 0 \\ \Sigma_{zR} & 0 & \Sigma_{zZ} \end{bmatrix} = \begin{bmatrix} \sigma_{rr} & 0 & \sigma_{rz} \\ 0 & \sigma_{\theta\theta} & 0 \\ \sigma_{rz} & 0 & \sigma_{zz} \end{bmatrix} \begin{bmatrix} \frac{r}{R}\frac{\partial z}{\partial Z} & 0 & -\frac{r}{R}\frac{\partial z}{\partial R} \\ 0 & \frac{\partial r}{\partial R}\frac{\partial z}{\partial Z} - \frac{\partial z}{\partial R}\frac{\partial r}{\partial Z} & 0 \\ -\frac{r}{R}\frac{\partial r}{\partial Z} & 0 & \frac{r}{R}\frac{\partial r}{\partial R} \end{bmatrix}; \tag{5.24}$$

in particular, the axial component of Piola-Kirchoff stress, $\Sigma_{zZ}$, is given by

$$\Sigma_{zZ} = \frac{r}{R}\left(\frac{\partial r}{\partial R}\sigma_{zz} - \frac{\partial z}{\partial R}\sigma_{rz}\right) = F_{\theta\Theta}(F_{rR}\sigma_{zz} - F_{zR}\sigma_{rz}). \tag{5.25}$$

Since there is no option to directly output the components of the deformation gradient from Presto, $\mathbf{F}$ may be computed from the mesh element stretch tensor, $\mathbf{V}$, and rotation tensor, $\mathbf{R}$, by means of polar decomposition, $\mathbf{F} = \mathbf{VR}$. The form of Eq. $(5.25)_2$, denoted as $\Sigma_{zZ}^{\mathrm{PK}}$, is used to sum Piola-Kirchoff nominal stresses over a referential cross-sectional area using the midpoint rule, resulting in the total force $F_z^{\mathrm{PK}}$,

$$-F_z^{\mathrm{PK}} = \sum_{n=1}^{n}\left(-\Sigma_{zZ}^{\mathrm{PK}}|_{[n]}\right)\left[\pi\left[\left(R|_{[n+1]}\right)^2 - \left(R|_{[n]}\right)^2\right]\right]. \tag{5.26}$$

The area-averaged nominal stress $\langle\Sigma_{zZ}^{\mathrm{PK}}\rangle$ is calculated by substituting $F_z^{\mathrm{PK}}$ into Eq. (5.19). Values obtained for the S-IB and S-TB interfaces are denoted by the subscripts [S-IB] and [S-TB].

## 5.4 Mesh Refinement Results

As the bar and specimen meshes are refined, an expected convergence in the axial stress is observed. The largest discrepancies between stresses are observed at the peak of the "inertial spike," as Figs. 5.7–5.9 show in detail. The labels in these figures correspond to the plotted nominal stress values at $e_z = 0.06$, which are summarized in Table 5.3.

Perhaps the most noticeable trend in the plots of Figs. 5.7–5.9 is the convergence of the nominal stresses at the S-IB interface. While the relative error between $\langle\Sigma_{zZ}^{\text{corr}}\rangle_{[\text{S-IB}]}$, $\langle\Sigma_{zZ}^{\text{corr}}\rangle_{[\text{S-IB}]}^{\text{spatial}}$, and $\langle\Sigma_{zZ}^{\text{PK}}\rangle_{[\text{S-IB}]}$ incrementally decreases during the progression of mesh refinement, the values calculated from the Piola-Kirchoff stress estimate $\langle\Sigma_{zZ}^{\text{PK}}\rangle_{[\text{S-IB}]}$ do not appear to

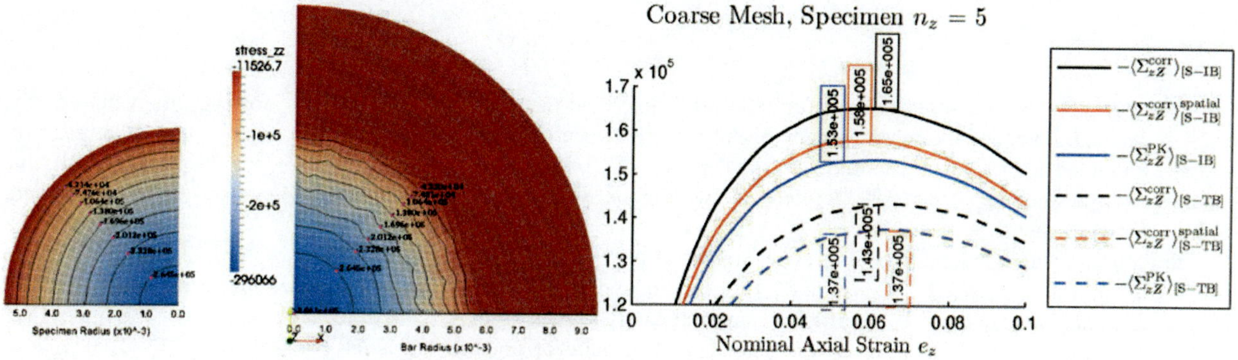


**Fig. 5.7** Shaded contour plots of uncorrected axial stress, $\sigma_{zz}^{\text{uncorr}}$, in coarse mesh, and the computed forces from $\sigma_{zz}^{\text{uncorr}}$

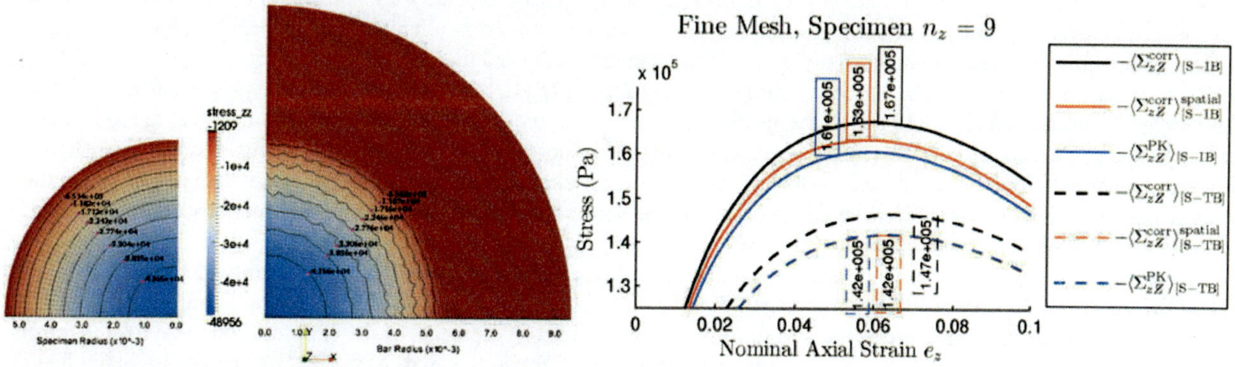


**Fig. 5.8** Shaded contour plots of uncorrected axial stress, $\sigma_{zz}^{\text{uncorr}}$, in fine mesh, and the computed forces from $\sigma_{zz}^{\text{uncorr}}$

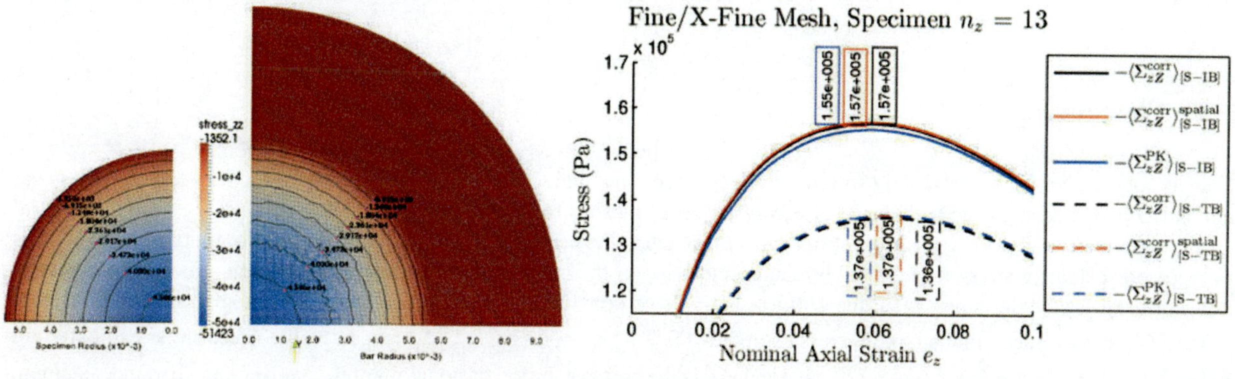


**Fig. 5.9** Shaded contour plots of uncorrected axial stress, $\sigma_{zz}^{uncorr}$, in fine/x-fine mesh, and the computed forces from $\sigma_{zz}^{uncorr}$

**Table 5.3** Cross-sectional Average Nominal Stresses at S-IB Interface

| | Coarse | Fine | Fine/X-Fine |
|---|---|---|---|
| $\langle \Sigma_{zZ}^{\text{spatial}} \rangle_{[\text{S-IB}]}^{\text{corr}}$ | 1.58 MPa | 1.63 MPa | 1.57 MPa |
| $\langle \Sigma_{zZ}^{\text{corr}} \rangle_{[\text{S-IB}]}^{\text{corr}}$ | 1.65 MPa | 1.67 MPa | 1.57 MPa |
| $\langle \Sigma_{zZ}^{\text{PK}} \rangle_{[\text{S-IB}]}^{\text{corr}}$ | 1.53 MPa | 1.61 MPa | 1.55 MPa |

completely converge to the other two estimates. In theory, this seems incorrect for stresses computed in elements adjacent to the S-IB interface, where it is assumed that $\frac{\partial z}{\partial R} = \sigma_{rz} = 0$ due to the frictionless boundary condition imposed. If $\frac{\partial z}{\partial R} = \sigma_{rz} = 0$, then Eq. (5.25) simplifies to

$$\Sigma_{zZ}^{\text{PK}}\big|_{\text{frictionless}} = \frac{\sigma_{zz}}{(1 - e_z)} \tag{5.27}$$

for a homogeneous, volume-preserving deformation. It is expected that $\Sigma_{zZ}^{\text{PK}}\big|_{\text{frictionless}} = \Sigma_{zZ}^{\text{spatial}}$. However, the values of $\frac{\partial z}{\partial R}$ and $\sigma_{rz}$ are negligible, and the reason for the discrepancy between $\langle \Sigma_{zZ}^{\text{PK}} \rangle$ and $\langle \Sigma_{zZ}^{\text{spatial}} \rangle$ is not exactly clear. The result of stresses computed at the element centroid, an axial distance of $[L_S/(2n_z)]$ from the specimen face, may be partly at fault; the relative error between $\langle \Sigma_{zZ}^{\text{PK}} \rangle$ and $\langle \Sigma_{zZ}^{\text{spatial}} \rangle$ decrease with higher values of $n_z$, so more accurate stresses may be computed within smaller elements.

The variation among nominal axial stress estimates at the S-IB interface for the mesh cases considered ($\langle \Sigma_{zZ}^{\text{corr}} \rangle_{[\text{S-IB}]}^{\text{spatial}}$: 3–4%; $\langle \Sigma_{zZ}^{\text{corr}} \rangle_{[\text{S-IB}]}$: 1–6.5%; $\langle \Sigma_{zZ}^{\text{PK}} \rangle_{[\text{S-IB}]}$: 5–6%) appears to be due to contact-dependent stress computations in Presto. The bar and specimen faces at the S-IB interface are approximately to scale in Figs. 5.7–5.9 for the coarse, fine, and fine/x-fine mesh cases, respectively. The surfaces are shaded by the value of uncorrected axial true stress, $\sigma_{zz}^{\text{uncorr}}$. The highest compressive stresses are located at the interior of the bar and specimen ($R = 0$). Note that shaded values of axial stress ($\sigma_{zz}^{\text{uncorr}}$) on the free surface of the incident bar face ($R > 5.48$ mm) actually corresponds to axial stress values computed at the centroid of the mesh element adjacent to the bar face.

Grids of mesh elements are overlaid on the specimen and incident bar cross-sections, and contour lines indicate gradients in the axial stress. The contours of each specimen mesh display a circumferential uniformity expected for an axisymmetric deformation, however the computed stresses at the IB faces in the coarse and fine meshes are less circumferentially uniform, particularly at bar radius values higher than half the specimen radius ($R_S/2 = 2.74$ mm). In the fine/x-fine mesh, stresses are less circumferentially uniform within half the specimen radius, $R_S/2$, but more circumferentially uniform at radial locations past this mark. This suggests sensitivity between contacting nodes located outside the finer "rhombus-shaped" interior mesh area of the bar, where the element sizes of the specimen mesh are slightly larger than the bar mesh elements to which they would map. The stresses computed within these bar mesh elements are consequently less circumferentially uniform, which is reflected in the average stresses computed from the total force, $\langle \Sigma_{zZ}^{\text{corr}} \rangle_{\text{S-IB}}$.

In contrast to the coarse and fine meshes, for the fine/x-fine case there is circumferential uniformity of computed stresses in the bar mesh at radial locations *greater* than half the specimen radius, $R_S/2$, and less circumferential uniformity among its interior elements. Despite circumferential non-uniformities in the interior elements of the bar mesh, minimal relative errors exist between stresses computed from bar and specimen mesh data ($\langle \Sigma_{zZ}^{\text{corr}} \rangle_{[\text{S-IB}]}$ and $\langle \Sigma_{zZ}^{\text{corr}} \rangle_{[\text{S-IB}]}^{\text{spatial}}$, respectively) in the fine/x-fine mesh case. This can be explained partly because circumferential non-uniformities in the outer portion of the bar mesh ($R_{\text{IB}} > R_S/2$) for the fine/x-fine case are not as severe as the coarse or fine mesh cases; the interior portion of the mesh where these circumferential non-uniformities exist also hold less weight in the total force calculation, $F_z^{\text{corr}}\big|_{[\text{S-IB}]}$.

## 5.5 Conclusions

The computed stresses for extremely soft, nearly-incompressible materials in a SHPB simulation are very sensitive to nodal contact, which was explored with successive refinement of meshes for the bar and specimen. To closely estimate the axial forces at the S-IB and S-TB interfaces, stress data computed at the centroid of the mesh elements adjacent to these interfaces must be corrected for stresses generated by the acceleration of the mesh volume between these points. The circumferential non-uniformity of axial stress resulting from sensitivity of contact at nodes on the bar and specimen mesh surfaces is reflected in the average nominal axial stress estimates $\langle \Sigma_{zZ}^{\text{corr}} \rangle_{[\text{S-IB}]}$ and $\langle \Sigma_{zZ}^{\text{corr}} \rangle_{[\text{S-IB}]}^{\text{spatial}}$, respectively. These stress estimates are in best agreement in the fine/x-fine mesh, where there is the best resolution of nodal mapping between the bar and specimen at radial coordinates in the bar beyond $R_S/2$.

While comparison of $\langle\Sigma_{zZ}^{\text{corr}}\rangle_{[S-IB]}^{\text{spatial}}$ and $\langle\Sigma_{zZ}^{\text{corr}}\rangle_{[S-IB]}$ illustrate the improved agreement between nominal stress estimates upon mesh refinement in the radial direction, the Piola-Kirchoff nominal stress estimate $\langle\Sigma_{zZ}^{\text{PK}}\rangle$ demonstrates the importance of refining the axial mesh length. Even in a fine/x-fine mesh, $\langle\Sigma_{zZ}^{\text{PK}}\rangle_{[S-IB]}$ differs from $\langle\Sigma_{zZ}^{\text{corr}}\rangle_{[S-IB]}^{\text{spatial}}$ and $\langle\Sigma_{zZ}^{\text{corr}}\rangle_{[S-IB]}$ at the peak of the inertial spike by 1.2%.

These comparisons of stress estimates for several grades of refined meshes encourage the use of fine/x-fine meshes in future simulations and the application of $\langle\Sigma_{zZ}^{\text{corr}}\rangle_{[S-IB]}^{\text{spatial}}$ to sensitive calculations of $\bar{s}_{zz}$. As we infer from the scale of the fine/x-fine mesh, a suitable radial element length $l_r$ in the bar mesh that is approximately 11% of the specimen length $L_S$, and a specimen mesh element length $l_z$ that is less than 8% of the specimen length $L_S$.

## References


1. Scheidler MJ, Fitzpatrick J, Kraft R (2011) Optimal pulse shapes for SHPB tests on soft materials. In: Proust T (ed) Dynamic behavior of materials. Conference proceedings of the society for experimental mechanics series 99, vol 1. The Society for Experimental Mechanics Inc/Springer, New York, pp 259–268. doi: 10.1007/978-1-4614-0216-9_37
2. Warren TL, Forrestal MJ (2010) Comments on the effect of radial inertia in the Kolsky bar test for an incompressible material. Exp Mech 50:1253–1255
3. Gray GT III (2000) Classic Split-Hopkinson pressure bar testing. In Kuhn H, Medlin D (eds) Mechanical testing and evaluation. ASM handbook, vol 8. American Society for Metals, Materials Park, pp 462–476
4. Gama BG, Lopatnikov SL, Gillespie JW Jr (2004) Hopkinson bar experimental technique: a critical review. Appl Mech Rev 57:223–250
5. Ramesh KT (2009) High strain rate and impact experiments. In: Sharpe WN (ed) Springer handbook of experimental solid mechanics, chap. 33. Springer, New York, pp 1–30
6. Chen W, Song B (2011) Split hopkinson (Kolsky) bar. Springer, New York
7. Pervin F, Chen WW (2009) Dynamic mechanical response of bovine gray matter and white matter brain tissues under compression. J Biomech 42(6):731–735
8. Moy P, Weerasooriya T, Juliano TF, VanLandingham MR, Chen W (2006) Dynamic response of an alternative tissue simulant, physically associating gels (PAG). In: Proceedings of the 2006 SEM annual conference, Saint Louis

# Chapter 6
# Identification of the Heterogeneous Elasto-plastic Behaviour of FSW Welds at High Strain Rates

**G. Le Louëdec, Fabrice Pierron, M.A. Sutton, and A.P. Reynolds**

**Abstract** Friction stir welding introduces spatial variations in local thermo-mechanical history, resulting in heterogeneity in both the local grain structure and elasto-plastic behavior. Such spatial heterogeneities make it difficult to extract the entire distribution of local material behavior. This information is required for effective and accurate simulations of structural response (e.g., localized yielding, fracture initiation, failure) under both quasi-static and dynamic loading conditions. At higher strain rates, the problem is even more challenging because of the need to use ultra-high speed (UHS) cameras.

The present work is a first attempt at identifying local elasto-plastic parameters within an aluminum friction stir weld that has undergone nominally in-plane dynamic loading. An IMACON200 UHS camera was used to acquire speckle images which were analyzed using two-dimensional digital image correlation to obtain full-field displacement, strain, strain rate and acceleration maps. The Virtual Fields Method is developed for dynamic loading conditions and coupled with the measurement data to identify a profile of yield stress along the weld. Results from our studies indicate that the impact force does not need to be measured, provided that the accelerations are measured within a volume to act as a form of "distributed load sensor"

**Keywords** Friction stir welding • FSW • Virtual fields method • Split hopkinson bar • SPHB • Digital image correlation • DIC

## 6.1 Introduction

Since its invention in 1991, the friction Stir Welding (FSW) process has allowed the use of large aluminium structures in various fields due to the high resistance of the welds produced. In various fields such as automotive and aeronautics these welds hold an important place. However, the high strain-rate mechanical properties used in numerical simulations are still estimates. Indeed, experimental set-ups used for high-speed measurements such as the Split-Hopkinson Pressure Bar (SHPB) are not suitable for heterogeneous materials. The Virtual Fields Method was introduced by Grédiac [1] in order to solve inverse problems in the materials' constitutive parameter identification from full-field measurements. It has already been successfully applied to the identification of constitutive parameters of heterogeneous materials in quasi-static experiments [2]. This work focuses on the identification of elasto-plastic properties of AL5456 FSW welds at high strain rate.

G. Le Louëdec
Department of Mechanical Engineering, Center for Mechanics, Material and NDE, University of South Carolina, 300 Main Street, Columbia, SC 29208, USA

Laboratoire de Mécanique et Procédés de Fabrication, Arts et Métiers ParisTech, rue Saint Dominique, BP 508, 51006 Châlons en Champagne, France

F. Pierron (✉) • A.P. Reynolds
Laboratoire de Mécanique et Procédés de Fabrication, Arts et Métiers ParisTech, rue Saint Dominique, BP 508, 51006 Châlons en Champagne, France
e-mail: fabrice.pierron@ensam.fr

M.A. Sutton
Department of Mechanical Engineering, Center for Mechanics, Material and NDE, University of South Carolina, 300 Main Street, Columbia, SC 29208, USA

V. Chalivendra et al. (eds.), *Dynamic Behavior of Materials, Volume 1: Proceedings of the 2012 Annual Conference on Experimental and Applied Mechanics*, Conference Proceedings of the Society for Experimental Mechanics Series, DOI 10.1007/978-1-4614-4238-7_6, 

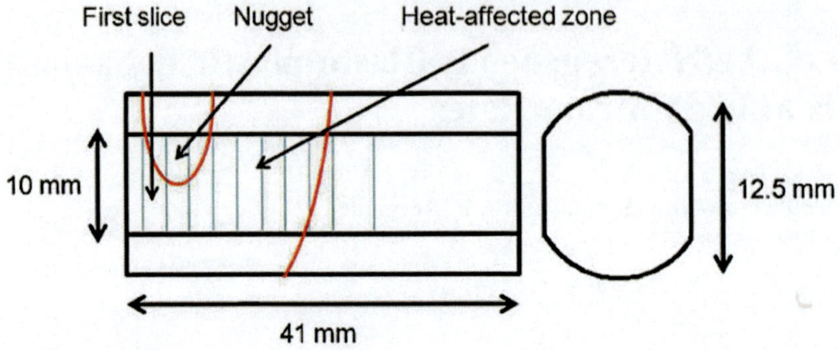


**Fig. 6.1** Test specimen, with field of view represented by the area containing the vertical lines where the elastic–plastic parameters are identified

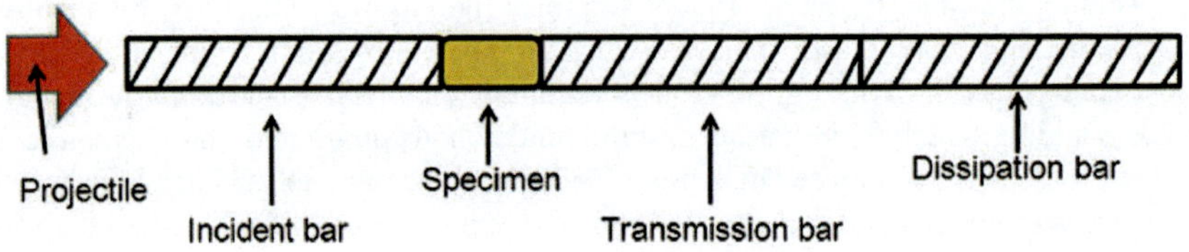


**Fig. 6.2** Diagram of the SHPB test

## 6.2 Experiment and Specimen

FSW was performed on an AL5456 12.7 mm thick plate specimen, at a rotational speed of 480 rpm and a welding speed of 3.4 mm/s. Transverse compressive specimens have been extracted by cutting across the weld. These are 41 mm long cylinders with a diameter of 12.7 mm. Flat surfaces have been machined on each side of the specimen to allow the use of 2D digital image correlation [3]. As shown in Fig. 6.1, the area of interest is mostly focused on the retreating side of the weld. The SHPB experiment (Fig. 6.2) was performed to obtain a global maximum strain rate of 1,500 $s^{-1}$. The specimen was impacted on the nugget side. The camera used for the image recording is a DRS Imacon 200 (10 bit 1,360 × 1,024 array of square pixels) with a 200 mm Nikon lens at a speed of 250,000 fps. It results in an area of interest of 25 × 10 mm.

## 6.3 Post-processing

Flat field correction was applied to the images, in order to reduce the noise between the different sensors of the camera that is caused by the technology of the camera (light amplifiers) [4]. Then images were processed using 2D-DIC using the software VIC-2D© with a 55 pixels subset and a step of 20 pixels. From the displacement fields, VIC-2D© was used to calculate strain fields using a least square quadratic fit to each set of 5 × 5 measurement points. Due to the complex evolution of the displacement fields throughout the weld zone, gradients determined by extrapolation beyond the data field were not used in this study. Acceleration was calculated by time derivation of the displacement field. Both strain fields and acceleration fields have been smoothed using an iterative least-square smoothing convolution (Savitzky-Gray) method [5] with a second order polynomial over a moving window containing seven data points (Figs. 6.3 and 6.4).

## 6.4 The Virtual Fields Method

The virtual fields method [1] makes use of the principle of virtual work. Due to the non-linearity of the stress–strain relationship in plasticity, the stress field is written as a non-linear function of the strain field which is dependent on the mechanical parameters to identify. To solve this problem a cost function dependent on the unknown parameters is built up and minimized [6]. Here special virtual fields are used in order to nullify the contribution of the load. Thus only the strain

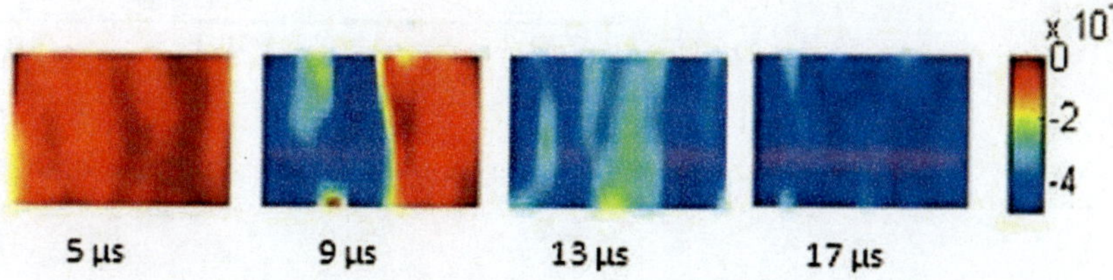


**Fig. 6.3** Evolution of the smoothed longitudinal strain field over time

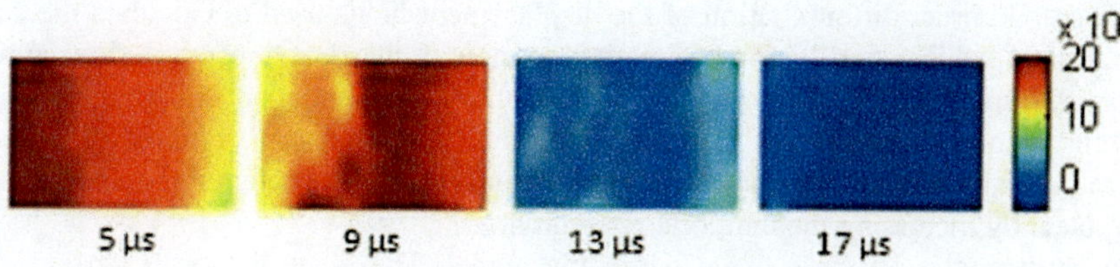


**Fig. 6.4** Evolution of the smoothed longitudinal acceleration field (in m.s$^{-2}$) over time

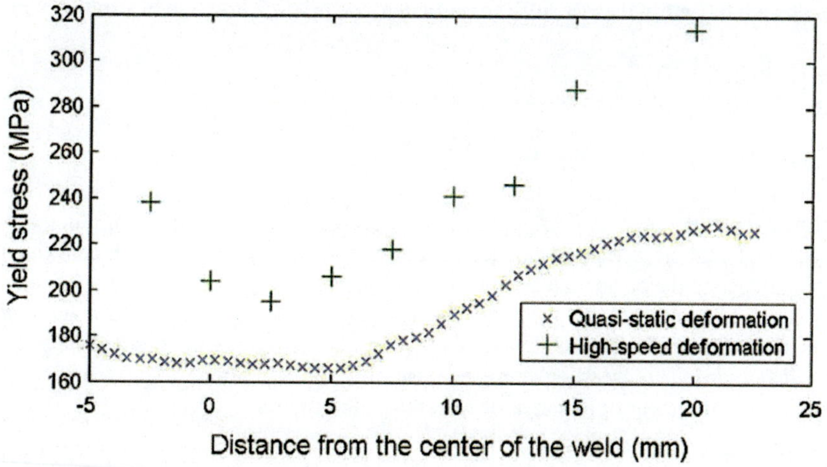


**Fig. 6.5** Identification of the yield stress along the weld

fields and acceleration fields are needed to proceed with the identification. This is a very recent and innovative approach that was already validated in [7] but in elasticity only. This is the first time that such a procedure is applied in elasto-plasticity, and even more challenging, in heterogeneous elasto-plasticity. It should be noted also that though the system to identify is non-linear, there is no need to iterate direct finite element solutions, leading to a very computationally efficient solution compared to finite element model updating.

Table 6.1 Elastic parameters identified by the VFM

| | Reference | Identified |
|---|---|---|
| E (GPa) | 70.0 | 15.0 |
| ν | 0.33 | 0.36 |

## 6.5 Results

It has been shown by Sutton and al. [2] that the elastic properties remain constant through the weld. Thus the identification of Young's modulus and Poisson's ratio is performed on the whole specimen in the early stage (2 first images) of the experiment. Results are shown in Table 6.1. While the identification of the Poisson's ratio is within the expected range, the identified Young's modulus is much lower than expected. This could be explained by the fact that this method makes use of the acceleration. The second order differentiation of the displacement fields used to calculate the acceleration tends to increase the noise significantly, which contribution is already important in the elastic response (low strains), while a loss of information results from the smoothing. On the other hand, Poisson's ratio is independent from the acceleration field and only results from the information gathered from the strain field. The fact that Poisson's ratio is well identified suggests that the main contribution to the badly identified Young's modulus comes from the acceleration computation. Future tests will have to address this problem by increasing the temporal resolution.

The model used for hardening is a simple linear hardening model. The identification is carried out on slices of the specimen, independently from each other (Fig. 6.1). The results concerning the yield stress are presented in Fig. 6.5. It is interesting to notice that the area with the lowest yield stress is the beginning of the heat-affected zone on the retreating side of the specimen, and not the nugget. The identification of the hardening coefficient proved to be very sensitive to noise, thus, so far, accurate identification of it has not been successful. The results show an evolution of the yield stress consistent with data obtained on quasi-static tests [8], with values about 20 % higher. These are only the first results obtained and future tests are presently being conducted to confirm these results. In particular, the grid method will be used in place of DIC to improve the spatial resolution.

## 6.6 Conclusion

In this work, the identification of the high strain rate mechanical properties of an FSW aluminium weld was investigated. The noise on the displacement field measurement remains a major problem, especially when it comes to the identification of elastic parameters, which explains the error on the identification of the Young's modulus. Nevertheless, it allows a more steady identification when it comes to plastic parameters which involves larger strain values and smoother acceleration profiles. Moreover, this method does not require the use of the load which measurement remains a major issue in dynamic. Further experiments will be needed to ascertain these results. And the continuous improvement of high-speed camera technology will support the development of this area of research in the future.

## References


1. Pierron F, Grédiac M (2012) The virtual fields method. Springer, New-York
2. Sutton MA, Yan J, Avril S, Pierron F, Adeeb SM (2008) Identification of heterogeneous constitutive parameters in a welded specimen: uniform stress and virtual fields methods for material property estimation. Exp Mech 48:451–464
3. Sutton MA, Orteu J, Schreier HW (2009) Image correlation for shape, motion and deformation measurements. Springer, New-York
4. Pierron F, Sutton MA, Tiwari V (2010) Ultra high-speed DIC and virtual fields method analysis of a three point ending impact test on an aluminium bar. Exp Mech 51:537–563
5. Gorry PA (1990) General least-squares smoothing and differentiation by the convolution (Savitzky-Golay) method. Anal Chem 62:570–573
6. Grédiac M, Pierron F (2006) Applying the virtual fields method to the identification of elasto-plastic constitutive parameters. Int J Plast 22:602–627
7. Moulart R, Pierron F, Hallett SR, Wisnom MR (2011) Full-field strain measurement and identification of composites moduli at high strain rate with the virtual fields method. Exp Mech 51:509–536
8. Le Louëdec G, Sutton MA, Pierron F, Reynolds AP (2011) Dynamic local elasto-plastic identification of the behaviour of aluminium FSW welds with the virtual fields method. In: Burguete RL, Lucas M, Patterson EA, Quinn S (eds) Applied mechanics and materials, vol 70, Advances in experimental mechanics VIII. Trans Tech Publications, Durnten-Zurich, joint BSSM/SEM ISEV conference, 7–9 September 2011 in Edinburgh (UK)

# Chapter 7
# Initiation Fracture Toughness of Human Cortical Bone as a Function of Loading Rate

**C. Allan Gunnarsson, Brett Sanborn, Mark Foster, Paul Moy, and Tusit Weerasooriya**

**Abstract** During injury causing events such as impact, blunt trauma, penetration or blast, the human body is subjected to high rate loading. These events will deform and damage human tissue, frequently causing tearing in soft tissues and cracking/fracture in hard tissue (bone) due to the inherent brittleness of bone. For the development of accurate fracture models for computer simulation of bone fracture and also to understand the failure mechanisms of hard tissues during blast and impact events, it is necessary to investigate the failure behavior (fracture) of hard tissues at different loading rates, including high loading rates. In this study, the Mode I fracture behavior of cortical bone from the human femur, perpendicular to the long axis, is investigated as a function of loading rate, including dynamic loading rates which are accomplished with a modified Kolsky bar. The bone fracture specimens were extracted from three different human donors, ages 36, 50, and 43. For all three donors, the fracture toughness increases as loading rate increases from quasi-static to intermediate. As loading rate increases from intermediate to dynamic rates, the initiation fracture toughness decreased on average, while remaining higher than at quasi-static rate. The observed variations in initiation fracture toughness with donor age for each loading rate did not show any specific trends. The initiation fracture toughness of cortical bone is dependent on the path of the crack relative to the internal bone structure.

## 7.1 Introduction

A material's fracture toughness represents the resistance to crack propagation of the material and must be accurately determined to allow for prediction of failure initiation from defects in the material. ASTM Standard C 1421-10 [1] specifies the standardized procedure to determine the initiation fracture toughness in opening mode (or Mode I), of advanced ceramic materials at quasi-static loading rates. In this standard, pre-cracked beam specimens are loaded in three or four-point bending configurations. The pre-crack can be a straight-through crack, a semi-elliptical crack, or a Chevron notch. Quasi-static loading rates are specified in this standard to ensure that the peak load measurements correspond to the local fracture toughness value at the crack tip.

There are no standard methods to determine the initiation fracture toughness at high loading rates. However, several techniques to determine the fracture toughness at high loading rates have been proposed over the past several decades, which can be categorized into three groups: high rate bending, high rate tension, and dynamic wedging. These techniques have attempted to extend the quasi-static ASTM standard procedure into the dynamic loading range through various approaches, resulting in a series of high rate bending techniques of pre-cracked beams [2–9]. In these experiments, the dynamic loading was applied using a modified split Hopkinson pressure bar (SHPB), a drop weight tower, or a modified Charpy tester. Most of these methods used the quasi-static relationships from [1] to relate far-field peak load to the initiation fracture toughness at the crack tip. Two conditions are necessary for valid high loading rate initiation fracture toughness experiments: the specimen must be in a state of dynamic equilibrium, and the loading rate must be constant so the data can be used for material studies. To accomplish these conditions for dynamic experiments, the input loading pulse is shaped to allow the specimen to reach a state of dynamic equilibrium while loading is applied at a constant rate [9–15]. Pulse shaping also

C.A. Gunnarsson (✉) • B. Sanborn • M. Foster • P. Moy • T. Weerasooriya
Weapons and Materials Directorate, Army Research Laboratory
e-mail: carey.a.gunnarsson.civ@mail.com

V. Chalivendra et al. (eds.), *Dynamic Behavior of Materials, Volume 1: Proceedings of the 2012 Annual Conference on Experimental and Applied Mechanics*, Conference Proceedings of the Society for Experimental Mechanics Series, DOI 10.1007/978-1-4614-4238-7_7, 

removes high frequency components of the loading pulse, preventing them from exciting resonance in the specimen which could cause the specimen to lose contact with the supports.

In SHPB experiments on low mechanical impedance specimens, such as pre-cracked beams, the incident and reflected pulses are typically close in magnitude, making dynamic equilibrium difficult to verify using traditional resistive strain gage analysis. Direct measurement of axial forces on both sides of the specimen during SHPB experiments is possible using quartz crystal force transducers, as reported by Chen et al. [11] for soft materials. This allows a direct verification that the specimen is in dynamic equilibrium. Weerasooriya et al. [12] developed a high loading rate four-point bending initiation fracture toughness experimental method, utilizing a modified SHPB with embedded quartz force transducers and pulse shaping to precisely control the loading input pulse. The high loading rate fracture toughness technique was used to measure the dynamic initiation fracture toughness of brittle materials such as ceramics (SiC) [12] and glassy polymers (PMMA) [13]. The method has been further extended to study the fracture behavior, such as fracture in an aluminum-epoxy interface by Syn et al. [14], and fracture in adhesive bond interfaces by Weerasooriya et al. [15]. Embedded quartz transducers are susceptible to errors caused by acceleration of the mass of the loading fixture between the specimen and quartz transducer. Such inertia effects are correctable, as noted by Casem et al. [16, 17].

There are many previous studies on the fracture behavior of cortical bone at lower loading rates [18]. There have been few studies conducted to understand the fracture behavior of animal cortical bone at high loading rates. Tanabe et al. [19] studied the dynamic fracture toughness of bovine cortical bone across the range of $10^{-4}$–$10^{6}$ MPa*m$^{1/2}$ s$^{-1}$ and found that the initiation fracture toughness is dependent on the specimen orientation with respect to the bone axis (0° and 90°) as well as the loading rate. Adharapurapu et al. [20] investigated the dynamic fracture toughness of bovine cortical bone (both dry and re-wetted) using a three point bend technique, with fracture toughness rates of $10^{-2}$–$10^{5}$ MPa*m$^{1/2}$ s$^{-1}$. The fracture toughness decreased with loading rate for dried bones (air dried for 15 or more days). For re-wetted bone, Adharapurapu et al. found that the fracture toughness increased as the loading rate increased from quasi-static to intermediate rate, but at high loading rates, the fracture toughness is lower than the quasi-static rate. A study by Currey [21] concluded that the effect of drying and re-wetting on the mechanical properties of bone is minimal. Kulin et al. [22] also investigated the fracture toughness of equine cortical bone up to dynamic loading rates using four-point bend techniques, although the dynamic fracture toughness rate was not given in this study. The fracture toughness was found to decrease from quasi-static to dynamic loading rates. Kulin et al. also found that the equine cortical bone fracture toughness decreases at quasi-static loading rate as the donor age increased. However, at the dynamic loading rate, no measurable age effect was found on the fracture toughness. These studies were focused on the dynamic fracture toughness of animal cortical bone.

The authors of this study are not aware of any prior studies on the dynamic fracture toughness of cortical bone at high loading rates from human donors. In this study, the high loading rate initiation fracture toughness experimental method utilized to measure the dynamic fracture toughness of brittle materials [10] was further developed to investigate the initiation fracture toughness of human femoral cortical bone from three donors, normal to the osteon micro-structure axis (normal to length axis of the femur). The focus of this study was on the effect of high loading rates on the initiation fracture toughness of human cortical bone. These high loading rates are experienced when a human is exposed to blast and ballistic impact. A modified SHPB was used to conduct high loading rate four point bend experiments on human cortical bone. Quartz transducers embedded in the SHPB were used to make direct force measurements at the specimen-bar interface, allowing the verification of dynamic equilibrium and the presence of a constant loading rate. High speed imaging was performed on the experiments to allow for analysis of crack propagation path and effect on loading rate. The imaging was also used to measure the full strain field at the crack tip, to correlate the strain field with fracture behavior, the results of which will be reported in a later publication.

## 7.2 Human Cortical Bone Structure

Bones in the human body, such as the leg bones (femur, tibia, fibula) and arm bones (humerus, ulna, radius), are comprised of two types of bone materials: a hard, brittle outer shell material that provides stiffness and load bearing capacity, known as cortical bone, and a soft, spongy inner material mostly containing blood vessels and marrow, known as cancellous or trabecular bone. In this study, the failure behavior of the cortical bone is investigated. Cortical bone is complex and anisotropic, as shown in Fig. 7.1. The cortical bone sub-structure consists of osteons, cylindrical structures comprised of concentric lamella sheets. The axes of osteons are aligned parallel to the bone axis. These osteons are on the order of 100–200 μm in diameter and a few millimeters long. Each lamella is further comprised of ever-smaller length scale components. The center of the osteon is hollow, and is known as the Haversian or osteonic canal, through which run blood vessels and nerves. Gaps between these osteons are occupied by interstitial lamellae. Interstitial lamellae are parts of

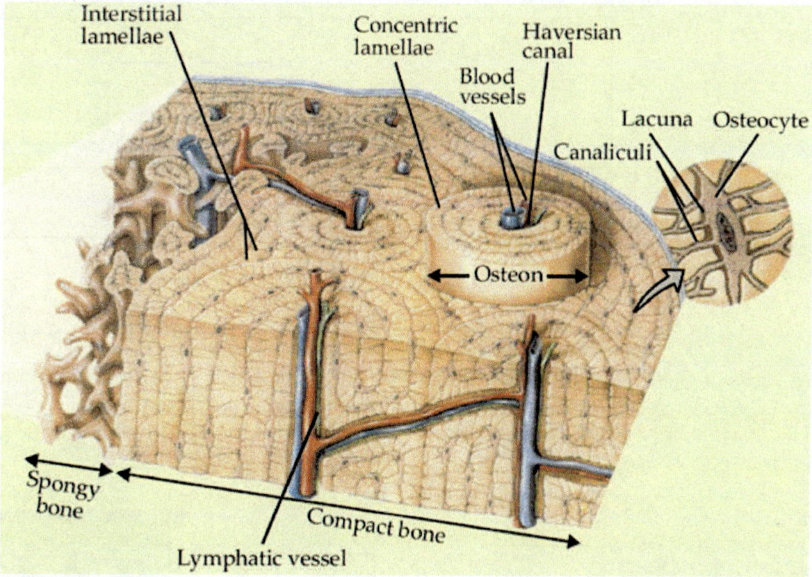


Fig. 7.1 Cortical bone sub-structure details [18]

former osteons that are broken down as the bone rebuilds itself with new osteons and are relatively weak. This remodeling process has a large influence on the mechanical properties of bone. A detailed illustration of the bone hierarchical substructure is shown in Fig. 7.1 [18].

Due to the anisotropic nature of cortical bone, the material properties are dependent on the orientation of the loading relative to the bone structure. In this study, the fracture behavior of cortical bone is investigated in the plane normal to the axis of the bone and the osteons. Based on the thickness of cortical bone in the human body, techniques have not yet been developed to extract specimens long enough to study the fracture behavior parallel to the osteon axis. Methods to perform high loading rate fracture experiments on these small specimens also are yet to be developed. The fracture toughness along the osteon axis would be at least an order of magnitude smaller than normal to the osteon axis, as the crack would follow the weaker osteon boundaries.

## 7.3 Specimen Fabrication

Six human lower extremities (legs) were obtained from three separate cadaveric donors. The three donors were all male and middle-aged (36–50 years old). The femurs were extracted from the donors; the upper leg was disarticulated, and the soft tissues (skin, muscle, etc.) were dissected and disposed of. Once most of the soft tissue was removed and the full femur was exposed, strips running along the axis of the bone were cut from the four sides: anterior, posterior, medial, and lateral. Each of these strips was long enough to make four or five femoral bend specimens. These individual specimens were then sanded to obtain dimensions of 3 mm thickness (B), 4 mm width (W), and approximately 50 mm long. The specimens were notched at the center; the notch length was set to be nominally 1.9 mm, and the thickness of the notch was approximately 200 μm. Once fabricated, the specimens were refrigerated in individual containers in a buffered saline solution (Hank's balanced salt solution, or HBSS) to prevent any mineral loss during storage. The nominal specimen dimensions and loading geometry are shown in Fig. 7.2 (dimensions are in mm).

## 7.4 Experimental Setup

To determine the loading rate effect on the fracture toughness of human cortical bone, four-point bending experiments on pre-cracked (notched) beam specimens were conducted at quasi-static, intermediate, and dynamic loading rates. The bone specimens were removed from wet storage and tested immediately. Approximately 9–11 experiments were conducted at each loading rate for each donor. These were randomized among locations around the femur, locations along the femur,

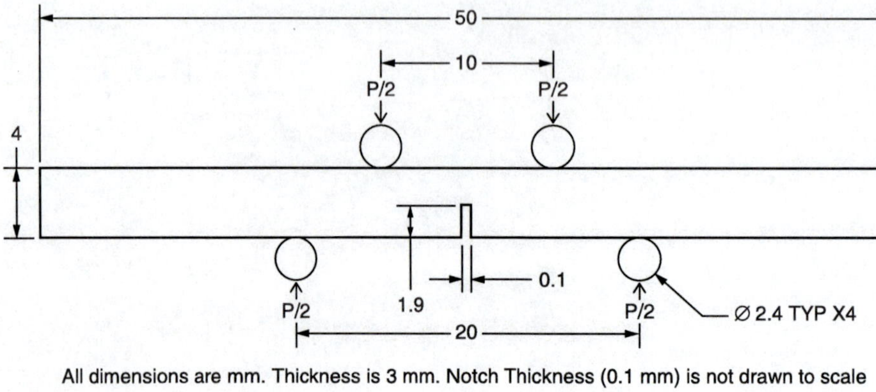


**Fig. 7.2** Specimen dimensions and loading geometry

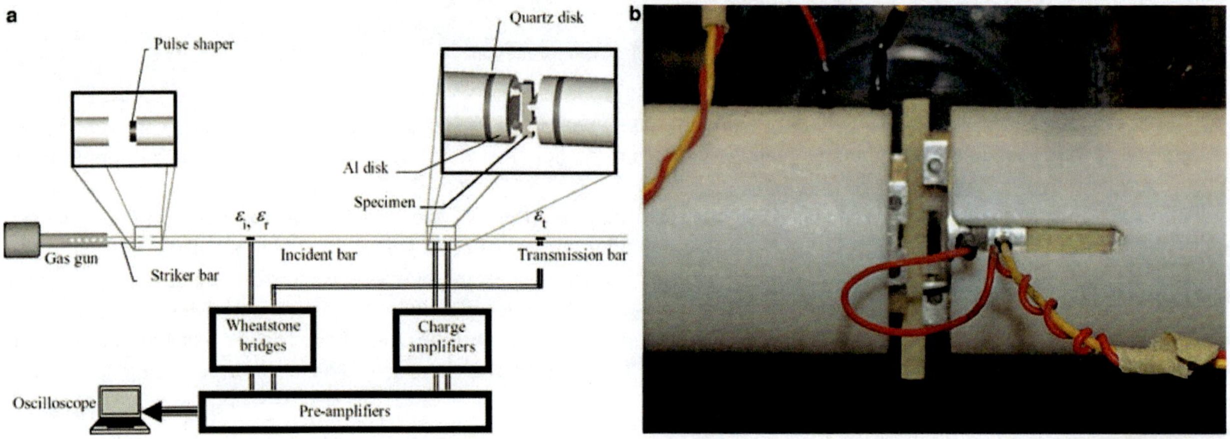


**Fig. 7.3** (**a**) SHPB system layout (**b**) experimental setup with specimen

and among the two different femurs from each donor. For the dynamic loading rate experiments, an aluminum SHPB was used with specially designed loading fixtures. The bars had a diameter of 31.75 mm (1.25 in.) and a length of 3.7 m (12 ft). For the dynamic experiments, the input pulse was shaped to ensure dynamic equilibrium and a constant loading rate on the specimen. The bar end velocity for the incident bar was approximately 550 mm/s after this pulse shaping was implemented.

To ensure that the specimen is in dynamic equilibrium, the force histories on both sides of the specimen were measured using quartz transducers embedded near the specimen/input-bar and specimen/transmitted-bar interfaces. These embedded quartz transducers provided high signal with little noise for the low loads inherent in experiments on low impedance material. The modified SHPB experimental setup with quartz transducers is shown in Fig. 7.3a. The loading fixtures on each bar consisted of two steel pins, with a center to center distance of 10 mm (incident fixture) and 20 mm (transmission fixture). This loading geometry replicates one of the loading configurations in the quasi-static ASTM standard [1], and was also used in the quasi-static and intermediate experiments. These customized force transducers were calibrated through separate calibration experiments using mating fixtures to allow the input stress pulse to pass through to the transmission bar. An input pulse was sent through the bars and the transducer. The transducer output was calibrated to the stress measured in the SHPB by traditional bar strain gages. Figure 7.3b shows a picture of the experimental setup with specimen just prior to testing.

Typical data from a dynamic experiment is shown in Figs. 7.4 and 7.5. Figure 7.4a, shows strain gage signals from the incident and transmission bar. As expected, there is no signal from the transmission bar strain gage, as the transmitted signal is too weak to be measured by traditional resistive gages. Although not shown here, a set of semi-conductor strain gages (which are about 100 times more sensitive than resistive strain gages) mounted on the transmission bar were able to record

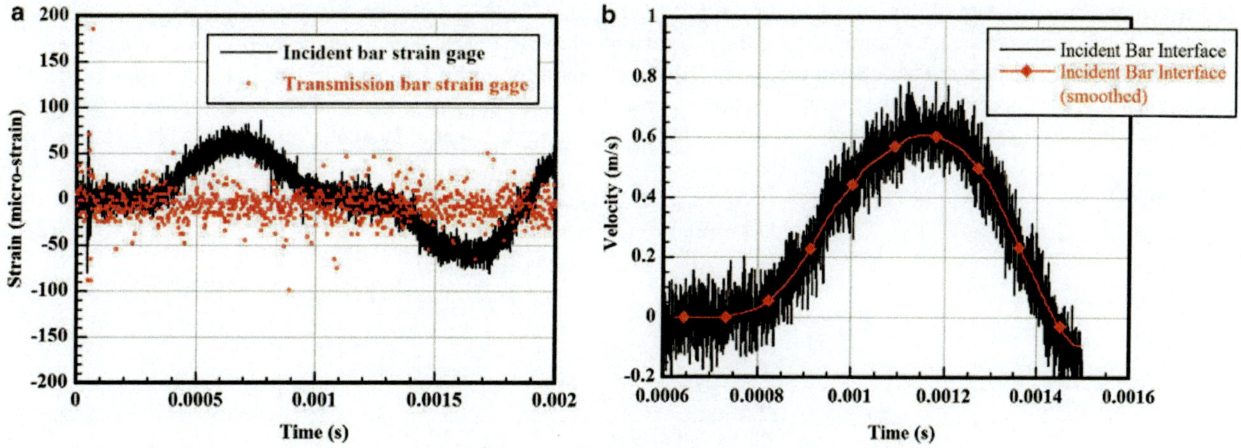


Fig. 7.4 (a) Typical incident and transmission bar strain gage signals (b) incident bar-end velocity profile

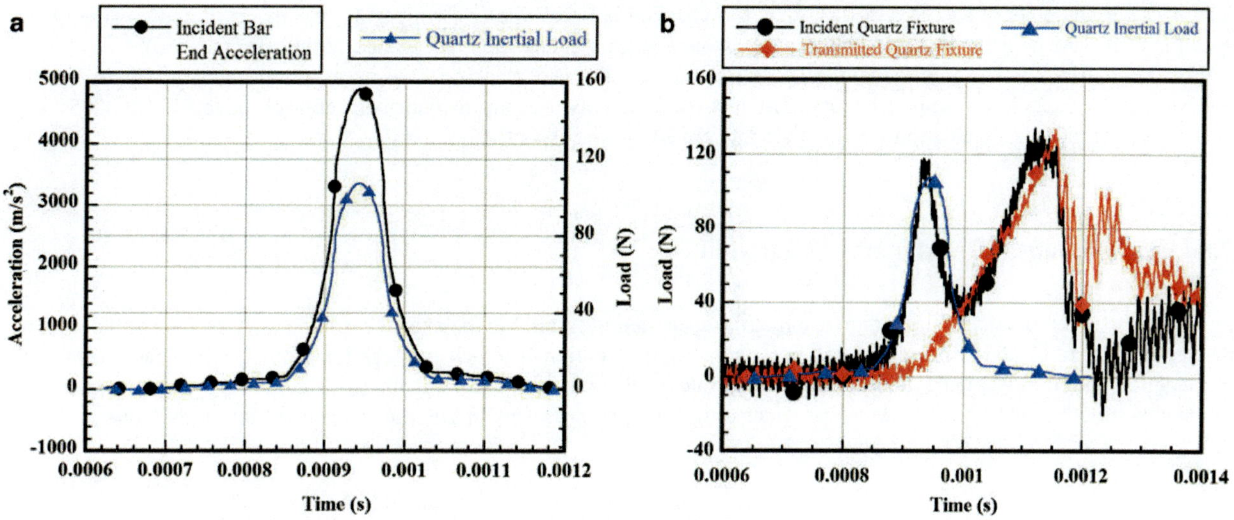


Fig. 7.5 (a) Acceleration of the incident bar end and corresponding quartz fixture inertial force and (b) typical incident and transmission quartz transducer signals

the transmission bar stress history, and provided an independent verification of the loading history. Figure 7.4b shows the velocity profile at the incident bar-end after the incident and reflected pulse strain data have been time resolved and summed using Eq. 7.1:

$$v_{INC\ BAR\ END}(t) = c_0[\varepsilon_I(t) - \varepsilon_R(t)] \tag{7.1}$$

where $c_0$ is the wave speed of the bar material, $\varepsilon_I(t)$ and $\varepsilon_R(t)$ are the incident and reflected strain histories, respectively.

Once the incident bar-end velocity profile is known, the data is smoothed and differentiated to obtain acceleration of the bar-end, and therefore acceleration of the incident quartz transducer (Fig. 7.5a). By knowing the mass of the loading fixture (which for this setup was 22.2 g), the incident quartz transducer signal can be corrected to remove the apparent force caused by inertia of the transducer during acceleration, a technique documented by Casem et al. [16, 17]. Figure 7.5b shows typical load histories during a dynamic experiment recorded by the incident quartz transducer and the transmission quartz transducer. Figure 7.5b also shows the apparent load at the incident interface caused by transducer inertia. Note that after removal of the apparent load from the incident transducer, the load histories at both sides of the specimen were nearly the same, demonstrating a state of dynamic equilibrium for the specimen. Additionally, the loading profile has a nearly linear slope, demonstrating a constant loading rate for the experiment which is a necessary criterion for dynamic experiments.

When the specimen is in dynamic equilibrium, the ASTM standard for quasi-static experiments can be used to relate the loading history of the specimen to the fracture toughness. From [1], Sect. A2.4, Eqs. A2.5 and A2.6, for four-point flexure of a pre-cracked beam specimen, with a crack length to height ratio [a/W] between 0.35–0.60:

$$K_{I_{PB}} = f\left[\frac{P_{max}(S_0 - S_1)10^{-6}}{BW^{\frac{3}{2}}}\right]\left[\frac{3[\frac{a}{W}]^{\frac{1}{2}}}{2[1-\frac{a}{W}]^{\frac{3}{2}}}\right] \tag{7.2}$$

with:

$$f = 1.9887 - 1.326\frac{a}{W} - \frac{\left\{3.49 - 0.68\left[\frac{a}{W}\right] + 1.35\left[\frac{a}{W}\right]^2\right\}\left[\frac{a}{W}\right]\left\{1 - \left[\frac{a}{W}\right]\right\}}{\left\{1 + \left[\frac{a}{W}\right]^2\right\}} \tag{7.3}$$

where $K_{IPB}$ (MPa$\sqrt{m}$) is the fracture toughness of a brittle four-point pre-cracked bending beam specimen, $P_{\max}$ is the measured peak axial force (N) by the transmission quartz transducer, $B$ is the specimen width (m) and $W$ the specimen height (m), $a$ is the notch (or crack) length (m), $S_O$ is the distance between the two supporting points on the notch side of the specimen (m), and $S_I$ is the distance between the two loading points on the loading side of the specimen (m).

For the quasi-static and intermediate loading rates, an Instron test machine was used with an integrated load cell for measurement of specimen loading history. The test machine was used in displacement control, at rates of 0.00051 and 0.51 mm/s for the quasi-static and intermediate rate experiments, respectively.

## 7.5 Experimental Results and Discussion

For all of the experiments, the long dimension of the specimen is parallel to the axis of the bone (and the axis of the osteons in the sub-structure) and is normal to the pre-crack. After initiation of crack propagation from the pre-crack, the crack propagation direction was dependent on the loading rate. Deflection of the crack propagation from the pre-crack axis is due to the sub-structural morphology, which provides toughening mechanisms in the transverse direction of the bone.

### *7.5.1 Quasi-static Experiments*

At the quasi-static loading rate, the bone specimens did not fail completely. The load on the specimen increased at a constant rate, reached the peak load and started to decrease. The experiment was stopped some arbitrary time after the peak load. Image analysis of the quasi-static experiments indicated that the crack begins to open up to the point of peak load, and after peak load it propagates along the specimen in a direction almost parallel to the osteon axis (Fig. 7.6). However, the specimen does not fail suddenly during the crack propagation, retaining a degree of load capacity, as shown by a typical quasi-static load history with corresponding images as shown in Fig. 7.7.

### *7.5.2 Intermediate Rate Experiments*

From the intermediate rate experiments, two failure behaviors were observed in the loading history, as shown in Fig. 7.8. The blue data shows abrupt specimen failure as the crack immediately propagated after maximum load (corresponding to Fig. 7.9a and f). The red data in Fig. 7.8 (corresponding to Fig. 7.9b, c, d, and f) showed stable crack propagation with reduced load capacity in the specimen after maximum load was reached; then, after some time, abrupt failure. At the intermediate rate, the crack propagation after initiation was more varied than the quasi-static experiments. Figure 7.9 shows a random sample of six intermediate rate experiments after failure, along with the fracture toughness value. The crack path is

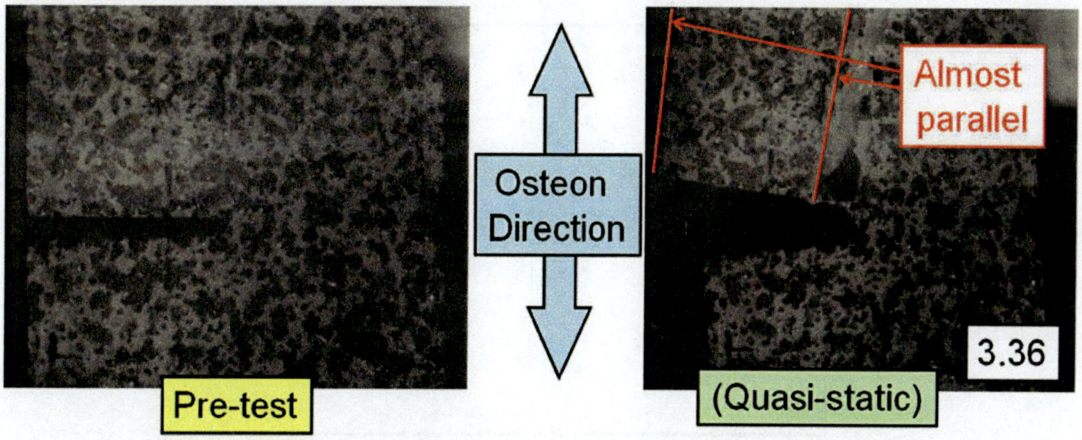


**Fig. 7.6** Typical crack propagation of cortical bone specimens in four point bending for quasi-static experiments

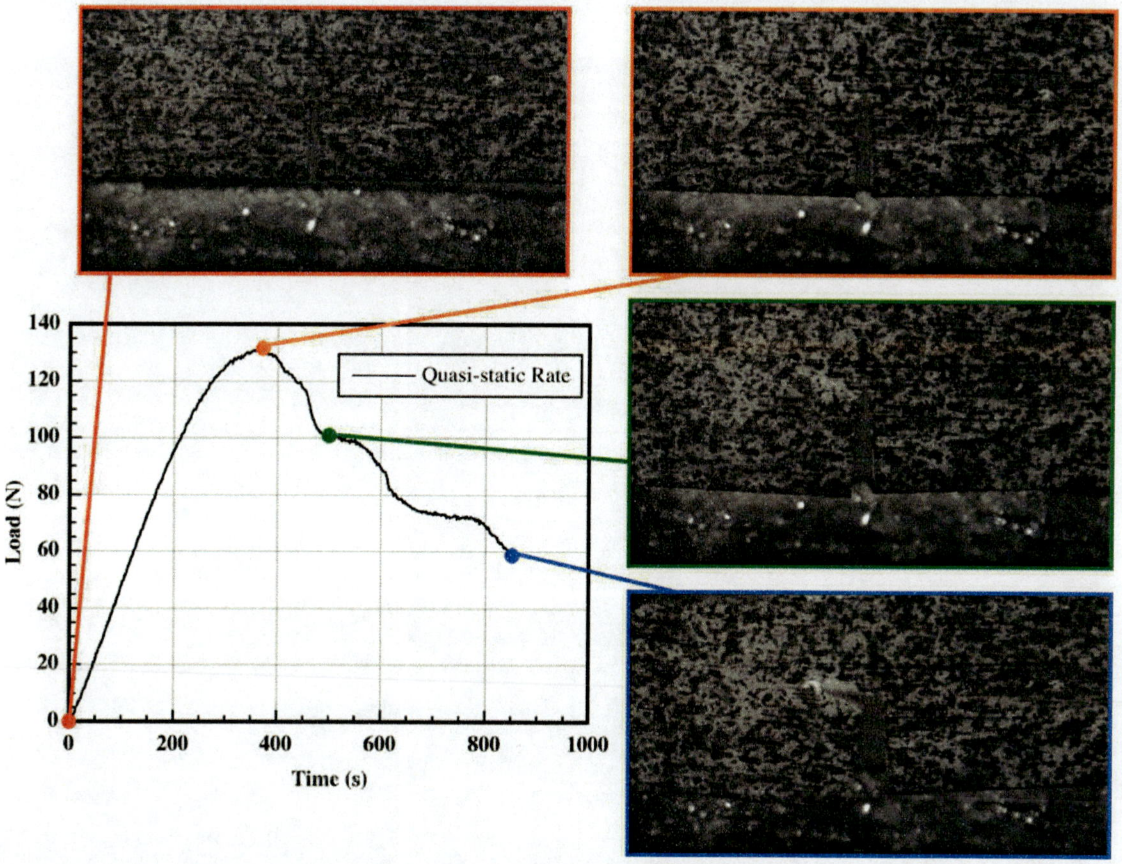


**Fig. 7.7** Loading history for quasi-static experiment

illustrated in red for the pictures difficult to view. Most of the intermediate rate experiments (Fig. 7.9a–c, e, and f) had crack propagation along a tortuous path across the specimen, changing directions (sometimes several times). A minority of the intermediate experiments (Fig. 7.9d) had crack propagation at 45° from the pre-crack; these experiments had lower fracture toughness. Generally, when the crack propagation was more parallel to the pre-crack and included changes in direction, the fracture toughness was higher.

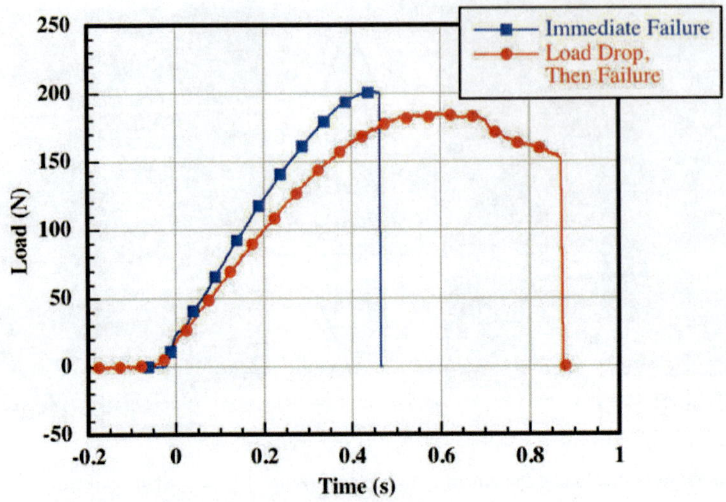


**Fig. 7.8** Typical load histories of specimens at intermediate rate

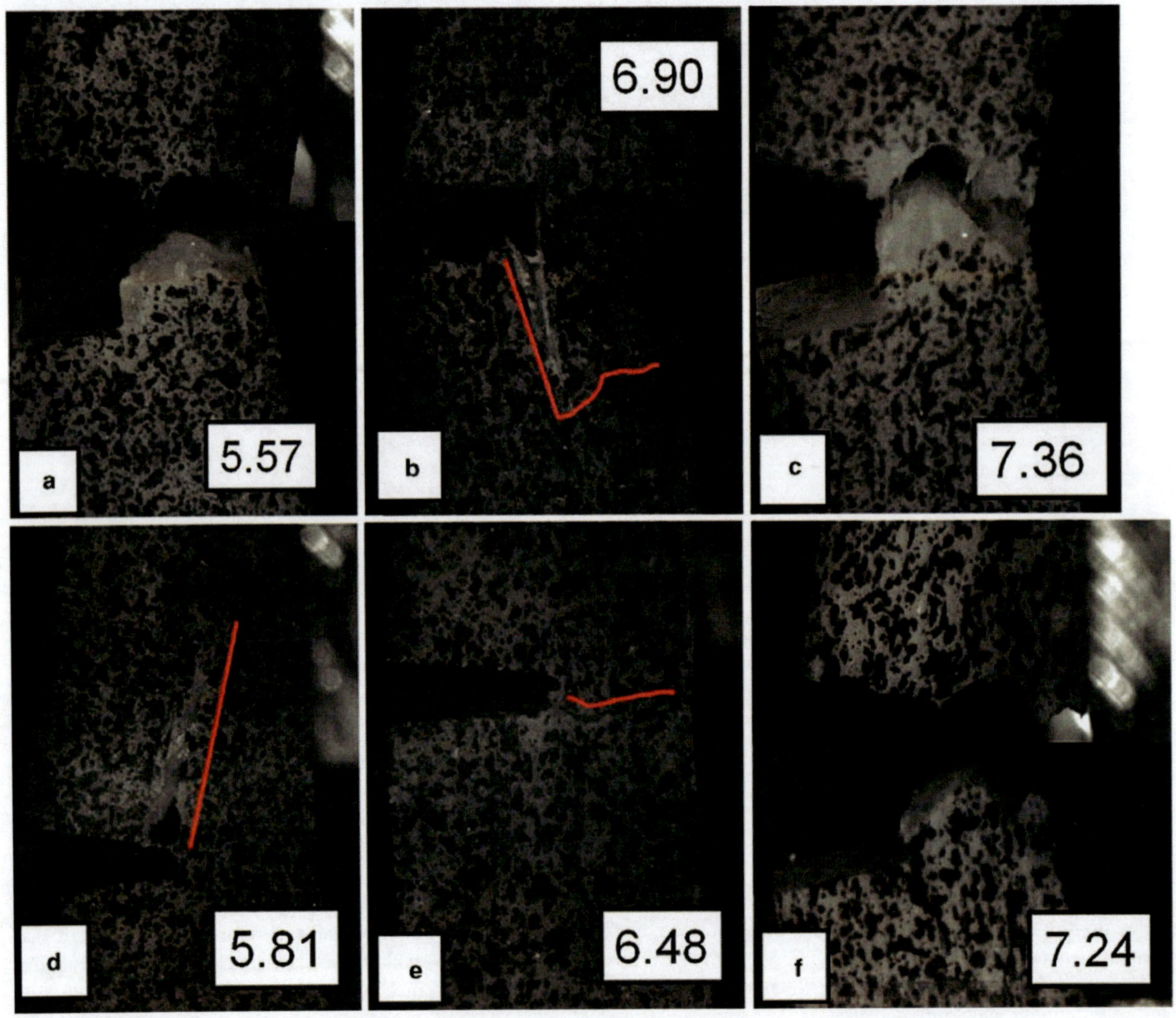


**Fig. 7.9** (a–f) Typical crack paths and fracture toughness values of cortical bone specimens for intermediate rate

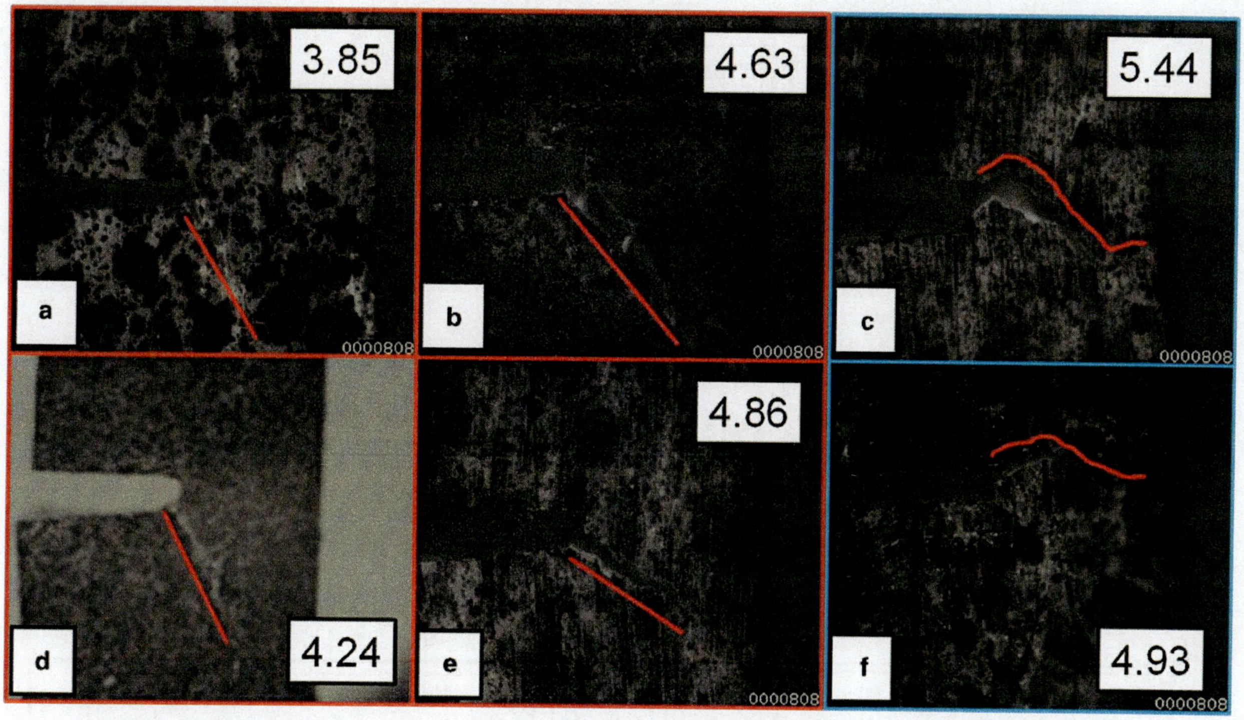


**Fig. 7.10** (a–f) Typical crack paths and fracture toughness values of cortical bone specimens for dynamic experiments

### *7.5.3 High Rate Experiments*

At high loading rates, the specimen load increased linearly with abrupt failure occurring at the peak load. The dynamic experimental results were more scattered compared to the quasi-static and intermediate rates. A typical loading history for a dynamic experiment is shown in Fig. 7.5b. Figure 7.10a–f shows crack propagation in six randomly selected dynamic experiments. The crack path has been roughly illustrated in red to make viewing easier.

The fracture toughness of the bone varied depending on the path of crack propagation. Most of the dynamic experiments had crack propagation in a straight line at an angle of approximately 45° from the corner of the pre-crack, shown in Fig. 7.10a, b, d, and e. This 45° crack propagation corresponded to lower initiation fracture toughness than the minority of dynamic experiments that had crack propagation in a general direction across the specimen (Fig. 7.10c and f). Note from the figure that the crack propagation in b and e are more parallel to the pre-crack (approximately 30°) than in a and d (approximately 45°) and have higher fracture toughness. The crack propagations in Fig. 7.11c and f are generally parallel to the pre-crack and have higher fracture toughness than any of a, b, d, and e. Figure 7.11c showed more changes in vertical direction (zigzagging) than f, resulting in the highest fracture toughness of these six experiments.

One possible mechanism causing this behavior involves relative osteon strength. The 45° crack propagation specimens may have had stronger osteons that forced the crack to propagate along the weak cement lines (osteon boundaries). The specimens with crack propagation across the specimen might have had weaker osteons, allowing individual osteons to fail in bending mode. This would seem counter-intuitive that the specimens with stronger osteons had lower initiation fracture toughness. Further research is needed to relate bone chemistry (mineral composition) and micro-structure to the fracture toughness.

### *7.5.4 Effect of Loading Rate on Fracture Toughness for Three Donors*

Figure 7.11 shows the initiation fracture toughness of human cortical bone, normal to the osteon axis, as a function of fracture toughness rate for all specimens from the three donors. Figure 7.12a shows the average fracture toughness for each donor, as well as the overall average of all three donors, at each rate with error bars representing ± one standard deviation (±σ). The initiation fracture toughness increases as the fracture toughness rate increases from quasi-static to intermediate

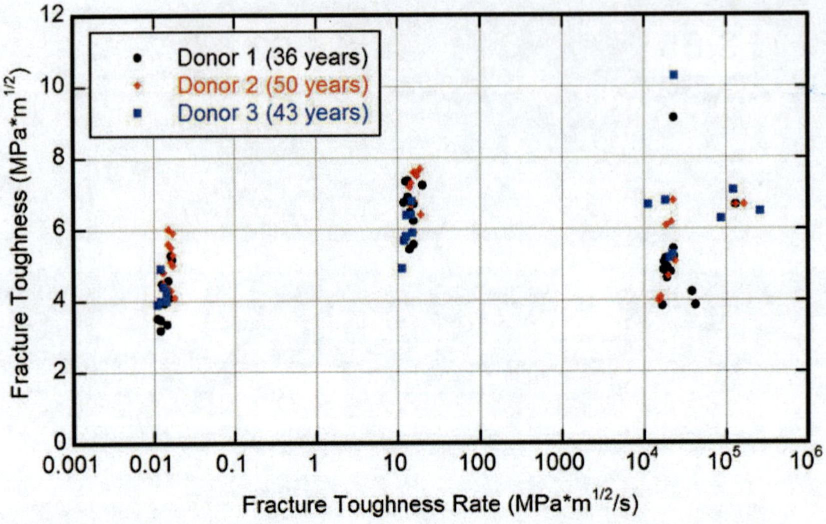


**Fig. 7.11** Fracture toughness of cortical bone (perpendicular to osteon axis) for individual donors

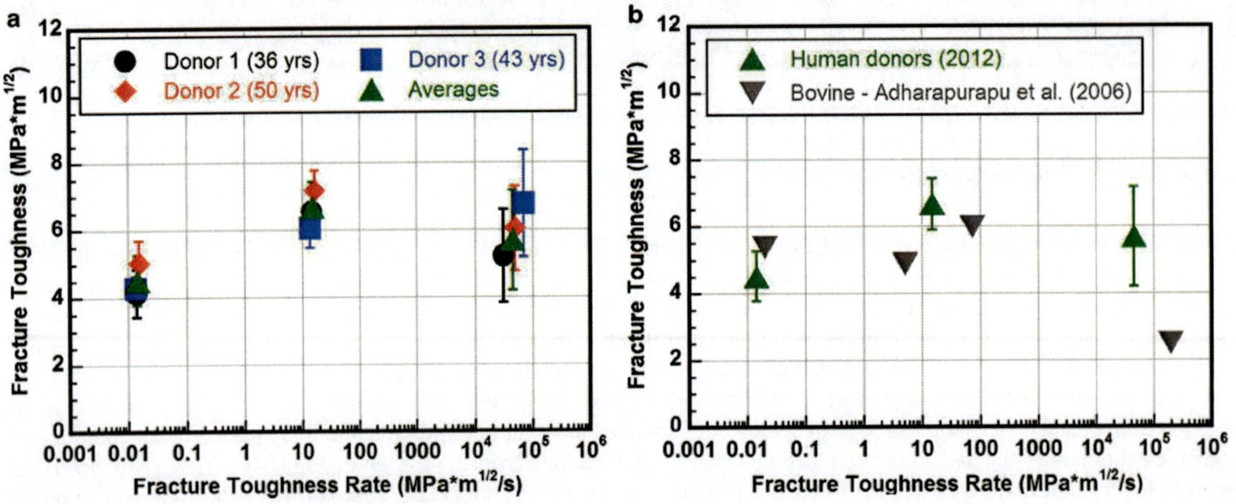


**Fig. 7.12** Fracture toughness of cortical bone (perpendicular to osteon axis) as a function of rate for (**a**) individual and overall averages as and (**b**) overall averages for this study (human) and Adharapurapu et al. (Bovine – wet) [20]

rate for all three donors. From intermediate to dynamic rates, the average initiation fracture toughness showed a decrease for two of the donors (donor 1, 36 years and donor 2, 50 years), but was higher than the quasi-static rate. For donor 3 (43 years), the fracture toughness increased slightly with the increase in loading rate from intermediate to dynamic. The overall average fracture toughness increased from quasi-static to intermediate and then decreased at high rate, but remained higher than the quasi-static. In general, the decrease of fracture toughness at high rate from intermediate rate is less than the observed increase from quasi-static to intermediate rate.

These trends are due to the path the crack takes during propagation, discussed with Figs. 7.9 and 7.10. At the quasi-static rate, crack initiation and growth is predominantly in the direction of the osteon axis, along osteon boundaries (cement lines), which is the weakest direction: This is reflected in the lower fracture toughness values for the quasi-static rate. For the intermediate rate, macroscopic failure generally progresses across the specimen, frequently changing direction. Some intermediate rate experiments had failure progress in a single direction, approximately 45° from the osteon axis, and these had lower fracture toughness values. This zigzag behavior indicates that the crack was crossing weaker osteons and then being redirected by stronger osteons. At high loading rates, macroscopic fracture typically followed a path 45° to the osteon axis; some experiments fractured across the specimen and had higher fracture toughness.

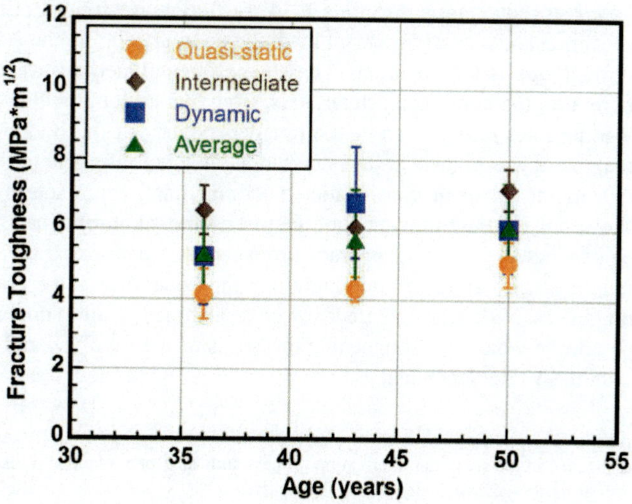


**Fig. 7.13** Fracture toughness of cortical bone (perpendicular to osteon axis) averaged for individual loading rates and overall average as a function of age

Figure 7.12b shows the fracture toughness data averaged across all three human donors as a function of loading rate. For comparative purposes, the figure also includes the averaged results from Adharapurapu et al. [20], showing the fracture toughness dependency of wet bovine cortical bone (normal to the osteon axis) as a function of loading rate. The data scatter was not reported. Adharapurapu et al. air dried all specimens for 15 or more days, creating "dry" bone. The "wet" bone was then stored in HBSS for 30 or more hours prior to testing to re-wet the specimens. The "wet" bone is more comparable to the bone experiments in this study (where the specimens were stored in HBSS) than the "dry" bone. The rate effect on the fracture toughness of equine cortical bone from Kulin et al. [22] was not included for comparison as the fracture toughness rates were not directly reported.

For bovine cortical bone, as well as human cortical bone, the intermediate rate fracture toughness was the highest. The high rate fracture toughness decreased significantly from the intermediate fracture toughness for the bovine cortical bone whereas the human cortical bone decreased slightly. The bovine high rate fracture toughness was lower than the quasi-static fracture toughness, by about half, which was consistent with the equine results in Kulin et al. The human cortical bone, by contrast, has dynamic fracture toughness higher than the quasi-static fracture toughness.

### 7.5.5 *Effect of Age on Fracture Toughness at Different Loading Rates*

Initiation fracture toughness values as a function of fracture toughness rate and grouped by donor (and therefore age) are shown in Fig. 7.11. There do not appear to be any significant age effects on the measured fracture toughness over the limited age range. The fracture toughness measurements from all three donors are in the same general range at each of the three loading rates. However, when averaged across all three loading rates, as shown in Fig. 7.13 (*green data*), a slight trend is observed; the initiation fracture toughness increases as age increases over this limited age range. However, when broken down into the separate loading rates in Fig. 7.13, it can be seen that there are no common trends. The quasi-static data increases directly with age; the intermediate data decreases with age first, then increases; and the dynamic data displays the opposite trend, increasing with age at first and then decreasing with age. These three donors have many unknown environmental variables (such as diet, general fitness and health, exercise history, genetics, etc.) that make age related conclusions difficult. Due to this variability, and the lack of any consistency in age dependence for the different loading rates, no age effects are concluded.

## 7.6 Conclusions

An existing experimental technique has been further developed to study the Mode I fracture behavior and to determine the fracture toughness for brittle bio-materials. This experimental method is based on a modified split-Hopkinson pressure bar, with embedded high sensitivity piezoelectric quartz transducers, using the quasi-static fracture toughness ASTM standard for advanced ceramic materials (C1421-10). Precise control of the loading pulse profile ensures dynamic equilibrium in the specimen and a nearly constant loading rate, thus relating the fracture toughness at the crack tip to the far-field peak loading through quasi-static relations. These experimental methods were used to study the fracture behavior of human cortical bone, normal to the bone (and osteon) axis, from human femurs. Fracture experiments were conducted at quasi-static to dynamic loading rates to demonstrate the rate dependency of the fracture toughness of human cortical bone. The initiation fracture toughness was highest at the intermediate loading rate. The initiation fracture toughness increased as the fracture toughness rate increased from quasi-static to intermediate. From intermediate to dynamic rates, the initiation fracture toughness showed a decrease in two out of the three donors but remained above quasi-static values. The fracture toughness appears to be dependent on the path of crack propagation; the more parallel to the pre-crack the propagation was, the higher the fracture toughness. Changes in direction along the bone axis as the crack propagated across indicated a higher initiation fracture toughness. No conclusive trends were determined relating donor age and bone fracture toughness. Future work is needed to determine what bone toughening mechanisms are involved and how bone chemistry and micro-structure affect the fracture toughness rate dependency.

**Acknowledgments** The authors would like to acknowledge Mr. Jeff Gair of the Oak Ridge Institute for Science and Education (ORISE) for assistance in the specimen fabrication process and Mr. Ronn Wade and his team at the Maryland State Anatomy Board (MSAB) for assistance in procuring the cadaveric donors for this study.

## References

1. ASTM C1421-10 (2010) Standard test methods for determination of fracture toughness of advanced ceramics at ambient temperatures. Annual book of ASTM standards, ASTM, West Conshohocken
2. Böhme W, Kalthoff J (1982) The behavior of notched bend specimens in impact testing. Int J Fracture 20:R139–R143
3. Rittel D, Mairge H, Bui H (1992) A new method for dynamic fracture toughness testing. Scripta Metall Mater 26(10):1593–1598
4. Bacon C, Färm J, Lataillade J (1994) Dynamic fracture toughness determined from load-point displacement. Exp Mech 34(3):217–223
5. Sharpe W, Böhme W (1994) Dynamic fracture toughness measurements on small Charpy specimens – A preliminary study. J Test Eval 22 (1):14–19
6. Anderson C, Popelar C, Nagy A, Walker D (2000) A novel method for determining dynamic fracture toughness. In: American Institute of Physics (ed) Shock compression of condensed matter. American Institute of Physics, Woodbury, pp 505–508
7. Weisbrod G, Rittel D (2000) A method for dynamic fracture toughness determination using short beams. Int J Fracture 104(1):89–103
8. Martins C, Prakash V (2002) Dynamic fracture characteristics of LMDPE utilizing a modified split Hopkinson bar. In: Proceedings of 2002 society of experimental mechanics annual conference. Milwaukee
9. Deobald L, Kobayashi A (1992) A bar impact tester for dynamic fracture testing of ceramic and ceramic composites. Exp Mech 32(2):109–116
10. Jiang F, Vecchio KS (2007) Experimental investigation of dynamic effects in a two-bar/three-point bend fracture test. Rev Sci Instrum 78:063903
11. Chen W, Lu F, Zhou B (2000) A quartz crystal imbedded split Hopkinson bar for soft materials. Exp Mech 40(1):1–6
12. Weerasooriya T, Moy P, Casem D, Cheng M, Chen W (2006) A four point bending load technique for determination of dynamic fracture toughness for ceramics. J Am Ceram Soc 89(3):990–995
13. Weerasooriya T, Moy P, Casem D, Cheng M, Chen W (2006) Fracture toughness for PMMA as a function of loading rate. In: Proceedings of the 2006 society of experimental mechanics annual conference
14. Syn C, Chen W (2008) Surface morphology effects on high-rate fracture of an aluminum epoxy interface. J Compos Mater 42:1639–1658
15. Weerasooriya T, Gunnarsson CA, Jensen R, Chen W (2011) Strength and failure energy for adhesive interfaces as a function of loading rate. Dynamic behavior of materials, volume 1, conference proceedings of the society for experimental mechanics series 99
16. Casem D, Weerasooriya T, Moy P (2003) Acceleration compensation of quartz transducers embedded in a split Hopkinson pressure bar. In: Proceedings of the SEM Annual Conference on Experimental Mechanics, Charlotte
17. Casem D, Weerasooriya T, Moy P (2005) Inertial effects of quartz force transducers embedded in a split Hopkinson pressure bar. Exp Mech 45 (4):368–376
18. Ritchie R, Buehler M, Hansma P (2009) Plasticity and toughness in bone. Phys Today 62(6):41–47
19. Tanabe Y, Tanner KE, Bonfield W (1998) Effect of loading rate on fracture toughness of bovine cortical bone. In: Proceedings of the 11th conference of the Eurpoean society of Biomechanics. J Biomech 31(Supplement 1): 121
20. Adharapurapu RR, Jiang F, Vecchio KS (2006) Dynamic fracture of bovine bone. Mater Sci Eng C 28(6):1325–1332
21. Currey J (1988) The effects of drying and re-wetting on some mechanical properties of cortical bone. J Biomech 21(5):439–441
22. Kulin R, Jiang F, Vecchio K (2011) Effects of age and loading rate on equine cortical bone failure. J Mechan Behav Biomed Mater 4:57–75

# Chapter 8
# Measurements of Particle Velocity Within a Kolsky Bar with Applications to Wave Separation

**Daniel T. Casem and Michael B. Zellner**

**Abstract** Several methods of measuring particle velocity at points along a Kolsky bar are investigated. Direct particle velocity measurements are of interest because they can be used in conjunction with strain gages to permit wave separation; this can be used to extend the lower strain-rate range accessible with the Kolsky bar technique. Two methods are evaluated with a PMMA Kolsky bar. They include (1) strain-gradient based measurements of acceleration, which is then integrated to obtain particle velocity, and (2) laser extensometer measurements of displacement, which is then differentiated to obtain velocity. Each of these methods has its limitations, and it is suggested that they can be used together to provide a velocity diagnostic useful for practical testing. Finally, preliminary experiments are presented with a Photon Doppler Velocimeter to measure particle velocity with a steel pressure bar.

## 8.1 Introduction

Over the past few decades, the Kolsky bar, or Split Hopkinson Pressure Bar, has become the standard method for characterizing the mechanical response of materials at high strain-rates, i.e., in the range of $10^3$/s–$10^4$/s [1–3]. The most common arrangement is shown in Fig. 8.1. The specimen is placed between two elastic bars, called the incident bar and the transmitter bar, as shown in the figure. A projectile impacts the incident bar and a compressive stress wave, the incident pulse, propagates into the bar. When this wave arrives at the specimen the specimen begins to deform. A reflected wave then propagates back into the incident bar while a third wave, called the transmitted pulse, propagates into the transmitter bar. These three waves are measured using strain gages at the mid-point of each bar, i.e., the incident bar strain gage, SG1, is used to measure the strain due to the incident and reflected pulses and the transmitter bar strain gage, SG2, is used to measure the transmitted pulse. Provided the bars remain elastic, the wave propagation is known and the forces applied to the specimen and its resulting deformation can be determined using an appropriate wave theory (i.e., forces $F_1$, $F_2$, and interface velocities $v_1$, $v_2$, in the figure).

In the simplest case, the wave propagation can be described by basic linear elastic one-dimensional bar wave propagation (uniaxial stress). In this case, the waves propagate through the bars without change in form at a fixed bar wave speed given by $c_0 = (E/\rho)^{1/2}$, where E is the elastic modulus of the bar material and $\rho$ is it's density. This description is adequate provided the wavelengths of the relevant stress waves are large in comparison to the diameter of the bar, i.e., relatively low frequencies. When high frequencies are significant, dispersion effects may become noticeable and more sophisticated theories [4–6] may be required. Furthermore, some researchers use polymeric bars to increase sensitivity and these materials may be viscoelastic. Theories have also been developed to account for this more general case [7–11].

D.T. Casem (✉)
U.S. Army Research Laboratory, RDRL-WMP-B, Aberdeen Proving Ground, Aberdeen, MD 21005-5069, USA
e-mail: daniel.t.casem.civ@mail.mil

M.B. Zellner
U.S. Army Research Laboratory, RDRL-WMP-D, Aberdeen Proving Ground, Aberdeen, MD 21005-5069, USA

V. Chalivendra et al. (eds.), *Dynamic Behavior of Materials, Volume 1: Proceedings of the 2012 Annual Conference on Experimental and Applied Mechanics*, Conference Proceedings of the Society for Experimental Mechanics Series,
DOI 10.1007/978-1-4614-4238-7_8, 

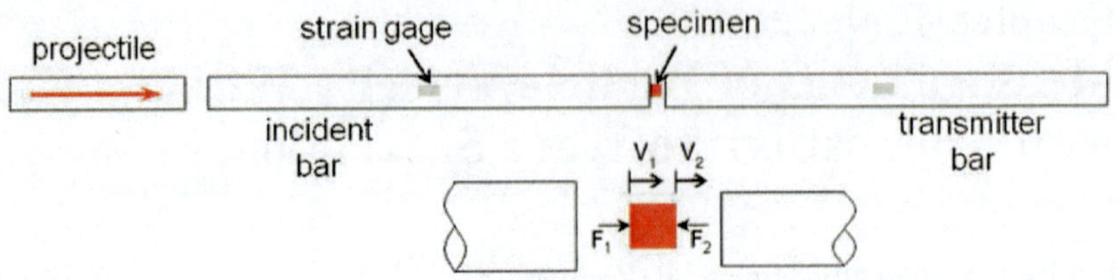


**Fig. 8.1** Standard Kolsky bar used for high rate compression

One shortcoming of the arrangement of Fig. 8.1 is that the waves need to be measured independently at the gage locations. For example, the strain measured at SG1 must be separable into the incident pulse and reflected pulse, i.e., they must not be superimposed at the gage location and must not interfere with later reverberations within the bar. Similarly, the transmitted pulse must not interfere with any other waves or reverberations in the transmitter bar at the gage location. This effectively limits the duration of the incident pulse, and ultimately the maximum duration of the experiment.

$$\Delta t = 2L/c_0 \tag{8.1}$$

Here L is the distance between the incident bar strain gage and the interface between the bar and the specimen. This can become a problem when low-strain-rate experiments are desired. Simply put, low-rate tests require longer test durations, and this may not be available from the standard system.

Two methods are used to overcome this limitation and permit low rate experiments to be performed with a Kolsky bar. The first is to simply use longer bars. By increasing bar length, and therefore L, the available test duration is increased; see for example [12]. The second method involves so-called "wave-separation." The fundamental problem in the standard analysis is that relevant stress-waves superimpose at the gage location, and a single measurement of strain at a point is unable to discern the waves without some additional information. In some cases, it is possible to obtain this information from the boundary conditions of the problem. For example, [13] has performed wave separation with essentially the standard arrangement of Fig. 8.1 by utilizing the knowledge that one end of the transmitter bar is stress-free. Alternately, this information can be obtained from additional instrumentation. For example, [14–16] have used multiple-point strain measurements to separate the relevant waves.

In this work, we expand on a technique used by [17], where co-located strain and particle velocity measurements are used to perform wave separation in a polymeric Kolsky bar. That research used standard electrical resistance strain-gages to measure strain and electromagnetic particle velocity gages to measure velocity. As noted in their paper, the latter instrumentation was less than ideal because of errors introduced by lead wire motion. The objective of the present work is to investigate alternate velocity measuring diagnostics that can be used with the Kolsky bar for the purpose of wave-separation. Preliminary results are presented using strain-gradient based accelerometers [18], line laser extensometers [19], and interferometry using Photon Doppler Velocimetry (PDV) [20]. For the first two techniques, a PMMA bar is used. The PDV experiments are performed with a single steel pressure bar.

## 8.2 Theory

The basic theory behind wave separation using single-point measurements of strain and particle velocity is described in [17]. It can include the effects of dispersion and attenuation due to viscoelastic behavior; [17] uses the method of [10]. However, the underlying concept can be illustrated with the simple one-dimensional theory. Assume strain is measured at a point x within a linear elastic bar. The measured strain is the super-position of two waves, F, a wave-train propagating in the direction of increasing x, and G, a wave-train propagating in the direction of decreasing x.

$$\varepsilon(x,t) = \frac{1}{c_0}\left[F\left(\frac{x}{c_0} - t\right) - G\left(\frac{x}{c_0} + t\right)\right] \tag{8.2}$$

Consider for example the case of the incident bar of Fig. 8.1, where F includes the incident pulse and G includes the reflected pulse. Provided the waves do not overlap at the gage, they are separable. If they do overlap, there is no way to uniquely determine the individual components.

Now assume particle velocity is also measured at x. It is written as follows.

$$v(x,t) = \left[F\left(\frac{x}{c_0} - t\right) - G\left(\frac{x}{c_0} + t\right)\right] \tag{8.3}$$

F and G can now be directly determined by solving Eqs. 8.2 and 8.3 simultaneously.

$$F\left(\frac{x}{c0} - t\right) = \frac{1}{2}[\varepsilon(x,t) + v(x,t)/c_0] \tag{8.4}$$

$$G\left(\frac{x}{c0} - t\right) = \frac{1}{2}[\varepsilon(x,t) - v(x,t)/c_0] \tag{8.5}$$

Thus the waves are separable without any additional assumptions or limitations.

## 8.3 Experiments

### 8.3.1 *Strain Gradient Accelerometer*

[18] used a strain-gradient based accelerometer to measure longitudinal acceleration within a Kolsky bar to compensate for inertial effects in quartz crystal force transducers. The method is based on the one-dimensional wave equation

$$\frac{\partial^2 u}{\partial t^2} = c_0^2 \frac{\partial^2 u}{\partial x^2} \tag{8.6}$$

where u is displacement. The term on the left hand side is acceleration, and the right hand side can be written in terms of a strain-gradient.

$$a = c_0^2 \frac{\partial \in}{\partial x} \tag{8.7}$$

Consider a set of strain-gages separated by a distance Δx as shown in Fig. 8.2. The acceleration, at least for wavelengths large in comparison to Δx, can be approximated from the difference in the readings of these two strain gages.

$$a \approx c_0^2 \frac{\Delta \in}{\Delta x}. \tag{8.8}$$

Velocity is then determined by time integration.

The acceleration measurement system of Fig. 8.2 generally works well for large accelerations. The sensitivity can be increased by increasing Δx, although this affects the rise-time. The numerical integration can also be a problem, resulting in "drift" of the final velocity signal needed for wave separation, especially if both gages are not precisely calibrated.

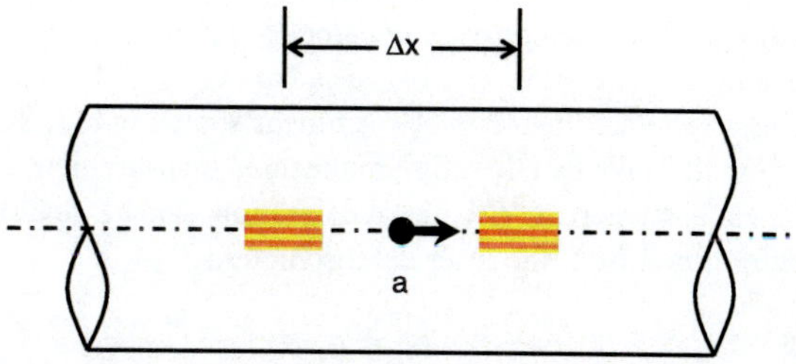


**Fig. 8.2** Two strain gages measure the strain-gradient; this can be related to acceleration

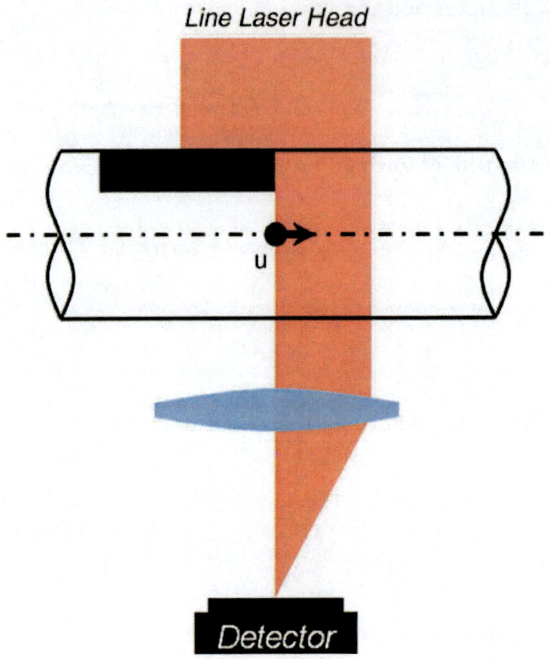


**Fig. 8.3** A laser extensometer (LORD) used to measure the displacement of a transparent bar

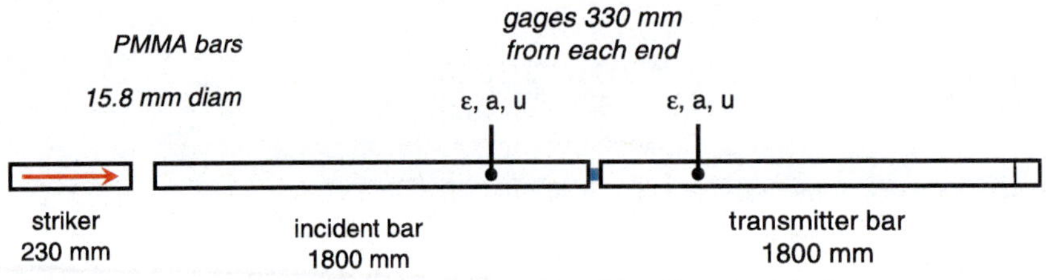


**Fig. 8.4** A PMMA Kolsky bar used for wave separation

### 8.3.2 *Laser Extensometer*

[19] developed a line laser extensometer to monitor the diameter of Kolsky bar specimens during high-rate compression (Line Occlusive Radius Detector, or LORD). It can be applied to a transparent bar, such as PMMA, as shown in Fig. 8.3.[1] The beam emitted from the line laser head is a sheet of uniform intensity along its width. This beam passes through the bar and is focused to a detector. A section of the bar (painted black) is opaque to the beam. As the bar moves, the amount of the beam passing through to the detector varies and the measured photo-detector output can be related to the bar displacement. Particle velocity is then obtained by time differentiation.

The laser extensometer system works well with low-accelerations. The difficulties arise in the differentiation required to determine velocity which makes it difficult to make measurements of rapidly changing velocity profiles.

*Example #1 – Comparison of velocity and acceleration measurements made with various methods* The bar used for our first example is shown in Fig. 8.4. The bars are PMMA and 15.8 mm in diameter. Each bar is instrumented with a set of strain-gages separated by $\Delta x = 19.1$ mm, as shown in Fig. 8.2. A laser extensometer passes through the same locations, as shown in Fig. 8.3. We can therefore make three measurements at each gage locations: (1) strain can be determined from the average of each gage pair, (2) the acceleration can be determined from the strain gradient, and (3) the displacement can be determined from the laser extensometer.

[1] A reflection based mode could perhaps be used with an opaque bar.

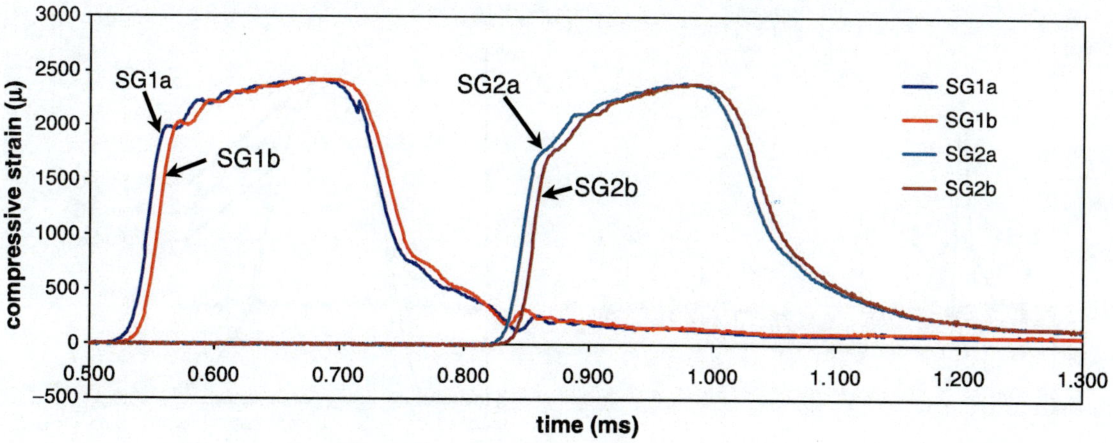


**Fig. 8.5** Strain measured from the strain gages on the incident and transmitter bars

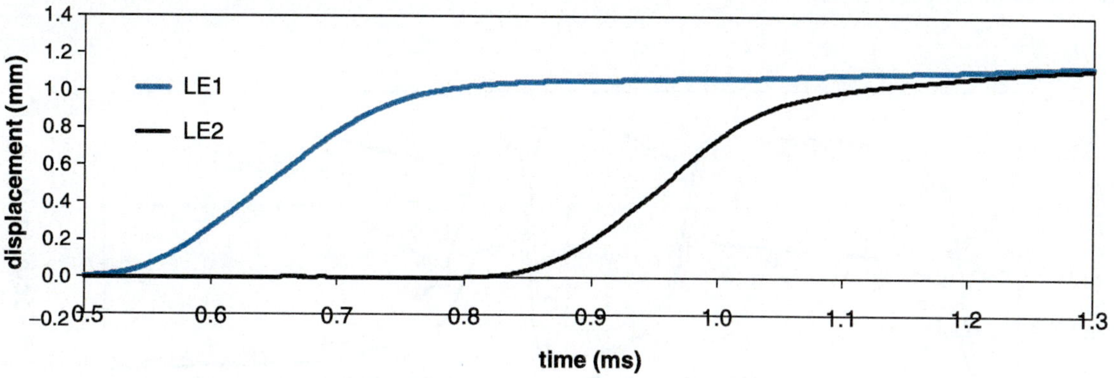


**Fig. 8.6** Displacement measured with the two laser extensometers

In this example, there is no specimen. The projectile impacts the incident bar at about 11 m/s. The bars are initially in contact and there is no reflected pulse. There are also no overlapping pulses at either gage, and no need for wave separation. The measurements are redundant so that comparisons can be made between the different techniques. Figure 8.5 shows the strain measured by each strain gage. SG1a and SG1b are the gages on the incident bar, and SG2a and SG2b are the gages on the transmitter bar (i.e., each pair is separated by $\Delta x$). The laser extensometer outputs (LE1 and LE2 for the incident and transmitter bars, respectively) are shown in Fig. 8.6.

From the raw data of Figs. 8.5 and 8.6, acceleration and velocity at the "gage location" on the incident bar can be determined in multiple ways. This is shown in Fig. 8.7. Figure 8.7a shows the acceleration measured by the strain-gradient accelerometer, AG1, in green. This is obtained by applying Eq. 8.8 to the incident bar data shown in Fig. 8.5. Because there is no wave separation in this case, the individual strain measurements can also be used to determine the velocity at this point using either of those same strain measurements, or the average, using a basic bar analysis ($v = c_0\varepsilon$). This is shown in red in Fig. 8.7b. It can be differentiated to obtain acceleration, shown in red in Fig. 8.7a. The acceleration from AG1 can of course be integrated to obtain velocity; this is shown in green in Fig. 8.7b. The agreement between velocity and acceleration measured by SG1 and AG1 can now be compared, and is quite good. Also shown in Fig. 8.7 is the velocity and acceleration determined by successive differentiations of the LE1 data of Fig. 8.6. In this case, the agreement with the other two measurement systems is poor, especially for the case of the acceleration. This is a result of the numerical differentiation required by the displacement measurement. What is important to note is the drift of the velocity determined by integrating the AG1 measurement. This can be seen for times larger than 0.9 ms in Fig. 8.7b. As mentioned above, this is a consistent problem with the strain-gradient measurements of acceleration. The laser extensometer does not have this problem, and it is suggested that both systems could be used in such a way as to compensate for the other's weakness.

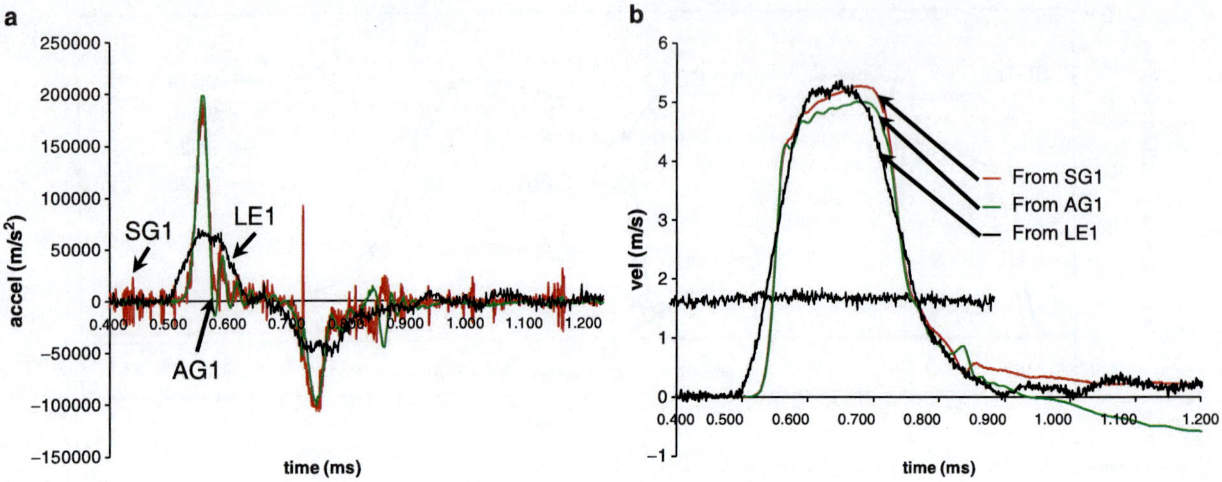


**Fig. 8.7** (**a**) Acceleration at the incident bar gage location using the strain gage, the acceleration gage, and the laser extensometer, (**b**) the velocity at the same location from the same gages

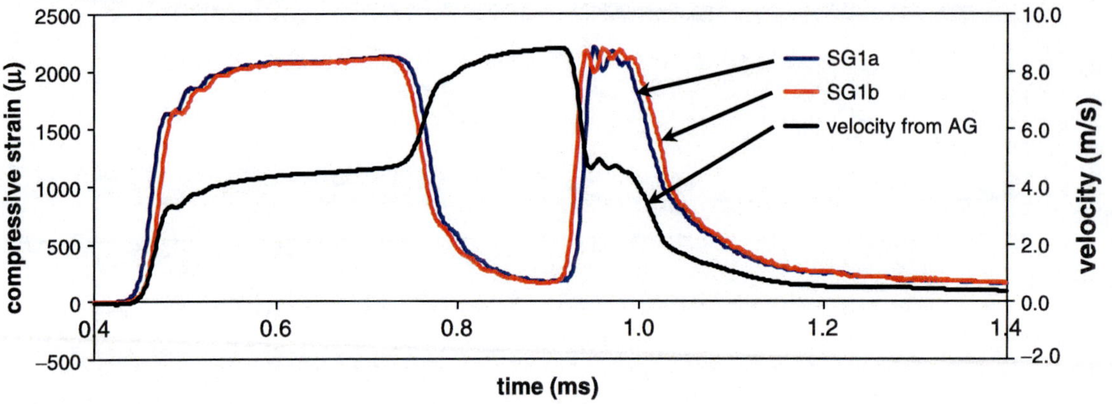


**Fig. 8.8** Strain and velocity used to perform wave separation

*Example #2 – Wave separation in a PMMA bar* This second example uses the same bar shown in Fig. 8.4 except a longer 610 mm projectile is used. Again there is no specimen, but in this instance a small gap is left between the bars. The effect of this is that the incident pulse initially reflects from the free-end of the incident bar until the gap closes. This creates a situation where wave separation is necessary because the refection superimposes with the incident pulse at the gage location. The impact speed is approximately 9 m/s, and the gap closes about 1/3 the way through the incident pulse. At this time, the reflected pulse ends and there is complete transmission into the transmitter bar.

Figure 8.8 shows the strain measured by SG1a and SG1b. The strain at the gage location (essentially the midpoint of $\Delta x$) is determined by averaging these two signals. The difference is then used in Eq. 8.8 to determine acceleration and integrated to obtain velocity; this velocity is shown in Fig. 8.8. The wave can now be separated by applying Eqs. 8.4 and 8.5 to the strain and velocity signals of Fig. 8.8. This is shown, in terms of F and G, in Fig. 8.9. It is now clear what happens in this experiment. The incident pulse (F) is a roughly square pulse[2] approximately 0.6 ms in duration. The reflected pulse is equal in magnitude but opposite in sign for about 0.2 ms while the gap is closing. At this time, the reflection ends.

[2] The incident pulse's deviation from a square pulse is in large part due to viscoelastic effects in the PMMA.

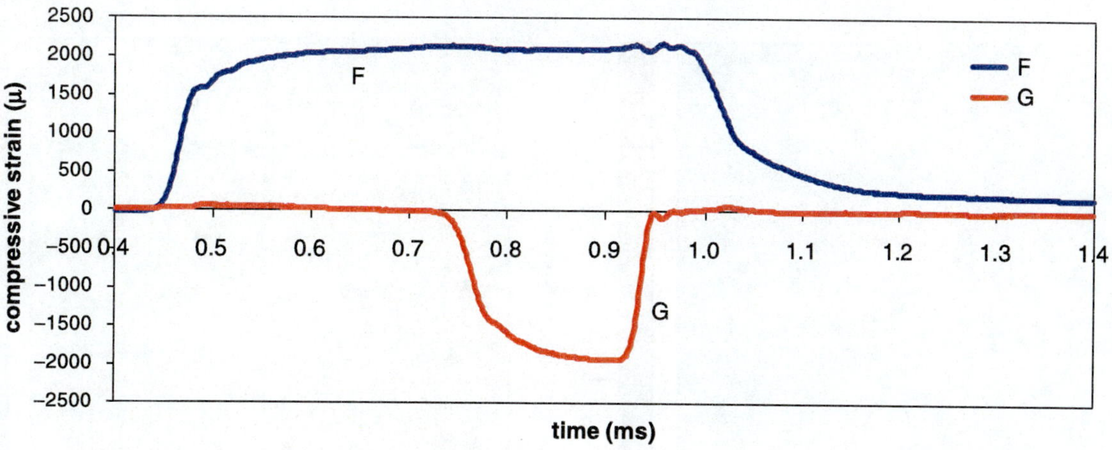


**Fig. 8.9** Separated waves F (incident pulse) and G (reflection)

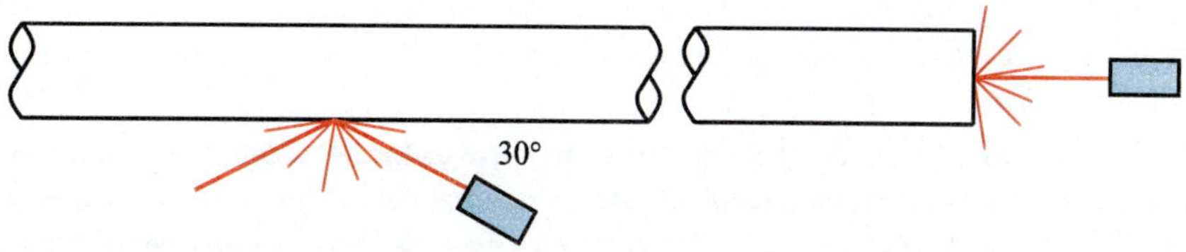


**Fig. 8.10** PDV applied to a Kolsky bar by reflecting from the side and free-end of the bar

## *8.3.3 Photon Doppler Velocimetry*

Interferometry is an extremely useful technique for displacement and velocity measurements in general, and can be applied to the Kolsky bar. For example, [21–23] document the development of Normal Displacement Interferometers (NDI) and Transverse Displacement Interferometers (TDI) [24] for measurements with small diameter Kolsky bars, bars which are too small to effectively use strain gages. Another interferometric technique that is gaining popularity is the PDV. [25] has used PDV with a standard Kolsky bar as an alternative to strain gages and obtained good results. We have performed similar preliminary experiments designed to investigate the application of the PDV system to wave separation.

The difficulty in using a PDV for wave separation using Eqs. 8.4 and 8.5 is that velocity must be measured at the location of the strain gage, typically at the mid-point of the bar. From the perspective of the PDV, this is a transverse displacement measurement, i.e., particle motion in the plane of the bar must be measured. It is assumed that the PDV measures the component of velocity along the path of the incident beam. Transverse velocity can therefore be measured by aligning the PVD probe at an angle to the lateral surface of the bar. This is shown in Fig. 8.10. The figure also shows a more direct application of the PDV where the probe is oriented to measure the longitudinal motion of the free-end of the bar, a configuration that is useful for practical testing provided at least one end of the bar is stress-free.

Figure 8.11 shows PDV spectrograms, which plots the spectral density of recorded particle velocities as a function of time, for two experiments with a 9.5 mm steel pressure bar (i.e., a single bar). In each experiment, the bar is impacted by a 230 mm long projectile. Figure 8.11a shows a free-end measurement. The PDV operates by measuring the beating of the Doppler shifted return beam with an unshifted reference beam. The reference beam is collected by the same PDV probe. In this case, the return beam is strong and the expected wave form is clear with good resolution both temporally and spatially. Figure 8.11b shows the transverse velocity measurement where the probe angle is 30°. Note the bar is sanded randomly (320 grit) at the measurement point to provide a diffuse reflection, however, the majority of the beam reflects away from the probe. This results in the relatively poor resolution seen in the figure. In spite of this, a useable signal is obtained.

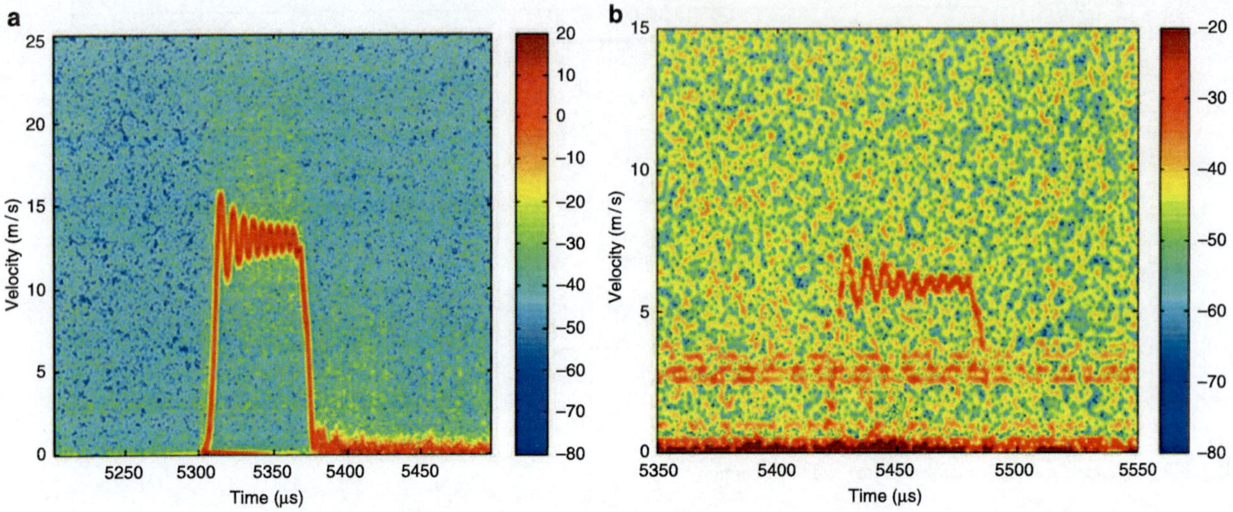


**Fig. 8.11** Velocity measured by the PDV system (**a**) directly from the free end (**b**) reflected from the side

## 8.4 Conclusion

Several methods of obtaining particle velocity at points within a Kolsky bar have been investigated. These include strain-gradient measurements of acceleration, displacement measurements with a laser extensometer, and velocity measurements with a PDV. Shortcomings of the acceleration and displacement measurements have been identified and are in part related to the numerical integration and differentiation that must be performed to determine velocity. Results obtained with the PDV are much more promising; the primary obstacle here is to devise an orientation in which the measured velocity is sufficiently resolved. These techniques can provide a method to make distinct measurements of individual stress waves in a Kolsky bar when the waves are superimposed at the measurement location. This permits low rate experiments to be performed with the Kolsky bar, extending the strain-rate range that can be achieved.

## References


1. Kolsky H (1949) Proc Phys Soc B 62:676–700
2. Follansbee PS (1985) The Hopkinson Bar, vol 8, 9th edn, Metals handbook. American Society for Metals, Metals Park, pp 198–217
3. Chen W, Song B (2011) Split Hopkinson (Kolsky) Bar. Springer, New York/Dordrecht/Heidelberg/London
4. Gorham DA (1983) A numerical method for the correction of dispersion in pressure bar signals. J Phys E Sci Instrum 16:477–479
5. Follansbee PS, Franz C (1983) Wave propagation in the Split-Hopkinson Pressure Bar. J Eng Mater Technol 105:61
6. Gong JC, Malvern LE, Jenkins DA (1990) Dispersion investigation in the Split-Hopkinson Pressure Bar. J Eng Mater Technol 112:309–314
7. Lundberg B, Blanc RH (1988) Determination of mechanical material properties from the two-point response of an impacted linearly viscoelastic rod specimen. J Sound Vib 126(1):197
8. Zhao H, Gary GA (1995) Three dimensional analytical solution of the longitudinal wave propagation in an infinite linear viscoelastic cylindrical bar, application to experimental techniques. J Mech Phys Solids 43(8):1335
9. Zhao H, Gary G, Klepaczko JR (1997) On the use of a viscoelastic Split Hopkinson Bar. Int J Impact Eng 19:319
10. Bacon C (1998) An experimental method for considering dispersion and attenuation in a viscoelastic Hopkinson Bar. Exp Mech 38(4):242–249
11. Sawas O, Brar NS, Brockman RA (1998) Dynamic characterization of compliant materials using an all-polymeric Split Hopkinson Bar. Exp Mech 38:205
12. Gilat A, Matrka T (2010) A new compression intermediate strain rate testing apparatus. In: Proceedings of the SEM annual conference, Indianapolis, 7–10 June 2010
13. Park SW, Zhou M (1999) Separation of elastic waves in Split Hopkinson Bars using one-point strain measurements. Exp Mech 39(4):287–294
14. Lundberg B, Henchoz A (1977) Analysis of elastic waves from two-point strain measurement. Exp Mech 17:213
15. Zhao H, Gary GA (1997) A new method for the separation of waves. Application to the SHPB technique for an unlimited duration of measurement. J Mech Phys Solids 45(7):1185
16. Othman R, Gary G (2007) Testing aluminum alloy from quasi-static to dynamic strain-rates with a modified Split Hopkinson Bar method. Exp Mech 47:295–299

17. Casem DT, Fourney W, Chang P (2003) Wave separation in viscoelastic pressure bars using single-point measurements of strain and velocity. Polym Test 22(2)
18. Casem D, Weerasooriya T, Moy P (2005) Inertial effects of quartz force transducers embedded in a Split Hopkinson Pressure Bar. Exp Mech 45(4):368–376
19. Ramesh KT, Narasimhan S (1996) Finite deformations and the dynamic measurement of radial strains in compression Kolsky bar experiments. Int J Solids Struct 33(25):3723–3738
20. Strand OT, Goosman DR, Martinez C, Whitworth TL (2006) Compact system for high-speed velocimetry using heterodyne techniques. Rev Sci Instrum 77:083108
21. Casem DT (2009) A small diameter Kolsky bar for high-rate compression. In: Proceedings of the 2009 SEM annual conference and exposition on experimental and applied mechanics, Albuquerque, 1–4 June 2009
22. Casem DT, Grunschel SE, Schuster BE (2010) Interferometric measurement techniques for small diameter Kolsky Bars. In: Proceedings of the 2010 SEM annual conference and exposition on experimental and applied mechanics, Indianapolis, pp 973–977, 6–10 June 2010
23. Casem DT, Grunschel SE, Schuster BE (2012) Normal and transverse displacement interferometers applied to small diameter Kolsky bars. Exp Mech 52(2):173–184
24. Kim KS, Clifton RJ, Kumar P (1977) Combined normal-displacement and transverse-displacement interferometer with an application to impact of y-cut quartz. J Appl Phys 48(10):4132–4139
25. Avinadav C, Ashuach Y, Kreif R (2011) Interferometry-based Kolsky bar apparatus. Rev Sci Instrum 82:073908

# Chapter 9
# Mechanical Response of Friction Stir Welded Aluminum 2139-T8 as a Function of Loading Rate and Stress-State

**C. Allan Gunnarsson, Tusit Weerasooriya, and Bryan Love**

**Abstract** Friction stir (FS) welding is a solid state process that does not melt the base metal. This has the benefit of changing the mechanical characteristics of the original material as little as possible. This welding process is used frequently on large size aluminum plates, where the size of the welded structure prohibits heat treating of it afterward to recover pre-weld material characteristics. In this study, two plates of aluminum 2139 are FS welded together, and then machined into experimental specimens from various locations (base metal, weld material). In addition, specimens are fabricated that span from one plate to the other, including the whole weld. These are used to experimentally characterize the mechanical response and failure behavior of the weld to allow for incorporation into simulation methodology to predict the deformation and failure of the weld. In addition, these experimental observations will be used to understand the mechanisms of weld failure for FS welds under various types of high rate loading. The characterization is performed using two types of experiments: uni-axial tensile testing and a modified three-point bending experiment with distributed loading over the weld. These experiments are performed at low and high loading rates to investigate the effect of loading rate on the deformation response and failure behavior. The high rate experiments are accomplished using a tensile Kolsky bar to conduct tensile experiments, and a modified compressive Kolsky bar to perform bending experiments. Digital image correlation is used in the experiments to measure the strain field on the specimen and observe how it differs between the bulk material, heat-affected zone, and weld zone. In this paper, the experimental techniques are discussed and results are presented.

**Keywords** Friction Stir Weld (FSW) • Digital Image Correlation (DIC)

## 9.1 Introduction

Aluminum 2139-T8 is a relatively new aluminum alloy, registered with the Aluminum Association in 2004. This alloy has high strength, fracture toughness, fatigue life, and better ballistic properties than other high performance aluminum alloys, including Aluminum 2519, which is currently the aluminum alloy used in modern armored vehicles. The mechanical behavior of this alloy has been studied by various researchers including Vural et al. [1].

Aluminum 2139-T8 plates are frequently welded together using the Friction Stir Weld (FSW) process in applications where the size of the welded structure prohibits heat-treating afterward to recover pre-weld material characteristics. FSW is a solid-state process that does not melt the base metal completely. Therefore, this joining process does not completely alter the mechanical characteristics of the original material at and near the weld. The FSW process does alter the microstructure, creating different zones in and around the weld. These different zones are subjected to different thermal-mechanical processes and have varying mechanical properties [2]. Based on the thermo-mechanical processes that occur during the friction stir welding, there are four separate material zones created: the bulk material zone (BZ), heat-affected zone (HAZ), thermo-mechanically affected zone (TMAZ) and weld or stir zone (WZ). The HAZ experiences micro-structural changes due to thermal effects (heating) only. The TMAZ undergoes plastic deformation due to the stirring process in addition to thermal-induced micro-structural changes. The thermo-mechanically affected zone and heat-affected zone are visually indistinguishable in an etched micrograph of a FSW plate. Therefore, this study considers only three zones, combining the

C.A. Gunnarsson (✉) • T. Weerasooriya • B. Love
Army Research Laboratory, Aderdeen Proving Ground, Weapons and Materials Directorate, RDRL-WMP-B, MD 21005
e-mail: carey.a.gunnarsson.civ@mail.mil

V. Chalivendra et al. (eds.), *Dynamic Behavior of Materials, Volume 1: Proceedings of the 2012 Annual Conference on Experimental and Applied Mechanics*, Conference Proceedings of the Society for Experimental Mechanics Series,
DOI 10.1007/978-1-4614-4238-7_9, 

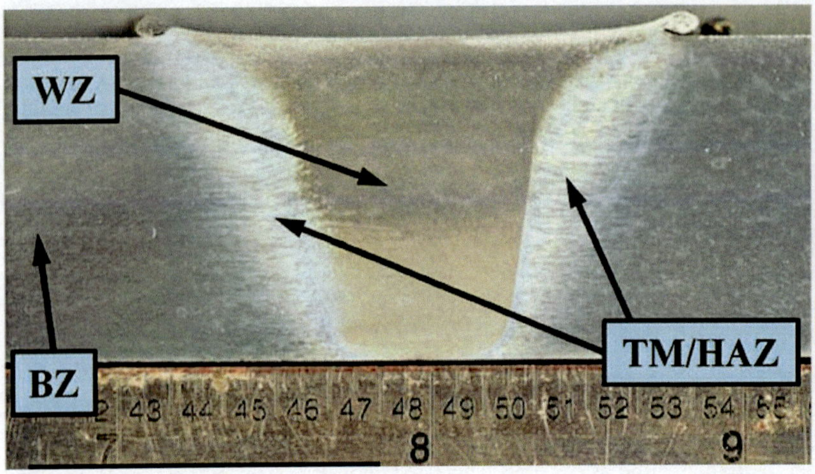


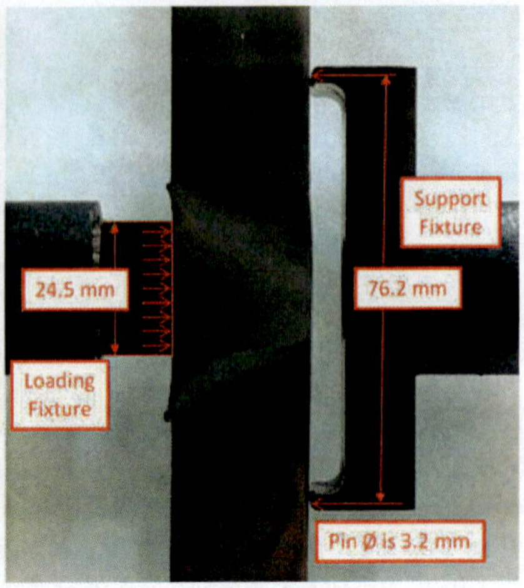


**Fig. 9.1** (a) Etched cross-section showing weld area detail for fsw 2139 plate and (b) bend experimental setup

TMAZ and HAZ into a single zone, labeled as TM/HAZ. Figure 9.1a shows a tensile specimen that has been chemically processed to illustrate the different zones (polished, then etched with a 10% NaOH/90% water solution). Other studies have been conducted by Sano et al. [3] to obtain the mechanical response of materials from the individual zones. The main focus of this study is to investigate the response of the welded plate as a whole (all three zones) at different loading rates and stress-states.

In this study, the mechanical response (deformation and failure) of an aluminum 2139-T8 friction stir welded plate is experimentally characterized. The deformation and failure response obtained from these experiments are being used to validate simulation methodologies. Two plates of aluminum 2139-T8 were friction stir welded together, and specimens were created from the plate that spanned across the weld and included all material affected by the FSW process. Two types of experiments were performed with these specimens: uni-axial tensile and a modified three-point bending with distributed loading over the weld zone. These experiments were performed at three different loading rates to investigate the effect of loading rate on the response. The dynamic bending experiments utilized a modified compressive split-Hopkinson bar. Digital image correlation (DIC) is used to measure the strain field along the length of the specimen and to understand the relative response of the three zones created by the FSW process.

## 9.2 Experiments

Two types of experiments were performed: uni-axial tensile and modified three-point bend. The tensile experiments were performed using an Instron test machine at a quasi-static rate (0.127 mm/s) and an intermediate rate (5,912 mm/min) at the machine's velocity limit. Tensile experiments were not performed at high loading rates due to the larger size of the plate (25.4 mm thick). The tensile specimens were fabricated using EDM (electro-discharge machining) to cut a 6.35 mm (0.25 in.) thick slice from the end of the plate, perpendicular to the direction of the weld. The specimen gage area was approximately 250 mm (10 in.) long, contained the weld area (which is approximately 50 mm long) and bulk material, and was thinned down to 3.2 mm, leaving the thicker ends for gripping. The specimen gage area was speckled for full-field strain measurement.

The modified bending experiments were conducted with 6.35 mm thick specimens that were sliced from the FSW plate. In these experiments, the loading devices were integrated into a steel compression Kolsky bar apparatus. The specimens were supported by a two pin fixture attached to the end of the transmission bar. The pin diameter was 3.2 mm and the pins were spaced 76.2 mm (3 in.) apart. The specimen was loaded using a fixture attached to the incident bar that had a flat rectangular loading face. The loading face was 6.35 mm thick (same as the specimen), with a loading height of 24.5 mm (Fig. 9.1b). This loading setup was designed to provide distributed loading over most of the weld zone length, instead of point loading. The specimen was loaded by the fixture at a velocity of approximately 25 m/s, which corresponded to a loading rate of 1.06 kN/μs to peak load.

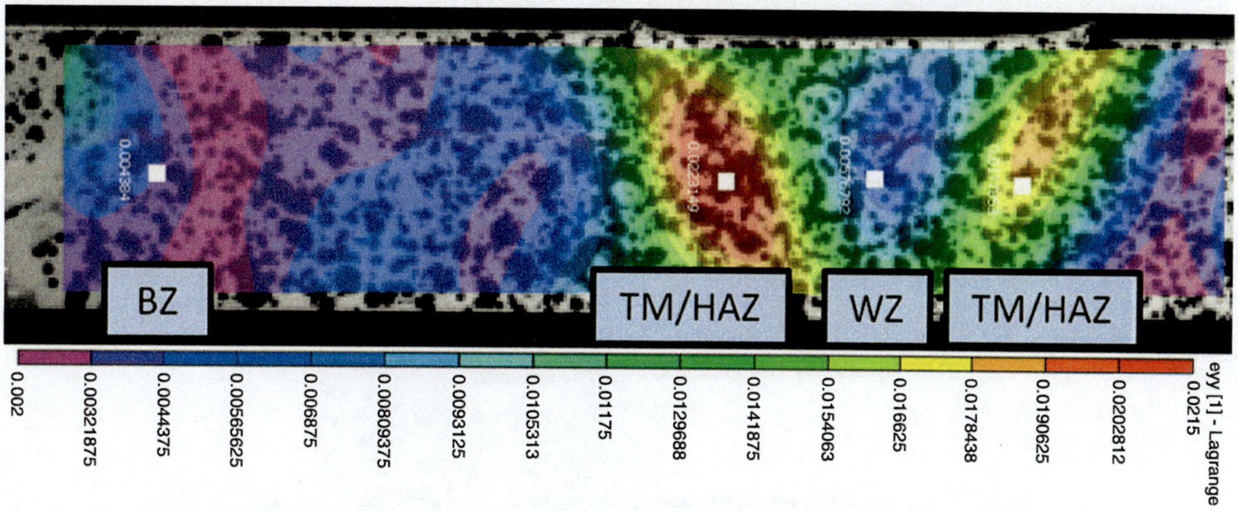


Fig. 9.2 Strain concentrations and locations of data extraction points for 5,912 mm/min rate

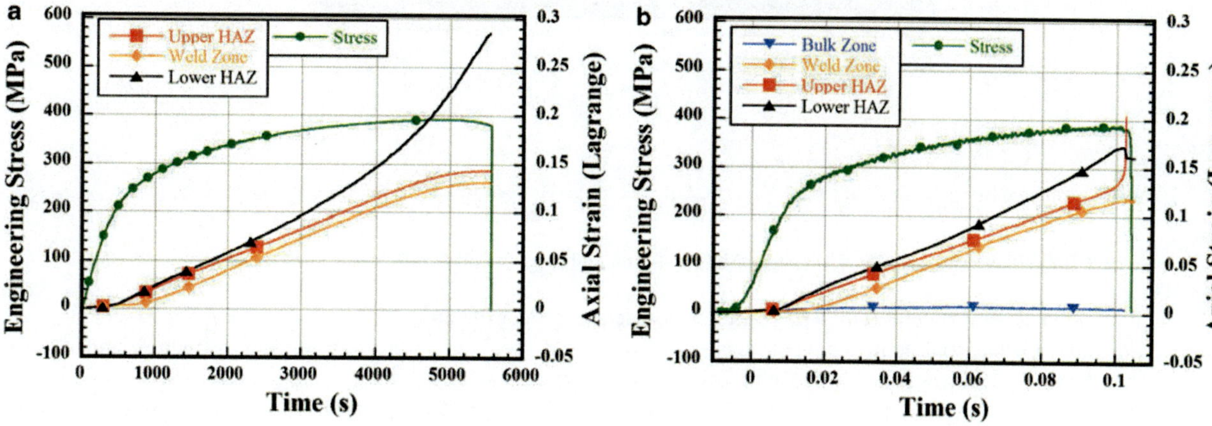


Fig. 9.3 Stress and strain histories for tensile experiments at (**a**) 0.127 mm/min and (**b**) 5,912 mm/min

## 9.3 Results

### *9.3.1 Tensile Experiments*

In the tension experiments, the specimen failed at the HAZ. During tensile loading, strain increased and localized in the two HAZs, and to a lesser extent the WZ, as shown in Fig. 9.2. The color map in Fig. 9.2 shows axial strain (along the length) on the specimen surface. Figure 9.2 also shows the locations of the data points where strain history data was extracted from each zone for the intermediate rate. Figure 9.3 shows a comparison of axial strain vs. time for the different zones, as well as far-field uni-axial engineering stress history for the (a) 0.127 mm/min (quasi-static) and (b) 5,912 mm/min (intermediate) displacement rate experiments. The quasi-static experiments were not able to provide strain in the BZ, which was outside of the images. The intermediate rate experiment (Fig. 9.3b) shows that the BZ is never plastically strained, with the maximum stress never reaching the yield stress of the bulk 2139-T8. All of the axial strain is localized in the material zones affected by the FSW process. It is assumed that this was the case for the quasi-static experiments as well. From Fig. 9.3, the strain histories of the different zones as well as the engineering stress history are very similar for both the quasi-

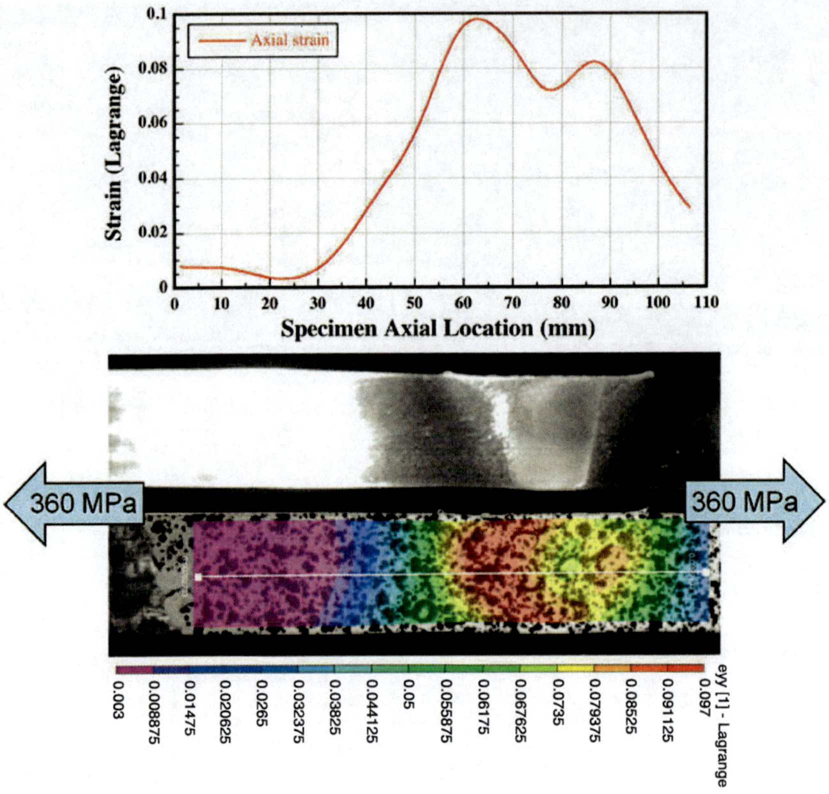


**Fig. 9.4** Strain distribution for 5,912 mm/min rate at 360 mpa. Strain in the graph was extracted along the *white-line*

static and intermediate loading rates. It is concluded that the plate as a whole does not exhibit notable rate dependent behavior.

In traditional tensile experiments with a uniform microstructure in the gage area, the strain distribution is expected to be uniform along the loading axis until localized necking. These weld specimens have varying microstructure and it was expected that the strain field would not be constant over the gauge length, but vary depending on proximity to the weld zone. As pointed out previously, strain concentrated in the HAZs and WZ. Figure 9.4 shows the strain distribution along the specimen at an engineering stress of 360 MPa for the 5,192 mm/min rate. The figure gives the axial strain as a function of location; the x-axis of the graph corresponds to the location in the specimen image and strain field shown below the graph. From the data in Fig. 9.4, the strain is concentrated in the two HAZs (shown as the two peaks), with significant strain in the WZ (the slight dip between the peaks). Additionally, strain concentration is pronounced some distance from the weld center; it does not decrease to the bulk material constant strain until approximately 40 mm from the HAZ. This is why the BZ strain was not captured at quasi-static rate; from the etched image, it appeared the bulk material zone was much closer to the weld center. This constant strain in the BZ is shown in Fig. 9.4 as the flat region under 30 mm. The axial strain in the BZ reaches only 0.005, meaning the BZ remained elastic.

## 9.3.2 High Rate Bend Experiments

The specimens did not fail during the modified high rate bend experiments. The specimens bent and bottomed out against the supporting structure. The peak load reached was approximately 40 kN. The maximum strain field (Fig. 9.5a), which occurred well after peak load, shows the expected tensile strain at the support side and compressive strain at the loading side. However, the tensile strain is concentrated at the HAZ/WZ interface, and is not uniform across the FSW-affected area. Figure 9.5b shows the loading history and strain history for both axial (tensile) and principal strain from the same bending experiment; the location of the strain data extraction point is shown in Fig. 9.5a (the *white point*). The principal and axial strains are identical, demonstrating that the stress-state at that point is purely tensile.

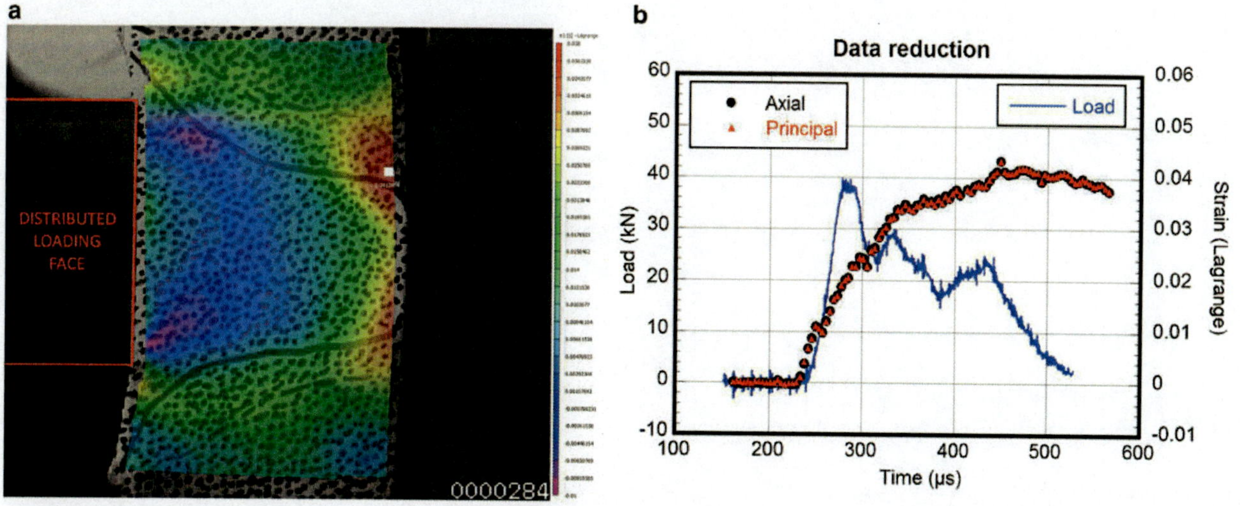


**Fig. 9.5** (**a**) strain field for bending experiments at 25 m/s showing distributed loading face (*left*) and strain data extraction location; (**b**) load history and peak strain point history

## 9.4 Conclusions

Two experimental techniques were developed to subject a FSW aluminum plate to loading across the entire area affected by the FSW process. DIC strain measurement method was used to obtain the full-field strain of the plate specimens during both tensile and bending loading. Under tensile loading, axial-strain concentrates in the HAZ, and to a lesser extent the WZ, and this is where tensile failure will occur. The bulk material in the far-field does not strain significantly, and is still elastic during failure at the TM/HAZ; however, part of the BZ in the vicinity of the weld is exposed to significant plastic strains. There is a significant strain gradient until reaching a distance of 60 mm from the weld center. FSW plate does not exhibit any significant rate effects in tensile loading for rates from quasi-static to intermediate. Under distributed bend-loading, the largest strain is tensile and is concentrated in the HAZ. This type of non-typical experimental data will be used to validate simulation methods that are used to understand the deformation and failure of welds. The experimentally determined full-field strain, up to failure, can be compared to that obtained from different simulation methodologies for different stress-states.

## References

1. Vural M, Caro J (2009) Experimental analysis and constitutive modeling for the newly developed 2139-T8 alloy. Mater Sci Eng A520:56–65
2. Fonda R, Bingert J (2004) Microstructural evolution in the heat-affected zone of a friction stir weld. Metall Mater Trans 35A:1487–1499
3. Sano T, Yu J, Medintz J, Yen C, Doherty K (2011) The effect of friction stir welding on the microstructure and tensile properties of Al 2139-T8 alloys. ARL report

# Chapter 10
# Characterization and Modeling of Polymeric Matrix Under Static and Dynamic Loading

**Brian T. Werner and Isaac M. Daniel**

**Abstract** A polymeric matrix (3501-6) used in composite materials was characterized under multi-axial loading at strain rates varying from quasi-static to dynamic ones. Tests were conducted under uniaxial compression, tension, pure shear and combinations of compression and shear. Quasi-static and intermediate strain rate tests were conducted in a servo-hydraulic testing machine. High strain rate tests were conducted using a split Hopkinson Pressure Bar (Kolsky bar) system built for the purpose. This SHPB system was made of a glass/epoxy composite (Garolite) bars having an impedance that is closer to that of the test polymer than metals. The typical stress-strain behavior of the polymeric matrix exhibits a linear elastic region up to a yield point, a nonlinear elastoplastic region up to an initial peak or "critical stress," followed by a strain softening region up to a local minimum and finally, a strain hardening region up to ultimate failure. A general three-dimensional elasto-viscoplastic model was developed incorporating strain rate effects, and including the large deformation region. The model was formulated in strain space unlike most models in the literature. The stress-strain curves obtained were used to develop and validate the new elasto-viscoplastic constitutive model.

**Keywords** Polymer characterization • Multi-axial testing • Dynamic testing • Strain rate effects • Elasto-plastic behavior • Constituive modeling

## 10.1 Introduction

Recent and ongoing research in fiber reinforced polymer composites (FRPC) has shown that there is a significant rate effect on the resin dominated stiffness and strength properties ($E_2$, $G_{12}$, $F_{2c}$, $F_{2t}$, $F_6$) of these materials [1, 2]. Fabrication and testing of these composites is costly due to the expense of the material itself, and its anisotropic behavior that requires time consuming multi-axial testing. In the case of carbon/epoxy composites, the fiber itself shows no rate dependence in its mechanical response. This suggests that all of the rate effects on the composite can be traced to the matrix. Since the polymer matrix is basically isotropic, a much less costly evaluation of a composite can be achieved by characterizing the bulk matrix at various strain rates.

The constitutive and strain rate behavior of epoxies under various loading conditions has been studied by many researchers [3–13]. Some studies focus on characterization of the resin at various rates [3, 4]. Many investigators have focused on viscoelastic analysis and the determination of shift factors used in temperature-time superposition modeling [5–8]. Some have attempted to link the monomer structure to the bulk properties [9]. Goldberg et al. have published works on both testing various resins under multiaxial conditions at various rates as well as modeling with a state variable approach [10–12].

The objective of this study was to characterize the matrix resin under multi-axial loading at different strain rates and develop a general three-dimensional elasto-viscoplastic model that incorporates rate effects and differentiates between dilatational and deviatoric loading. Emphasis was placed on development of a relatively simple model that would not require extensive testing for evaluation of a given matrix.

B.T. Werner • I.M. Daniel (✉)
Northwestern University, Evanston, IL, USA
e-mail: bwerner@u.northwestern.edu; imdaniel@northwestern.edu

V. Chalivendra et al. (eds.), *Dynamic Behavior of Materials, Volume 1: Proceedings of the 2012 Annual Conference on Experimental and Applied Mechanics*, Conference Proceedings of the Society for Experimental Mechanics Series,
DOI 10.1007/978-1-4614-4238-7_10, 

## 10.2 Material Characterization

### *10.2.1 Material and Specimen Preparation*

The polymer matrix investigated is a high stiffness, high strength epoxy (3501-6). It has a highly crosslinked structure that provides stiffness and strength but also reduces its ductility which leads to a fairly brittle polymer. The B-staged resin (supplied by Applied Poleramic Inc.) is normally stored in a freezer. The partially cured and frozen resin was chipped out, weighed, and placed in the mold. A small amount of acetone was added to reduce the resin viscosity and facilitate casting of the partially cured resin into the mold. The temperature was raised at a rate of 4 F/min up to 250°F, and held at that level for 1 h. During this stage the resin viscosity was low and vacuum was drawn at 30 in. Hg to remove any air that may have been trapped during the resin melt and to boil off any acetone that may remain in the resin. This causes the resin to foam considerably. Subsequently, the vacuum was removed and air was slowly bled back into the mold to collapse the bubbles. For more complex casting geometries a small amount of solvent (acetone) is used to reduce the viscosity of the partially cured epoxy and facilitate casting into the mold. For large volumes of resin, the amount of solvent is restricted to prevent the acetone from igniting as it boils off during the exothermic stage of the cure. The material was cast into closed molds to produce thin (3 mm) plates for tensile coupons, thick blocks for prismatic compression coupons, and thin-wall cylinders for specimens to be tested under torsion and combinations of torsion and axial tension or compression. In casting the cylinders, an aluminum tube was used as the mold with a Teflon rod inserted at its center as the core.

The geometry and dimensions of the specimens used are shown in Fig. 10.1. Specimens for uniaxial tensile testing were thin dogbone coupons with a gage section of 12.7 × 50.8 mm machined from 3 mm thick sheets using a router with an abrasive bit. Specimen edges were hand polished and a thin layer of mounting epoxy was used at the fillets to fill in any microcracks formed from machining. Uniaxial compressive tests were conducted on thick prismatic coupons 12.7 mm long with a cross section of 7.62 × 8.89 mm. It was important to maintain an aspect ratio of around 1.5 as anything shorter than that would lead to end effects inhibiting a uniaxial stress condition which would result in an increased apparent modulus and critical stress. A higher aspect ratio would cause buckling of the specimen and a decrease in the critical stress. Aspect ratios from 0.125 to 2.7 were tested and it was found that the small aspect ratios yielded stiffness values approaching the plane strain stiffness value ($C_{11}$) and the larger aspect ratios yielded the Young's modulus E of the material. An aspect ratio of 1.5 was chosen for the determination of Young's modulus E (Fig. 10.2). Tests under pure shear and combinations of shear and normal tensile or compressive stress were conducted on thin-wall cylindrical specimens. Since the resin used has a shear strength higher than that of available mounting epoxies, the cylindrical specimens were designed and machined with enlarged ends and fillets as shown in Fig. 10.1. The cylinders were machined using a lathe to produce a 1.27 mm wall thickness in the gage section and a fillet radius of 11.43 mm. The resulting dogboned cylinder was mounted on a steel post which was used to apply compression and torsion.

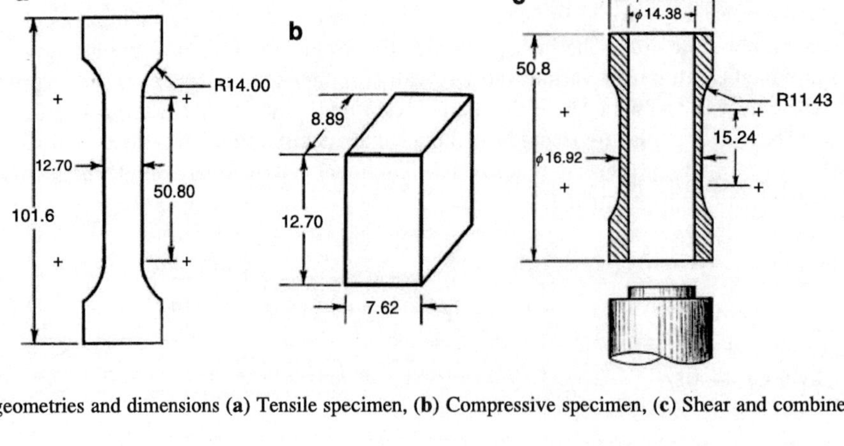


**Fig. 10.1** Specimen geometries and dimensions (**a**) Tensile specimen, (**b**) Compressive specimen, (**c**) Shear and combined stress specimen

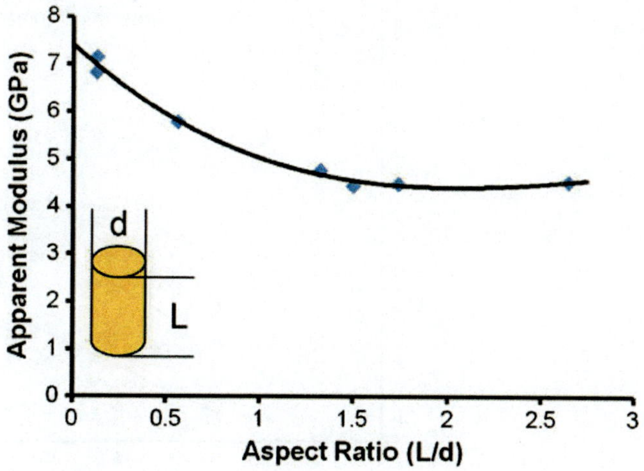


**Fig. 10.2** Effect of aspect ratio on compressive response

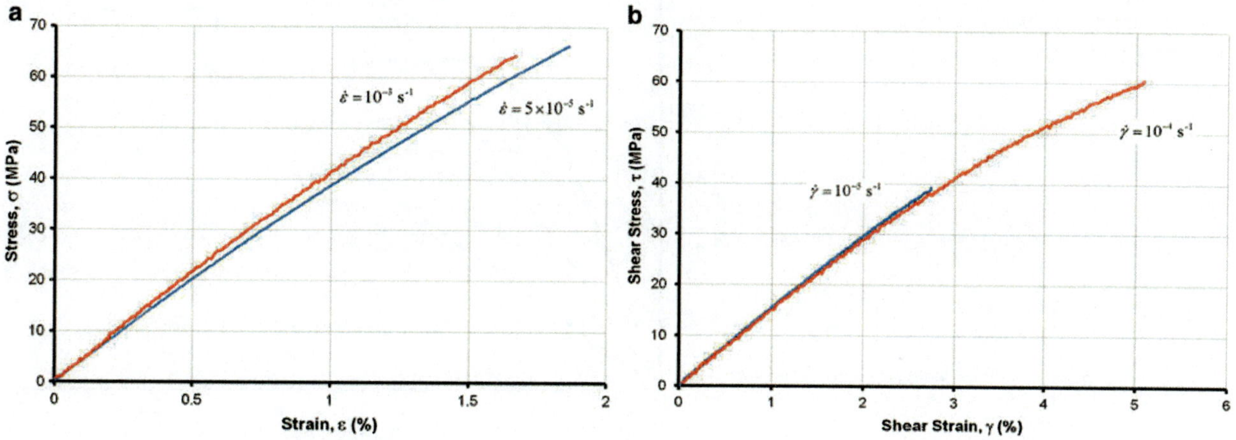


**Fig. 10.3** Tension and shear stress-strain curves at various strain rates

### *10.2.2 Testing and Results*

Experiments were conducted at various strain rates ranging from $10^{-5}$ to 1,500 $s^{-1}$. Lower rate experiments at less than 1 $s^{-1}$ were conducted in a servo-hydraulic testing machine while the higher rate tests were conducted in a Split Hopkinson Pressure Bar system using composite (G-10) bars. Tension and pure shear tests at lower rates, $10^{-4}$ to 1 $s^{-1}$, did not show much rate dependence in stress-strain behavior. Similarly, the brittle failure under tensile loading did not show much variation (Fig. 10.3). The rate effect was much more pronounced under uniaxial compression.

Dynamic testing at high strain rates was conducted in a Split Hopkinson Pressure Bar (SHPB) system with G-10 bars. Typically either polymeric bars such as polycarbonate or metallic bars such as steel or aluminum are chosen for SHPB testing. However, the resin being tested is stiffer and stronger than most polymers so it would damage the ends of the bars during testing. For metallic bars, not only is there a very large impedance mismatch, the resin is expected to reach ultimate failure at strains higher than 20%. The wave speed in metallic bars is very high and the length of the bars needed to produce a long enough stress pulse to achieve the required ultimate strain was impractical. A compromise was reached with G-10 composite rods made of stacked layers of woven glass/epoxy. The impedance of the bars was not too different from that of the resin, which allowed stress equilibration within the specimen to occur much sooner than in the steel or aluminum Hopkinson bar systems (Fig. 10.4). The wave speed within the bars was much lower than that in the metallic bars, so that a 1.83 m incident bar was sufficient to capture most of the resin behavior. Because the glass fibers in the G-10 bars are oriented along the axis of the

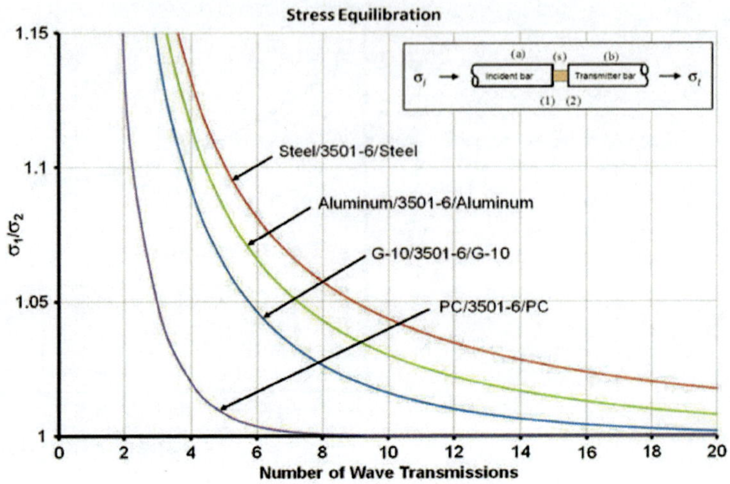


**Fig. 10.4** Stresses at ends of resin specimen for different Hopkinson bar materials

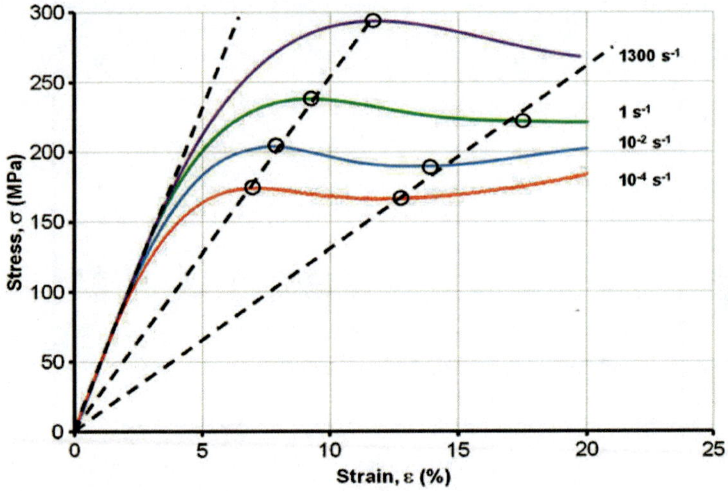


**Fig. 10.5** Compressive stress-strain curves at various rates showing radial alignment of characteristic features

bar, the stress wave did not show much dispersion or attenuation, and thus, there was no need for a more complex viscoelastic analysis of the signals. The only concern was the lack of axisymmetry in the rods, but no significant difference was observed in the propagating pulse monitored by strain gages along the principal planes of the bar cross section.

Stress-strain curves were obtained over a wide range of strain rates. The compressive stress-strain curves shown in Fig. 10.5 have the same overall shape and share many key features Each curve has a linear elastic region up to the yield point where the primary bonds in the polymer chains shorten to store the elastic energy. This is followed by a plastic region where the chains between crosslinks begin to flex up to a critical point of material instability, the "critical stress." Next, the chains between crosslinks start to buckle until the free volume within the resin reaches a minimum at the plateau point (local minimum) of the stress-strain curve. From that point on, the resin begins to harden as it is forced to flow transversely to the loading direction. This causes the chains to begin straightening again in the transverse to the loading direction up to a critical stress when cracking starts due to tension in the transverse direction. All curves share the same initial modulus. They only differ in the nonlinear behavior that shows a rate dependence. What is notable is that the key features on the curves expand radially from the origin with increasing strain rate. This is highlighted in Fig. 10.5 where it can be seen that the yield point, the critical point, and the plateau point all fall along these radial lines.

These characteristic properties vary linearly with the logarithm of strain rate as shown in Fig. 10.6. The resulting rate dependence can be expressed as

$$P(\dot{\varepsilon}) = P(\dot{\varepsilon}_0)\left[m\log_{10}\frac{\dot{\varepsilon}}{\dot{\varepsilon}_0} + 1\right] \tag{10.1}$$

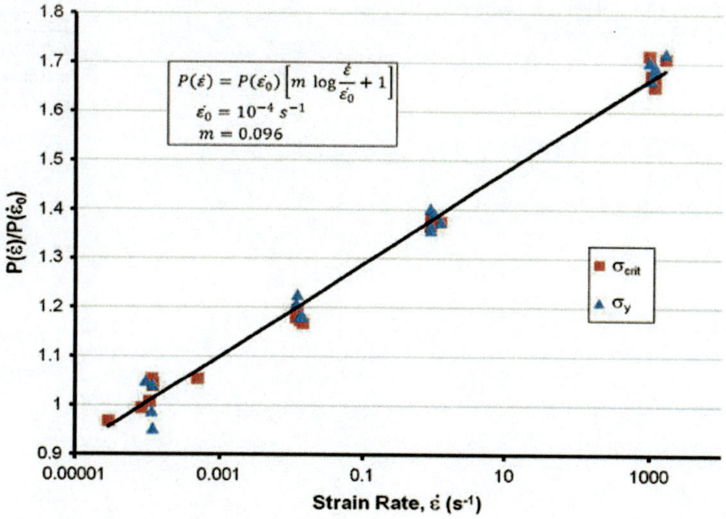


Fig. 10.6 Variation of normalized critical and yield stresses with strain rate

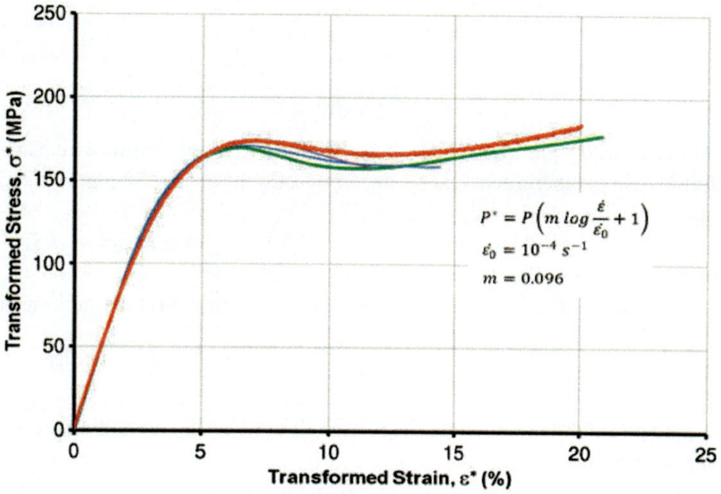


Fig. 10.7 Master compressive stress-strain curve at reference strain rate

where $P(\dot{\varepsilon})$ is the rate dependent property, $P(\dot{\varepsilon}_0)$is that property at the reference strain rate, and $m$ is the slope of the linear logarithmic relation. In the present case, a reference strain rate of $10^{-4}\ s^{-1}$ was used and $m$ was found to be 0.096. Using this relation, the stress-strain curves were transformed into one master curve at the reference strain rate as shown in Fig. 10.7.

## 10.3 Constitutive Modeling

No rate effect was observed in the shear and tensile stress-strain behavior of the material. Thus, constitutive modeling was based and focused on the master curve and the rate effect observed in the compressive response of the material. The measured elastic properties for the resin are tabulated in Table 10.1.

Most models for polymeric resins are expressed in stress space. However, because of the softening and subsequent hardening behavior of the stress-strain curves, the strain dependence of the stress is not unique but multi-valued. For this reason an approach formulated in strain space was used. It is also valuable to break up the constitutive response into two regions, one pre-critical stress and the other post-critical stress region. Here the emphasis was placed on the material

**Table 10.1** Elastic properties of 3501-6 epoxy resin

| Property | Value |
|---|---|
| Young's modulus, E | 4.65 GPa |
| Poisson's ratio ,ν | 0.35 |
| Shear modulus, G | 1.72 GPa |

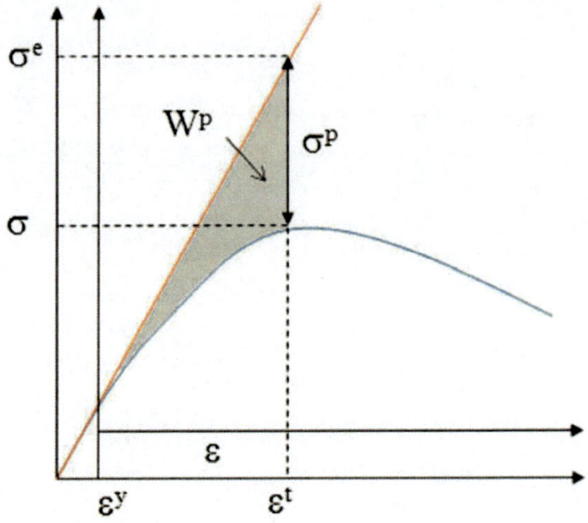


**Fig. 10.8** Definition of plastic stress and plastic work

behavior in the pre-critical stress region. In a general nonlinear elastic-plastic stress-strain response shown in Fig. 10.8, the total stress increment can be decomposed into an elastic part and a plastic part as

$$d\sigma_i = d\sigma_i^e - d\sigma_i^p \tag{10.2}$$

The elastic stress increment is linearly related to the total strain increment by the stiffness $C_{ij}$ as

$$d\sigma_i^e = C_{ij} d\varepsilon_j \tag{10.3}$$

The plastic stress increment is related to a potential (or loading) function through the associated flow rule and normality rule as follows:

$$d\sigma_i^p = d\lambda \frac{\partial f}{\partial \varepsilon_i} \tag{10.4}$$

where $i = 1, 2, \ldots, 6$; $f$ is a plastic potential function; and $d\lambda$ is a scalar function of proportionality. The constitutive model sought is obtained from the plastic potential and scalar functions.

A plastic potential function was proposed based on a modified Drucker-Prager [14] yield criterion consisting of deviatoric and dilatational deformation components. The proposed model describes the multi-axial stress-strain behavior and accounts for the sign of the normal stress (tension or compression). A similar stress-based approach was used to model a transversely isotropic composite [2]. Here a strain based potential function for an isotropic material is used as follows:

$$f = \left\{ a\left[ (\varepsilon_1 - \varepsilon_2)^2 + (\varepsilon_1 - \varepsilon_3)^2 + (\varepsilon_2 - \varepsilon_3)^2 + 6\left(\varepsilon_4^2 + \varepsilon_5^2 + \varepsilon_6^2\right) \right] + a_v (\varepsilon_1 + \varepsilon_2 + \varepsilon_3)^2 \right\}^{\frac{1}{2}} + b(\varepsilon_1 + \varepsilon_2 + \varepsilon_3) = \bar{\varepsilon} \tag{10.5}$$

where $a$, $a_v$, and $b$ are plastic anisotropy parameters, $\bar{\varepsilon}$ is the effective strain that contributes to plastic work, and all strains are elastic-plastic strains defined as:

$$\varepsilon_i = \left\langle \varepsilon_i^t - \varepsilon_i^y \right\rangle \tag{10.6}$$

where $\varepsilon_i^t$ is the total strain and $\varepsilon_i^y$ is the yield strain. For this material, under most loading conditions the yield strain is effectively zero as the material shows nonlinear behavior from the start. However, under compression, the normal strain is linear up to a certain point. This region does not contribute to the plastic work performed, as shown in Fig. 10.8. For the master compressive stress-strain curve, this normal yield strain is 1.6%.

The incremental plastic strain energy per unit volume is given in terms of the effective plastic stress as

$$dW_p = \varepsilon_i d\sigma_i^p = \bar{\varepsilon} d\bar{\sigma}^p \tag{10.7}$$

Where $d\bar{\sigma}^p$ is the increment in effective plastic stress. Replacing the plastic stress increment with the gradient of the potential function by the flow rule of Eq. 10.4 we obtain,

$$\bar{\varepsilon} d\bar{\sigma}^p = \varepsilon_i \frac{\partial f}{\partial \varepsilon_i} d\lambda = \bar{\varepsilon} d\lambda \tag{10.8}$$

Therefore, the scalar function of proportionality is given by

$$d\lambda = d\bar{\sigma}^p \tag{10.9}$$

Assuming a power law relationship between the effective plastic stress and effective strain as

$$\bar{\sigma}^p = A\bar{\varepsilon}^n \tag{10.10}$$

we obtain the following expression for the scalar function of proportionality:

$$d\lambda = nA(\bar{\varepsilon})^{n-1} d\bar{\varepsilon} \tag{10.11}$$

The incremental effective strain, as defined in Eq. 10.4, is

$$d\bar{\varepsilon} = \frac{\partial f}{\partial \varepsilon_j} d\varepsilon_j \tag{10.12}$$

Using Eqs. 10.2, 10.3, 10.4, and 10.9, 10.10, 10.11, 10.12 we obtain the constitutive elasto/viscoplastic model for the total stress and strain increments as

$$d\sigma_i = \left[ C_{ij} - nA\bar{\varepsilon}^{n-1} \frac{\partial f}{\partial \varepsilon_i} \frac{\partial f}{\partial \varepsilon_j} \right] d\varepsilon_j \tag{10.13}$$

In order to account for the rate effect, compressive normal stresses and strains must be transformed to the reference strain rate using Eq. 10.1.

## 10.4 Model Parameters

The constitutive law derived above is expressed in terms of three parameters associated with the potential function, $a$, $a_v$, and $b$, and two power law parameters, $A$ and $n$, which relate the effective plastic stress with the effective strain. However, the three parameters in the potential function are not independent. They are used to transform any given state of strain into an effective strain. For this reason, the potential function can be normalized by one of the parameters, or any loading condition can be chosen to represent a strain equivalent to the effective strain. Since the pure shear and tension tests produced relatively low strains before failure, the uniaxial compression test was chosen to yield the effective strain. In this test, $\varepsilon_1$ denotes the axial strain, the shear strains being zero, and

$$\varepsilon_2 = \varepsilon_3 = -\nu\varepsilon_1 \tag{10.14}$$

Substituting these into the potential function, we obtain the following relation

$$f = \varepsilon_1 \left\{ \left[ 2a(1+v)^2 + a_v(1-2v)^2 \right]^{\frac{1}{2}} - b(1-2v) \right\} = \bar{\varepsilon} \tag{10.15}$$

Since, under uniaxial compression, the axial strain is chosen as the effective strain

$$\left[ 2a(1+v)^2 + a_v(1-2v)^2 \right]^{\frac{1}{2}} - b(1-2v) = 1 \tag{10.16}$$

This provides a direct relation for the sign dependent dilatational constant $b$ in terms of $a$, and $a_v$

$$b = \frac{\left[ 2a(1+v)^2 + a_v(1-2v)^2 \right]^{\frac{1}{2}} - 1}{(1-2v)} \tag{10.17}$$

A direct expression for parameter $a$ can be found by comparing the pure shear and uniaxial compression tests at the critical point. At the critical point the slope of the stress-strain curve is zero

$$\frac{d\sigma_1}{d\varepsilon_1} = \frac{d\tau_6}{d\varepsilon_6} = 0 \tag{10.18}$$

Using the incremental constitutive law in Eqs. 10.13 and 10.18, we obtain the following relations

$$C_{11} = nA\bar{\varepsilon}^{n-1} \left( \frac{\partial f}{\partial \varepsilon_1} \right)^2 \frac{(1-v)}{(1+v)(1-2v)} \tag{10.19}$$

$$C_{66} = nA\bar{\varepsilon}^{n-1} \left( \frac{\partial f}{\partial \varepsilon_6} \right)^2 \tag{10.20}$$

By definition, from Eq. 10.15 for a uniaxial compression test,

$$\frac{\partial f}{\partial \varepsilon_1} = 1 \tag{10.21}$$

and for a pure shear test

$$\frac{\partial f}{\partial \varepsilon_6} = (6a)^{\frac{1}{2}} \tag{10.22}$$

Thus, by combining Eqs. 10.19, 10.20, 10.21 and 10.22, an expression for parameter $a$ can be found

$$a = \frac{C_{66}(1-v)}{6C_{11}(1+v)(1-2v)} = \frac{G}{3E} = 0.123 \tag{10.23}$$

The final potential function parameter, $a_v$, can be found from the tension tests. It is sensitive to the value of $a_v$, so that it is not difficult to fit it to the remaining data. The parameter $a_v$ was found to have a value of 5.8. The power law parameters $A$ and $n$ were determined by using the critical point under uniaxial compression. Equation 10.19 can be rewritten in terms of the product $nA$ at the critical point $\bar{\varepsilon}_c$

$$nA = C_{11}{\bar{\varepsilon}_c}^{1-n} \frac{(1+v)(1-2v)}{(1-v)} = E{\bar{\varepsilon}_c}^{1-n} \tag{10.24}$$

**Table 10.2** Charcteristic properties at reference strain rate of $10^{-4}$ $s^{-1}$

| Property | Value |
|---|---|
| Yield strain, $\varepsilon_y$ | 0.016 |
| Critical strain, $\varepsilon_c$ | 0.07 |
| Critical stress, $\sigma_c$ | 176 MPa |

**Table 10.3** Model parameters

| Parameter | Value |
|---|---|
| a | 0.123 |
| $a_v$ | 5.4 |
| b | −0.108 |
| A | 20,123 |
| n | 1.68 |

Then, substituting in the total strain value from Eq. 10.6 we obtain

$$nA = E\left(\varepsilon_{1c}^{t} - \varepsilon_{1}^{y}\right)^{1-n} \tag{10.25}$$

The critical stress is then found by integrating the incremental constitutive law, Eq. 10.13, from zero to the total critical strain, $\varepsilon_c^t$

$$\sigma_{1c} = \int_{0}^{\varepsilon_{1c}^{t}} \left[E - nA\left\langle\varepsilon_{1}^{t} - \varepsilon_{1}^{y}\right\rangle^{n-1}\right] d\varepsilon_1 \tag{10.26}$$

Substituting in Eq. 10.25 and integrating, we obtain an expression for $n$

$$n = \frac{E\left(\varepsilon_{1c}^{t} - \varepsilon_{1}^{y}\right)}{E\varepsilon_{1c}^{t} - \sigma_{1c}} \tag{10.27}$$

Finally, rearranging Eq. 10.25 provides an expression for $A$

$$A = \frac{E\left(\varepsilon_{1c}^{t} - \varepsilon_{1}^{y}\right)^{1-n}}{n} \tag{10.28}$$

All five model parameters can be obtained using the elastic constants listed in Table 10.1, the characteristic properties of the stress-strain curve shown in Table 10.2, and by fitting the stress-strain curve under uniaxial tension.

A summary of the resulting model parameters is shown in Table 10.3.

Using these parameters and the experimental data from uniaxial tension and compression, pure shear, and combined loading tests, as well as the transformed stress-strain curves at various strain rates, all of the effective strain-effective plastic stress curves collapse into a master curve as shown in Fig. 10.9. The heavy solid black curve is the power law approximation for this relationship obtained as discussed above.

## 10.5 Model Validation

The model was formulated using uniaxial tension and compression and pure shear tests at different strain rates. Figures 10.10 and 10.11, show that the predictions of the proposed constitutive model are in excellent agreement with experimental results.

All of the tests and model predictions reported thus far have been uniaxial stress or strain tests, but the model is valid for combined loadings as well. Two minor changes must be made to the model derivation described above. First, to account for

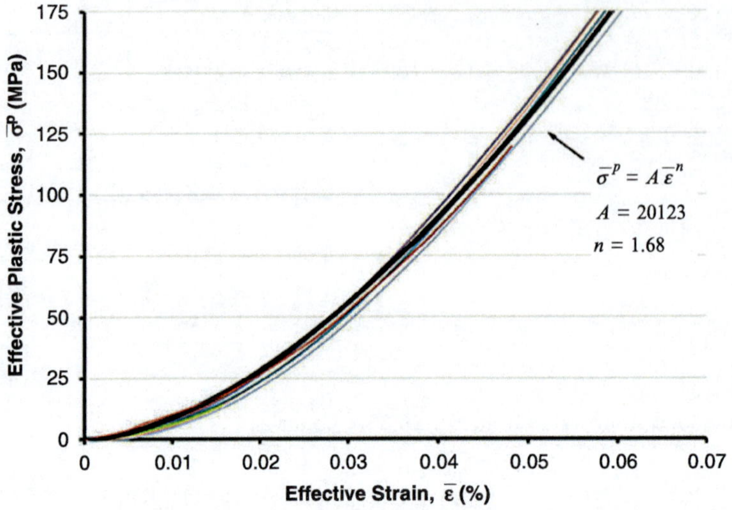


**Fig. 10.9** Collapsed curves of tension, shear, compression and combined loadings at various strain rates with power law model

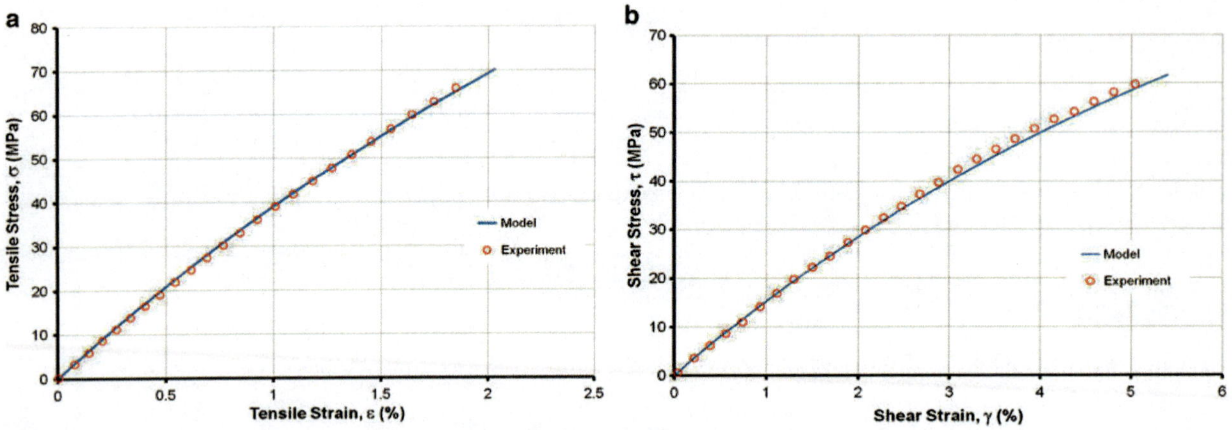


**Fig. 10.10** Comparison between model predictions and experimental results for (**a**) uniaxial tension and (**b**) shear tests

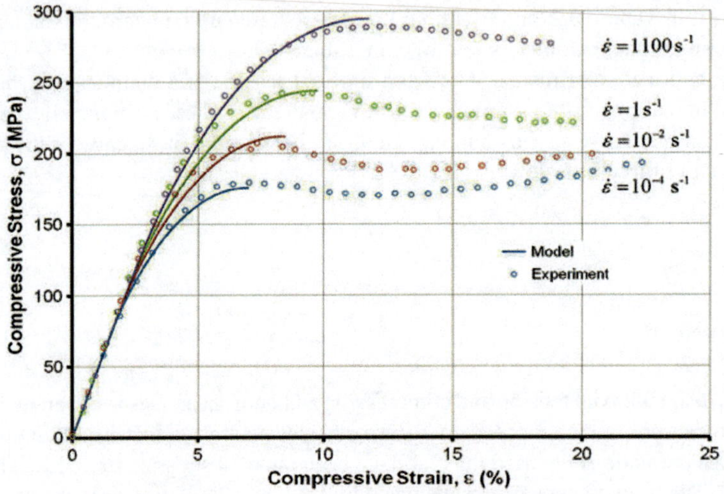


**Fig. 10.11** Comparison between model predictions and experimental results for uniaxial compression tests at different strain rates

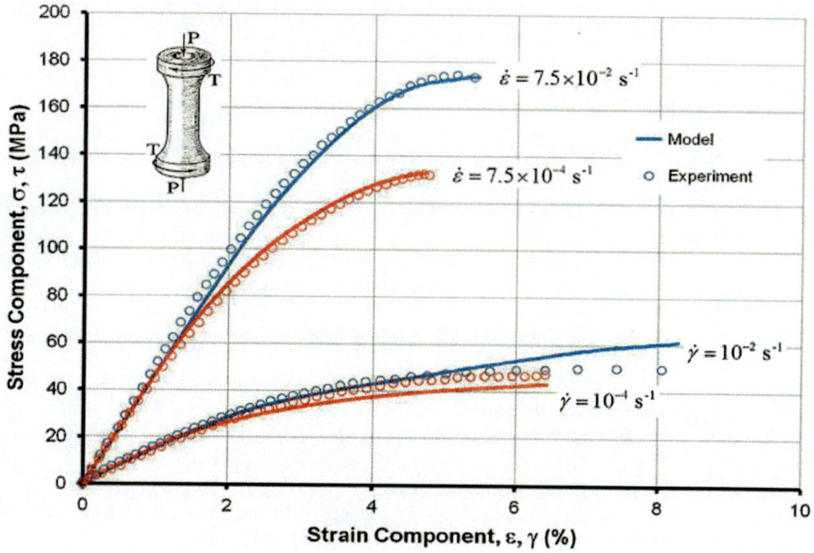


**Fig. 10.12** Combined loading experiments and model predictions at different strain rates

the rate effect, the strain rates used in Eq. 10.1 must be converted to effective strain rates. This extends the strain rate portion of the model to include multi-axial stress and strain effects

$$P(\bar{\dot{\varepsilon}}) = P(\bar{\dot{\varepsilon}}_0)\left[m\log_{10}\frac{\bar{\dot{\varepsilon}}}{\bar{\dot{\varepsilon}}_0} + 1\right] \tag{10.29}$$

This change does not affect any of the results presented earlier, applied only for uniaxial compression tests and under those loading conditions, the effective strain is equivalent to the axial strain. The other modification is that the yield strain reported in Table 10.2 is a principal yield strain and it must be changed to the axial yield strain under the given loading condition.

For combined loading tests, the same specimen type and procedure were used as for the pure shear tests. The servo-hydraulic testing machine was controlled to apply a constant displacement rate in compression and rotation to provide the desired loading condition. For these tests, the compressive strain rate was specified to be 75% of the shear strain rate. This would provide a loading condition that would exhibit nonlinearity in both compression and shear. The test was run at two different strain rates which would serve to verify whether the model could take into account rate effects under combined loading. Experimental results and model predictions are shown in Fig. 10.12.

As can be seen from the experimental data, most of the rate effects are shown in the compressive stress-strain component of the test. The model appears robust in describing the resin response to combined loadings and various strain rates. Some further verification is needed at different biaxialities and strain rates to further reinforce the above observations.

## 10.6 Conclusions

It was shown in this study that a constitutive law that can account for rate effects in an epoxy matrix can be formulated based on a limited number of relatively simple tests. It was demonstrated that the elastic-viscoplastic constitutive model can be determined from tests under three separate loading conditions; uniaxial tension, compression, and pure shear. To determine the strain rate effect, a limited number of tests at various rates must be performed to produce an expression for the rate dependence. This model was verified by comparing predictions and experimental results of the behavior under a combined compression and shear loading at different strain rates. Excellent agreement was observed.

**Acknowledgement** The work described in this paper was sponsored by the Solid Mechanics Program of the Office of Naval Research (ONR). We are grateful to Dr. Y. D. S. Rajapakse of ONR for his encouragement and cooperation.

## References

1. Daniel IM, Cho J-M, Werner BT, Fenner JS (2011) Characterization and constitutive modeling of composite materials under static and dynamic loading. AIAA Journal 49(8):1658–1682
2. Daniel IM, Werner BT, Fenner JS (2011) Strain-rate-dependent failure criteria for composites. Compos Sci Technol 71(3):357–364
3. Liang YM, Liechti KM (1996) On the large deformation and localization behavior of an epoxy resin under multiaxial stress states. Int J Solids Struct 33(10):1479–1500
4. Buckley CP et al (2001) Deformation of thermosetting resins at impact rates of strain. Part I: Experimental study. J Mech Phys Solids 49(7):1517–1538
5. Schapery RA (1969) On characterization of nonlinear viscoelastic materials. Polym Eng Sci 9(4):295
6. Padmanabhan K (1996) Time-temperature failure analysis of epoxies and unidirectional glass/epoxy composites in compression. Compos Part A 27A:585–596
7. Mayr AE, Cook WD, Edward GH (1998) Yielding behaviour in model epoxy thermosets – I. Effect of strain rate and composition. Polymer 39(16):3719–3724
8. Behzadi S, Jones FR (2005) Yielding behavior of model epoxy matrices for fiber reinforced composites: effect of strain rate and temperature. J Macromol Sci Phys 44(6):993–1005
9. Yamini S, Young RJ (1980) The mechanical properties of epoxy resins: Part 1 Mechanisms of plastic deformation. J Mater Sci 15:1814–1822
10. Goldberg RK, Roberts GD, Gilat A (2005) Implementation of an associative flow rule including hydrostatic stress effects into the high strain rate deformation analysis of polymer matrix composites. J Aerosp Eng 18(1):18–27
11. Goldberg RK et al (2008) Approximation of nonlinear unloading effects in the strain rate dependent deformation analysis of polymer matrix materials utilizing a state variable approach. J Aerosp Eng 21(3):119–131
12. Littell JD et al (2008) Measurement of epoxy resin tension, compression, and shear stress-strain curves over a wide range of strain rates using small test specimens. J Aerosp Eng 21(3):162–173
13. Buckley CP et al (2004) Deformation of thermosetting resins at impact rates of strain. Part 2: constitutive model with rejuvenation. J Mech Phys Solids 52(10):2355–2377
14. Drucker DC, Prager W (1952) Soil mechanics and plastic analysis or limit design. Q Appl Math 10:157–165

# Chapter 11
# Study on Penetration Equation by Experimental Investigation on Penetration into Wax

**Hwun Park and Weinong Chen**

**Abstract** Penetration of a projectile into a target is a complicated phenomenon where both materials deform with very high strain rate. Most of experiments on penetration were to obtain resultant quantities such as depth of penetration and residual speeds of projectiles. To verify conventional penetration equations based on dynamic cavity expansion, the histories of strain and stress inside target are required. In this study, we used wax as the target material because the penetration equation of wax is relatively simple due to its hydrodynamic behavior in low speed penetration. We installed specially designed piezoelectric gages in a wax target and then launched metal projectiles into the target. The stress variations in the target were obtained with the gages and the motions of the projectile were recorded with flash X-rays.

**Keywords** Dynamic cavity expansion • Radial stress

## 11.1 Extended Abstract

Analysis with dynamic cavity expansion is the main approach to understand high speed penetration. Motions of a projectile inside a target produce cavity expansion in the radial direction [1]. To improve penetration equations, radial stress need to be measured experimentally. We employed a high speed gas gun and instrumented wax targets to measure radial stress during penetration by steel projectiles as shown in Fig. 11.1. The projectile penetrates into the wax having PVDF pressure gage [2]. The gages are strategically located to observe the change of radial stress by motion of the projectile. The speed of the projectile inside the wax is obtained with the shadow graphics by flash X-rays.

Figure 11.2 shows speeds of the projectile versus the positions of the projectiles striking with the speed of 167 m/s and 170 m/s. The speeds gradually decrease and suddenly drop. The speed of projectile passing by PVDF gages can be obtained by interpolation with the points.

The measured stress by PVDF gage is presented in Fig. 11.3. The gage locating 205 mm away from the impacting surface of the target has lower peak stress because the projectile has lower speed. The correlation between the projectile speed and peak stress obtained from the experiments will be compared with the analytic results [1].

H. Park (✉) • W. Chen
School of Aeronautics and Astronautics, Purdue University, 701 West Stadium Avenue West Lafayette, Indiana 47907–2045, USA
e-mail: wan.park@gmail.com; wchen@purdue.edu

V. Chalivendra et al. (eds.), *Dynamic Behavior of Materials, Volume 1: Proceedings of the 2012 Annual Conference on Experimental and Applied Mechanics*, Conference Proceedings of the Society for Experimental Mechanics Series,
DOI 10.1007/978-1-4614-4238-7_11, 

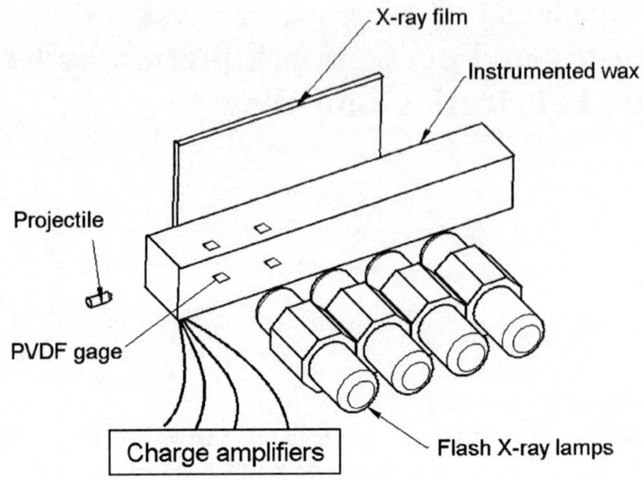


**Fig. 11.1** Configuration of experimental setup

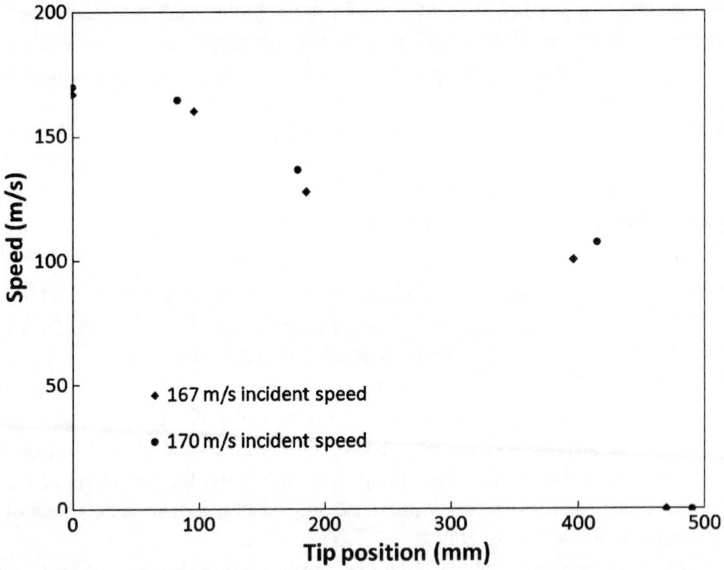


**Fig. 11.2** Positions of projectile tips versus time and speed versus tip position

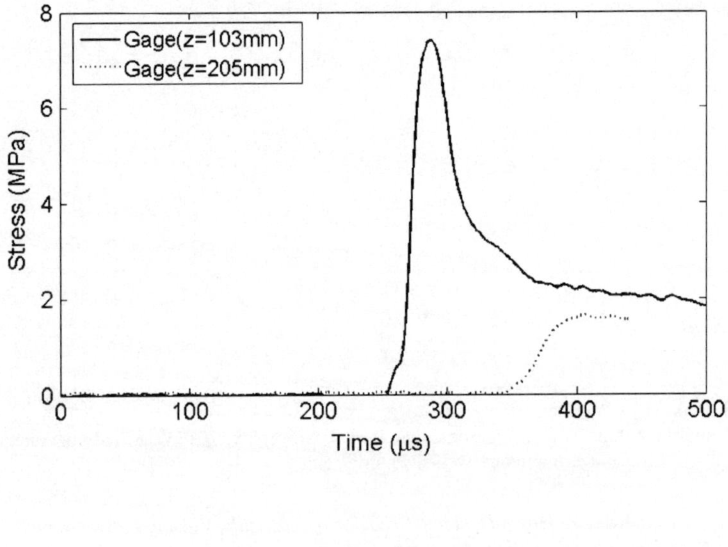


**Fig. 11.3** Positions of projectile tips versus time and speed versus tip position

## References

1. Forrestal MJ, Luk VK (1988) Dynamic spherical cavity expansion in a compressible elastic–plastic solid. J App Mech 55:275–279
2. Park H, Chen W, Piezoelectric gages to measure three-dimensional dynamic stress propagation in granular materials, (in preparation)

# Chapter 12
# Micro-Tensile Tests of Aluminum Thin Foil with the Variation of Strain Rate

**J.B. Kwon, J. Huh, Hoon Huh, and J.S. Kim**

**Abstract** This paper is concerned with the rate-dependent material properties of an aluminum thin foil. In the micro-scale tensile tests were conducted for Al1100 thin foils with thickness of 96 μm at various strain rates ranged from 1/s to 100/s using a High-Speed Micro Material Testing Machine (HSMMTM). The dimension of the test specimen was 1,500 μm in the gauge length and 300 μm in the gauge width. Test equipments designed and constructed for micro-tensile tests as well as a new jig system for easy installation and accurate alignment of a specimen between a specimen and the jig system. The jig mechanism developed enhances the reproducibility of experiments. Experimental results show that the flow stress of an aluminum thin foil increases more remarkable as the strain rate increases than the flow stress of an aluminum sheet with moderate thickness.

**Keywords** Aluminum thin foil • Rate-dependent material properties • Micro-testing machine

## 12.1 Introduction

Thin films of silicon, polysilicon, and some metals have been widely used in micro-electromechanical systems (MEMS) and other micro-devices such as micro-formed parts and micro-accelerometers. The reliable mechanical properties of these thin films are essential to accurately evaluate the structural reliability of micro devices and MEMS products. Tensile tests of thin films that are only a few microns thick require special techniques and procedures for specimen preparation and handling. Several techniques have been developed to evaluate the mechanical properties of these thin films. The tensile testing methods of thin films at high strain rate, however, are not yet thoroughly established on the contrary to those at quasi-static strain rate. For micro-specimens, material tests are mostly conducted at quasi-static strain rates of less than 0.01/s using mechanical [1], piezoelectric [2, 3], or electro-magnetic actuators [4–6].

Kim and Huh [7] developed a high-speed micro-material testing machine, HSMMTM, to investigate the high strain rate material properties of an OFHC (Oxygen Free High Conductivity) copper film. The present paper investigates the material properties of Al1100 thin foil with the thickness of 96 μm using the HSMMTM. A new gripping mechanism was proposed for easier installation and accurate alignment of a micro-specimen. A method to generate a speckled pattern on a specimen was suggested to measure the strain with digital image processing technique. Flow curves of the material were obtained at the strain rate of 0.001/s and 100/s.

J.B. Kwon • J. Huh • H. Huh (✉)
School of Mechanical, Aerospace and Systems Engineering, Korea Advanced Institute of Science and Technology, Daeduk Science Town, Daejeon 305–701, South Korea
e-mail: hhuh@kaist.ac.kr

J.S. Kim
Railway Structure Research Department, Korea Railroad Research Institute, 360–1, Woram-Dong, Uiwang-Si, Gyeonggi-Do 437–757, South Korea

V. Chalivendra et al. (eds.), *Dynamic Behavior of Materials, Volume 1: Proceedings of the 2012 Annual Conference on Experimental and Applied Mechanics*, Conference Proceedings of the Society for Experimental Mechanics Series, DOI 10.1007/978-1-4614-4238-7_12, © The Society for Experimental Mechanics, Inc. 2013

## 12.2 Improvement of Gripping System of High-Speed Micro-material Testing Machine

To obtain stress–strain curves of 1100 aluminum micro-specimen, High-Speed Micro-Material Testing Machine (HSMMTM) developed by Kim and Huh in 2009 [7]. The HSMMTM consists of a loading actuator, a gripper, a strain measurement system and a load measurement system as shown in Fig. 12.1. The nominal strain rate range for the HSMMTM is from 1/s to 1,000/s with regard to a micro-specimen proposed with the gauge length of 1.5 mm

The gripping mechanism suggested by Kim and Huh [7] is a modified slack adapter type, as shown in Fig. 12.2. This gripping mechanism was used for its simplicity. However, a lot of efforts are needed to obtain reliable test results with this mechanism due to the following reasons. The micro-specimen could be damaged during the mounting procedure due to comparably heavy clipper attached to the specimen. It also is difficult to maintain the alignment of the specimen since the moving jig can rotate during a high-speed tensile tests as in Fig. 12.3. Misalignment of specimens is a critical defect in micro-material testing. Hence, a new gripping system was developed to increase the reliability of micro-tests. Figure 12.4 shows the proposed gripping system. The new gripping system has an additional guide part, which enables easy and accurate alignment of a specimen and prevents damage of a specimen during the specimen mounting procedure. Furthermore, effect of the moving jig rotation can be cancelled out using cylindrical rod in acceleration region. Figure 12.5 shows the new gripping system after mounting a specimen.

In order to verify the performance of the new gripping system, tensile tests were carried out at the strain rate of 100/s and were recorded with the high-speed camera. Figure 12.6 compares deformed shapes of micro-specimens after tensile tests

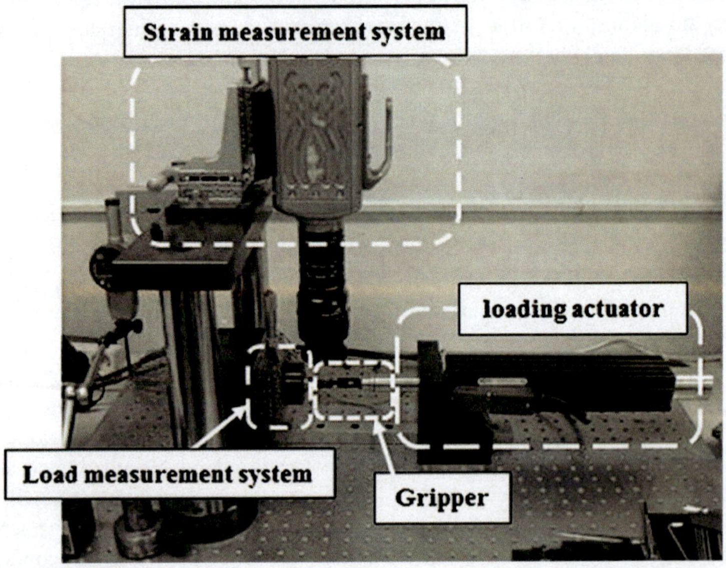


**Fig. 12.1** A full view of High-Speed Micro-Material Testing Machine (HSMMTM)

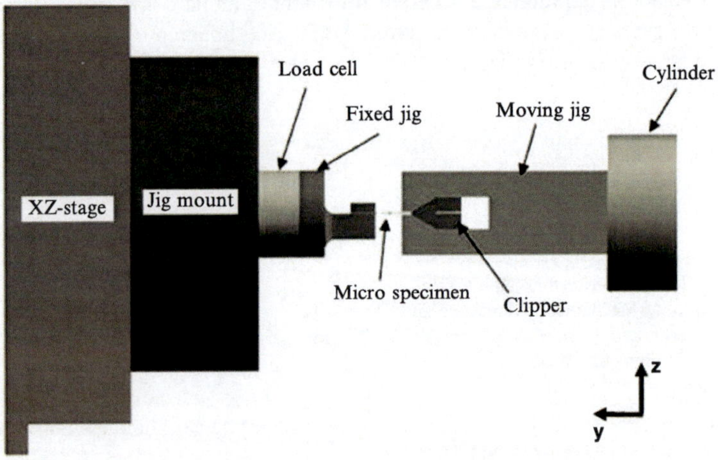


**Fig. 12.2** A schematic view of the previous gripping mechanism (Modified slack adapter type)

**Fig. 12.3** The previous moving jig and cylinder

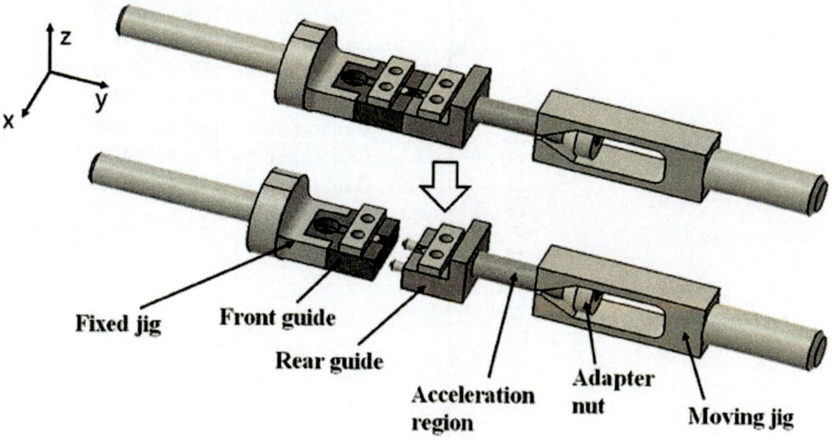


**Fig. 12.4** The new gripping system proposed

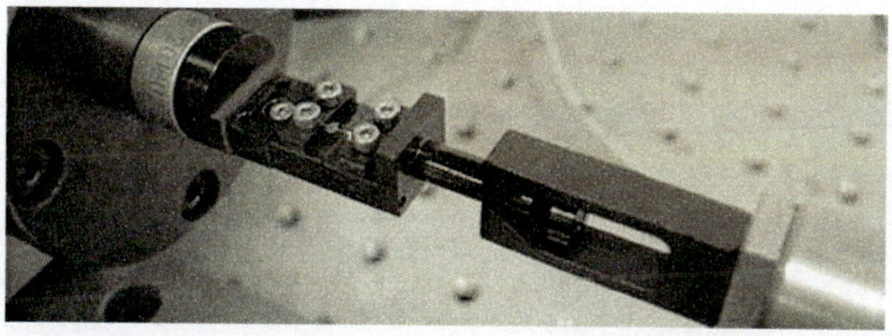

**Fig. 12.5** The new gripping system after mounting a specimen

with the previous gripping system and the new gripping system, respectively. Misaligning of a micro-specimen occurred frequently when the previous gripping mechanism was used as shown in Fig. 12.6a. The new gripping mechanism solved this problem, and alignment of a specimen was maintained throughout the test as shown in Fig. 12.6b.

For accurate measurement of the load during the tensile tests, the load scale factor for the new gripping system should be obtained. The load scale factor was calibrated by comparing the load of a reference static UTM with the output voltage of the load cell ten times. The obtained scale factor is 110.04 mV/N as shown in Fig. 12.7 and the standard uncertainty of the load scale factor is 0.53 mV/N, which was obtained following ISO/IEC standard GUM 98-3 [8].

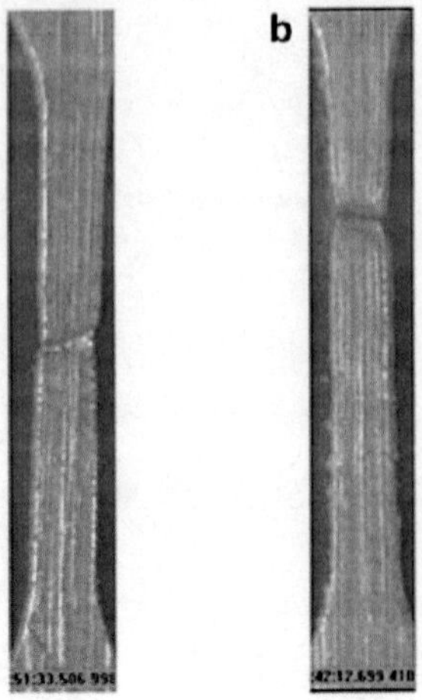


**Fig. 12.6** Deformed shapes of micro-specimens at the strain rate of 100/s: (**a**) the result obtained with the previous gripping system; (**b**) the result obtained with the new gripping system

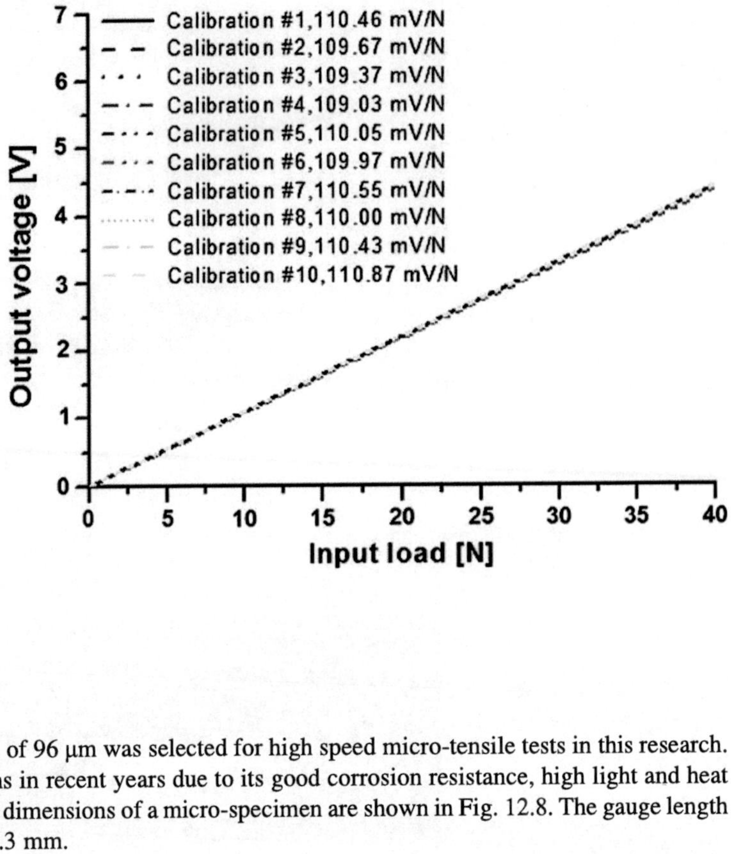


**Fig. 12.7** Obtained load scale factor

## 12.3 Preparation of Micro-specimens

Al1100 (99.0% purity) thin foil with the thickness of 96 μm was selected for high speed micro-tensile tests in this research. The material is widely used in MEMS applications in recent years due to its good corrosion resistance, high light and heat reflexibility, and high forming characteristics. The dimensions of a micro-specimen are shown in Fig. 12.8. The gauge length is 1.5 mm and the width of the gauge section is 0.3 mm.

The micro-specimens were fabricated with micro-photo etching, which is a process used in micro-fabrication to remove parts selectively from a thin film or from a bulk substrate. SEM images of a fabricated micro specimen confirm fine surface quality, as shown in Fig. 12.9.

An optical microscope was used in measuring the width of the micro-specimens. The reference dimension was measured with stage micro-meter whose uncertainty is 0.25 μm in Fig. 12.10. The average gauge width of 30 micro-specimens was measured five times at evenly spaced location and tabulated in Table 12.1 along with the standard deviation and the standard uncertainty. The standard uncertainty was calculated with Eq. 12.1 following ISO/IEC standard GUM 98-3 [8].

$$u_m{}^2 = u_r^2 + u_e^2 \tag{12.1}$$

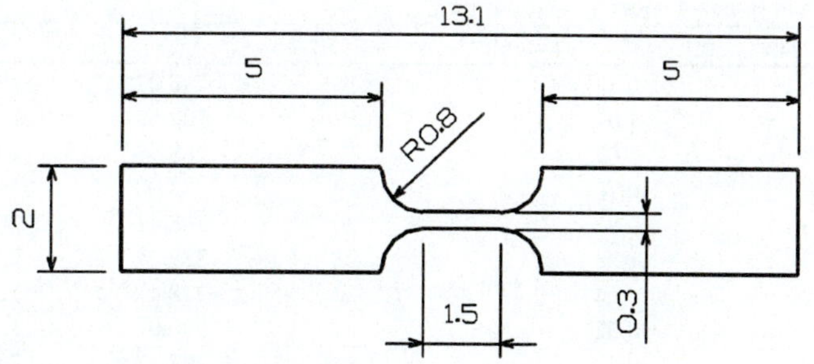


**Fig. 12.8** Dimensions of a micro-specimen for high speed micro-tensile tests [unit: mm]

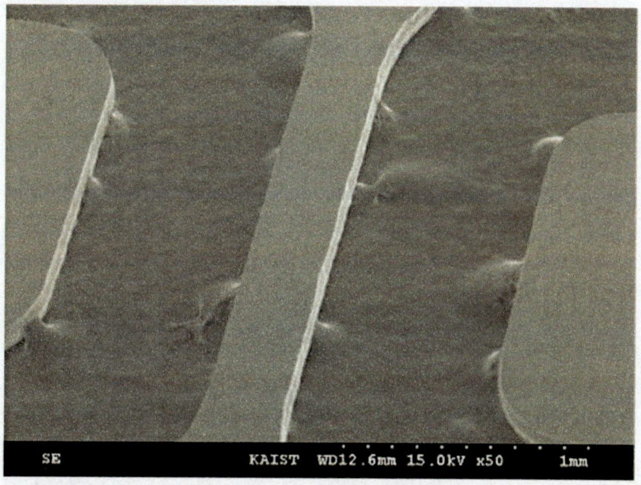


**Fig. 12.9** Micro-specimen with the gauge width of 300 μm machined with micro-photo etching

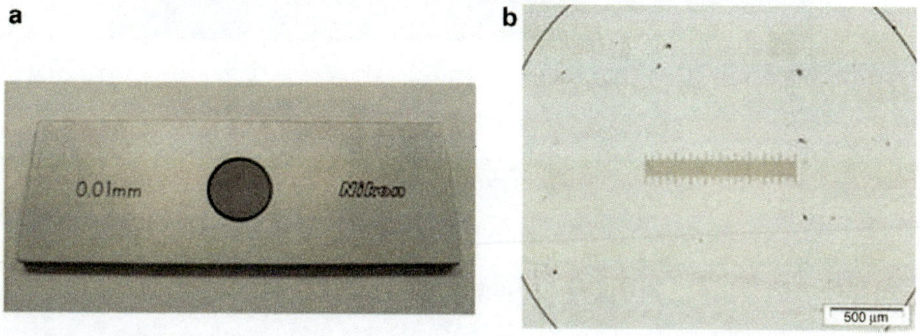


**Fig. 12.10** Stage micrometer: (**a**) a full view; (**b**) a magnified view

Where, $u_m$ is the uncertainty of measurement, $u_r$ is the uncertainty by repetition of gauge width and $u_e$ is the uncertainty of measuring equipment. The uncertainty by repetition of the gauge width, $u_r$ can be obtained from Eq. 12.2 and $u_e$ is 0.25.

$$u_r = \frac{s}{\sqrt{n}} \tag{12.2}$$

where, s is the standard deviation, n is the total number of repetition. The micro-specimens which have the standard uncertainty of less than 1.0 were only used for reliable test results.

**Table 12.1** Average gauge width and standard uncertainty [unit: μm]

| No. | Avg. | Std. dev. | Std. uncertainty | No. | Avg. | Std. dev. | Std. uncertainty |
|---|---|---|---|---|---|---|---|
| 1 | 289.65 | 0.93 | 0.48 | 16 | 290.07 | 1.85 | 0.86 |
| 2 | 287.59 | 2.26 | 1.04 | 17 | 287.59 | 1.85 | 0.86 |
| 3 | 289.65 | 1.73 | 0.81 | 18 | 288.41 | 1.85 | 0.86 |
| 4 | 291.31 | 1.46 | 0.70 | 19 | 288.00 | 1.13 | 0.56 |
| 5 | 295.85 | 1.73 | 0.81 | 20 | 288.41 | 1.13 | 0.56 |
| 6 | 291.72 | 2.69 | 1.23 | 21 | 289.24 | 2.53 | 1.16 |
| 7 | 287.17 | 1.31 | 0.64 | 22 | 288.41 | 1.13 | 0.56 |
| 8 | 290.07 | 1.01 | 0.52 | 23 | 290.07 | 1.13 | 0.56 |
| 9 | 295.02 | 0.92 | 0.48 | 24 | 287.17 | 2.07 | 0.96 |
| 10 | 287.99 | 1.14 | 0.57 | 25 | 288.41 | 1.13 | 0.56 |
| 11 | 295.02 | 1.73 | 0.81 | 26 | 288.83 | 1.73 | 0.81 |
| 12 | 288.00 | 1.13 | 0.56 | 27 | 290.48 | 1.85 | 0.86 |
| 13 | 288.41 | 1.13 | 0.56 | 28 | 290.48 | 1.13 | 0.56 |
| 14 | 288.83 | 1.73 | 0.81 | 29 | 290.07 | 1.13 | 0.56 |
| 15 | 289.24 | 2.07 | 0.96 | 30 | 289.24 | 1.46 | 0.70 |

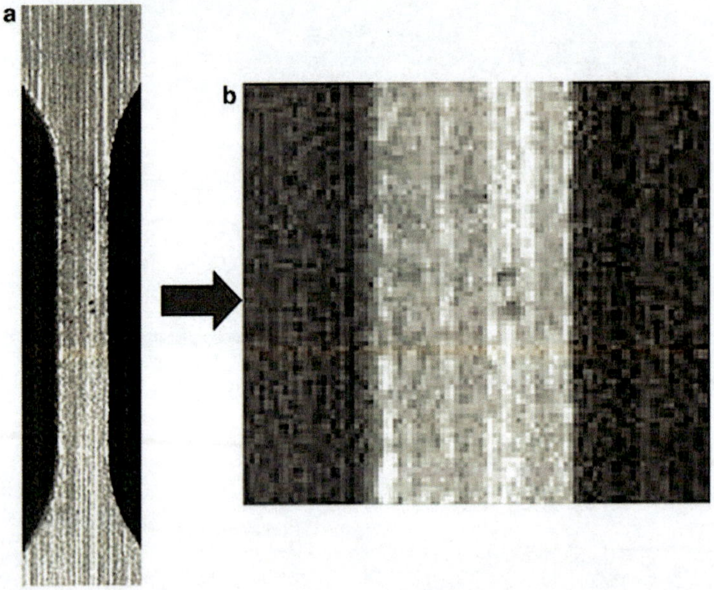


**Fig. 12.11** An aluminum micro-specimen with 0.3 mm in gauge width and 1.5 mm in gauge length: (**a**) a full view; (**b**) a magnified view (×4) of the aluminum specimen surface

## 12.4 Measurement of Strain by a Digital Image Processing Technique

The strain along a specimen was measured by a noncontact digital image processing technique using a high-speed camera, Phantom v9.0. A distinct pattern or markers are needed to correlate particular points on a series of images and measure the axial strain by a digital image processing technique. Kim and Huh [7] used as-received specimen without any processing since the material they used, OFHC copper film, has low light reflexibility so that light and dark spots on a specimen are easily distinguishable. Those spots, however, are not clear in case of aluminum micro-specimen due to its high reflexibility as shown in Fig. 12.11. The digital image processing technique does not correlate particular points on the bare aluminum specimen surface, so a speckled pattern should be generated on the specimen surface.

A speckled pattern was generated on the Al1100 micro-specimen surface with spray paint and a fan. The average size of a paint particle is 0.02 ~ 0.3 mm when sprayed directly to a specimen. It is too large compared to the gauge length of a specimen, 1.5 mm. A fan, therefore, was used to blow paint particles toward the specimen surface instead of spraying paint directly to a specimen. Using this method, only minute particles of spray paint can reach the specimen surface.

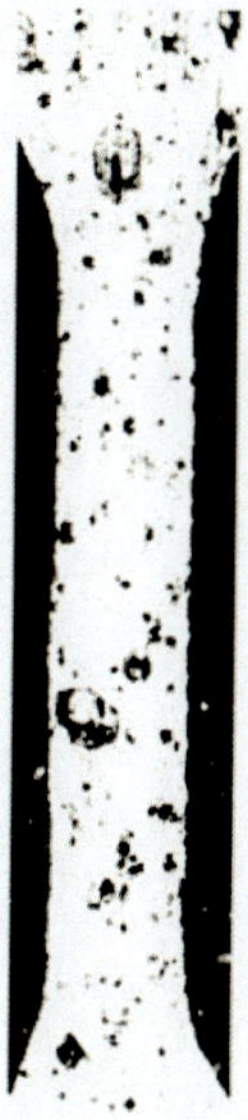

**Fig. 12.12** High speed camera image of a micro-tensile specimen with speckled patterns

The size of the paint particles made by this method is about 0.012 ~ 0.025 mm when tested with black spray paint. Figure 12.12 shows a sample image of a specimen with minute particles of spray paint on recorded by the high speed camera. The captured image was adjusted using contrast value to make similar effects of white layer on the specimen surface. This procedure makes black particles distinct. An image of a specimen have about 400 pixels in the gauge region along the axial direction, and the number of pixels of the smallest particle of black spray paint is about 2 ~ 3. It gives a minimum detectable strain of 0.005 ~ 0.0075.

## 12.5 Rate-Dependent Material Properties of Al1100 Thin Foil

Table 12.2 shows the tensile testing conditions, including the materials, testing machines, specimen dimensions, strain measurement techniques and strain rates. Two materials with different thicknesses and sizes were tested using a quasi-static testing machine and a high-speed material testing machine at the strain rates that ranged from 0.001/s to 100/s. A Static Micro Material Testing Machine (SMMTM) was developed for the tensile testing of Al1100 film with a thickness of 96 μm at the strain rates ranging from 0.001/s to 0.1/s. The modified HSMMTM was used for the tensile tests of micro-specimens at the strain rates ranging from 1/s to 100/s. For comparison of the rate-dependent material properties of Al1100 thin film with those of a bulk Al1100 sheet, the tensile tests have been carried out at the strain rates ranging from 0.001/s to 100/s using Instron 5583 and high speed material testing machine (HSMTM) [9] for A1100 with the thickness of 1.0 mm. The dimensions of macro-specimen are shown in Fig. 12.13. The stress-strain curves of both micro-and macro-specimen can be obtained, as shown in Fig. 12.14. It is observed that both the flow stress and the strain hardening increase as the strain rate increases. The cause of the strain rate hardening becomes larger as the strain rate increases for both specimens was explained by Follansbee and Kocks [10] and Tong et al. [11]. They demonstrated that this strain rate dependence of strain hardening arises due to the increasing difficulty of dislocations to overcome obstacles at high strain rates, leading to an increased multiplication rate and to an increase in the strain-rate sensitivity of strain hardening. The fracture elongation is also increased as the strain rate increase. But the slope of the stress-strain curve of Al1100 macro-specimen with thickness of 1.0 mm is lower than that of Al1100 micro-specimen with thickness of 96 μm. The fracture elongation of macro-specimen is also lower than that of micro-specimen. To quantify theses material properties, the variation of the strain rate sensitivity which indicates the variation of the flow stress with respect to the strain rate with the variation of the strain were illustrated in Fig. 12.15. The lowest curve marked with square symbols in Fig. 12.15 denotes the initial yield stress curve. The interval between the symbols under the same strain rate indicates the amount of strain hardening. The yield stress of Al1100 micro-specimen at a strain rate of 0.001/s was 127.15 MPa, which is 10.9% higher than that of Al1100 macro-specimen, 114.66 MPa. The yield stress of Al1100 micro-specimen at a strain rate of 100/s was 147.06 MPa, which is 10.4% higher than that of Al1100 macro-specimen, 133.22 MPa. Figure 12.16 shows the fracture elongation of Al1100 micro-and macro-tensile specimen with respect to the logarithmic scale of the strain rate. The fracture elongation of Al1100

**Table 12.2** Tensile testing conditions for Al1100

| Material | Machine | Strain rate[/s] | Strain measurement techniques | Dimension of gauge section Length[mm] | Width[mm] |
|---|---|---|---|---|---|
| Al1100 thickness 0.096 mm | SMMTM | 0.001 | Digital image correlation | 1.5 | 0.3 |
| | | 0.01 | | | |
| | | 0.1 | | | |
| | HSMMTM | 1 | | | |
| | | 10 | | | |
| | | 100 | | | |
| Al1100 thickness 1.0 mm | Instron 5583 | 0.001 | Extensometer | 30 | 6 |
| | | 0.01 | | | |
| | HSMTM [9] | 0.1 | Digital image correlation | | |
| | | 1 | | | |
| | | 10 | | | |
| | | 100 | | | |

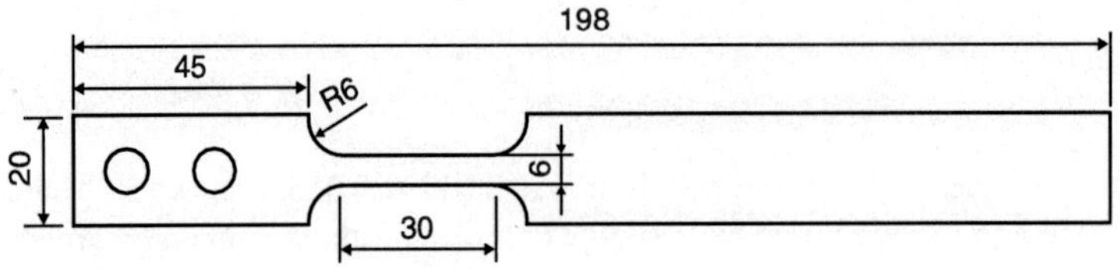


**Fig. 12.13** Dimension of a conventional specimen of a sheet metal for high speed tensile tests [unit: mm]

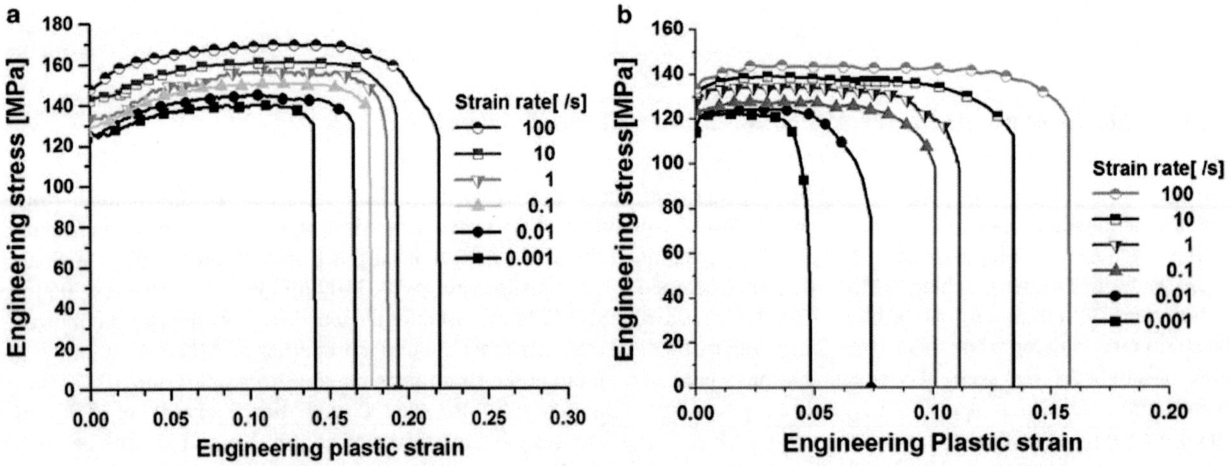


**Fig. 12.14** Rate-dependent stress/strain curves of A1100: (**a**) with a thickness of 0.096 mm; (**b**) with a thickness of 1.0 mm

micro-specimen at a strain rate of 0.001/s is 13.89% while the fracture elongation at a strain rate of 100/s is 21.88%, which is an increase of 57.5% compared to the quasi-static case. The fracture elongation of Al1100 macro-specimen at a strain rate of 0.001/s is 4.61% while the fracture elongation at a strain rate of 100/s is 15.94%, which is an increase of 245.8% compared to the quasi-static case.

Through the tensile tests with the variation of strain rate, it is observed that initial flow stress, strain hardening and fracture elongation of Al1100 micro-specimen is larger than that of Al1100 macro-specimen. To examine the effect of grain size and specimen size on the tensile properties of A1100, the grain sizes of Al1100 micro-and macro-specimen were measured using an electron back-scattered diffraction(EBSD) machine, as shown in Fig. 12.17a, b. In the Grain Boundary (GB) map and the Inverse Pole Figure(IPF) map, lines indicate the grain boundary and each color shows an individual grain. The statistical grain size distribution of Al1100 micro-and macro-specimen in Fig. 12.18 indicates the average grain size of each specimen, which was obtained following ASTM E1382 [12]. The average grain size of Al1100 micro-specimen is

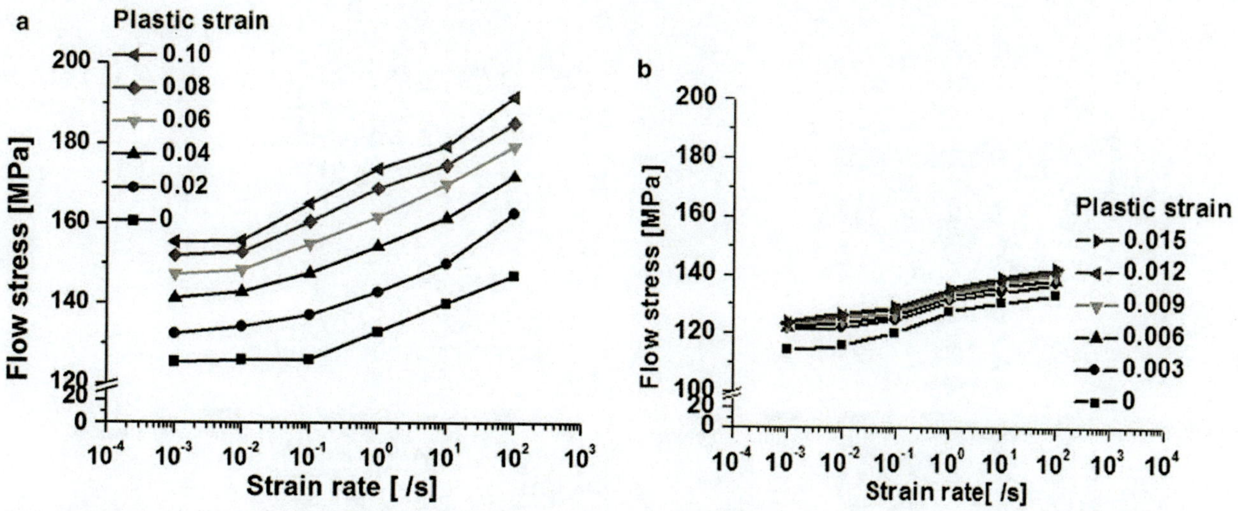


**Fig. 12.15** Strain rate sensitivity of A1100: (**a**) with a thickness of 0.096 mm; (**b**) with a thickness of 1.0 mm

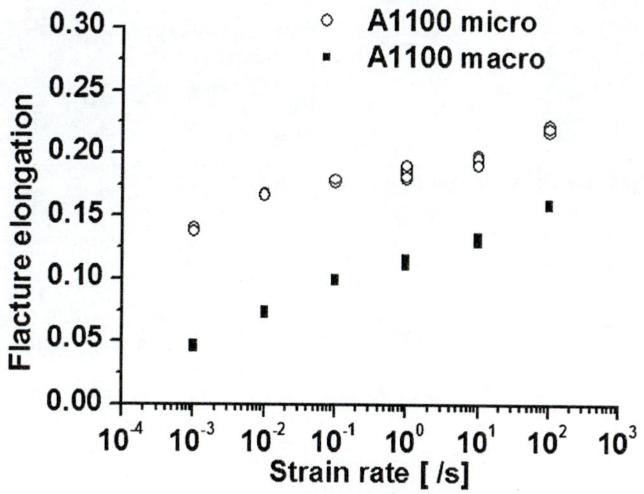


**Fig. 12.16** Fracture elongation of A1100 micro-and macro-tensile specimen with respect to the strain rate

8.2 μm. For Al1100 macro-specimen, this is 12.6 μm. Kim et al. [13] demonstrated that the Hall-Petch relationship describes the dependence of the flow stress on the grain size. They suggest and confirm that the material flow stress depends not only on the grain size but also on the interactions with the specimen/feature size. They explained that the specimen/feature size effect involves geometrical scaling of the workpiece. In this study, comparison of flow stresses with respect to the average grain size between Al1100 and OFHC copper [7] can be used to confirm the relationship between grain size effect and specimen/feature size effect. Figure 12.19a, b illustrates the flow stress curves of Al1100 and OFHC copper [7] fitted by a power-law hardening formular for strain rates of 0.001/s and 100/s, respectively. The flow stress of the Al1100 micro-specimen with smaller grains was higher than the Al1100 sheet with larger grains. This result shows the opposite result with those of OFHC cooper film. This penomenon can be explained by specimen/feature size effect [13]. Hansen [14] suggested that the grain number, $n$, which can be expressed the width of the gauge, $w$, thickness of specimen, $t$ and the average grain size, $d$, in the follwing form

$$n = \sqrt{\frac{4wt}{\pi d^2}} \tag{12.3}$$

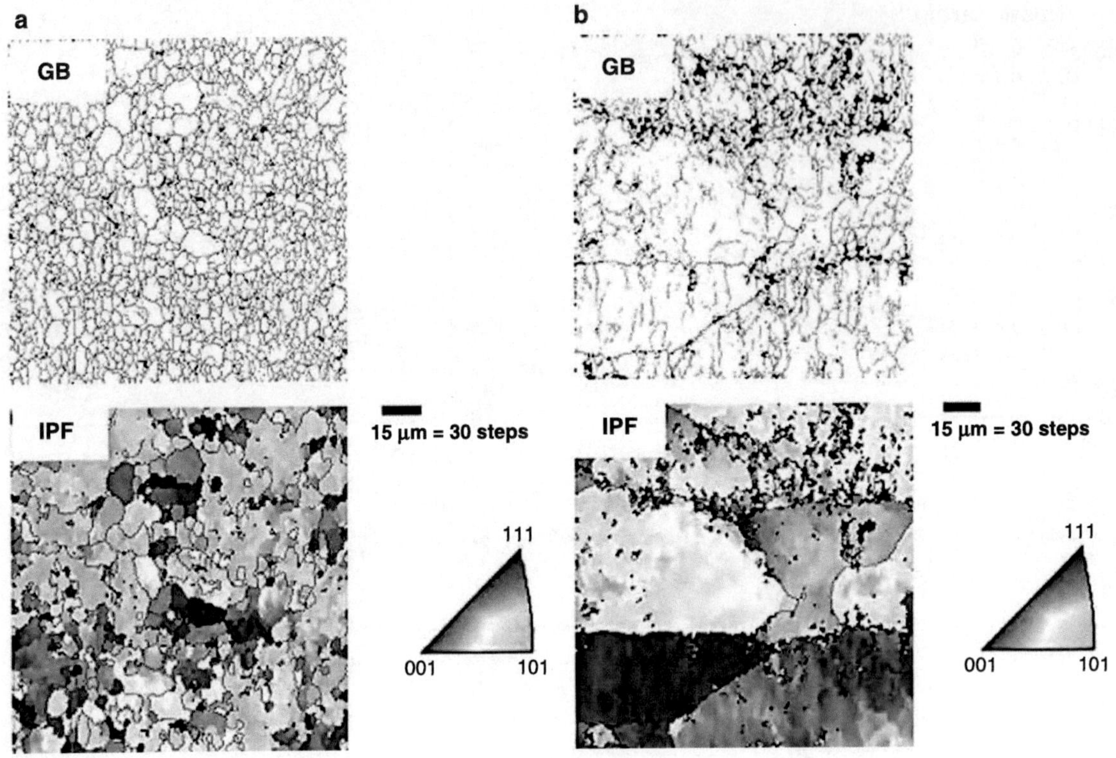


**Fig. 12.17** GB and Inverse Pole Figure maps of A1100: (**a**) A1100, 0.096 t; (**b**) A1100, 1.0 t

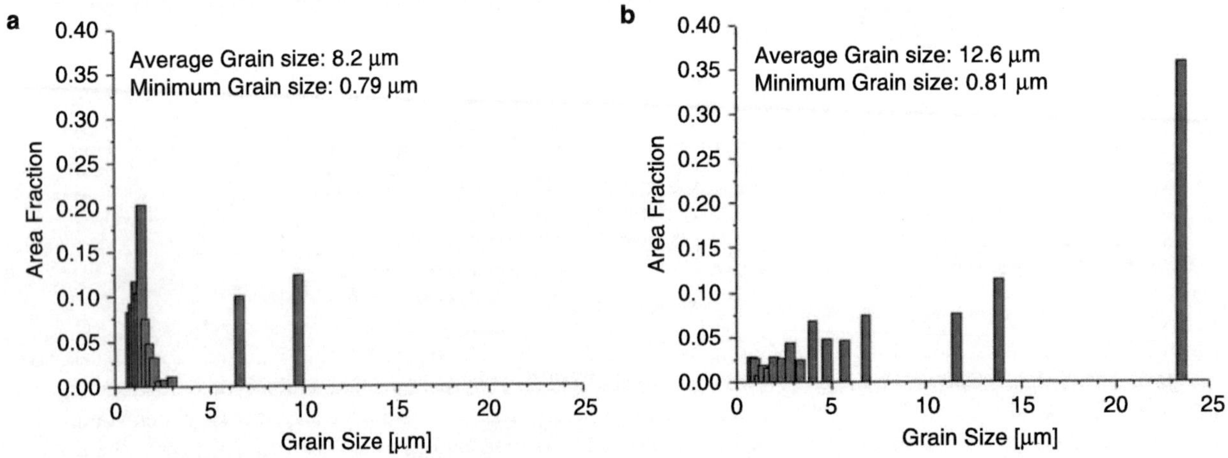


**Fig. 12.18** Statistical grain size distribution: (**a**) A1100, 0.096 t; (**b**) A1100, 1.0 t

The limit of $n = 15$ was suggested by Hansen as when a specimen may start to show specimen/feature size effect [14]. Table 12.3 shows the grain number of each material, including the dimension of gauge section, average grain size and the percentage of the ratio of the number of the grain at the specimen surface to that of the internal grain. The grain number of OFHC copper film with thickness of 0.1 mm is 17 while the grain number of Al1100 micro-specimen with thickness of 96 μm is 23.4, which is close to the limit of grain number, 15, suggested by Hansen [14]. Thus, specimen/feature size effect affected the flow stress of OFHC copper film. Engel et al. [15] proposed the surface layer model to qualitaivey describe the feature size effect. In the surface layer model, a specimen is subdivided into a free surface and an internal portion. The model argued that the free surface did not represent a boundary comparable to a grain boundary, and thus the dislocation movement

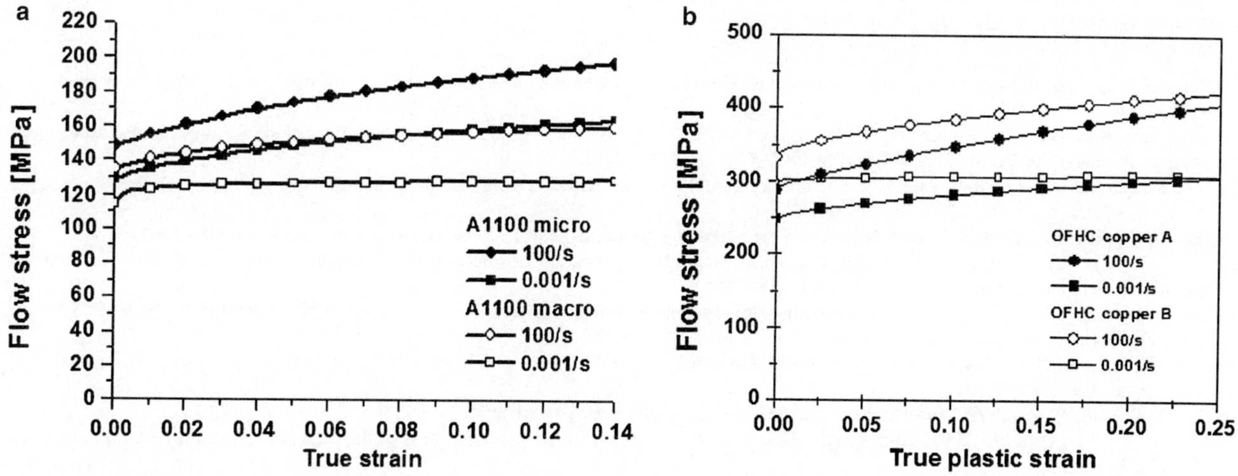


**Fig. 12.19** Comparison of the flow stress curves at the strain rate of 0.001/s and 100/s: (**a**) A1100 micro-and macro-tensile specimen; (**b**) OFHC copper A and B [7]

**Table 12.3** Comparison of grain number w.r.t grain size between A1100 and OFHC copper

| Material | Gauge section *w*[mm] | *t*[mm] | Average grain size [μm] | *n* | #of surface grain/# of internal grain [%] |
|---|---|---|---|---|---|
| OFHC copper 0.1 t | 0.2 | 0.1 | 9.4 | 17 | 31.8 |
| OFHC copper 1.0 t | 6.0 | 1.0 | 57.7 | 47.9 | 13.4 |
| Al1100 0.096 t | 0.3 | 0.096 | 8.2 | 23.4 | 17.2 |
| Al1100 1.0 t | 6.0 | 1.0 | 12.6 | 219.4 | 2.4 |

in surface grains was not restricted in the same way as in internal grains, yielding a less distinct hardening. Therefore, it was concluded that the flow stress of the surface portion is weaker than that of the internal portion of a material.

## 12.6 Conclusion

In this study, a new gripping mechanism was proposed for easier installation and better alignment of a micro-specimen. An additional guide part in the jig mechanism developed enhances the reproducibility of experiments. Digital image correlation (DIC) method is employed to measure the axial strain of micro-specimens. To enhance the accuracy of the strain measurement, a novel technique is introduced such that very fine speckle patterns can be made. Al1100 micro-specimens with the gauge length of 1.5 mm were fabricated by a micro-photo etching technique showing a good dimensional accuracy. The standard uncertainty of the average gauge width of micro-specimens was adopted for reliable test results. The material properties of A1100 film with a thickness of 96 μm has been evaluated at strain rates ranging from 0.001/s to 100/s using Static Micro Material Testing Machine (SMMTM) and High-Speed Micro Material Testing Machine (HSMMTM). It is observed that Al1100 thin foil has considerable strain-rate sensitivity showing that the flow stress, strain hardening and the fracture elongation increase as the strain rate increases. To examine the effect of grain size and specimen size on the tensile properties of A1100, the tensile tests have been carried out at the strain rates ranging from 0.001/s to 100/s using Instron 5583 and a conventional high-speed material testing machine (HSMTM) for A1100 with the thickness of 1.0 mm. The Hall-Petch equation was concluded that the flow stress of the A1100 film with smaller grains is higher than the A1100 sheet with larger grains. The flow stress of the OFHC copper film [7], however, indicated different behavior of Al1100. The size effects on the flow stress were influenced by the ratio of the number of surface grain to that of internal grain. The portion of surface grain relates to the grain number, which suggested by Hansen [14]. The grain number of OFHC copper film with thickness of 0.1 mm is closed to the limit of grain number, 15, as when a specimen may start to show specimen/feature size effect [14]. Therefore, not only the grain size but also the feature/specimen size effect should be considered in order to simulate the microscale forming process accurately.

## References


1. Sharpe WN, Yuan B, Edwards RL (1997) A new technique for measuring the mechanical properties of thin films. J Microelectromech Syst 6:193–199
2. Sharpe WN, Pulskamp J, Gianola DS, Eberl C, Polcawich RG, Thompson RJ (2007) Strain measurements of silicon dioxide microspecimens by digital imaging processing. Exp Mech 47:649–658
3. Lee SJ, Han SW, Hyun SM, Lee HJ, Kim JH, Kim YI (2009) Measurement of Young's modulus and Poisson's ratio for thin Au films using a visual image tracing system. Curr Appl Phys 9:S75–S78
4. Emery RD, Povirk GL (2003) Tensile behavior of free-standing gold films. Part I. Coarse-grained films. Acta Mater 51:2067–2078
5. Huh YH, Kim DL, Kim DJ, Park P, Kee CD, Park JH (2004) Application of micro-ESPI technique for measurement of micro-tensile properties. Adv Nondestruct Eval 270–273(1–3):744–749
6. Huh YH, Kim DL, Kee CD (2005) Measurement of continuous micro-tensile strain using micro-ESPI technique. Adv Fract Strength 297–300(1–4): 53–58
7. Kim JS, Huh H (2011) Evaluation of the material properties of an OFHC copper film at high strain rates using a micro-testing machine. Exp Mech 51:845–855
8. ISO/IEC Guide 98–3 (2008) Guide to the expression of uncertainty in measurement
9. Huh H, Lim JH, Park SH (2009) High speed tensile test of steel sheets for stress-strain curve at the intermediate strain rate. Int J Automot Technol 10:195–204
10. Follansbee PS, Kocks UF (1988) A constitutive description of the deformation of copper based on the use of the mechanical threshold stress as an internal state variable. Acta Metall 36:81–93
11. Tong W, Clifton RJ, Huang S (1992) Pressure-shear impact investigation of strain rate history effects in oxygen-free high-conductivity copper. J Mech Phys Solids 40:1251–1294
12. ASTM E1382-97 (2010) Standard test methods for determining average grain size using semiautomatic and automatic image analysis
13. Kim GY, Ni J, Koc M (2007) Modeling of the size effects on the behavior of metals in microscale deformation processes. J Manuf Sci Eng-Trans ASME 129:470–476
14. Hansen N (1977) The effect of grain size and strain on the tensile flow stress of aluminum at room temperature. Acta Metall 25:863–869
15. Engel U, Messner A (1998) Numerical simulation of metal forming processes for the production of microparts. Wire 2:94–97

# Chapter 13
# Mechanical Behavior of SS 304LN at High Strain Rates in Compression

**Sumit V. Prasad and Krishna N. Jonnalagadda**

**Abstract** Stainless steel 304LN is used in various industries, where it is subjected to both static and dynamic loads. However, the austenitic phase of the steel is not stable, which upon loading partly transforms into martensite. Although the martensitic phase is stronger, the ductility of 304 series steels reduces with increasing volume fraction of the martensite. Besides, it is known that austenitic SS 304 grade steels show strain rate sensitivity and increased flow stress at high strain rates, similar to other FCC metals. In this work, quasi-static and dynamic compression experiments were performed on fully annealed SS 304LN between the strain rates of 0.001–4,800 $s^{-1}$ using an UTM and Kolsky bar setup, to understand their mechanical response over a wide range of strain rates. The experimental results showed that the flow stress increased with increasing accumulated strain and strain rate. The work hardening rate decreased with increasing strain rate and strain. To investigate strain rate sensitivity at constant structure of SS 304LN, interrupted rate jump experiments were performed with a change of strain rates under dynamic loading. It was observed that the constant structure rate sensitivity was much higher compared to that of the constant strain rate sensitivity.

**Keywords** Stainless steel 304LN • High strain rates • Kolsky bar • Strain rate sensitivity

## 13.1 Introduction

Austenitic stainless steels represent a large family of alloys with FCC crystallographic structure and are one of the most commonly used in the industry because of their superior mechanical properties such as high strength, ductility, toughness and work hardening, good corrosion resistance and excellent weldability [1]. The mechanical responses of the austenitic steels under different types of loading, such as, quasi-static and dynamic conditions are dissimilar [2]. The dynamic or high strain rate behavior of austenitic steels is of great interest in a wide range of engineering applications like forming, automotive industry, space industry, etc. Also, it is important to study the basic mechanical behavior at different strain rates to understand and predict deformation, ductility and failure. For designing structures and components, subjected to dynamic loading, the current literature data available for the specific SS 304 LN is inadequate in terms of total stain, strain rate range and constant structure rate sensitivity.

In austenitic stainless steels family, SS 304 is the most commonly used steel and a lot of work has been done on this material under quasi-static & dynamic loads. Andrade-Campos et al. [1] investigated the interdependencies between strain rate, plastic anisotropy and strain hardening for SS 304 in the strain rate range of $10^{-4}$ –$10^{-1}$ $s^{-1}$ at room temperature and found that changes in the mechanical behaviour due to increasing strain rate depend strongly on the stability of austenite phase. Kundu et al. [3] studied the effect of strain rate and strain rate sensitivity on quasi-static ($5 \times 10^{-4}$ and $1 \times 10^{-1}$ $s^{-1}$) tensile flow behaviour of a commercial SS 304 and found that with increase in strain rate, yield strength increases, but tensile strength, ductility and strain rate sensitivity decreased.

Austenitic steels are metastable and promptly transform phase during plastic deformation, commonly referred to as strain-induced martensitic transformation [4], except at high temperatures where the austenitic phase is stable. Talonen et al. [5] have studied the effect of strain rate between $3 \times 10^{-4}$ and 200 $s^{-1}$ on the mechanical behavior and strain-induced phase

S.V. Prasad • K.N. Jonnalagadda (✉)
Indian Institute of Technology Bombay, Powai, Mumbai, India
e-mail: krishnajn@iitb.ac.in

V. Chalivendra et al. (eds.), *Dynamic Behavior of Materials, Volume 1: Proceedings of the 2012 Annual Conference on Experimental and Applied Mechanics*, Conference Proceedings of the Society for Experimental Mechanics Series, DOI 10.1007/978-1-4614-4238-7_13, © The Society for Experimental Mechanics, Inc. 2013

transformation of two commercial SS 304 and SS 301LN steels and it was observed that these steels show significantly different behavior at high strain rates compared to quasi-static tensile test. The effect of the strain rate was found to be the adiabatic heating that suppresses the strain induced-martensite transformation and the work-hardening behavior affecting the ultimate tensile strength and ductility of the studied steels. Similar studies for strain rates between $1.25 \times 10^{-4}$ and $400\ s^{-1}$ were done on material SS 304L and SS 309 by Lichtenfeld et al. [6] and they found that the phase transformation was strongly dependent on the alloy composition, strain, strain rate, stress state and temperature. Chen et al. [7] explored the effect of strain rate and temperature of SUS 304 under dyanmic loading at $1{,}200\ s^{-1}$ and the experimental results showed that the flow stress increases with increasing strain rate, but decreases with increasing temperature and martensitic volume fraction depended not only on strain and temperature but also on strain rate. Lee et al. [2] characterized the mechanical propeties of SS 304L plastic deformation under high strain rates (800– $4{,}800\ s^{-1}$) at room temperature and discussed the dislocation density, shear band formation and $\alpha$ martensite nucleation base on observed deformed sample microstructure. Comparative studies with and without prestrain [8] and at high temperatures [9] were also done under high strain rates for SS 304L.

In general, dynamic loading not only affects the macroscopic behavior of the material but also has a strong effect on the evolution of the microstructure. Many researchers have studied the effect of strain rate on the microstructure structure [2, 6, 8]. Microscopic observations, such as, dislocation density, shear bands and the amount of martensite fraction were correlated with macroscopic response of 304L stainless steel under high impact loading by Lee et al. [2]. They used the data to analyze the variation in strength and the work hardening behaviour both of which are important from a practical point of view. When these materials are plastically deformed, part of the austenite is converted to $\alpha$ and/or $\varepsilon$-martensite; the latter forming predominantly at low stress values. With gradual increase in the deformation, there was an increase in the amount of martensite transformed, imparting a higher mechanical strength to the material [10].

Material SS 304LN studied in this work, is a low carbon and nitrogen added version of SS 304. It is well known that SS 304 is susceptible to inter-granular corrosion [11] and to achieve excellent intergranular corrosion resistance, carbon content is reduced in it and designated as SS 304L, which has lower mechanical strength. In order to achieve superior mechanical strength, nitrogen is added and this material is known as SS 304LN. However, to date a lot of experimental work has been conducted on SS 304 and SS 304L but no such studies have been performed on the stainless steels 304LN under dynamic loading, even though it has superior quasi-static mechanical properties and widely used in sea water systems, chemical and other industries. The main objective of this work is to investigate the dynamic behaviour of stainless steel 304LN over a wide range of strain rates and accumulated strains by using Kolsky bar setup.

## 13.2 Experimental Method

### *13.2.1 Specimen Material for Quasi-Static and Dynamic Experiments*

The material used in this work was commercially available stainless steel 304LN pipe fully annealed with an average grain size of 22.6 μm. The grain size was measured via metallography using optical microscope. The chemical composition of SS 304LN used in this study is given in Table 13.1. Cylindrical compressive specimens of 5 mm diameter and a L/D ratio of 0.6 and cylindrical specimens of 5 mm diameter with a L/D ratio of 1 were prepared for dynamic and quasi-static experiments, respectively. All the specimens' flat faces were surface grinded/polished and have a maximum flatness defect of 0.01 mm. The material obtained was in the form of a pipe and therefore to study any anisotropic effects both radial and longitudinal direction cuts were made for metallographic inspection. The microstructure as observed under optical microscope showed no difference in grain structure. Experiments were conducted on both categories of specimens and negligible difference in mechanical response was observed under quasi-static as well as dynamic loading. From the results, it was concluded that the material was isotropic, only radial cut samples were chosen for all the experiments in this work, and only those results are presented.

**Table 13.1** Chemical composition (in wt%) of SS 304LN

| Steel | Cr | Ni | Mo | Mn | Si | C | N | P | S | Fe |
|---|---|---|---|---|---|---|---|---|---|---|
| 304LN | 18.77 | 9.47 | 0.3 | 1.63 | 0.42 | 0.022 | 0.16 | 0.024 | 0.005 | balance |

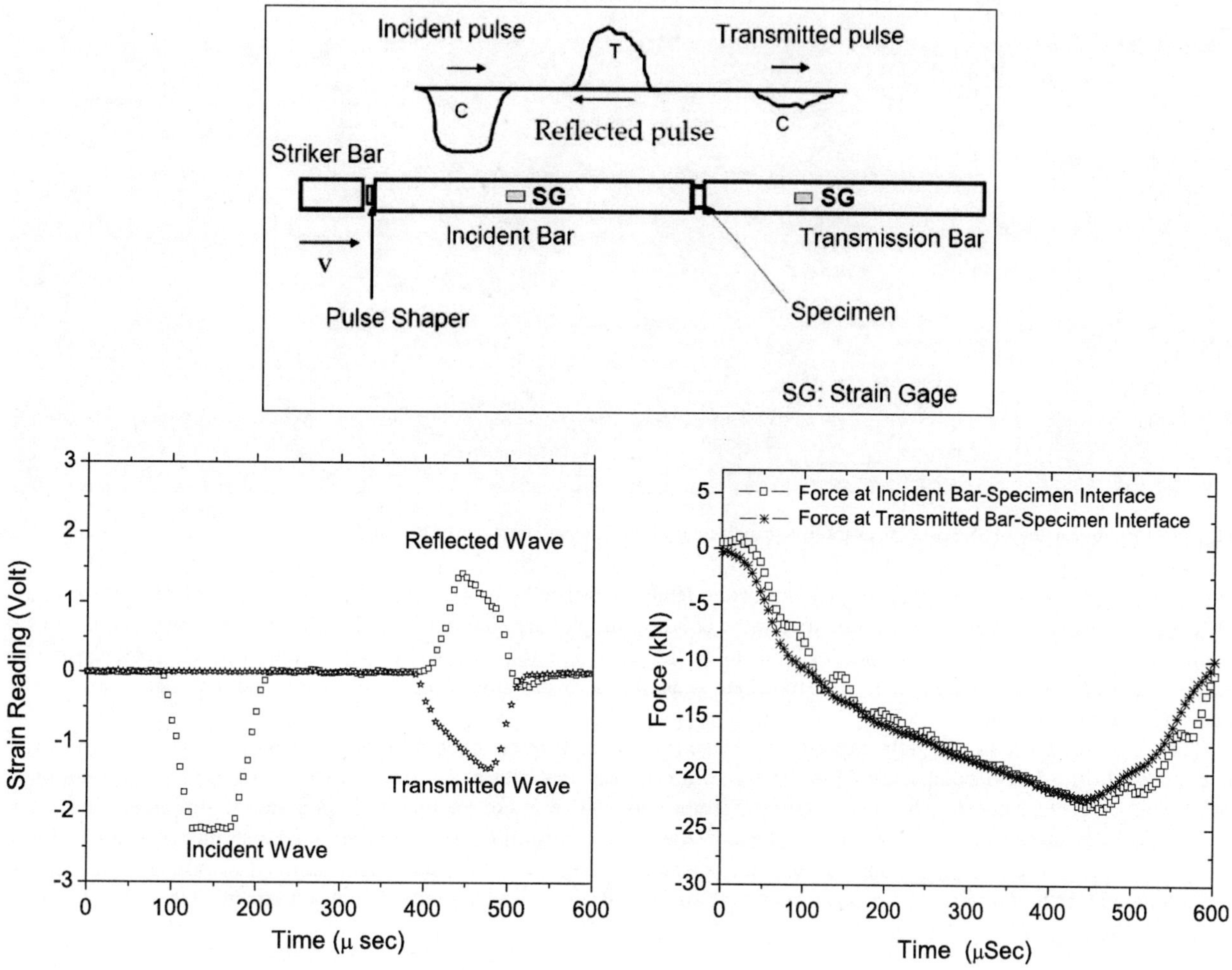


**Fig. 13.1** (**a**) A schematic of Kolsky bar setup [13], (**b**) typical strain gauge signals from the experiment, and (**c**) force equilibrium plot showing the forces acting on either side of the specimen during loading

### 13.2.2 Experimental Method

Quasi-static experiments were conducted at a strain rate of 0.001 $s^{-1}$ at room temperature on MTS compression testing machine (100 kN capacity). Dynamic compression experiments were conducted for the strain rate range of 1,700–4,800 $s^{-1}$ at room temperature by using Kolsky bar setup. In a Kolsky compression bar, the compressive stress wave is generated by the impact of the striker on the incident bar. When the compression wave in the incident bar arrives at the interface between the incident bar and the specimen, part of it is reflected back into the incident bar while rest transmits into the specimen. When the wave reaches the interface of the specimen and transmission bar, part of wave is reflected back into specimen and rest propagates into the transmission bar. Due to impedance mismatch between specimen and the bars, several reverberations of the loading wave occur within the specimen, and it builds up the stress level gradually and compresses the specimen. From one-dimensional elastic wave propagation theory and the recorded strain histories on the bars, the specimen stress, strain and strain rate were calculated by using the following relations:

$$\acute{\varepsilon} = -2(C_0/L_s)\varepsilon_R;\ \varepsilon = \int \acute{\varepsilon} dt \text{ and } \sigma = (A_b/A_s)E_b\varepsilon_T \tag{13.1}$$

where, $C_0$ is the longitudinal wave velocity in the bars, $L_s$ is the length of the specimen, $E_b$ is the Young's modulus of the bars, while $A_b$ and $A_s$ are the cross-sectional areas of the bar and the specimen, respectively. A critical review on the Kolsky

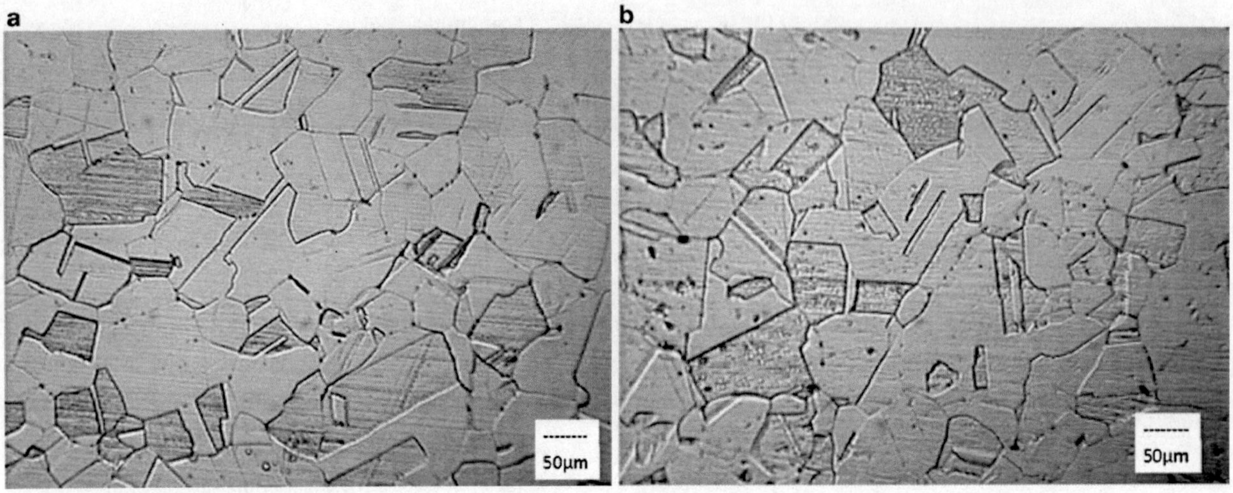


**Fig. 13.2** Microstructure of SS 304LN in as received condition (**a**) Longitudinal direction (**b**) Radial direction

bar setup and experimental procedures have been completely described by Gama et al. [12]. A schematic of Kolsky bar setup [13], typical strain gauge signals and obtained force equilibrium plot are shown in Fig. 13.1. As can be observed, the force equilibrium was achieved for most part of the loading duration. The reflected and the transmitted signals show the strain hardening behavior of the specimen. The strain rate peaks in the beginning of the deformation and then decreases due to material strain hardening.

The interrupted strain-rate jump experiments in compression were conducted on Kolsky bar with a change of strain rates to study the rate-dependence of flow stress in SS 304LN material at constant structure. Unlike the conventional strain rate jump experiments where instantaneous strain rate change at constant strain is imparted, in the interrupted rate jump experiments the sample is loaded at particular strain rate controlling the total accumulated strain either by limiting the pulse width or by using a collar around the specimen. Then the fully unloaded, strained sample is reloaded at room temperature at a different (higher or lower) strain rate. Strain rate sensitivity at constant structure is then defined as $(\partial \ln\sigma/\partial \ln\dot{\varepsilon})_{\text{structure}}$.

### *13.2.3 Microstructural Observations*

Samples of SS 304LN were polished up to diamond finish and microstructures were examined under optical microscopy after electrochemical etching in freshly prepared 10% oxalic acid etchant as per ASTM Standard A262, Practice A [14]. Optical micrographs of SS 304LN are shown in Fig. 13.2, which shows the grains have similar morphology and size distribution is uniform both in the longitudinal and radial directions. Therefore, the macroscopic material behavior should be isotropic.

## 13.3 Results and Discussion

### *13.3.1 Stress–Strain Characteristics*

The true stress–true strain curves for SS 304LN deformed at room temperature over a wide range of strain rates are shown in Fig. 13.3a. From quasi-static experimental result, the flow stress of SS 304LN at 0.2% strain ($\sigma_{0.2}$) was observed as 315 MPa which is much higher when compared to SS 304 (205 MPa [15]) and SS 304L (170 MPa [16]). There is remarkable difference in flow stress between quasi-static and dynamic high strain rates experiments. The flow stress ($\sigma_{0.2}$) from quasi static to highest strain rate loading increased from 315 to 612 MPa, an increase of 297 MPa. The experimental results show that the flow stress increased with increasing strain and strain rates. It is noticed that, at 15% true strain, the flow stress

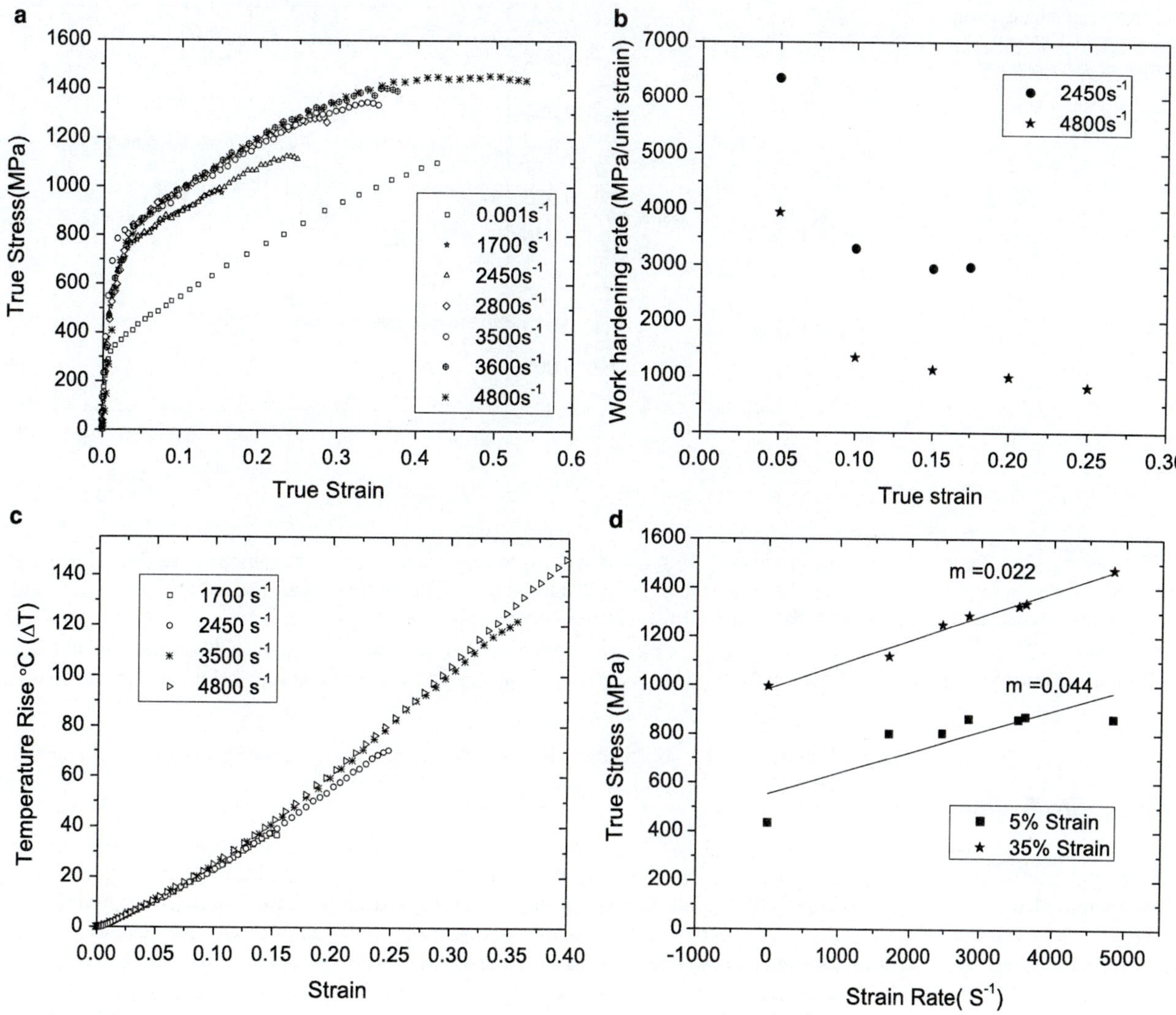


**Fig. 13.3** SS 304LN (**a**) Stress–strain curves at room temperature at different strain rates, (**b**) variation in work-hardening rate as a function of true strain, (**c**) temperature (ΔT) rise due to plastic deformation as a function of true strain, and (**d**) effect of strain rate on flow stress at different accumulated strains

increased from 976 to 1,096 MPa, an increase of 120 MPa from lowest (1,700 $s^{-1}$) to highest (4,800 $s^{-1}$) strain rates. Therefore, the effect of strain rate on flow stress is significant in 304LN stainless steel. It is observed that flow stress for 4,800 $s^{-1}$ strain rate at 20% strain is 1,200 MPa which is 300 MPa higher that reported flow stress of SS 304L (900 MPa) [2]. During high strain rate experiments, no macroscopic shear bands were observed in deformed sample and from experimental results at the highest strain rate (4,800 $s^{-1}$), flow stress saturation is observed after 40% strain.

Figure 13.3b shows the variation of work hardening rate ($\partial\sigma/\partial\varepsilon$) obtained from the stress–strain curves. It was observed that the work-hardening rate decreases with increasing strain rate and total accumulated strain. The key reason for this is with increasing strain, grains gradually get saturated with dislocation network and this effect leads to reduction of work hardening rate. In addition plastic deformation induces heat, leading to thermal softening, which further reduce the work hardening effect. Similar trends of work hardening rate variation with strain rate and plastic strain for SS 304L was observed in literature [8].

The heat generated during high rate deformation was calculated assuming that 95% of the work done during deformation is converted into heat. This assumption is valid for adiabatic loading conditions related to lower thermal conductivity of specimen under deformation or the total time of loading (i.e., high strain rate). The rise in temperature can be calculated for frictionless deformation by $\Delta T = \alpha[\int \sigma d\varepsilon/\rho C_p]$, where ΔT is the temperature rise, ρ is the density of the speicmen

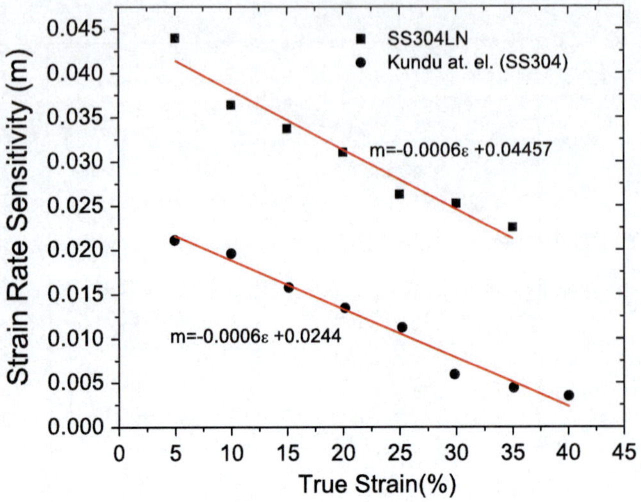


**Fig. 13.4** Variation in strain rate sensitivity with as a function of true strain and comparison with earlier work on SS 304

(7,790 kg/m$^3$), $C_p$ is the specific heat capacity (477 J/kg/K), $\sigma$ is the true stress and d$\varepsilon$ is the increment of strain and $\alpha$ is the fraction of deformational work that is converted to heat, assumed to be 0.95. The temperature increase with strain rate and total strain is shown in Fig. 13.3c. At 25% strain and for 4,800 s$^{-1}$ strain rate, the homologous temperature ($T/T_{melt}$) of SS 304LN was 0.23 where as at same strain and strain rate the homologous temperature of SS 304L [2] was calculated 0.21. This small difference can have an effect on the phase transformation, as austenite is stable at higher temperatures.

### 13.3.2 *Strain Rate Sensitivity*

The strain rate effect is evident from the plot of flow stress vs. strain rate, as shown in Fig. 13.3d. It is observed that flow stress increased with increasing strain rate. The strain rate sensitivity (m) at constant strain was evaluated by finding the slope of flow stress and the strain rate on a log-log scale. Figure 13.4 shows the variation of strain rate sensitivity with true strain. From the Fig. 13.4, strain rate sensitivity is decreasing with true strain as $m = -0.0006\varepsilon + 0.04457$ and this may be due to the accumulation of dislocations leading to saturation and thermal softening. Kundu et al. [3] also observed that the strain rate sensitivity linearly decreases with increasing true strain for SS 304 at room temperature as $m = -0.0006\varepsilon + 0.0244$. SS 304LN has higher strain rate sensitivity compared to SS 304 although both have same slope but SS 304LN has higher intercept value.

### 13.3.3 *Interrupted Rate Jump Experiments*

The interrupted strain-rate jump experiments in compression provide the direct evidence of rate-dependence of flow stress in metals at constant structure. To investigate strain rate sensitivity at constant structure of SS 304LN, interrupted rate jump experiments were performed with a change of strain rates at 1,900s$^{-1}$, 2,100 s$^{-1}$, 2,350 s$^{-1}$, 3,000 s$^{-1}$, 3,550 s$^{-1}$ and at 16.5%, 18%, 21%, 28.5% and 30% corresponding accumulated strains, respectively. The experiments conducted in this work, although do not correspond to the traditional rate jump experiments, but we can use the data to calculate the constant structure rate sensitivity with the following assumptions: (a) microstructure during unloading and reloading remain the same, (b) effect of volume expansion related to the strain induced martensite transformation on specimen dimension is small and can be neglected, and (c) dimensions of the deformed specimen used for successive experiments has no effect on the mechanical response with similar microstructures. The maximum temperature rise at strain rates 1,900s$^{-1}$, 2,100 s$^{-1}$, 2,350 s$^{-1}$, 3,000 s$^{-1}$, 3,550 s$^{-1}$ are 40 °C, 66 °C, 90 °C, 68 °C, 119 °C, respectively, which are well below the recrystallization temperature (520 °C [17]) and full anealing temperature (1,040 °C [18]) of steel. So, the assumption of constant microstructure is reasonable during interrupted strain-rate jump experiments.

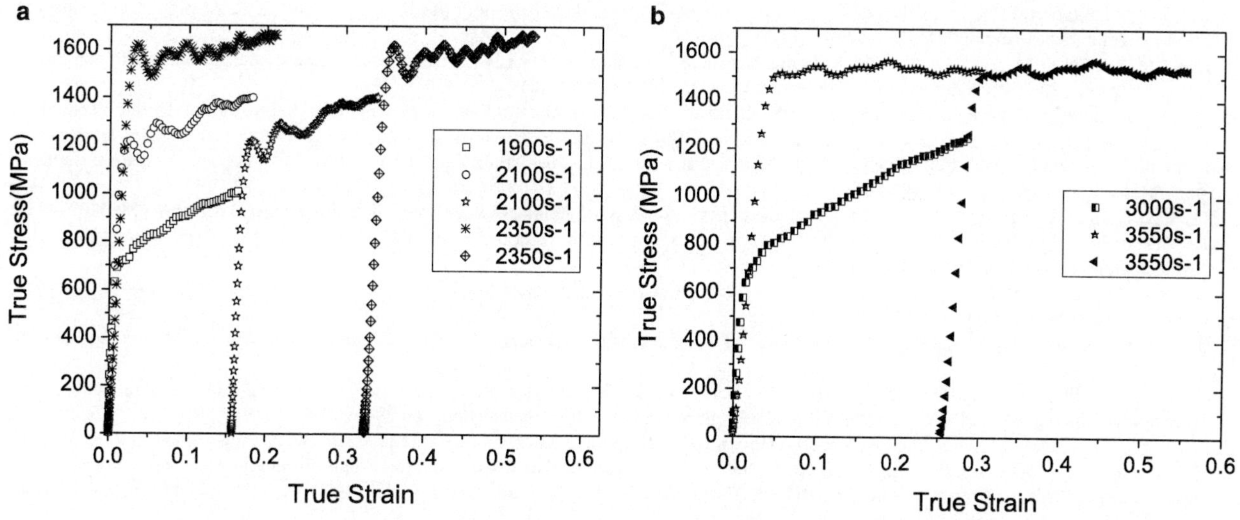


**Fig. 13.5** Interrupted rate Jump experiments at (**a**) strain rate 1,900s$^{-1}$, 2,100 s$^{-1}$ and 2,350 s$^{-1}$ (**b**) strain rate 3,000 s$^{-1}$ and 3,550 s$^{-1}$

Figure 13.5a show the true stress–true strain curves for the rate jump experiments performed at strain rates 1,900s$^{-1}$, 2,100 s$^{-1}$ and 2,350 s$^{-1}$. With change in strain rate at constant structure, flow stress increased considerably. The strain rate sensitivity at constant structure was calculated and found to be 0.276 for strain rate jump from 1,900 to 2,100 s$^{-1}$ and 0.331 for strain rate from 2,100 s$^{-1}$ to 2,350 s$^{-1}$. Similarly, Fig. 13.5b shows the true stress–strain curves for the strain rate jump tests performed at higher strain rates of 3,000 and 3,550 s$^{-1}$ and the strain rate sensitivity is 0.793 for this strain rate jump. At higher strain rate (3,550 s$^{-1}$), stress saturation is observed. From the results, it is concluded that SS 304LN material is showing positive strain rate sensitivity at constant strain and constant strucuture. It was observed that flow stress of SS 304LN increased upon reloading compare to single loading at the same strain rate because the previous plastic loading increased the dislocation density and possibly martensite transformation during the deformation process [9] and therefore, increased the barriers to dislocation movement resulting in the increased flow stress [8]. Similar observation was reported for SS 304L by Lichtenfeld et al. [6] at lower strain rates (0.00125 and 0.125 s$^{-1}$).

## 13.4 Conclusions

The mechanical behaviour of SS 304LN material was investigated by performing quasi-static and dynamic compression experiments in the strain rate range of 0.001–4,800 s$^{-1}$ at room temperature using UTM and compression Kolsky bar setup. The dynamic behaviour of SS 304LN is strongly affected by strain rate. It was clearly observed that flow stress increases with increasing strain rate at any specific accumulated strain and experiments showed that the flow stress of 304LN was higher than that of 304L both at quasi-static and dynamic strain rates. Work hardening rate decreased with increasing strain rate and accumulated strain. It was observed that strain rate sensitivity at constant strain was linearly decreasing with increasing true strain. Interrupted strain rate jump experiments were performed with change of strain rates and it was observed that strain rate sensitivity at constant structure was positive for SS 304LN and much higher than constant strain sensitivity.

**Acknowledgement** The authors wish to thank Prof. Parmeswaran, IIT Kanpur for his help to conduct the experiments on Kolsky bar setup and Dr. R.K. Sinha, Dr. P.K.Vijayan, Dr.A.K Nayak and Mr. Kamlesh Chandra for their valuable help.

## References

1. Andrade-Campos A, Teixeira-Dias F, Krupp U, Barlat F, Rauch EF, Grácio JJ (2010) Effect of strain rate, adiabatic heating and phase transformation phenomena on the mechanical behaviour of stainless steel. Strain 46:283–297
2. Lee W-S, Lin C-F (2000) Impact properties and microstructure evolution of 304L stainless steel. Mater Sci Eng A308(2001):124–135

3. Kundu A, Chakraborti PC (2010) Effect of strain rate on quasistatic tensile flow behaviour of solution annealed 304 austenitic stainless steel at room temperature. J Mater Sci 45:5482–5489
4. Zaera R, Rodríguez-Martínez JA, Casado A, Fernández-Sáez J, Rusinek A, Pesci R (2012) A constitutive model for analyzing martensite formation in austenitic steels deforming at high strain rates. Int J Plast 29:77–101
5. Talonen J, Nenonen P, Pape G, Hänninen H (2005) Effect of strain rate on the strain-induced – martensite transformation and mechanical properties of austenitic stainless steels. Metal Mater Trans A 36A:421–432
6. Lichtenfeld JA, Mataya MC, Van Tyne CJ (2006) Effect of strain rate on stress–strain behavior of Alloy 309 and 304L Austenitic stainless steel. Metall Mater Trans A 37A:147–161
7. Chen X, Wang Y, Gongand M, Xia Y (2001) Dynamic behavior of SUS304 stainless steel at elevated temperatures. J Mater Sci 39:4869–4875
8. Lee W-S, Lin C-F (2002) Comparative study of the impact response and microstructure of 304L stainless steel with and without prestrain. Metall Mater Trans A 33A:2801–2810
9. Lee W-S, Linb C-F, Chenc T-H, Yanga M-C (2010) High temperature microstructural evolution of 304L stainless steel as function of pre-strain and strain rate. Mater Sci Eng A 527:3127–3137
10. da Rocha MR, de Oliveira CAS (2009) Evaluation of the martensitic transformations in austenitic stainless steels. Mater Sci Eng A 517(2009):281–285
11. Singh R, Dey PK, Kumar A, Das SK, Kumar R, Chattoraj I (2001) Intergranular corrosion of deformed SS304. In: National seminar on corrosion and its prevention (NSCP-2001), National Metallurgical Laboratory, Jamshedpur, pp 29–33
12. Gama BA, Lopatnikov SL, Gillespie JW Jr (2004) Hopkinson bar experimental technique: a critical review. Appl Mech Rev 57(4):223–250
13. Abotula S, Chalivendra V (2011) Effect of aspect ratio of cylindrical pulse shapers on force equilibrium in Hopkinson pressure bar experiments. In: Proceedings of the SEM Annual Conference, June 7–10, 2010, Indianapolis, Indiana USA, © 2010 Society for Experimental Mechanics Inc., vol 1, pp 453–461
14. A 262 (1994) Annual book of ASTM standards, vol 3.01. ASTM, West Conshohocken
15. http://masteel.co.uk/type-304-stainless-steel.htm
16. http://masteel.co.uk/type-304l-stainless-steel.htm
17. George ET (2006) Steel heat treatment: metallurgy and technologies. Steel treatment hand book, 2nd edn., p 352
18. A480/A480M (2011) Standard specification for general requirements for flat-rolled stainless and heat-resisting steel plate. Sheet, and strip, ASTM Standards

# Chapter 14
# Plasticity and Ballistic Characterization of a New Armour Steel

**Jacques N. Bester and Waldo E. Stumpf**

**Abstract** Some results are presented on an ongoing experimental, ballistic and numerical characterization of a novel new armour steel with comparison to a leading commercial benchmark alloy. The apparent better resistance to plugging failure of the new steel is shown with comparative ballistic tests over a range of plate thicknesses for two types of projectiles. Split Hopkinson Pressure Bar tests in compression and tension as well as a series of quasi-static tensile tests were performed to characterize the dynamic plasticity and failure of the materials.

**Keywords** Ballistic impact • Steel armour plate

## 14.1 Introduction

The problem of terminal ballistics can broadly be defined as the study of the interaction of a projectile with a target and an interpretation of the outcome. Fig. 14.1 shows the difference in interaction of a lead filled projectile and hardened steel core projectile on a steel plate. Exact test conditions of the projectiles, impact velocities, target material and thickness could not be found for these figures [1], however the general projectile interaction and failure is clear.

For a steel plate of sufficient thickness to resist a lead filled projectile at standard impact velocity, the projectile shows catastrophic deformation with apparent phase transformation to liquid state and shock waves. The interaction of a hardened steel core projectile at standard impact velocity with a similar thickness steel plate results in failure of the target with little damage to the core of the projectile [2]. This relatively small change of different projectile material, reveals some of the extensive complexities involved in this field. Some of the target perforation mechanisms are defined by [3] as pedalling, fragmentation, fracture due to the initial stress wave, radial brittle failure, ductile hole enlargement, spall failure [4, 5] and plugging.

This paper forms part of an ongoing investigation into the armour material design and optimization for a level 1 kinetic energy threat [6]. The 5.56 × 45 mm M193 projectile is the limiting calibre for this level and the typical target perforation mechanism is plugging (Fig. 14.2) with adiabatic shear bands as well as inter-granular cracks (Fig. 14.3).

Numerical algorithms with specialized material plasticity and damage models [7–10] yield promising correlation with impact events. The decomposition of complex material behaviour into sets of simplified relationships, allows for almost infinite differentiation [11–13]. A continuous challenge for researchers is the balance between accuracy of models and range of validity for practical implementation. Numerical simulation of lead filled projectiles on steel plates requires the full coupling of Euler and Lagrangian meshes or the use of hybrid techniques [14]. This type of problem is therefore still a numerical challenge and requires further attention. However this paper represents the initial steps towards a better understanding of ballistic performance.

At present commercial armour plate data sheets only focus on the following mechanical properties; hardness, yield strength, tensile strength, minimum elongation and low temperature impact toughness. The hardness of the material is often included in the commercial product name and seems to be the only required mechanical property distinguishing ballistic resistant plate from structural steel. Hardness is an indicator of tensile strength and as extensive analytical work [15–17]

J.N. Bester (✉) • W.E. Stumpf
Department of Materials Science and Metallurgical Engineering, University of Pretoria, Pretoria, South Africa
e-mail: jacques.bester@megchem.co.za

V. Chalivendra et al. (eds.), *Dynamic Behavior of Materials, Volume 1: Proceedings of the 2012 Annual Conference on Experimental and Applied Mechanics*, Conference Proceedings of the Society for Experimental Mechanics Series,
DOI 10.1007/978-1-4614-4238-7_14, 

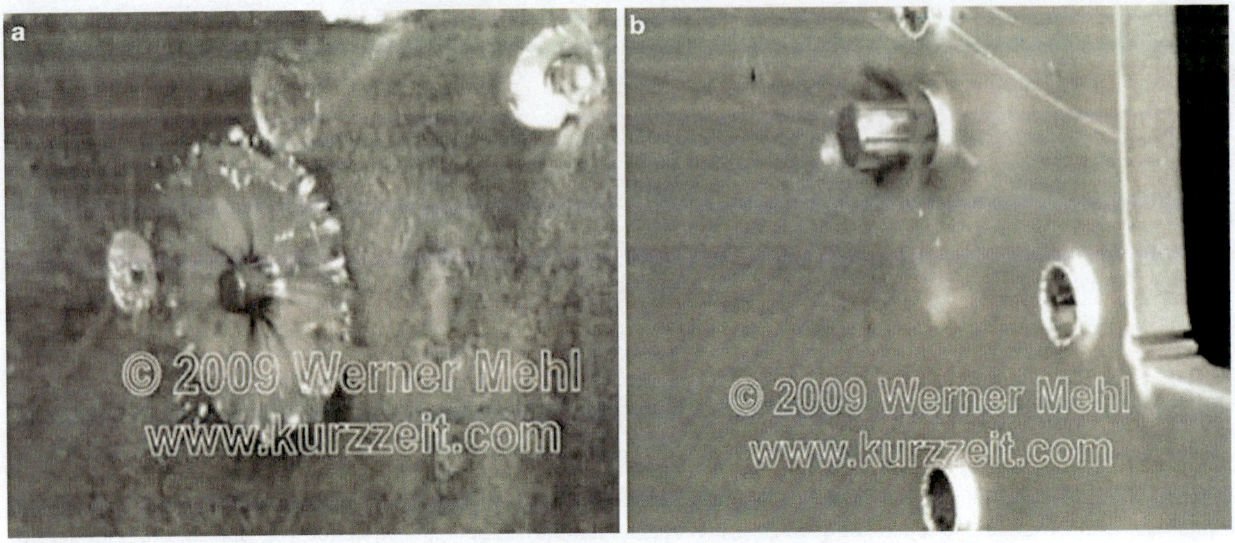


**Fig. 14.1** (**a**) Lead filled projectile on steel plate (**b**) Hardened steel core projectile on steel plate [1]

**Fig. 14.2** Micrograph of ballistic plug formation and partial expulsion on the newly developed armour steel (5.56 × 45 mm of 7.5 mm plate)

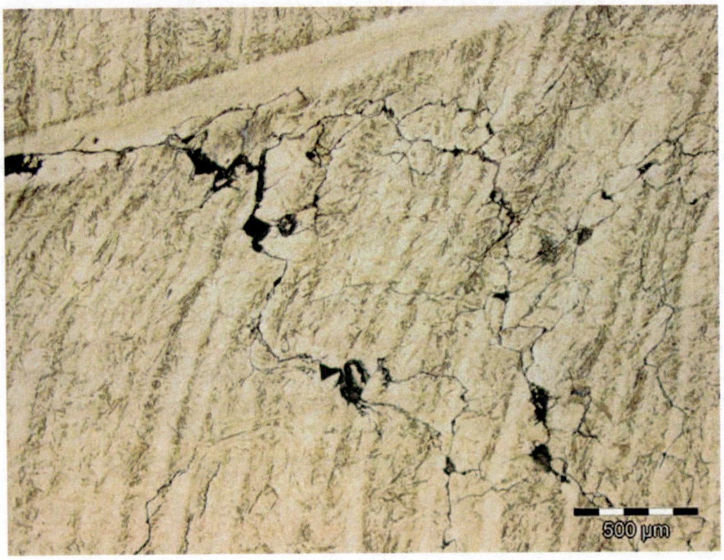


**Fig. 14.3** High magnification of micrograph in Fig. 14.2 (area close to partial plug formation)

relies almost entirely on the tensile strength of the material, the use of material hardness as a ballistic performance indicator is entirely valid.

Recent work [18] has indicated that material hardness alone might not be the only required mechanical property for optimal ballistic performance. The development of Dual-Phase, TRIP and especially TWIP alloys [19] (with tensile strength of ≈1,700 MPa at 45 % strain), indicates that enhanced plastic energy absorption could be attained with higher work-hardening rates. Thermal softening delays, dynamic plasticy, damage and phase transformations should ideally also be included in the development of optimal ballistic performance.

The new experimental alloy was developed as part of a group of alloys, aimed at providing superior ballistic resistance for small-arms projectiles at lower plate thickness levels, primarily for use in light armoured vehicles. The chemical composition of the alloys was based on high manganese content, as well as moderate levels of chromium, nickel and molybdenum. In the as-rolled condition, the plate produced to these alloy formulations produced a predominantly martensitic microstructure, with some retained austenite evident. Alloy 1 was the alloy in the group aimed at the semi-structural 500 HB application, where the armour plate is fabricated into the vehicle chassis and where the 7–8 mm armour plate would provide adequate resistance to the NATO STANAG 4569 Level 1 ballistic threat [6].

The materials included in this testing program are the new experimental Alloy 1 [20], a tempered variant Alloy 1 T200 (200°C for 1 h) and a leading commercial ballistic benchmark material. The ballistic, normal tensile test, high temperature tensile test and notched tensile test results are summarized with the high strain-rate SHPB test results discussed in detail.

## 14.2 Ballistic Performance

A high pressure ballistic test barrel with standard cartridge charge was used in this investigation. Each cartridge was loaded before the shot to control the impact velocity. A total number of 17 plates were tested with the 5.56 × 45 mm M193 caliber round. The results are presented in Fig. 14.4 and data range bars have been removed from Alloy1 data points to clarify the results. Alloy1 is positioned above both the benchmark and Alloy1 T200. The top limit of 937 + 20 = 957 m/s for 5.56 × 45 mm (M193) projectile at 10 m [6] is shown by the dotted line which can be used to determine the minimum required thickness to defeat the 5.56 × 45 mm (M193) projectile. A plate thickness of 9 mm is required by both the benchmark and Alloy1 T200 materials with only 7.3 mm (lower bound value) required by Alloy1.

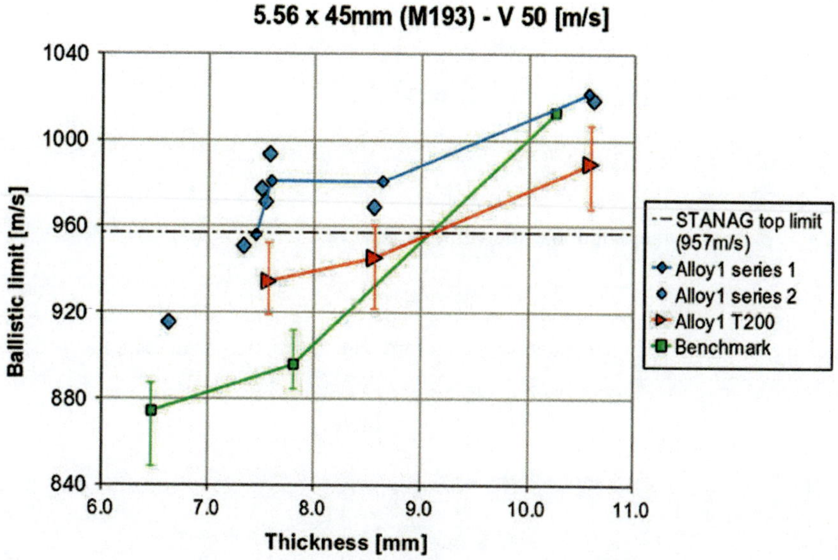


**Fig. 14.4** Ballistic performance of Alloy1, Alloy1 T200 and the Benchmark material against 5.56 × 45 mm (M193) projectile at 10 m

## 14.3 Split Hopkinson Pressure Bar Evaluation

### 14.3.1 Tensile SHPB Equipment

The tensile configuration [21] of the Split Hopkinson Pressure Bar (THPB) used in this experimental investigation is seen in Fig. 14.5. A tubular striker is used with reversed gas gun to strike the transfer flange and send a tensile stress wave down the input bar. The striker as well as the input and output bars, all aluminium, are supported with several Teflon bush assemblies along the length to reduce any axial reaction forces during testing. Strain gages located in the middle of both bars are used to record the stress wave at 10 MHz.

The specimen is connected to the bars using a split taper threaded adapter, with a special cut-out on the inside shown in Fig. 14.6. Longitudinal slots in the adapter allow for a tight fit to the specimen, allowing the adapter to be snap-fitted to the specimen before fitment. Similarly the use of a tapered thread simplifies the assembly by reducing the number of rotations to tighten the connection. The compact design and use of titanium in the adapter minimizes inertial effects on the test results.

### 14.3.2 Specimen Results from Measured Bar Stress Waves

The stress waves measured in the pressure bars are related to force and displacement values next to the specimen as shown in the next section. The dynamic plasticity response of the material is then calculated using Fig. 14.7 as reference.

The general one dimensional elastic wave equation [21],

$$\frac{\partial^2 u}{\partial x^2} = \frac{1}{c^2}\frac{\partial^2 u}{\partial t^2} \tag{14.1}$$

can be written as follows for incident and reflected waves [22].

$$u_1 = f(x - c_{b1}t) + g(x + c_{b1}t) \tag{14.2}$$

Differentiating Eq. 14.2 with respect to x, the strain in the bar is given by,

$$\varepsilon_{b1} = \frac{\partial u}{\partial x} = f' + g' = \varepsilon_i + \varepsilon_r \tag{14.3}$$

and differentiating Eq. 14.2 with respect to t and using Eq. 14.3 gives,

$$v_1 = \frac{\partial u}{\partial t} = c_{b1}(f' + g') = c_{b1}(\varepsilon_i + \varepsilon_r) \tag{14.4}$$

As no reflected wave is present in the output bar the velocity is,

$$v_2 = c_{b2}\varepsilon_t \tag{14.5}$$

Therefore the instantaneous increase in specimen length, strain and strain rate is calculated as follows:

$$\Delta u = \int v_1 dt - \int v_2 dt \tag{14.6}$$

$$\varepsilon_s = \frac{\Delta u}{l_0} \tag{14.7}$$

$$\dot{\varepsilon} = \frac{v_1 + v_2}{l_{si}} \tag{14.8}$$

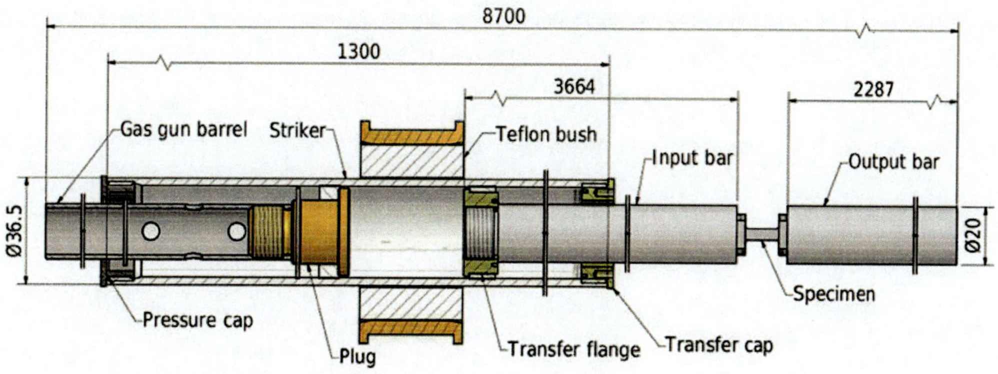


**Fig. 14.5** THPB configuration – in section view

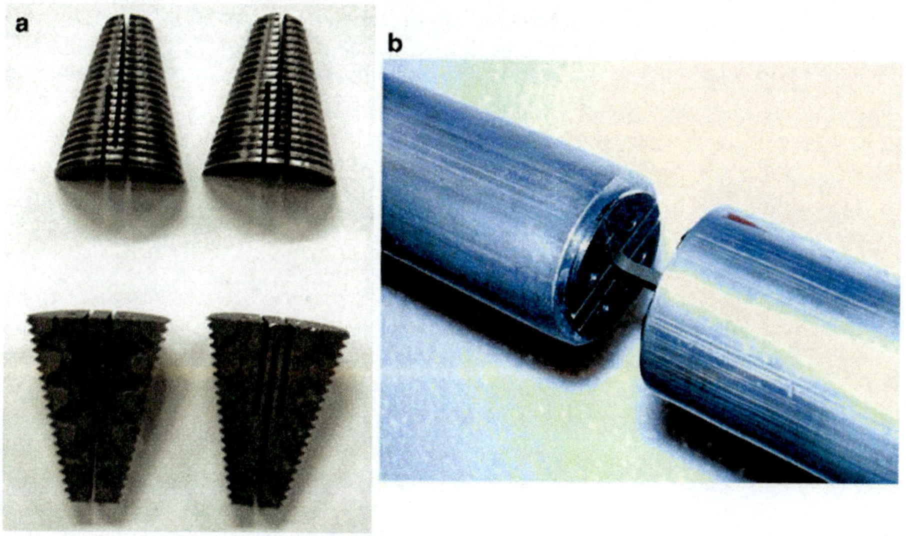


**Fig. 14.6** (**a**) Snap on taper threaded specimen connection adaptor (**b**) Specimen and adaptor assembly in THPB

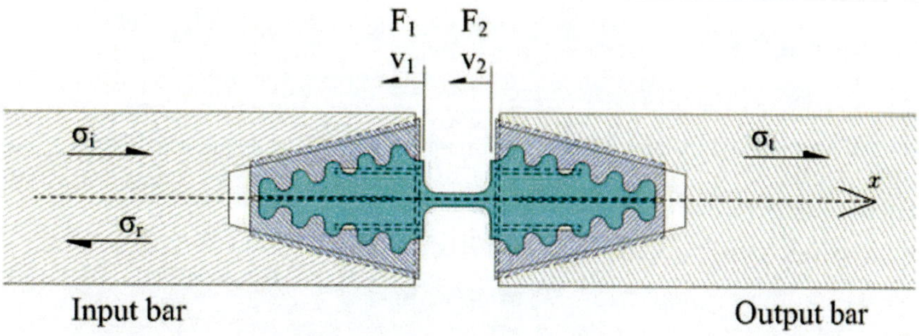


**Fig. 14.7** THPB stress wave diagram with specimen attachment adaptor shown

**Fig. 14.8** Comparative stress results of curves using $F_1$ and $F_2$ (Eqs. 14.9 and 14.10) respectively

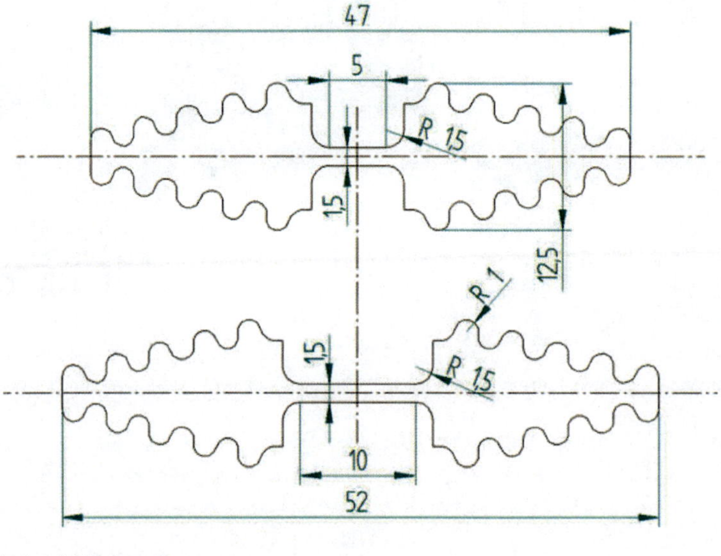


**Fig. 14.9** Tensile specimen design for high strain rate tests

with $l_0$ and $l_{si}$ the specimen original and instantaneous lengths respectively. As the bars remain elastic, the forces in each bar are calculated as,

$$F_1 = A_{b1}E_{b1}(\varepsilon_i + \varepsilon_r) \tag{14.9}$$

$$F_2 = A_{b2}E_{b2}\varepsilon_{b2} \tag{14.10}$$

Using these forces and Eqs. 14.9 and 14.10, the dynamic plasticity response can be calculated for each force signal as shown in Fig. 14.8.

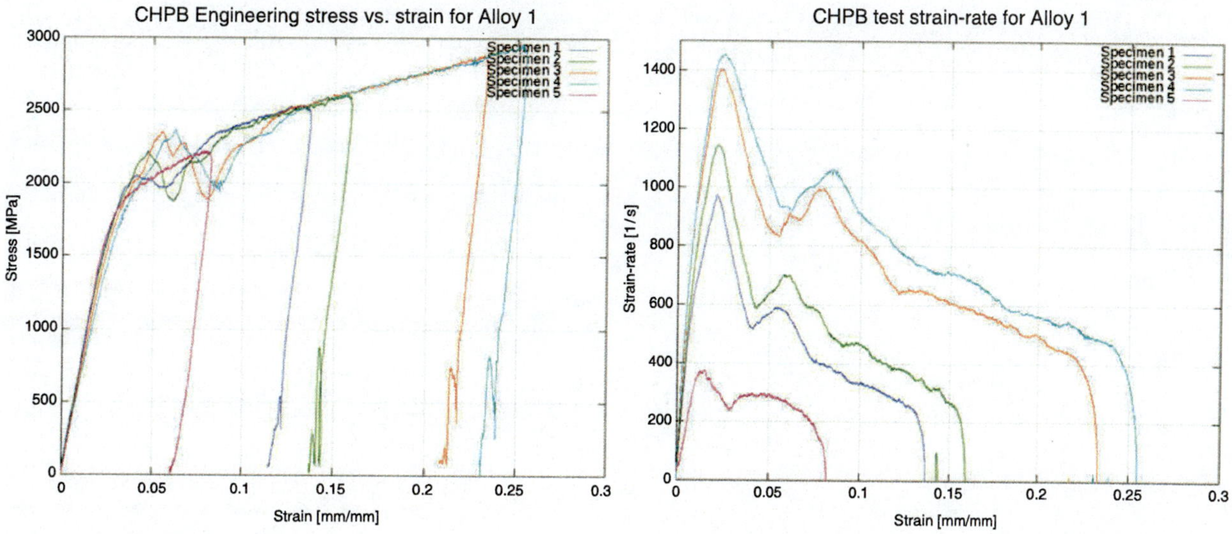


**Fig. 14.10** CHPB stress and strain rate results for Alloy 1

The force signal calculated using Eq. 14.9 seems to have some additional dynamic noise component superimposed on the signal. This was observed in all the tensile tests and is assumed to be related to the experimental setup. The force signal from the output bar was therefore used to generate all the dynamic plasticity results.

$$\sigma_s = \frac{F_2}{A_s} \tag{14.11}$$

The strain rate in the specimen is related to the inverse of the specimen length, Eq. 14.8. Two different lengths of the tensile specimen were accordingly used as shown in Fig. 14.9 to increase the range of strain rates during testing (Fig. 14.15b) and therefore also the range of accuracy in implementation of the material model.

### *14.3.3 SHPB in Compression*

The Split Hopkinson Pressure Bar in compression (CHPB) configuration is similar to the THPB shown in Fig. 14.5, as the same gas gun and vertical support structure are used. However a conventional striker bar [22] is used to impact the input bar and the compression specimen is simply held in place before the test by friction with a thin layer of high pressure grease as this also reduces the lateral friction, therefore reducing barrel formation. The design of the bars is limited by contact stresses to the specimen and should therefore be harder than the material tested and hardened steel bars were used. The same procedure (Sect. 3.2) is used to calculate dynamic plasticity in compression; however as the stress wave is in compression for CHPB tests, the sign of the wave equations is reversed.

### *14.3.4 SHPB Results*

All the SHPB results are presented with engineering stress strain as well as strain-rate strain curves for each tested specimen. Results are grouped together with two figures per material in a final summary plot of all the stress results for each of the tensile and compression configurations. These summary plots show each material as a single colour to simplify interpretation.

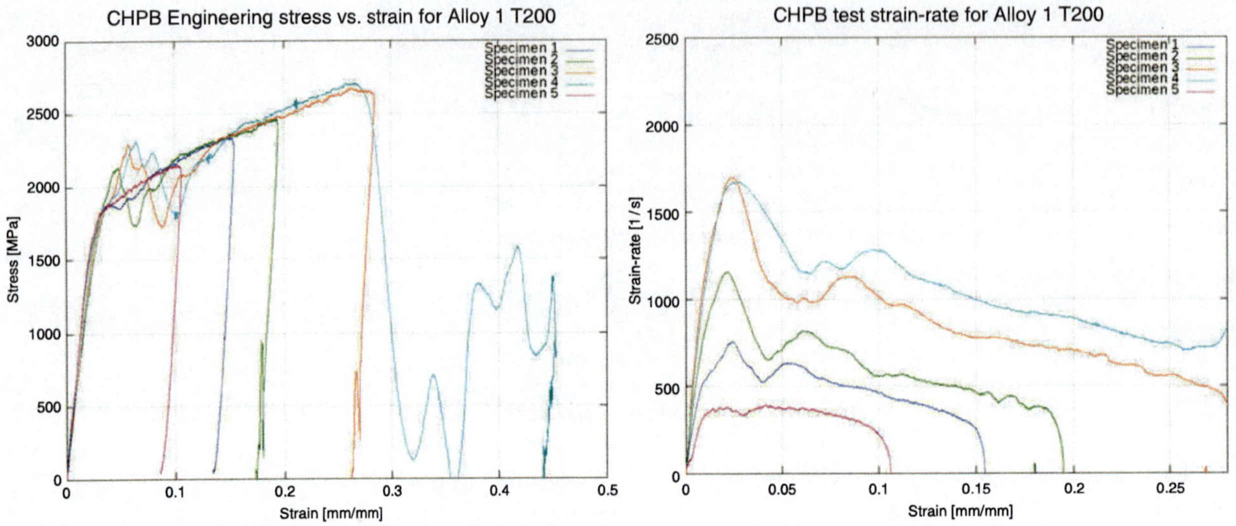


**Fig. 14.11** CHPB stress and strain rate results for Alloy 1T200

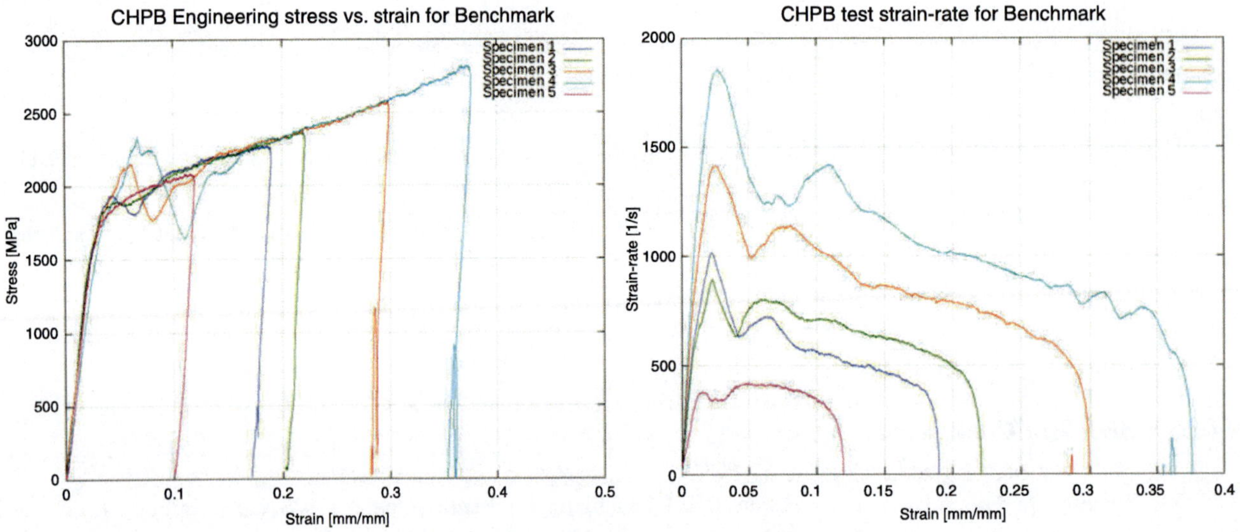


**Fig. 14.12** CHPB stress and strain rate results for the Benchmark alloy

#### 14.3.4.1 CHPB Results

The initial oscillation after the yield point, seen in all the compression results, is due to the dynamic instability as force equilibrium from both sides has not been reached. This is also clearly seen in the strain-rate results (Fig. 14.10). Each successive test was performed at higher pressures in the gas gun charge cylinder to increase the strain-rate and deformation. Only one of the Alloy 1 T200 specimens fractured during the test (Fig. 14.11), however these tests will be used to fit the high strain-rate yield point and not damage or failure parameters. Strain-rate sensitivity in all three materials can be observed as the plasticity curves form slightly offset parallel lines in the overlapping regions (Fig. 14.12).

A clear trend is seen in Fig. 14.13 with Alloy 1 above Alloy 1 T200 and the benchmark material below both. The energy absorption capacity in compression, as well as the work-hardening rate of the experimental Alloy 1 is therefore higher than Alloy 1 T200 and the benchmark alloy, within the strain-rate range of 300–1,000 $s^{-1}$.

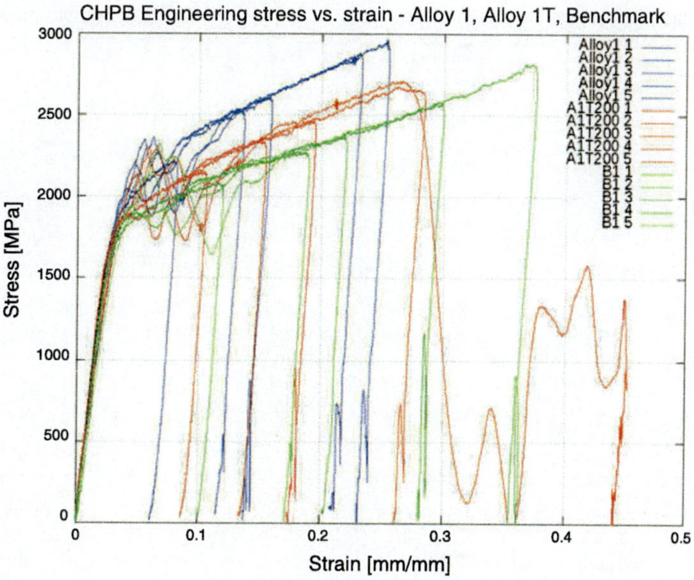


**Fig. 14.13** CHPB stress result summary of Alloy 1, Alloy 1T200 and the Benchmark alloy

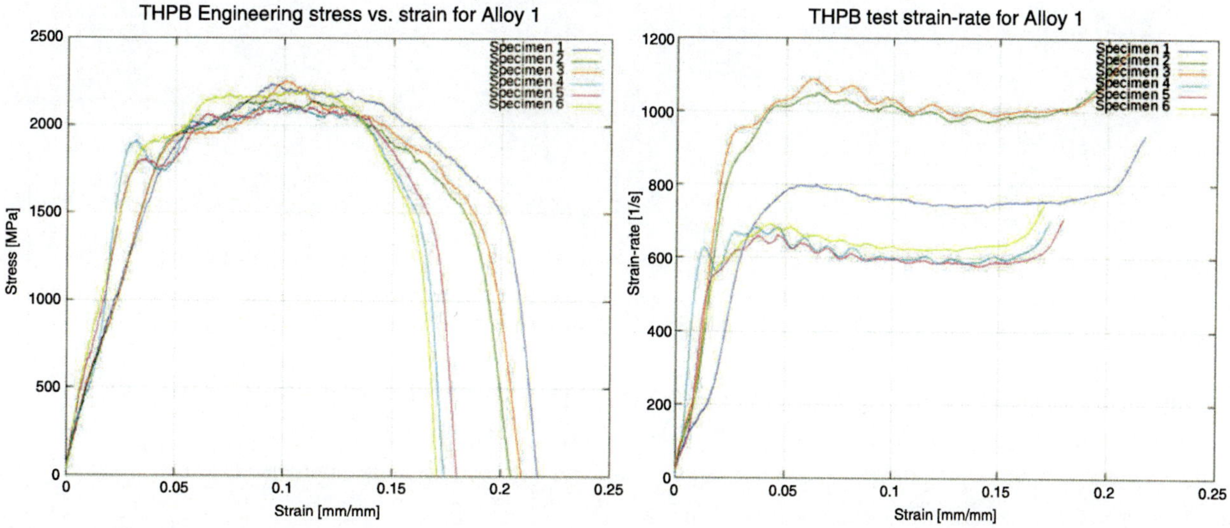


**Fig. 14.14** THPB stress and strain rate results for Alloy 1

#### 14.3.4.2 THPB Results

The aforementioned post yield dynamic instability is again observed in all the tensile results with an apparent increase in severity for the benchmark material (Figs. 14.14, 14.15, and 14.16). Strain-rate control by specimen design (Sect. 3.2) is an effective procedure to define and increase the testing range (Fig. 14.9). Elongation results are also grouped according to the initial specimen length as seen in all the stress result figures and could be misleading as the elongation value could be shifted around by using different initial specimen lengths. Reporting values for elongation past the point of necking is also not technically accurate; nevertheless it is informative to directly compare the results.

True fracture strain values used in the model fitting algorithm are calculated as follows. A set of photographs (Fig. 14.18) was taken of the neck area of an Alloy 1 specimen from all four sides with a stereo-microscope to accurately measure the total average reduction in area.

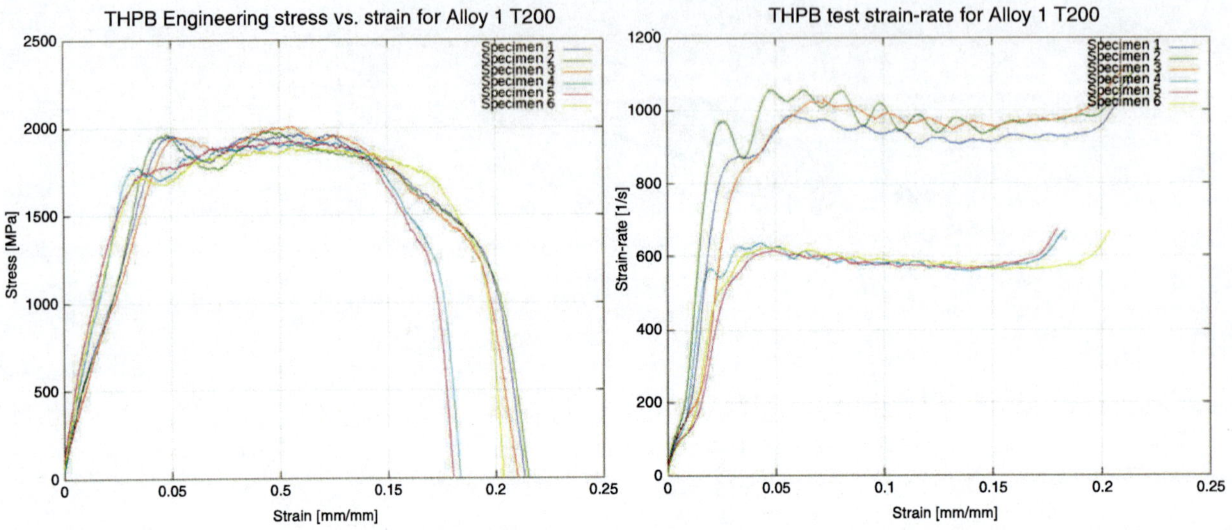


**Fig. 14.15** THPB stress and strain rate results for Alloy 1T200

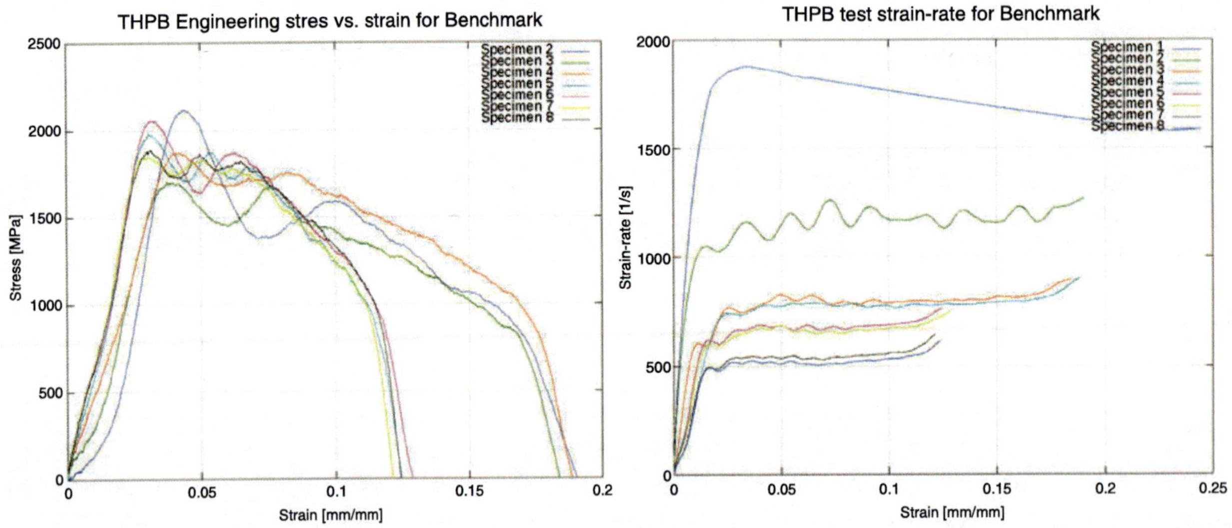


**Fig. 14.16** THPB stress and strain rate results for the benchmark alloy

$$A_{av} = \left(\frac{1}{4}(w_{n1} + w_{n2} + w_{n3} + w_{n4})\right)^2 = \left(\frac{1}{4}(1.151 + 1.162 + 1.176 + 1.151)\right)^2 = 1.346mm^2$$

Therefore the true fracture strain is,

$$\varepsilon_f = \ln\left(\frac{A_0}{A_{av}}\right) = \ln\left(\frac{1.5^2}{1.3456}\right) = 51.4\%$$

In addition the identification of the necking point or the tensile instability point should be performed by fitting the data to a plasticity model. Nevertheless the initiation of localized deformation seems to be delayed in the experimental Alloy 1 and Alloy 1 T200 and as seen in Fig. 14.17, the curves for Alloy 1 are above Alloy 1 T200 with the benchmark material below. The energy absorption capacity in tension, as well as an apparent delay in localized deformation of the experimental Alloy 1 are therefore higher within the strain-rate range of 500–1,200 $s^{-1}$ (Fig. 14.18).

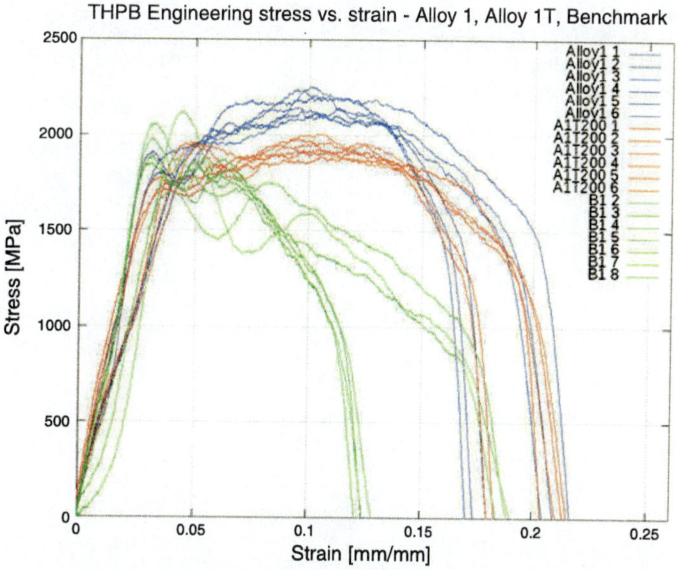


**Fig. 14.17** THPB stress results summary of Alloy 1, Alloy 1T200 and the benchmark alloy

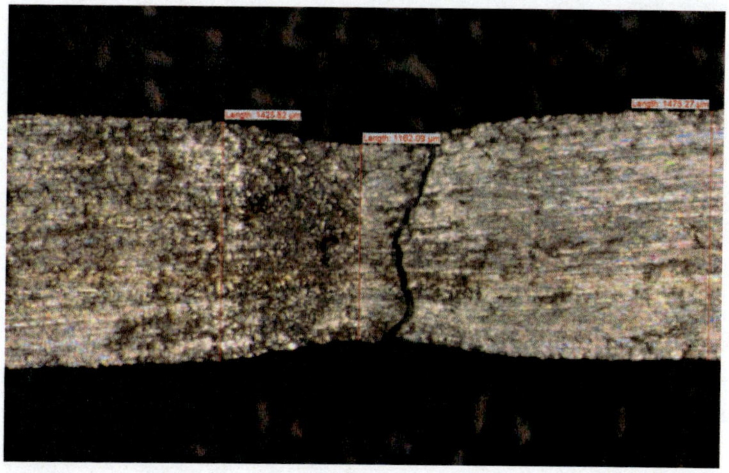

**Fig. 14.18** Microscope view of necked area for an Alloy 1 THPB specimen

**Table 14.1** Quasi-static tensile result summary from 35 tensile tests of Alloy 1

| Value | Tensile strength [MPa] | Elongation [%] |
|---|---|---|
| Minimum | 1,954.59 | 4.06 |
| Maximum | 4,216.40 | 17.77 |
| Mean | 2,104.95 | 13.58 |
| Standard deviation | 72.13 | 3.20 |

## 14.4 Standard Tensile Tests

Some of the completed tensile results are reported in this section for direct comparison of the materials.

### *14.4.1 Normal Tensile Tests*

A number of standard tensile tests were performed on Alloy 1 using EDM profiled and face ground specimens, yielding geometric accuracy of 0.010 mm on the width and 0.050 mm on the thickness. The results are summarized in Table 14.1.

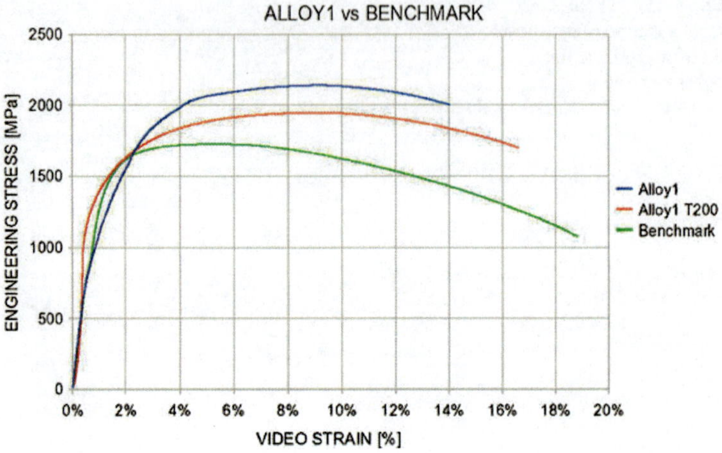


**Fig. 14.19** Quasi-static tensile test result comparison of Alloy 1, Alloy 1T200 and the benchmark alloy

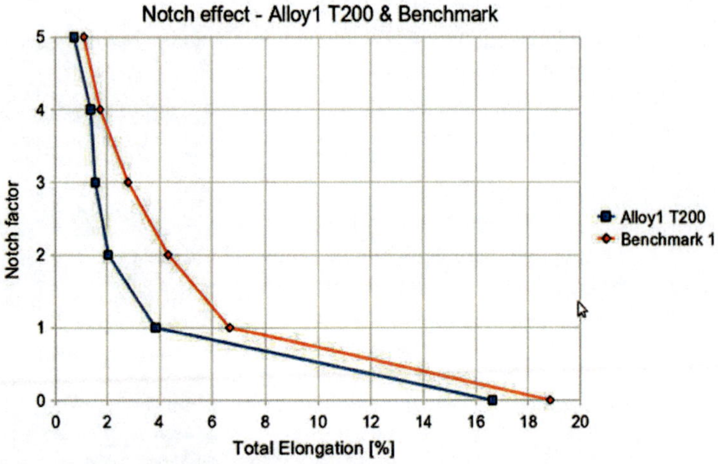


**Fig. 14.20** Comparison of the notch effect on Alloy 1T200 and the benchmark alloy

Tensile behaviour of all the materials, as tested with a video extensometer, is shown in Fig. 14.19. The energy absorption capacity of the materials is not clear, nevertheless direct comparison shows the experimental Alloy 1 to have a higher tensile strength with Alloy 1 T200 lower and the benchmark material the lowest. The delay in localized deformation for Alloy 1 is also evident.

### 14.4.2 *Notched Tensile Tests*

Some of the notched tensile tests on Alloy 1 were performed on a different machine and the direct comparison with the video extensometer is not possible at this stage. Only results for the Alloy 1 T200 and benchmark materials are therefore shown. A decrease in notch radius is used to increase the stress triaxiality and reduce the total elongation. The stress triaxiality effect on the fracture strain can be compared using the video strain in Fig. 14.20, with the notch factor simply a number assigned to the specimen as the local triaxiality value still needs to be calculated. Alloy 1 T200 failed at a lower strain for each of the notch factors and is clearly more severely affected by triaxiality than the benchmark material.

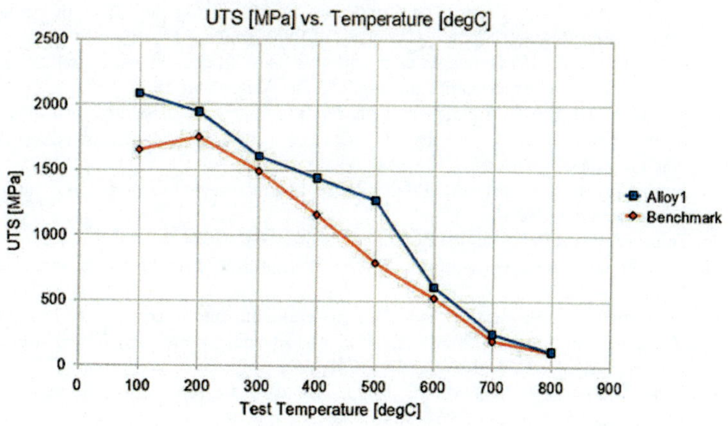


**Fig. 14.21** High temperature tensile result comparison for Alloy 1 and the benchmark alloy

### *14.4.3 High Temperature Tests*

Only Alloy 1 and the benchmark materials were tested at elevated temperatures as heating of the specimen past 200°C would negate any tempering effects on the Alloy 1 T200. The comparative results in Fig. 14.21 show a clear delay in thermal softening of Alloy 1 up to 500°C if compared to the benchmark alloy. After this point the tensile strength of the two materials converges.

## 14.5 Summary and Conclusions

Some results of an ongoing experimental, ballistic and numerical characterization of a novel new armour steel with comparison to a current commercial benchmark alloy, are presented. A new experimental Alloy 1 [20], a tempered variant Alloy 1 T200 (200°C for 1 h) and a commercial ballistic benchmark material were tested.

The ballistic test indicates that Alloy 1 requires a minimum thickness of 7.5 mm for STANAG Level 1 [6], while both the tempered variant and the benchmark seem to require a 9 mm thickness. SHPB tests show an apparent higher energy absorption capacity of Alloy 1 at strain-rates of 500–1,000 $s^{-1}$. Standard tensile tests reveal the higher tensile strength of Alloy 1. Alloy 1 T200 is more severely affected by triaxiality than the benchmark material, as shown in the notched tensile tests (results from Alloy 1 could not be compared at this time). High temperature tensile tests show a clear delay in thermal softening of Alloy 1 up to 500°C if compared to the benchmark alloy.

**Acknowledgments** The authors thank Dr. R. J. Mostert and Dr. P.H.G. Pistorius for all the informative discussions as well as T. Cloete of the Blast Impact Survivability Research Unit at the University of Cape Town for the help with specimen design and SHPB tests. The very kind help of Mr. J. Calitz of Eskom Research and Innovation Centre in the tensile tests was exceedingly valuable. Partial funding by the NRF through the Technology and Human Resource for Industry Program is appreciated. Partial funding by the Damascus Armour Development Consortium through the Innovation Fund (currently the Technology Innovation Agency) (Project no. T70044) is also appreciated. Finally, permission to publish from the University of Pretoria and the Damascus Armour Development Consortium is acknowledged.

## References

1. Images from Flash video. www.kurzzeit.com
2. Bester JN (2010) Private communication with Dr. R. J. Mostert, Dr. P. G. H. Pistorius. Level 2 STANAG test
3. Backman ME, Goldsmith W (1978) The mechanics of penetration of projectiles into targets. Int J Engng Sci 16:1–99
4. Grady DE (1988) The spall strength of condensed matter. J Mech Phys Solids 36(3):353–384
5. Chevrier P, Klepaczko JR (1999) Spall fracture: mechanical and microstructural aspects. Eng Fracture Mech 63:273–294
6. Protection Levels for Occupants of Logistic and Light Armoured Vehicles. STANAG 4569. 2004
7. Borvik T, Hopperstad OS, Berstad T, Langseth M (2001) A computational model of viscoplasticity and ductile damage for impact and penetration. Eur J Mech A/Solids 20:685–712
8. Johnson GR (2011) Numerical algorithms and material models for high-velocity impact computations. Int J Imact Eng 38:456–472
9. Bao Y, Wierzbicki T (2004) A comparative study on various ductile crack formation criteria. J Eng Mat Tech 126(3):314–324

10. Bowden AS (2009) Experimental and numerical study on the effect of strain rate to ductile damage. M.Sc. thesis, Blast Impact and Survivability Research Center, University of Cape Town
11. Roters F et al (2010) Overview of constitutive laws, kinematics, homogenization and multiscale methods in crystal plasticity finite-element modeling: theory, experiments and applications. Acta Mater 58:1152–1211
12. McDowell DL (2010) A perspective on trends in multiscale plasticity. Int J Plas 26:1280–1309
13. Klein PA et al (2001) Physics-based modeling of brittle fracture: cohesive formulations and the application of meshfree methods. Theor Appl Fracture Mech 37:99–166
14. Borvik T, Dey S, Clausen AH (2009) Perforation resistance of five different high-strength steel plates subject to small-arms projectiles. Int J Imact Eng 36:948–964
15. Zukas JA et al (1982) Impact dynamics. ISBN 0-471-08677-0
16. Corbeitt GC, Reid SR, Johnson W (1996) Impact loading of plates and shells by free-flying projectiles: a review. Int J Imact Eng 18(2):141–230
17. Goldsmith W (1999) Review: non-ideal projectile impact on targets. Int J Imact Eng 22:95–395
18. Maweja K, Stumpf WE (2008) The design of advanced performance high strength low-alloy martensitic armour steels. Part 1: mechanical property considerations. Mat Sci Eng A 485:140–153
19. Irdissi H, Renard K, Schryvers D, Jacques PJ (2010) On the relationship between the twin internal structure and the work-hardening rate of TWIP steels. Scr Mater 63:961–964
20. High Ballistic Strength Martensitic Armour Steel Alloy. PCT/IB2010/052675
21. Kolsky H (1953) Stress waves in solids. Oxford University Press
22. Gray GT III (1999) Classic split-Hopkinson pressure bar technoque. ASM Int 8:1–36

# Chapter 15
# Inertia Effects in High-Rate Compression Experiments of Soft Materials

**Xu Nie, Brett Sanborn, Tusit Weerasooriya, and Weinong Chen**

**Abstract** Inertia effects in Kolsky compression bar experiments have long been studied ever since the establishment of this high-rate characterization technique. As the materials under investigation become softer, new uncertainties arise in relation to the inertia stress induced by acceleration of specimen radial deformation rate. When the soft specimen is compressed at constant axial strain rate, the radial deformation rate gradually ramps up as the volume of the material needs to be conserved. This acceleration introduces an extra radial pressure term which superimposes on the intrinsic material response. In this study, we presented some recent experimental evidences on how this pressure affected the radial deformation behavior of annular soft specimens under dynamic compression at large strains. By incorporating high-speed photography to a modified Kolsky compression bar setup, the radial expansion of several soft rubbers was investigated. Inhomogeneous deformation was identified as the inner boundaries of several annular specimens exhibited counter-intuitive motions.

**Keywords** Kolsky bar • Soft materials • Inertia effects

## 15.1 Introduction

Dynamic compressive responses of soft materials have received continuous attention in the past decade [1]. Up to date, the Kolsky compression bar technique serves as the most popular high-rate characterization method for the soft materials except for a very few extensions to dynamic lap shear [2, 3]. During the entire compressive deformation process, in most of the cases the disk specimen is accelerated from rest to a certain high strain rate, and then maintains at this constant rate until the specimen is unloaded. The initial acceleration of axial strain rate inevitably leads to the acceleration in radial deformation. A radial inertia induced pressure is therefore created and this pressure term is believed to superimpose on the intrinsic high-rate material behavior [4]. In this case the stress history signal recorded by the strain gages on the transmission bar is composed of the radial pressure; and the material constitutive response under this inhomogeneous pressure. The early solution to alleviate the interference from this undesired pressure is to use the annular specimen geometry [4]. By creating a stress-free inner surface, the inertia stress brought out by the initial axial strain rate acceleration has been successfully reduced. However, as the axial strain continues to accumulate, the radial inertia again dominates over the intrinsic material response even though the axial strain rate may still remain constant. The purpose of this study was to investigate the possible inertia effects of soft materials at high rates and large deformations. Annular specimens made from several rubber materials were subjected to uniaxial compression at high strain rate. High-speed photography was used to observe the radial deformation of these specimens.

X. Nie (✉) • W. Chen
AAE&MSE schools, Purdue University, 701 W. Stadium Ave, West Lafayette, IN 47907–2045, USA
e-mail: xnie@purdue.edu

B. Sanborn • T. Weerasooriya
U.S. Army Research Laboratory, 701 W. Stadium Ave, West Lafayette, IN 47907–2045, USA

V. Chalivendra et al. (eds.), *Dynamic Behavior of Materials, Volume 1: Proceedings of the 2012 Annual Conference on Experimental and Applied Mechanics*, Conference Proceedings of the Society for Experimental Mechanics Series, DOI 10.1007/978-1-4614-4238-7_15, © The Society for Experimental Mechanics, Inc. 2013

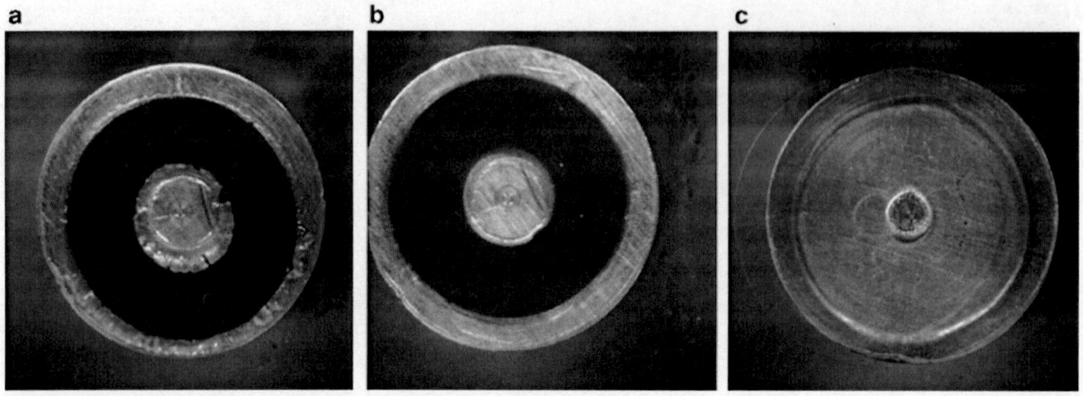


**Fig. 15.1** High-speed camera images revealing the radial deformation of annular specimens made from several rubber materials. (**a**) EPDM rubber, (**b**) Polyurethane rubber, and (**c**) PDMS rubber (Note the transition of the inner boundary motion)

## 15.2 Experiments and Results

The three materials being studied are EPDM rubber, polyurethane rubber, and PDMS rubber. For the purpose of comparison, in all the experiments a 10 mm O.D. and 5 mm I.D. specimen geometry was adopted, and the axial compression strain rate was maintained at 2,000/s. To facilitate high-speed imaging of the specimen radial deformation, the transmission bar in a conventional Kolsky compression bar setup was replaced by a transparent polycarbonate plate which was fixed to a rigid mount. A Cordin 550 high-speed camera was positioned behind the fixed plate to record the radial deformation of the specimen while it was axially compressed by the incident bar at high strain rate.

Figure 15.1 shows the typical radial deformation modes of different rubber materials. These images represent the cross-section geometry of the annular specimens at approximately 50% of axial compressive strain. While compared with the undeformed specimens, both the inner and outer boundaries of the deformed EPDM rubber specimen moved away from the center at roughly the same proportion. This behavior agrees with the solid mechanics theory based on the Poisson's ratio effect. For the relatively softer polyurethane rubber, it was found that the inner boundary of the annular specimen retained almost stationary during the compressive loading. PDMS rubber is the softest among the three rubbers and Fig. 15.1c shows that its inner boundary unimaginably collapsed towards the center. The transition of the inner boundary motion of the three rubber materials could not be solely explained by friction at the specimen-bar interface, because friction would only restrict the radial motion instead of actively pushing the inner boundary to collapse. If the annular specimens are subjected to uniaxial stress loading, it is expected that both boundaries should move outward at the same proportion. Evidences presented in Fig. 15.1 suggested the existence of internal pressure in the solid rubber materials deformed at high strain rate. The effect of this pressure on the specimen radial deformation becomes more explicit while the stiffness of the material decreases. Once the inertia induced pressure is significant enough to compromise the assumption of homogeneous radial deformation, the Kolsky compression bar technique may need to be further modified for high-rate soft material characterization.

## References

1. Song B, Chen W (2005) Split Hopkinson pressure bar techniques for characterizing soft materials. Lat Am J Solids Struct 2(2):113–152
2. Saraf H, Ramesh KT, Lennon AM, Merkle AC, Roberts JC (2007) Measurement of the dynamic bulk and shear response of soft human tissues. Exp Mech 47:439–449
3. Subhash G, Kwon J (2010) High rate shear response of gelatin. In: SEM Annual Conference & Exposition on Experimental & Applied Mechanics, IN, 7–10 June 2010
4. Song B, Chen W, Ge Y, Weerasooriya T (2007) Radial inertia effects in Kolsky bar testing of extra-soft materials. Exp Mech 47(5):659–670

# Chapter 16
# Challenges for Implementing Polymer Gels in Defense Applications

**Joseph L. Lenhart, Randy A. Mrozek, Kenneth R. Shull, and Kathryn J. Otim**

**Abstract** Polymer gels are soft, lightly crosslinked polymers that are highly swollen with solvent. The gel properties can be tuned by manipulating the polymer and solvent chemistry, solvent loading, polymer and solvent chain architecture, and the incorporation of various fillers and additives. This tunability provides broad utility in military applications including electronic devices, sensors, robotics, multi-functional textiles, responsive coatings, combat medical care, and tissue surrogates for ballistic testing. While potentially useful, a number of challenges can hinder gel utility for the Army. This paper describes recent efforts that offer promise to overcome these obstacles, including improving operational temperature performance and gel toughness.

**Keywords** Polymer gel • Gelatin • Toughness

## 16.1 Introduction

A polymer gel is a physically or chemically cross-linked polymer that is highly swollen by solvent. Mechanically, the solvent creates a "soft solid" that is easily deformable, yet still recovers from applied strain due to the elastic nature of the cross-links in the polymer [1]. The gel properties can be tuned by varying the polymer chemistry, solvent type, polymer-solvent architecture and molecular weight, and solvent loading. In addition, small molecule additives and fillers can be incorporated into the gel formulation to enhance the properties further. This tunability offers the potential to implement polymer gels in an array of Army related technologies ranging from combat casualty care and tissue surrogates, to robotics and electronics devices. A significant amount of work has been focused on water-based hydrogels for applications including biomedical technology [2–5], optical materials [6], and sensors [7–9]. A classic example is a hydrogel based on poly-(N-isopropylacrylamide) (PNIPAM), which exhibits a volume shrinkage transition near 33°C [29]. Because of the near room temperature responsiveness of PNIPAM hydrogels, they have been exploited in the biological community for many applications including controlled release [10], cellular adhesion [11], and nucleic acid purification [12]. While potentially versatile materials, several obstacles hinder widespread deployment of gel-based technologies including: (1) limited operational temperature windows and material lifetimes; (2) poor toughness and durability; (3) unstable performance in other harsh environmental conditions (i.e. extreme mechanical, electrical, chemical, and radiation environments); and (4) limited multi-functional capability. This proceedings paper will summarize some of our recent research focused on overcoming these challenges and enabling gels to be exploited for Army applications.

Our research program is focused on several areas including (a) improving operational temperature windows, (b) increasing gel toughness, (c) developing computational tools and measurement approaches to facilitate materials design, and (d) incorporating fillers/additives into the gel formulation to impart multi-functional properties. This paper will only discuss the first two areas (improving operational temperatures and increasing toughness). However, it is important to justify the need for multi-functional soft materials. As the name implies, multi-functional materials and devices must perform more

J.L. Lenhart (✉) • R.A. Mrozek
US Army Research Laboratory, Aberdeen Proving Ground, Aberdeen, MD 21005, USA
e-mail: joseph.l.lenhart.civ@mail.mil

K.R. Shull • K.J. Otim
Northwestern University, Evanston, IL 60208, USA

V. Chalivendra et al. (eds.), *Dynamic Behavior of Materials, Volume 1: Proceedings of the 2012 Annual Conference on Experimental and Applied Mechanics*, Conference Proceedings of the Society for Experimental Mechanics Series, DOI 10.1007/978-1-4614-4238-7_16, 

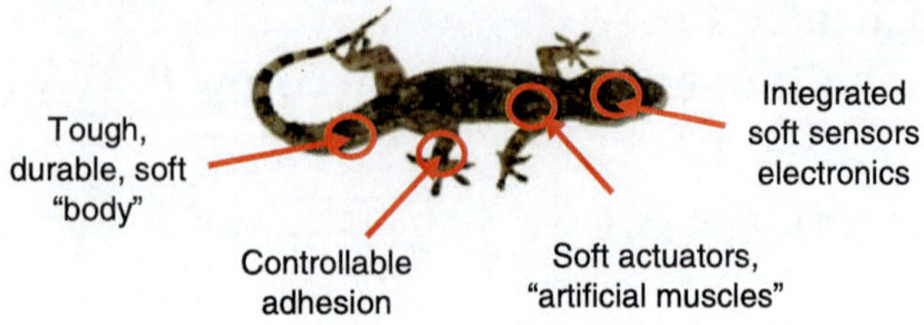


**Fig. 16.1** An example of how multi-functional soft polymers and gels would enable future robotics

than one function. For example, composites composed of continuous fiber reinforcements, to meet structural requirements, and impregnated with nano-particulate reinforced resins to control electrical or thermal transport properties represent an example of a multi-functional material. The need for multi-functional materials and devices is driven by increasingly stringent military requirements to reduce size, weight, power consumption, and cost of systems, while still providing improved capabilities. A specific example, illustrating the benefits of multi-functional soft polymers and gels is in the area of robotics. Figure 16.1 presents a schematic of a small, biomimetic robot, composed predominantly of flexible materials. The soft robots could obtain covert access to spaces, through confined openings and over difficult terrain. Soft materials advances are critical for enabling new robotics including (1) the development of a flexible and elastic body that is tough and durable, (2) the incorporation of controlled adhesion mechanisms for traversing difficult terrain, (3) the design of soft actuators or artificial muscles that can move the robot reliably with low power input, and (4) the integration of flexible sensors and electronics to provide information gathering capability, environmental recognition, and stimuli-responsiveness. The complex, multi-scaled structure of the robotic soft materials would enable a broad range of functionality including durable body; controlled transport; chemical sensing; integrated communications, sensors, and power technology; etc. While the schematic in Fig. 16.1 is conceptual and has not been fully realized to date, it provides a useful framework to guide the development of multi-functional soft polymers and gels, which can be realized through the judicious selection of materials and processing approaches.

## 16.2 Operational Temperature Window

While hydrogels are pursued extensively in the biomedical community, their use is limited in defense related applications due to the limited temperature range and high volatility of the aqueous solvent. In principle, this limitation can be circumvented by utilizing non-aqueous polymer gel chemistries.

Past research has focused on expanding operational temperature windows and materials lifetimes by designing non-aqueous gel materials from commercial precursors, and investigating processing effects [13, 14, 30]. In general the criteria for gel performance over a broad range of temperatures includes: a loosely crosslinked elastomer with a flexible polymer backbone; a low volatility polymer plasticizer as a solvent; an absence of thermal transitions (such as crystallinity or glass transition) over the operational temperature range, except in the case where responsive materials are desired; and polymer-solvent miscibility over the desired operational temperature range. Figure 16.2 shows an example of a polybutadiene (PBD) based gel that meets these criteria.

Figure 16.3 plots the shear storage modulus, G′, as a function of temperature for a cross-linked polybutadiene gel with various mass loadings of BEHS. In these gels, the maleic anhydride functionalized PBD, MA5, was used which has an average functionality of 2.5 groups per chain. At low temperatures the gel is glassy with a modulus approaching $10^9$ Pa. As the temperature increases the gel goes through a glass transition temperature (Tg), where the modulus drops several orders of magnitude. At higher temperatures the modulus reaches a rubbery plateau region and changes minimally with temperature. Both the Tg and rubbery modulus for the crosslinked polybutadiene (without solvent) are low due to the small functionality of the maleic anhydride and hydroxyl groups, which are about 2.5 functional groups per monomer for the MA5 and R45, respectively. As the BEHS loading increases, the plateau value for the modulus in the rubbery region decreases by approximately an order of magnitude, and the gel Tg, decreases by approximately 20°C. With the higher solvent loadings of (50 and 60) %, the modulus exhibits a plateau value at temperatures above −40°C. However, the gel material is still flexible with a modulus near $10^5$ Pa even at −70°C. At elevated temperatures the low volatility of BEHS slows solvent evaporation, leading to gel materials that are stable across a typical military operational temperature window of approximately (−55 to 70) °C. Similar broad temperature performing gels have been developed with epoxy based resins that are swollen with a polymer plasticizer, as well as crosslinked silicone based resins with non-reactive silicone oil as the solvent.

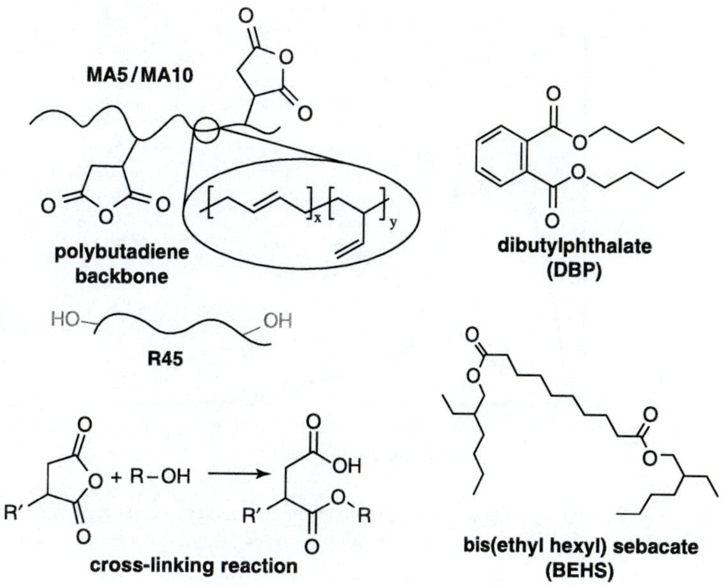


**Fig. 16.2** Schematic of a polybutadiene based gel where maleic anhydride functionalized polybutadiene (MA5/MA10) is crosslinked with hydroxyl functionalized polybutadiene (R45) in a low volatility polymer plasticizer like bisethyl-hexyl-sebacate (BEHS) or dibutylphthalate (DBP)

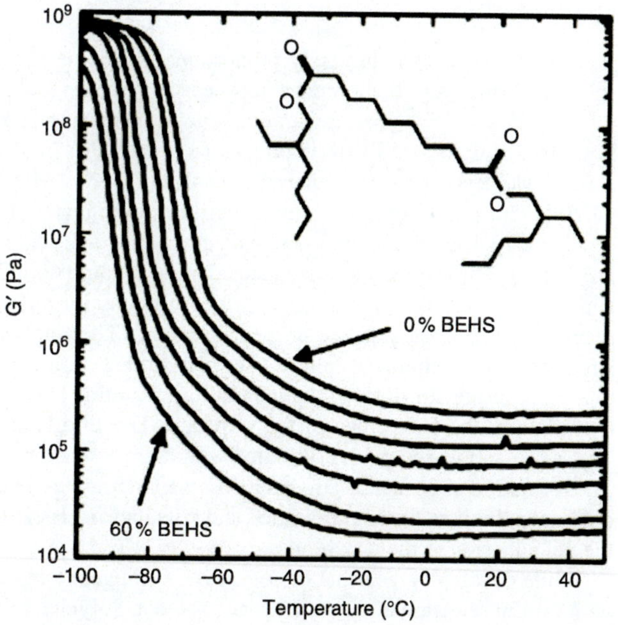


**Fig. 16.3** Shear storage modulus, G′, as a function of temperature for polybutadiene based gels with mass loadings of BEHS (inset) ranging from (0 to 60) mass%

While it is conceptually simple to design a polymer gel that will meet military temperature requirements, additional factors can complicate the materials selection. For example maintaining polymer-solvent miscibility over a broad range of temperatures with a crosslinked resin can be difficult. The thermodynamics of the solvent miscibility in a crosslinked polymer can be expressed using the Flory-Rehner equation (16.1) which describes the equilibrium swelling of a polymer network [1]. Where $V_1$ is the molar volume of the solvent, n is the concentration of elastic crosslinking chains, $\Phi_2$ is the polymer volume fraction at equilibrium, and $X$ is the polymer-solvent interaction parameter. Equilibrium swelling is a balance of three terms: (1) an enthalpic term with the parameter $X$, that is generally slightly positive, opposing miscibility; (2) entropy of mixing terms that are favorable for miscibility; and (3) a term that accounts for stretching of cross-linked chains, opposing polymer-fluid miscibility. Several factors thermodynamically drive polymer solvent phase separation in a polymer gel including: (a) increasing the number of cross-linked chains (n), (b) increasing the molecular weight of the solvent, and (c) increasing $X$ by changing the polymer-solvent interactions or decreasing temperature.

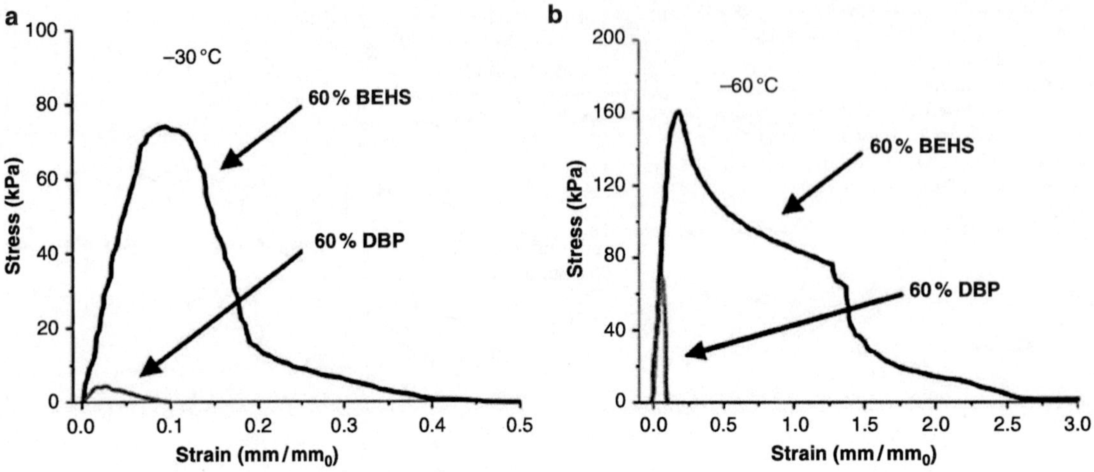


**Fig. 16.4** Typical stress-strain curves from tack adhesion measurements on an MA5/R45 stoichiometric gel formulation with either 60 mass% DBP or 60 mass% BEHS at (**a**) −30 °C and (**b**) −60 °C

$$\underbrace{X\Phi_2^{\,2}}_{1} + \underbrace{\ln(1-\Phi_2)+\Phi_2}_{2} + \underbrace{V_1 n[\Phi_2^{\,1/3} - \Phi_2/2]}_{3} = 0 \qquad (16.1)$$

The $X$ parameter is inversely proportional to temperature and can be estimated as proportional to the square of the solubility parameter, δ, difference between the polymer and solvent. At low temperatures, even small solubility parameter differences can drive phase separation, due to increasing $X$. For example, with the MA5/R45 polybutadiene based gels in Figs. 16.2 and 16.3, if BEHS is utilized (δ ~ 18 $MPa^{1/2}$), phase separation does not occur even at temperatures as low as −80°C. However, if DBP is utilized as a solvent (δ ~ 19 $MPa^{1/2}$), phase separation occurs just below room temperature [13, 14, 30]. This phase separation can be detrimental to materials performance such as tack adhesion.

Figure 16.4 shows the stress-strain diagram for a tack adhesion test on the MA5/R45 gels with 60 mass% loading of either DBP or BEHS. The tests were conducted at −30 ° and −60°C where BEHS is miscible with the crosslinked polybutadiene and DBP is not. The BEHS gels exhibit substantially higher tack adhesion (essentially the integral of the stress-strain diagram as described in the experimental section). The gels containing 60 mass% DBP exhibit substantially lower maximum stress and lower strains at failure than gels loaded with BEHS. In addition, at −60°C the BEHS-loaded gel exhibits an extension in the stress-strain diagram at larger strains [13]. The extension is the result of the formation of gel fibrils that extend from the tack probe to the substrate. The fibrillation process was visibly observable for the BEHS loaded gels at −60°C, and in general is associated with large extensions in the stress-strain diagram during tack adhesion testing [15]. The formation of fibrils has also been observed in pressure sensitive adhesives [16, 31]. In contrast, the gel loaded with DBP does not exhibit extension/fibrillation and fails like an elastic-type soft adhesive, due to the DBP phase separation degrading the gel adhesion to the tack probe.

While developing gels that perform over a stringent temperature range has challenges, it can generally be accomplished with careful selection of the polymer and solvent. Polymer-solvent miscibility is often overlooked, which can have profound consequences on performance, particularly when gel adhesion is critical.

## 16.3 Exploiting Chain Effects

Chain level effects, such as dangling chain ends and physical chain entanglements can be exploited to tune the strain rate dependent material response. To study these effects in a systematic manner, model silicone gels were investigated. Figure 16.5 shows the chemistry of the silicone systems where a vinyl-terminated polydimethyl siloxane (v-PDMS) is crosslinked with a tetra-functional silane crosslinker. The solvent in these gels is a non-reactive methyl-terminated polydimethy siloxane (m-PDMS). The entanglement molecular weight ($M_{ent}$) for PDMS is ~30 K g/mol. Both the v-PDMS and m-PDMS are available in chain sizes well above and below $M_{ent}$.

One variable that can impact the properties of gels is the size of the solvent molecules. Figure 16.6 plots the frequency dependent shear storage modulus for silicone gels as a function of the molecular weight of the m-PDMS solvent.

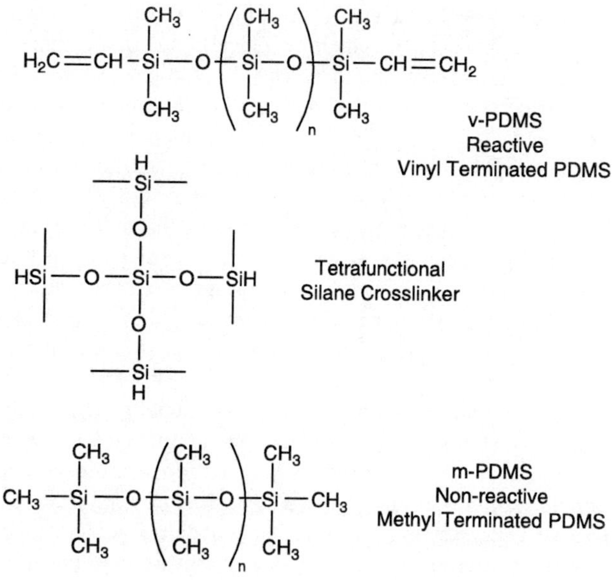


**Fig. 16.5** Schematics of the silicone based gel components

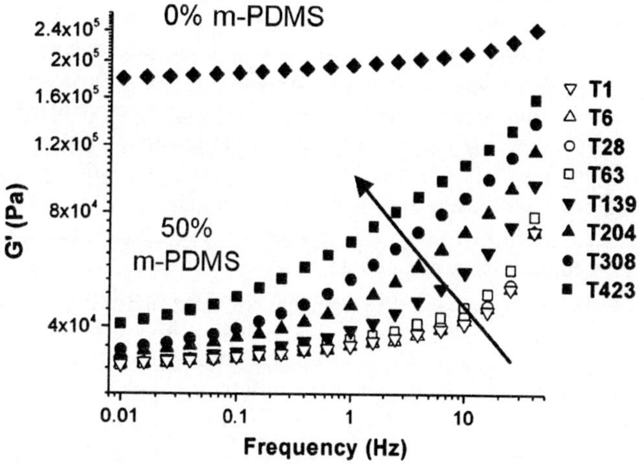


**Fig. 16.6** Shear storage modulus, G′, is plotted as a function of frequency for silicone gels with 50 vol% loading of m-PDMS solvent. The *arrow* indicates the direction of increasing m-PDMS molecular weight. The legend numbers represent the molecular weight of the m-PDMS solvent

The samples are composed of a 117 K g/mol v-PDMS crosslinked with the tetra-functional silane hardener. For each sample, the silane to vinyl functional group ratio was 4:1, which provided the optimum (stiffest) network at zero loading of the m-PDMS. The black diamonds (stiffest sample) represent the crosslinked silicone with 0 % m-PDMS solvent, where the modulus only exhibited small frequency dependence. The remaining samples contained 50 Vol% of the m-PDMS solvent, with a molecular weight varying from 1 K to 400 K g/mol. The frequency dependence increased dramatically when the m-PDMS molecular weight was greater than $M_{ent}$. The arrow shows the direction of increasing m-PDMS solvent molecular weight.

The molecular weight of the solvent in polymer gels can be exploited to manipulated the strain rate dependent mechanical response as well as the adhesive performance of the gel, which is beneficial to a host of Army relevant applications.

## 16.4 Gel Toughness

Understanding the fracture behavior of soft, compliant materials such as polymer gels is an area of growing importance for a host of Army applications. For example, the Army and its collaborators are designing bio-inspired robotics concepts [17]. These future robotics will be required to perform task similar to their biological inspiration, such as fly like an insect, jump like a bug, crawl like a lizard, or manipulate objects with the dexterity of the human hand. One challenge is that synthetic soft tissues are not as tough as their biological counterparts, therefore, these future robotics will not exhibit the same performance as biological systems. Our research has a strong focus on improving the fracture toughness of soft materials. Crack growth in these materials cannot be accurately modeled using traditional linear elastic fracture mechanics because chains are highly stretched near the crack tip. Additionally, polymer gels can undergo crack blunting, which increases fracture resistance. Gels that are highly viscoelastic in nature have additional stress dissipation mechanisms which result in a greater fracture resistance when compared to elastic materials [18].

Knowledge of the stress state and chain behavior at the crack tip is fundamental to understanding crack propagation through polymer gels. Lake and Thomas proposed that the fracture energy is determined by the length of chains spanning the fracture plane [19]. Network chains at the crack tip must be fully stretched before failure occurs. Each chain is therefore capable of storing much more energy than is required to break a single bond. Mazich et al. found that the primary mechanisms of fracture resistance in lightly cross-linked systems were chain pullout and viscous dissipation due to chain deformation [20].

Soft materials typically undergo crack blunting. The blunting process dissipates a significant amount of energy that could account for the increase in fracture energy that typically exceeds the value predicted by Lake and Thomas [21]. Because material near the crack tip does not conform with linear elastic fracture mechanics it is difficult to accurately predict the sample response in that region. Krishnan et al. used finite element analysis to model crack blunting in a cohesive zone in front of the crack tip [22]. Non-linear models of the crack tip are being developed based on experimental data.

Baumberger et al. compared the energy release rate (Gc) as a function of velocity for gelatin samples of different concentrations and viscosities (Baumberger et al. 2006; [25, 26]). The resulting Gc values scaled quasi-linearly with the gelatin viscosity and crack velocity. The strong viscosity dependence highlights its importance in the fracture process. Recently, double network hydrogels, in which a highly cross-linked network is interpenetrated by a more loosely cross-linked one, have gained attention for their high fracture energy and mechanical toughness despite their high water content. Gong et al. synthesized double network gels with fracture energies ranging from 30–700 J/m$^2$ (Tanaka et al. 2009; [24]). The group is still investigating the nature of the high fracture resistance in their gels. While double network gels are promising, multi-step processing methodology hinders practical utility.

Our research has focused on exploiting dangling chain ends [13] and solvent entanglement to improve gel toughness with single step processing approaches. Silicone gels represent a model system wher entanglements can be varied in a systematic way via the m-PDMS molecular weight. These model systems are needed to help increase understanding of the relationship between the energy release rate, crack blunting, and the viscoelastic behavior of polymer gels. Next, we discuss the energy release rates for these model silicone networks as a function of the solvent fraction and molecular weight.

Fracture experiments were conducted on Instron machines using a method similar to those described previously [25, 26]. Rectangular strips with lengths ranging from 40 to 70 mm were cut from the silicone samples. The average thickness was 5 mm. Samples were aligned in the Instron grips so that a 5 mm pre-crack, made with a razor blade, was in the center of one side and the sample height between the grips was 20 mm. The samples were stretched to a fixed displacement and held in place while a crack propagated through the mid-line of the sample. The measured load and displacement data were converted to nominal stress, elongation, and extension ratios, while the crack velocity was calculated from images taken during the experiment. The energy release rate (Gc) was calculated from the integral of the stress-elongation curve according to Eq. 16.2:

$$G_c = 2h\int P d\lambda = 2h_o\int \sigma d\lambda \tag{16.2}$$

where, h is the distance between the grips, $h_o$ is the half height of the sample, P is the measured load, $\sigma$ is the stress, and $\lambda$ the extension ratio.

The energy release rates for the silicone gels are quite different from values obtained from physical gels, such as triblock copolymers, which show a constant dependence of Gc on crack velocity [27]. The velocity dependence of the silicones changes depending on composition, where gels with higher molecular weight solvent exhibit fracture energies much higher than those typically observed in physically cross-linked gels. These fracture energies are comparable to those exhibited by double network hydrogels [23].

The ratio Gc/E defines a length scale that is useful for determining the large strain response of an elastic material (Baumberger et al. 2006; [28]). Dividing this value by the initial height of the sample provides a dimensionless parameter, Gc/E$h_o$, which can be used to compare the behavior of our samples with different molecular weight and loading of m-PDMS. A comparison of the velocity dependence of the samples that have been normalized in this manner reveals that the neat gels

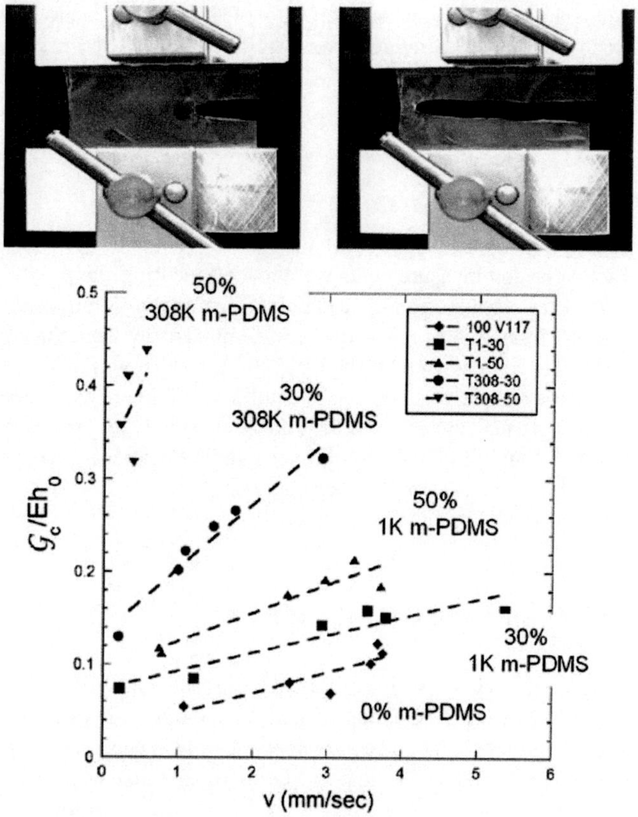


**Fig. 16.7** Normalized energy release rate versus crack velocity. *Dashed lines* are drawn as guides for the eye

have the lowest $Gc/Eh_o$ values (Fig. 16.7). Since all of the rates have been normalized by their respective elastic moduli, the differences in behavior highlight the viscoelastic contributions of the solvent. Increasing the solvent fraction raises the normalized values. Addition of the short T1 (1 K g/mol m-PDMS) solvent results in a moderate increase in $Gc/Eh_o$, while addition of the longer T308 (308 K g/mol m-PDMS) greatly increases $Gc/Eh_o$. Thus, the presence of the long sol chains enhances the fracture toughness of the gels by providing additional mechanisms for energy dissipation through chain disentanglement and pullout.

## 16.5 Experimental

### *16.5.1 Polybutadiene Gel Preparation*

Stoichiometric amounts of maleic anhydride and hydroxyl functionalized polybutadienes were mixed along with 0–60 wt% solvent and 0.5 wt% didecylmethylamine catalyst for 10 min. The mixture was then cured for 6 days at 75°C. The extent of cure was monitored by rheological measurements that determined after 6 days at 75°C the value of the modulus in the plateau region remained unchanged indicating complete cure. The cured gels had an extractable content equal to the solvent loadings along with an additional 2–3 wt% of unreacted polymer precursors.

### *16.5.2 Rheological Characterization*

Rheology measurements were made with rectangular gel samples in a torsion geometry. The gel samples had dimensions of approximately 12 × 4.5 × 28 mm. The measurements were made on a Rheometric Scientific ARES instrument at a frequency of 1 Hz and a scan rate of 2°C/min. An environmentally-controlled oven permitted determination of the modulus over temperatures from (−100 to 70)°C. Strain sweeps were conducted at various temperatures to ensure that the modulus was independent of strain at a frequency of 1 Hz.

### *16.5.3 Tack Adhesion*

Tack adhesion measurements were made by curing the polybutadiene gel in a thin film (~0.5 mm thick) on an aluminum disc. After plasma cleaning, a stainless steel probe (8 mm diameter) was brought into contact with the gel film and held for 60 s at a force of 500 g. The probe was pulled away from the gel film at 0.02 mm/s, while measuring the force-displacement curve. The temperature for the tack measurements was held constant utilizing an environmental chamber. Unless otherwise noted, the sample was allowed to equilibrate at the measurement temperature for 10 min prior to each measurement. To get an effective tack adhesion energy, the stress (force/gel cross-sectional area) – strain (displacement/film thickness) curve was integrated and multiplied by the gel sample thickness. The gel cross-sectional area was calculated, assuming a Poisson's ratio of 0.5, typical for a rubbery polymer.

### *16.5.4 Silicone Gel Preparation*

Trimethylsiloxy – terminated polydimethylsiloxane (PDMS) sol fractions were added to a vinyl-terminated PDMS base network in weight fractions of 0.30 and 0.50. Two different molecular weight sols were used: 1,250 g/mol (T1) and 308,000 g/mol (T308). The base network (V117) had a molecular weight of 117,000 g/mol. The vinyl-terminated PDMS (v-PDMS) and methyl-terminated PDMS (m-PDMS) were mixed with 800 ppm of a platinum-cyclovinylmethylsiloxane complex by mechanical mixing for 15 min. A tetrafunctional silane cross-linker, tetrakis(dimethylsiloxy)silane, was added at 4.0 molar equivalents of hydride to vinyl groups and stirred for an additional 10 min. The samples were degassed by vacuum and cured for 72 h at 82°C. In most samples the modulus did not appreciably change after 24 h. Eight different m-PDMS molecular weights were used: 1,250 g/mol (T1); 5,970 g/mol (T6); 28,000 g/mol (T28); 62,700 g/mol (T63); 139,000 g/mol (T139); 204,000 g/mol (T204); 308,000 g/mol (T308); and 423,000 g/mol (T423).

### *16.5.5 Silicone Tensile Tear Tests*

Fracture experiments were conducted on Instron machines using a method similar to those described previously [17, 25, 26, 27]. Rectangular strips with lengths ranging from 40 to 70 mm were cut from the silicone samples described above. The average thickness was 5 mm. Samples were aligned in the Instron grips so that a 5 mm pre-crack, made with a razor blade, was in the center of one side and the sample height between the grips was 20 mm. The samples were stretched to a fixed displacement and held in place while a crack propagated through the mid-line of the sample. The measured load and displacement data were converted to nominal stress, elongation, and extension ratios, while the crack velocity was calculated from images taken during the experiment. The energy release rate was calculated from the integral of the stress-elongation curve.

## 16.6 Conclusions

Polymer gels are chemically or physically crosslinked polymers that are highly swollen with solvent. The properties of gels can be tuned by varying polymer and solvent chemistry, solvent loading, chain architecture for the polymer and solvent, and the incorporation of fillers or additives. This tunability offers potential for gel implementation in a variety of defense related applications. Implementation requires the advancement of gel materials including operational temperature windows and toughness. Operational temperature windows can be expanded by utilizing non-aqueous based gels as replacements for hydrogel materials. Incorporating entangled solvent can be exploited to improve gel toughness.

**Acknowledgements** K. Otim was funded by the U.S. Army Research Laboratory (ARL) and the National Physical Sciences Consortium Fellowship program. R. A. Mrozek was funded at ARL through a contract with the Oak Ridge Institute of Science and Engineering (ORISE). Certain commercial equipment and materials are identified in this paper in order to specify adequately the experimental procedure. In no case does such identification imply recommendations by the Army Research Laboratory nor does it imply that the material or equipment identified is necessarily the best available for this purpose.

## References

1. Flory PJ (1953) Principles of polymer chemistry, vol 1. Cornell University, Ithaca, p 576
2. Kwon IC, Bae YH, Kim SW (1991) Electrically erodible polymer gel for controlled release of drugs. Nature 354:291–293
3. Miyata T, Asami N, Uragami T (1999) A reversibly antigen-responsive hydrogel. Nature 399:766–769
4. Murdan S (2003) Electro-responsive drug delivery from hydrogels. J Control Release 92:1–17
5. Peppas NA, Huang Y, Torres-Lugo M, Ward JH, Zhang J (2000) Physicochemical, foundations and structural design of hydrogels in medicine and biology. Annu Rev Biomed Eng 2:9–29
6. Jiang H, Su W, Mather PT, Bunning TJ (1999) Rheology of highly swollen chitosan/polyacrylate hydrogels. Polymer 40:4593–4602
7. Hu Z, Chen Y, Wang C, Zheng Y, Li Y (1998) Polymer gels with engineered environmentally responsive surface patterns. Nature 393:149–152
8. Holtz JH, Asher SA (1997) Polymerized colloidal crystal hydrogel films as intelligent chemical sensing materials. Nature 389:829–832
9. Li J, Hong X, Liu Y, Li D, Wang Y-W, Li J-H, Bai Y-B, Li T-J (2005) Highly photoluminescent CdTe/Poly(N-isopropylacrylamide) temperature-sensitive gels. Adv Mater 17:163–166
10. Park TG (1999) Temperature modulated protein release from pH/temperature-sensitive hydrogels. Biomaterials 20:517–521
11. Schmaljohann D, Oswald J, Jorgensen B, Nitschke M, Beyerlein D, Werner C (2003) Thermo-responsive PNiAAm-g-PEG films for controlled cell detachment. Biomacromolecules 4:1733–1739
12. Elarssari A, Rodrigue M, Meunier F, Herve C (2001) Hydrophilic magnetic latex for nucleic acid extraction, purification and concentration. J Magn Magnetic Mater 225:127–133
13. Lenhart JL, Cole PJ (2006) Adhesion properties of lightly crosslinked solvent-swollen polymer gels. J Adhes 82:945–971
14. Lenhart JL, Cole PJ, Unal B, Hedden R (2007) Development of nonaqueous polymer gels that exhibit broad temperature performance. Appl Phys Lett 91:061929
15. Zosel A (1998) The effect of fibrilation on the tack of pressure sensitive adhesives. Int J Adhes Adhes 18:265–271
16. Lakrout H, Sergot P, Creton C (1999) Direct observation of cavitation and fibrillation in a probe tack experiment on model acrylic pressure-sensitive-adhesives. J Adhes 69:307–359
17. Mrozek R, Otim K, Shull K, Lenhart JL (2011) Influence of solvent size on the mechanical properties and rheology of polydimethylsiloxane-based polymeric gels. Polymer 52:3422–3430
18. Gent AN, Lai SM (1994) Interfacial bonding, energy dissipation, and adhesion. J Polym Sci B 32:1543–1555
19. Lake GJ, Thomas AG (1967) Strength of highly elastic materials. Proc R Soc London A Math Phys Sci 300:108
20. Mazich KA, Samus MA, Smith CA, Rossi G (1991) Threshold fracture of lightly crosslinked networks. Macromolecules 24:2766–2769
21. Hui CY, Jagota A, Bennison SJ, Londono JD (2003) Crack blunting and the strength of soft elastic solids. Proc R Soc Lond A Math Phys Eng Sci 459:1489–1516
22. Krishnan VR, Hui CY, Long R (2008) Finite strain crack tip fields in soft incompressible elastic solids. Langmuir 24:14245–14253
23. Tanaka Y, Kuwabara R, Na YH, Kurakawa T, Gong JP, Osada Y (2005) Determination of fracture energy of high strength double network hydrogels. J Phys Chem B 109:11559–11562
24. Gong JP, Katsuyama Y, Kurokawa T, Osada Y (2003) Double-network hydrogels with extremely high mechanical strength. Adv Mater 15:1155–1158
25. Baumberger T, Caroli C, Martina D (2006) Solvent control of crack dynamics in a reversible hydrogel. Nature Mater 5:552–555
26. Baumberger T, Caroli C, Martina D (2006) Fracture of a biopolymer gel as a viscoplastic disentanglement process. Eur Phys J E 21:81–89
27. Seitz ME, Martina D, Baumberger T, Krishnan VR, Hui CY, Shull KR (2009) Fracture and large strain behavior of self-assembled triblock copolymer gels. Soft Matter 5:447–456
28. Creton C, Hu GJ, Deplace F, Morgret L, Shull KR (2009) Large-strain mechanical behavior of model block copolymer adhesives. Macromolecules 42:7605–7615
29. Hirokawa Y, Tanaka T (1984) Volume phase transition in a nonionic gel. J Chem Phys 81(12):6379–6380
30. Mrozek RA, Cole PJ, Cole SM, Schroeder JL, Schneider DA, Hedden RC, Lenhart JL (2010) Design of nonaqueous polymer gels with broad temperature performance: impact of solvent quality and processing conditions. J Mater Res 25:1105–1117
31. O'Connor AE, Willenbacher N (2004) The effect of molecular weight and temperature on tack properties of model polyisobutylenes. Int J Adhes Adhes 24:335–346

# Chapter 17
# Mode I Fracture Response of Polymer Based Gelatins as a Function of Loading Rate

**Mark Foster, Paul Moy, C. Allan Gunnarsson, Tusit Weerasooriya, Joseph Lenhart, and Randy Mrozek**

**Extended Abstract** Traditional collagen-based ballistic gelatin has been used extensively to determine the effectiveness of bullets and firearms because of its similar density to natural tissue. In these investigations, the gel is subjected to a projectile penetration experiment, during which temporary and permanent cavities are observed in the gel medium. The tensile failure mode dominates the damage mechanisms in the formation of these cavities. However, the mechanical response of ballistic gelatin is sensitive to temperature and the gel has a limited shelf life. Synthetic polymer-based gelatins alleviate these issues and thus are good candidates for an alternative tissue simulant. The ability to closely control the synthesis of these polymer gels gives them a distinct material structure, which governs mechanical characteristics on the macro-scale (i.e. strength, compliance). This ability to tune mechanical properties deems them suitable for many other applications such as in sensors, robotics, chemical, and biomedical uses. In order to determine the tensile failure properties of these materials, a Mode I experimental method was developed. The present work builds upon prior studies to incorporate styrene-ethylene-butylene-styrene (SEBS) gelatins at low and high loading rates. Using a digital image correlation (DIC) technique to quantify the full-field surface strains around a crack tip, the critical strain required for initiation of failure and crack growth was determined as a function of loading rate. The experimental methods are presented with results from Mode I fracture experiments, including energy and strain based criteria for failure initiation and growth and their corresponding loading rate dependence.

**Keywords** Fracture • Loading rate • Gelatin • Polymer

The low rate test methodology was adapted from Mrozek et al. [1]. A wider specimen than the dogbone shaped configuration used by Moy et al. [2] provides a more consistent crack initiation by satisfying a plane-stress condition. The SEBS triblock copolymer gelatin was mixed in a polymer to solvent ratio of 1:4 and poured into 5 mm thick sheets. Specimens were sliced to a width of 70 mm and a razor blade was used to precrack the gelatin to a depth of 15 mm. The specimens were then fitted into uniquely designed flat grips for compliant materials at a gage length of 50 mm and were tested in an Instron 8871 load frame at constant extension rates of 0.127, 12.7, and 127 mm/s to full failure. Data from these experiments is included in Fig. 17.1. As expected, load increases with extension rate with the maximum load occurring when the crack begins to propagate. The strain perpendicular to the crack tip remains constant throughout propagation for the two slower rates, with a critical strain of 0.25, 0.80, and 1.0 required for initiation at extension rates of 0.127, 12.7, and 127 mm/s respectively. Video of the deforming specimen was recorded for DIC measurements and a typical screenshot is included in Fig. 17.2. The hyperelastic region around the crack tip resembles models of mode I fracture and shows a similar strain field to the hyperelastic softening material investigated by Buehler et al. [3].

High rate testing was performed using a 7075 aluminum direct tension Hopkinson bar similar to that of Nie et al. [4]. The material tested was the commercially available polymer gelatin Perma-Gel$^{TM}$. A tubular striker was fired into a flange on the incident end of the bar, and the corresponding incident stress wave was recorded with semiconductor strain gages. The low stiffness of the gelatin material requires a long rise time pulse, so annealed copper pulse shapers were used. Through pulse shaping a rise time of 390 μs was achieved. A typical strain signal from the strain gages is included in Fig. 17.3. In order to

M. Foster (✉) • P. Moy • C.A. Gunnarsson • T. Weerasooriya • J. Lenhart • R. Mrozek
US Army Research Laboratory, Weapons and Materials Research Directorate, Bldg 4600 Deer Creek Loop, Aberdeen Proving Ground, MD 21005–5069, USA
e-mail: mark.foster12.ctr@mail.mil; paul.moy.civ@mail.mil; carey.a.gunnarsson.civ@mail.mil; joseph.l.lenhart.civ@mail.mil

V. Chalivendra et al. (eds.), *Dynamic Behavior of Materials, Volume 1: Proceedings of the 2012 Annual Conference on Experimental and Applied Mechanics*, Conference Proceedings of the Society for Experimental Mechanics Series, DOI 10.1007/978-1-4614-4238-7_17, 

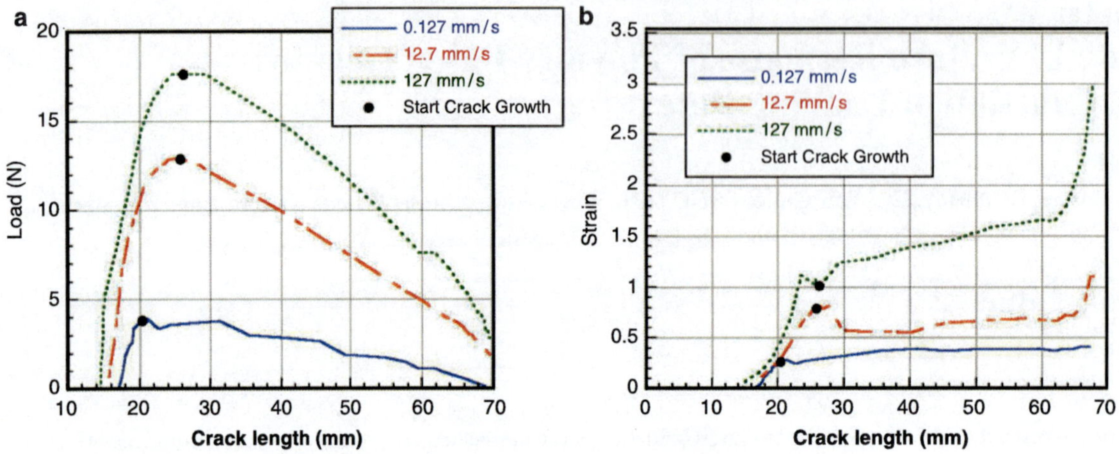


**Fig. 17.1** (**a**) Load versus crack length and (**b**) Crack tip strain versus crack length for SEBS polymer gelatin

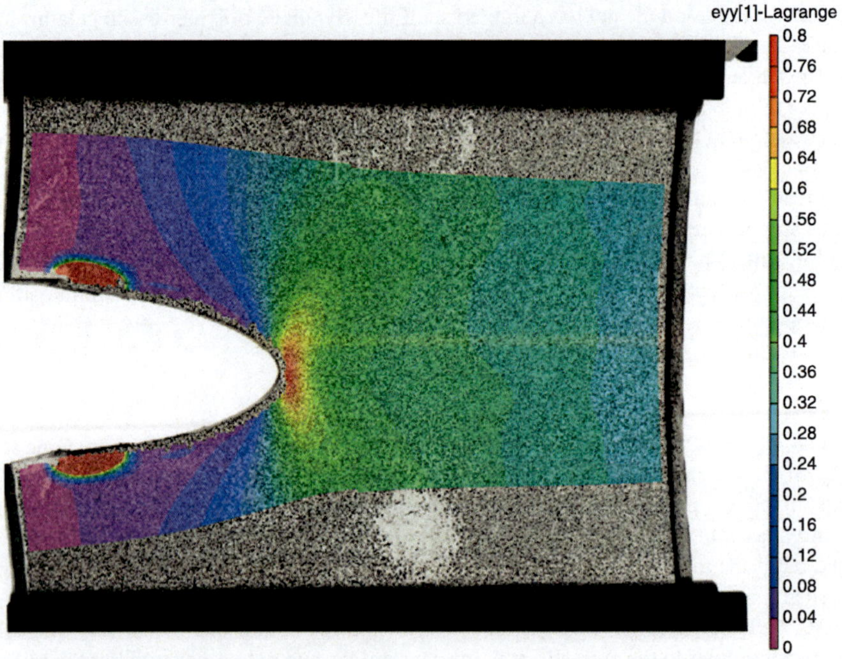


**Fig. 17.2** DIC measurement of Lagrangian strain field around crack tip

measure a small transmitted signal due to an impedance mismatch between the gelatin and aluminum, a 5 lb capacity piezoelectric load cell was rigidly mounted on the far (i.e. output) side of the specimen from the impact end. This technique was adapted from Cheng et al. [5] who developed a miniature split Hopkinson bar to test single fibers in tension at high rates. The specimen was glued to flats on the bar ends at a gage length of 1.93 mm. In order to minimize through-thickness shear, the specimen thickness was reduced to approximately 1.27 mm. Like the low rate studies, a precrack was added using a razor blade and the strain perpendicular to the crack tip was recorded throughout the experiment using high speed DIC at 125,000 frames per second. This data is displayed in Fig. 17.4. While a strain level of 0.6 was reached, the crack did not propagate in the first loading pulse. The corresponding time histories for the crack tip strain and load cell are shown in Fig. 17.5. Using the input bar end displacement data also included in Fig. 17.5, an average bar end velocity was found to be 1,953 mm/s.

A SEBS triblock polymer gelatin and Perma-Gel$^{TM}$ were tested at different rates of loading to determine surface strains surrounding the crack tip. Further work will focus on the high rate test technique and explore different gripping methods. In addition, gels of different polymer composition will be investigated.

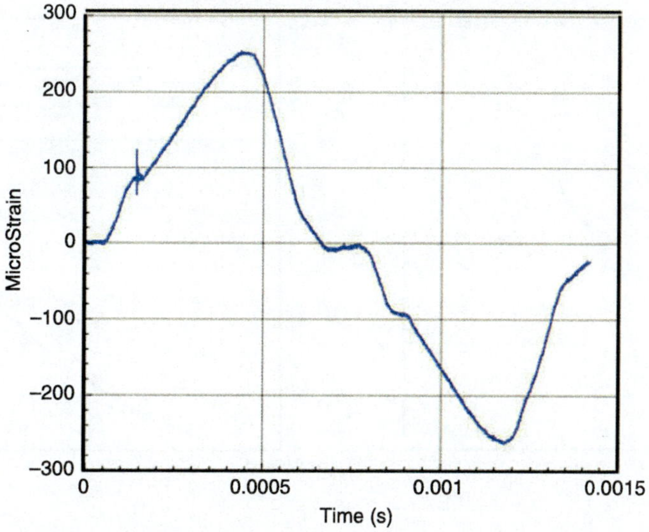


Fig. 17.3 Time history of incident and reflected pulse from semiconductor strain gage on input bar

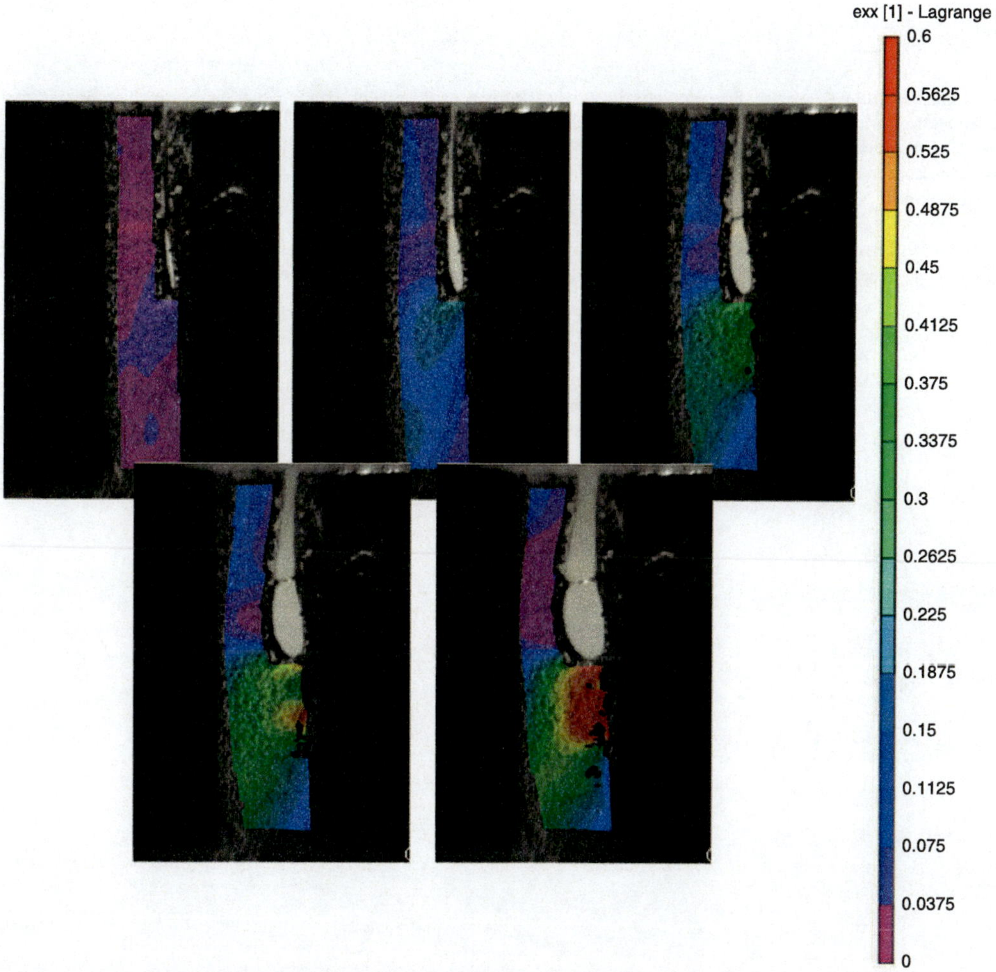


Fig. 17.4 DIC measurement of strains perpendicular to the crack tip in Hopkinson bar test

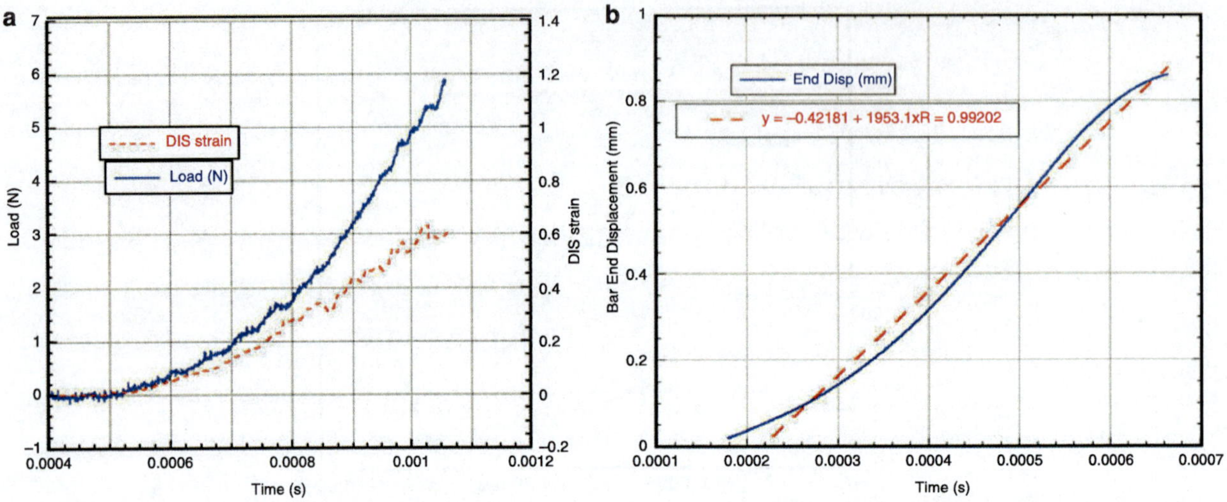


**Fig. 17.5** (**a**) Specimen load and strain (perpendicular to the crack tip) versus time and (**b**) average input bar end velocity

## References

1. Mrozek RA, Cole PJ, Otim KJ, Shull KR, Lenhart JL (2011) Influence of solvent size on the mechanical properties and rheology of polydimethylsiloxane-based polymeric gels. Polymer 52:3422–3430
2. Moy P, Foster M, Gunnarsson CA, Weerasooriya T (2010) Loading rate effect on tensile failure mode of gelatins under mode I. In: Proceedings of the SEM annual conference, Indianapolis
3. Buehler MJ, Abraham FF, Gao H (2003) Hyperelasticity governs dynamic fracture at a critical length scale. Nature 426:141–146, 13 November 2003
4. Nie X, Song B, Ge Y, Chen WW, Weerasooriya T (2009) Dynamic tensile testing of soft materials. Exp Mech 49:451–458
5. Cheng M, Chen WW, Weerasooriya T (2005) Mechanical properties of Kevlar KM2 single fiber. J Eng Mater Technol 127:197–203

# Chapter 18
# Determination of Simple Shear Material Properties of the Brain at High Strain Rates

**Kyle A. Ott, R.S. Armiger, A.C. Wickwire, A.S. Iwaskiw, and Andrew C. Merkle**

**Abstract** Split Hopkinson Pressure Bars are used to study dynamic material response in uniaxial compression, tension or torsion. Modifications of traditional test techniques have permitted the successful measurement of highly elastic soft biological tissue properties. Testing biological materials, specifically brain material, using the modified Split Hopkinson Pressure Bar method is effective in characterizing the tissue response at strain levels and strain rates that are greater than traditional tissue testing modalities. The high strain rate material properties are of particular interest for use in constitutive models that govern material response and injury predictions for computational models of the human body exposed to dynamic impact events. Since the use of viable brain material is critical, and the viability largely depends on post mortem interval, harvesting fresh, non-frozen brain tissue is critical. In this paper, we present the shear response of fresh human brain tissue under high rate loading. The investigated strain rates ranged from 25 to 248 strain/s. Specimen were prepared from various orientations and locations within the brain across two specimen. The average elastic modulus was 13.0 ± 10 kPa computed across all trials. The response was also characterized based on specimen, loading rate, and strain level to analyze contributing factors. This information is critical for targeting surrogate material properties and generating parameters for computational model constitutive equations.

**Keywords** Brain • High rate • Shear • Modulus • Material properties • Tissue characterization • Split hopkinson pressure bar • Computational modeling • Finite element

## 18.1 Introduction

Accurately characterizing the mechanical response of biological tissues is required to understand the response and injury potential that results from high-rate loading events. These types of events range from automotive crashes to ballistic impacts and blast overpressure loading. It has been shown that biological materials respond differently at these high loading rates. Therefore, accurate modeling of these events through computational or physical modeling necessitates the understanding of how the response of these materials change under these unique loading conditions. For physical surrogates, knowledge of biological material response at high loading rates allows the development and verification of more accurate simulant materials. For computational models, material response data allows for the development of constitutive equations that model the response of these biological materials. Specifically, many of the commonly utilized viscoelastic material models require an understanding of the bulk and shear material response in both the quasi-static and high strain rate loading regime.

The requirement of understanding how a particular material responds at both low and high strain rates, however, presents a unique and interesting challenge due to the non-standard nature of these materials, both on the acquisition and preparation side as the physical testing and data collection. Biological tissues exhibit a greater compliance than those materials typically tested in a Split Hopkinson Pressure Bar (SHPB) setup, such as metals and ceramics. The characteristics of these biological tissues results in long stress equilibration times during experiments involving a single loading pulse (i.e., non-oscillatory), and special efforts to adapt the test setup must be made to verify uniform deformations of the samples [1–3]. As a result of these and other experimental obstacles such as tissue acquisition and preservation prior to testing, there currently is a lack of

K.A. Ott • R.S. Armiger • A.C. Wickwire • A.S. Iwaskiw • A.C. Merkle (✉)
The Johns Hopkins University Applied Physics Laboratory, 11100 Johns Hopkins Road, Laurel, MD 20723, USA
e-mail: Andrew.Merkle@jhuapl.edu

V. Chalivendra et al. (eds.), *Dynamic Behavior of Materials, Volume 1: Proceedings of the 2012 Annual Conference on Experimental and Applied Mechanics*, Conference Proceedings of the Society for Experimental Mechanics Series, DOI 10.1007/978-1-4614-4238-7_18, 

test data at the loading rates of interest. Recently, techniques have been developed and shown to be successful in testing biological material under simple shear utilizing a modified SHPB technique [1–5]. This technique differs from traditional high strain rate material characterization, such as Ultrasound or Dynamic Mechanical Analysis (DMA), in that it is not limited to small strain levels and avoids the oscillatory motion, which may not be indicative of high-rate impact loading, while still achieving high strain rates. This modified SHPB method has been shown successful in measuring the response of soft biological simulants as well as human tissue under high loading rates.

Another challenge in testing these biological tissues, especially tissues that have been shown to rapidly degenerate post expiration, relates to the acquisition and preparation of the tissue itself. The effect of freezing or fixing tissue on the material properties of biological tissue has long been studied. These studies indicate a varying level in the change of response based on tested time post mortem [6]. In the specific case of brain tissue, it has been shown that freezing the tissue severely affects the mechanical properties [7]. This presents the need for robust acquisition and testing procedures that ensure tissue is tested prior to degradation. As a result of the inherent difficulties with these tests, both in tissue acquisition and testing, the available data that employs high strain-rate, non-oscillatory methods applied to fresh specimen is limited.

The objective of this study was to devise a procedure by which fresh, never frozen human brain tissue could be acquired with a Post Mortem Interval less than 36 h and tested over a period of 12 h utilizing a modified SHPB technique. Preliminary test results indicate varying levels of stiffness in simple shear from brain tissue at strain rates of 10–200 strain/s.

## 18.2 Material and Methods

### *18.2.1 Human Brain Tissue*

Human brain tissue was obtained from the NICHD Brain and Tissue Bank for Developmental Disorders at the University of Maryland, Baltimore MD. A novel tissue acquisition protocol was established whereby never-frozen specimen were acquired in approximately 3 cm coronal slices and placed in short term storage prior to testing. While awaiting testing, the tissue was immersed in Minimal Essential Medium (MediaTech Inc.) in order to maintain proper hydration and stored in a refrigerator at 38°F. The specimen acquired would capture large coronal sections (~3 cm thick, see Fig. 18.1), this allows for the tissue type, orientation and location to be experimentally controlled. Tissue acquisition and testing would be completed within 12 h of notification of specimen availability. Slices were taken at different portions of the brain to determine differences in response based on location, if any. The size of the slices also allowed for multiple samples to be produced for each specimen obtained, Table 18.1.

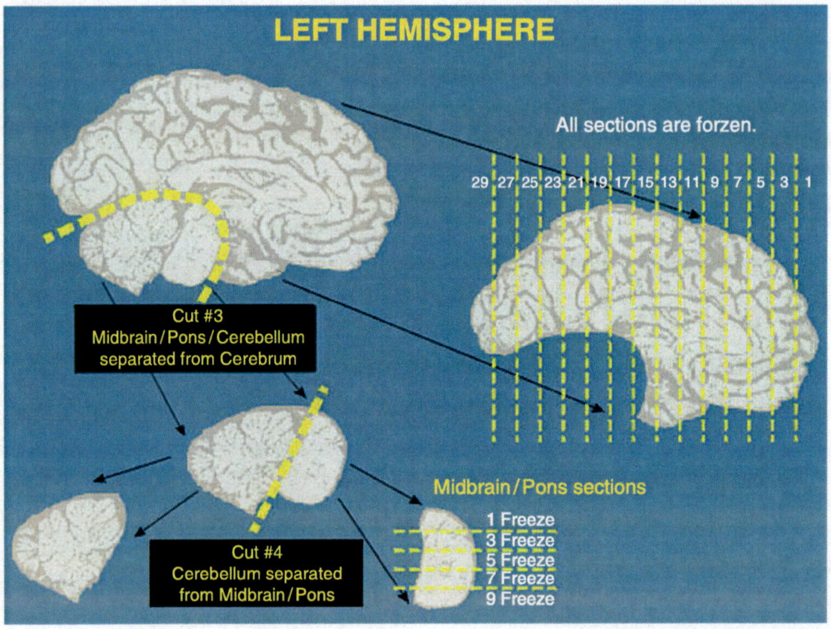


**Fig. 18.1** Coronal slice diagram for tissue acquisition (Image c/o NICHD BTB)

**Table 18.1** Fresh human brain donor information

| Specimen ID | Cause of death | Age | Race | Sex | Brain mass (g) | PMI (years) |
|---|---|---|---|---|---|---|
| HB01 | Myocardial Infarction (Pending Confirmation) | 37 | Caucasian | Male | 1,312 | 34 |
| HB02 | Complications of diabetes | 45 | African American | Male | 1,234 | 25 |

**Fig. 18.2** Human brain fixture and tissue preparation, **(a)** Agarose base poured to form initial layer in base of slice fixture, **(b)** Brain encased in Agarose in slice fixture, **(c)** Blade inserted into slice fixture with assist fixture to maintain straight blade configuration

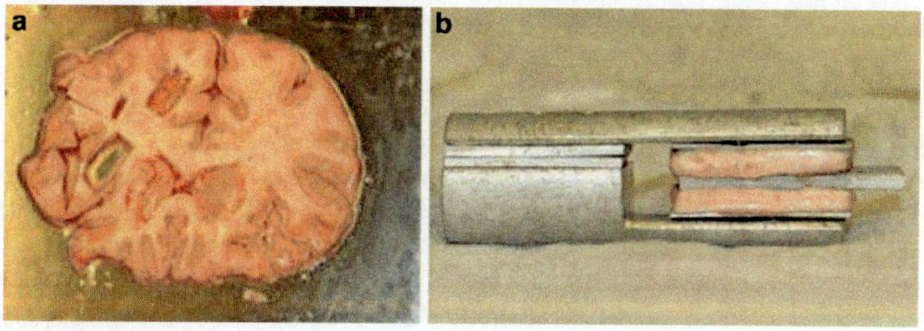

**Fig. 18.3** Removal of human brain tissue from fixture, example of tissue punches and specimen mounting into test fixture

A custom designed slice fixture with 0.3 mm cutting guides spaced 1.5 mm apart combined with an Agarose solution that surrounded the sample were used to generate the specimen used in the SHPB test. This procedure was designed to produce an even slice thickness that could be used later in preparation of test samples and produced slices that had an average thickness of $1.6 \pm 0.6$ mm for HB01 and $1.7 \pm 0.5$ mm for HB02. An initial layer of Agarose was poured to form a base and cooled before the brain then inserted on top of this Agarose layer. After insertion, more Agarose was poured over the top of the brain until the specimen was completed covered. Blades were then inserted into the guides at every location or every other location depending on the desired material thickness. An additional inserting fixture was utilized to ensure that the blades were inserted in a straight configuration, Fig. 18.2.

### *18.2.2 Sample Preparation and Mounting*

Prior to mounting of the samples into the test fixture, dissected tissue slices remained hydrated by immersion in Hank's Buffer Solution (GIBCO, Life Technologies™). Pictures of each individual 1.5 mm slice were taken prior to building the integrated test specimen so that characterization of the amount of white and gray material could be determined as well as details about punch location within the brain (Fig. 18.3). A custom punch was used to create $9 \times 18$ mm rectangular specimen samples from the prepared slices. Two of these punched samples were subsequently bonded to aluminum shims in a double lap shear configuration using a thin layer of cyanoacrylate tissue adhesive (3 M™ Vetbond™). Microscopy performed on prepared samples revealed that the adhesive layer used to bond the specimen had an average thickness of approximately 50 μm. The specimen was mounted in a double lap shear configuration (Fig. 18.3). The tested specimen samples were chosen from areas that were parallel to one another in the in vivo condition.

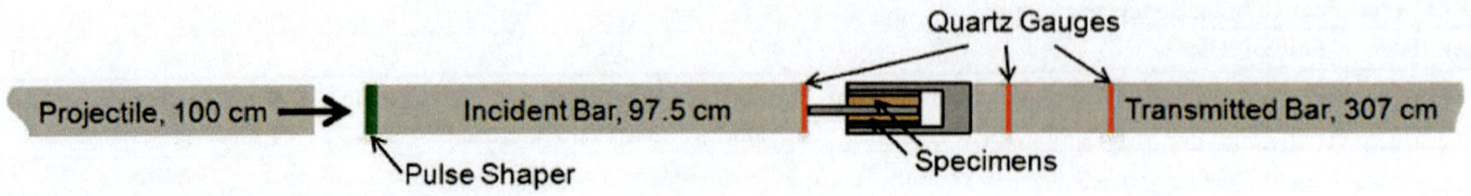


**Fig. 18.4** Schematic of the modified SHPB setup showing the aluminum bars, specimen, pulse shaper and instrumentation

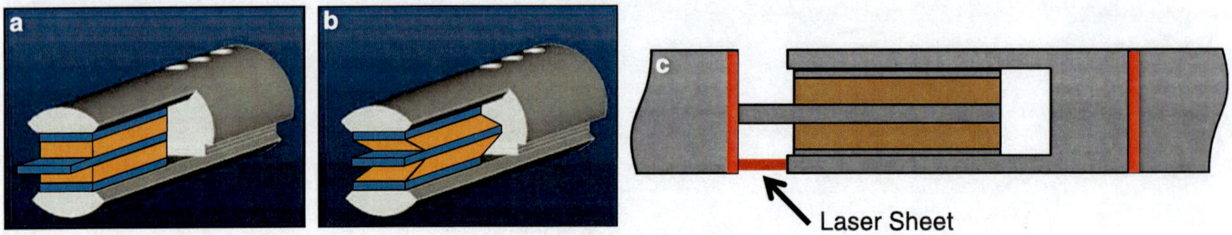


**Fig. 18.5** Schematic of specimens in double lab shear configuration (**a**) Unstressed and (**b**) During Impact (Fixtures were Adjustable for Individual Specimen Height), (**c**) Schematic showing measured specimen thickness and location of line laser measurement, Saraf 2007

### *18.2.3 Split Hopkinson Pressure Bar (SHPB) Setup*

The test setup used was described by Trexler et al. [3], and utilizes a modified SHPB setup to induce shear deformation in biological specimens. In brief, this setup uses an air pressurized breech to generate an impact of a projectile on an incident bar, creating a stress wave which propagates down the bar until reaching the specimen. The specimen was mounted in a double lap shear configuration as described earlier, the central plate translated as a result of the compressive incident stress pulse which develops symmetric shear in the samples during specimen deformation. The double lap configuration uses two opposing specimens for symmetry and this configuration cancels the bending stresses. This specimen configuration is commonly used for DMA shear testing [8, 9] and resists changes in specimen thickness, creating a state of simple, uniform shear. Force was measured using piezoelectric quartz gauges (Boston Piezo-Optics Inc.) [10] which have a 2.3 pC/N piezoelectric contant and 10 MHz frequency response and are able to detect small forces (~5 mN). These quartz gages were used in conjunction with Kistler dual mode amplifiers (Type 5010), which converted and amplified the signal produced by the quartz crystals into the recorded voltage response. Specimen deformation was measured directly by measuring the change in the input bar relative to the test fixture using a line laser (Stocker Yale Lasiris 5 mW Laser and Detector) [11] which was calibrated before each test by measuring the change in voltage at known displacements of the input bar. A schematic of the test setup can be seen in Fig. 18.4. Schematics of the double lap shear configuration prior to and during testing can be seen in Fig. 18.5. Additionally, a schematic illustrating where the specimen thickness and line laser measurement is acquired is shown in Fig. 18.5c. Post-processing of results was completed using Matlab (MathWorks, Inc.) for calculation of stress and strain response from measured voltage as well as creating graphical representation of the results. Statistical analysis of the results utilized JMP (SAS Institute, Inc.) software. In addition to the instrumentation on the bar itself, high speed video was recorded of each test using a Phantom V10 high speed camera at 8,000 fps (Phantom Research, Inc.).

An example of a characteristic test result is found in Fig. 18.6, which shows the response of the input and output gauges as well as the line laser in volts. Although the entire recorded response lasted 150 ms, this response has been truncated to illustrate where the signals start to diverge. Agreement between the two sensors is shown to be acceptable for a large amount of displacement, as observed by the change in the line laser response from approximately $-7.0$ to $-3.5$ ms, the negative values a result of the triggering technique applied. The voltage response for the output and input quartz gauges are scaled by a gain factor as determined by amplifier settings to obtain a force response, which is then converted into a shear stress through knowledge of the punch area. The approximated error in the calculation of this stress value was found to be 10.9 % using methods described in Taylor (1997) [12] based on an assumed error in the quartz gauge measurement and measurement in the variation of the test samples. The line laser response was scaled by a calibration factor to obtain the shear strain and strain rate response respectively. An example of the calibration curve used to scale the line laser response is shown in Fig. 18.6. The output shear stress-strain curve, combining the recorded force and displacement data, Fig. 18.6a, b this process repeated for all samples tested.

In addition to checking the correspondence of the input and output gages, the high speed video for each test was reviewed to verify that the center shim stayed level throughout the test until either the tissue failed or the input gage contacted the

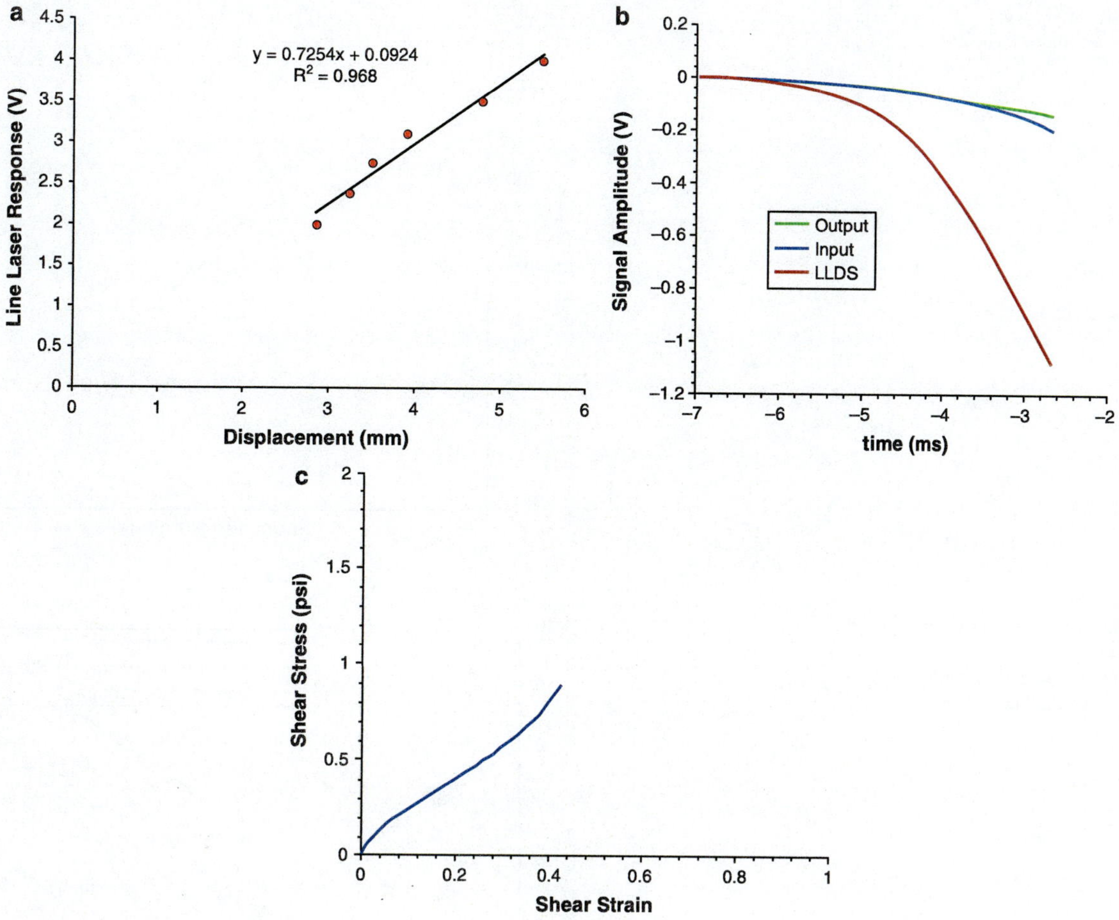


**Fig. 18.6** Example of line laser calibration curve (**a**) Example of gauge and line laser data response (**b**) Example stress – strain response from single test (**c**)

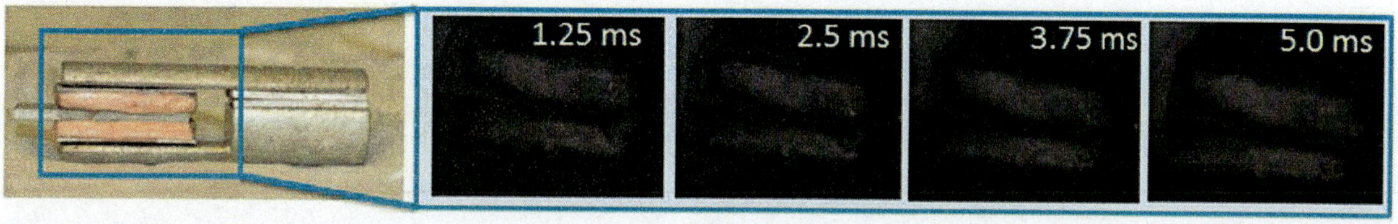


**Fig. 18.7** Stills of brain shear SHPB test

test fixture. Selected images from one of the tests can be found in Fig. 18.7. From this figure, it is clear that the center shim maintained its initial orientation with respect to the tissue during displacement. Additionally, this data was used to confirm when the calculation of the shear response should end, corresponding to failure of the tissue or the adhesive layer.

## 18.3 Results

A total of 69 shear tests were completed, 32 tests from the first specimen and 37 tests from the second specimen. Of the tests completed, 60 tests had results that were found to be acceptable based on post-processing quality checks, as determined by stress equilibrium being achieved through analysis of the quartz gauge input and output response curves. All tests were

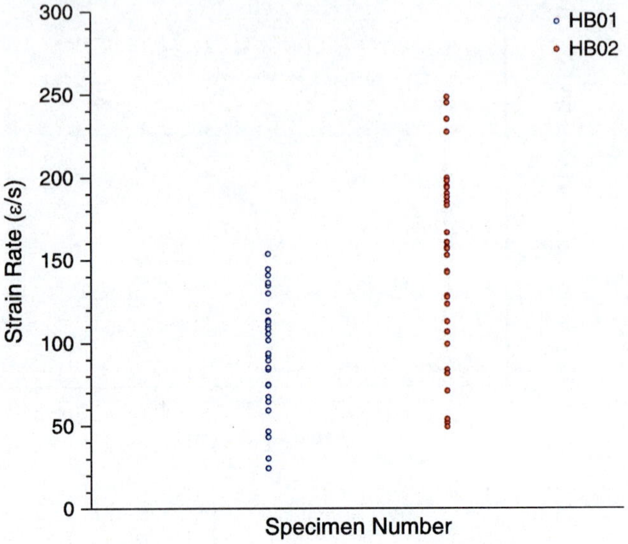


**Fig. 18.8** Strain rate distribution for each specimen tested

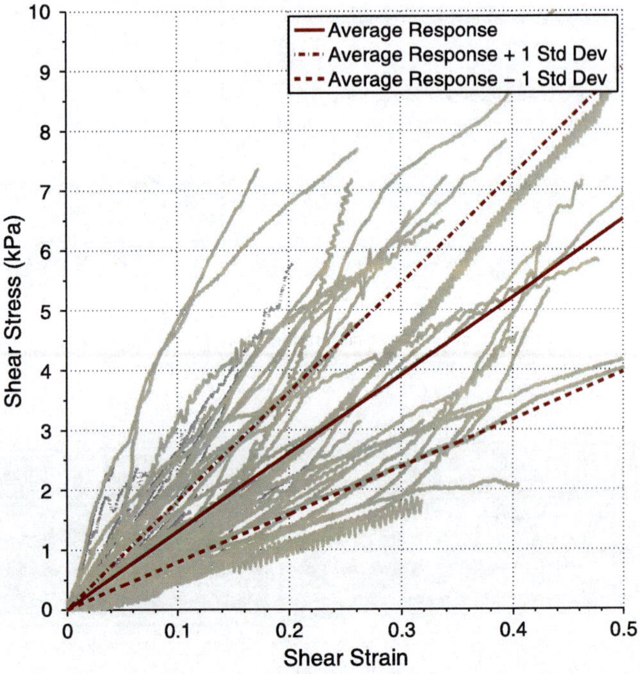


**Fig. 18.9** Shear stress – strain response for all samples, linear fit of average response with ±1 standard deviation

executed within a 36 h post-mortem interval. Specimen HB02 was tested at slightly higher input pressures, which resulted in higher achieved peak strain rates. The distribution of achieved strain rate values for each specimen is shown in Fig. 18.8. The peak strain rate levels tested have been indicated to be relevant to computational modeling of brain response to dynamic impact events.

The complete results for the human brain tests are found in Fig. 18.9 where each shear stress-strain response is represented. The linear fit of the data results in a shear modulus value of 13.0 ± 10 kPa (R = 0.981). This value was determined by taking the average shear stress response at each strain level and then fitting a linear curve through the resulting points. Additionally, linear fit approximations were made for each of the individual test results with an average R value of 0.988 and a minimum R value of 0.950.

To investigate the variance observed in Fig. 18.9, a few of the many different possible parameters were investigated to determine the effect on the fitted linear slope values. The results of this investigation are found in Fig. 18.10. The linear slope

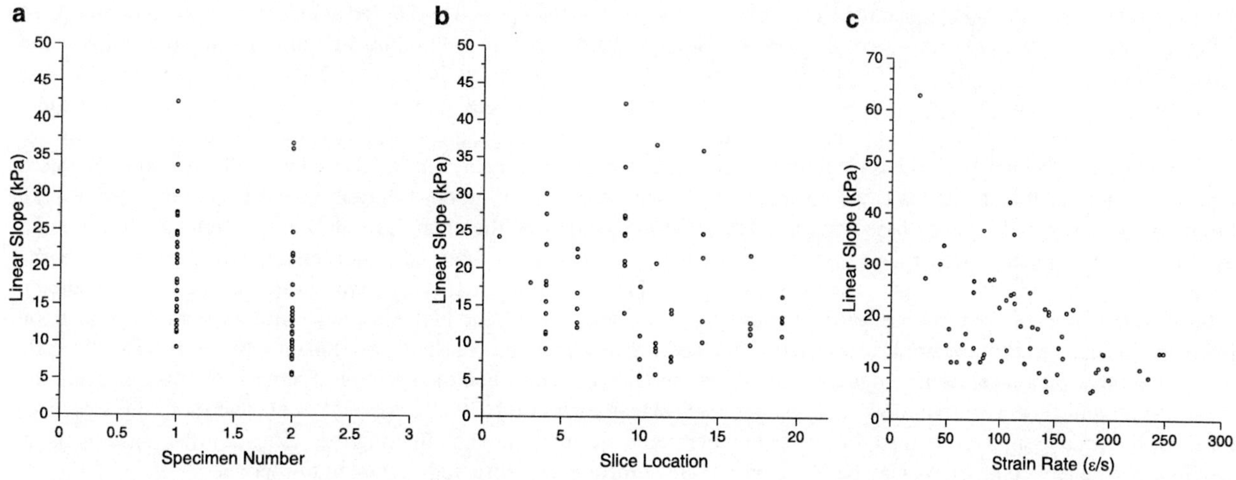


**Fig. 18.10** Linear slope as a function of tested specimen (*Left*), Slice location arranged anterior to posterior (*Middle*), tested strain rate (*Right*), * indicates a significant difference

was found to vary significantly (*t*-test, $p = 0.0011$) based on tested specimen (Fig. 18.10, *left*) but did not appear to vary with slice location (Fig. 18.10, *Middle*). A weak decreasing trend with increasing strain rate was observed (R = 0.238), however it is unclear if this trend would be affected by other parameters not represented.

## 18.4 Discussion

To our knowledge, this is the first study to involve testing fresh human brain tissue within 36 h post-mortem using high rate testing at large strain (> 20 % deformation). The objective of this study was to develop a procedure by which fresh, never frozen human brain tissue could be acquired and tested utilizing a modified Split Hopkinson Pressure Bar technique in order to determine the simple shear material properties of brain tissue at high strain rates. This procedure was applied to two different specimen and a total of 69 tests were completed. Of the 69 completed tests, 60 were confirmed to have developed stress equilibrium during the testing through comparison of the input and output response characteristics. High speed video confirmed the displacement of the center shim and confirmed timing for ending of shear stress response. From these response curves, the tangent modulus was calculated at various strain values and strain rates as well as a function of percent white matter and compared within each specimen to determine any differences in the result.

The test setup was shown to be able to achieve stress equilibrium in the brain tissue for a majority of the tests. Given the relative compliance of the samples and the high rates of testing, the ability to achieve a high rate of success, as defined by the previously described quality checks, in measuring a relevant response is seen as non-trivial. During the execution of these tests it was found that the thickness of the samples was critical in determining if stress equilibrium would be developed within the sample. Although this relationship is expected with thinner samples expected to perform better in both developing equilibrium as well as higher strain rates, due to the compliance of the brain tissue it was found to be difficult to produce slices that were thinner than 3 mm. Additionally, the amount and coverage of the cyanoacrylic was found to have a large effect on the results by determining how long the adhesion to the various shims would be maintained. If the sample was bonded only on a portion of the aluminum shims due to incomplete coverage of adhesive, the input and output force responses would not equilibrate. Further, from the recorded video data it was clear during post-processing when a sample was not properly bonded as indicated by the observed failure pattern.

Across the two tested specimen, HB01 had a higher mean elastic modulus (22.2 ± 10.8 kPa) than HB02 (14.0 ± 7.7 kPa). These averaged linear slope values are slightly different than the overall linear estimation of all the response values (Fig. 18.9) and is a reflection of some of the non-linear aspects of the response. Overall, the high R values resulting from the linear fits of both the individual results as well as the average response indicates that this approximation is a decent initial representation of the data. Additionally, while there did not appear to be any change in the linear stiffness with respect to slice location anterior to posterior, there did appear to be a weak trend of decreasing linear slope with increasing strain rate. This trend of decreasing stiffness with increasing strain rate is unexpected and requires further investigation. Overall, the stiffness values reported here appear to be higher than those previously reported and tested at lower shear strain rates [13, 14] although the test methods employed in previous studies were different and thus difficult to directly compare too. An additional study by Manduca 2001 reported a shear modulus in white brain tissue of 14.2 kPa at 100 Hz using an in vivo experimental technique [15]. This value is similar to the values reported here, however direct comparison is again difficult since the test method is dissimilar, the strain rates tested here representing a much higher level of strain.

As mentioned previously, variation in the stress-strain response can be attributed to several factors beyond those reported. One parameter not investigated was the change in the response based on the post mortem interval. The response of brain tissue has been shown to change based on this interval [16] and therefore there may be trends in the data that correspond to the time from acquisition that may explain some of the relative scatter observed. Additionally, it has been shown that different areas of the brain have different stiffness response [17, 18] and where these samples were taken could be studied to determine the effect on the response. While initially there was not a clear trend in the response with respect to slice location, investigation as to where the sample was tested with respect to location within a slice may explain some of the variance in the results. Another parameter to investigate is the difference observed specimen to specimen. From Fig. 18.10 it appears that the linear modulus was generally lower for specimen HB02 than it was for HB01. However, it is unclear if this difference is a result of the specimen, the tested strain rate or another parameter that may be affecting the results. Further specimens are required before any conclusions may be made based on the differences observed specimen to specimen.

## 18.5 Conclusions

A novel test technique to measure the shear properties of soft materials was applied using a modification to the traditional Split Hopkinson Pressure Bar test setup. This test setup allowed for measurements to be taken at greater strains and strain rates than was previously possible. A procedure for fresh human brain tissue acquisition, preparation and testing was also developed and executed utilizing two specimen and two different locations in the brain. Force and displacement response was collected from over 60 samples and the results were used to determine the shear strain response. The strain rates tested range from 25 to 248 strain per second. The elastic modulus of HB01 and HB02 was 13.0 $\pm$ 10 kPa. The results of these tests will prove useful for both computation and surrogate model development.

**Acknowledgement** This effort was partially funded by contract # W81XWH-09-2-0168 with the U.S. Army Medical Research Acquisition Activity, 820 Chandler Street, Fort Detrick MD 21702-5014 as the awarding and administering acquisition office as well as contract # W911QY-09-C-0058 awarded by the Office of Naval Research and the Naval Health Research Center. The content included in this work does not necessarily reflect the position or policy of the U.S. government.

## References

1. Saraf H et al (2007) Measurement of the dynamic bulk and shear response of soft human tissues. Exp Mech 47:439–449
2. Saraf H et al (2007) Mechanical properties of soft human tissues under dynamic loading. J Biomech 40(9):1960–1967
3. Trexler MM et al (2011) Verification and implementation of a modified split Hopkinson pressure bar technique for characterizing biological tissue and soft biosimulant materials under dynamic shear loading. J Mech Behav Biomed Mater 4(8):1920–1928
4. Lennon A et al (2007) Modifications to the compression kolsky bar for characterizing soft biomaterials at impact loading rates, in SAMPE 2007: Society for the Advancement of Material and Process Engineering, M and P. Baltimore
5. Pervin F, Chen WW (2009) Dynamic mechanical response of bovine gray matter and white matter brain tissues under compression. Biomechanics 42:731–735
6. Garo A et al (2007) Towards a reliable characterisation of the mechanical behaviour of brain tissue: the effects of post-mortem time and sample preparation. Biorheology 44(1):51–58
7. Zhang J et al (2011) Effects of tissue preservation temperature on high strain-rate material properties of brain. J Biomech 44(3):391–396
8. Cognard JY et al (2005) Development of an improved adhesive test method for composite assembly design. Compos Sci Technol 65(3–4):359–368
9. Yang H et al (2007) Multi-layer in-situ for evaluation of dynamic mechanical properties of pressure sensitive adhesives. Int J Adhes Adhes 27(7):536–546
10. Chen W et al (2002) Dynamic compression testing of soft materials. Appl Mech Trans ASME 69(3):214–223
11. Ramesh KT, Kelkar N (1995) Technique for the continuous measurement of projectile velocities in plate impact experiments. Rev Sci Instrum 66(4):3034–3036
12. Taylor JR (1997) An introduction to error analysis: the study of uncertainties in physical measurements, 2nd edn. University Science, Sausalito
13. Hrapko M et al (2006) The mechanical behaviour of brain tissue: large strain response and constitutive modelling. Biorheology 43(5):623–636

14. Nicolle S, Lounis M, Willinger R (2004) Shear properties of brain tissue over a frequency range relevant for automotive impact situations: new experimental results. Stapp Car Crash J 48:239–258
15. Manduca A et al (2001) Magnetic resonance elastography: non-invasive mapping of tissue elasticity. Med Image Anal 5(4):237–254
16. McIlwain H, Bachelard HS (1985) Biochemistry and the central nervous system, 5th edn. Edinburgh: Churchill Livingstone
17. Arbogast KB, Margulies SS (1998) Material characterization of the brainstem from oscillatory shear tests. J Biomech 31(9):801–807
18. Prange MT, Meaney DF, Margulies SS (2000) Defining brain mechanical properties: effects of region, direction, and species. Stapp Car Crash J 44:205–213

# Chapter 19
# Injury Potential of Shock Induced Compressive Waves on Human Bone

**Ann M. Bailey, S. Boruah, J.J. Christopher, B.C. Bennett, M. Shafieian, D.S. Cronin, and R.S. Salzar**

**Abstract** Recent military conflicts have provided evidence of significant lower extremity injury resulting from under-vehicle blasts. Improved vehicle design elements have decreased the frequency of hull breech into the occupant compartment, yet catastrophic injuries to the lower extremities continue to occur. In under-vehicle blast, the lower extremities, with intermediate to low acoustic impedance, are loaded by a high speed compression wave resulting from the blast load on the lower impedance vehicle armor. The dynamic stresses created in the tissues are significant and may exceed the dynamic strength of the bone. Numerical analysis of this steel plate/lower extremity interaction reveals the potential for the accumulation of stresses in the hard tissues, which has the potential to induce injuries separate from the more traditional inertia-based injuries. To investigate this potential injury mechanism, an experimental test was developed to apply high rate loading to a bone sample.

**Keywords** Acoustic • Impedance • Shock • Bone • Injury

## 19.1 Introduction

Efforts to improve the design of armored military vehicles by decreasing the amount of floor intrusion from under-vehicle blasts have revealed the existence of a previously uninvestigated lower extremity injury mechanism. While previous lower extremity injuries were attributed to the inertia-based loading caused by breaching of the occupant compartment or severe deflection of the floor plate, the more recent injuries cannot be explained in the same manner [1]. Relatively small deflections in the vehicle floor panels suggest the injuries could be a result of another mechanism.

Attempts to reproduce these under-vehicle blast injuries have focused mainly on the inertial loading of the lower extremity and have neglected to consider alternative injury mechanisms such as impedance mismatch or the destruction that could be caused by a high speed compression wave traveling through bone. Considering the medium through which the blast energy is delivered to the foot could provide insight into the reason for seeing injury at very small floor plate deflections.

Quenneville, et al. concluded that "factors in addition to force" such as impulse and energy have potential for contributing to tibia injury criteria under impulse loading conditions [2]. Using shorter duration impulses than previous experiments of a similar nature such as Yoganandan et al. [3] the Quenneville et al. experiment resulted in lower peak forces prior to injury. While such results may be artifacts of the lack of skin and other soft tissues between the bone and steel plate or other differences in the experimental set-up, these results could also be evidence of a separate injury mechanism which causes bone to fail in a different manner.

Landmine explosion injuries are classified into four categories based on the aspect of the explosion that causes the injury. These types include primary (injuries caused by the shockwave), secondary (injuries caused by fragments and other projectiles), tertiary (injuries due to the acceleration of the vehicle or floor pan deformation), and quaternary (injuries due to increased temperature) [4]. While most underbody blast injuries are classified as tertiary [4], some injuries from

A.M. Bailey (✉) • S. Boruah • J.J. Christopher • B.C. Bennett • M. Shafieian • R.S. Salzar
Center for Applied Biomechanics, University of Virginia, 4040 Lewis and Clark Drive, Charlottesville, VA 22911, USA
e-mail: amb9um@virginia.edu

D.S. Cronin
University of Waterloo, 200 University Avenue West, Waterloo, ON N2L 3G1, Canada
e-mail: dscronin@mecheng1.uwaterloo.ca

V. Chalivendra et al. (eds.), *Dynamic Behavior of Materials, Volume 1: Proceedings of the 2012 Annual Conference on Experimental and Applied Mechanics*, Conference Proceedings of the Society for Experimental Mechanics Series,
DOI 10.1007/978-1-4614-4238-7_19, 

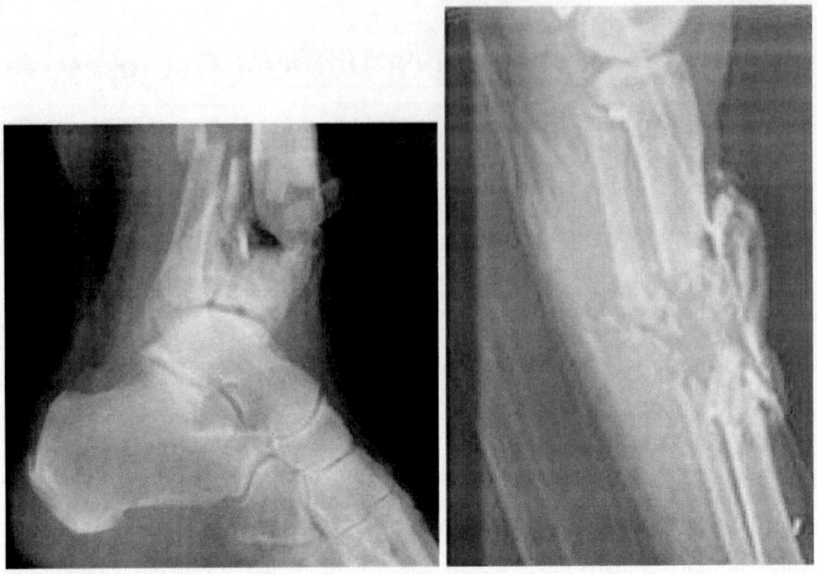

**Fig. 19.1** Typical inertia-based distal tibia fracture (*left*) [5] vs. an under-vehicle blast mid-tibia fracture (*right*) [4]

underbody blast are characteristically different than those deemed to be a result of inertial loading. Figure 19.1 provides an example of a fracture produced from high rate inertial loading and one produced by under-vehicle blast. Notice the more comminuted state of the under-vehicle blast fracture compared to the fracture produced by pure inertial loading.

The goal of this study was to separate out the inertial loading and concentrate on the effect of shock wave propagation through the vehicle floor and into the lower extremities. Arepally, et al. divided the injury mechanisms into three categories: elastic, viscous, and inertial [6]. Of these categories, the elastic injury mechanism was of most interest to the current study due to its focus on the compression and tension of a body rather than the inertial loads caused by accelerations. While the inertial loading is still an important factor, this experiment focused on the influence of the high speed compression wave on injury, separate from the global accelerations the body experiences.

When considering the possibility of acoustic impedance as a potential contributor to this mechanism of injury, it is important to understand the properties of the materials involved. The floor panels in military vehicles are made of metals such as steel, which have a large speed of sound, approximately 5,800 m/s. According to various research articles, the speed of sound in bone (specifically the tibia) is between 3,000 and 4,000 m/s [7–9]. When an object or shock wave hits the outer surface of the vehicle floor pan, a high speed compression wave, traveling at the speed of sound of the material through which it is moving, will form and propagate through the material. In this case, the wave is traveling at around 5,800 m/s. The wave will travel through the steel and then reach a material on its opposite surface, the occupant's lower extremity. At this interface, part of the wave will be transmitted through the boot/ankle and into the skeleton, and part will be reflected back into the steel. Despite the fact that part of the wave will be reflected, even a small fraction of the shock wave's energy transmitted has enough energy to produce injuries in the bone.

The objective of our research was to demonstrate the influence acoustic impedance has on failure in bone. To prove this concept, a series of finite element simulations and experimental tests were developed examining the injury potential of a high speed compression wave traveling from a low impedance material into a higher impedance bone material. By varying both the amount of energy input and the amount of impedance mismatch between the armor material and the bone, the potential for significant energy accumulation was demonstrated in the finite element simulations based on the experimental setup. Additional analysis of this phenomenon also revealed the importance of capturing the strain-rate dependent material properties of bone. The results of this study provide a foundation for understanding the nature of some under-vehicle blast injuries as well as insight into potential lower extremity injury mitigation strategies.

## 19.2 Methodology

The objective of this study was to demonstrate that loading a material using a high speed compression wave through a material of a higher speed of sound (lower impedance around 14 MN-s/m$^3$) into a material with a lower speed of sound (higher impedance around 5 MN-s/m$^3$) will produce failures in the lower speed of sound material prior to the onset of inertial loading.

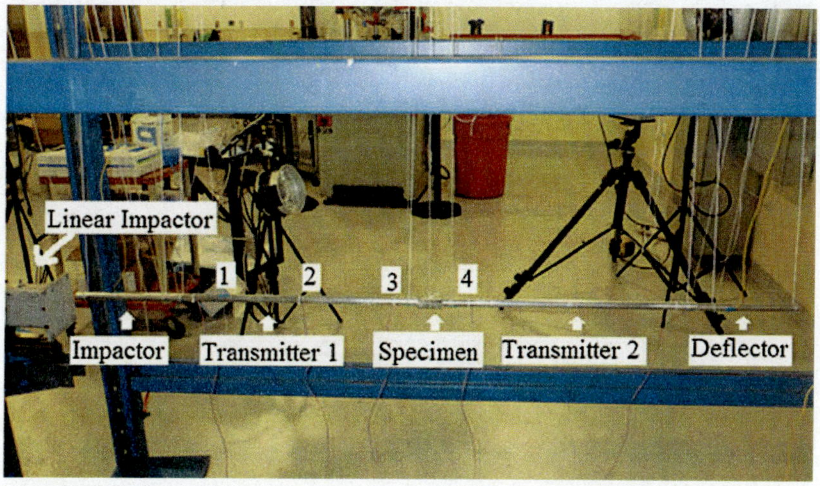


**Fig. 19.2** Experimental setup denoting strain gage position numbers

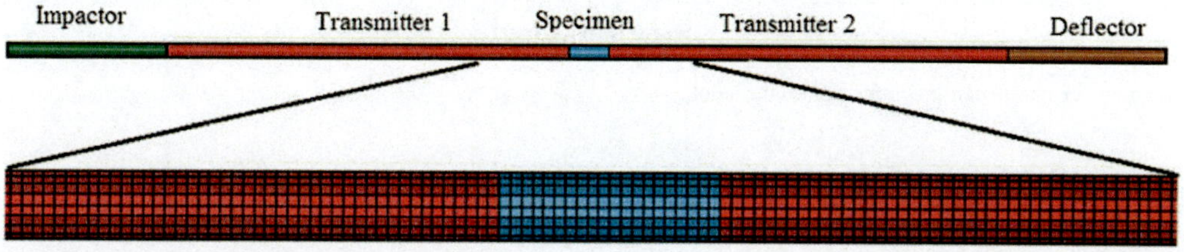


**Fig. 19.3** FE model showing common nodes between transmitters and specimen

An experimental setup using the concepts of Newton's cradle and the split-Hopkinson bar was developed in order to create a loading scenario that eliminated inertial loading via global accelerations. The setup consisted of an impactor, a transmitter bar placed on either side of the specimen, and a deflector. Each of the bars had a diameter of 19 mm. The transmitters, specimen, and deflector were 603, 241, and 60 mm in length and were made of 2024 aluminum. Aluminum was used in place of steel because of its higher speed of sound, which would enhance the difference in speed of sound with the specimen material.

Each bar was suspended from a frame using nylon string (Fig. 19.2). The bars were aligned such that each end was in direct contact with its neighboring bars. The impactor was given an initial separation from the first transmitter and then hit with a pneumatic linear impactor causing it to hit the first transmitter with a velocity of approximately 5 m/s. The impactor velocity was measured using an optoCONTROL 1,200 laser micrometer. To track the high speed compression wave, three strain gages were placed on the first transmitter at 40, 304, and 653 mm along the bar length, and one was placed 40 mm from the sample on the second transmitter. All were sampled at 2 MHz. A high speed camera was used to check for translational motion of the specimen and for crack detection. Two different materials were used as the specimen material for these tests: sidewalk chalk, and 2024 aluminum. Chalk was chosen for its brittle behavior since the goal was to produce fracture in the specimen with the energy levels achievable by the setup.

Additionally, a 3-D finite element model was created in LS-Dyna based on this experimental setup. However, to prevent errors associated with artificial contact stiffening, the surface nodes of the specimen in contact with the transmitter bars were made common. The impactor was placed 5 mm from the end of the first transmitter bar and the deflector was placed in direct contact with the end of the second transmitter bar (see Fig. 19.3). An automatic surface-to-surface contact was placed between the impactor and transmitter 1 and between the deflector and transmitter 2.

Supplementary investigation of the hypothesized injury mechanism was performed using the finite element model. For purposes of ensuring that the mass was balanced on either side of the specimen, the transmitters, impactor, and deflector were assigned the same material properties. The nodes of the impactor were given an initial velocity and allowed to impact transmitter 1; this generated a high speed compression wave that carried energy through the transmitter and into the specimen. After traveling through the specimen, the wave entered transmitter 2, and eventually the deflector. The deflector, because of symmetry in the design, then obtained a velocity, separating it from transmitter 2, and allowing the remaining parts to stay stationary and free of inertial loading as occurs in the Newton's cradle.

**Table 19.1** Material properties for FE models

| Material name | Young's modulus (GPa) | Density (kg/m$^3$) | Speed of sound (m/s) | Impedance (MN-s/m$^3$) |
|---|---|---|---|---|
| Aluminum 2024 | 73 | 2,780 | 5,124 | 14.24 |
| Modified material 1 (MM1) | 18.765 | 2,780 | 2,598 | 7.22 |
| Modified material 2 (MM2) | 40 | 2,780 | 3,793 | 10.56 |
| Rate dependent bone | 13.5 | 2,000 | 2,598 | 5.20 |

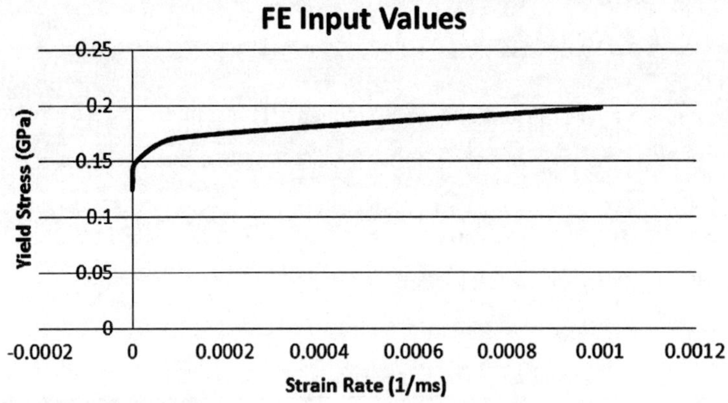


**Fig. 19.4** Yield stress for rate-dependent bone material model

Several simulations were run using this finite element model in order to observe trends associated with the relationship between the material properties of the transmitter bars and the specimen. Initially, the impactor, transmitters, and deflector were assigned the properties of linear-elastic 2024 aluminum, and the specimen was given rate-dependent bone properties, outlined in this study. Two additional sets of simulations were run in which the material properties for the impactor, deflector, and transmitters were changed in order to modify the acoustic properties without changing the mass of these parts. To accomplish this, only the Young's modulus was altered, while the density remained equal to that of 2024 aluminum since the speed of sound for these models is calculated using the equation $c = \sqrt{\frac{E}{\rho}}$, where c is the speed of sound in the material, E is Young's modulus, and ρ is the density. The properties of all materials used are listed in Table 19.1.

While bone is often modeled as an elastic material, studies such as the one done by McElhaney [10] on human femur cortical bone show that bone properties exhibit strain-rate dependence. Current finite element biomechanical material models may incorporate a strain rate dependent material model, for example a rate-dependent plasticity model [11] (*MAT_019, LS Dyna) based on material testing results for femur cortical bone where applicable. This material model was used in the current study to model the constitutive response of the bone material, where the tangent modulus was considered to be 5% of the Young's modulus and independent of strain rate [12, 13]. The failure criterion for the material model was set to a plastic strain of 0.0088. A graph of the strain rate versus yield stress curve used for the bone material model is shown in Fig. 19.4.

For each of the three sets of material properties, a series of simulations were run with different initial velocities assigned to the impactor. Impactor velocities were 5, 15, 30, 40, and 50 m/s. For each, the longitudinal compressive strain wave was tracked through the bone specimen. Peak strain and internal energy were then plotted as a function of the distance along the bone specimen. This was important for determining the relationship between speed of sound and location of injury.

## 19.3 Results

The initial results of experimentation using the physical setup revealed minimal translational motion of the specimen during impact, and thus negligible inertial loading. The tests performed using chalk as the specimen were deemed inconclusive due to defects and voids in the material causing unpredictable fracture behavior. Tests were then performed using a 2024 aluminum specimen and the strain-time history was recorded for the four locations. The finite element model was run using

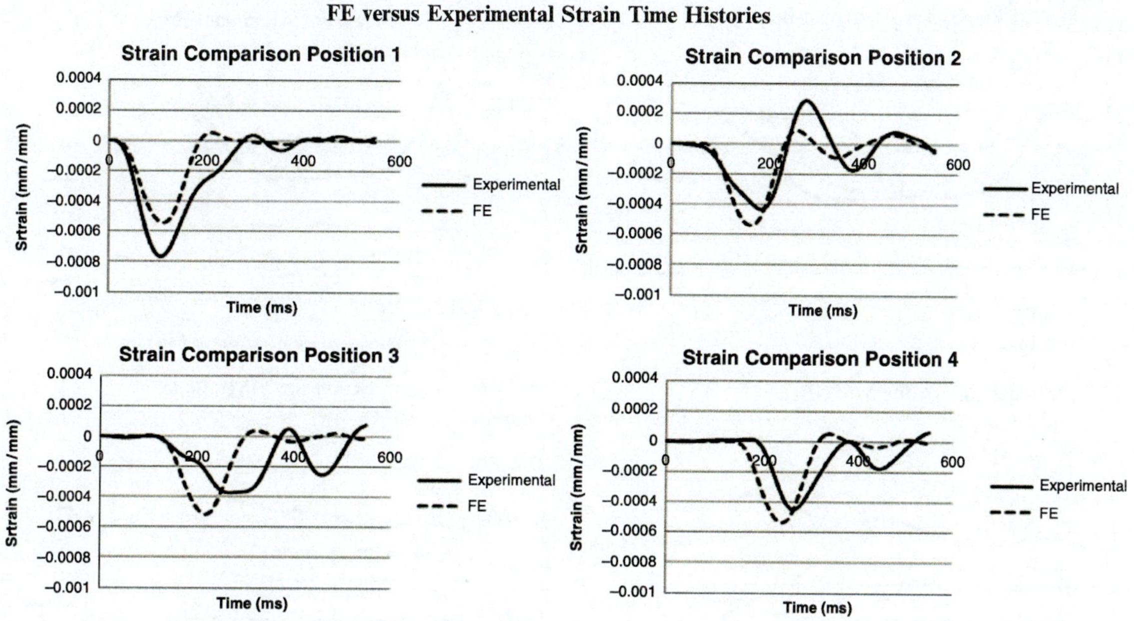


**Fig. 19.5** Strain-time history plots using a 40 kHz Butterworth filter for FE and experimental results

the same conditions from the experimental test and strains were recorded for the locations where strain gages were placed in the experiment. Strain-time history plots comparing the experiment and the FE simulation are shown in Fig. 19.5.

Results in Fig. 19.5 show that for the case in which the specimen was of the same material as the transmitters, the maximum strain measured in the transmitters on either side of the specimen was similar for both the FE and experimental cases. This was not found to be the case for the chalk, due to failure of the material during the impact which clipped or cut off the transmitting wave. Differences observed in the experimental and FE strain-time history plots may be due to assumptions made for the material properties. Despite these differences, the experimental results verify the finite element model concept for the conditions in which the material is constant throughout the setup. However, additional work is necessary to validate the FE model for impactor velocities high enough to induce fracture in bone.

From the finite element simulations, strain data was recorded for elements along the length of the specimen as well as at selected points within the transmitters. First, the strain time history was plotted for each of the elements in the bone specimen. The peak strain corresponding to the initial arrival of the compression wave at each element was then recorded for specific elements along the length of the bone specimen. This data is shown in Fig. 19.6. Note that the strain peaks are consistent with those in the internal energy plots.

For the case where the acoustic properties of the transmitters are equal to those of bone (material MM1), the strain varies almost linearly along the length of the bone with a very gradual slope. As the speed of sound in the transmitter was increased as for MM2 and then aluminum, the strain developed peaks along the length of the bone. These peaks provide evidence that the speed of sound in the transmitters affects the amount of strain the bone material experiences.

It is also important to note that these strain peaks do not form for lower impactor velocities, which correspond to lower input energies; thus, a certain amount of energy is required to produce these strain elevations in the material. Additional simulations were run in order to determine the impactor velocity required to generate these drastic strain peaks. This velocity was determined to be around 35 m/s for the aluminum-bone interaction case. Figure 19.7, in which the time history for various elements is plotted, illustrates the onset of these strain spikes. Figure 19.7 also displays similar plots for the case where the transmitters have the same speed of sound as the bone. Note the lack of strain peaks in this case.

The material model used for bone allowed for observance of failure. All failures occurred upon the initial arrival of the high speed compressive wave and these locations are described in Table 19.2. The fact that fractures did not occur in the specimen where the bone and the transmitters had the same speed of sound (MM1) does not seem to be coincidental. Since the transmitter and the specimen had the same speed of sound, the high speed compressive wave created by the impact was able to move smoothly through the transmitter and then through the specimen. In the cases where an impedence mismatch existed, the strain (and stress) accumulated in the material due to the bone specimen's inability to move the wave through at the same rate it was being loaded by the transmitter. Since the wave speed was forced to decrease because of its moving into a material with a higher impedance, the wave was forced to increase in amplitude to adjust, particularly in the cases where more energy was introduced into the system by increasing the impactor velocity.

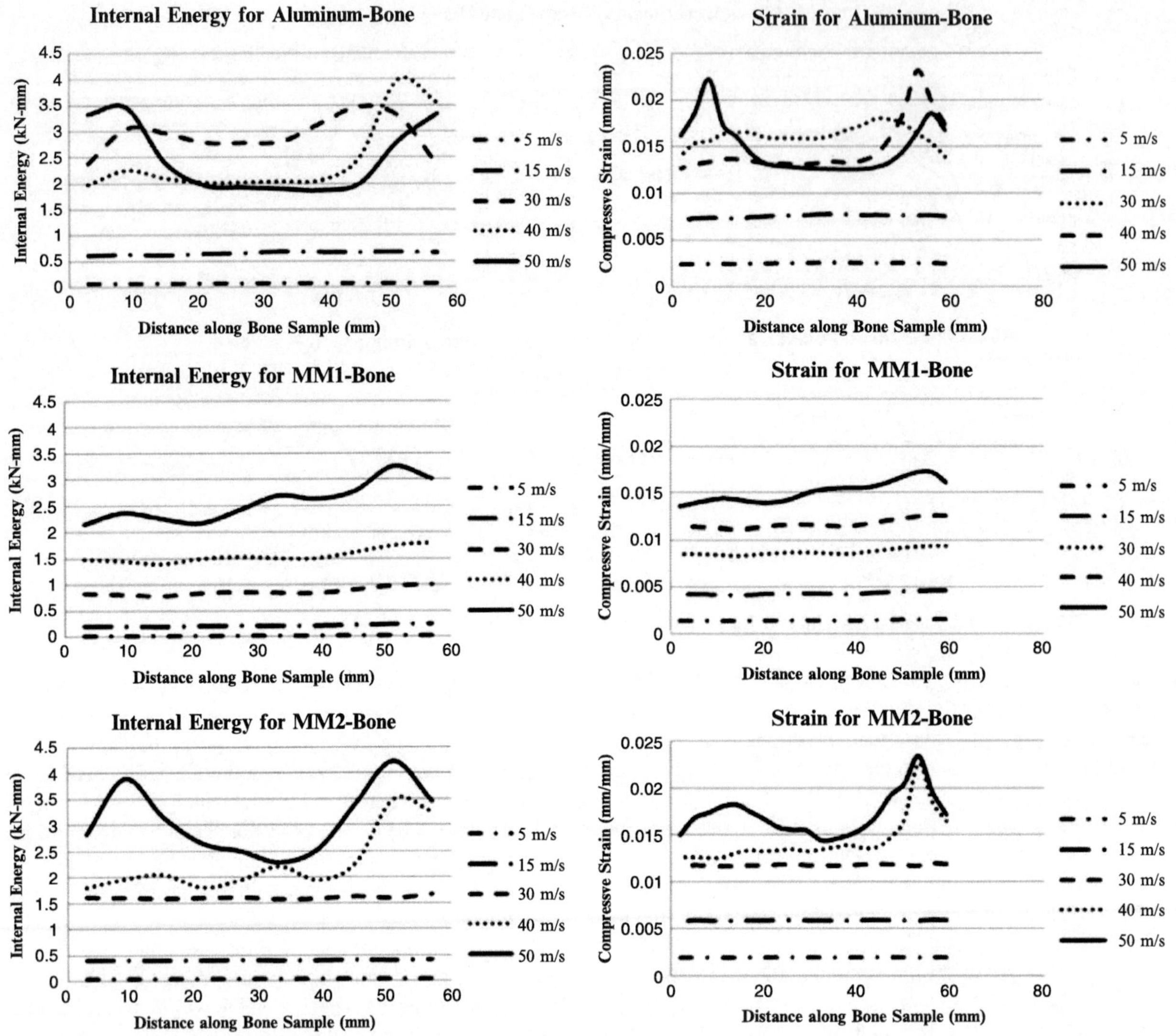


**Fig. 19.6** Strain and internal energy plots showing changes along length of specimen

This is similar to what occurs in the proposed injury mechanism for the underbody blast scenario. A shock wave, carrying a large amount of energy, loads the underside of the vehicle, a steel plate. The plate allows the wave to travel through and into the lower extremities. Because of the mis-match in acoustic properties, a build-up of strain occurs in the lower extremity, and eventually reaches a point where the bone fractures catastrophically, producing injuries similar to that seen in Fig. 19.1. Additionally, in cases where inertial loading is prominent, one would expect injury to occur near the impacted surface; however, certain underbody blast injuries appear to occur at mid-shaft of the tibia and more proximal (Fig. 19.1). This would require peak strains or energies away from the contact surface of the bone sample. Figure 19.6 provides evidence of this phenomenon.

## 19.4 Conclusion and Discussion

Despite the inability of our experimental setup to validate our finite element model, much progress was made in determining the behavior of materials loaded at rates higher than their speed of sound. The finite element model using rate-dependent bone as the specimen material concluded that impactor velocities exceeding 35 m/s, which is comparable to the 7–30 m/s

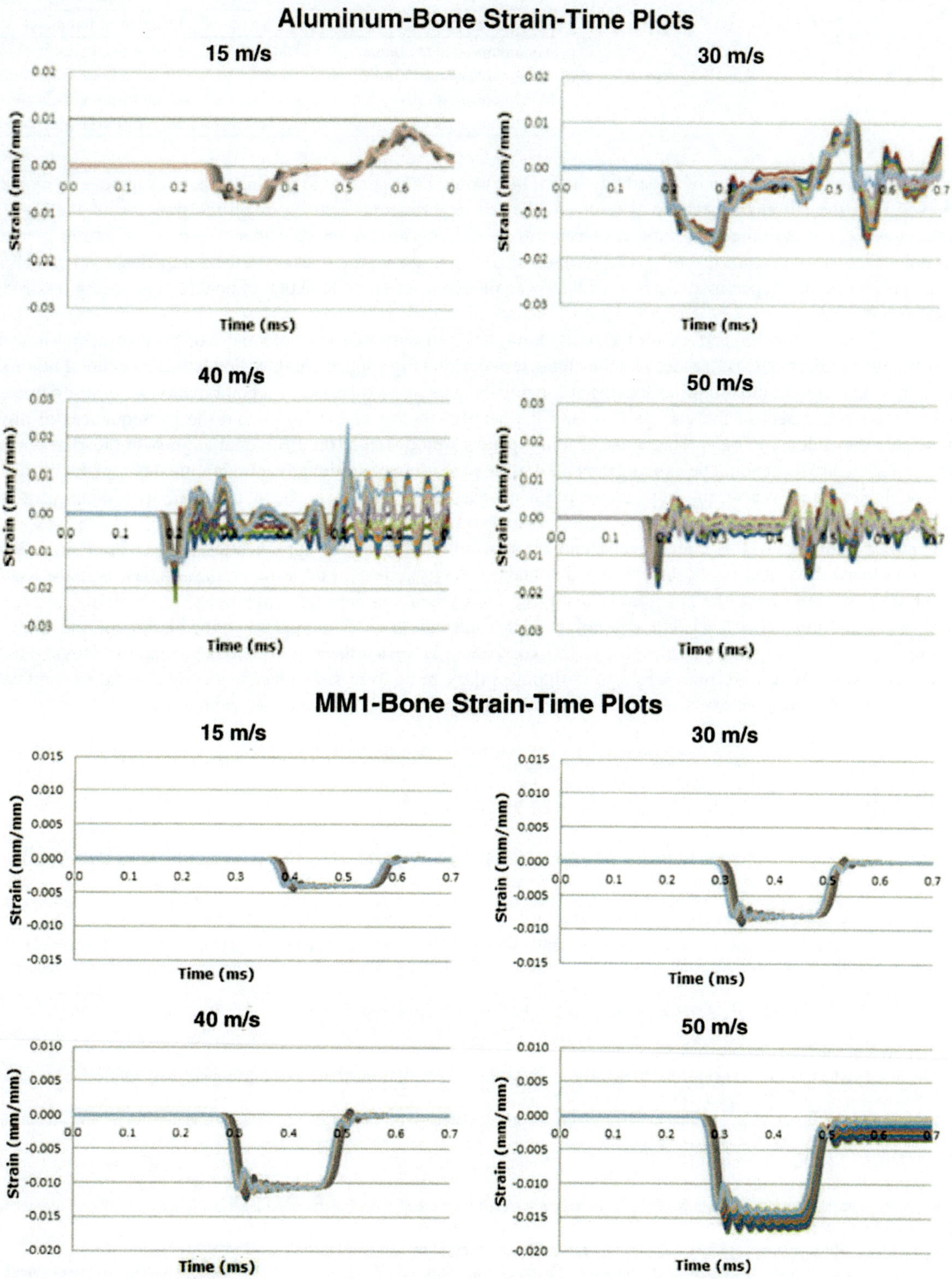


**Fig. 19.7** Strain plots for aluminum and MM1 transmitter simulations (Note: Individual lines represent different elements along the length of the bone specimen)

Table 19.2 Element failure locations for FE models

| Model description | Location of failure |
|---|---|
| Aluminum-bone at 40 m/s | 54 mm from impact site |
| Aluminum-bone at 50 m/s | 12 and 54 mm from impact site |
| MM2-bone at 40 m/s | 54 mm from impact site |
| MM2-bone at 50 m/s | 54 mm from impact site |

floor plate velocities characteristic of underbody blast [14], would be necessary to produce fractures in bone, which cannot be achieved with the current experimental setup. While the experimental setup enabled the tracking of the high speed compression wave, it was limited in its ability to transmit enough energy into the specimen to produce fractures in materials such as bone. The key to future use of this experimental set up to prove the existence of an impedance mismatch injury mechanism is identifying a specimen material with a speed of sound comparable to that of bone, but requiring less energy to produce fracture.

The finite element modeling effort, on the other hand, has demonstrated the potential existence of a high-rate injury mechanism related to acoustic properties and the characteristics of a high speed compression wave. Anecdotal reports from theater suggest that comminuted injuries located at the mid-tibia can be explained by a combination of acoustic impedance mismatch between the steel of the vehicle floor and the bone of the lower extremity, and the consequences of high-rate loading of rate-dependent biological materials. Fracture patterns produced in the FE model align with those of underbody blast field injuries, and that cannot be explained by inertia-based loading. Additionally, eliminating the impedance mismatch in the model eliminated the strain and energy peaks which occur within the bone material, thus demonstrating the dependence of injury on the material through which a shock wave transmits energy.

While the key concepts of this injury mechanism are shown by the modeling approach taken, there are many additional factors to consider such as the tibial geometry and the inhomogeneity of bone. Furthermore, the soft tissues in the foot/ankle and materials in the boot have the potential to damp the high speed compression wave before it enters the bone. Future modeling and experimental efforts will determine their contribution in damping the transmission of the high speed compression wave. Future research will attempt to experimentally verify these finite element models. Moving forward, the results of this research may play a role in designing injury mitigation strategies for preventing under vehicle blast injuries, and should provide an explanation for some of the injuries reported from under vehicle blasts.

## References

1. Nelson T, Clark T, Stedje-Larsen ET, Lewis CT, Grueskin JM, Echols EL, Wall DB, Felger EA, Bohman HR (2008) Close proximity blast injury patterns from improvised explosive devices in iraq: a report of 18 cases. J Trauma 65:212–217
2. Quenneville CE, McLachlin SD, Greeley GS, Dunning CE (2011) Injury tolerance criteria for short-duration axial impulse loading of the isolated tibia. J Trauma 70(1):E13–E18
3. Yoganandan N, Pintar FA, Boynton M et al (1996) Dynamic axial tolerance of the human foot-ankle complex. SAE, Albuquerque
4. Ramasamy A, Hill AM, Hepper AE, Bull AMJ, Clasper JC (2011) Blast mines: physics, injury mechanisms, and vehicle protection. Philos Trans Roy Soc 155(4):689–698
5. Funk JR, Crandall JR, Tourret LJ, MacMahon CB, Bass CR, Khaewpong N, Eppinger RH (2001) The effect of active muscle tension on the axial injury tolerance of the human foot/ankle complex. In: Proceedings International Technical Conference on the Enhanced Safety of Vehicles, Amsterdam, Netherlands, 237
6. Arepally S, Gorsich D, Hope K, Gentner S, Drotleff K (2008) Application of mathematical modeling in potentially survivable blast threats in military vehicles, Army RDECOM-TARDEC unclassified
7. Weiss M, Ben-Shlomo AB, Hagag P, Rapoport M (2000) Reference database for bone speed of sound measurement by a novel quantitative multi-site ultrasound device. Osteoporos Int 11:688–696
8. Falk B, Galili Y, Zigel L, Constantini N, Eliakim A (2007) A cumulative effect of physical training on bone strength in males. Int J Sports Med 28:449–455
9. Miyatake N, Muta H, Murota C, Haga M, Fujii M (2002) Bone status assessment in Japanese subjects using speed of sound along the tibia. Chin Med J 115(2):254–257
10. McElhaney J (1966) Dynamic response of bone and muscle tissue. Journal of Applied Physiology 21(4):1231–1236
11. Untaroiu C, Kerrigan J, Kam C, Crandall J, Yamazaki K, Fukuyama K, Kamiji K, Yasuki T (2007) Correlation of strain and loads measured in the long bones with observed kinematics of the lower limb during vehicle-pedestrian impacts. Stapp Car Crash J 51
12. Bayraktar HH, Morgan EF, Niebur GL, Morris GE, Wong EK, Keaveny TM (2004) Comparison of the elastic and yield properties of human femoral trabecular and cortical bone tissue. J Biomech 37:27–35
13. Krone R, Schuster PS (2006) An investigation on the importance of material anisotropy in finite element modeling of the human femur. SAE International
14. Ramasamy A et al (2011) In-vehicle extremity injuries from improvised explosive devices: current and future foci. Philos Trans Roy Soc 366:160–170

# Chapter 20
# Indentation and Dynamic Compression Experiments on Microcrystalline and Nanocrystalline Nickel

**S. Rajaraman, Krishna N. Jonnalagadda, and P. Ghosh**

**Abstract** A lot of work has been done on nickel and its alloys in the past over a wide range of strain rates. In this work, the dynamic behaviour of microcrystalline (mc) and nanocrystalline (nc) nickel was studied through quasi-static and dynamic compression Kolsky bar experiments over the strain rate range of 0.004–6,450 $s^{-1}$. Nanoindentation experiments were performed on nc nickel to determine the strain rate sensitivity over a range of strain rates from 0.0125 to 0.1 $s^{-1}$. The effect of grain size on mechanical response was investigated via stress vs. strain curves. From the quasi-static experiments, it was observed that the yield strength of nc nickel was around four times that of mc nickel and this can be attributed to the Hall-Petch effect. The ductility of both the materials remained similar under compression (greater than 35%). Nanocrystalline nickel at high strain rates showed strain softening compared to strain hardening in mc nickel. Strain rate sensitivity at constant plastic strain for nc nickel was found to be higher by a factor of 2.37 compared to that of mc nickel over the strain rate range of 0.004–6,450 $s^{-1}$.

**Keywords** Nanocrystalline nickel • High strain rate experiments • Nanoindentation • Strain rate sensitivity (SRS)

## 20.1 Introduction

Nickel and its alloys are one of the most widely used materials because of its high strength at elevated temperatures. Mechanical properties measured from high strain rate experiments (strain rates $>10^2$ $s^{-1}$) differ drastically when compared to the properties measured from conventional quasi-static experiments (strain rates $<10^{-1}$ $s^{-1}$), as flow stress of the material depends on the strain, strain rate and temperature. This emphasizes on the need to conduct high strain rate experiments to determine the material behaviour at high rates of loading which are commonly encountered in impact, vehicle collision, forging, etc.

Muller [1] conducted the first set of high strain rate experiments on nickel (mean grain size 70 μm), both at room and high temperatures using split-Hopkinson pressure bar (SHPB) at strain rates ranging from 500 to 6,400 $s^{-1}$. He observed that flow stress increased with increase in strain rate and decreased with increasing temperature. Gray et al. [2] conducted high strain rate experiments on ultrafine-grained nickel at strain rates from 0.001 to 3,000 $s^{-1}$. They determined the strain rate sensitivity (SRS) at constant strain ($\partial \ln \sigma / \partial \ln \dot{\varepsilon}$, $\sigma$ being the flow stress at constant strain and $\dot{\varepsilon}$ being the strain rate) as 0.006 over the range of strain rates. Dirras et al. [3] investigated the microstructure evolution during high strain rate testing on nickel (mean grain size 3.9 μm) using direct impact Hopkinson bar. They reported dynamic recrystallization at the highest velocity of impact and reduction in percentage of the high angle grain boundaries (HAGB) and Σ3 CSL (coincidental site lattice) boundaries compared with the initial structure. Jonnalagadda [4] reported increase in twin density in coarse grained nickel (mean grain size 32 μm) when deformed at high strain rates and reported a SRS of 0.066 over the strain rates ranging from 2,600 to 8,400 $s^{-1}$. Gurao et al. [5] conducted high strain rate and quasi-static experiments on nickel (mean grain size 75 μm). They found an increase in the fraction of high angle grain boundaries (HAGB) at the expense of decrease

S. Rajaraman • K.N. Jonnalagadda (✉)
Indian Institute of Technology Bombay, Powai, Mumbai, India
e-mail: krishnajn@iitb.ac.in

P. Ghosh
Indian Institute of Science, Bangalore, India

V. Chalivendra et al. (eds.), *Dynamic Behavior of Materials, Volume 1: Proceedings of the 2012 Annual Conference on Experimental and Applied Mechanics*, Conference Proceedings of the Society for Experimental Mechanics Series, DOI 10.1007/978-1-4614-4238-7_20, 

in the low angle grain boundaries (LAGB). However, no reason was provided for the change in microstructure. They also observed that the volume fraction of <101> direction of orientation of atomic planes decreasing with increase in strain rate with saturation at higher strain rates. They also found microband formation and grain fragmentation at higher strain rates. In all these studies, it has been shown that coarse grained nickel exhibits various active deformation mechanisms at high strain rates and was found to be moderately rate sensitive, with its dependence on grain size and initial work hardening.

Over the past two decades, nanocrystalline metals have gained much interest because of their higher strength and high hardness compared with their coarse or ultrafine grained counterparts. Though they have better mechanical properties, the ductility was low in tension [6] owing to the fact that limited dislocation plasticity was seen in them and plastic instability in the form of shear bands or micro cracks created during specimen preparation (e.g., severe plastic deformation) causes early failure. Some authors reported high strength and high ductility in nc nickel [7, 8] by having a bimodal distribution of mc and nc grains. Dalla torre et al. [6] conducted high strain rate experiments in tension on nc nickel (mean grain size 21 nm) over the strain rates ranging from$10^{-5}$ to $10^{3}$ $s^{-1}$. They reported an ultimate tensile strength (UTS) ranging from 1,250 to 2,500 MPa over the range of strain rates, with UTS of nc nickel being 70% more than the UTS of coarse grained nickel and the SRS of nc nickel as 0.01–0.03 over the strain rates of $5.5 \times 10^{-5}$ to $5.5 \times 10^{-2}$ $s^{-1}$. They observed the presence of shear bands in nc nickel when deformed at high strain rates and observed no dislocations in the grains in transmission electron microscope (TEM). Wang et al. [9] performed shock experiments on nc nickel (grain size 30–100 nm) at strain rates greater than $10^{7}$ $s^{-1}$. They reported an increase in dislocation density with increase in shock pressure and the absence of deformation twins in the deformed sample. They also found an increase in hardness of the material after subjecting the material to shock loads. Wu et al. [10, 11] observed the inverse effect of grain size on twinning for nc nickel (mean grain size 25 nm). They found that with increasing strain rate from quasi-static tension ($3 \times 10^{-3}$ $s^{-1}$) to high strain rate compression by SHPB (2,600 $s^{-1}$), twin fraction increased at the expense of decrease in stacking faults and showed that for grains with diameter less than 16 nm, only stacking faults are created without twin formation in compression. They showed that for grain size smaller than 50 nm, there exists a particular grain size at which twinning as a mode of deformation is predominant.

Due to difficulties in fabricating bulk specimens of nc metals, nanoindentation experiments were alternatively used to measure the SRS ($\partial \ln H / \partial \ln \dot{\varepsilon}_i$, H being the hardness in GPa and$\dot{\varepsilon}$being the indentation strain rate) by measuring the hardness at various indentation strain rates. Schwaiger et al. [12] performed constant indentation strain rate and constant load rate experiments on nc nickel (mean grain size 20 nm) and found that the hardness increased with increase in strain rate. Gu et al. [13] reported a SRS of 0.033 for nc nickel (mean grain size 20 nm) and compared with the SRS obtained from uni-axial tension experiments. Maier et al. [14] performed nanoindentation experiments on nc nickel (mean grain size 25 nm) and reported a SRS of 0.052 from constant indentation strain rate experiments, as against a SRS of 0.019 from indentation strain rate jump experiments.

The current work focuses on determining the strain rate sensitive behaviour of nc nickel (mean grain size 17 nm) under compression over strain rates ranging from 0.007 to 5,215 $s^{-1}$ using an UTM and Kolsky bar setup. A comparison of results with that of mc nickel (mean grain size 48.44 μm) was performed to understand the operative mechanisms during plastic deformation. Nanoindentation experiments were also performed to measure the rate sensitivity of nc nickel at strain rates less than $10^{-1}$ $s^{-1}$.

## 20.2 Experimental Procedure

Commercially available coarse grained pure nickel was used for the experiments. It was produced through vacuum arc melting process and extruded into a rod of 12 mm diameter. For quasi-static experiments, cylindrical samples of diameter 5 mm and length 5 mm were used and for Kolsky bar experiments cylindrical samples of diameter 5 mm and length 2.7 mm were used. The samples were cut with its axis parallel to the extrusion direction using wire-cut electric discharge machine (EDM) and then polished. Nanocrystalline nickel in the form of plate of thickness 2.5 mm and area of 30 × 35 mm was produced using pulsed electro deposition technique with on/off time of pulse being 5/40 ms at 60 °C. The pH of electrolyte was maintained at 4 and saccharin was used as an additive in the electrolyte. For nanoindentation and quasi-static experiments, square samples of length 4.7 mm and thickness 2.15 mm were used and for Kolsky bar experiments cylindrical samples of diameter 4 mm and length 2 mm were used, with the samples cut in the direction parallel to the direction of deposition.

Grain size measurements for mc nickel were carried out using EBSD and bulk XRD was used for nc nickel. Mean grain size of mc nickel was found to be 48.44 μm and that of nc nickel in the range of 14–20 nm. Conventional Kolsky bar setup with bar diameter of 12.74 mm was used for dynamic compression experiments. The incident and the transmitted bar lengths were 1,511 and 1,351 mm, respectively. Pulse shaping and desired rise time of pulse were achieved using a pulse shaper and to ensure that specimen is loaded with a single compression pulse, momentum trap was used. A thin layer of molybdenum sulphide grease ($MoS_2$) was used to reduce friction at the specimen-bar interface. Strain gage signals were captured using digital data acquisition system and were corrected for wave dispersion effects [15, 16] using a Fast Fourier Transform (FFT) code written in Matlab$^{TM}$. Frequencies up to Nyquist frequency was taken for dispersion correction [17]. Data analysis from Kolsky bar experiments can be referred from the literature [18].

Quasi-static experiments were performed using a MTS universal testing machine with 100 KN load capacity. During the experiment a thin layer of lubricant was used at the specimen/platen interface to reduce friction. For nc nickel, two tungsten carbide inserts of diameter 10 mm and thickness 3 mm were used to sandwich the specimen between the compression platens (a small amount of lubricant at the specimen interface was also used), to avoid damage to platens due to punching effect. Compliance correction was performed using a maraging steel specimen for which Young's modulus was known a priori. Nanoindentation specimens were prepared by mechanical polishing using SiC paper of decreasing grit size up to 4,000 grit paper, followed by diamond polishing using 3 μm diamond paste and finally colloidal silica polishing using a suspension containing 0.5 μm particles with the surface of the specimen cleaned with acetone using lint free paper. Nanoindentation experiments were performed on Hysitron Triboindentor using Berkovitch indentor (tip radius 500 μm) for a depth of 1 μm. Indentations were performed in high load – constant displacement mode and indentation depth rates [19] were varied from 12.5 to 100 nm/s to attain indentation strain rates ranging from 0.0125 to 0.1 $s^{-1}$. Two indentations were performed for each strain rate and the average of measurements was used for the calculation of SRS.

## 20.3 Results and Discussion

The compression stress vs. strain curves for mc and nc nickel are shown in Fig. 20.1 over the strain rate range of 0.004–6,450 $s^{-1}$. It is clear that flow stress increased with increase in strain rate, which shows the dependence of flow stress on strain rate.

In case of mc nickel, one can easily observe the increase in strain hardening from quasi-static to high strain rate loading due to an increase in the dislocation density. Strain hardening rate defined by $\partial\sigma/\partial\varepsilon$ was found to decrease with increasing strain, indicating that increase in strain reduced the generation/motion of dislocations. Saturation of flow stress was seen for stress vs. strain curve at 5,637 $s^{-1}$for strains greater than 50%, which indicates saturation of dislocation density at large strains and higher strain rates. The observed behaviour could also be due to thermal softening at high accumulated strains.

In case of nc nickel, we observed a transition from limited strain hardening to strain softening with increase in strain rate from quasi-static to high strain rate deformation. Offset yield strength (at 0.2% strain) at 1,173 $s^{-1}$was found to be 1,990 MPa which was 13% higher than the offset yield strength of 1,755 MPa obtained at 0.004 $s^{-1}$. Reported offset yield strength at 0.004 $s^{-1}$ was found to be higher than the compression yield strength of 1,210 ± 20 MPa reported by

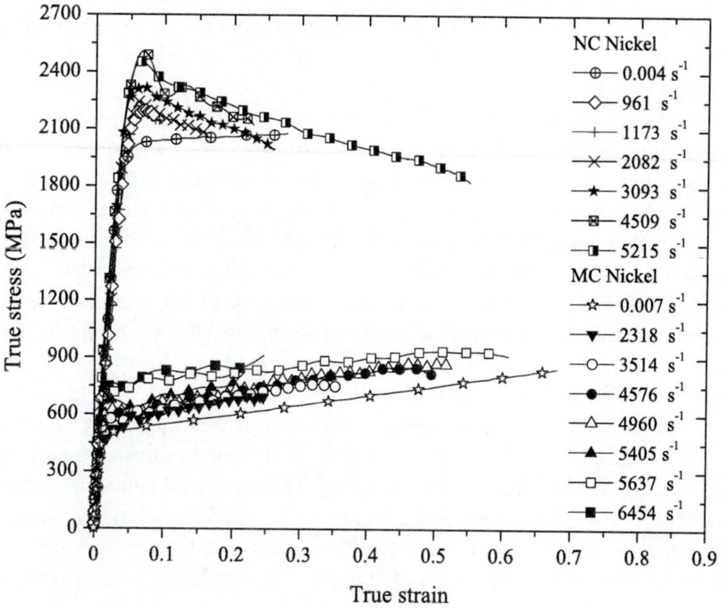


Fig. 20.1 Stress vs. strain curves of mc and nc nickel at various strain rates

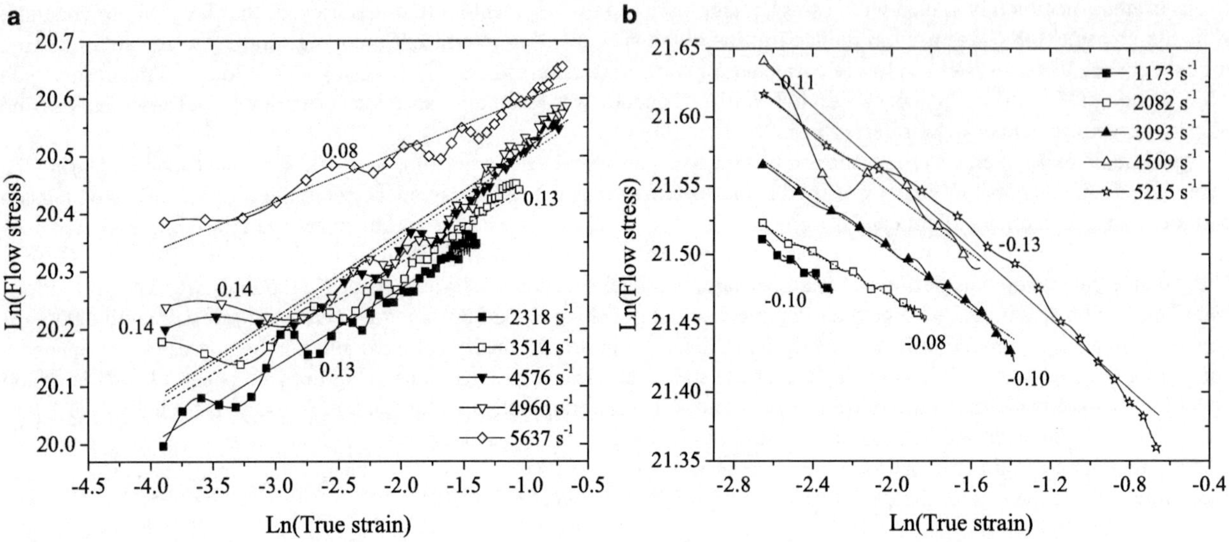


**Fig. 20.2** Strain hardening exponent for **(a)** mc and **(b)** nc nickel, at various strain rates

Schuster et al. [20] from their micro pillar compression experiments on nc nickel (mean grain size 29 nm). The difference in value was due to smaller grain sized nickel used in our experiments and also the larger size of the specimen. Observed yield strength under compression for nc nickel at high rate of loading (2,000–2,100 MPa) was comparable to the yield strength of nc nickel at high rate of loading under tension [6]. All the specimens deformed uniformly under compression exhibiting ductility as high as 55% at the highest strain rate (5,215 $s^{-1}$) achieved in this work. A similar behaviour of high compression ductility was reported in nc copper at high rates of loading [21] due to presence of higher fraction of LAGB in electrodeposited material. It was clearly revealed that nc nickel possess higher dynamic yield strength, which was around 4–5 times that of mc nickel. All the stress vs. strain curves were obtained till the end of loading and failure of sample was not observed in any experiment reported.

Strain hardening exponent ‘n’ was calculated for mc and nc nickel at various strain rates and are shown in Fig. 20.2a, b. It was found that strain hardening exponent showed a decreasing behaviour with increase in the strain rate for mc nickel. Strain hardening exponent ($n = 0.13$) for stress vs. strain curves at 2,318 $s^{-1}$ was comparable to the strain hardening exponent reported for mc nickel at 2,000 $s^{-1}$ [5]. At high strain rates, strain softening was seen in nc nickel compared to strain hardening in mc nickel. A similar behaviour of strain softening was reported by Jia et al. [22] in their high strain rate experiments on nc iron (mean grain size 80 nm), where it was found that formation of deformation shear bands caused instability, which caused the strain softening behaviour. No such deformation bands have been observed in this work. Further investigations with the aid of TEM are required to understand the microstructural variations causing this strain softening behaviour.

The measure of hardness is related to the flow stress of the material [23]. Therefore, nanoindentation experiments were performed on nc nickel to determine its hardness, which was subsequently used to determine the strain rate sensitivity of nc nickel over a range of indentation strain rates from 0.0125 to 0.1 $s^{-1}$. Since the deformation zone produced by nanoindentation experiments was very small compared to bulk compression, nanoindentation experiments offer the simplest way to determine the strain rate sensitivity for strain rates below $10^{-1}$ $s^{-1}$. Load vs. displacement curves obtained in nanoindentation experiment are shown in Fig. 20.3. Rate sensitivity of nc nickel was evident from the increase in load required to achieve the required indentation depth. Hardness value of nc nickel ranged from 4.71 to 5.58 GPa, with highest hardness attained for highest indentation strain rate (0.1 $s^{-1}$). Figure 20.4 shows the plot of hardness vs. indentation strain rate. After indentation, the sample was observed in SEM to observe the indent for pile-up or sink-in. SEM image of the indented region is shown as an inset in Fig. 20.4, which doesn’t show sink-in or pile-up near the indented region. SRS value of 0.021 is obtained in the current work, which is lesser than the value reported by Gu et al. [13].

The Hall-Petch relation ($\sigma = \sigma_0 + Kd^{-0.5}$, d being grain size in nm) constants for nickel were calculated and found to be, $\sigma_0 = 426$ MPa and $K = 5{,}479$ MPa$\sqrt{nm}$. The calculated values were more than those reported in the literature [24] because of prior cold working done on mc nickel. Figure 20.5 shows the logarithmic plot of true stress (at 3% plastic strain) vs. strain rate for mc and nc nickel. From the Fig. 20.1, it can be clearly seen that nc nickel exhibited strain softening compared to limited strain hardening observed at quasi-static strain rates. This limited the calculation of strain rate sensitivity only up to

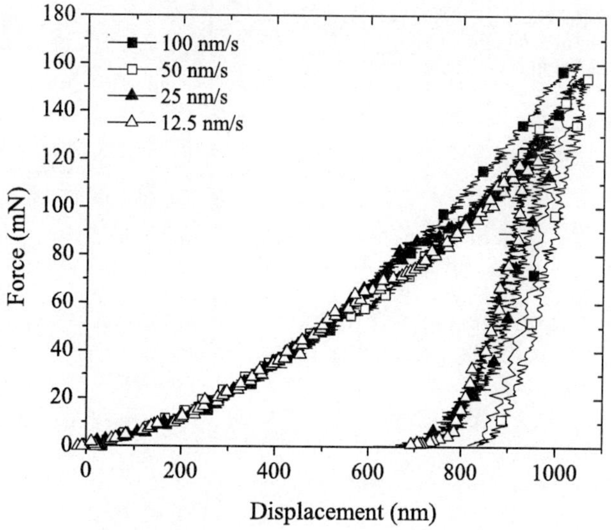


**Fig. 20.3** Load vs. displacement curves for nc Nickel obtained for various depth rates from indentation experiments

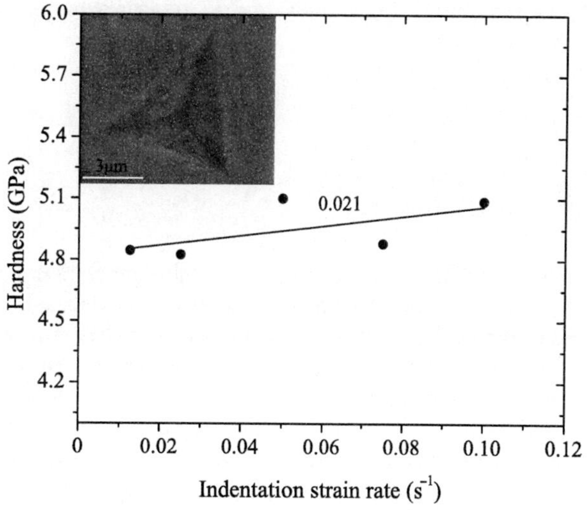


**Fig. 20.4** Plot of hardness vs. indentation strain rate for nc Nickel with the inset showing the SEM image of indent

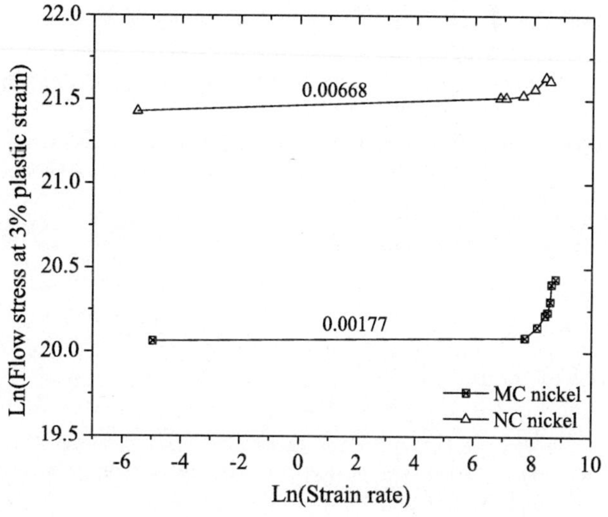


**Fig. 20.5** Plot of flow stress at 3% plastic strain vs. strain rate for mc and nc nickel in compression

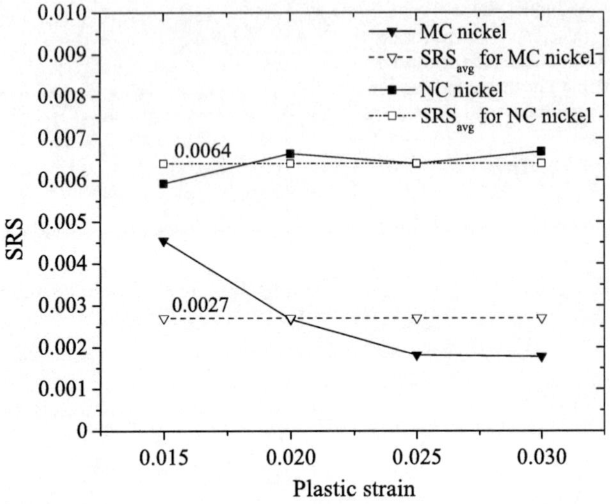


**Fig. 20.6** Plot of SRS vs. plastic strain for mc and nc nickel in compression

3% plastic strain where both mc and nc nickel showed strain hardening behaviour. Figure 20.6 shows the variation of SRS with plastic strain, with average SRS shown in dashed lines. From Fig. 20.6, it is clear that rate sensitivity of nc nickel was found to be higher by a factor of 2.37 compared to mc nickel over the strain rate regime of 0.004–6,450 $s^{-1}$. A similar behaviour was reported in nc aluminium by Khan et al. [25], with the SRS of nc aluminium being more than that of mc aluminium by a factor of 1.37, over the strain rate range of $10^{-4}$ to 2,470 $s^{-1}$.

## 20.4 Conclusion

Material response of mc and nc nickel subjected to high strain rate loading under compression was discussed in this paper. Effect of grain size strengthening was clearly established with nc nickel possessing higher strength than mc nickel. Both mc and nc nickel were found to be strain rate dependent with nc nickel being more strain rate sensitive (by a factor of 2.37) than mc nickel over the strain rate range investigated in this work. Although, poor ductility is usually reported from tensile experiments conducted on nc materials, current work shows good compression ductility of nc nickel comparable to mc nickel. Yield strength of nc nickel at high strain rates was found to be 4–5 times that of mc nickel, with nc nickel showing strain softening behaviour at high rates of loading. Reason for this strain softening behaviour, besides thermal softening could be attributed to the formation of adiabatic or deformation shear bands and the possibility of twinning and formation of stacking faults, which has to be investigated using TEM. There was a transition from strain hardening to strain softening behaviour for nc nickel over the strain rates of 0.004–5,215 $s^{-1}$. The temperature rise at high strain rates may have caused the formation of adiabatic shear bands, although no evidence was obtained through SEM imaging. This has to be however confirmed with the help of TEM. Sudden increase in flow stress of mc nickel around 5,000 $s^{-1}$ shows the possibility of dynamic recrystallization, which has to be further investigated.

**Acknowledgements** We would like to thank Prof. V. Parameswaran, IIT Kanpur for allowing us to use their Kolsky bar setup for dynamic compression experiments. We would like to thank Prof. A. H. Chokshi, IISc Bangalore for providing the materials for the work. We would like to thank the staff in high speed experimental mechanics lab (IIT Kanpur), Nanoindentation lab (IIT Bombay), FIST UTM lab (IIT Bombay), and OIM lab (IIT Bombay) for providing us the much needed support in this work. Finally, we would like to thank all the members of Mechanics of Materials Lab, IIT Bombay for the valuable discussions and support during the course of our work.

## References

1. Muller T (1972) High strain rate behaviour of iron and nickel. J Mech Eng Sci 14(3):161–167
2. Gray GT III et al (1997) Influence of strain rate & temperature on the mechanical response of ultrafine grained Cu, Ni and Al-4Cu-0.5Zr. Nanostruct Mater 9:477–480

3. Dirras G et al (2010) Fine-grained nickel deformed by direct impact at different velocities: microstructure and mechanical properties. Mater Sci Eng A 527:4128–4135
4. Jonnalagadda K (2011) Rate sensitivity in pure Ni under dynamic compression. Conf Proc Soc Exp Mech Ser 99:439–442
5. Gurao NP et al (2010) Effect of strain rate on evolution of the deformation microstructure and texture in polycrystalline copper and nickel. ASM Int Metall Mater Trans A 41A:2794–2804
6. Dalla Torre F et al (2002) Nanocrystalline electrodeposited Ni: microstructure and tensile properties. Acta Mater 50:3957–3970
7. Shen X et al (2008) High strength and high ductility of electrodeposited nanocrystalline Ni with a broad grain size distribution. Mater Sci Eng A 487:410–416
8. Gu C et al (2006) Enhanced tensile ductility in an electrodeposited nanocrystalline Ni. Scripta Mater 54:579–584
9. Wang YM et al (2006) Deforming nanocrystalline nickel at ultrahigh strain rates. Appl Phys Lett 88:061917
10. Wu XL, Zhu YT (2008) Inverse grain-size effect on twinning in nanocrystalline Ni. Phys Rev Lett 101:025503
11. Wu XL et al (2009) Strong strain hardening in Nano crystalline nickel. Phys Rev Lett 103:205504
12. Schwaiger R et al (2003) Some critical experiments on the strain-rate sensitivity of nanocrystalline nickel. Acta Mater 51:5159–5172
13. Gu CD et al (2007) Experimental and modelling investigations on strain rate sensitivity of an electrodeposited 20 nm grain sized Ni. J Phys D Appl Physs 40:7440–7446
14. Maier V et al (2011) Nanoindentation strain-rate jump tests for determining the local strain-rate sensitivity in nanocrystalline Ni and ultrafine-grained Al. J Mater Res 26(11):1423–1430
15. Gong JC et al (1990) Dispersion investigation in the Split Hopkinson pressure bar. J Eng Mater Technol 112(3):309–314
16. Follansbee PS, Frantz C (1983) Wave propagation in Split Hopkinson pressure bar. J Eng Mater Technol 105(1):61–66
17. Li Z, Lambros J (1999) Determination of the dynamic response of brittle composites by the use of the split Hopkinson pressure bar. Compos Sci Technol 59:1097–1107
18. Gray III GT (2000) Classic split-Hopkinson pressure bar testing. In: Mechanical testing and evaluation. ASM Handbook, vol 8. ASM International, Materials Park
19. Lucas BN, Oliver WC (1999) Indentation power-law creep of high purity indium. ASM Int Metall Mater Trans A 30:601–610
20. Schuster BE et al (2006) Microcompression of nanocrystalline nickel. Appl Phys Lett 88:103112
21. Jia D et al (2001) Compressive behaviour of an electrodeposited nanostructured copper at quasistatic and high strain rates. Scripta Mater 45:613–620
22. Jia D et al (2003) Effects of nanocrystalline and ultrafine grain sizes on constitutive behavior and shear bands in iron. Acta Mater 51:3495–3509
23. Tabor J (1951) The hardness of metals. Oxford University Press, New York
24. Thompson AW (1975) Yielding in nickel as a function of grain or cell size. Acta Mater 23:1337–1342
25. Khan AS et al (2008) Effect of grain refinement on mechanical properties of ball-milled bulk aluminium. Mater Sci Eng A 489:77–84

# Chapter 21
# Plastic Deformation of Ti-6Al-4V Plate over a Wide Range of Loading Conditions

**Jeremiah T. Hammer, Ravi S. Yatnalkar, Jeremy D. Seidt, and Amos Gilat**

**Abstract** Plastic deformation of Ti-6Al-4V plate stock is investigated under multiple loading conditions. The objective of the test program is to generate experimental data that can be used for the development and calibration of constitutive and failure models for numerical simulations of dynamic events. Uniaxial tension, compression and pure shear experiments are conducted at strain rates ranging from 1E-4 to 3,000 $s^{-1}$. Specimens are fabricated from both 6.35 mm (0.25 in.) and 12.7 mm (0.5 in.) thick plate stock. Compression and tension tests are conducted on specimens oriented in several different directions within the plate. The data shows significant strain rate sensitivity in tension, compression and shear. Both plates exhibit anisotropic plastic deformation behavior in tension and compression. The response of 6.35 mm plate is significantly different than the 12.7 mm plate in tension, compression.

**Keywords** Ti-6Al-4V • Plastic deformation • Strain rate sensitivity • Anistropy

## 21.1 Introduction

Jet engines are extremely complex mechanical systems which need to have certain precautions taken to ensure safe operation. Failure events such as blade-off and disk fracture can be catastrophic to the vehicle operation and can result in loss of life. In these events, a mechanical failure causes parts, typically a blade or disk fragments, to be propelled outward from the center of the engine. Since these machines are operating at high rotational velocities the energy released is extremely high and can cause immense damage to the vehicle, Fig. 21.1. Prior to an engine entering service it must be rigorously tested and analyzed.

Advancements in computational mechanics have allowed for the initial validation of an engineering design through numerical simulation. This is a cost efficient way to test complex designs prior to fabrication and has been implemented in most engineering design processes. Finite element codes, such as LS-DYNA, offer various options for rate and temperature dependent constitutive equations for deformation and failure. The well known Johnson-Cook [1] and the recently introduced MAT224 models are examples of such equations which both require accurate experimental data for calibration.

Titanium alloys are often used in the aerospace industry because of their high strength to weight ratio, thermal properties, and their resistance to corrosion. Ti-6Al-4V (Ti64) accounts for 50 % of all titanium alloys sold and is primarily used in aerospace applications [2]. This project studies the mechanical properties of Ti64 under various loading conditions in order to develop an accurate material model suitable for simulations of dynamic events, such as engine failure (Fig. 21.1).

Many researchers have previously tested and analyzed plastic deformation properties of Ti64 under different loading conditions, strain rates, and orientations. Results from the work by Follansbee, et al. [4] showed high strain-rate sensitivity for both yield and flow stresses for several heat treated microstructures. The work by Wulf [5] displayed linear strain rate dependence for Ti64 in compression at strain rates of $3 \times 10^3\ s^{-1}$ to $3 \times 10^4\ s^{-1}$. Lee et al. [6] show the behavior of Ti64 at strain rates of 0.02, 0.1, and 1.0 $s^{-1}$ in compression. The researchers found that yield strength increased with strain rate at similar temperatures. Another work by Lee et al. [7] investigates the effects of temperature at high strain rates on Ti64 which

J.T. Hammer • R.S. Yatnalkar • J.D. Seidt (✉) • A. Gilat
Department of Mechanical and Aerospace Engineering, Scott Laboratory, The Ohio State University,
201 W 19th avenue, 43210 Columbus, OH, USA
e-mail: seidt.2@osu.edu

V. Chalivendra et al. (eds.), *Dynamic Behavior of Materials, Volume 1: Proceedings of the 2012 Annual Conference on Experimental and Applied Mechanics*, Conference Proceedings of the Society for Experimental Mechanics Series,
DOI 10.1007/978-1-4614-4238-7_21, 

**Fig. 21.1** Damage to turbofan engine and subsequent fuselage damage following a disk failure [3]

show similar results. They also found that the work hardening of the material decreased considerably as strain rate increased. Lee et al. [8] also investigates the shear properties of Ti64 at strain rates of $1.8 \times 10^3$ to $2.8 \times 10^3$ s$^{-1}$. This work once again demonstrated the material has a significant temperature and strain rate dependence.

Titanium alloys are known to have different mechanical properties depending on the material processing. These variations can cause substantial errors in numerical simulations if assumed constitutive and fracture models are employed. It is critical to calibrate these material models with experiments conducted on the actual material used in the eventual application. The goal of this work is to characterize the strain rate and anisotropic strength behavior for two separate plate stocks.

## 21.2 Material, Specimens, and Testing Plan

The plate stocks listed in Table 21.1 come from two different sources both within specified Aerospace Material Standards (AMS) parameters. Tests are conducted on a 6.35 and 12.7 mm plate (plate 1 and 2, respectively) at several strain rates and in various orientations within the plates. Tension, compression, and torsion tests are conducted according to the tabulated text matrix seen in Table 21.2.

Strain Rate characterization includes strain rates from 1.0E-4 to 1.5E3 s$^{-1}$ for tension and torsion and up to 3.8E3 s$^{-1}$ for compression. Quasi-static tests (<1.0E0 s$^{-1}$) are conducted on a servohydraulic biaxial load frame. Load data is taken from a load cell mounted in series with the specimen. High strain rate tests are performed on a compression Split-Hopkinson bar [9], direct tension Split-Hopkinson bar [10], or direct torsion Split-Hopkinson bar [11]. Torsion tests are only performed on the plate 2 stock due to difficulties that arise from manufacturing spool samples from thin plate. The Anisotropy test series consists of tension and compression tests on specimens oriented in rolled direction as well as 45°, 90° (transverse), and 135° from rolled directions. The anisotropic test series is conducted at a strain rate of 1.0 s$^{-1}$. Torsion samples are fabricated such that the axis of symmetry of the spool specimen was oriented in the thickness direction of the plate. Three dimensional digital image correlation (DIC) is used to measure the full field surface strains in all the tests.

## 21.3 Experimental Results

Although both plates are within AMS specifications, they show notably different plastic deformation characteristics. The material shows significant strain rate dependence in tension, compression, and shear as seen in Fig. 21.2. Plate 2 shows higher yield and flow stresses compared to plate 1 in tension and compression at similar strain rates. In tension, the specimens from both plates have similar fracture strains and strain hardening characteristics. In compression the

**Table 21.1** Material composition of the various plate stocks tested

| Plate No. | Plate thickness | Material source | Ingot source | Rolling source | Chemical makeup Al | V | Fe | O | C | N | Ti |
|---|---|---|---|---|---|---|---|---|---|---|---|
| 1 | 6.35 mm | Tri-tech | RTI | Allegheny Ludium | 6.13 | 3.97 | 0.18 | 0.173 | 0.016 | 0.006 | Bal |
| 2 | 12.7 mm | Titanium Ind. | Timet | Timet | 6.64 | 4.04 | 0.13 | 0.190 | 0.011 | 0.006 | Bal |

**Table 21.2** Experimental Outline for the Plastic Deformation Testing

| Test no. | Loading mode | Testing apparatus | Strain rate (1/s) | Specimen orientation | Temp (°C) | Plate stock no. |
|---|---|---|---|---|---|---|
| 1 | Tension | Hydraulic load frame | 1.0E-4 | Rolled | Room temp | 1, 2 |
| 2 | | | 1.0E-2 | | | |
| 3 | | | 1.0E0 | | | |
| 4 | | Kolsky bar | 5.0E2 | | | |
| 5 | | | 1.5E3 | | | |
| 6 | | Hydraulic load frame | 1.0E0 | 45° from rolled | | |
| 7 | | | | Transverse | | |
| 8 | | | | 135° from rolled | | |
| 9 | Compression | Hydraulic load frame | 1.0E-4 | Rolled | Room temp | 1, 2 |
| 10 | | | 1.0E-2 | | | |
| 11 | | | 1.0E0 | | | |
| 12 | | Kolsky bar | 1.5E2 | | | |
| 13 | | | 3.0E3 | | | |
| 14 | | Hydraulic load frame | 1.0E0 | 45° from rolled | | |
| 15 | | | | Transverse | | |
| 16 | | | | 135° from rolled | | |
| 17 | | | | Through | | |
| 18 | Torsion | Hydraulic load frame | 1.0E-4 | Through | Room temp | 2 |
| 19 | | | 1.0E-2 | | | |
| 20 | | | 1.0E0 | | | |
| 21 | | Kolsky bar | 5.0E2 | | | 2 |
| 22 | | | 1.5E3 | | | 2 |

hardening behavior of the two plates is similar in all tests. At low strain rates (<1.0E-2 $s^{-1}$) after an approximate strain of 0.28, plate 2 fails in shear at a 45° angle. Plate 1 continues to deform and does not fail below a strain of 0.45.

The materials show significant anisotropic effects in tension and compression as seen in Fig. 21.3. For tension both plates have noticeably lower flow stresses for 45° and 135° from rolled directions. Plate 1 has comparable rolled and transverse flow stresses, while plate 2 shows a slight difference. Both plates have similar fracture strains in tension.

Compression data from these plates indicate a much more complex anisotropic effect. Plate 2 shows a comparably isotropic properties with only slight deviation in flow stress for all tests performed in the rolled, 45°, transverse, and 135° directions. Similar hardening curves are seen from the rolled and transverse as well as the 45° and 135° directions. Plate 1 has similar flow stresses for the rolled, 45°, and 135° directions up to true strains of 0.3. The transverse direction has significantly higher flow stresses and lower fracture strains then the other directions tested. Through direction stresses for plate 1 are higher than those in the rolled direction unlike plate 2 where they are lower. Plate 2 transverse specimens sustain much larger strains than those from plate 1 which fracture in shear at a true strain of 0.24.

## 21.4 Summary and Conclusions

A test series was designed and executed to determine the strain rate sensitivity and anisotropic strength characteristics of two Ti-6Al-4V plate products from different manufacturers (6.35 and 12.7 mm thick). Both plate stocks are strongly sensitive to strain rate in tension and compression. The 12.7 mm plate also exhibits rate sensitivity in shear (torsion). Higher flow

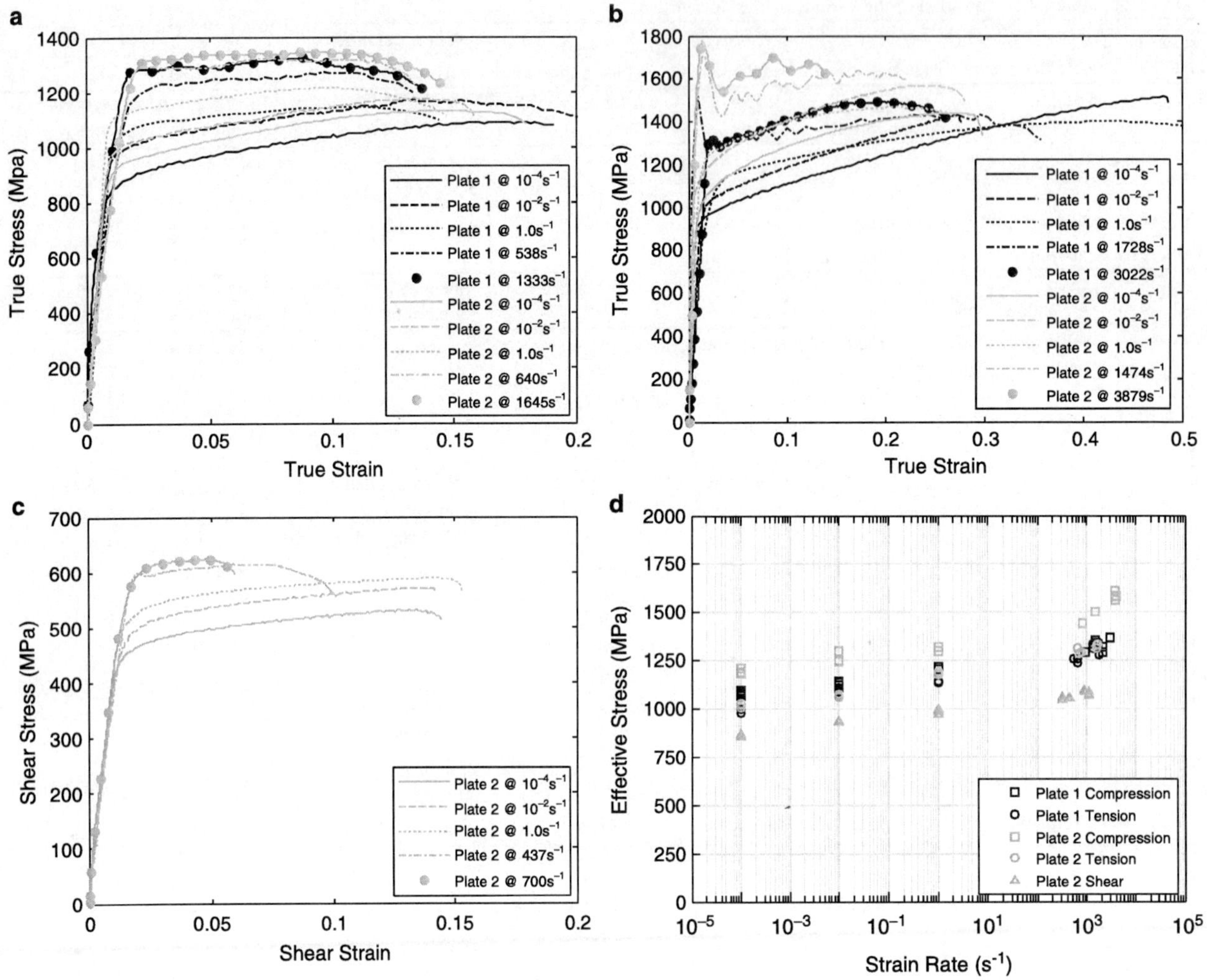


**Fig. 21.2** Experimental data from the strain rate dependence test series: (**a**) tension, (**b**) compression, (**c**) shear (**d**) effective stress for each loading condition at various strain rates at an equivalent strain of 0.05

stresses are seen in plate 2 as compared to plate 1 in tension and compression. Both plates showed anisotropic behavior with regard to both plasticity and fracture. The two plates showed slight differences in anisotropic behavior in tension, however, the compression properties are considerably different.

The results of these experiments demonstrate that this material is an anisotropic material which is also strain rate sensitive. Ti-6Al-4V is a complex material that can have drastically different material properties depending on its processing. These differences, however, may go unnoticed if standard tensile testing is conducted in only one or two directions. Dynamic finite element models of this material must employ rate sensitive and anisotropic material models that are calibrated from tests on the actual material used in the application. Otherwise, numerical results may contain significant errors.

**Acknowledgements** The research was supported by the U.S.A. Federal Aviation Administration, Grant No. 2011-G-003. The Authors are grateful to Mr. Don Altobelli, Mr. William Emmerling, and Dr. Chip Queitzsch for their support and involvement.

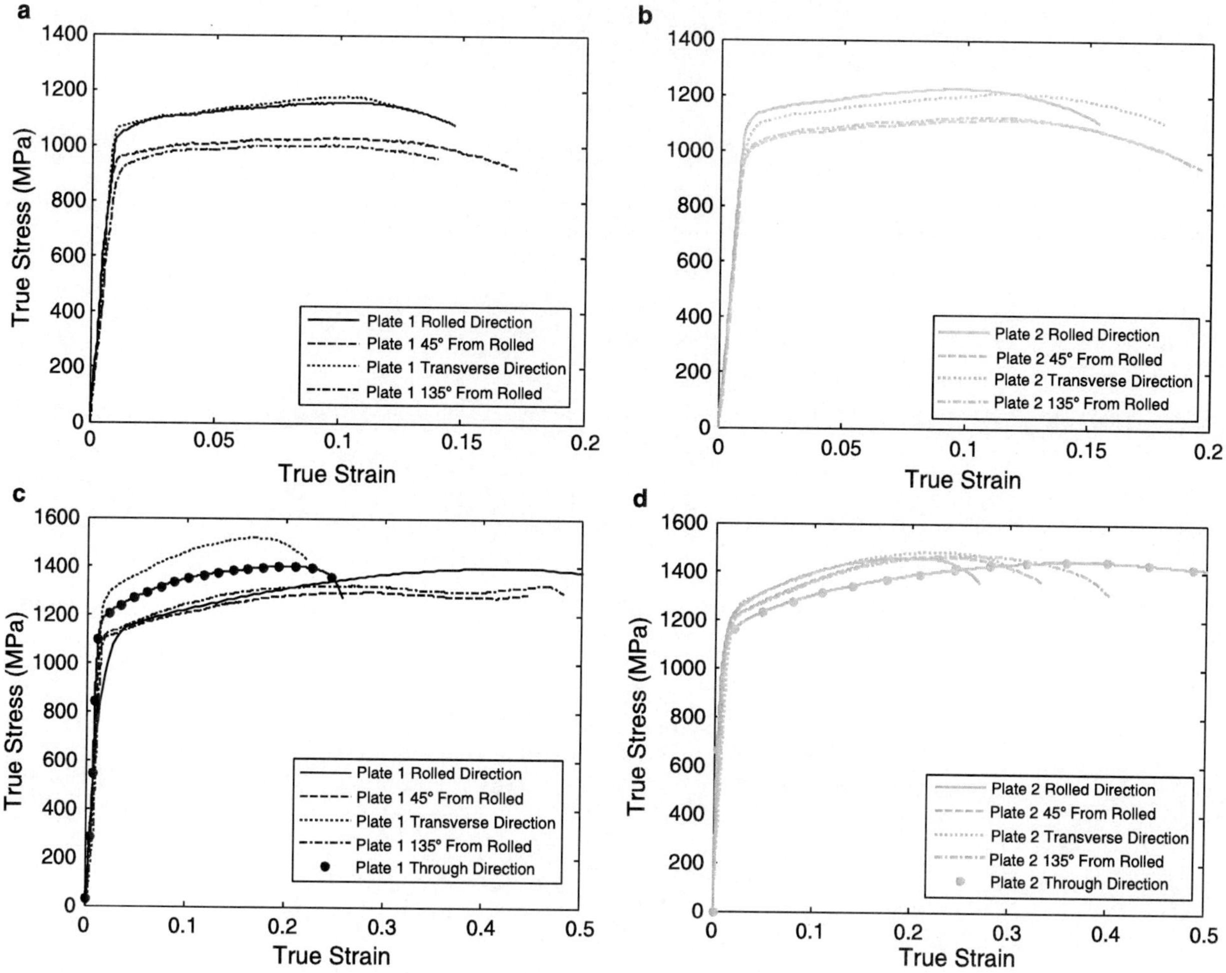


**Fig. 21.3** Experimental data from the anisotropic test series: (**a**) plate 1 tension, (**b**) plate 2 tension, (**c**) plate 1 compression, (**d**) plate 2 compression

# References


1. Johnson G, Cook W (1983) A constitutive model and data for metals subjected to large strains, high strain rates, and high temperatures. In: Proceedings of the 7th international symposium on ballistics
2. Boyer R, Welsch G (1994) Material properties handbook – titanium alloys. ASM International, Metals Park
3. NTSB (2004) Fail safe design requirements and features, regulatory requirements. AIAA Conference, July
4. Follansbee P, Gray G III (1989) An analysis of the low temperature, low and high strain-rate deformation of Ti-6Al-4V. Metall Trans A 20A:863–874
5. Wulf G (1976) High strain rate compression of titanium and some titanium alloys. Int J Mech Sci 21:713–718, July
6. Lee W-S, Lin M-T (1995) The effects of strain rate and temperature on the compressive deformation behaviour of ti-6al-4v alloy. J Mater Process Technol 71:235–246
7. Lee W-S, Lin C-F (1998) Plastic deformation and fracture behavior of ti-6al-4v alloy loaded with high strain rate under various temperatures. Mater Sci Eng 241:48–59
8. Lee W, Lin C, Huang S (2005) Effect of temperature and strain rate on the shear properties of ti-6al-4v alloy. J Mech Eng Sci 220:127–136
9. Gray III GT (2000) Classic Split-Hopkinson pressure bar testing. In: ASM Handbook, vol 8. ASM International located in Materials Park, Ohio
10. Staab GH, Gilat A (1991) A direct-tension Split Hopkinson bar for high strain-rate testing. Exp Mech 31:232–235
11. Gilat A (2000) Torsional Kolsky bar testing. In: Mechanical testing. ASM Handbook, vol 8. ASM International located in Materials Park, Ohio

# Chapter 22
# A Miniature Impact Tester for Dynamic Testing of Small Specimens

**J.F. Liu, V.P.W. Shim, and V.B.C. Tan**

**Abstract** A novel dynamic testing methodology was conceptualized and a portable miniature impact tester was fabricated for obtaining the dynamic force-displacement or stress-strain response of small specimens such as single solder joint. The tester comprises an input and an output bar; a tubular striker impinges an impact plate initially in contact with the specimen to generate a tensile or compressive load. Careful alignment is required only for the striker, the impact plate and the ends of the two bars adjacent to the specimen. The mechanical response of a specimen is derived from the stress waves in bars, and wave dispersion or attenuation in bars has less influence on the results compared to a split Hopkinson bar system. Theoretical derivation and interpretation of the governing equations was presented, with respect to calculation of results from recorded wave signals. FEM simulation was performed to examine the loading wave generated at the impact plate and its planar nature during propagation along the bars was verified. Experimental tests on cylindrical aluminum specimens showed that the results obtained using the miniature impact tester are comparable to those obtained from a split Hopkinson bar system, validating its performance.

**Keywords** Dynamic testing • Miniature impact tester • Stress-strain response • Small specimen

## 22.1 Introduction

Experimental measurements of the force-displacement or stress–strain responses of specimens are necessary for characterizing the mechanical properties of a material, and to formulate constitutive relationships to serve as inputs to FEM modeling. Appropriate equipment is needed for accurate data measurements at different loading speeds. The response of a specimen subjected to quasi-static loading can be determined using a universal testing machine; inertial and wave propagation effects in specimens or the testing device need not be considered. In contrast, the testing of a specimen under impact or dynamic loading is generally based on stress wave concepts; for example, the commonly used split Hopkinson bar system was based on one-dimensional wave theory [1–4].

A split Hopkinson bar system comprises a cylindrical striker and two straight slender bars that are supported such that they are coaxial so that one-dimensional wave propagation in the bars can be assumed. The specimen is sandwiched between the two bars and usually has size similar or comparable to the bar diameter. However, there are small specimens such as solder joints in IC packages that have aroused interest in understanding their dynamic mechanical behavior [5–9]. A single solder joint can be measured only about 0.5 mm or less in diameter, posing potential difficulties in dynamic testing. Even if thin bars (ϕ5 mm) are employed, their cross-sectional area is still significantly larger than that of a solder joint. Testing of bulk solder alloy specimens is easier, as such specimens can be made large enough to fit the split Hopkinson bar system used. However, the stress-strain curves obtained cannot be used to describe the mechanical response of a solder joint. This is because, solder joint is a complex structure and its mechanical properties are determined by many factors such as the properties of the solder bump, copper pad, substrate and intermetallic compound (IMC), as well as the strengths of

J.F. Liu (✉)
Temasek Laboratories @ NTU, Nanyang Technological University, 50 Nanyang Avenue, Singapore 639798, Singapore
e-mail: jianfeiliu@ntu.edu.sg

V.P.W. Shim • V.B.C. Tan
Impact Mechanics Laboratory, Department of Mechanical Engineering, National University of Singapore, Singapore 119260, Singapore

V. Chalivendra et al. (eds.), *Dynamic Behavior of Materials, Volume 1: Proceedings of the 2012 Annual Conference on Experimental and Applied Mechanics*, Conference Proceedings of the Society for Experimental Mechanics Series,
DOI 10.1007/978-1-4614-4238-7_22, 

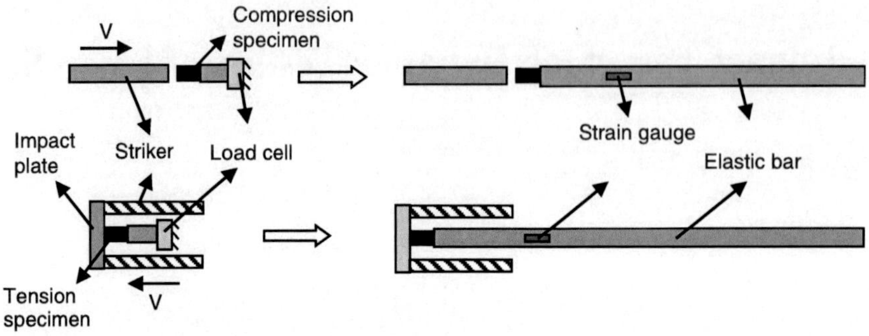


**Fig. 22.1** Replacement of load cell by an elastic bar to record the loading pulse

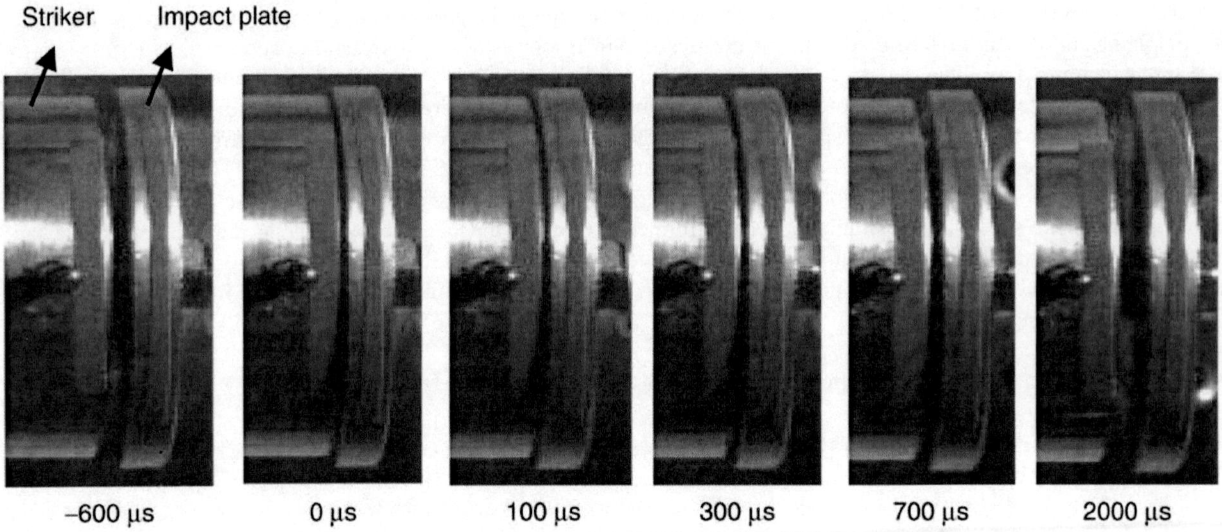


**Fig. 22.2** High speed camera footage at 10,000 fps showing impact between the striker and the impact plate

the interconnections between these components. Therefore, dynamic testing of actual single solder joint specimens is necessary to obtain their force-deformation responses.

There are some problems associated with tests using direct impact approach. From a review of literature, it is noted that some devices used for dynamic tests on solder joint specimens employ load cells [10, 11]. However, the loading rate is limited because of the frequency response of such devices [12]. To overcome this limitation, a method is to replace the load cell by a solid cylindrical bar bonded with strain gauges. A schematic drawing is illustrated in Fig. 22.1, which shows a direct impact approach for compression or tension. A striker that impacts a specimen directly corresponds to a compression test, while impacting against an impact plate generates tension in the specimen which is connected with the impact plate.

The loading history experienced by the specimen can be derived from the strain gauge readings – similar to that for a split Hopkinson bar system. The measurable duration of the loading wave is determined by the length of the bar; a long bar can accommodate a long loading pulse which is required for dynamic tests at relatively low strain-rates. The overall average deformation of a specimen can be calculated from the speeds of the two contact interfaces between the specimen and the bar/plate; one is equal to the speed of the striker/plate, while the other can be derived from the strain gauge readings. Obviously, accuracy of the deformation history derived will depend on accurate measurement of the striker/plate speed.

In this direct impact mode, the initial velocity of a striker/plate (just before impact commences) is used to calculate the deformation history. However, for hard specimens, the speed of the contact interface between specimen and striker/plate, will be varies and decreased from its initial value due to resistance from the specimen; even for a compliant specimen, this also may not be accurate. To demonstrate this, a tensile test was performed whereby a 570 g steel striker was propelled to impact a 40 g steel impact plate at a speed of 2.69 m/s (no specimen was mounted). A high speed camera was used to capture the impact between the striker and the impact plate at a framing rate of 10,000 fps (Fig. 22.2). It shows that soon after the striker makes contact with the plate, the plate separates and moves away from the striker at a higher speed. From the images, there appears to be no discernible separation between the striker and the plate within the initial 100 μs after commencement of impact. However, separation is obvious at around 300 μs. From careful examination of the images, it was observed that the separation gap is about 0.96 mm at 700 μs, and about 2.4 mm at 2,000 μs. The residual speeds after separation were measured and found to be 2.25 m/s for the striker and 3.44 m/s for the plate. This can be explained from the law of conversation of momentum of the striker and the plate, from which the residual speed is 2.33 m/s for the striker and 5.02 m/s for the impact plate. The small discrepancy is acceptable, as friction was not taken into account in the calculations.

This phenomenon reveals that when a striker impinges an impact plate (such as in tension configuration) instead of the specimen directly, the speed of the specimen-plate interface is different from the initial speed of striker and varies according to the compliance of the specimen. Hence, this configuration involving an impact plate is not suitable for dynamic tensile tests; it is also unsuitable for tests where the deformation history of a specimen needs to be measured accurately.

When a specimen is very small and experiences brittle fracture under tension (such as solder joint), it requires the test equipment to have a high sensitivity and frequency response to capture short loading pulses as well as small deformations. Therefore, in this study, a novel dynamic testing methodology is proposed and a miniature impact tester is fabricated for testing of small specimens. It is capable of both compressive and tensile testing, and the deformation or strain history is derived from the wave signals recorded by strain gauges, instead of the velocity of the striker. The configuration of this tester is somewhat different from a conventional split Hopkinson bar system. It contributes an alternative approach to impact testing and has some special features that make it convenient for extension to an ultra-long-thin bar system for testing very small specimens.

## 22.2 Principles Governing the Miniature Impact Tester

In a split Hopkinson bar system, a specimen is mounted between an input and an output bar. A striker impacts the other end of the input bar and generates a loading wave which travels along the entire input bar before loading the specimen. Similarly, the miniature impact tester also comprises an input and an output bar with a specimen sandwiched between them. However, as shown in Fig. 22.3, an impact plate is connected to the end of the input bar; a hollow striker is propelled to impinge the impact plate.

With this configuration, two loading waves are initiated at the impact plate and propagate in opposite directions along the two bars. Therefore, the strain gauges on the bars can be mounted close to the specimen once the stress in the bar cross

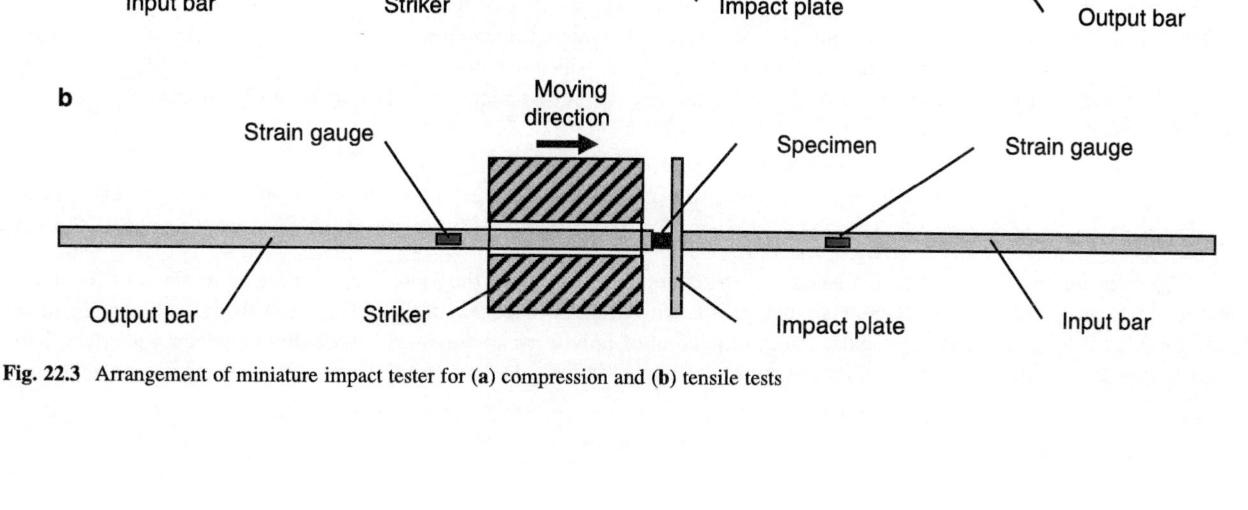


**Fig. 22.3** Arrangement of miniature impact tester for (a) compression and (b) tensile tests

section becomes uniform, because only the input and output waves are used in calculation of the specimen load and deformation. Compared to a conventional split Hopkinson bar system, the short distance between the specimen and the strain gauges decreases the degree of discrepancy between recorded signals and the actual loading pulse applied on the specimen (e.g. by wave dispersion or attenuation [13, 14]). In a split Hopkinson bar, straightness and alignment between the striker and the bars is a strict requirement. However, with the miniature impact tester, waves that propagate away from the strain gauges are not involved in calculations. This implies that good alignment between the striker and the bars is only strictly required within a small vicinity of the specimen, i.e., between the positions of strain gauges. These characteristics make the miniature impact tester more convenient for use with long and thin bars. Another advantage of this tester is that it can induce both compression and tension via a simple rearrangement of the impact plate; i.e. by detaching the impact plate from input bar and connecting it to the output bar, a compressive mode is converted to a tensile one, as shown in Fig. 22.3.

A conventional split Hopkinson bar system is based on one-dimensional wave theory and assumes equilibrium of stress in the specimen, thus making it possible to employ only the input and output waves to calculate the stress, strain and strain-rate history of the specimen. These principles are also adopted for the miniature impact tester. The governing equations for this tester are derived with reference to the schematic diagram in Fig. 22.3a for a compression configuration, where tensile wave is recorded as positive signal on oscilloscope. Denoting the speeds of the specimen-bar interfaces at position $x_1$ and $x_2$ as $v(x_1, t)$ and $v(x_2, t)$, the average strain rate of a specimen with initial length $L_s$ is,

$$\dot{\varepsilon}_s(t) = (v(x_2, t) - v(x_1, t))/L_s \tag{22.1}$$

where $v(x_2, t)$ corresponds to the output strain signal $\varepsilon_t(t)$recorded by the strain gauges on the output bar,

$$v(x_2, t) = -C_0\varepsilon_t(t) \tag{22.2}$$

where $C_0$ is the elastic wave velocity in the bars. The particle velocity $v(x_1, t)$on the input bar-specimen interface is the resultant of two components ($v(x_1, t) = v_1(x_1, t) + v_2(x_1, t)$) corresponding to two sources, i.e. the impact of striker on the impact plate and the resistance of the specimen. When the striker makes contact with the impact plate (without taking into account the resistance of the specimen), it causes the plate to attain a common particle velocity $v_1(x_1, t)$. $v_1(x_1, t)$ corresponds to the tensile wave traveling towards the free end of the input bar, and is recorded by the strain gauge which yields the signal $\varepsilon_{i1}(t)$. On the other hand, resistance from the specimen is generated when compression commences. This induces the particle velocity $v_2(x_1, t)$ at the interface with the impact plate, and this corresponds to a compressive pulse propagating in the same direction as $\varepsilon_{i1}(t)$. The wave is defined as $\varepsilon_{i2}(t)$ and is also recorded by the strain gauge on the input bar. Thus, the input wave $\varepsilon_i(t)$recorded by the strain gauge on the input bar is actually a superposition of the two pulses, i.e. $\varepsilon_i(t) = \varepsilon_{i1}(t) + \varepsilon_{i2}(t)$. As the impact plate is thin and positioned at the input bar–specimen interface, it is acceptable to make the assumption that $v_1(x_1, t)$and $v_2(x_1, t)$ are generated simultaneously at the interface. Therefore, the velocity $v(x_1, t)$at the input bar-specimen interface is obtained from the superposition of these two waves:

$$v(x_1, t) = v_1(x_1, t) + v_2(x_1, t) = C_0[\varepsilon_{i1}(t) + \varepsilon_{i2}(t)] = C_0\varepsilon_i(t) \tag{22.3}$$

By substituting Eqs. 22.2 and 22.3 into Eq. 22.1, the average strain rate in the specimen can be obtained, and thus the corresponding average specimen strain is derived from integration of the strain rate.

$$\dot{\varepsilon}_s(t) = (v(x_2, t) - v(x_1, t))/L_0 = -C_0(\varepsilon_t(t) + \varepsilon_i(t))/L_s \tag{22.4}$$

Since it is difficult to separate$\varepsilon_{i2}(t)$ from$\varepsilon_{i1}(t)$, the force at the input bar-specimen interface (=$A_0E_0\varepsilon_{i2}(t)$, where $A_0$ is the bar cross sectional area) cannot be obtained. This is not a problem, since stress equilibrium in the specimen is assumed. Thus, the average stress history $\sigma_s(t)$experienced by the specimen with cross sectional area$A_s$ can be obtained solely from the output wave $\varepsilon_t(t)$,

$$\sigma_s(t) = \frac{A_0}{A_s}E_0\varepsilon_t(t) \tag{22.5}$$

If the material and cross sectional area of the striker are same as that of the input bar, then the component of the input wave $v_1(x_1, t)$ should correspond to a rectangular profile with a length twice that of the striker. As a sufficiently long loading duration is necessary to deform the solder joint to the point of failure for loading at relatively low impact speeds (less than 1 m/s), another technique is employed for the miniature impact tester – use of a heavy striker. This makes it convenient for

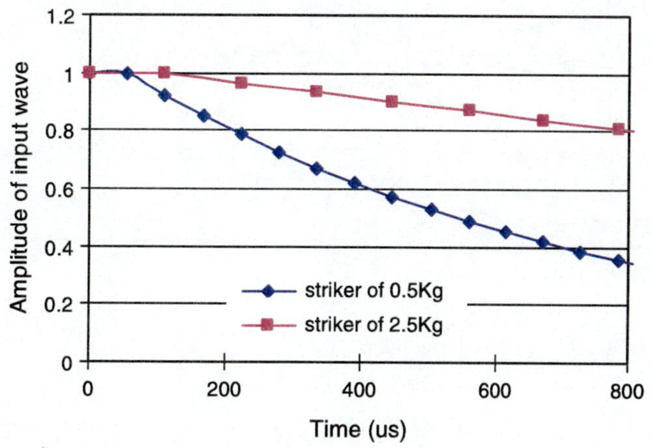


**Fig. 22.4** Predicted profile of the input wave generated by two heavy strikers of different mass

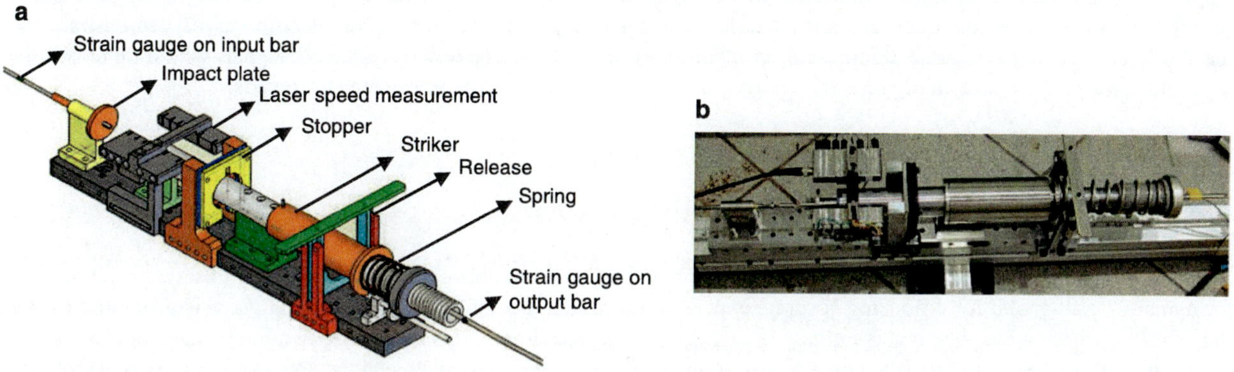


**Fig. 22.5** (**a**) schematic assembly and (**b**) picture of the miniature impact tester for dynamic compression and tension testing

the miniature tester to be constructed as an ultra-long thin bar system for moderate speed testing, whereby a long loading wave is generated by a short thick heavy striker with a large radius-length ratio made of a dense material, as illustrated in Fig. 22.3b.

This technique of employing a heavy striker to impact a light input bar had been used previously to generate an extended input wave, and the wave profile can be theoretically estimated as described by [15]. In a split Hopkinson bar, the mechanical impedances of the striker and the input bar are usually made equal through similarity in cross-sectional area and material. This results in the striker coming to rest after the stress wave generated upon impact travels to the other end of the striker and back. In contrast, if a thick, heavy tubular striker is used to impact a light, thin input bar, a long loading wave could be generated only by a short striker. Two steel strikers of 28 cm long but with different thickness were fabricated for the miniature impact tester. The corresponding input waves in elastic bar up to 800 μs were predicted as illustrated nondimensionalizedly in Fig. 22.4. It is seen that the amplitude of input wave generated by the 2.5 kg striker is higher and decrease less than that by 0.5 kg striker. Currently a 2 m long input bar is employed in the miniature impact tester – this corresponds to an input wave of about 800 μs in duration before superposition with the reflected wave occurs.

Figure 22.5 shows the schematic assembly designed and a picture of the miniature impact tester that was fabricated. A series of laser emitter-receiver pairs are sited along the axis of the bar to measure the speed of the striker before impact. The two ends of the input and output bars are supported through two precisely aligned supports, to ensure coaxiality during impact. A linear bearing is incorporated into the support of the output bar; this minimizes friction with the output bar end to improve test accuracy. Two typical input waves generated by this tester are illustrated in Fig. 22.6 for a low and a high speed impact test, showing the almost constant amplitude. The input wave (700 μs) was generated by a 28 cm long 2.5 kg steel striker against a $\phi$5 mm $\times$ 2 m steel bar; the decrease at the fall of the wave is caused by superposition of the input and reflected waves due to limited input bar length. As the cross sectional area of the striker is much larger than that of the bar,

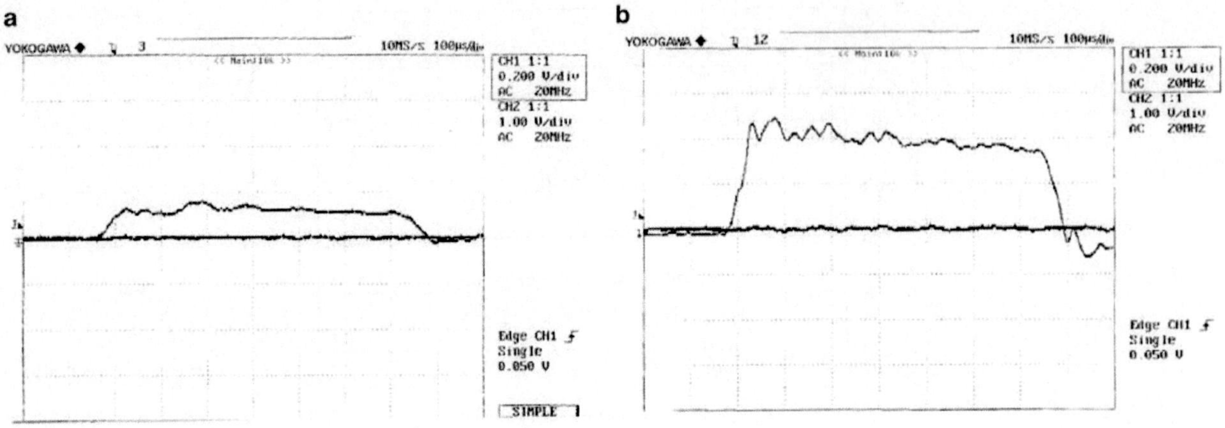


**Fig. 22.6** Input waves corresponding to (**a**) low and (**b**) high speed impact respectively

the particle velocity induced in the input bar is close to the value of the striker velocity. This is different from a split Hopkinson bar system, whereby the amplitude of the input wave is approximately half the striker velocity when the bar and striker have same cross sectional area and material. The relatively constant amplitudes of both curves demonstrate the feasibility of imposing a constant deformation speed on small specimens. The relatively smooth signals with little noise also shows that waves generated in the thin bar can be obtained, even for very low striker impact speeds.

## 22.3 Numerical and Experimental Validation

As a new experimental technique, analysis and validation of the miniature impact tester are necessary to assess its feasibility for dynamic testing, and for producing accurate experimental results. For this tester, the loading wave is generated by the impact of a striker on the plate that is initially in contact with the specimen. This may cause two immediate concerns about the validity of the test. The first is whether planar contact between the impact plate and the specimen is maintained during testing, as this will affect the uniformity of loading on the specimen. The second is whether the stress wave generated at the bar/specimen interface is uniform across the bar cross section and maintains its profile during propagation. To investigate these issues, an experiment may be not able to elicit sufficient information. Hence numerical modeling using ABAQUS was performed to examine some details.

FEM modeling was carried out to simulate an input steel bar (diameter 5 *mm* and length 2 *m*) connected to a steel plate (diameter 40 *mm* and thickness 5 *mm*) that is subjected to an impact. The magnitude of the stress wave in an elastic bar is directly related to the particle velocity $v(t)$, the bar density $\rho_0$ and elastic wave speed $C_0$ (i.e., $\sigma(t) = \rho_0 C_0 v(t)$). A velocity boundary condition comprising a 400 $\mu s$ trapezoidal pulse with a 20 $\mu s$ rising slope to a velocity of 2.5 m/s was imposed on the steel plate to simulate the impact of a striker against it. As shown in Fig. 22.7, the application of the velocity boundary condition to the annular area is between a diameter of 10 and 40 *mm* on the steel plate. The specimen and output bar are not composed in the simulation because they do not influence the two concerns about the validity of the test, as mentioned above. The wave studied in this simulation corresponds to $v_1(x_1, t)$ as denoted by Eq. 22.3.

Thirteen nodes along a radius of the steel plate (Fig. 22.7) were studied to examine how deformation of the plate affects the contact with a specimen. The first six nodes near the center are 0.5 mm from each other and the rest are spaced at intervals of 2.5 mm. The respective nodal displacements relative to the center of the plate, as well as the nodal velocities along the bar axis are plotted in Fig. 22.8 for all the 13 nodes. The relative nodal displacements reflect the profile of the impact plate surface when it is loaded. The positive means in front of the plate surface. It shows that all the relative displacement curves (Fig. 22.8a) quickly converge and there is less vibration; the profile of the plate remains constant during loading. With regard to the flatness of the impact plate during loading, the most important region is within the $\phi$5 mm circle (diameter of input bar). This is because the cross-sectional area a specimen is required to be less than that of the input bar. From the relative displacement values of the first six nodes in each curve, it is found that the maximum unevenness is only 0.266 μm. These small values should have negligible effects on the test results. The velocity profiles of all the 13 nodes on the impact plate (Fig. 22.8b) are consistent and represented by a trapezoid with a 20 μs rising slope, which is similar to the input boundary condition prescribed to the steel plate.

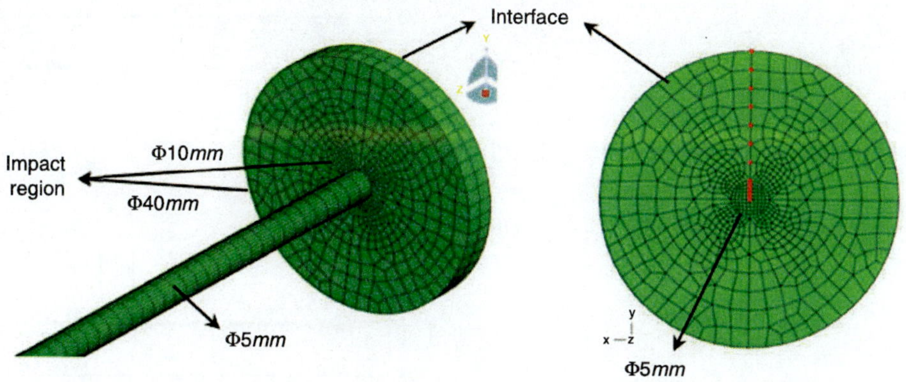


**Fig. 22.7** Three impact contact areas in numerical simulations

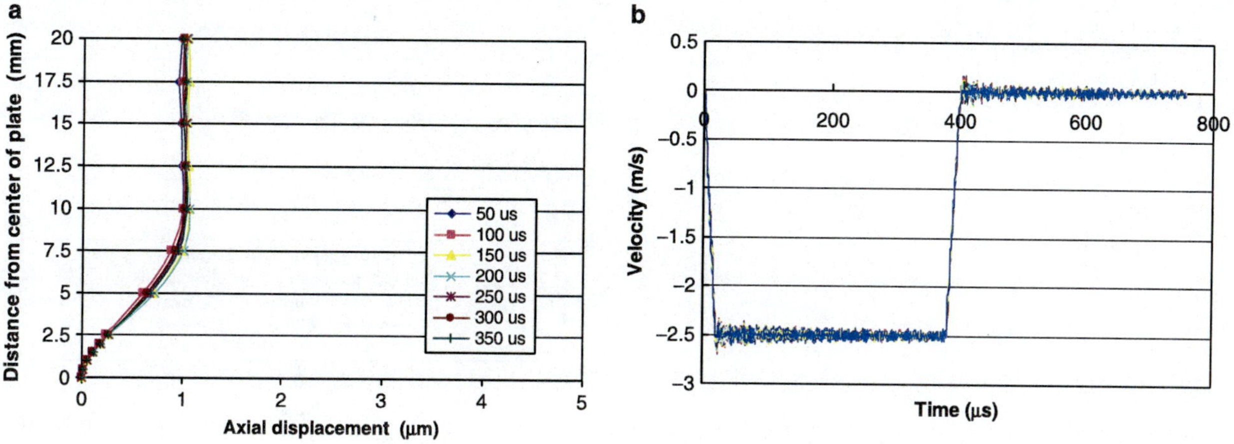


**Fig. 22.8** (**a**) nodal displacement relative to center and (**b**) nodal velocity histories, of the thirteen nodes along radius of impact plate at the interface

The next objective is to examine whether the stress generated is uniform across the cross section of the bar and propagates without distortion. Nodal velocities of the first six nodes along a 2.5 mm radius at the interface surface of the impact plate are plotted in Fig. 22.9, together with the corresponding radius at bar cross sections 25 and 50 cm away from the impact plate. Coincidence of the six curves at each position clearly shows that the particle velocity in cross-sections along the bar is quite uniform, even at the interface between the impact plate and the specimen. This is maintained as the wave travels along the bar, demonstrating that the input waves are acceptable in dynamic testing.

To experimentally validate the applicability of the miniature impact tester, tests on commercially available aluminum alloy were conducted, to verify whether the miniature tester can produce results with comparable accuracy to that from a traditional Hopkinson bar system. Small cylindrical specimens of aluminum alloy (2.4 mm in diameter and length) were fabricated and subjected to dynamic compression using a traditional Hopkinson bar system (with ϕ12 mm steel bars) at two striker speeds. Dynamic tests were also carried out using the miniature impact tester (with ϕ5 mm steel bars) at a striker speed of around 0.45 m/s (corresponding to a test on a split Hopkinson bar system with striker speed around 1 m/s). Static compression tests were also performed. Figure 22.10 shows that the results obtained using the miniature impact tester are comparable to those from the traditional Hopkinson bar system, and are more consistent among each test than the SHPB results. Demonstrating that this novel testing approach is capable of effecting dynamic testing. The discrepancy in the initial portion of these curves could have been caused by variation in the specimen-bar contact, as well as limitations in deriving early response from the Hopkinson bars. The results from the traditional Hopkinson bar system showed slightly larger oscillations, especially in the region of yielding; this could be due to the large difference in the cross sectional area between the bar (ϕ12 mm) and the specimen (ϕ2.5 mm), demonstrating the need for a thin bar for small specimens [16].

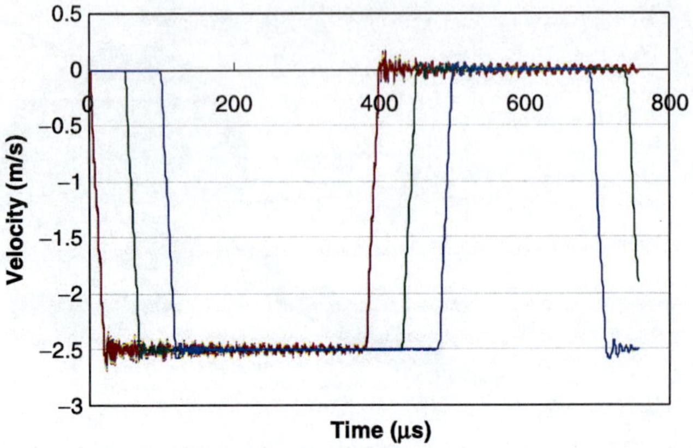


**Fig. 22.9** Velocity histories extracted from nodes along a radius at bar cross-sections located at the interface of impact plate, and 25 and 50 cm away from the interface

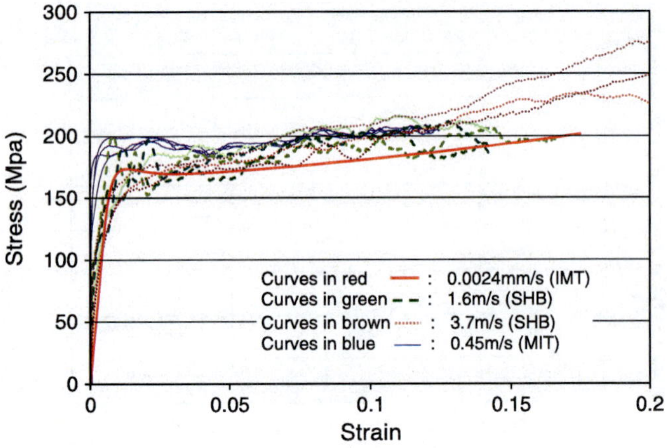


**Fig. 22.10** Results of tests on aluminum specimens (IMT – Instron micro tester; SHB – Split Hopkinson bar; MIT – Miniature impact tester)

## 22.4 Conclusion

A novel dynamic test methodology has been proposed. The motivation for this comes from the need to handle small specimens, whereby the load and deformation of the specimen must be measured accurately because of their small amplitudes. This is not easy with a traditional split Hopkinson bar system, where straight alignment between the striker, input and output bars is critical. With the miniature impact tester, the region that requires precise coaxial alignment is between the striker, the impact plate and the two surfaces in contact with the specimen; this is a short region compared to a split Hopkinson bar system. The strain gauges on both the input and output bars can be mounted close to the specimen-bar interfaces. Consequently, supporting the very slender bars is less of a concern and wave dispersion or attenuation in the bars has minimal influence on the results for the miniature impact tester. Stress waves in bars are used to derive the force and deformation response of a specimen for accurate results. The proposed miniature impact tester also has the advantage that it can impose either compression or tension simply through attachment of the impact plate to the input or output bar. These features facilitate establishment of a portable dynamic compression cum tension tester which has test accuracy comparable to a split Hopkinson bar system.

Theoretical derivation and interpretation of the governing equations have been presented, with respect to calculation of results from recorded wave signals. FEM simulation was also performed to evaluate the validity for dynamic testing. Experimental tests on aluminum specimens showed results comparable with that from a split Hopkinson bar system. The portable miniature impact tester fabricated in this study is relatively compact and based on a spring to propel a tubular striker. Alternative designs that use compressed gas to drive cylindrical striker are possible. An important requirement for achieving a good input wave is that the striker cross-section should cover as much of the impact plate as possible.

## References


1. Kolsky H (1949) An investigation of the mechanical properties of materials at very high rates of loading. Proc Phys Soc B62:676–701
2. Ogawa K (1984) Impact-tension compression test by using a split-Hopkinson bar. Exp Mech 24(2):81–86
3. Li M, Wang R, Han MB (1993) A Kolsky bar: tension, tension-tension. Exp Mech 33(1):7–14
4. Bragov AM, Lomunov AK (1995) Methodological aspects of studying dynamic material properties using the Kolsky method international. J Impact Eng 16(2):321–330
5. Tan VBC, Tong MX, Lim KM, Lim CT (2005) Finite element modeling of electronic packages subjected to drop impact. IEEE Trans Compon Pack Technol 28(3):555–560
6. Lai YS, Yang PF, Yeh CL (2006) Experimental studies of board-level reliability of chip-scale packages subjected to JEDEC drop test condition. Microelectron Reliab 46(2–4):645–650
7. Yeh CL, Lai YS (2006) Transient fracturing of solder joints subjected to displacement-controlled impact loads. Microelectron Reliab 46 (5–6):885–895
8. Luan J, Tee TY, Pek E, Lim CT, Zhong ZW (2007) Dynamic responses and solder joint reliability under board level drop test. Microelectron Reliab 47(2–3):450–460
9. Siviour CR, Walley SM, Proud WG, Field JE (2005) Mechanical properties of SnPb and lead-free solders at high rates of strain. J Phys D Appl Phys 38:4131–4139
10. Gonzalez M, Vandevelde B, Hoof V, Beyne E (2004) Characterization and FE analysis on the shear test of electronic materials. Microelectron Reliab 44(12):1915–1921
11. Chen KM, Chiang KN (2003) Impact of probing procedure on flip chip reliability. Microelectron Reliab 43:123–130
12. Yeh CL, Lai YS (2007) Design guideline for ball impact test apparatus. J Electron Pack 129(1):98–104
13. Gong JC, Malvern LE, Jenkins DA (1990) Dispersion investigation in the split Hopkinson pressure bar. J Eng Mater Technol 112(3):309–314
14. Wang LL, Labibes K, Azari Z, Pluvinage G (1994) Generalization of split Hopkinson bar technique to use viscoelastic bars. Int J Impact Eng 15(5):669–686
15. Shim VPW, Liu JF, Lee VS (2006) A technique for dynamic tensile testing of human cervical spine ligaments. Exp Mech 46(1):77–89
16. Liu JF, Shim VPW, Tan VBC, Lee TK (2007) Dynamic testing of solder joint strength under compression, tension and shearing. In: Proceedings of the 9th electronics packaging technology conference, Singapore, pp 380–385

# Chapter 23
# Kolsky Tension Bar for Samples with Very Low Strength

**Matthew Hudspeth, Xu Nie, and Weinong Chen**

Following the inception of the original Split-Hopkinson bar apparatus, being first designed to load specimens in a high strain rate compression environment [1], the need for testing materials in high-rate tension has led to several modifcations of the typical Kolsky bar setup. Of the many possible configuration geometries, two main arrangements have proven to be quite adept at impinging a tension stress wave upon the incident bar while ensuring test setup robustness. The first and less widely used mechanism involves storing elastic energy into a section of the incident bar via stretching a portion of the bar end and clamping it into place. This stored energy is then released via unclamping the specimen thereby sending a stress pulse down the length of the incident bar and impinging upon the sample. This mode of operation was performed by Staab et al. in efforts to produce a high-rate tension experiment wherein a tensile stress was imposed on a section of the incident bar and then clamped so as to store the energy used to produce the desired stress pulse [2]. This method, while quite productive, lacks the ability of instigating the adequate pulse shaping needed to generate a constant strain rate in the sample. The second approach, which promotes a much more repetitive result and modifiable stress wave, involves striking a flange which is attached to the end of the incident bar, thereby generating a tensile stress pulse which then runs the length of the bar and imposes on the sample. There are various fashions of performing the impact, with the most common involving a tube which is concentric to the incident bar being fired at the flange from either a gas chamber or spring firing mechanism. This experimental setup, in tandem with an open-loop styled pulse control, allows for adequate shaping of the incident tensile wave [3]. Another mode of flange-styled pulse generation incorporates a rotating disk, which when a certain rotational speed is met, an arm extends from the disk and strikes the flange at the end of the incident bar. This type experimental setup was used by Yuanxin when testing AAL laminate, wherein a hammer would strike the end of an aluminum bar attachment causing breakage, thereby sending a rectangular pulse through the incident bar towards the specimen [4]. Both of these mechanisms, rotating disk and tube firing, have proved to be adept at pulse shaping and thus quite successful at generating a uniform stress pulse in the sample at a constant strain rate.

One of the most imperative, yet tricky aspects of performing a high-rate Kolsky bar tension experiment involves the mounting/attachment of the sample to the incident and transmission bars. Unlike a high rate compression experiment, the sample cannot simply be sandwiched between the aforementioned bar ends, rather, it must be fixed to both bars by a constraining mechanism making sure to minimize the edge effects evolving from the sample interaction with the gripping device (e.g. glue, clamps). For example, Wang et al. glued several wraps of E-glass fiber bundle into the ends of the incident and transmission bars with a high shear strength adhesive [5]. The selection of glue was imperative in this setup in order to ensure that slippage of the fibers at the bar surfaces was negligible. Another attachment mechanism performed by Nie et al.,

M. Hudspeth (✉)
School of Materials Engineering, Purdue University, West Lafayette, IN, USA
e-mail: mhudspet@purdue.edu

X. Nie
School of Aeronautics and Astronautics, Purdue University, West Lafayette, IN, USA
e-mail: xnie@purdue.edu

W. Chen
School of Materials Engineering, Purdue University, West Lafayette, IN, USA

School of Aeronautics and Astronautics, Purdue University, West Lafayette, IN, USA

V. Chalivendra et al. (eds.), *Dynamic Behavior of Materials, Volume 1: Proceedings of the 2012 Annual Conference on Experimental and Applied Mechanics*, Conference Proceedings of the Society for Experimental Mechanics Series, DOI 10.1007/978-1-4614-4238-7_23, © The Society for Experimental Mechanics, Inc. 2013

involves wrapping the specimen around the outside of the incident and transmission bars and then clamping the sample on both sides [6]. In many cases, the sample geometry requires special mounting grips, which inherently require both bar ends to be tapped in order to allow for possible sample removal via screw in and out attachments [7]. Usage of these types of grips may alter the stress experienced by the sample due to bar-attachement-sample impedance mismatch, along with imposing a possible stress spike due to poor contact between the bar and grip threads [8]. Therefore, the attachment mechanism used in a tension environment must be fully understood in efforts to minimize misappropriation of measured response to actual material behavior.

During the testing of weak materials using a Kolsky bar approach, it is quite common for the majority of the incident pulse to be reflected leaving just a minute portion of the pulse to travel through the transmission bar, leading to a very high noise to strain ratio measured with the resistive strain gauge on the transmission bar. As used in typical Kolsky bar testing, the 1-D wave mechanics solution requires that an accurate representation of the strain experienced by the transmission bar be gathered in order to ensure correct determination of the stress measured in the specimen. As the failure stress of the sample reduces, this transmission bar strain becomes very minimal, thus leaving it impossible for an accurate determination of the strain in the transmission bar to be resolved. Therefore, two typical approaches have resulted, both being quite adept at determining bar strain. The first, being the simpler of the two, is performed by solely replacing the typically used resistive strain gauge attached to the transmission bar with a much more sensitive semi-conductor strain gauge. The second, which is a bit more complicated and requires more in depth analysis, consists of strategic placement of quartz crystals so as to more directly determine the force experienced by the sample [9]. This technique was further improved by Chen et al., who reported a three orders of magnitude increase in testing sensitivity as opposed to the typical Kolsky bar approach when embedding the quartz crystal into the transmission bar, yielding the Kolsky bar capable of testing materials with both extremely low compressive strengths and low mechanical impedance [10]. More recently, when testing EDPM rubber in dynamic tension, Nie et al. placed quartz force transducers at the sample contact ends of both the incident and transmission bars so as to make direct measurements of the stress on each end of the sample [6]. With the introduction of force detection on both bar-sample contact points, verification of the assumption regarding dynamic equilibrium is possible. During the high rate testing period, it is imperative the sample undergoes a dynamic equilibrium deformation cycle, i.e. both sides of the sample experience identical force levels. As it is possible to embed quartz crystals into both the incident and transmission bars, only remove the inertial portion of detected force signal, which may mask the true instantaneous force seen by the sample-bar contact zone. This is especially true on the incident bar side which experiences a greater acceleration than the transmission bar [6].

As the failure stress of a sample decreases, it becomes necessary to decrease both the incident and transmission bar testing diameter so as to decrease this acceleration term experienced by the quartz discs, which can become astronomical compared to the true sample failure force. Of coarse there reaches a point wherein the bar geometry can no longer be reduced if a robust testing device is to be designed, thus leading to the entire removal of the transmission bar, which at such low transmitted force levels, can be replaced with a stationary force detection device [3]. Typically, a quartz load cell is used due to its high degree of sensitivity and its extremely fast response time. Furthermore, at such low failure force levels exhibited by the sample, the introduction of various noise levels may creep into to the force history during testing. It is of upmost importance to isolate the three main experimental components, which consist of the incident bar, firing mechanism, and force detection system. During testing with single fibers, it has been found that placing the transmission load cell on its own optical table can suppress mechanical vibrations which may arise due to the firing mechanism and because of bar sliding. Additionally, it is imperative the mounting fixture for the load cell be extremely rigid in efforts to reduce possible deflection which could result in an acceleration term being experienced by the load. It is also recommended that the firing mechanism require minimal contact with the incident bar, preferably none, leaving solely the final contact of the firing mechanism with the bar end flange, as seen in Fig. 23.1. Finally, when testing extremely low force level samples such as spider fiber, the introduction of acoustical noise may arise due to the inevitable contact of the striker tube into bar end flange. This ringing noise then travels down the length of the incident bar and traverses across the expanse between the bar end and load cell, possibly creating a noise level on the same order of magnitude as the sample testing force regime. In efforts to quell this noise culprit, some sort of blocker plate must be put in between the bar end and load cell, leaving only a small aperture for the passage of the single fiber sample. Though the introduction of this blocker plate may prove to be quite tricky, it is of upmost importance in order to render the testing sequence free from appreciable noise.

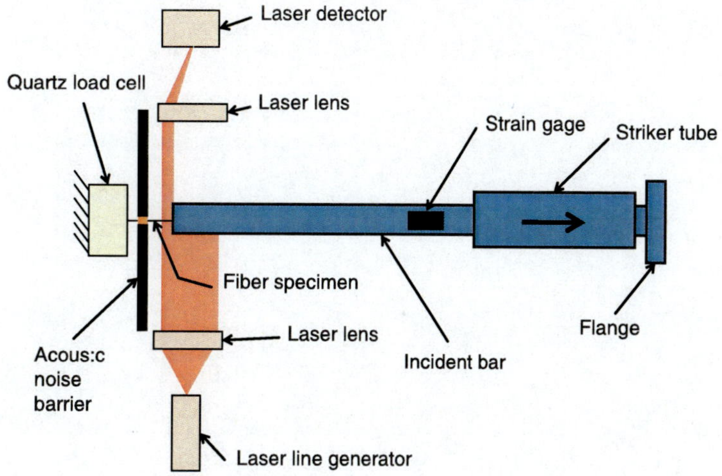


**Fig. 23.1** Typical miniature kolsky bar showing acoustic noise blocker plates

# References


1. Kolsky H (1949) An investigation of the mechanical properties of materials at very high rates of loading. Proc Phys Soc B 62:676–700
2. Staab GH, Gilat A (1991) A direct-tension split hopkinson bar for highstrain-rate testing. Exp Mech 31:232–235
3. Cheng M, Chen W, Weerasooriya T (2005) Mechanical properties of kevlar R km2 single fiber. J Eng Mater Technol 127:197–203
4. Yuanxin Z, Yuanmin X (1998) The of effect of strain rate and prestress on the mechanical behaviour aal. Compos Part A 29A:517–524
5. Wang Z, Xia Y, Yang B (1996) A statistical model and experimental study of the strain rate and temperature dependence of the strength of fibers. Appl Compos Mater 3(2):89–101
6. Nie X, Song B, Ge Y, Chen W, Weerasooriya T (2009) Dynamic tensile testing of soft materials. Exp Mech 49:451–458
7. Chen W, Lu F, Cheng M (2002) Tension and compression tests of two polymers under quasi-static and dynamic loading. Polym Test 21(2):113–121
8. Song B, Antoun B (2011) Pseudo stress response in kolsky tension bar experiments. Exp Mech 1–4. doi:10.1007/s11340-011-9509-9
9. Wasley R, Hoge K, Cast J (1969) Combined strain gauge—quartz crystal instrumented hopkinson split bar. Rev Sci Instrum 40(7):889–894
10. Chen W, Lu F, Zhou B (2000) A quartz-crystal-embedded split hopkinson pressure bar for soft materials. Exp Mech 40:1–6. doi:10.1007/BF02327540

# Chapter 24
# Preloaded High-Rate Tension Techniques

**Bo Song and Wei-Yang Lu**

**Abstract** A current Kolsky tension bar has been implemented with pre-tension-load capability to investigate the effect of preload on the high-rate response in tension of materials and structures. In this study, fully threaded brass studs have been experimentally investigated in terms of pre-tension-load effect on the tensile stress-strain response at the same high strain rate. The preload is not observed to significantly influence the plastic flow stress. The failure responses are quite different, however, when different pre-tension loads are applied.

**Keywords** Kolsky bar • High rate • Tensile response • Preload

Engineering parts are usually assembled together with joint units. For example, bolts are very common components to connect parts together. During assembly, a certain amount of torque is applied to the bolts, producing a pre-tension load in the bolts, to maintain the structural integrity under various loading conditions in applications. One should be very careful of the amplitude of the preload applied to the bolts as either over-loaded or under-loaded bolts may result in unexpected failure or malfunction of the engineering structures. The effect of preload on the mechanical response including failure and fracture of the bolts in tension is desired to be thoroughly understood. Particularly, for the cases where the bolts are subject to impact loading, the dynamic tensile response of the preloaded bolts is necessary to be determined. This requires the development of new preloaded high-rate tension techniques. In this study, we properly modified the current Kolsky tension bar to characterize the high-rate tensile properties of bolts under various amount of preload.

Figure 24.1 shows the schematic of the modified Kolsky tension bar with preload capability. As shown in Fig. 24.1, two bar stoppers were installed on the ends of the incident and transmission bars, respectively. The first bar stopper (#1) is used to prevent the bar system from moving when the bar system is preloaded in tension from the transmission bar side. The threaded far end of the transmission bar passes through the second bar stopper (#2). Tightening the lock nut against the bar stopper #2 provides a tension load in the bar system. The preload can be controlled and measured with the load cell installed between the lock nut and the bar stopper (#2). Alternatively, the preload can be measured with the strain gages on the incident bar or the transmission bar, which is the case described in this paper. Figure 24.2 shows actual photographs of both bar stoppers.

In this study, we employed this preloaded Kolsky tension bar system to characterize brass alloy fully threaded studs. The studs are 1″ (25.4 mm) long with ¼″-20 threads. Before dynamic test, the stud specimen was directly threaded into the ends of incident and transmission bars via adapters (Fig. 24.1), making a gage length of ¼″ (6.35 mm). Experimental procedure is the same as the regular Kolsky tension bar experiments [1, 2]. In order to eliminate strain-rate effect from preload effect, the striking speed was changed appropriately to maintain the same strain rate for different pre-tension load experiments. A Ø1/4″ × 0.02″ annealed copper disk was used as the pulse shaper to facilitate constant plastic strain rate and dynamic stress equilibrium [1, 2]. Figure 24.3 shows oscilloscope records of three experiments with different pre-tension loads: 0, 1,140, and 3,420 N. The baseline offsets shown in Fig. 24.3 indicate such different pre-tension loads. In this study, we used a MTI 2000 Fotonic sensor to replace the Micro-Epsilon laser system [2] to directly measure the displacement of the incident bar end. This Fotonic sensor has a sufficiently high resolution in measuring small displacement of the bar end,

B. Song (✉) • W.-Y. Lu
Sandia National Laboratories, Livermore, CA 94551-0969, USA
e-mail: bsong@sandia.gov; wlu@sandia.gov

V. Chalivendra et al. (eds.), *Dynamic Behavior of Materials, Volume 1: Proceedings of the 2012 Annual Conference on Experimental and Applied Mechanics*, Conference Proceedings of the Society for Experimental Mechanics Series, DOI 10.1007/978-1-4614-4238-7_24, 

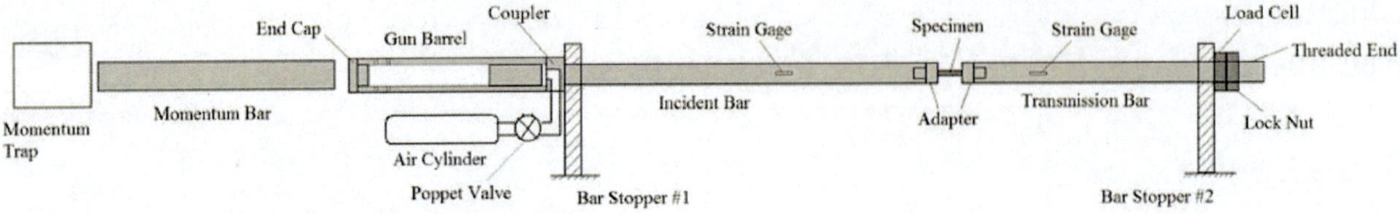


**Fig. 24.1** Schematic of preloaded Kolsky tension bar

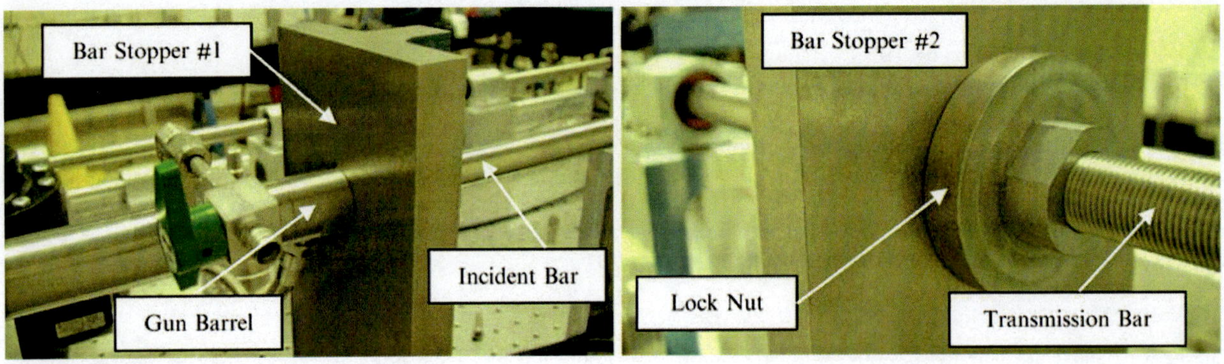


**Fig. 24.2** Photographs of bar stoppers

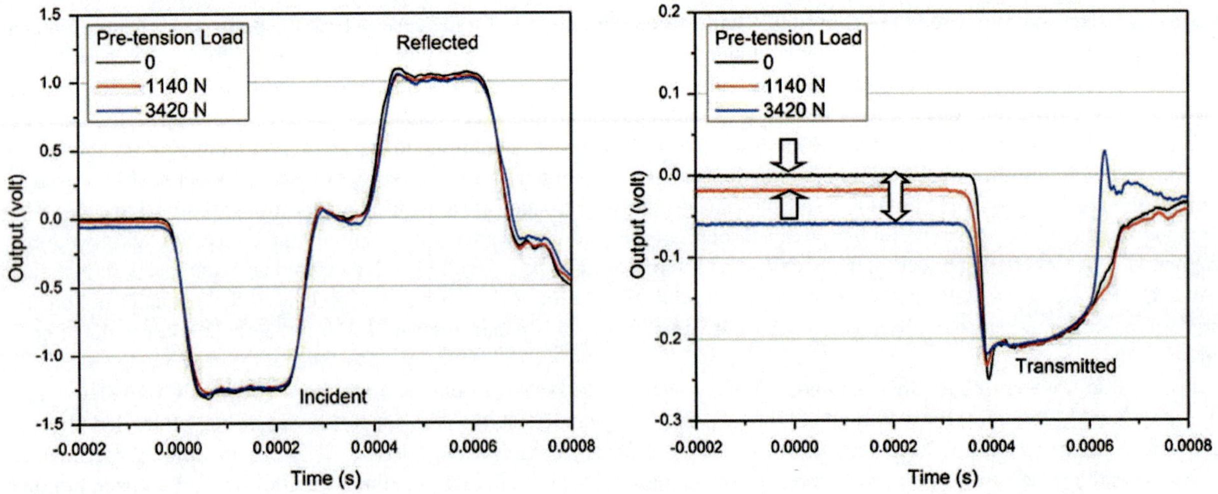


**Fig. 24.3** Incident, reflected, and transmitted pulses at different pretension loads

enabling precise small-strain measurement of the specimen. The strain rates for all three dynamic experiments were maintained identical, which is approximately 1,350 $\pm$ 50 $s^{-1}$. The early stress peaks ridden on the stress-strain curves are possibly pseudo due to threads and the amplitudes decreased with increasing pre-tension load [2, 3]. Figure 24.4 shows the resultant tensile stress-strain curves of the brass threaded studs at different pre-tension loads. The stress-strain curves shown in Fig. 24.4 have been shifted according to different preloads (1,140 and 3,420 N). The brass studs exhibit nearly perfectly plastic flow response. With the strain is over 10 %, the flow stress then decreases with increasing strain, which is a precursor of failure. The results show the preload does not significantly influence the plastic flow stress. However, the failure responses are quite different for different pre-tension loads. The mechanism of preload effect on the failure behavior is still under study.

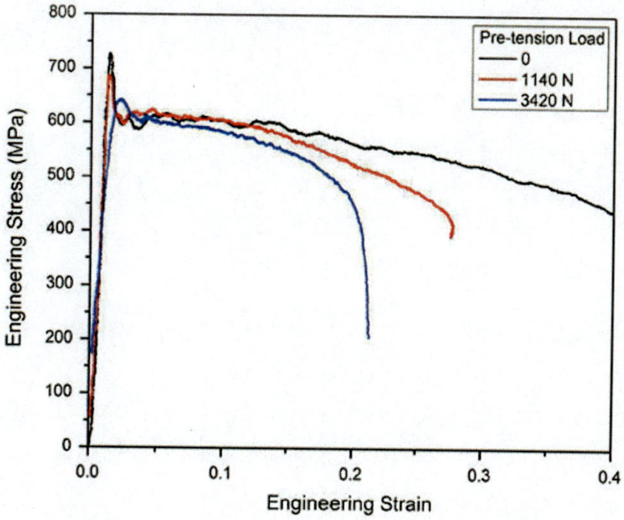


**Fig. 24.4** Tensile stress-strain curves of threaded studs at different pre-tension loads

**Acknowledgements** Sandia National Laboratories is a multi-program laboratory managed and operated by Sandia Corporation, a wholly owned subsidiary of Lockheed Martin Corporation, for the U.S. Department of Energy's National Nuclear Security Administration under contract DE-AC04-94AL85000.

## References

1. Song B, Antoun BR, Connelly K, Korellis J, Lu W-Y (2011) Improved Kolsky tension bar for high-rate tensile characterization of materials. Meas Sci Technol 22:045704 (1–7)
2. Song B, Jin H, Antoun BR (2011) Kolsky tension bar techniques for characterization of alloys. In: Proceedings of 2011 SEM annual conference and exposition on experimental and applied mechanics, Mohegan Sun, Uncasville, June pp 13–16
3. Song B, Antoun BR (2012) Pseudo stress response in Kolsky tension bar experiments. Exp Mech 52:525–528

# Chapter 25
# Spinel Modeling Embedded with Small Diameter Surface Transparent Material Particles

**Costas G. Fountzoulas**

**Abstract** Ceramic technology has greatly enhanced the ballistic performance of transparency systems for future tactical platforms. To continue the rapid advancements, numerical approaches to investigating material response to dynamic loading have been implemented. A recent effort reported the success of strategically placed surface and near-surface voids on the dynamic response of a transparent magnesium aluminate spinel (spinel) backed by polycarbonate. It was shown that the shape, the distance from the line of impact of the defects affected the ballistic performance of the laminate. However, fabrication of pristine systems and understanding damage sensitivity of the laminates remain critical factors for brittle materials. Fountzoulas et al. studied by dynamic simulation the effect of additional parametric variables including shape, size, number, location and type of defect filled with air at atmospheric pressure [1]. The present effort reports on the effect of surface glass fillet of circular, rectangular and elliptical cross section on the dynamic response of spinel ceramic subjected to high rate loading. The numerical study will continue to refine the synergistic relationship between ballistic behavior and composite designs of transparent laminates. For the numerical analysis, the existing or modified materials strength and failure models of ANSYS/AUTODYN commercial software are used. The report offers a survey of inclusions types and locations and reports the estimated impact on dynamic performance.

**Keywords** Modeling • Spinel • Embedded • Small glass particles • Surface defects

## 25.1 Introduction

Transparent armor systems using ceramics as the striking face have been explored since the early 1970s because they potentially provide superior ballistic protection to conventional glass based transparent armor systems [2]. The U.S. Army has invested heavily in the development of next generation materials, including ceramics, for military systems [3]. The result of the on-going investments is a critical understanding of ceramics strengths and weaknesses for military platforms. Among the potential ceramic materials considered for armor-sapphire, edge-form-growth sapphire, magnesium aluminate spinel, aluminium oxynitride ($MgAl_2O_4$)- magnesium aluminate spinel was selected for the current pursuit. Although individually, single impact testing has shown small variations in ballistic efficiencies ($<10\%$) of ceramics, to achieve multi-hit performance, all of the ceramics are effectively equivalent. Technology Assessment and Transfer (Annapolis MD, USA) is providing ceramic spinel plates produced via hot-pressing in sizes up to $28 \times 36 \times 1.5$ cm for this report [4].

Finite element modeling has progressed substantially in the ability to predict failure of materials under extreme dynamic loading conditions. One of the limitations of predictive models is lack of a complete dynamic materials properties database which is needed for materials models for each of the materials in the simulations. In order to compensate for parameters whose dynamic values were extrapolated from their static or quasi-static properties, baseline experiments are often used to recalibrate the models [5, 6]. However, the recalibration method of modeling lacks many of the physical properties and failure mechanisms associated with real-world materials. Therefore, often recalibrated models lack the ability to predict within statistical error future failures over any substantial ranges due to the existence of defects, and materials substitutions often lead to new calibration requirements. The desired approach is to validate a fully characterized materials database with

C.G. Fountzoulas (✉)
U.S. Army Research Laboratory, WMRD, RDRL-WMM-B Aberdeen Proving Ground, Aberdeen, MD 21005-5069, USA
e-mail: Constantine.Fountzoulas.civ@mail.mil

V. Chalivendra et al. (eds.), *Dynamic Behavior of Materials, Volume 1: Proceedings of the 2012 Annual Conference on Experimental and Applied Mechanics*, Conference Proceedings of the Society for Experimental Mechanics Series,
DOI 10.1007/978-1-4614-4238-7_25, 

one calibration model, and subsequently apply the model to modified designs. However, despite its apparent problems, recalibration of existing materials models has been proven to be an effective tool in the hands of the modeler by minimizing the number of simulation iterations and resulting in more successful predictions. Regardless of methodology, finite element tools can be applied effectively to reduce the variability between impact tests and can be used to validate designs with fewer experimental failures, when robust models are created [6].

## 25.2 Experimental

Due to the sensitive nature of ballistic test results, surrogate materials were assembled that do not represent current or future armored designs. Experimental coupons for ballistic testing consisted of laminated layers of spinel bonded using Huntsman 399 polyurethane adhesive to Bayer polycarbonate. The laminate targets consisted of the spinel varied between 9 and 15 mm thick, depending on the density of flaws introduced the bonding layer typically 1–2 mm and the backing polycarbonate of 12 mm thickness. Experimental samples were evaluated only to attain penetration velocity to confirm the strength and failure material models parameters. In addition, square cuts of 1.5 mm width and 5 mm height, and cones of 4 mm diameter and 4 mm height were introduced into the surface of the spinel. The density of the surface defects varied and it represented a 2% and 4% mass loss of the solid spinel. Otherwise, all results and discussion, including the internal defects architecture, are based on simulation analysis.

## 25.3 Modeling

The ballistic behavior of an identical to the actual target geometry, which was consisted of spinel, polyurethane and polycarbonate, and impacted by a surrogate projectile, was simulated using the non-linear ANSYS/AUTODYN commercial package [7]. The material models used were obtained from the AUTODYN library. The modeled laminated target consisted of panels of spinel, polyurethane, and polycarbonate of 900 $cm^2$ cross sectional area (30 cm in 2D models). In our previous study [1] the defects were filled with air at one atmospheric pressure. Some of the strength material models of the laminated target materials had been previously recalibrated using existing ballistic data. The numerical modeling of undamaged armor coupons, and the coupons with surface defects, was carried out in two and three dimensions. The projectile applied in the models was a 3.0 cm long, 1,095 steel projectile, of conical frustum geometry, (6 mm large base, 1.0 mm small base), using two dimensional axi-symmetric models. Depending on the geometric complexity of the laminated target and the existing pre-processing capability of the solver, smooth particle hydrodymamics (SPH) solver was used for spinel with no defects, and rectangular and circular cross section defects, Euler solver was used for elliptical cross section defects. The projectile, polyurethane and polycarbonate were discretized by SPH solver with particle size of 0.2 mm for increased modeling accuracy. The element size used for the Euler solver was 0.2 mm. Since each of those solvers has its own characteristics, to ensure result compatibility a target containing no defects was simulated by the two solvers. All solvers produced similar results. Results were obtained by simulating projectiles impacting the targets at the experimental velocity of 975 m/s.

## 25.4 Defect Models

One of the advantages of modeling methods is the ability to create physically challenging architectures to investigate effects of point defects on failure. The sensitivity of ballistic measurement tools is typically less than $\pm 10\%$ due to the range of available failure modes invoked in the high energy exchange between projectile and target. Additionally, capturing the real-time failure modes in the impact event requires highly specialized video equipments. These factors contribute to a very difficult and expensive set of experiments for investigating small flaws and the impact on performance in the experimental realm. By using modeling tools, however, the effects of macroscopic flaws and the location of these flaws can easily be investigated in a model that correctly captures the physics of failure in the materials. Therefore, to enhance the understanding of flaws and the behavior of spinel with defects, the modeling approach, which had been validated previously by experimental data for the case of surface introduced discontinuities, defined also as defects, is employed. Figures 25.1 and 25.2 show surface defects of rectangular and elliptical cross-sectional area respectively introduced in the spinel.

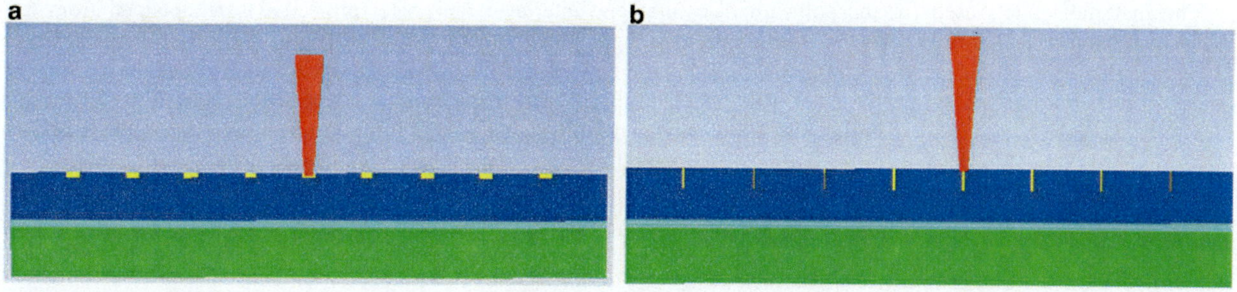

**Fig. 25.1** Target containing rectangular defects with width (**a**) parallel; and (**b**) perpendicular to the line of impact

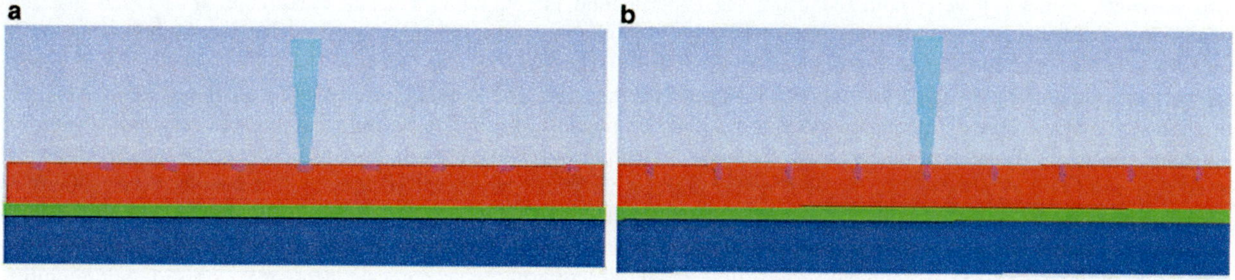

**Fig. 25.2** Target containing elliptical defects with the minor axis (**a**) parallel; and (**b**) perpendicular to the line of impact

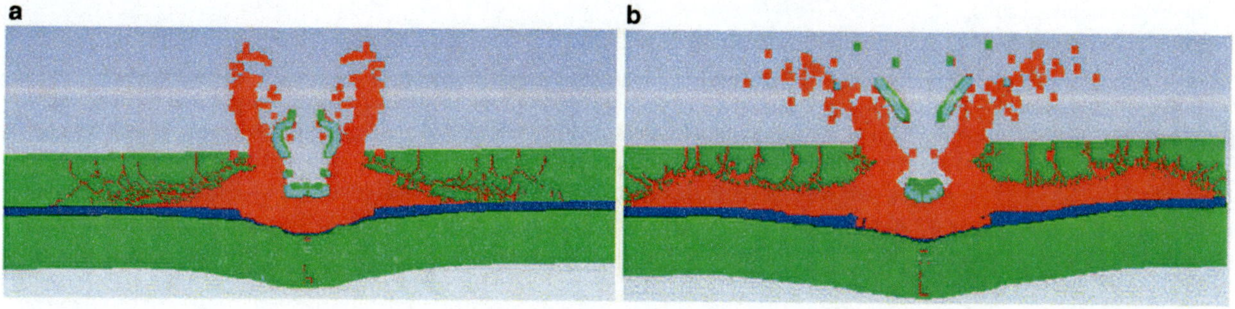

**Fig. 25.3** Snapshot images of simulations of damage propagation into rectangular cross section laminate after 100 μs with slug moving 975 m/s with the width (**a**) parallel to line of impact and (**b**) perpendicular to line of impact

Consideration of defects in calculating the failure probability of ceramic articles with short-term loading has been reported by Gorbatsevich et al. [8]. The effect of various shapes and locations of defects of equal cross sectional area was studied by simulation of the impact. Nine (9) various cross section defects of equal total area were introduced in the spinel. Due to limitations of the SPH solver preprocessor, for this study the Euler solver was also used. The 2-D Euler solver produced results for all defect shapes. To ensure compatibility of our previous results produced by both solvers, we compared the results of the impact simulation for a target without defects which were run under identical impact and boundary conditions. Both solvers produced similar results. In the interest of computational consistency, defect results were compared using residual velocities based on a constant initial impact velocity. To keep simulation times reasonable, all specimens are impacted at an initial velocity of 975 m/s.

The effect of surface flaws on the ballistic performance of the target was studied by introducing square cuts into the surface of the spinel of 1.5 mm width and 5 mm height and comparing the results against a target without defects. The effect of the defects on areal density of the coupons is less than 1%. Simulation images after impact are shown in Fig. 25.3. The defect closer to the outer edge of the spinel was placed at a distance equal to the surface defect from the outer edge of the target. The projectile impacted the target perpendicular to the center of middle defect. Analysis of the experimental tests and the simulations included evaluation of the extent of the damage, residual velocity after impact, and extent of deflection.

The material models used for the polycarbonate, urethane interlayer, and steel projectile were obtained from the AUTODYN material library [7]. The PC was modeled using a shock equation of state (EOS), piecewise Johnson-Cook (JC) strength model, and a plastic strain failure criterion. The polyurethane was modeled using a linear EOS and a principle stress failure criterion. The projectile steel was modeled using a shock EOS and a JC strength model [7]. The spinel, however, due to lack of existing material model, was modeled using a recalibrated form of the existing AUTODYN library $Al_2O_3$ material model, produced from existing experimental data and consequently validated many times by predicting within 3% results of new target design, which included a polynomial EOS and Johnson-Holmquist (JH2) strength and failure models [7].

## 25.5 Results

Our previous results [1] were validated using the experimental target, with excellent agreement. It is noted that the simulation results were completed more than 30 days prior to experimental results and agreement was within 3%. Therefore, the failure criteria and confidence in the material parameters were excellent. The extent of damage in the simulation result did not coincide precisely with the experimental coupons for this case. The simulation results showed potential edge effects that did not appear in the experimental system. It should be noted, however, that during the impact event, failure modes appeared consistent. Therefore, the 2D model showed an effective and rapid method of producing laminate constructions for interrogation of failure criteria. The baseline results allowed confident investigation of the defect models. That prediction disagreement may be attributed to the fact that a 2D defects, when expanded in 3D is not a localized 3D defect, but rather a groove. Therefore, more material is removed at a 2D simulation than the 3D respective simulation. Currently performed 3D simulation on targets containing conical surface defects tend to confirm this hypothesis. Moreover, the simulation almost duplicated the experimental surface cracking. The authors believe that a very accurate reproduction of a surface cracking is possible by using a SPH particle size smaller than the currently used of 1 mm. Figure 25.4 shows a summary of our previous study [1] for surface and internal defects filled with air and placed strategically in the hard face of the laminate and compared to the experimental results.

Figure 25.5 shows the exit velocity of the projectile from the back face of the spinel hard face. The simulations showed that the exit velocity from the back face of the spinel was smaller for defects filled with glass for all the rectangular cross sections, circular cross section, and for the elliptical cross section with minor axis parallel to the line of impact. However, for the elliptical cross section with the minor axis perpendicular to the line of impact was smaller for the defect filled with glass. The resulting exit velocities of the projectile are dramatic. As it appears in Fig. 25.6, the final exit velocity of the projectile from the laminate was larger for targets with targets filled with glass for rectangular cross section with the width perpendicular to line of impact defects, and all the elliptical defects. The exit velocity from the laminate containing rectangular cross section defects with the width parallel filled with glass was to the line of impact was smaller than the laminate with unfilled defects. Both circular defects, filled and unfilled with glass appear to have captured the impactor

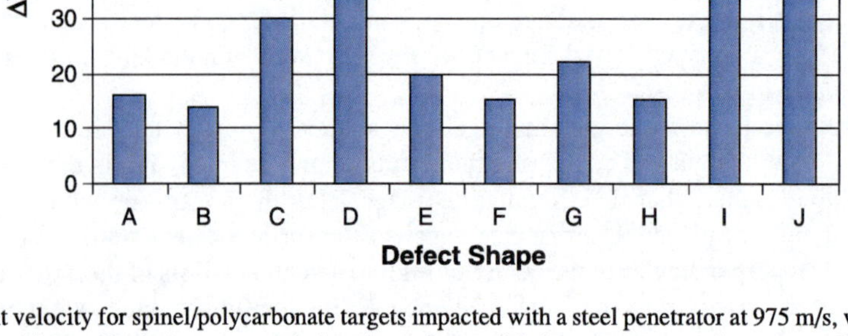


| | |
|---|---|
| A | Circle 9 |
| B | Circle 10 |
| C | Ellipse 9 |
| D | Ellipse 9r |
| E | Ellipse 10 |
| F | Rectangle 9 |
| G | Rectangle 9r |
| H | Rectangle 10 |
| I | Rectangular Surface Groove (5) |
| J | Rectangular Surface Groove (9) |

**Fig. 25.4** Exit velocity for spinel/polycarbonate targets impacted with a steel penetrator at 975 m/s, where defects are purposefully included in the ceramic layer with specific design parameters keeping volume of defects constant; and compare effect of internal defects to surface defects [1]

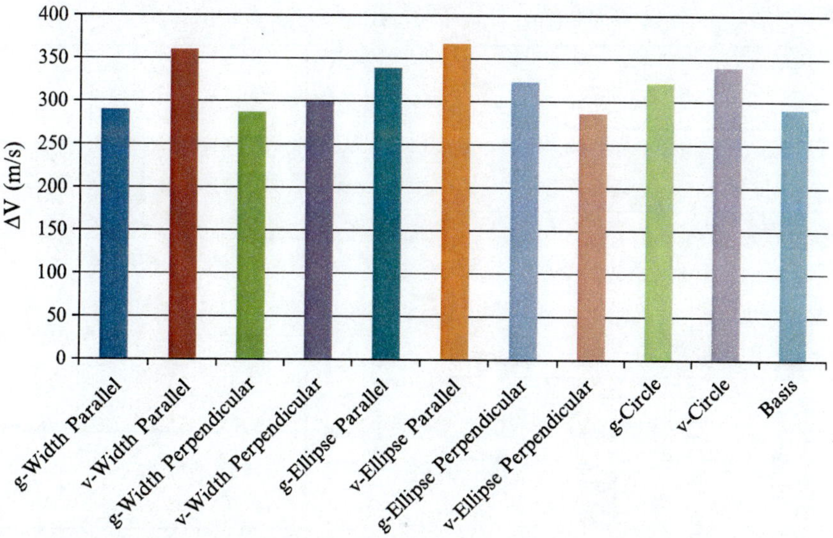


**Fig. 25.5** Exit velocity from the back face of the spinel hard face (50 μs simulation time): g = glass filled, v = no filling, Perpendicular and Parallel denote width of the rectangle and ellipse minor axis with respect to line of impact

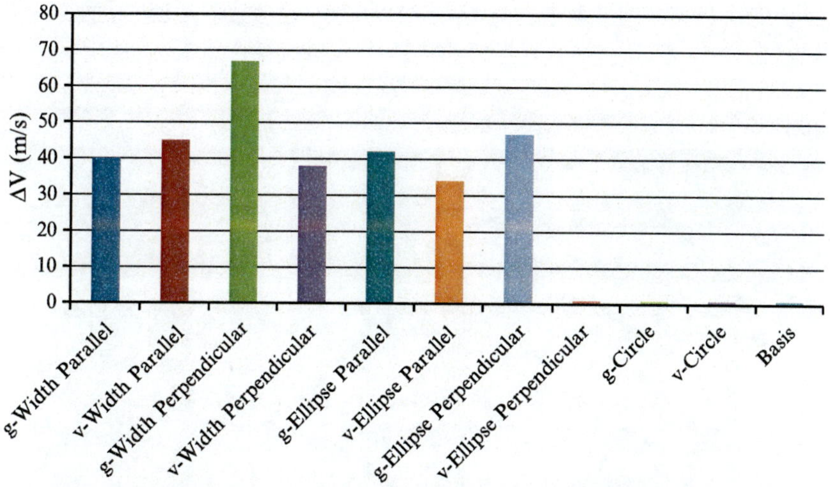


**Fig. 25.6** Exit velocity from laminated targets containing various cross-sectional surface defects

(Fig. 25.6). Figure 25.7a, b show the difference of the exit velocity between the unfilled and filled with glass, same cross section defect exiting the back face of the spinel and the PC respectively. With the exception of the elliptical defect with the minor axis perpendicular to the line of impact, defects filled with glass provide greater resistance to penetration of the spinel hard face (Fig. 25.7a). However, the simulations showed that, the glass filling only of defects of circular cross section and rectangular cross section with the width parallel to the line of impact resulted in smaller exit velocity of the projectile from the laminate (Fig. 25.7b), although not much smaller than the unfilled defects. Figure 25.8a–c show damage progression through the spinel hard face 40, 43 and 46 μs after impact. The damage appears to start at the area of the defect, where higher stress concentration is anticipated.

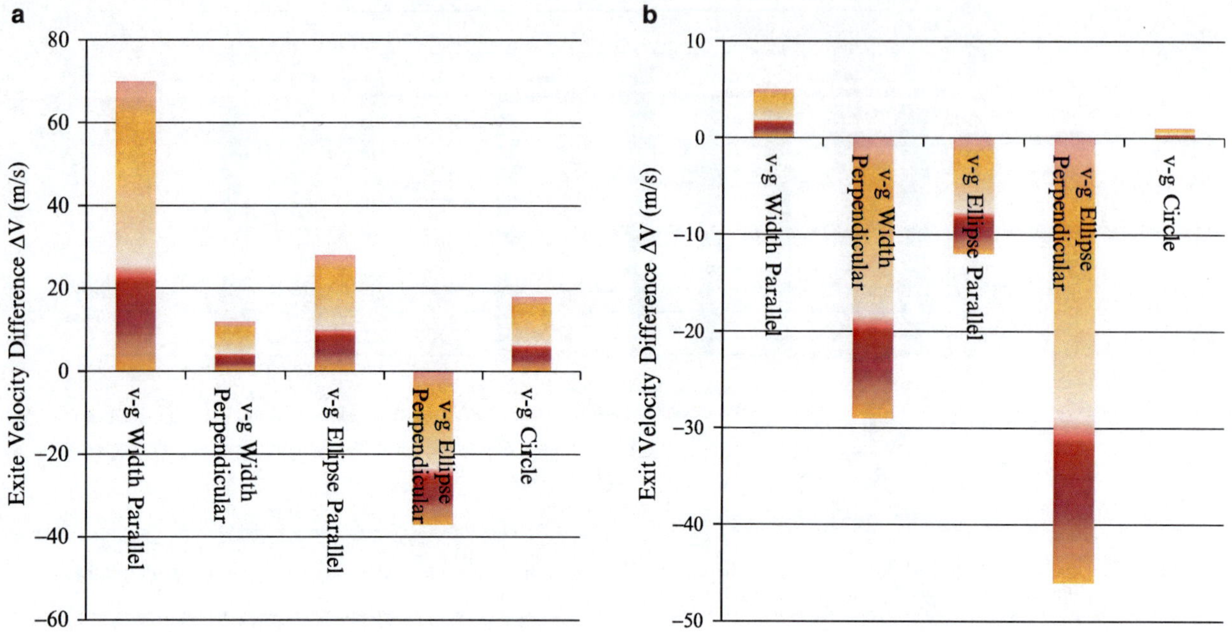


**Fig. 25.7** Exit velocity difference between unfilled and filled (v–g) defects of same cross section: (**a**) just exiting the back face of the spinel hard face, and (**b**) exiting polycarbonate backing layer

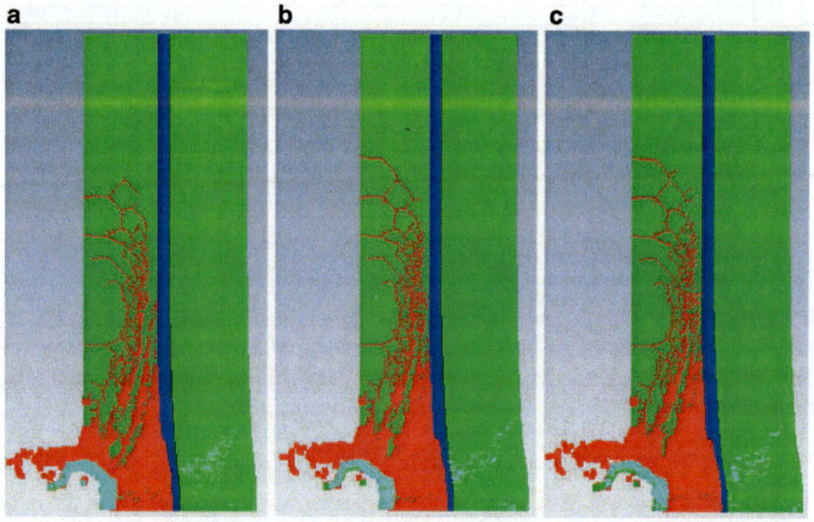


**Fig. 25.8** (**a**) Damage progression in spinel hard face containing rectangular cross section defects with width parallel to the line of impact after 40 μs; (**b**) damage progression after 43 μs; (**c**) damage progression after 46 μs

## 25.6 Conclusions

While the need for advanced materials solutions for protection of vehicles from ballistic threats continues to grow, the ability to predict material performance using advanced modeling tools increases. The current efforts underway in the U.S. Army include the use of ballistic modeling, ballistic testing, and historic knowledge of ballistic design to create structural armors for the transparent armor needs of the U.S. military [9]. The current paper has demonstrated the powerful use of computational modeling to predict the effects of defects on failure in ceramic materials. The current numerical analysis showed that various cross sectional surface defects filled with glass, with the exception of circular defects and rectangular defects with width parallel to the line of impact, did not behave better than unfilled defects of similar cross section. Although, the glass filled defects appear to initially slow down the exit of the steel projectile from the spinel strike face of the laminate more than the unfilled ones, eventually the projectile, for most of the defect cross sections exited the laminate at higher velocity. Unlike the unfilled defects, this is attributed to the fact that the projectile continues pushing the glass filled defect through the polycarbonate at high enough velocities to penetrate it. The combined kinetic energy of the projectile and the glass bead ahead of it deteriorated the resistance of the polycarbonate backing more compared to only the projectile kinetic energy. Defect shapes, such as rectangles and ellipses, favor the damage progression and reduce the resistance of the ceramic to penetration by the projectile. In agreement with our prior analyses the numerical study also showed that for the same cross sectional area, the orientation of the defect with respect to the line of impact can affect the ballistic behavior of the ceramic. Orientations of same area defects which result in reduced mass ahead of the penetrating projectile, such as rectangular cross sections with the width parallel or perpendicular to the line of impact, affect the resistance of the ceramic to penetration less for the same initial impact velocity.

## References

1. Fountzoulas CG, Sands JM, Gilde GA, Patel PJ (2009) "Applying modeling tools to predict performance of transparent ceramic laminate armors", 32 ICACC, advances in ceramic armor IV. Ceram Eng Sci Proc 29(6):45–53
2. Gatti A, Noone MJ (1970) Feasibility study for producing transparent spinel (MgAl2O4), AMMRC-CR-70-8, February 1970
3. Annex D (2006) Army modernization plan, "building, equipping, and supporting the modular force," March 2006
4. http://www.techassess.com/tech/spinel/spinel_prop.htm. Accessed 20 Sep 2007
5. Fountzoulas CG, Cheeseman BA, Dehmer PG, Sands JM (2007) A computational study of laminate transparent armor impacted by FSP. In: Proceedings of the 23rd international ballistic symposium, Tarragona, 14–19 April 2007
6. Fountzoulas CG, LaSalvia JC, Cheeseman BA (2007) Simulation of ballistic impact of a tungsten carbide sphere on a confined silicon carbide target. In: Proceedings of the 23rd international, ballistic symposium, Tarragona, 14–19 April 2007
7. ANSYS/AUTODYN Vol 11.0, Manual, Century Dynamics Inc., Concord
8. Gorbatsevich MI, Ginzburg AE (1991) Strength of materials, Springer ISSN 1573-9325, 23(4):392–397
9. Sands JM, Patel PJ, Dehmer PG, Hsieh AJ, Boyce MC (2004) Transparent materials safeguard the army's vision. AMPTIAC Quart 8(4):28–36
10. Fountzoulas CG, Cheeseman BA, Dehmer PG, Sands JM (2007) A computational study of laminate transparent armor impacted by FSP. In: Proceedings of the 23rd international ballistic symposium, Tarragona, 14–19 April 2007

# Chapter 26
# Effect of Particle Size Distribution on the Response of Metal-Matrix Composites

**Brandon McWilliams, K.T. Ramesh, and C. Yen**

**Abstract** An enhanced continuum model for ceramic particle reinforced metal matrix composites (MMCs) is used to explore the effect of particle size distribution on the variability in deformation response of heterogeneous microstructures. The model incorporates particle size dependent strengthening through a "punched" zone around the particles that is the result of an increase in dislocation density due to geometrically necessary dislocations generated by the mismatch in coefficients of thermal expansion of the particle and matrix. In this work, these zones are explicitly accounted for in mesoscale finite element simulations of representative heterogeneous composite microstructures consisting of randomly distributed particles in a metal matrix. Additionally, particle-matrix interface decohesion is incorporated through the use of cohesive zones. The results demonstrate that in the absence of material failure, the mean particle size of a distribution is sufficient to predict the elastic–plastic response with nominal variance in the response of the composite. The effect of interface strength on particle stresses is quantified and shown to reduce particle fracture in distributions containing large particles.

**Keywords** Metal matrix composite • MMC • Modeling • Particle reinforcement

## 26.1 Introduction

Ceramic particle reinforced metal matrix composites (MMCs) offer the potential for enhanced stiffness and yield strength over conventional lightweight structural alloys. A tradeoff in properties typically develops because high stiffness and/or strength is achieved by increasing the volume fraction of reinforcement particles, which often results in a significant decrease in ductility of the composite. This reduced ductility has thus limited the use of MMCs in structural applications. Therefore it is desirable to identify and understand the effect of heterogeneous microstructure design variables such as particle size, particle size distributions, particle morphology, and spatial distributions of particles, to narrow the design space and achieve higher strength/stiffness along with enhanced ductility.

The main strengthening mechanism in MMCs is load transfer from the matrix to the reinforcement and can be successfully modeled using classical continuum homogenization techniques [1]. However, for a given volume fraction of reinforcement it has been experimentally shown that the magnitude of strengthening is inversely proportional to the reinforcement size [2, 3]. Particle size dependent strengthening arises from several sources, which include the inherent size dependence of the strength of the ceramic particles due to the higher probability of larger particles more likely to have flaws of critical size, and from quench hardening (dislocation punching) due to mismatch between the coefficients of thermal expansion of the particles and matrix [2]. The dominant mechanism for particle size dependence at small strains is dislocation punching which is formed by geometrically necessary dislocations (GNDs) which develop to accommodate thermal strains as the composite is cooled from elevated processing temperatures and results in a hardened matrix zone (dislocation punching) surrounding the reinforcement. The size of the GND zone and the magnitude of the increase in dislocation density and subsequent increase in yield strength are dependent upon the size and volume fraction of particles

B. McWilliams (✉) • C. Yen
US Army Research Laboratory, Weapons and Materials Research Directorate, Aberdeen Proving Ground, MD 21005, USA
e-mail: brandon.a.mcwilliams.ctr@mail.mil

K.T. Ramesh
Center for Advanced Metallic and Ceramic Systems, Johns Hopkins University, Baltimore, MD 21218, USA

V. Chalivendra et al. (eds.), *Dynamic Behavior of Materials, Volume 1: Proceedings of the 2012 Annual Conference on Experimental and Applied Mechanics*, Conference Proceedings of the Society for Experimental Mechanics Series, DOI 10.1007/978-1-4614-4238-7_26, 

[4, 5]. Particle fracture contributes to the overall strengthening, but this mechanism is generally only active at larger strains at which stresses are high enough to fracture particles [6].

Classical plasticity modeling approaches do not include such size dependent mechanisms and have been modified to incorporate these effects, for example the work of Nan and Clarke [6]. Another approach is a direct numerical approach which explicitly accounts for the GND zones such as the periodic unit cell finite element model used by Suh et al. [7], and more recently by the current authors to explore the effect of size dependent strengthening mechanisms on the variability in response of multi-particle heterogeneous MMC microstructures with uniform size reinforcements [8]. In this paper, we extend this work to explore the effect of particle size distribution on the response and variability in response of heterogeneous multi-particle MMC microstructures. This method captures the probabilistic response of the material which is linked to stochastic variations in the local microstructure of the material. This microstructure dependent effective response can be used to generate variations in global response in macro scale simulations [9]. Although we choose to explore MMCs consisting of circular particles in the current model system, this approach can easily be extended to simulate additional particle morphologies and/or real microstructures without the assumptions on geometry and arrangement required by continuum homogenization and strain gradient plasticity techniques.

## 26.2 Method

We consider a 2-dimensional representative volume element (RVE) of an idealized heterogeneous MMC microstructure containing randomly distributed circular cross section particles. A given microstructure is defined by the particle size distribution and volume fraction of particles. The volume fraction of particles is fixed at 0.20 for all microstructures. Each particle is surrounded by a finite GND zone of higher strength matrix material. The size of the GND zone and the amount of strengthening in these regions are dependent upon the particle size and particle volume fraction (mean particle spacing). The GND zones were not allowed to overlap to minimize particle clustering effects on strain localization. A RVE with a length of 27 times the mean particle diameter was chosen.

Microstructures were generated with particles having a normal distribution of sizes, a bounded Pareto (power law) distribution, and uniform (mono) particle size. In all cases the mean particle diameter was 3 μm. For the normal distribution the standard deviation was 0.5 μm, and for the Pareto distribution the bounds were $1.3 < D_p < 8.0$ μm. For all cases, 5 realizations of each microstructure (Fig. 26.1) were generated to explore the effects of size dependent strengthening and particle size distribution on the variability of the mechanical response of the heterogeneous composites.

A 2D plane strain finite element model was used to conduct analysis of the microstructures using the commercial code ABAQUS/Explicit. The object oriented finite element program (OOF2) was used to generate 4-noded quad element meshes on each microstructure. A mesh density that gave at least two to three elements across the GND zone (smallest feature of interest) was found to give converged values of modulus for the microstructures studied in this paper, and typically resulted in ~80,000 elements per microstructure realization. We note that a fully 3-dimensional RVE would be the ideal system for analysis, however for the present systems which include size dependent GND zones this would require a prohibitive amount of CPU time to run multiple realizations of each microstructure with a sufficient RVE size to be statistically relevant. Therefore we choose to make the tradeoff between computational efficiency and accuracy and go with the 2D RVE. Further, it has been previously demonstrated that 2D models are fully capable of capturing the response of MMCs with little sacrifice in accuracy compared to fully 3D models [10].

A symmetry boundary condition is applied to the bottom, the left boundary is fixed, and a uniform displacement boundary condition is applied to the right boundary to deform the composite in tension. The effective true stress ($\bar{\sigma}_t$) and true strain ($\bar{\varepsilon}_t$) of the volume element are calculated from the reaction force ($F_x$) and displacement ($u_x$) at the right boundary using the following equations:

$$\bar{\varepsilon}_t = \ln\left(1 + \frac{u_x}{w_0}\right) \tag{26.1}$$

$$\bar{\sigma}_t = \frac{1}{w} \int_0^{h_0} F_x|_{x=w} dy \tag{26.2}$$

where $w$ is the current width, $w_0$ and $h_0$ are the initial width and height of the 2D RVE respectively.

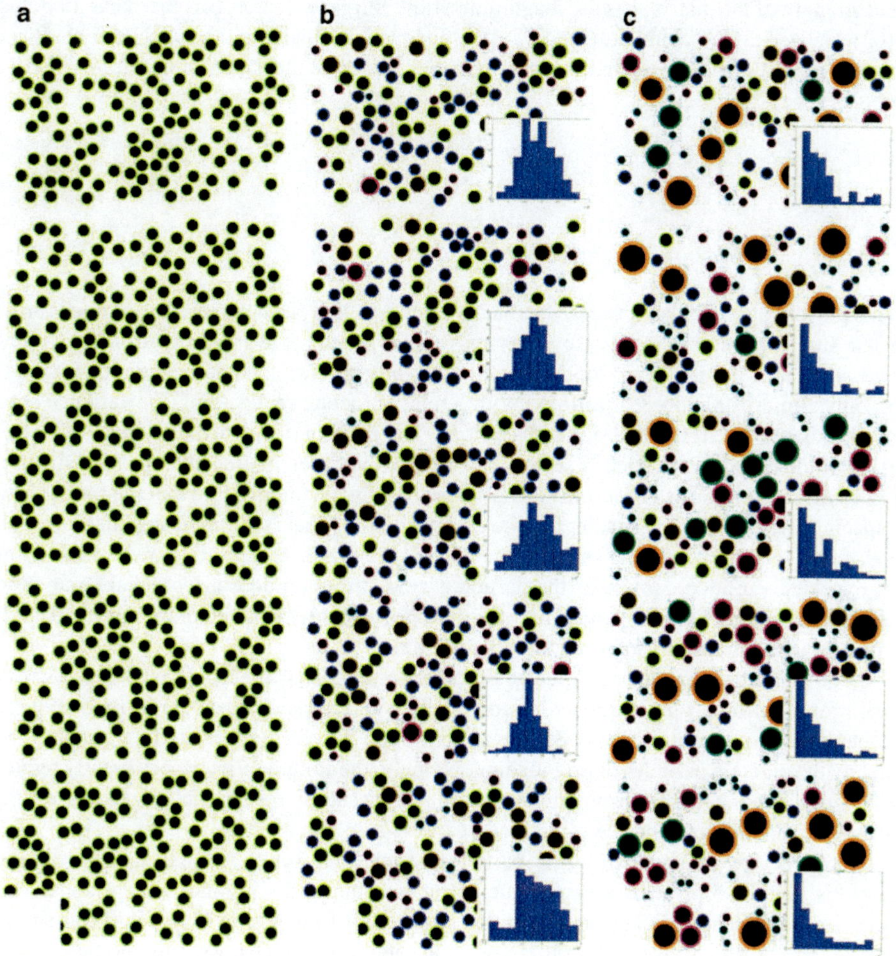


**Fig. 26.1** Uniform particle size (**a**), normal particle size distribution (**b**), and Pareto particle size distribution (**c**) microstructure realizations modeled in this paper

**Table 26.1** Summary of material properties used for matrix and reinforcement particles in simulations

| Particle properties | | Matrix properties | |
|---|---|---|---|
| $E^*$ | 427 GPa | $E$ | 76 GPa |
| $v^*$ | 0.17 | ν | 0.33 |
| $a^*$ | $4.3 \times 10^{-6}$ K$^{-1}$ | $a$ | $23.63 \times 10^{-6}$ K$^{-1}$ |
| | | $b$ | 0.28 nm |
| | | $\sigma^y$ | 208 MPa |
| | | $\tau$ | 104 MPa |

The ceramic reinforcement particles are assumed to be isotropic elastic and the aluminum alloy matrix and GND zones were modeled as isotropic elastic–plastic. A summary of material properties used in the simulations is provided in Table 26.1. Strain hardening of the matrix and GND zones were included using the relation $\sigma = 464(0.00274 + \varepsilon_p)_{0.136}$ from [7] with the GND zones having higher initial yield strength calculated using Eq. 26.3, the formulation of which is presented elsewhere [7] based on particle size, volume fraction, and a thermal quenching of $\Delta T = 474$ K. The strength of the GND zone $\sigma$ is described by:

$$\sigma = \sigma_y + aGb\sqrt{\rho_g} \tag{26.3}$$

where $\sigma_y$ is the yield strength of the matrix, $b$ is the magnitude of the Burger's vector, $G$ is the shear modulus of the matrix, $a$ is a geometric factor which is 1.25 for aluminum, and $\rho_g$ is the dislocation density in the punched zone. It is noted that in this work no material damage or failure is considered, which limits the validity of the data to small strains, particularly for tensile loading conditions.

## 26.3 Results

Figure 26.2a presents the average calculated stress at a macro strain of 1 % versus the number of microstructure realizations for each of the particle size distributions. In all cases, the stress converges to the average very quickly which can also be seen in the variance of the stress at 1 % strain (Fig. 26.2b). For all cases the variance (standard deviation divided by mean) is less than 5 % even with only 2 microstructure realizations. With 5 microstructure realizations, the variance appears to converge to an acceptable level of 0.3 % for all particle size distributions. The particle size distribution has little effect on the overall stress–strain response of the composite, which is seen in Fig. 26.3 which presents the average (5 realizations) stress–strain response for each microstructure. This is consistent with prior work which demonstrated that the mean particle size controls the mean particle spacing and hence strain gradient of the composite [6]. This result indicates that in the absence of failure the mean particle size and volume fraction is sufficient to predict the elastic–plastic response of the composite within an acceptable level of expected variation. However, the particle size distribution does affect local strain gradients which are expected to have significant impacts on the variability of failure processes which are controlled by local damage initiation processes such as ductile void nucleation and growth in the matrix.

As an example of how failure mechanisms affect the response, Fig. 26.4 presents the average stress in the ceramic particles versus strain for a uniform 3 μm particle size composite with cohesive zones used to model interface separation. It can be seen that the degree of interface failure which was controlled by varying the interface strength, greatly affects the stresses experienced by the ceramic particles during loading. Namely, the failure of interfaces greatly reduces the stresses in the particles. This has implications for microstructures containing a distribution of particle sizes. For example, a composite with a Pareto particle size distribution, and the same mean particle size, the average stresses in the particles will be equal except the Pareto microstructure will contain larger particles which have lower strength and have more propensity for failure. To illustrate this, Fig. 26.5 presents a Pareto microstructure modeled with and without cohesive zones (interface failure) with particle size dependent particle strength. In this case the Griffith criteria $\sigma_c = K/\sqrt{d}$ was used to determine the particle strength with $K = 2.4$ MPa $\sqrt{\text{m}}$ [6]. Here it is seen that interface failure lowers the particle stresses and prevents particle fracture at a given level of strain.

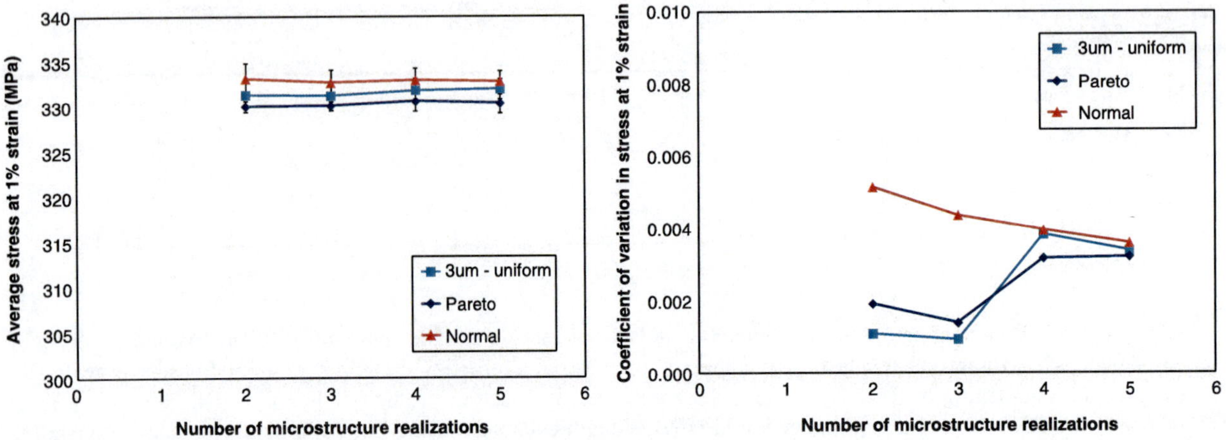


**Fig. 26.2** Simulation results for the average stress at 1 % overall strain versus the number of microstructure realizations (**a**), and the coefficient of variation (standard deviation dived by mean) versus the number of microstructure realizations (**b**) for uniform, normal, and Pareto distributions with mean particle size of 3 μm. *Error bars* in (**a**) represent standard deviation

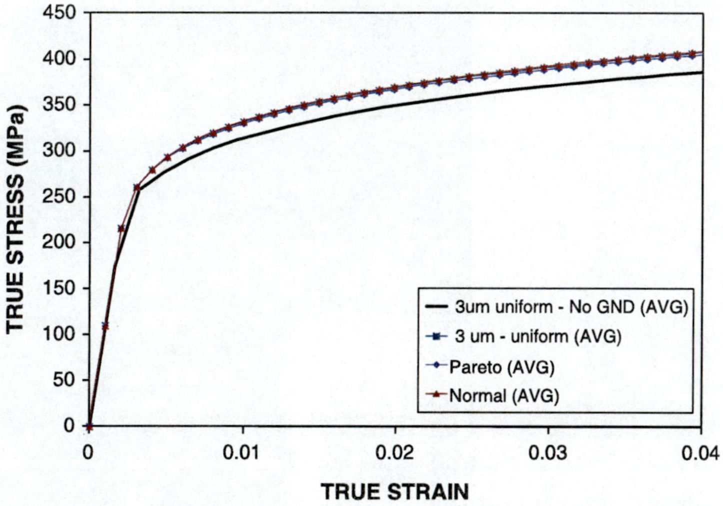


**Fig. 26.3** Average stress–strain curves for 5 simulated MMC microstructures with 3 μm mean particle size with different particle size distributions

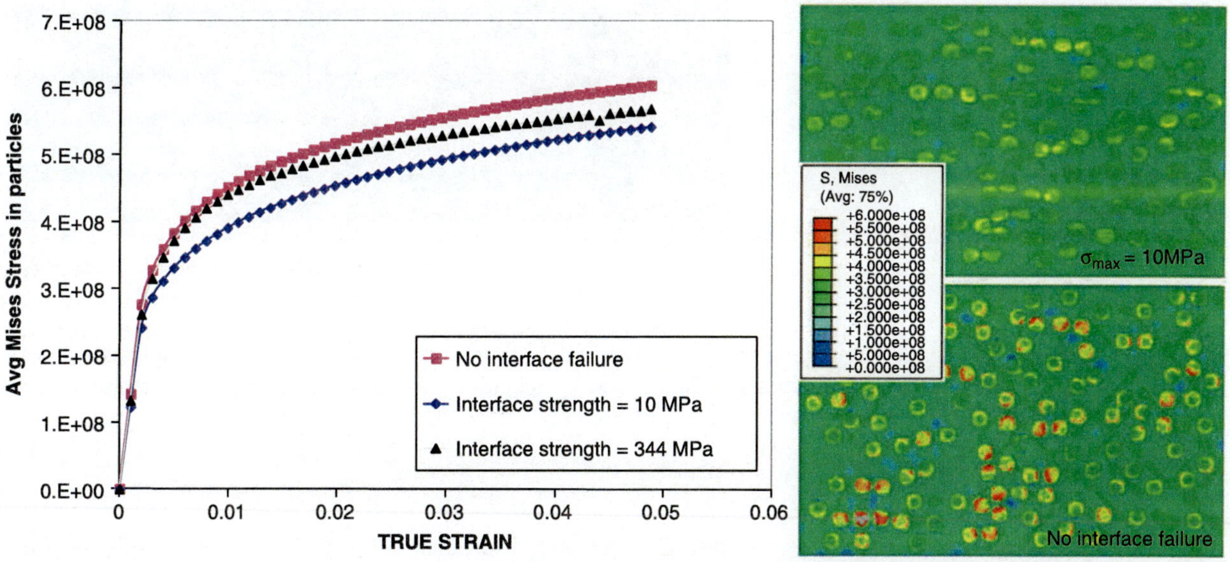


**Fig. 26.4** Average stress in ceramic particles versus strain for a microstructure with uniform 3 μm particles and three level of interface strength and Mises stress contours showing reduction in particle stresses when particle-matrix interface separation is considered

## 26.4 Conclusions

This paper presents an enhanced continuum approach for modeling particle size dependent strengthening of MMCs with particle size distributions. The numerical results are consistent with previous results using modified classical plasticity modeling, which show that in the absence of material failure, the mean particle size is sufficient to predict the elastic–plastic response of an MMC with a given volume fraction of reinforcement within an acceptable level of variance when accounting for the heterogeneous nature of the materials. Particle-matrix interface separation was shown to reduce the magnitude of stresses

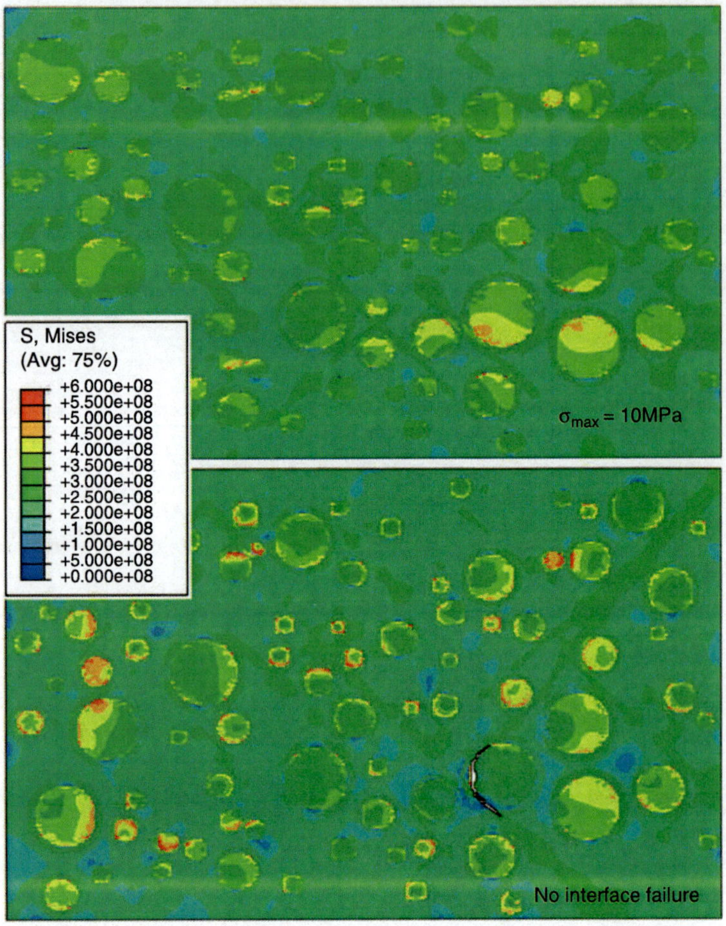


**Fig. 26.5** Mises stress contours for a microstructure with a Pareto particle size distribution with (*top*) and without (*bottom*) interface failure. Particle fracture can be observed in the case without interface failure

present in the ceramic particles, which implies that for a given mean particle size, volume fracture, and applied strain, a microstructure with a wider distribution of sizes (e.g. Pareto) will be more likely to have damage initiation through particle fracture. The affect of failure mechanisms on the variability in heterogeneous MMC response will be addressed in future work.

## References

1. Christman T, Needleman A, Suresh S (1989) An experimental and numerical study of deformation in MMCs. Acta Met 37:3029–3050
2. Lloyd DJ (1994) Particle reinforced aluminum and magnesium matrix composites. Int Mater Rev 39:1–23
3. Ling Z (2000) Deformation behavior and microstructure effect in 2124Al/SiCp composite. J Comp Mat 34:101–115
4. Shibata S, Taya M, Mori T, Mura T (1992) Dislocation punching from spherical inclusions in a metal matrix composite. Acta Met Mater 40:3141
5. Dai LH, Ling Z, Bai YL (2001) Size-dependent inelastic behavior of particle-reinforced metal-matrix composites. Comp Sci Tech 61:1057
6. Nan CW, Clarke DR (1996) The influence of particle size and particle fracture on the elastic/plastic deformation of metal matrix composites. Acta Mat 44(9):3801–3811
7. Suh YS, Joshi SP, Ramesh KT (2009) An enhanced continuum model for size-dependent strengthening and failure of particle-reinforced composites. Acta Mat 57(19):5848–5861
8. McWilliams B, Ramesh KT, Yen C (2010) Mesoscale modeling of size dependent strengthening of particle reinforced metal matrix composites. In: Proceedings of the 16th US National Congress of theory and applied mechandics, State College, 2 June 27-July 2010
9. Acton K, Graham-Brady L (2009) Meso-scale modeling of plasticity in composites. Comp Methods Appl Mech Eng 198:920–932
10. Chawla N, Chawla KK (2006) Microstructure based modeling of the deformation behavior of particle reinforced metal matrix composites. J Mater Sci 41:913–925

# Chapter 27
# Molecular Dynamics Simulation on Plastic Deformation of Nanocrystalline Copper

**Lu Guo, Wang Shuaichuang, and Zhang Guangcai**

**Abstract** Molecular dynamics (MD) simulation is used to simulate the plastic deformation of a nanocrystalline model Cu sample with an average grain size of 5 nm containing 27 grains under high strain rate shear loading. Results show that the plastic deformation of nanocrystalline copper can be divided into two parts: grain sliding dominated part, dislocation motion and grain sliding together dominated part. When the shear strain is less than 4%, the macro plastic deformation is the result of relative sliding of grains, grain rotation does not contribute to the overall plasticity, but it plays a supporting role in grain network adjustment. In this part, the deformation process can be described by viscous flow model. When the strain is greater than 4%, a large number of partial dislocations nucleate in grain boundaries and then slip into grains, most of them finally annihilate in corresponding grain boundaries and some deformation twins emerge near triple junctions. Therefore, the plastic deformation is the result of grain sliding, dislocation slipping and twinning. During shear deformation, shear planes involving several grain boundaries which are sub parallel to the loading plane have been observed.

**Keywords** Molecular dynamics • Nanocrystalline copper • Plastic deformation • Grain boundary • Dislocation

## 27.1 Introduction

The mechanical properties of metals depend on the characteristics and properties of the internal defect structures. The plastic deformation mechanics of microcrystal is closely related to the behavior of dislocations, such as slip, reaction, proliferation and aggregation. A reduction in the average grain size results in an increase in the yield strength of metals, which is known as the Hall–Petch relationship [1]. When the average grain size is less than 100 nm, the percentage of grain boundary atoms is gradually increasing and dislocation behavior is suppressed, the relationship between yield stress and average grain size no longer meets the Hall–Petch formulation. Recent studies indicates that a further decrease in grain size (less than ~10 nm) can results in the weakening of metal and this weakening is referred to as the inverse Hall–Petch behavior [2].

In nanocrystalline metals, grain boundaries become a major defect structure and the size, type and network characteristics of grain boundaries determine their special mechanical properties, such as high yield strength, high fracture strength, superplasticity and so on. A quantitative clarification of the relationship between macroscopic mechanical properties and microscopic structures of nanocrystalline is a challenging work, because the details on grain size and its distribution, grain boundary type, grain boundary network characteristics, impurities and porosity are not well obtained experimentally, not only that, the mechanical properties of nanocrystalline is also affected by temperature, loading methods and rate. However, experimental and theoretical studies in recent years have made some interesting progresses. The structure analysis of grain boundaries shows that grain boundaries are not highly disordered amorphous liquid-type but with certain regular patterns and structure units [3]. Swygenhoven points out by simulation that grain boundary sliding contains atomic shuffling and stress-assisted free-volume migration, the corresponding grain boundary network adjustment process includes the grain boundaries and triple junctions migration and dislocation activities [4]. At high temperature, the plastic deformation process of nanocrystalline is different from that of the room temperature. Shear planes involving several grain boundaries are observed during the deformation and they are results of grain boundary migration, intragranular slip, rotation and

L. Guo (✉) • W. Shuaichuang • Z. Guangcai
Institute of Applied Physics and Computational Mathematics, 2 Fenghao East Rd, 100094 Beijing, China
e-mail: lu_guo@iapcm.ac.cn; stallenlu@gmail.com

V. Chalivendra et al. (eds.), *Dynamic Behavior of Materials, Volume 1: Proceedings of the 2012 Annual Conference on Experimental and Applied Mechanics*, Conference Proceedings of the Society for Experimental Mechanics Series, DOI 10.1007/978-1-4614-4238-7_27, © The Society for Experimental Mechanics, Inc. 2013

coalescence of grains [5]. These results give lots of useful microscopic images of defect structures and the evolution characteristics, which lays the foundation for proposing a reasonable model of nanocrystalline structure. We must point out that the microstructure of nanocrystalline depends on the way it was prepared and the detail characteristics are not accurately measured in experiments, the simulation sample can only be constructed according to the known characteristics of structures. In many studies, the samples are highly idealized, so some of their conclusions are not general and need to be carefully examined.

Many microscopic activities which are interrelated exist during the deformation of nanocrystalline with smaller grain size (3 ~ 10 nm), such as grain sliding, grain rotation, grain migration, deformation twinning and activities of dislocations. To develop a scientific polycrystalline model, it is necessary to obtain the characteristics of grain boundaries, dislocations and other defects and to understand the relationship among these defects. The high-resolution TEM experiments can be used to study defect structure of deformed crystal, but cannot track the microstructure evolution. In recent decades, with the rapid development of computer hardware and large-scale parallel computing frameworks, atomic-level molecular dynamics simulations have been wildly used to study the structure evolution of sub-micron materials in several picoseconds. With the development and application of different loading methods, molecular dynamics simulation can mimic more loading phenomena, such as constant strain rate loading, which is suitable for the case of overall uniform deformation. Due to the inherent limit of molecular simulation, the high deformation rate ($>10^7$/s) loading must be used to obtain results in an acceptable simulation time, but the rate must be suitable to avoid unphysical behavior, the deformation velocity is usually less than the sound speed of sample. Swygenhoven studies the dislocation emission mechanism of nanocrystalline nickel with average grain size of 5, 12 and 20 nm. The results show that there are no partial dislocations in 5 nm samples. However, for the total deformation is small (<4%) and the effect of size and shape of each grain is not carefully taken into account, the conclusion is not general. Studies on the high temperature deformation of nanocrystalline indicate that grains can rotate during deformation, but the characteristics and evolution of rotation and its relationship with grain sliding are lack of detailed analysis. In addition, the most commonly used loading methods are shock loading, uniaxial/triaxial tension and compression loading, but these methods cannot avoid the body effect (volume expansion or compression) effects.

Aiming at these issues, molecular dynamics simulation is used to study the deformation of nanocrystalline copper under high strain rate shear loading. The paper focuses on analyzing the evolution of defect structures, characteristics of grain sliding and rotation, contribution of grain sliding and rotation to the overall plastic deformation and properties of shear plane.

## 27.2 Simulation Procedure and Analysis Techniques

The grain structure of nanocrystalline depends on the specific preparation process. We create samples that have similar grain structure of nanocrystalline metals generated by inert gas condensation. All grains are almost equiaxed, dislocation free and separated by narrow grain boundaries. In our simulation, the sample is generated using a Voronoi construction: a cubic whose side length is 15 nm is created, a set of 27 grain centers are chosen at random and then the Voronoi tessellation is constructed with periodic boundary condition, perfect fcc copper atoms with randomly selected orientation are created in each Voronoi cell to avoid texture structure, and one atom of the atom pair whose distance is smaller than 0.2 nm is deleted. The initial sample contains about 274 thousand atoms and the average grain size is about 5 nm. In the limit of a large number of grains, the Voronoi construction will generate a grain-size distribution close to a log-normal distribution [6].

The Mishin Cu EAM potential [7] which can well describe the twin formation energy and stacking fault energy is chosen in the simulation using LAMMPS. The initial sample is relaxed to reach the state of minimum potential energy by conjugate gradient method, and then it is relaxed in the isobaric-isothermal ensemble under both the pressure 0 bar and temperature 300 K for 200 ps. In equilibration process, the periodic boundary condition is used and the timestep is 0.1 fs. The final density of the sample is about 98% of the perfect crystal value and the percentage of defect atoms which locate in the grain boundaries is about 29.3%. Figure 27.1 shows the relaxed sample. The relaxed sample is sheared in xz plane (yx deformation) and the strain rate is $2 \times 10^8$/s. During the shear loading simulation, simulation is carried out in the canonical ensemble at 300 K for 1,000 ps, the timestep is 1 fs and the final shear deformation is 20%.

The total and local stress is needed to analyze the mechanical properties of deformation of metal. In this paper, the per-atom stress is calculated by the following formula:

$$\sigma_{i,\alpha\beta} = \frac{1}{V_i}\left(-m_i v_{i,\alpha} v_{i,\beta} + \frac{1}{2}\sum_{j\neq i} \frac{\partial E_p}{\partial r_{ij}} \frac{r_{ij,\alpha} r_{ij,\beta}}{r_{ij}}\right) \tag{27.1}$$

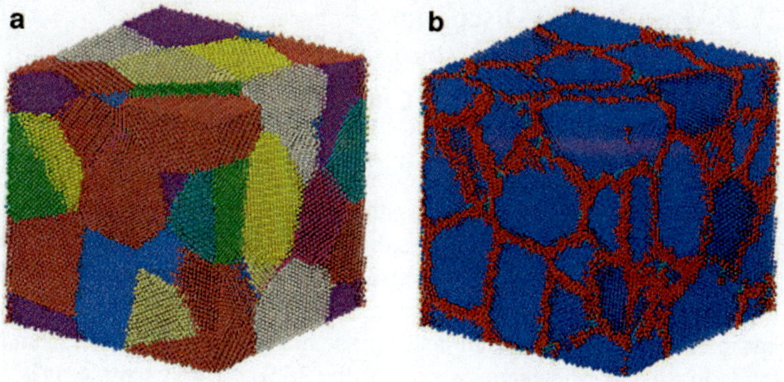


**Fig. 27.1** (Color Online) The relaxed sample of nanocrystalline copper. (**a**) Different color represents different grain. (**b**) The *red* symbols refer to boundary atoms, the *blue* ones represent atoms with local fcc order and the *cyan* ones represent atoms with local hcp order

where $\sigma_{i,\alpha\beta}$ is the $\alpha\beta$ component of the stress tensor for atom $i$, $V_i$ is the volume occupied by atom ($\Sigma_i V_i = V$, where $V$ is the total volume of the sample), $m_i$ is the mass of atom $i$. $v_{i,\alpha}$ is the $\alpha$ component of the velocity of atom $i$, $r_{ij}$ is the distance between atom $i$ and $j$, and $r_{ij,\alpha}$ is a component of the vector from $i$ to $j$. For a given region, the average stress tensor is defined as follow:

$$\bar{\sigma}_{\alpha\beta} = \frac{\sum_i \sigma_{i,\alpha\beta} V_i}{\sum_i V_i} \tag{27.2}$$

In nanocrystalline sample, the orientation of each grain is different and there are a high percentage of boundary atoms. For such as a system, the deformation under shear loading is non-uniform. The constant rate loading [8] is a commonly used method to control the total deformation rate of a system in molecular simulation. However, it lacks an efficient way to analysis the details of deformation of each grain. The statistical moment method [9] is a newly proposed one to the issue. In this method, the local strain distribution is calculated according to the moment of atom displacements and positions before and after deformation, the moment links the macroscopic strain and the microscopic physical quantities (position and displacement). For a system containing N particles, the displacement gradient tensor $\boldsymbol{\beta}$ is as follow:

$$\boldsymbol{\beta} = \mathbf{M}_s^{-1} \cdot \mathbf{M}_b \tag{27.3}$$

where $\mathbf{M}_s$ is the second central moment of the initial atom positions, $\mathbf{M}_b$ is the second mixed central moment, they are defined as follow:

$$\mathbf{M}_s = \frac{1}{N} \sum_i^N (\mathbf{X}_i - \overline{\mathbf{X}})(\mathbf{X}_i - \overline{\mathbf{X}}) \tag{27.4}$$

$$\mathbf{M}_b = \frac{1}{N} \sum_i^N (\mathbf{X}_i - \overline{\mathbf{X}})(\mathbf{U}_i - \overline{\mathbf{U}}) \tag{27.5}$$

where $\overline{\mathbf{X}} = \sum_i^N \mathbf{X}_i/N$, $\overline{\mathbf{U}} = \sum_i^N \mathbf{U}_i/N$. The strain tensor $\boldsymbol{\varepsilon}$ is the symmetric part of $\boldsymbol{\beta}$ and the rotation tensor $\boldsymbol{\Omega}$ is the antisymmetric part. For the strain analysis of each grain, the statistical region is set as the region occupied by the grain. To evaluate the overall equivalent strain, all atoms in the deformed sample should be involved.

In order to identify the defect structures during the deformation of nanocrystalline, the common neighbor analysis (CNA) developed by Honneycutt and Andersen is used [10]. According to connection between each atom and its neighbors, all atoms can be classified into three types: fcc-type, hcp-type and other-type. The use of CNA makes dislocations, stacking faults and grain boundaries visible in the simulation. Two adjacent 111 planes of hcp atoms are the intrinsic stacking faults, if they are separated by a single 111 plane of fcc atoms, those are the extrinsic stacking faults, whereas a single 111 plane of hcp atoms is just the twin boundary. The majority of dislocation cores and grain boundaries are other-type atoms.

## 27.3 Results and Discussions

### 27.3.1 A. Microstructural Changes

Figure 27.2 show the stress–strain curve of nanocrystalline copper under shear loading. The point A in stress-stain curve refers to 4% shear strain. According to the shape of the curve, the deformation can be divided into two parts: an increasing part in which the shear strain is less than 4% and a plateau part where the shear strain is greater than 4%. The average flow stress of the plateau part is about 1.043 GPa. By tracking the evolution of dislocations, grain boundaries and stacking faults, it indicates that, in the increasing part, several deformation twins nucleate around triple junctions and no dislocations emit, but in the plateau part, partial dislocations firstly emit from grain boundaries and then gliding into grains leaving stacking faults in. Figure 27.3a show a slice of the 3.2% sheared sample, it is parallel to the y-plane and its thickness is 2.5 nm.

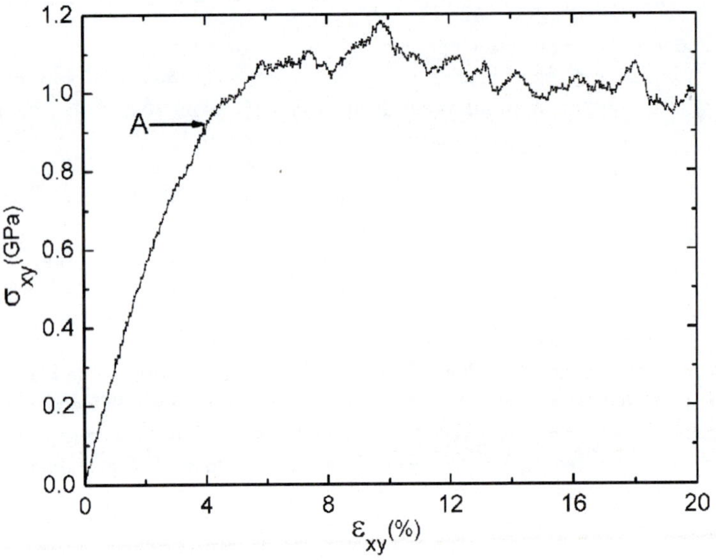


**Fig. 27.2** Stress ($\sigma_{xy}$) – strain ($\varepsilon_{xy}$) cuves for the system in Fig. 27.1, deformed under shear loading at room temperature

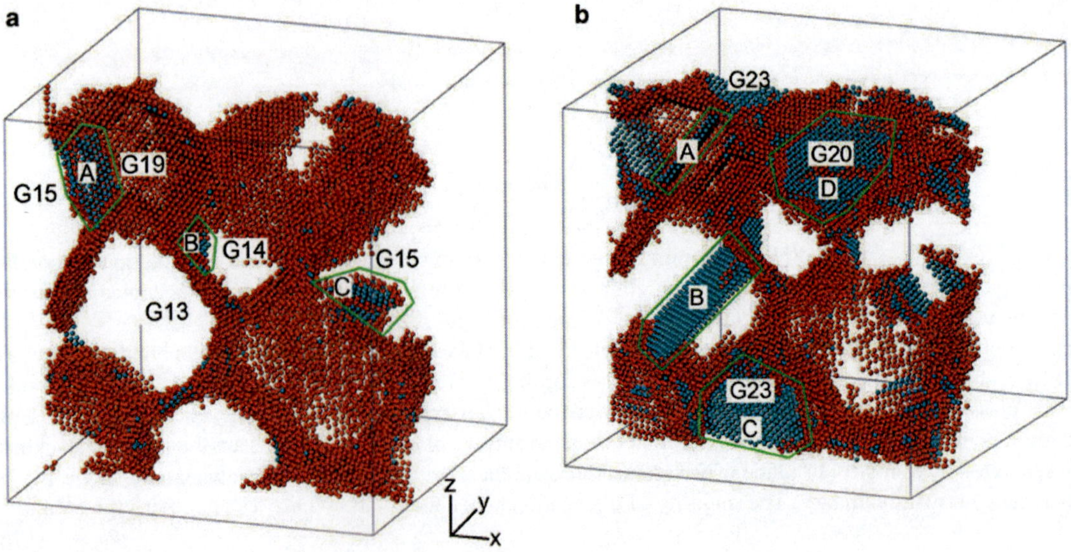


**Fig. 27.3** (Color Online) Cross section of the deformed sample. The shear strain is 3.2% in (**a**) and 12% in (**b**)

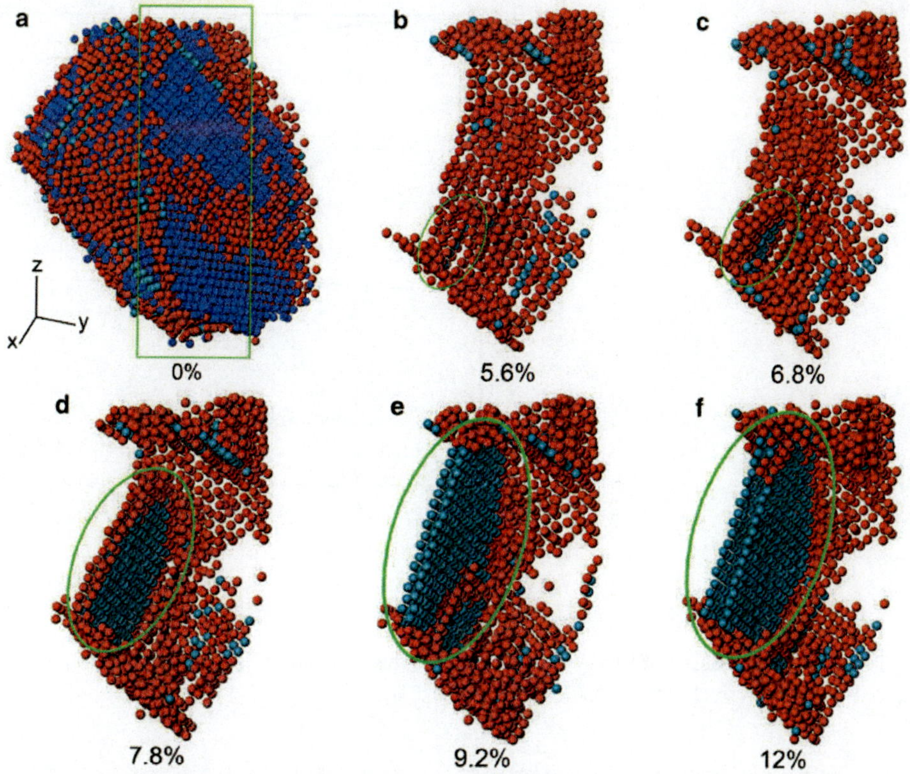


**Fig. 27.4** (Color Online) Snapshots of the formation of extrinsic stacking fault in grain 23. (**a**) is grain 23, the *blue* atoms are fcc type, hcp atoms are colored as cyan, the *red* ones are atoms located in grain boundary and dislocation core. All other pictures ((**b**) ~ (**c**)) are corresponding to the rectangle region of picture (**a**) and perfect atoms are deleted

The single hcp plane in region A of Fig. 27.3a is a twin plane between grain 15 and 19. In region B of Fig. 27.3a, the single hcp plane is a deformation twin in grain 14, which is close to the triple junction of grain 13, 14 and 19. In region C, the two adjacent hcp planes generated in the relaxation process is in grain 15 and the edge of the two planes is partial dislocation. Figure 27.3b is a slice of the 12% sheared sample, the two adjacent hcp planes in grain 19 of Fig. 27.3b is generated by the gliding of the partial dislocation which nucleates in the grain boundary of grain 13 and 19, and similarly, the stacking fault in grain 13 is the result of gliding of partial dislocation which is emitted from the grain boundary of grain 13 and 14. In grain 23, the two hcp planes is separated by a single fcc plane and it is extrinsic stacking fault, but the one in grain 20 is intrinsic stacking fault.

By analyzing the distribution and generation of stacking faults during shear deformation of nanocrystalline copper, it indicates that almost all stacking faults are located in grains and they are the results of dislocation gliding. The partial dislocations are generated in grain boundaries, most of them annihilate in the opposite grain boundaries, only a few ones are retained in grain interior. Figure 27.4 shows the defects evolution of grain 23 during shear deformation. Figure 4b–f corresponds to the green rectangle region of Fig. 27.4a, and fcc atoms are deleted for the convenience of observing. There are several hcp atoms located around triple junctions and grain boundaries (in Fig. 27.4a), the first partial dislocation nucleates in grain boundary when the shear strain is 5.6% and then it glides into grain 23 leaving intrinsic stacking fault behind, further gliding enlarges the area of stacking fault until it reaches the opposite grain boundary. The second partial dislocation appears when the strain is about 9.2%, and it nucleates in the region of grain boundary which is just adjacent to left plane of the stacking fault. Due to its gliding, the former stacking fault changes to extrinsic one.

The CNA technology is used to track the evolution of all types of defect atoms and Fig. 27.5 shows the change of the percentage of defect atoms with strain. In the figure, it is found that when the shear strain is less than 4%, due to absence of dislocation activities and little deformation twinning, the percentage of each type of defect atom has no major change and they are respectively 69.3%, 29.3% and 1.4%. In the plateau part, partial dislocations are generated in grain boundaries and then glide into grains along 111 planes, due to the enhanced dislocation activities, part of fcc-type atoms are change to

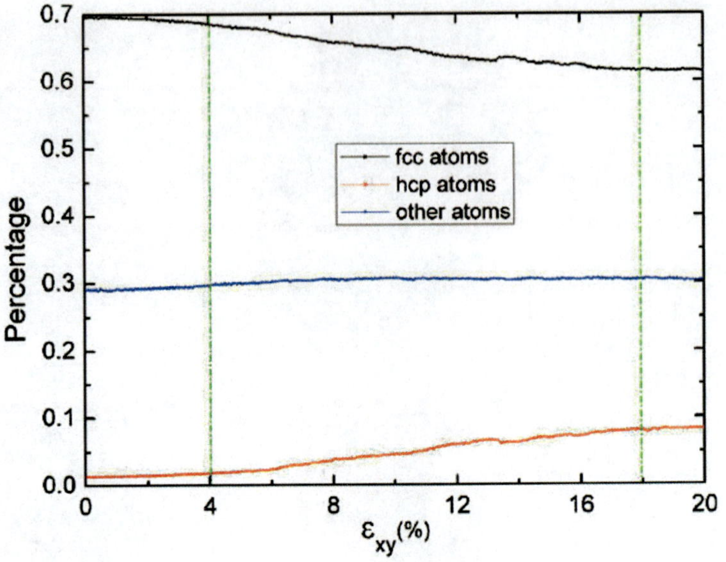


**Fig. 27.5** (Color Online) Percentage of three types of atoms

hcp-type resulting in the rapid increasing of hcp-type atoms. When the shear strain is larger than 18%, the percentages reach new stable value, and they respectively are 61.5%, 30.4% and 8.1%.

### 27.3.2 *B. Grain Slip and Rotation*

The relative slip between grains is observed by calculating the displacement of atoms. In Fig. 27.6, the direction of displacement of adjacent grain 6 and 12 is opposite, the relative displacement increases as the deformation increases. By comparing the overall spatial distribution of atoms of each grain, small rotation is a very popular grain activity. Figure 27.7 shows the overlapped picture of grain 23, the red atoms represent the initial grain and the green ones refer to the 8% sheared grain. Rotation of grain 23 can be easily seen. The rotation axis is close to z-axis, the direction of rotation is counterclockwise and the rotation angle is about 2°.

The local characteristics and atomic process on grain slip and rotation have been discussed in detail in lots of literatures [4, 5]. However, few work focus on the overall analyzing of the slip and rotation of all grains and the relationship between slip and rotation. For detailed description of grain slip and rotation, a spherical region is selected within each grain and then tracks the motion of atoms in the sphere. According to the spatial distribution of atoms in the selected spherical region before and after deformation, the least square method is used to calculate the deformation gradient matrix **F** and the translation vector **t** of the spherical region (in center-of-mass coordinate system of each grain). The matrix **F** can be decomposed into a multiplication of a rotation matrix **R** and deformation matrix **U** (**F** = **R.U**). The net rotation matrix $\mathbf{R}_0$ of each grain is obtained via deducting the rotation part imposing by external shear. Thus the rotation angle and axis can be computed using matrix $\mathbf{R}_0$. Figure 27.8 show the transitional vectors of all grain in the 4% sheared sample. It can be seen that the translation of grains in the middle part of the sample is smaller than the upper and lower ones and it's consistent with the continuous model on shear process. Figure 27.9 shows rotation of each grain in the 4% sheared sample. Most rotation angles are around 1° (0.018 in radian), the rotation angle of grain 23 is the largest and it is about 1.9°. The instantaneous rotation axis of each grain changes during deformation, but its variation ranges is small. Thus, it is possible determined by the geometrical configuration of grain boundaries. Due to further analyzing which shows that there is no obvious consistency orientation among the rotation direction of all grains, it indicates that grain rotation has little contribution to the overall plastic deformation.

Due to the relative slip of grain, grain boundary network needs to adjust and the possible mechanisms involve grain boundary migration, triple junction migration and grain rotation [4]. For two adjacent grains, we can analyze the relationship between grain slip and rotation by investigating the activity of the grain boundary. Figure 27.10 is a schematic on the motion of two adjacent grains, point $O_1$ and $O_2$ are grain centers and A is the intersection point of line $O_1O_2$ and grain boundary,

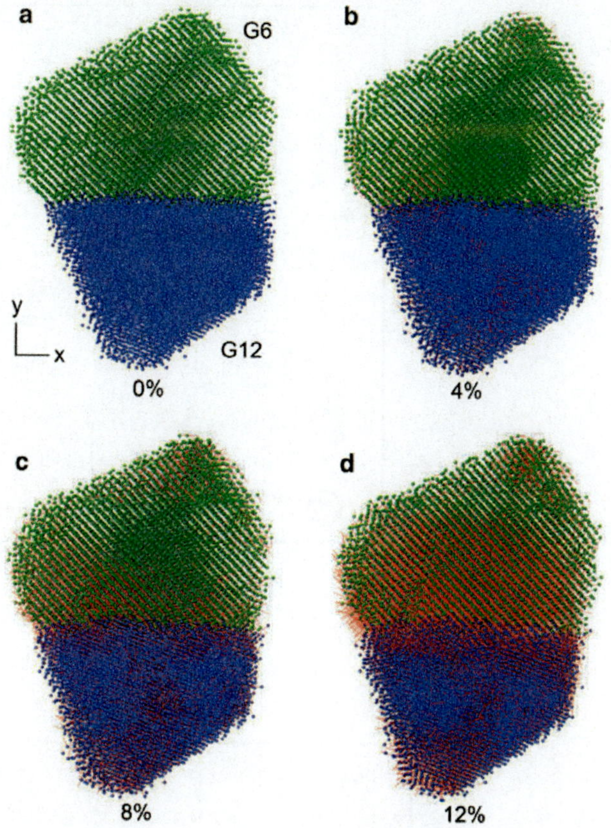


**Fig. 27.6** (Color Online) Relative slipping between grain 6 and 12. The *vector lines* represent displacements of atoms

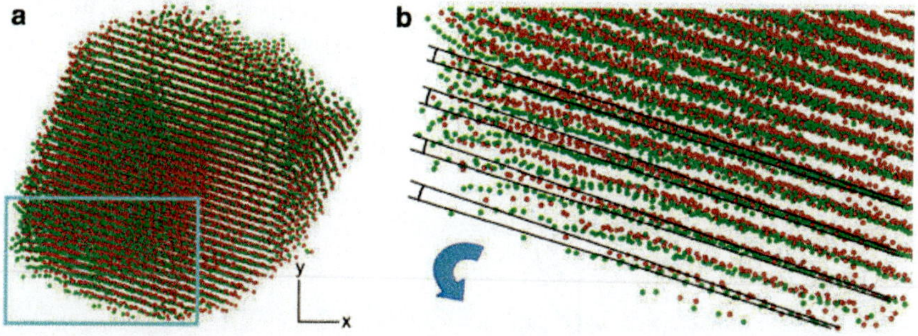


**Fig. 27.7** (Color Online) Rotation of grain 23. The *red* atoms represent the initial grain and the *greens* refer to the grain after 8% shear loading. In (**a**) the viewpoint direction is z and (**b**) is the enlarged picture of the *blue rectangle* region of (**a**)

$\mathbf{r}_1$, $\mathbf{r}_2$ and $\mathbf{r}$ present vector $OO_1$, $OO_2$ and A respectively, $\omega_1$ and $\omega_2$ are instantaneous angular velocities, $\mathbf{v}_1$and $\mathbf{v}_2$ are translational velocities of two grains. The total velocity of point A is

$$\mathbf{v}_A = \mathbf{v}_1 + \boldsymbol{\omega}_1 \times (\mathbf{r} - \mathbf{r}_1) - \mathbf{v}_2 - \boldsymbol{\omega}_2 \times (\mathbf{r} - \mathbf{r}_2) \tag{27.8}$$

To measure the incompatibility of rigid body motion between grains, a parameter $C_A$ is defined as $|v_A|/|v_1 - v_2|$. If grain motion is compatible, $C_A$ is equal to zero. If there are no grain rotation ($\omega_1 = \omega_2 = 0$), $C_A$ is equal to 1, it refers to fully incompatibility of grain motion. Due to the $C_A$ of all grain boundaries are near 1, it shows the incompatibility of grain motion and further indicates that grain rotation is not a dominated mechanism of nanocrystalline deformation.

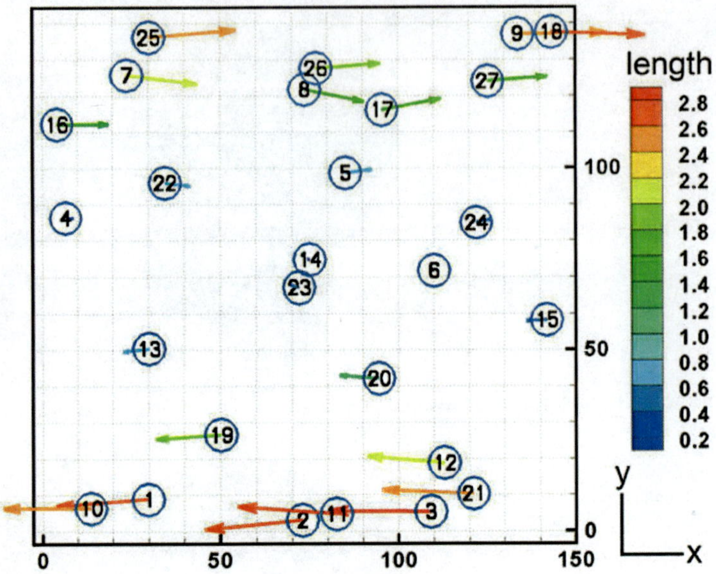


**Fig. 27.8** (Color Online) Distribution of translation vectors of grains in 4% sheared sample

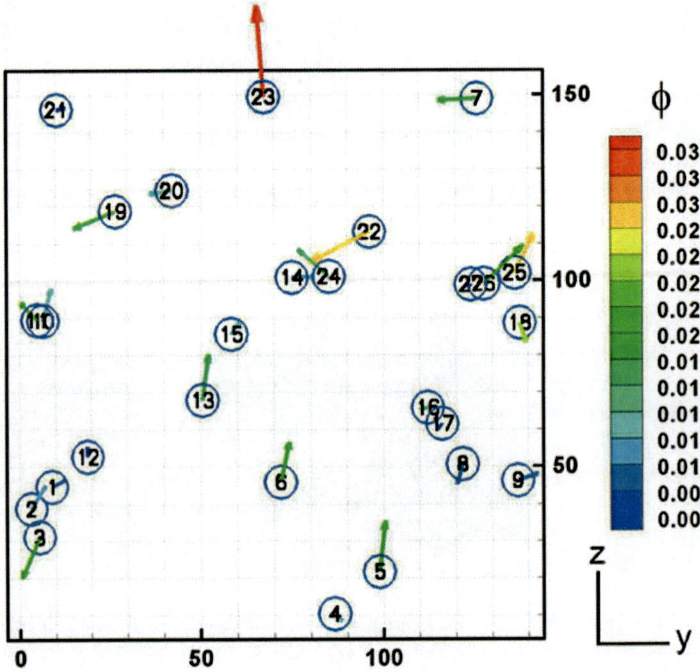


**Fig. 27.9** (Color Online) Grain rotations in 4% sheared sample. The *blue circles* refer to grain centers and the number in it marks the grain ID, the vectors represent the rotation axis and the rotation angle φ is in radian

According to the above conclusion, we propose a plastic deformation model of nanocrystalline, in which, assuming plastic deformation of nanocrystalline is fully the result of relative slip of grains and grain boundaries are viscous layers. Based on this model, the total equivalent plastic strain $\varepsilon^p_{xy}$ can be evaluated by statistical moment method [9]. By subtracting the plastic part from the total shear strain, the total elastic strain $\varepsilon^e_{xy}$ is obtained, and then we can calculate shear stress $\sigma_{xy}$ via Hook's law $\sigma_{xy} = G\varepsilon^e_{xy}$, where $G$ is shear modulus, it is about 48 GPa and it is equal to the slope of the straight part of the curve in Fig. 27.1. The red curve in Fig. 27.11 is the calculated stress–strain curve base on the above model. The black curve is the simulation result and the extension of its straight part is the blue line (shown in inlet). The blue line refers to a pure elastic deformation model. In Fig. 27.11, the red and black curves are qualitatively close to each other in the increasing part,

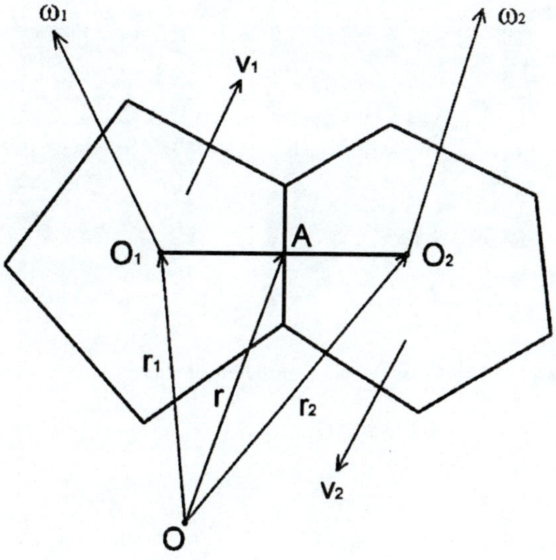


**Fig. 27.10** Schematic on the movement of two neighbor grains, $O_1$ and $O_2$ represent grain centers

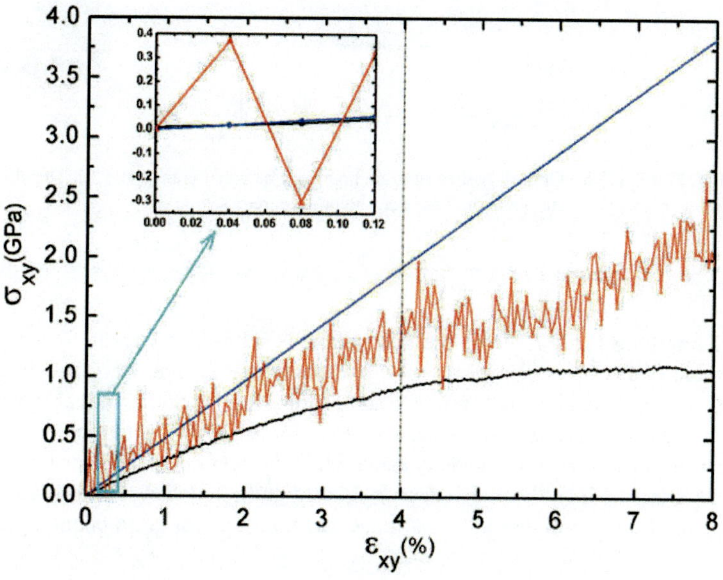


**Fig. 27.11** (Color Online) Stress–strain curve. The *blue line* is a reference line corresponding to pure elastic model, the *red curve* is the evaluated curve of a elastic–plastic model in which plastic deformation is attributed to grain slip, the *black curve* is the result of MD simulation

and in plateau part, the red cure is closer to the black cure than the blue line. It suggests that the relative slip of grains is the dominated mechanism of plastic deformation in the increasing part and it is also an important mechanism in the plateau part. Due to the large percentage of grain boundary atoms and the complexity of the atomic motion in grain boundaries, the contribution of grain boundary activities is difficult to accurately evaluate. In statistical moment method, the grain translation is set as the translation of its selected spherical part, which is the cause of the oscillation of the red curve.

### 27.3.3 Shear Plane

The existence of shear plane containing several grain boundaries is a typical characteristic of nanocrystalline under shear loading. Figure 27.12 shows the distribution of atoms whose x component of displacement is larger than 1.5 Å after shearing 8% (Taking into account the periodic boundary condition, the observation area can be set as the initial cube region).

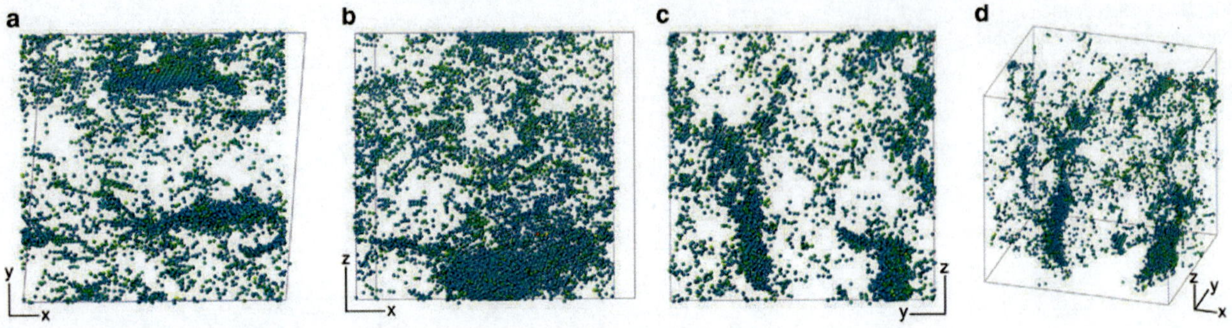


**Fig. 27.12** (Color Online) Distribution of atoms whose x component of displacement is larger than 1.5 Å after shearing 8%

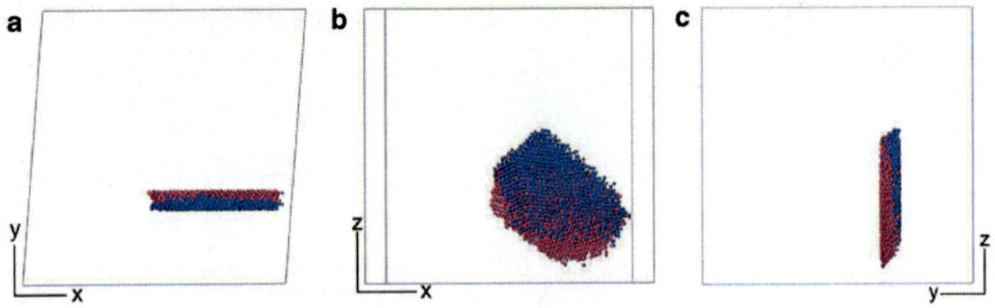


**Fig. 27.13** (Color Online) A slice of grain boundary between grain 6 and 12. The slice is parallel to y plane and its thickness is 1 nm, the *purple* atoms belong to grain 6 and the *blue* ones belong to grain 12

Comparing Fig. 27.12a, b, the shear plane is sub-parallel to the y-plane, grain 2, 6, 10, 11, 12 and 13 are involved in the shear plane. The main part of the plane is the grain boundary between grain 6 and 12. Figure 27.13 is a slice of grain boundary of two adjacent grains 6 and 12, the slice is parallel to the y-plane, its thickness is 1 nm and the normal direction is (0.0537,-0.9901,0.1289), which is close to $[0\bar{1}0]$, the rotation axis of the gain boundary is close to $[2\bar{5}2]$ and the corresponding rotation angle is about 23.3°. Further detail analysis indicates that shear plane is more likely to form around the area in which there are several adjacent grain boundaries whose normal direction is close to the external shear loading plane. In the formation of shear plane, the migration of grain boundary and triple junctions are observed.

## 27.4 Conclusions

In this paper, molecular dynamics simulation is used to study the deformation process of nanocrystalline copper under high strain rate shear loading at room temperature. According to the dominated deformation mechanism, the plastic deformation can be divided into two parts: increasing part and the plateau part. In the increasing part, the relative slipping of grains is the dominated activities, but the dislocation activities take a great role in the plateau part. The plastic deformation of the increasing part is all from the relative slip of grains, the rotation of each grain is small and has no contribution to plasticity, grain rotation is a kind of mechanism of grain boundary adjustment and the rotation direction of each grain is determined by the grain boundary network characteristics, detailed strain analysis indicates that the viscous flow model is suitable for describing the plastic deformation of the increasing part. However, the plastic deformation of the plateau part is dominated by grain sliding, dislocation gliding and deformation twining. As the deformation increases, the relative slip between grains is not enough to carry all the plastic deformation. The partial dislocations nucleate in grain boundaries under local stress and then glide into the grains, while some deformation twins emerge near the triple junctions. When the orientation of adjacent grain boundary is close to the normal direction of the external shear plane, the overall slip used to occur in those regions to form shear planes containing several grain boundaries.

**Acknowledgements** Our thanks are due to Profs. Haifeng Liu, Haifeng Song, Guicun Ma, Hui Zheng, and Drs. Bo Sun, Yanhong Zhao, Shuaichuang Wang, Gongmu Zhang, Hongzhou Song, Qili Zhang, Mingfeng Tian, for helpful comments and discussions. This work was supported by the National Natural Science Foundation of China (Grant No. 11075021) and the Science Foundations of the Laboratory of Computational Physics and China Academy of Engineering Physics (Grant Nos. 2009A0102005 and 2011A0201002).

## References

1. Wolf D, Yamakov V, Phillpot SR, Mukherjee A (2005) Deformation of nanocrystalline materials by molecular-dynamics simulation: relationship to experiments? Acta Mater 53(1):1
2. Kumar KS, Swygenhoven HV, Suresh S (2003) Mechanical behavior of nanocrystalline metals and alloys. Acta Mater 51(19):5743
3. Swygenhoven HV, Farkas D, Caro A (2000) Grain-boundary structures in polycrystalline metals at the nanoscale. Phys Rev B 62(2):831
4. Swygenhoven HV, Derlet PM (2000) Grain-boundary sliding in nanocrystalline fcc metals. Phys Rev B 64(22):224105
5. Hasnaoui A, Swygenhoven HV, Derlet PM (2002) Cooperative process during plastic deformation in nanocrystalline fcc metals: a molecular dynamics simulation. Phys Rev B 66(18):184112
6. Kumar KS, Kurtz SK, Banavar JR, Sharma MG (1992) Properties of a three-dimensional poisson-voronoi tessellation: a monte carlo study. J Stat Phys 67(3–4):523
7. Mishin Y, Mehl MJ, Papaconstantopoulos DA, Voter AF, Kress JD (2001) Structural stability and lattice defects in copper: Ab initio, tight-binding, and embedded-atom calculations. Phys Rev B 63(22):224106
8. Schiøtz J, Jacobsen KW (2003) A maximum in the strength of nanocrystalline copper. Science 301(5638):1357
9. Guo Lu, Shuaichuang Wang, Guangcai Zhang, Aiguo Xu (2012) Moment method for strain analysis and its application in molecular dynamics. Acta Phys Sin 61(7):073101
10. Honeycutt JD, Andersen HC (1987) Molecular dynamics study of melting and freezing of small Lennard-Jones clusters. Phys Chem 91(19):4950

# Chapter 28
# Nano-scale Investigation of Microstructural Phenomenon Contributing to Toughening of Nanoparticulate Reinforced Polymers

**Ryan L. Karkkainen, T. Walter, and A. Bujanda**

**Abstract** The objective of the current study is to employ precise and detailed nano-scale investigation to determine the physical mechanisms that contribute to the mechanical performance benefits of nano-particulate inclusions in polymer and polymer composite materials. A significant amount of work by the research community has experimentally illustrated the effective property benefits of nano-inclusions, thus the current work has elucidated some of the phenomenological causes behind these enhancement observations in a controlled fashion through a combination of micromechanical modeling and nano-scale experimentation under SEM observation. Nano-scale Mode I and Mode II fracture specimens have been designed to allow direct observation of crack growth in a nanoparticulate-reinforced epoxy system. Modeling employing the X-FEM node enrichment scheme for investigation of arbitrary-path crack propagation has been performed. Considerations critical to the accuracy of representation as well as the tractability of computation have been developed.

**Keywords** Nano-inclusions • XFEM • SEM

## 28.1 Introduction

A significant amount of work has been done in this area to experimentally illustrate the effective property benefits of nano-inclusions in polymer and polymer composite materials. Yet, much remains to be done to shed light on the actual mechanisms and phenomenon behind these property improvements, in such a way that the benefits can be fully understood and exploited. Further, work in this area has generally been done in an Edisonian trial-and-error fashion, or efforts in which varying percentages of inclusions are added and property enhancements are observed.

Impact energy absorption mechanisms of polymers, and especially polymeric composite materials, are quite different from those of traditional metals which are employed for their ductility. Yet, even brittle composites which exhibit no plastic deformation possess superior specific energy absorption to metallic materials. Multiple mechanisms contribute to their performance, including delamination, reinforcing fiber breakage, or tortuous crack propagation through a particulate-infused polymer phase. Multiple factors have been shown to be relevant, including particulate (or second-phase) dispersion characteristics, size, volume fractions, and interface properties [1–4].

There is a great deal of experimental evidence to indicate the effect of microstructure or particulate additions on mechanical enhancement. In general, experimental studies are conducted using conventional characterization tests for impact and fracture, and they are evaluated in terms of macro-level property improvements. Note that although carbon nanotubes are sometimes considered in fracture studies, there are significant cost and supply considerations and thus study of these materials is not proposed herein. Impact and fracture studies including essential work of fracture calculations have been performed for ceramic nanoparticles such as alumina and $TiO_2$ with marked improvements [5, 6]. Particle diameter-to-spacing ratios have been seen to have an effect on fracture properties in an alumina reinforced epoxy [7]. Epoxy including both nano and micro sized silica particles was seen to exhibit fracture toughness increases of a factor of two [8]. Multiple studies have observed such improvements in properties across multiple polymers types, generally observing diminishing returns as inclusion percentage is increased, and giving some treatment to a controlled variation in particle properties

R.L. Karkkainen (✉) • T. Walter • A. Bujanda
U.S. Army Research Laboratory, Aberdeen Proving Ground, MD 21005, USA
e-mail: ryan.l.karkkainen.civ@mail.mil

V. Chalivendra et al. (eds.), *Dynamic Behavior of Materials, Volume 1: Proceedings of the 2012 Annual Conference on Experimental and Applied Mechanics*, Conference Proceedings of the Society for Experimental Mechanics Series, DOI 10.1007/978-1-4614-4238-7_28, 

measured against subsequent property improvements [9–11]. Several experimental studies have attempted to infer the relative contribution of specific mechanisms towards the toughening of nano-reinforced polymers [12, 13]. This has proven difficult with a lack of understanding of microscale phenomenon and thus has pointed to a need for computational analysis which can quantify the exact effects of morphology and properties, as well as the relative effects of phenomena which contribute to toughening.

Comparatively less research has been accomplished in terms of effective modeling of material systems with micron and nano scale morphology effects. Inclusion of all relevant phenomena and simultaneous incorporation of effects at both the nano/micron and continuum levels has been elusive for many approaches due to the large difference in length scales. Molecular dynamic simulation can reveal much about the underlying nature of interface properties (inter-phase or particle-matrix interfaces), but at current is often not amenable to long time or length scales [14]. Furthermore, the results are often dependent on the initial boundary conditions [15]. In finite element based simulations, phase interfaces are often represented with perfect bonding or with a thin elastic interphase [16–18]. Some papers are limited to basic mechanical properties [19, 20] outside the regime of impact or fracture. Often, plateletized or idealized geometries are considered, and particle size, dispersion, and mechanical properties are varied, sometimes with observation of crack-path-influencing stress fields [21–23]. Detailed fracture mechanics studies employing cutting-edge fracture simulations such as cohesive zones [24], XFEM [25], or AFEM [26] have yet to be considered. Many issues have yet to be addressed with regards to nanoscale composites, including test methods [27], lack of models to predict and exploit such small-scale mechanisms, thorough evaluation of the tradeoffs and advantages to microstructural manipulation, and the full parametric study thereof.

The current ongoing work is designed to employ companion efforts in both modeling and experimental capacities to investigate and exploit the phenomenological causes behind the property enhancement observations seen with nano-particulate reinforced polymers in order to develop a material design strategy to improve performance and reduce material development lead times. Both modeling and experimental efforts take place at the micron and sub-micron level on the length scale of active energy absorbing mechanisms.

## 28.2 Design of Experimental Fracture Nanospecimens

Experimental work is based upon on micron sized specimens manufactured under a Focused Ion Beam (FIB) for actuation under SEM observation. This allows for direct observation of in situ crack propagation across a field of fully discernible embedded nanoparticles. The specimen is end-loaded with a nano-probe while under SEM observation, which enables visualization and measurement of crack propagation from a controlled pre-crack through nano-particulate enhanced polymer materials. Total force and displacement can be precisely applied and measured through use of a specialized intra-SEM loading stage. The FIB setup allows for accurate fabrication of specimens on the micron scale which are conducive to nano/micro scale observations, and these are milled to several desired shapes. The current study will include double cantilever beam specimens of two configurations to allow Mode I and Mode II loading, as well as simple cantilever bending, as shown in Fig. 28.1. Epon 828 with silica nanoparticles has been selected for initial study and proveout, as constituent properties are well known and the material is easily available. Specimens are approximately 5 μm in length in order to capture a particle field containing a representative distribution of fully resolvable 30 nm inclusions.

The particle-infused resin solution is mixed using a high speed shear mixing device. This procedure is short, simple, and industrially scalable. Previous investigations have shown the ability to adequately disperse particles and avoid agglomeration. Such observation is important as an agglomerized nanoparticle will not exhibit the same benefits as a well-dispersed nano-particle reinforcement system.

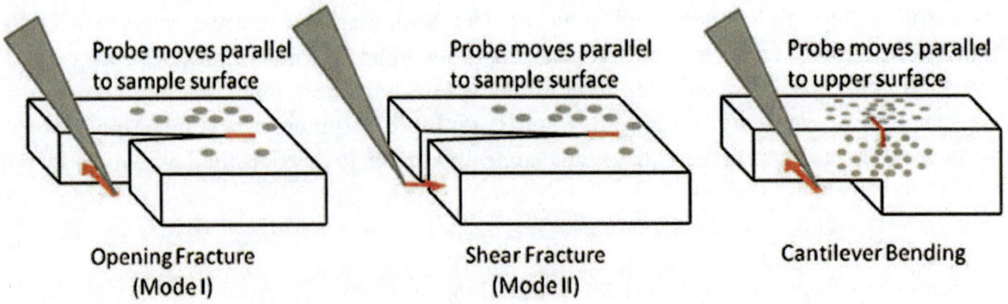


**Fig. 28.1** Nanospecimens for evaluation of crack propagation in mode I, mode II, and bending

At the component level, these tests characterize the overall performance of a material or adhesive. Analogous tests were performed at the micro/nano scale to investigate local phenomenon which contribute to the material level response, i.e. observe crack growth, branching, and interaction through a particulate-reinforced material field under various loading modes with various particulate reinforcement configurations (parameterized as described in the modeling section). These tests provide a quantification of material response, critical information explaining why certain materials will outperform others, and thus develop engineering knowledge enabling proactive design of nano-reinforced materials. Thus, with force, displacement, and crack growth information, along with visualization of the particulate crack arrest behavior, the fracture properties of each reinforced material can be determined, explained, and optimized.

One of the challenges of investigating composites with particulate reinforcement sizes on the order of nanometers is the inability to directly image and observe the particles during the loading process. Currently, it has been observed that imaging for 20 nm diameter particulate inclusions does not adequately resolve under SEM observation. Thus, the material system under study has been re-selected to include slightly larger silica particles on the order of 30 nm with a history of successful observation in this setup (previously employing different specimen types).

## 28.3 XFEM Based Fracture Modeling: Formulation and Methods

Modeling is performed within the Abaqus™ finite element software package using finite element based fracture micromechanics simulations applied to micron-sized representative material elements for which crack propagation across a field of particulates can be observed and quantified. A relatively newly developed node enrichment scheme known as the extended finite element method (XFEM) [25, 28] is employed to allow for fracture prediction with no required a priori knowledge of arbitrary crack paths and locations. Crack growth can be modeled without the presence of an initial flaw. The domain of the XFEM method can be a subset of the model, but in this case has been applied to the entire model. XFEM elements are enriched with additional terms to the base shape functions which are incorporated into an existing FEM framework. Displacements in an element ($u$) are interpolated from nodal displacements ($u_i$) according to:

$$u = \sum_{i=1}^{n} N_i(x)\left[u_i + H(x)a_i + \sum_{\alpha=1}^{4} F_\alpha(x)b_i^\alpha\right] \tag{28.1}$$

Where $N_i$ are the standard FEM shape functions, thus two additional terms are added to the standard FEM framework. The discontinuous jump function across crack surfaces (which can span an element) $H(x)$ and the enriched nodal degree of freedom vector $a_i$ comprise the first additional term. The second additional term involves elastic asymptotic crack-tip functions $F_\alpha(x)$ and another associated nodal enriched degree of freedom vector $b_i^\alpha$. These additional second and t hird terms are only active for nodes whose shape functions have been split by a crack or cut by a crack tip, respectively. The function $H$ is analogous to a Heaviside function which enforces the discontinuity across the crack and defines the sense of the contribution of the enrichment function $a_i$ as follows:

$$H(x) = \begin{cases} 1 & if \quad (x - x^*) \cdot n \geq 0 \\ -1 & if \quad (x - x^*) \cdot n < 0 \end{cases} \tag{28.2}$$

Where $x$ is a Gauss point, $x^*$ is the point on the crack closest to x, and $n$ is the unit outward normal to the crack at the point $x^*$. The function $F$ is designed to represent the singularity and process zone associated with a crack tip, which enriches the inter-element detail needed to define such a region. As per Eq. 28.1, this is accomplished with a series of four functions in a polar coordinate system with origin at the crack tip and the zero angle is defined as tangent to the crack tip:

$$F_\alpha(x) = \left[\sqrt{r}\sin\tfrac{\theta}{2}, \sqrt{r}\cos\tfrac{\theta}{2}, \sqrt{r}\sin\theta\sin\tfrac{\theta}{2}, \sqrt{r}\sin\theta\cos\tfrac{\theta}{2}\right] \tag{28.3}$$

Note that the asymptotic crack tip functions above are for an isotropic elastic material and must be modified to accommodate other constitutive laws employed in an element.

For initial study and for initial implementation of the XFEM computational scheme, a two-dimensional idealized field of nanoparticles has been analyzed. Figure 28.2 shows the evenly distributed field of 40 nm diameter particles, with inter-particle spacing of 120 nm edge-to-edge, in a regularly staggered array. The particle field size is 800 × 1,200 nm in order to capture a representative number of particles, and to ensure the model is large enough to fully contain the process zone.

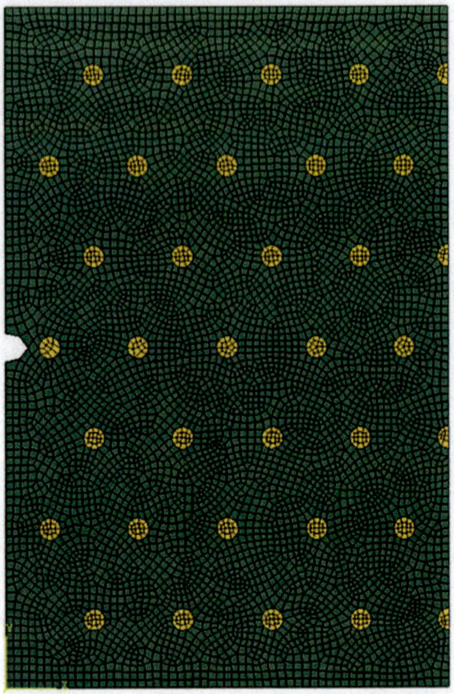

**Fig. 28.2** Nanoparticle field used for simulation, with a geometric pre-crack

**Table 28.1** Basic material properties used for simulation

| | Polycarbonate | Silica particles |
|---|---|---|
| *Stiffness (E)* | 2.40 GPa | 73.0 GPa |
| *Poisson's ratio (ν)* | 0.32 | 0.17 |
| *Strength* | 62.1 MPa | 1,100 MPa |

While the experimental material system is being finalized, polycarbonate with silica nanoparticles has been chosen for initial study. Material properties employed are shown in Table 28.1. Future microstructures employed in model simulations will represent direct matches to actual fabricated materials, as well as numerous simulated structures (such as in the current work) which will be investigated parametrically.

Loading has been applied via displacement boundary conditions on the leading edge of the fracture specimen sub-model to simulate opening. A displacement has been applied to each side, i.e. upward displacement is applied to the edge above the crack tip, and downward displacement has been applied below the crack tip. The applied displacements are each of a magnitude of 5% of the total specimen height. The actual displacement achieved will be somewhat less than this, once a simulation reaches a point of crack arrest or total through-section failure. The nodes on the right trailing edge of the specimen are fixed.

The Maximum Principal Stress Theory is used to determine material failure and crack initiation, as per Table 28.1. Damage evolution is specified to be total and immediate property degradation at the element level, once this criterion has been surpassed. The tolerance during solver iterations has been slightly relaxed from 5% to 7% to aid in rate of convergence.

Note that once the current model formulation is verified, additional simulations can be performed to rapidly iterate through various microstructural configurations with parameters spanning relevant categories such as: particle/phase geometry, distrubtion, adhesion (interface strength and interphase area), and constituent mechanical properties. The model is designed to provide explicit focus upon critical consideration of the interface between reinforcing particles and the bonding polymer matrix phase. Effects can be evaluated both in terms of the force and displacement of a simulated double cantilever beam, the crack propagation rate, as well as local computations such as J-integral, or smaller scale phenomenon such as total strain energy required to debond cohesive interface elements.

## 28.4 XFEM Based Fracture Modeling: Initial Results

Initial results have shown the potential for prediction of crack propagation at arbitrary locations, with no need for a priori knowledge of crack paths or locations. The XFEM formulation has been successfully employed, as a tortuous crack path across a particle field has been predicted successfully. This enables improved future simulations with increased sophistication and accuracy of representation. Some issues have been identified with regards to convergence of the FEM solver, and significant user alterations of solver parameters are required to obtain a solution.

An extremely small step size can be required, even in a standard or implicit FEM framework, especially given a small-scale micromechanics framework in which length scales and element sizes are very small with respect to the material properties, and analysis time scales are also very short. In the current analyses, step sizes of $10^{-8}$ or $10^{-10}$ (s) can be required. As a further issue, some default solver settings not appropriate to discontinuous analyses may result in premature cutbacks and termination. As such, the number of iterations in a time integration scheme preceding a time increment cutback have been changed from 4 to 8. Similarly, a slight increase in the allowance for estimated iterations required for convergence before reducing a time increment has been changed from 8 to 10 in order to avoid premature time increment cutbacks. Lastly, simply increasing the number of attempts to obtain convergence per increment has been increased to 10. Note that values of 50 or more can be employed, if needed, without great penalty to solution times.

Further, although the XFEM method is largely mesh-independent, an irregular mesh around key geometric features, such as a crack initiation point, can limit the accuracy of representation around these points. For example, in Fig. 28.3a, it can be seen that crack initiation does not occur directly at the pre-crack tip, as the XFEM computation is amenable to starting a crack front by splitting an element rather than splitting element bounds (subsequently, either can occur during the calculation of crack propagation). This can be solved either by mesh refinement at the crack tip or by pre-defining a crack initiation point within the XFEM methodology, as will be incorporated into future work. Figure 28.3b shows the Von-Mises stress field at a time point corresponding to the crack illustrated in Fig. 28.3a. The process zone can be observed in detail, and it is seen that the particle field size is adequate to capturing the fracture process while maintaining highly detailed features.

Future work will employ more sophisticated constitutive models incorporating hyperelastic behavior. Cohesive interface elements will be applied around each nanoparticle to simulate interface and interphase effects. These can be modified to investigate the importance of such properties on the fracture process and properties. Further parametric study will involve base material properties, particle size, and particle distribution.

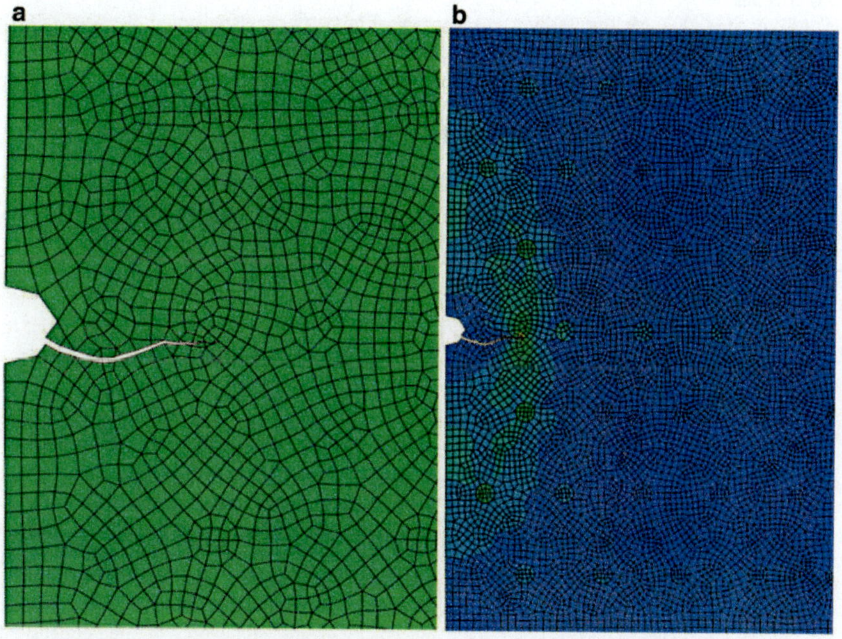


**Fig. 28.3** (a) Crack tip propagation simulation (b) Von-mises stress field

## 28.5 Conclusions

Intra-SEM fracture nanospecimens have been designed to provide direct observation and measurement of fracture processes in nanoparticulate reinforced polymers on the length scale of active mechanisms. Mode I, Mode II, and bending specimens of 5 μm in size have been employed. 20 nm silica particles in Epon 828 have presented difficulties in adequate image resolution, thus the specimens are currently being reprocessed with slightly larger silica particles known to resolve well under SEM observation. FEM modeling has successfully incorporated the XFEM methodology to predict arbitrary crack initiation and propagation with no a priori knowledge of crack paths or ubiquitous element insertion required.

## 28.6 Future Work

The experimental data set will be quickly expanded once specimens have been remanufactured with particulates more easily amenable to resolution under SEM. With the initial successful execution of the XFEM computational scheme, modeling will be expanded to include more sophisticated and accurate representation, as well as parametric study of interface properties, general mechanical properties, and particulate geometry and distribution. Further modeling will be performed to extend the above analysis into the dynamic regime, although physical testing under dynamic conditions within SEM observation is not currently feasible due to limitations on raster scan times which exceed impact event time scales by several orders of magnitude.

**Acknowledgements** The authors would like to express their appreciation to Dr. Brian Schuster and Dr. Jason Robinette of the Army Research Laboratory for their efforts and support.

## References

1. Viana JC (2006) Polymeric materials for impact and energy dissipation. Plast Rubber Compos 35(6–7):260–267
2. Zhang J, Wang X, Lu L, Li D, Yang X (2003) Preparation and performance of high impact polystyrene (HIPS)/Nano-TiO2 nanocomposites. J Appl Polym Sci 87(3):381–385
3. Tang W, Santare MH, Advani SG (2003) Melt processing and mechanical property characterization of multi-walled carbon nanotube/high density polyethylene (MWNT/HDPE) composite films. Carbon 41(14):2779–2785
4. Bartczak Z, Argon AS, Cohen RE, Weinberg M (1999) Toughness mechanism in semicrystalline polymer blends: II high-density polyethylene toughened with calcium carbonate filler particles. Polymer 40(9):2347–2365
5. Singh D, Jayasimha T, Rai KN, Kumar A (2007) Preparation of poly(methylmethacrylate) nanocomposites with superior impact strength. J Appl Polym Sci 105(6):3183–3194
6. Yang JL, Zhang Z, Zhang H (2005) The essential work of fracture of polyamide 66 filled with TiO2 nanoparticles. Compos Sci Technol 65(15–16):2374–2379
7. Wetzel B, Rosso P, Haupert F, Friedrich A (2006) Epoxy nanocomposites – fracture and toughening mechanisms. Eng Fract Mech 73(16):2375–2398
8. Han JT, Cho K (2006) Nanoparticle-induced enhancement in fracture toughness of highly loaded epoxy composites over a wide temperature range. J Mater Sci 41(13):4239–4245
9. Uddin MF, Sun CT (2008) Impact resistance and toughness of composite laminates with nanoparticle – enhanced matrix. In: Proceedings international SAMPEL symposium and expo 2008, Long Beach, CA
10. Brunner AJ, Necola A, Rees M, Gasser P, Kornmann X, Thomann R et al (2006) The influence of silicate-based nano-filler on the fracture toughness of epoxy resin. Eng Fract Mech 73(16):2336–2345
11. Mortezaei M, Famili M, Kokabi M (2011) The role of interfacial interactions on the glass-transition and viscoelastic properties of silica/ polystyrene nanocomposites. Comp Sci Tech 71(8):1039–1045
12. Hsieh TH, Kinloch AJ, Masania K et al (2010) The toughness of epoxy polymers and fibre composites modified with rubber microparticles and silica nanoparticles. J Mater Sci 45:1193, 2011, Vol:46, Pages:4092–4092
13. Hsieh TH, Kinloch AJ, Taylor AC et al (2011) The effect of carbon nanotubes on the fracture toughness and fatigue performance of a thermosetting epoxy polymer. J Mater Sci 46:7525–7535
14. Lau KT, Gu C, Hui D (2006) A critical review on nanotube and nanotube/nanoclay related polymer composite materials. Compos Part B: Eng 37(6):425–436
15. Zhou SJ, Beazley DM, Lomdahl PS, Holian BL (1997) Large-scale molecular dynamics simulations of three-dimensional ductile failure. Phys Rev Lett 78(3):479–482
16. Gawand AA, Whitney JM, Brockman RB, Tandon GP (2008) Interaction between a nanofiber and an arbitrarily oriented crack. J Compos Mater 42(1):45–68

17. Liu YJ, XU N, Luo JF (2000) Modeling of interphases in fiber-reinforced composites under transverse loading using the boundary element method. J Appl Mech Trans ASME 67(1):41–49
18. Liu H, Brinson LC (2006) A hybrid numerical–analytical method for modeling the viscoelastic properties of polymer nanocomposites. J Appl Mech Trans ASME 73(5):758–768
19. Hua L, Brinson LC (2008) Reinforcing efficiency of nanoparticles: a simple comparison for polymer nanocomposites. Compos Sci Technol 68 (6):1502–1512
20. Hbaieb K, Wang QX, Chia YHJ, Cotterell B (2007) Modelling stiffness of polymer/clay nanocomposites. Polymer 48(3):901–909
21. Sheng N, Boyce MC, Parks DM, Rutledge GC, Abes J, Cohen RE (2004) Multiscale micromechanical modeling of polymer/clay nanocomposites and the effective clay particle. Polymer 45(2):487–506
22. Thiagarajan G, Deshmukh K, Wang Y (2007) Nano finite element modeling of the mechanical behavior of biocomposites using multi-scale (virtual internal bond) material models. J Biomed Mater Res A 83(2):332–344
23. Wanga J, Pyrzb R (2004) Prediction of the overall moduli of layered silicatereinforced nanocomposites – part I: basic theory and formulas. Compos Sci Technol 64(7–8):925–934
24. Karkkainen R, Yen C (2012) Dynamic modeling for rate-dependent and mode dependent cohesive interface failure analysis, J Comp Mat 46:2193–2201
25. Belytschko T, Black T (1999) Elastic crack growth in finite elements with minimal remeshing. Int J Numer Methods Eng 45:601–620
26. Ling D, Yang Q, Cox B (2009) An augmented finite element method for modeling arbitrary discontinuities in composite materials. Int J Fract 156:53–73
27. Boesl B (2009) Fracture toughening mechanisms in nanoparticle and microparticle reinforced epoxy systems using multi-scale analysis. Dissertation, University of Florida
28. Abaqus Users Manual, Section 10.6

# Chapter 29
# Peridynamic Simulations of Infinite Regions Using a Perfectly Matched Layer

**Raymond A. Wildman and George A. Gazonas**

**Abstract** Absorbing boundary conditions for non-local methods (peridynamics) will be presented. Peridynamics is a non-local method in which the force term of the momentum equation is evaluated as the convolution of the displacement with a suitable micro-force function, or kernel. Though the method involves integration, it is based on a differential equation formulation and requires specialized absorbing boundary conditions to simulate infinite regions. Applications include simulation indentation experiments, which can provide verification/validation of the method. Perfectly matched layers (PML) are absorbing boundary regions that are reflection-less at their interface and decay impinging waves exponentially. Here, PML are applied to peridynamics by treating the kernel as the convolution of the displacement with the second derivative of a nascent Dirac delta distribution, such as the appropriately defined Gaussian distribution. In this sense, the peridynamics formulation can be split into a coupled auxiliary field formulation, suitable for a PML application.

## 29.1 Introduction

Peridynamics, a non-local formulation of elastodynamics, was introduced as a method that more easily incorporates discontinuities such as cracks and damage in a numerical simulation [1]. The formulation treats an elastic solid as a collection of nodes, connected by bonds that break if stretched past some predefined limit. While most peridynamics work has focused on simulating problems with free or fixed boundary conditions, there are applications in which the simulation of an infinite medium may be useful, such as wave or crack propagation in a half-space. Absorbing boundary conditions are a way of simulating an infinite medium by absorbing any impinging waves at the computational boundaries so they do not reflect back into the simulation. A PML is such an absorbing boundary, and was originally introduced for electromagnetic simulations [2, 3]. PMLs differ from traditional absorbing boundary conditions in that they are an absorbing layer, placed between the computational region of interest and the truncation of the grid or mesh. They can also be thought of as an artificial, anisotropic absorbing material, which is why the flexibility of a state-based peridynamics is necessary [4]. PMLs have two important qualities: First, waves in a PML decay exponentially, and second, in their analytic form, the interface of a PML and the computational region is reflectionless. Since their introduction, PMLs have been extended to many different types of media [5, 6], different numerical methods [7, 8], and different fields [9–11].

This paper implements a peridynamic formulation of elastodynamics in one (1D) and two dimensions (2D) and terminates the boundary with a PML. As is discussed, the use of a PML is facilitated with an auxiliary field formulation, derived from state-based peridynamics for 2D. In 1D, the auxiliary formulation of peridynamics is formulated using the results of [12]. This approach requires a matrix representation of the auxiliary field, which may be too memory prohibitive in higher dimensions.

R.A. Wildman (✉) • G.A. Gazonas
U.S. Army Research Laboratory, ATTN: RDRL-WMM-B, Aberdeen Proving Ground, MD 21005, USA
e-mail: raymond.a.wildman.civ@mail.mil

V. Chalivendra et al. (eds.), *Dynamic Behavior of Materials, Volume 1: Proceedings of the 2012 Annual Conference on Experimental and Applied Mechanics*, Conference Proceedings of the Society for Experimental Mechanics Series,
DOI 10.1007/978-1-4614-4238-7_29, 

## 29.2 Formulation

The continuum equation of motion in an elastic solid can be stated as

$$\rho \frac{\partial^2}{\partial t^2}\mathbf{u} = \nabla \cdot \bar{\sigma} + \mathbf{b}, \tag{29.1}$$

where $\rho(\mathbf{x})$ is the density, $\mathbf{u}(\mathbf{x}, t)$ is the displacement, $\bar{\sigma}(\mathbf{x}, t)$ is the stress tensor, and $\mathbf{b}(\mathbf{x})$ is a body force [13]. (Throughout, boldface type denotes a vector and a boldface variable with an overbar denotes a tensor.) Equation 29.1 is a local formulation because the divergence of the stress (and gradient of the displacement implied in its definition) represents a local operation on a variable. In problems involving discontinuities, such as cracks, the divergence at such discontinuities is not well defined, leading to numerical implementation problems.

In one dimension, by ignoring body forces and using Hooke's law, Eq. 29.1 can be simplified to

$$\frac{\partial}{\partial x}\left(a\frac{\partial}{\partial x}u\right) = \frac{1}{b}\frac{\partial^2}{\partial t^2}u, \tag{29.2}$$

with material properties Young's modulus, $a(x)$, and inverse density, $b(x)$.

### 29.2.1 *Perfectly Matched Layer*

The first step in formulating a PML is to construct an analytic continuation to the complex plane, $x^* = x + ig(x)$, where $g(x)$ is a given function describing the deformation of $x$ into the complex plane [14]. This mapping has the effect of transforming traveling waves of the form $e^{ikx}$, where $k = \omega/c$ is the wave number, into evanescent waves of the form $e^{ikx}e^{-kg(x)}$, thus attenuating such waves in the PML region. Next, we change variables back to the real part $x$, which requires a relation for the differential quantities, $\partial x^*/\partial x$. A convenient choice for $g(x)$ is

$$\frac{d}{dx}g(x) = \frac{\phi(x)}{\omega}, \tag{29.3}$$

because the $1/\omega$ factor creates a frequency independent attenuation rate in dispersionless materials [14]. (Peridynamic formulations are not dispersionless, as discussed in Ref. [15], though this standard choice is used for simplicity.) Finally, the substitution that must be made for any spatial derivative can be written

$$\frac{\partial}{\partial x} \to \frac{1}{1 + i\frac{\phi(x)}{\omega}}\frac{\partial}{\partial x}. \tag{29.4}$$

As will be discussed below, the PML transformation remains valid in peridynamics because we can substitute the spatial derivatives in Eq. 29.4 with those linear operators acting as spatial derivatives in peridynamics.

Wherever $\phi > 0$, $\mathbf{u}$ and $\bar{\sigma}$ will exponentially decay. Before discretization, any change in $\phi$ will not result in any reflections, so the region of interest would have $\phi = 0$, and the PML region could have a discontinuity in applying $\phi$. In practice, however, numerical reflections can result from discontinuous material parameters after discretization, so it is better to use a smooth transition for $\phi$. Here, we divide the PML region into two parts, one in which $\phi$ is a constant value, and $\phi$ in the other ramps up to that constant value following a Gaussian distribution.

### 29.2.2 *One-Dimensional Peridynamics*

Peridynamics proposes replacing $\nabla \cdot \bar{\sigma}$ in Eq. 29.1 with a nonlocal operator that nonetheless also represents a force

$$\rho \frac{\partial^2}{\partial t^2}\mathbf{u} = \int_{\mathcal{H}_\mathbf{x}} \mathbf{f}(\mathbf{u}' - \mathbf{u}, \mathbf{x}' - \mathbf{x})dV' + \mathbf{b}, \tag{29.5}$$

where $\mathbf{f}(\mathbf{u}' - \mathbf{u}, \mathbf{x}' - \mathbf{x})$ represents a micro-modulus force function (or kernel) that defines a force between two points and $\mathcal{H}_\mathbf{x}$ represents a horizon or maximum distance over which two points can influence each other [1]. The micro-modulus function becomes the constitutive response in the formulation, replacing, for example, Hooke's law in the continuum case. In its original form, the micro-modulus function was developed as a simple linear elastic response, which can be linearized as

$$\begin{aligned} \mathbf{f}(\eta, \xi) &= \bar{\mathbf{C}}(\xi)\eta, \\ \bar{\mathbf{C}}(\xi) &= c\frac{1}{|\xi|^3}\xi \otimes \xi H(\ell - |\xi|), \end{aligned} \tag{29.6}$$

where $c$ is some constant, $H(\cdot)$ is the Heaviside step function, and $\ell$ is the radius of the horizon region, which defines $\mathcal{H}_\mathbf{x}$, and $\otimes$ denotes an outer product [1]. Because we are only concerned with one dimension, Eq. 29.6 can be simplified to

$$\begin{aligned} f(\eta, \xi) &= C(\xi)\eta, \\ C(\xi) &= c\frac{1}{|\xi|}H(\ell - |\xi|). \end{aligned} \tag{29.7}$$

Equation 29.6 represents one choice of kernel, though there are other choices. Ref. [15] discusses a few alternatives, one of which is

$$C_{Gauss}(\xi) = \frac{4E}{\ell^3\sqrt{\pi}}e^{(\xi/l)^2}, \tag{29.8}$$

where $\ell$ is a given intrinsic width parameter similar to the horizon $\ell$ above, though here it does not define a strict cutoff, and $E$ is Young's modulus. Both kernels in Eq. 29.8 are derived from functions that converge to a Dirac delta function in the limit as $\ell \to 0$. As a result, they approach 0 as $\xi \to \pm\infty$, so that no Heaviside function is necessary.

#### 29.2.2.1 PML in One Dimension

Equation 29.2 can be split into two coupled equations by introducing an auxiliary field $\upsilon(x, t)$

$$\begin{aligned} -i\omega\hat{u} &= b\frac{\partial}{\partial x}\hat{\upsilon}, \\ -i\omega\hat{\upsilon} &= a\frac{\partial}{\partial x}\hat{u}, \end{aligned} \tag{29.9}$$

assuming an $e^{-i\omega t}$ time dependence, and where $\hat{u}(x, \omega)$ and $\hat{\upsilon}(x, \omega)$ are the Fourier transforms of $u(x, t)$ and $v(x, t)$, respectively. Now, using the PML substitution of Eq. 29.4 results in

$$\begin{aligned} b\frac{\partial}{\partial x}\hat{\upsilon} &= -i\omega\hat{u} + \phi\hat{u}, \\ a\frac{\partial}{\partial x}\hat{u} &= -i\omega\hat{\upsilon} + \phi\hat{\upsilon}. \end{aligned} \tag{29.10}$$

#### 29.2.2.2 Auxiliary Field Formulation and PML Application

The previous subsections reviewed the one-dimensional formulations of peridynamics and PMLs, and this subsection details a PML implementation for peridynamics. The main difficulty is deriving an auxiliary field formulation, after which applying a PML is simple. To begin, consider the following definition from Ref. [12], which formalizes the nonlocal point divergence operator

$$\mathcal{D}(\mathbf{u}) = -\int_{\mathcal{H}_\mathbf{x}} (\mathbf{u}' + \mathbf{u}) \cdot \alpha(\mathbf{x}', \mathbf{x}) dV', \tag{29.11}$$

where $\alpha(\mathbf{x}', \mathbf{x})$ is a kernel with specific properties, which are discussed below. Also necessary is the adjoint of the above operator, defined as

$$\mathcal{D}^*(\upsilon) = (\upsilon' - \upsilon)\alpha(\mathbf{x}', \mathbf{x}). \tag{29.12}$$

An important result from Ref. [12] is that the above operators are equivalent to their local counterparts, i.e.,

$$\begin{aligned} \mathcal{D}(\mathbf{u}) &= \nabla \cdot \mathbf{u}, \\ \int_{\mathcal{H}_\mathbf{x}} \mathcal{D}^*(\upsilon) dV' &= -\nabla \upsilon, \end{aligned} \tag{29.13}$$

with the choice

$$\alpha(\mathbf{x}', \mathbf{x}) = \nabla' \delta(\mathbf{x}' - \mathbf{x}), \tag{29.14}$$

where here $\delta(\cdot)$ is the Dirac delta function. Ultimately, the result we need is for the Laplace operator, which is present in all wave equations and implicit in the peridynamic formulation. In terms of the point divergence operator and its adjoint, the Laplace operator can be written as [12]

$$\begin{aligned} \mathcal{D}(\mathcal{D}^* v) &= \int_{\mathcal{H}_\mathbf{x}} (v' - v)|\alpha(\mathbf{x}', \mathbf{x})|^2 dV' \\ &= -\nabla^2 \upsilon, \end{aligned} \tag{29.15}$$

a relation that formalizes the results of Ref. [1], which are summarized in Sect. 29.2.2.

Of course, to obtain a viable numerical method we cannot choose $\alpha$ in direct terms of the Dirac delta function, but we can choose derivatives of functions that converge to the delta function, similar to the derivation of those in Eq. 29.8. Equation 29.14 also suggests important properties for $\alpha$: it should be spatially invariant, $\alpha(\mathbf{x}', \mathbf{x}) = \alpha(\mathbf{x}' - \mathbf{x})$ and it should be an odd function, $\alpha(\mathbf{x}' - \mathbf{x}) = -\alpha(\mathbf{x} - \mathbf{x}')$.

We are now ready to derive a peridynamic auxiliary field formulation, necessary for implementing a PML. Equation 29.2 can be rewritten in terms of the point divergence operator and its adjoint as discussed above in one dimension

$$\mathcal{D}[a\mathcal{D}^*(\hat{u})] = \frac{\omega^2}{b}\hat{u}. \tag{29.16}$$

Again, introducing an auxiliary field $\hat{\upsilon} = \frac{a}{-i\omega}\mathcal{D}^*(\hat{u})$ gives

$$\begin{aligned} b\mathcal{D}(\hat{\upsilon}) &= i\omega\hat{u}, \\ a\mathcal{D}^*(\hat{u}) &= -i\omega\hat{\upsilon}. \end{aligned} \tag{29.17}$$

Because the operators $\mathcal{D}$ and $\mathcal{D}^*$ are equivalent to the standard differential operators under certain conditions, we can use the same PML substitutions as given in Eq. 29.4, giving

$$\begin{aligned} i\omega\hat{u} &= b\mathcal{D}(\hat{\upsilon}) + \phi\hat{u}, \\ -i\omega\hat{v} &= a\mathcal{D}^*(\hat{u}) - \phi\hat{\upsilon}. \end{aligned} \tag{29.18}$$

Substituting in for $\mathcal{D}$ and $\mathcal{D}^*$ gives and converting back to the time domain gives

$$\frac{\partial}{\partial t}u = b\int_{\mathcal{H}_x}(\upsilon' + \upsilon)\alpha(x' - x)dx' - \phi u,$$
$$\frac{\partial}{\partial t}\upsilon = a(u' - u)\alpha(x' - x) - \phi\upsilon. \tag{29.19}$$

A final issue is the choice of $\alpha(\xi)$, which should either satisfy Eqs. 29.13 and 29.14 by using a function that converges to the derivative of the delta function, or be derived as discussed in Ref. [15] and restated in Eq. 29.8 by essentially choosing the kernel of the standard formulation and then using Eq. 29.15 to find $\alpha(\xi)$. For the latter, we can define $\alpha(\xi)$ as

$$\alpha^1_{Gauss}(\xi) = \text{sgn}(\xi)\sqrt{C_{Gauss}(\xi)}, \tag{29.20}$$

which demonstrates a limitation of this formulation in that the kernel function implicitly used to replace the Laplacian is always positive as it is the square of $\alpha(\xi)$, and consequently, any kernel $C(\xi)$ must be positive. We can also define $\alpha(\xi)$ as the derivative of distributions that converge to the Dirac delta function so that Eqs. 29.13 and 29.14 are satisfied. Again consider, e.g., a Gaussian distribution, defined as

$$\alpha^2_{Gauss}(\xi) = -\frac{2\xi}{\ell^3\sqrt{\pi}}e^{-\frac{\xi^2}{\ell^2}}, \tag{29.21}$$

As before, the parameter $\ell$ is not strictly a horizon, as this function does not have compact support in the strict sense, though it will be used interchangeably.

#### 29.2.2.3 Discretization

The discretization of Eq. 29.19 used here is relatively straightforward, and follows the method of Ref. [16]. First, temporal integration is performed using an explicit forward Euler approximation. The spatial discretization is performed using a simple point-wise approximation for the integral, and enforcing both equations in a point-wise fashion. The only complication here is that while $u_j$ is a vector as usual, $\upsilon_j$ is actually a matrix as it is a two-point function in the nomenclature of Ref. [12]. (Physically, $\upsilon_j$ is a matrix because of the nonlocality of the formulation: It is a function of $a\mathcal{D}^*(u)$, which, in elasticity, represents the stress on all individual points due to a single point.) Performing the spatial discretization gives

$$u^m_{j+1} = \Delta t\left[b\Delta x\sum_n\left(\upsilon^{nm}_j + \upsilon^{mn}_j\right)\alpha(x^n - x^m) - \phi^m_u u^m_j\right] + u^m_j,$$
$$\upsilon^{mn}_{j+1} = \Delta t\left[a\left(u^n_{j+1} - u^m_{j+1}\right)\alpha(x^n - x^m) - \phi^{mn}_\upsilon \upsilon^{mn}_k\right] + \upsilon^{mn}_k, \tag{29.22}$$

where $\Delta x$ is the grid spacing, $x^m$ is the $m$th node, $j$ is the time step, $\Delta t$ is the time step size, and $\phi_u^m$ and $\phi_\upsilon^{mn}$ are the PML functions for $u$ and $\upsilon$, respectively. For $\upsilon$, we require a matrix representation of the PML function, which is defined as

$$\phi^{mn}_\upsilon = \sqrt{\phi^m_u\phi^n_u}. \tag{29.23}$$

The limit on $n$ in the summation in Eq. 29.22 depends on the chosen kernel $\alpha$ and its decay rate. For small simulation regions, it may be most efficient to simply use the entire region for summation; however, for large regions some multiple of $\ell$ can be used depending on the decay rate of the kernel. The stability of this discretization is discussed in Ref. [16]. In contrast to differential equation-based methods, the time step size $\Delta t$ scales with $\ell$ and is only weakly dependent on $\Delta x$.

### 29.2.3 *Two-Dimensional, State-Based Peridynamics*

A PML application requires an auxiliary field formulation, as it is essentially an anisotropic absorbing material, if a non-physical one. Consequently, a state-based peridynamic formulation [4] is necessary to implement the required constitutive relations in the absorber. The state-based peridynamics uses a family of bonds to determine a given force rather than a single bond independently. This more general approach allows for inelastic behavior and more general elastic behavior, and is governed by

$$\rho \frac{\partial^2}{\partial t^2}\mathbf{u} = \int_{\mathcal{H}_\mathbf{x}} (\bar{\mathbf{T}}[\mathbf{x},t]\langle \mathbf{x}'-\mathbf{x}\rangle - \bar{\mathbf{T}}[\mathbf{x}',t]\langle \mathbf{x}-\mathbf{x}'\rangle) dV_{\mathbf{x}'} + \mathbf{b}, \tag{29.24}$$

where $\bar{\mathbf{T}}[\mathbf{x},t]\langle \mathbf{x}'-\mathbf{x}\rangle$ is a peridynamic vector state, with parameters in the square brackets indicating variables that act as arguments to any functions referenced in the vector state and variables in the angle brackets acting as arguments to the vector state itself. In the state-based formulation, the deformation gradient, given by

$$\bar{\mathbf{F}} = \bar{\mathbf{I}} + \mathbf{u}\nabla, \tag{29.25}$$

can be approximated as a vector state as

$$\bar{\mathbf{F}}[\mathbf{x},t] = \left[\int_{\mathcal{H}_\mathbf{x}} C(|\xi|)(\mathbf{Y}[\mathbf{x},t]\langle\xi\rangle \otimes \xi) dV_{\mathbf{x}'}\right]\bar{\mathbf{K}}^{-1}, \tag{29.26}$$

where $\bar{\mathbf{K}}$ is a shape tensor given by

$$\bar{\mathbf{K}}[\mathbf{x},t] = \int_{\mathcal{H}_\mathbf{x}} C(|\xi|)(\xi\otimes\xi) dV_{\mathbf{x}'}, \tag{29.27}$$

$\mathbf{Y}$ is a deformation vector state given by

$$\mathbf{Y}[\mathbf{x},t]\langle\xi\rangle = \eta + \xi, \tag{29.28}$$

with $\eta = \mathbf{u}[\mathbf{x}',t] - \mathbf{u}[\mathbf{x},t]$ and $\xi = \mathbf{x}' - \mathbf{x}$ [17].

The deformation gradient can now be substituted into Hooke's law and strain-displacement relations, giving a stress term $\bar{\sigma}$ in terms of $\mathbf{u}$ in plane strain

$$\rho \frac{\partial^2}{\partial t^2}\mathbf{u} = \nabla\cdot\bar{\sigma} = \nabla\cdot(\bar{\mathbf{c}}:\bar{\epsilon}), \tag{29.29}$$

where

$$\bar{\epsilon}[\mathbf{x},t] = \frac{1}{2}(\nabla\mathbf{u} + \mathbf{u}\nabla) = \frac{1}{2}\left(\bar{\mathbf{F}}[\mathbf{x},t] + \bar{\mathbf{F}}[\mathbf{x},t]^{\mathrm{T}} - 2\bar{\mathbf{I}}\right), \tag{29.30}$$

where $\bar{\mathbf{c}}$ is the plane strain stiffness matrix with defining Lamé parameters $\lambda$ and $\mu$. Ultimately, the peridynamic vector state $\mathbf{T}$ for plane strain elasticity is given by

$$\bar{\mathbf{T}}[\mathbf{x},t]\langle\xi\rangle = C(|\xi|)\bar{\sigma}[\mathbf{x},t]\bar{\mathbf{K}}^{-1}\xi. \tag{29.31}$$

#### 29.2.3.1 PML in Two Dimensions

Before applying a PML directly to the peridynamic equation, Eq. 29.1 will be treated so that the PML application to peridynamics will be clear. It is convenient to convert Eq. 29.1 to the Laplace domain, assuming $e^{-st}$ time dependence, giving

$$\rho s^2 \tilde{\mathbf{u}} = \nabla \cdot \tilde{\bar{\sigma}}, \tag{29.32}$$

where the Laplace transform of a variable is indicated by $\mathcal{L}\{f\} = \tilde{f}$. Next, we express the wave equation as two coupled first order partial differential equations, the first in $\tilde{\mathbf{u}}$ and the second in $s\tilde{\bar{\psi}} = \tilde{\bar{\sigma}}$

$$\begin{aligned} \rho s \tilde{\mathbf{u}} &= \nabla \cdot \tilde{\bar{\psi}} \\ s\tilde{\bar{\psi}} &= \bar{\mathbf{c}} : \tilde{\bar{\varepsilon}}. \end{aligned} \tag{29.33}$$

Expanding Eq. 29.33 into components gives five coupled equations

$$\begin{aligned} &\rho s \tilde{u}_x = \frac{\partial}{\partial x}\tilde{\psi}_x + \frac{\partial}{\partial y}\tilde{\psi}_\tau, \; \rho s \tilde{u}_y = \frac{\partial}{\partial x}\tilde{\psi}_\tau + \frac{\partial}{\partial y}\tilde{\psi}_y \\ &s\tilde{\psi}_x = (\lambda + 2\mu)\frac{\partial}{\partial x}\tilde{u}_x + \lambda \frac{\partial}{\partial y}\tilde{u}_y, \; s\tilde{\psi}_y = \lambda \frac{\partial}{\partial x}\tilde{u}_x + (\lambda + 2\mu)\frac{\partial}{\partial y}\tilde{u}_y \\ &s\tilde{\psi}_\tau = \mu\left(\frac{\partial}{\partial y}\tilde{u}_x + \frac{\partial}{\partial x}\tilde{u}_y\right). \end{aligned} \tag{29.34}$$

Here, we will make the substitution given in Eq. 29.4 for all spatial derivatives, written as

$$\frac{\partial}{\partial x} \to \frac{s}{s + \phi(x)} \frac{\partial}{\partial x} \tag{29.35}$$

in the Laplace domain, and later define $\phi_x$ and $\phi_y$ in the desired absorbing boundary locations. Using the first of Eq. 29.34 as an example, we get

$$\rho(s + \phi_x)\left(s + \phi_y\right)\tilde{u}_x = \left(s + \phi_y\right)\frac{\partial}{\partial x}\tilde{\psi}_x + (s + \phi_x)\frac{\partial}{\partial y}\tilde{\psi}_\tau. \tag{29.36}$$

The remaining components of Eq. 29.34 can be expanded in a similar way.

#### 29.2.3.2 Auxiliary Field Formulation and PML Application

Peridynamics is not typically stated in terms of Cartesian components as Eq. 29.34, but we can expand the state-based formulation into components and match terms to Eq. 29.34. Following this approach yields a viable method for performing PML substitutions.

First, the state-based peridynamic equations listed above in Eqs. 29.24–29.31 can be written explicitly as

$$\begin{aligned} \rho s \tilde{u}_x[\mathbf{x}, s] = &\int_{\mathcal{H}_\mathbf{x}} C(|\xi|)\left[\left(\tilde{\psi}_x[\mathbf{x}, s]k^{inv}_{xx} + \tilde{\psi}_\tau[\mathbf{x}, s]k^{inv}_{yx}\right)\xi_x + \left(\tilde{\psi}_x[\mathbf{x}', s]k^{inv}_{xx} + \tilde{\psi}_\tau[\mathbf{x}', s]k^{inv}_{yx}\right)\xi_x\right] dV_{\mathbf{x}'} \\ &+ \int_{\mathcal{H}_\mathbf{x}} C(|\xi|)\left[\left(\tilde{\psi}_x[\mathbf{x}, s]k^{inv}_{xy} + \tilde{\psi}_\tau[\mathbf{x}, s]k^{inv}_{yy}\right)\xi_y + \left(\tilde{\psi}_x[\mathbf{x}', s]k^{inv}_{xy} + \tilde{\psi}_\tau[\mathbf{x}', s]k^{inv}_{yy}\right)\xi_y\right] dV_{\mathbf{x}'}, \end{aligned} \tag{29.37}$$

where

$$\bar{\mathbf{K}}^{-1} = \begin{bmatrix} k^{inv}_{xx} & k^{inv}_{xy} \\ k^{inv}_{yx} & k^{inv}_{yy} \end{bmatrix}, \tag{29.38}$$

and only $\tilde{u}_x[\mathbf{x}, s]$ is shown for brevity. Though no derivatives appear in Eqs. 29.37, the correspondence of each term to those in 29.34 is apparent – with partial derivatives following the component of $\xi$ – and the PML substitutions can be made. Equation 29.37 can then be rewritten as

$$
\begin{aligned}
\rho(s+\phi_x)\left(s+\phi_y\right)\tilde{u}_x = &\left(s+\phi_y\right)\int_{\mathcal{H}_{\mathbf{x}}} C(|\xi|)\left(\tilde{\psi}_x[\mathbf{x},s]k_{xx}^{inv} + \tilde{\psi}_{xy}[\mathbf{x},s]k_{yx}^{inv}\right)\xi_x dV_{\mathbf{x}'} \\
&+ \left(s+\phi_y\right)\int_{\mathcal{H}_{\mathbf{x}}} C(|\xi|)\left(\tilde{\psi}_x[\mathbf{x}',s]k_{xx}^{inv} + \tilde{\psi}_{xy}[\mathbf{x}',s]k_{yx}^{inv}\right)\xi_x dV_{\mathbf{x}'} \\
&+ (s+\phi_x)\int_{\mathcal{H}_{\mathbf{x}}} C(|\xi|)\left(\tilde{\psi}_x[\mathbf{x},s]k_{xy}^{inv} + \tilde{\psi}_\tau[\mathbf{x},s]k_{yy}^{inv}\right)\xi_y dV_{\mathbf{x}'} \\
&+ (s+\phi_x)\int_{\mathcal{H}_{\mathbf{x}}} C(|\xi|)\left(\tilde{\psi}_x[\mathbf{x}',s]k_{xy}^{inv} + \tilde{\psi}_\tau[\mathbf{x}',s]k_{yy}^{inv}\right)\xi_y dV_{\mathbf{x}'},
\end{aligned}
\tag{29.39}
$$

with the remaining equations following similarly.

#### 29.2.3.3 Discretization

For the temporal discretization, forward Euler will be used to simplify the presentation and implementation. Higher order temporal discretizations can be used, though they lead to more terms in what follows. A forward Euler approximation can be stated as the substitution

$$
s \rightarrow \frac{z-1}{\Delta t}, \tag{29.40}
$$

where $z$ is the $z$-transform[1] variable representing a unit delay and $\Delta t$ is the time step size. Multiplying by $z^{-2}$ and rearranging gives an update equation in terms of $z$, which can be converted to a time-stepping method via the inverse $z$-transform[2] (assuming vanishing initial conditions and an appropriate region of convergence). Finally, the spatial discretization used here is a simple one-point integration and point match testing, giving, for the $x$-component of displacement

$$
\begin{aligned}
u_x[\mathbf{x}_i,l] = &-(\gamma_x+\gamma_y)u_x[\mathbf{x}_i,l-1] - \gamma_x\gamma_y u_x[\mathbf{x}_i,l-2] \\
&+\frac{\Delta t}{\rho}\sum_{j=1}^{N_i} C(|\xi_{ij}|)\left(\psi_x[\mathbf{x}_i,l-1]k_{xx}^{inv} + \psi_\tau[\mathbf{x}_i,l-1]k_{yx}^{inv}\right)\xi_x V_j \\
&+\frac{\Delta t}{\rho}\sum_{j=1}^{N_i} C(|\xi_{ij}|)\left(\psi_x[\mathbf{x}_j,l-1]k_{xx}^{inv} + \psi_\tau[\mathbf{x}_j,l-1]k_{yx}^{inv}\right)\xi_x V_j \\
&+\frac{\Delta t}{\rho}\gamma_y\sum_{j=1}^{N_i} C(|\xi_{ij}|)\left(\psi_x[\mathbf{x}_i,l-2]k_{xx}^{inv} + \psi_\tau[\mathbf{x}_i,l-2]k_{yx}^{inv}\right)\xi_x V_j \\
&+\frac{\Delta t}{\rho}\gamma_y\sum_{j=1}^{N_i} C(|\xi_{ij}|)\left(\psi_x[\mathbf{x}_j,l-2]k_{xx}^{inv} + \psi_\tau[\mathbf{x}_j,l-2]k_{yx}^{inv}\right)\xi_x V_j \\
&+\frac{\Delta t}{\rho}\sum_{j=1}^{N_i} C(|\xi_{ij}|)\left(\psi_x[\mathbf{x}_i,l-1]k_{xy}^{inv} + \psi_\tau[\mathbf{x}_i,l-1]k_{yy}^{inv}\right)\xi_y V_j \\
&+\frac{\Delta t}{\rho}\sum_{j=1}^{N_i} C(|\xi_{ij}|)\left(\psi_x[\mathbf{x}_j,l-1]k_{xy}^{inv} + \psi_\tau[\mathbf{x}_j,l-1]k_{yy}^{inv}\right)\xi_y V_j \\
&+\frac{\Delta t}{\rho}\gamma_x\sum_{j=1}^{N_i} C(|\xi_{ij}|)\left(\psi_x[\mathbf{x}_i,l-2]k_{xy}^{inv} + \psi_\tau[\mathbf{x}_i,l-2]k_{yy}^{inv}\right)\xi_y V_j \\
&+\frac{\Delta t}{\rho}\gamma_x\sum_{j=1}^{N_i} C(|\xi_{ij}|)\left(\psi_x[\mathbf{x}_j,l-2]k_{xy}^{inv} + \psi_\tau[\mathbf{x}_j,l-2]k_{yy}^{inv}\right)\xi_y V_j,
\end{aligned}
\tag{29.41}
$$

where $V_j$ is the volume of node $j$, $N_i$ is the number of nodes in the neighborhood of node $i$, and $\xi_{ij} = \mathbf{x}_j - \mathbf{x}_i$.

[1] The $z$-transform is defined here as $X(z) = Z\{x[n]\} = \sum_{n=0}^{\infty} x[n]z^{-n}$ for causal signals.

[2] The only necessary property is the delay: $x[n-k] \leftrightarrow z^{-k}X(z)$.

## 29.3 Results

The PML was tested on two types of problems, first a 1D wave propagation problem and second a 2D wave propagation problem.

### 29.3.1 *1D Wave Propagation*

In the first example, the PML was applied at both sides of the region with a Gaussian pulse initial condition. The second Gaussian kernel, $\alpha^2_{\text{Gauss}}$ (Eq. 29.21), was used and the following parameters: a time step size of $\Delta t = 2.5$ ms, node spacing of $\Delta x = 5$ mm, horizon of $\ell = 2.1$ mm, PML width of 0.4 m, $a = 1$ N, and $b = 1$ m/kg. Figure 29.1 shows a waterfall plot of the simulation progress, with the initial condition visible at the front of the plot. As the plot illustrates, the PML regions absorb the impinging pulse. Figure 29.2 shows the maximum absolute value of the displacement over a longer simulation time. As can be seen, the initial waves are reduced by about three orders of magnitude. Inspection of the simulation results show that the residual displacement is from reflection of the initial wave from the PML interface. Though effort was taken to smooth the transition, there are still numerical reflections; however, these reflections are continually reduced by three orders of magnitude each time they reach the PML. After about 6 s, the wave is reduced to very near machine precision.

Finally, the 1D formulation is verified against an exact solution for wave propagation in a finite layer on a halfspace (one interface). The half space is terminated with a PML to simulate an infinite medium and the input is a step stress, modeled by setting a Neumann boundary condition on $u$. The first layer has a length of 0.5 (all units are arbitrary), with $a = 1$ and $b = 1$. The second layer, starting at $x_0 = 0$, has an actual length of 2, with a PML of width 0.5 placed at the end. The reflection coefficient between the layers is 0.5, set by choosing $a = 1$ and $b = 9$ in the second layer. The kernel $\alpha^1_{\text{Gauss}}$ was used with a horizon $l = 10^{-3}$, node spacing $\Delta x = 2.5 \times 10^{-3}$, and time step $\Delta t = 0.8 \times 10^{-3}$. The stress, defined here as the first off-diagonal entry of $a\mathcal{D}(u) = a\left(u^{m+1}_{k+1} - u^{m}_{k+1}\right)\alpha(x_{m+1}, x_m)$, was recorded for all time at two locations: $x_1 = 0.25$ and $x_2 = -0.75$. Figure 29.3 shows the exact solution (solid line) at $x_1 = 0.25$ compared with the peridynamic solution (dashed line). Some overshoot is evident for early times and some dispersion for later times, but otherwise the solutions agree and demonstrate code verification for transient wave phenomena. Figure 29.4 shows both solutions at $x_2 = -0.75$ (which is in the second layer) with similar results.

### 29.3.2 *2D Wave Propagation*

The PML was first tested on a wave propagation problem with PML boundary layers and a Gaussian distribution, centered at the midpoint of the computational region and with a variance of 0.05 m, as an initial condition for the $x$-directed

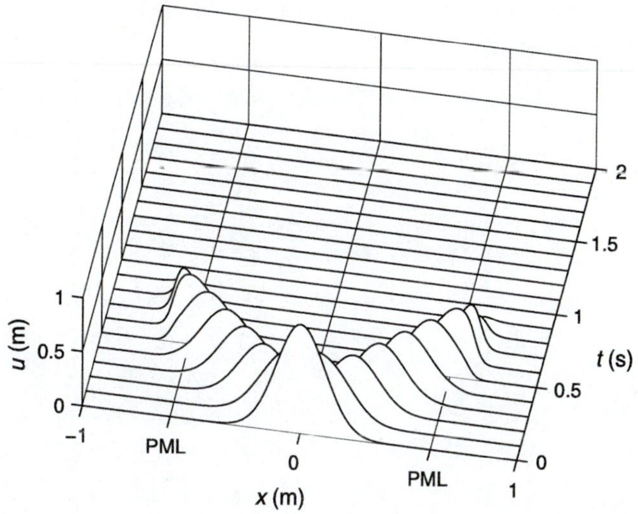


**Fig. 29.1** Gaussian initial condition with PML boundary layers at $x = \pm 0.6$ m for $\alpha^2_{\text{Gauss}}$

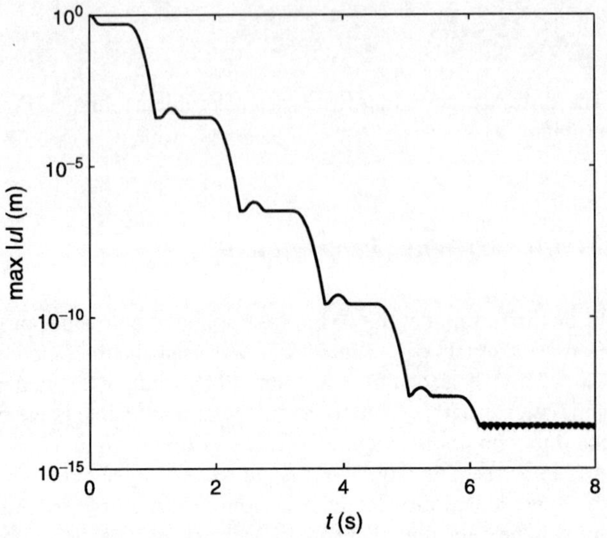


**Fig. 29.2** Maximum displacement for the first example

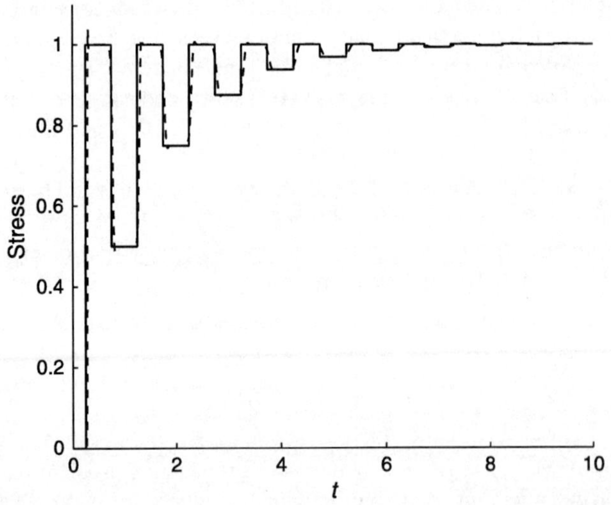


**Fig. 29.3** Finite layer and halfspace problem at $x = 0.25$. The *solid line* is the exact solution and the *dashed line* is the peridynamic solution

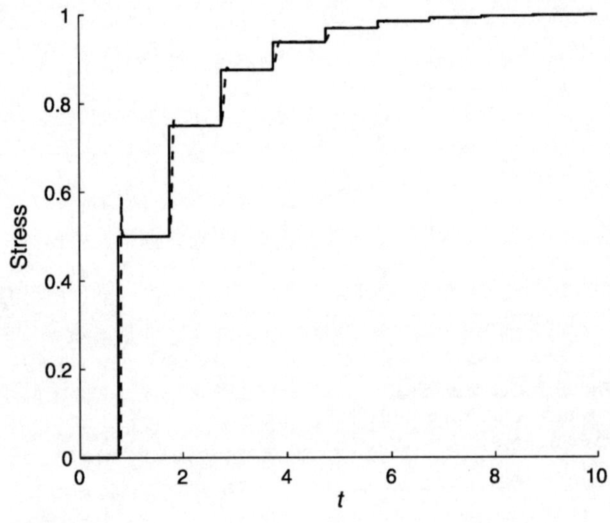


**Fig. 29.4** Finite layer and halfspace problem at $x = -0.75$

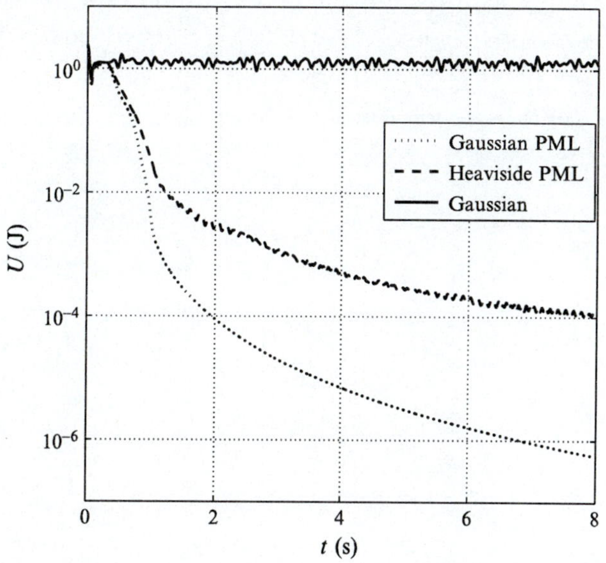


**Fig. 29.5** Total strain energy in a simulation terminated by a PML

displacement. The computational region in this example was defined as the square $0 \leq x, y \leq 1$ and discretized with $\Delta x = \Delta y = 0.01$ m. Young's modulus for the region was set to 1 Pa, Poisson's ratio was 1/4, and the density was 1 $kg/m^3$. The PML region was defined as the 0.3 m border around the 1 m-by-1 m region and used a Gaussian ramp with a width of 0.2 m, finally reaching a maximum of 50 $s^{-1}$ for the remaining 0.1 m. For the Gaussian kernel, a horizon size of $\ell = 1.1\Delta x$ was used, and for the Heaviside kernel, a horizon of $\ell = 3.1\Delta x$ was used.

The simulation was run with both the Heaviside and Gaussian kernels, with the total strain energy shown in Fig. 29.5. The Gaussian kernel (dotted line) shows the largest drop in energy, reaching a minimum of $5.6 \times 10^{-7}$, and the Heaviside kernel (dashed line) decreases to $1.0 \times 10^{-4}$. A bounded simulation is shown for reference as the solid line, which used a fixed displacement boundary condition and the Gaussian kernel.

For verification, the method was compared with an exact analytical solution. Consider a cylindrically symmetric wave propagating in an infinite elastic medium with the same constitutive parameters as the above example, and with an initial condition given by

$$u_0(r) = b\left(\frac{r}{a}\right)\left[1 + \left(\frac{r}{a}\right)^2\right]^{-\frac{3}{2}} \tag{29.42}$$

where here we take $b = 1$ and $a = 0.1$. The exact solution is given by Eringen and Suhubi [18]

$$\begin{aligned} u(r,t) &= \frac{br}{\sqrt{2}aR^6}\sqrt{R^2+\alpha}\left(2\alpha - R^2\right), \\ \alpha &= 1 + \frac{r^2 - c^2t^2}{a^2}, \\ R^2 &= \sqrt{\alpha^2 + \frac{4c^2t^2}{a^2}}, \end{aligned} \tag{29.43}$$

where $c$ is the longitudinal wave speed. This problem was simulated in a 2D region, 2 m-by-2 m and $\Delta x = \Delta y = 0.01$ m, terminated by a PML with the same dimensions and magnitude as the above problem. The Gaussian kernel was used with a horizon size of $\ell = 0.75\Delta x$, with an actual cutoff of 0.028 m. The results are shown in Fig. 29.6, with the exact solution shown as the solid line and the peridynamic solution shown as the dashed line. The peridynamic solution shows good agreement with the exact solution and minimal reflections from the PML boundary.

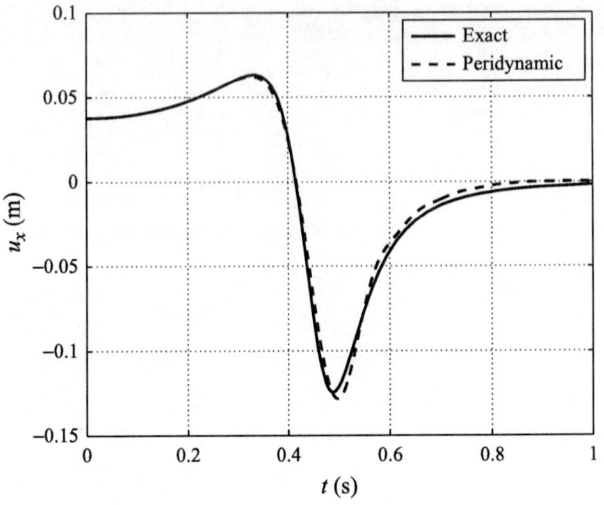


**Fig. 29.6** $x$-directed displacement at $x = 0.5$, $y = 0$ m. The *solid line* shows the exact solution and the *dashed line* results terminated by a PML

## 29.4 Conclusions

A perfectly matched layer was applied to peridynamics in one and two dimensions, allowing for the simulation of infinite regions. A matrix-based auxiliary formulation was used in one dimension and state-based peridynamics was used in two dimensions as more flexible constitutive relations are necessary to implement a perfectly matched layer, essentially an artificial anisotropic absorbing material. For simplicity, standard discretization techniques were used: one-point integration and point-matching for the spatial discretization and forward Euler for the temporal discretization. Results show that the PML absorbs incoming waves and results in minimal reflections at the boundary between the absorbing layer and the computational region. A Gaussian function was used as a ramp to avoid these numerical reflections.

## References


1. Silling S (2000) Reformulation of elasticity theory for discontinuities and long-range forces. J Mech Phys Solid 48(1):175–209
2. Berenger JP (1994) A perfectly matched layer for the absorption of electromagnetic-waves. J Comput Phys 114(2):185–200
3. Chew WC, Weedon WH (1994) A 3D perfectly matched medium from modified Maxwell's equations with stretched coordinates. Microw Opt Technol Lett 7(13):599–604
4. Silling SA, Epton M, Weckner O, Xu J, Askari E (2007) Peridynamic states and constitutive modeling. J Elast 88(2):151–184
5. Uno T, He Y, Adachi S (1997) Boundary condition for dispersive medium. IEEE Microw Guided Wave Lett 7(9):264–266
6. Teixeira FL, Chew WC (1998) A general approach to extend Berenger's absorbing boundary condition to anisotropic and dispersive media. IEEE Trans Antennas Propag 46(9):1386–1387
7. Alles EJ, van Dongen KW (2009) Frequency domain perfectly matched layers for acoustic scattering integral equation problems. 2009 IEEE International Ultrasonics Symposium, Rome, Big Sky, Montana pp 1610–1613
8. Pissoort D, Olyslager F (2003) Termination of periodic waveguides by PMLs in time-harmonic integral equation-like techniques. Antennas Wirel Propag Lett 2(20):281–284
9. Chew WC, Liu QH (1996) Perfectly matched layers for elastodynamics: a new absorbing boundary condition. J Comput Acoust 4(4):341–359
10. Liu QH, Tao JP (1997) The perfectly matched layer for acoustic waves in absorptive media. J Acoust Soc Am 102(4):2072–2082
11. Festa G (2003) PML absorbing boundaries. Bull Seismol Soc Am 93(2):891–903
12. Du Q, Gunzburger MD, Lehoucq RB, Zhou K (2010) A nonlocal vector calculus , nonlocal volume-constrained problems, and nonlocal balance laws. Tech. Report. SAND-2010-8353J, Sandia National Laboratories
13. Malvern LE (1969) Introduction to the mechanics of a continuous medium. Prentice-Hall, Englewood Cliffs
14. Johnson SG (2007) Notes on perfectly matched layers. http://math.mit.edu/~stevenj/18.369/pml.pdf. Accessed on 2007
15. Weckner O, Abeyaratne R (2005) The effect of long-range forces on the dynamics of a bar. J Mech Phys Solid 53(3):705–728
16. Silling S, Askari E (2005) A meshfree method based on the peridynamic model of solid mechanics. Comput Struct 83(17–18):1526–1535
17. Foster JT, Silling SA, Chen WW (2010) Viscoplasticity using peridynamics. Int J Numer Methods Eng 81(10):1242–1258
18. Eringen AC, Suhubi ES (1975) Elastodynamics. Academic, New York

# Chapter 30
# Dynamic Behavior and Numerical Modeling of Titanium 15-3-3-3 Alloy

**Mikko Hokka, Tõnu Leemet, Aviral Shrot, Martin Bäker, and Veli-Tapani Kuokkala**

**Abstract** Metastable beta titanium alloys combine excellent mechanical properties with low density, and are therefore very attractive in many mechanically demanding applications. The high strength and hardness, however, cause several challenges in the machining of these materials, and the machining costs of titanium components can be significant compared to the overall costs of the component. The cutting conditions can be optimized using finite element simulations, leading to reduced machining costs and improved machining quality. However, the simulations of the rather complex machining processes need reliable material models. The models can only be generated when the mechanical behavior of the material is well understood. In this work, the mechanical response of Ti-15-3-3-3 alloy has been characterized in a wide range of strain rates and temperatures. The Johnson-Cook material model was fit to the measurement data, and the model was used to simulate orthogonal cutting of the material. The simulation results were further compared to cutting experiments at high cutting speeds. The current model is able to simulate the serrated chip formation frequently observed in machining of titanium alloys at high cutting speeds. Also, the simulated cutting forces match well with the experimentally obtained forces. However, the model needs to be further developed to match also the fine details of the chip, such as the chip curl and thickness of the individual serrations.

**Keywords** Dynamic behavior • Numerical modeling • High speed machining • Titanium 15-3-3-3

## 30.1 Introduction

The extraordinary properties of titanium and its alloys have rapidly increased the use of these materials in various engineering applications, especially in the aerospace industry. Typical properties of titanium alloys include low density, excellent corrosion resistance, and high strength and hardness. Other properties typical to titanium alloys include relatively low heat conductivity and high chemical reactivity. On the other hand, the high strength and hardness combined with the overall chemical reactivity can lead to several difficulties in the machining of these alloys [1–3]. As the high strength and hardness are usually retained to high temperatures the machining forces are typically rather high. The hardness also increases the abrasive wear of the tools. Chemical reactions between the tool and the workpiece can lead to strong adhesion or galling of the tool and the chip leading to a rapid wear and even failure of the tool. Finally, the low heat conductivity and high strength of the material easily induce strong adiabatic heating during machining, resulting in higher machining temperatures and faster failure of the tool. For these reasons, most titanium alloys are considered hard-to-machine alloys.

The experimental research on machining and machinability is somewhat complicated by the vast number of different machining operations, e.g., turning, milling, drilling etc., and the number of different conditions or machining parameters, e.g., cutting speed and depth, rake angles, etc. Therefore, studying the material behavior in machining would require extensive testing with various techniques and machines. The testing can, however, be complemented by simulating the machining processes. On the other hand, the simulation results are reliable only when using accurate material models that

M. Hokka (✉) • T. Leemet • V.-T. Kuokkala
Department of Materials Science, Tampere University of Technology, P.O.B. 589, 33101 Tampere, Finland
e-mail: mikko.hokka@tut.fi

A. Shrot • M. Bäker
Technische Universität Braunschweig, Institute for Materials, Langer Kamp 8, 38106 Braunschweig, Germany

V. Chalivendra et al. (eds.), *Dynamic Behavior of Materials, Volume 1: Proceedings of the 2012 Annual Conference on Experimental and Applied Mechanics*, Conference Proceedings of the Society for Experimental Mechanics Series, DOI 10.1007/978-1-4614-4238-7_30, 

are able to reproduce the true material behavior at the cutting conditions. These models are typically generated based on experimental data acquired from rather simple experiments, such as tension or compression tests. However, the material properties can be measured only in certain and often quite narrow ranges of strain, strain rate, and temperature. In machining, the strains, strain rates, and temperatures are most likely to be much higher than what can be covered by the laboratory experiments, and therefore the material model must be able to predict the properties and behavior of the material far beyond the ranges where the model and its parameters were determined.

In this work, the behavior of titanium 15-3-3-3 alloy was extensively studied in compression, and a material model based on the Johnson-Cook equation was generated. This model was used to simulate orthogonal cutting experiments at high cutting speeds.

## 30.2 Experimental

In this work, the compression behavior of a commercial metastable beta titanium alloy Ti-15V-3Sn-3Al-3Cr was studied in a wide range of strain rates and temperatures. The alloy was studied in as-received condition without any aging treatments. Cylindrical compression specimens with a diameter of 8 mm and length of 8 mm and/or 6 mm were machined using electric discharge machining. At low strain rates the mechanical testing was performed using an MTS servohydraulic materials testing machine with an induction heating system for high temperature tests. The high strain rate compression tests, on the other hand, were done with a Split Hopkinson Pressure Bar (SHPB) device equipped with a high temperature test setup. The device used in this study consists of 22 mm diameter high strength Maraging steel incident, transmitted, and striker bars. Both incident and transmitted pressure bars were aligned along the same horizontal axis, and the bars were supported by three bearings on four stanchions, which can be moved in y- and z- directions to ensure proper alignment. The lengths of the incident, transmitted, and momentum trap bars were 1,200 mm each, and the length of the striker bar varied from 200 to 400 mm. A room temperature experiment is performed simply by placing the cylindrical specimen between the lubricated ends of the incident and transmitted bars, and then impacting the striker bar to the free end of the incident bar. The impact velocity is adjusted by changing the pressure and measured using three IR sender receiver pairs. The impact of the striker creates a stress pulse that propagates in the incident bar at the speed of sound (~4,900 m/s), and when the stress pulse reaches the bar-specimen interface, part of the pulse is transmitted through the specimen into the transmitted bar, whereas the remaining part of the incident stress pulse is reflected back to the incident bar as a wave of tension. The incident, reflected, and transmitted stress pulses are measured with strain gages bonded on the surface of the stress bars, and the signals are amplified and recorded using Kyowa CDV 700A series signal conditioner and a Yokogawa DL 708 high sampling rate digital oscilloscope. The stress, strain, and strain rate in the specimen are calculated from the dispersion corrected stress pulses using Eqs. 30.1, 30.2 and 30.3. The numerical dispersion correction used in this work was adopted from the work of Gorman [4].

Average stress in the specimen is given by:

$$\sigma(t) = \frac{A_b \, E\varepsilon_t(t)}{A_s} \tag{30.1}$$

Average strain in the specimen is given by:

$$\varepsilon(t) = \frac{2C_0}{L_s} \int_0^t \varepsilon_r(t) \, dt \tag{30.2}$$

Average strain rate in the specimen is given by:

$$\dot{\varepsilon} = \frac{2C_0\varepsilon_r(t)}{L_s} \tag{30.3}$$

In Eqs. 30.1, 30.2 and 30.3, $A_b$, E, and $C_0$ are the cross sectional area, Young's modulus, and the speed of sound of the bar material, $L_s$ and $A_s$ are the length and cross sectional area of the specimen, and $\varepsilon_r$, and $\varepsilon_t$ are the reflected and transmitted strain pulses. The Eqs. 30.1, 30.2 and 30.3 assume stress equilibrium, i.e., that the stresses on the front and back sides of the specimen are the same. For metallic materials this assumption is true at least for strains higher that a few percent.

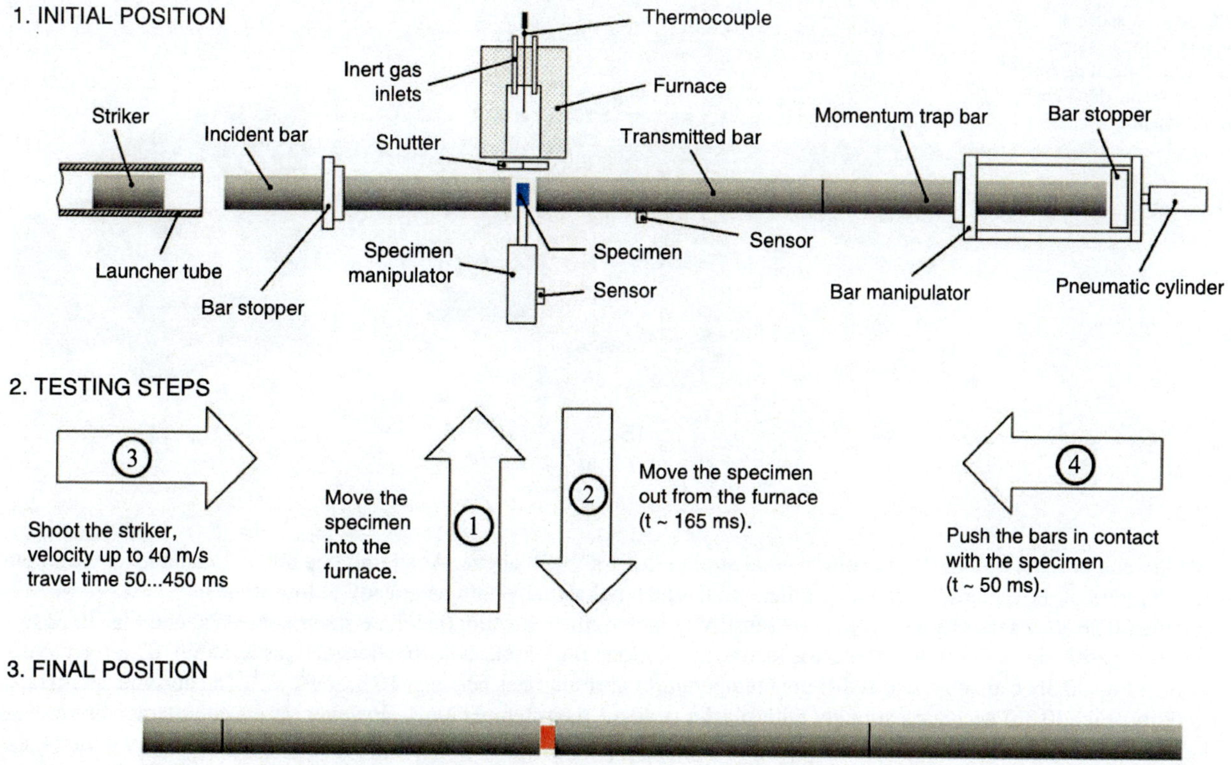


**Fig. 30.1** A schematic picture of the testing steps during a test with the high temperature setup

The high temperature compression test system at high strain rates is based on computer controlled pneumatic manipulation of the stress bars and the specimen. This system allows keeping the bars at room temperature while heating the specimen in a furnace located beside the bars. The specimen is inserted into a ceramic insulating wool ring exposing only the ends of the cylindrical specimen. The wool ring itself is very soft and does not mechanically constrain the deformation of the specimen. The wool ring is further placed into a specimen holder arm that is pneumatically pushed to the furnace. After heating the specimen holder arm quickly retracts the specimen to the centerline of the bars. The striker bar is shot, and just a fraction of a second before the impact, a second manipulator pushes the cold bars and the hot specimen into contact. The current high temperature setup limits the total time that the specimen spends outside the hot furnace before the dynamic loading to about 200 ms. More importantly the system limits the contact time of the hot specimen and the cold bars to less than 50 ms, which keeps the temperature of the specimen essentially constant. The heat flow from the specimen to its surroundings has been studied by numerical simulations, and the results show only a minor temperature drop of a few degrees at the depth of less than 150 μm for a test performed at the initial temperature of 800°C [5, 6]. Figure 30.1 shows the testing steps during a test with the high temperature system.

A material model was generated based on the results of the mechanical testing. The model was further used in the simulations of orthogonal cutting of the material. The results of the simulations were further compared to orthogonal cutting experiments with the U-shape specimens. A schematic picture of the U-shape test specimen is shown in Fig. 30.2. The test is performed by inserting the cutting tool between the arms of the 'U', placing the whole specimen-tool assembly between the bars, by impacting the striker bar to the incident bar, and reading the cutting forces from the transmitted signal.

## 30.3 Results and Discussion

The room temperature stress–strain curves for the titanium 15-3-3-3 alloy are shown in Fig. 30.3a. At low strain rates the yield behavior of the material is continuous, followed by moderate strain hardening. When the strain rate is increased from $10^{-3}$ $s^{-1}$ to 1 $s^{-1}$ the strain hardening rate decreases strongly with increasing strain rate. Also, the yield behavior changes

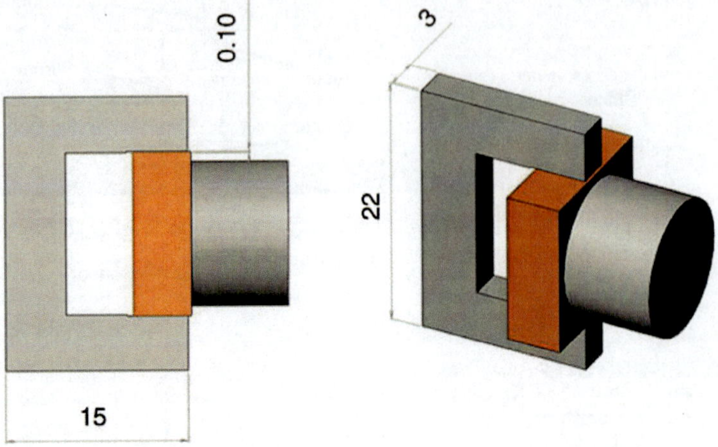


**Fig. 30.2** A schematic picture of the U-shape specimen for cutting force measurement. Units are in millimeters

gradually and pronounced yielding is observed at strain rates 1 $s^{-1}$ and above. At strain rates 600 $s^{-1}$ and above, the strain hardening rate of the material decreases further and the material actually softens already at low strains.

Figure 30.3b shows the absolute strain rate sensitivity factor, $m_a = d\sigma/d\log(\dot{\varepsilon})$ where stress is in MPa, and Fig. 30.3c the relative (logarithmic) strain rate sensitivity factor, $m_r = d\log(\sigma)/d\log(\dot{\varepsilon})$, as a function of plastic strain. The sensitivity factors in Fig. 30.3b, c are obtained at different temperatures at strain rates between $10^{-3}$ and 1 $s^{-1}$. The absolute strain rate sensitivity (Fig. 30.3b) decreases strongly with plastic strain at room temperature. However, at elevated temperatures the behavior is clearly different and the sensitivity factor higher than that at room temperature. The rate sensitivity increases with plastic strain at temperatures 800°–950°C. The highest absolute strain rate sensitivity is achieved at 800°C. On the other hand, if the strain rate sensitivity is calculated with respect to the material's strength, (Fig. 30.3c) the rate sensitivity increases strongly with increasing temperature and the highest rate sensitivity is achieved at 950°C. The relative strain rate sensitivity decreases with plastic strain at room temperature, is essentially constant at 600°C, slightly increases at 800°C, and clearly increases at 950°C.

The Jonhson-Cook plasticity model was used to describe the mechanical behavior of the titanium alloy. The model used for the adiabatic strength of the material is shown in Eq. 30.4. The parameters for the model were found using the following procedure: first the parameter 'm' was obtained using Eq. 30.5, where the yield strengths $\sigma_Y^{Tr}$ and $\sigma_Y^{T2}$ were obtained at the same strain rate at reference (room) and at an elevated temperature, respectively. Second, the adiabatic stress–strain curve obtained experimentally at the reference temperature and the reference strain rate was converted to an isothermal one using Eq. 30.6. Third, the parameters A, B, and n were obtained by fitting a straight line to the log(σ-A) vs. log(ε) curve. Finally, the parameter C was obtained by comparing the experimental stress–strain curves obtained at different strain rates and the curves calculated using Eq. 30.4. The parameter A was determined as the yield strength of the material at the reference strain rate. However, this value overestimated the strength of the material at higher strains, and therefore the value was decreased so that the strength calculated by the model better matches the experimental stresses at higher strains. Also, the value for the parameter m that was obtained using Eq. 30.5 did not produce a good match between the simulated and experimental stresses at higher strains. Therefore, the value was manually adjusted to improve the fit. The final values for the parameters are shown in Table 30.1. Figure 30.4 shows the comparison of the experimentally obtained adiabatic stress and the values calculated using the current material model. The fit is relatively good at strains higher than about 7%. The initial pronounced yielding was omitted when obtaining the model parameters as this behavior is very difficult to reproduce with the current material model and, more importantly, the observed yield behavior is not likely to have a significant effect on the chip formation process.

$$\sigma_a = (A + B\varepsilon^n)\left(1 + C_{JC}\ln\frac{\dot{\varepsilon}}{\dot{\varepsilon}_{ref}}\right)\left(1 - \left[\frac{T + \left(\frac{\beta}{\rho c}\int_0^{\varepsilon}\sigma d\varepsilon\right) - T_{ref}}{T_m - T_{ref}}\right]^m\right) \quad (30.4)$$

$$m = \log\left(1 - \frac{\sigma_Y^{T2}}{\sigma_Y^{Tr}}\right) / \log\left(\frac{T_2 - T_r}{T_m - T_r}\right) \quad (30.5)$$

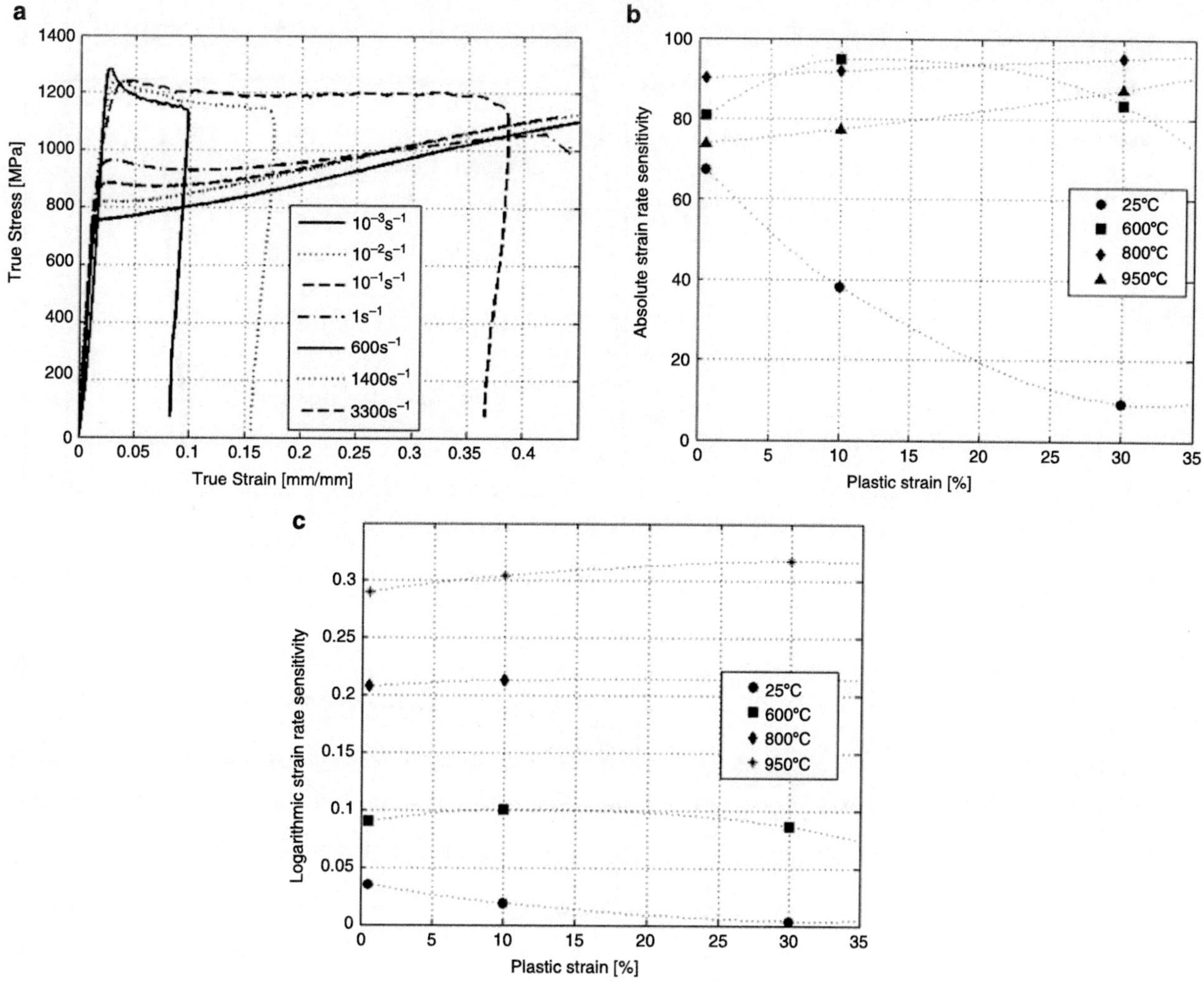


**Fig. 30.3** Results of the compression tests: (**a**) true stress versus true strain curves at different strain rates, (**b**) absolute, and (**c**) relative (logarithmic) strain rate sensitivity as a function of plastic strain at different temperatures

**Table 30.1** The parameters for the modified Johnson-Cook model used in this work

| Parameter | A | B | n | C | m | $T_r$ [°C] | $T_m$ [°C] | $\dot{\varepsilon}_{ref}$ [$s^{-1}$] | β | Cp [J/kgK] | ρ [$kg/m^3$] |
|---|---|---|---|---|---|---|---|---|---|---|---|
| Value | 984 | 380 | 0.3 | 0.069 | 1.1 | 23 | 1,669 | 1,400 | 0.9 | 500 | 4,760 |

$$\sigma_i = \frac{\sigma_a}{\left[1 - \left(\frac{T + \left(\frac{\beta}{\rho c}\int_0^{\varepsilon}\sigma d\varepsilon\right) - T_r}{T_m - T_r}\right)^m\right]} \tag{30.6}$$

Figure 30.5 shows the effect of the specific heat capacity on the isothermal stress–strain curves of the studied titanium alloy. In this work, both the modeling and simulations were done using heat capacity of 500 J/Kg*K, which is a value commonly found in literature, e.g. [7]. However, the heat capacity of the alloy is known to increase with temperature. In high speed cutting, the deformation induced heating therefore increases both the temperature and the heat capacity of the material. Figure 30.5 shows the stress strain curves calculated from the measured adiabatic curves at room temperature at the strain rate of 3,300 $s^{-1}$ using Eq. 30.6 with the specific heat capacities of 500 J/Kg*K and 650 J/Kg*K. At low strains the difference between the isothermal stress–strain curves is negligible, but the difference increases towards higher strains. The higher heat capacity leads to lower calculated strength values, as the temperature of the material increases less during deformation.

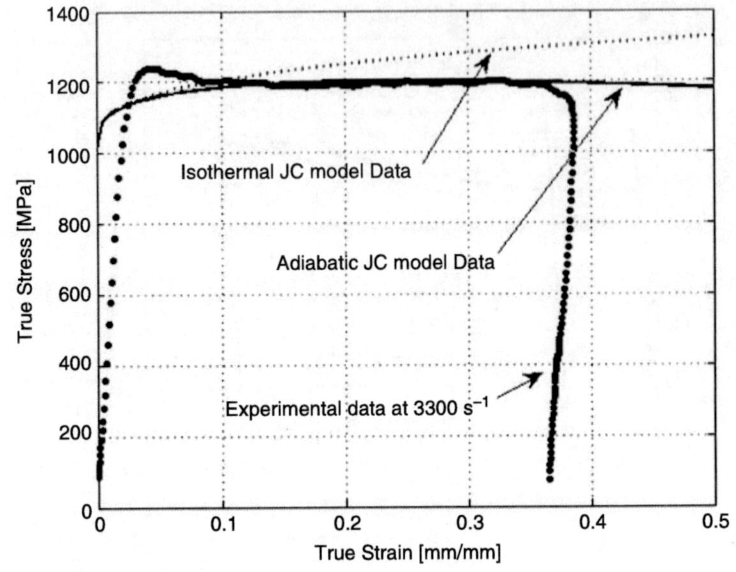


**Fig. 30.4** Comparison of the measured adiabatic stress data and the values calculated by the Johnson-Cook model at room temperature and at the strain rate of 3,300 s$^{-1}$

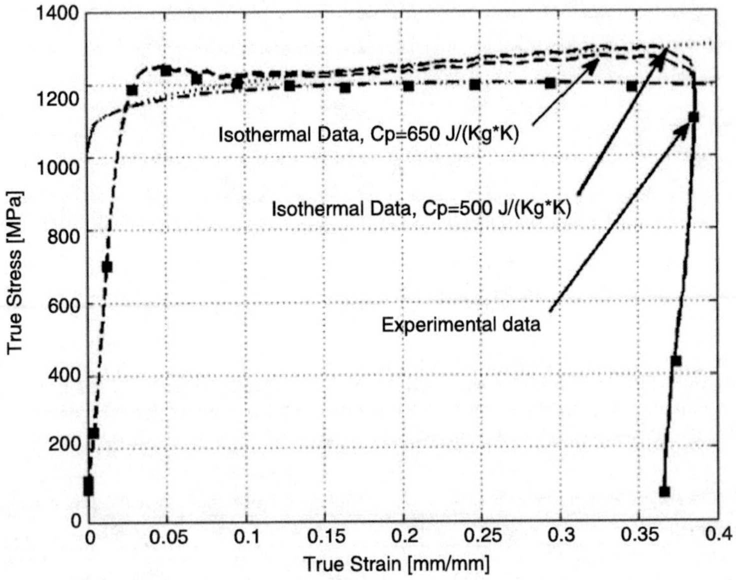


**Fig. 30.5** Effect of specific heat capacity on isothermal stress strain curves at room temperature. The *dotted* and *dash-dot lines* are stresses calculated using the Johnson-Cook model

In the simulations of high speed cutting, the results were found to be very sensitive to small changes in the heat capacity, and the serrated chips were only produced when using the lower heat capacity value. With $C_p = 650$ J/Kg*K, the deformation induced heat was not sufficient to localize the strain and to produce segmented chips.

A two dimensional explicit finite element model was created using the commercial software ABAQUS 6.9-1. The tool was assumed rigid and no heat conduction was allowed between the tool and the chip. A conservative friction coefficient of 0.1 was used between the tool and the deforming material. The workpiece was meshed very finely with CPE4R elements, and the model contained 101151 nodes and 100049 elements (Fig. 30.6). Since the alloy has a low thermal conductivity, the machining process was assumed adiabatic at the cutting speed of 15 m/s. The chip separation was modeled by using a sacrificial layer whose elements were deleted when the shear strain exceeded the value of 1.0.

The same cutting conditions, i.e., cutting speed of 15 m/s and cutting depth of 44 μm were used in both the simulations and experiments. The simulated (Fig. 30.7) and experimentally obtained chips show somewhat similar serrations, but the

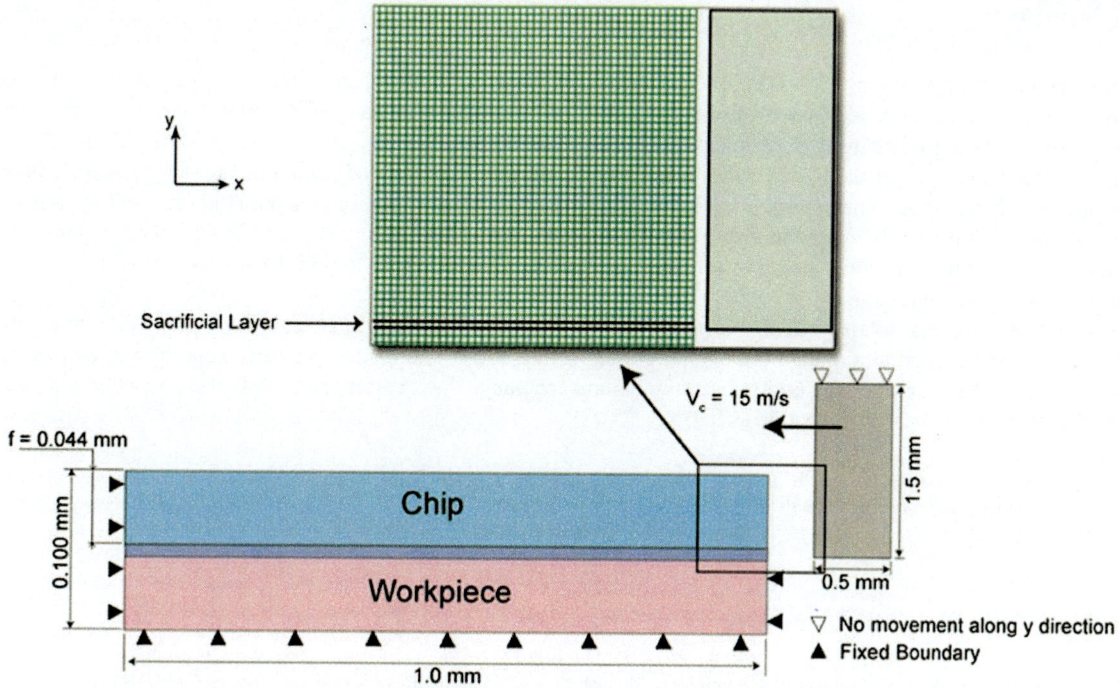


**Fig. 30.6** The finite element model used in the machining simulations

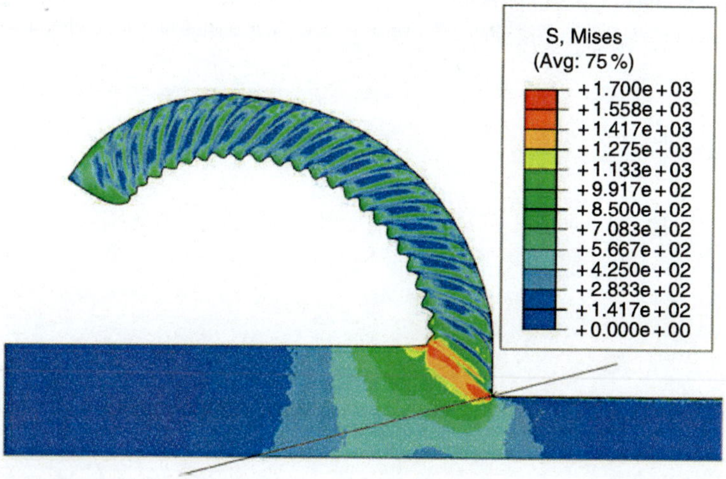


**Fig. 30.7** von Mises stress in the simulated chip

fine details such as the chip curvature and the dimensions of individual segments are somewhat different. On the other hand, the cutting stresses obtained from the experiments and from the simulations were 2,000 MPa and 2,200 MPa, respectively. The difference in the simulated and experimentally obtained cutting forces can be explained at least partly with the differences in the frictional conditions between the simulated and experimental setups. However, the model still needs further development to match also the chip shape. One possibility for the optimization of the parameters is the use of inverse methods [8].

## 30.4 Summary

Dynamic response of the titanium 15-3-3-3 alloy was studied at room and elevated temperatures. Based on the results of the mechanical tests, a material model based on the Johnson-Cook equation was generated. The model was further used in the simulations of orthogonal cutting experiments.

The strain hardening rate of the material was found to be a sensitive function of strain rate and temperature. The strain hardening rate decreases fast when strain rate is increased. Also, at higher temperatures the strain hardening rate at high strain rates was much higher compared to that observed at room temperature. The strain rate sensitivity of the material was also found to be a very sensitive function of temperature. In the low strain rate region the relative strain rate sensitivity increased rapidly with temperature.

The parameters for the Johnson-Cook plasticity model were determined from the isothermal stress strain curves. The model was further used to simulate orthogonal cutting of the alloy. The model is clearly able to produce segmented chips and to match the experimental cutting forces with good accuracy. However, more work is still needed to fine tune the model to match also the details of the chip shape.

**Acknowledgements** The research leading to these results has received funding from the European Union Seventh Framework Programme (FP7/2007-2013) under grant agreement No. PITN-GA-2008-211536, project MaMiNa.

## References

1. Rahman M, Wang Z, Wong Y (2006) A review on high-speed machining of titanium alloys. JSME Int J, Series C 49(1):11–20
2. Hartung PD, Kramer BM, von Turkovich BF (1982) CIRP Annals – Manufact Tech 31:75–80
3. Jawaid A, Che-Haron CH, Abdullah A (1999) J Mat Proc Tech 92–93:329–34
4. D. Gorman, J. Phys. E: Scientific Instruments, 16 (1983).
5. Apostol M, Kuokkala V-T, Vuoristo T (2004) In: Papalettere C (ed) Advances in experimental mechanics. McGraw-Hill, Bari
6. Apostol M, Vuoristo T, Kuokkala V-T (2003) J de Phys IV 110:459–64
7. Boyer R, Welsch G, Collings EW (1994) Materials properties handbook: titanium alloys. ASM International, Materials Park
8. Shrot A, Bäker M (2012) Determination of Johnson–Cook parameters from machining simulations. J Com Mat Sci 52:298–304

# Chapter 31
# Dynamic Load Measurement of Ballistic Gelatin Impact Using an Instrumented Tube

**Jeremy D. Seidt, J. Michael Pereira, Jeremiah T. Hammer, and Amos Gilat**

**Abstract** Bird strikes are a common problem for the aerospace industry and can cause serious damage to an aircraft. Ballistic gelatin is frequently used as a surrogate for actual bird carcasses in bird strike tests. Numerical simulations of these tests are used to supplement experimental data, therefore it is necessary to use numerical modeling techniques that can accurately capture the dynamic response of ballistic gelatin. An experimental technique is introduced to validate these modeling techniques. A ballistic gelatin projectile is fired into a strike plate attached to a 36 in long sensor tube. Dynamic load is measured at two locations relative to the strike plate using strain gages configured in a full Wheatstone bridge. Data from these experiments are used to validate a gelatin constitutive model. Simulations of the apparatus are analyzed to investigate its performance.

**Keywords** Dynamic force measurement • Ballistic impact • Gelatin

## 31.1 Introduction

Bird strikes can inflict serious damage to an aircraft. Photographs of bird strike damage to a Boeing 767–300 aircraft are presented in Fig. 31.1 [1]. This aircraft was damaged by large birds as it climbed from Charles de Gaulle Airport in Paris. The cockpit area of the aircraft was perforated by the birds, causing depressurization. Fortunately, the crew safely returned to the airport and no injuries were reported. Another recent and well-publicized example of bird strike damage is the January 15th 2009 incident involving US Airways Flight 1549. The captain of the Airbus A320-214 was forced to make an emergency landing on the Hudson River after both engines ingested Canadian geese resulting in a complete loss of thrust [2]. Because of many incidents like these, aircraft structures and engines must be certified to operate safely after a bird strike or ingestion event.

Certification tests are conducted using actual bird carcasses [3], however, ballistic gelatin is frequently used as a surrogate material for preliminary testing. The main advantage to using gelatin for testing is repeatability. No two birds are identical, even if they are of the same species. These natural variations make it difficult to perform repeatable experiments [4]. In addition, there are obvious sanitary benefits to testing with gelatin opposed to actual birds. Wilbeck and Rand [4] developed a bird substitute by mixing commercial gelatin with phenolic microballoons. The authors note that this surrogate projectile creates impact loads similar to those generated by actual birds.

Many researchers have conducted numerical simulations of bird impact with various aircraft structures, such as wing leading edges [5], windshields [6], flaps [7], and engine blades [8]. The majority of these simulations treat the bird as circular cylinders with rounded edges, and assume that the birds will behave hydrodynamically on impact. In these studies, various numerical treatments of the bird model are used, including Lagrangian, Arbitrary Lagrangian Eulerian (ALE) and Smoothed Particle Hydrodynamics (SPH). These simulations need to be validated by comparing the results to experimental data.

J.D. Seidt (✉) • J.T. Hammer • A. Gilat
Scott Laboratory, Department of and Mechanical and Aerospace Engineering, The Ohio State University, 201 W 19th Ave, +43210 Columbus, OH, USA
e-mail: seidt.2@osu.edu

J.M. Pereira
NASA Glenn Research Center 21000 Brookpark Road, MS 49-8 Cleveland, OH 44135

V. Chalivendra et al. (eds.), *Dynamic Behavior of Materials, Volume 1: Proceedings of the 2012 Annual Conference on Experimental and Applied Mechanics*, Conference Proceedings of the Society for Experimental Mechanics Series, DOI 10.1007/978-1-4614-4238-7_31, 

**Fig. 31.1** Photographs of bird strike damage to a Boeing 767–300 aircraft [1]: (**a**) below cockpit window, (**b**) underside fuselage, (**c**) aft of radome

An experimental measurement of the dynamic force imparted into a target by a projectile is very useful to validate simulations. However, impact force measurements are difficult to obtain accurately because of the short duration of the load. Even piezoelectric load cells with the highest frequency response are generally not sufficient to measure forces involved in ballistic impact events, and resonances in the supporting structure typically make results difficult to interpret and simulate. In the past, gelatin impact experiments have been conducted by firing a projectile into a flat plate instrumented with different types of pressure transducers [9, 10]. Typically, the diameter of the pressure transducer is less than the diameter of the gelatin projectile. The pressure on the impacting face of the projectile is certainly non-uniform, especially when the projectile begins to "mushroom". The pressure is maximum at the center and decreases radially outward. If it is assumed that the pressure measured by the transducer is applied to the entire impacting surface area, the impulse transmitted to the target by the projectile could be substantially overestimated. While pressure measurements can be very valuable, it is also useful to measure the overall impact force. In this paper, a tubular sensor is used to measure the transmitted force from an impacting gelatin projectile. These results are compared to simulations of the measurement device to determine how the measured response varies from the input loading function.

## 31.2 Experimental

The tube sensor (Fig. 31.2) is a 36 in long 2024-T3 aluminum drawn tube with an outside diameter of 3.5 in and a 0.125 in thick wall. The tube dimensions were chosen such that the waves inside the tube would remain elastic during the impact loading. Transmitted forces were anticipated to be roughly 20,000 lb, thus the peak dynamic stress in the tube is 15.1 ksi, 30% of the yield strength for 2024-T3 aluminum (50.0 ksi). A 0.5 in thick 7075-T6 aluminum strike plate is attached to end of the tube to serve as a contact surface for the gelatin projectile.

The tube is instrumented with Vishay Micro-Measurements EA-06-125BZ-350 electric resistance strain gages configured in full Wheatstone bridges at two locations: 6 in and 12 in from the edge of the tube. For each bridge, two strain gages are oriented in the axial (loading) direction and two are oriented in the transverse direction, a configuration commonly used for load cells. The sensitivity of each bridge is:

$$\frac{\Delta E}{V} = \frac{S_g F}{2AE}(1 + v) \tag{31.1}$$

where, $\Delta E$ is the signal voltage, $V$ is the excitation voltage, $S_g$ is the gage factor, $F$ is the force in the tube, $A$ is the tube's cross-sectional area and $E$ and $v$ are the Young's modulus and Poisson's ratio of the tube, respectively. The strain gages have a gage factor of 2.105, so for 10 V excitation, the sensitivity is 0.996 mV per kip. For comparison, a solid 3.5 in diameter rod with the same instrumentation would have a sensitivity of 0.137 mV per kip.

Bridge #1 can measure an impact event that is 297 μs long. This is the duration required for an elastic wave, traveling at 201,884 in/s, to traverse 60 in or twice the distance from the bridge to the end of the tube. After 297 μs, the measurement from the bridge is affected by the beginning of the reflected wave. Bridge #2 can only measure a 238 μs event, since this bridge is 6 in closer to the end of the tube.

The projectiles (Fig. 31.3) were nominally 1.25 in diameter, 3.0 in long cylinders, a mixture of gelatin, water and phenolic microballoons used to modify the density. A 20% gelatin mixture recipe was used with the appropriate volume of phenolic

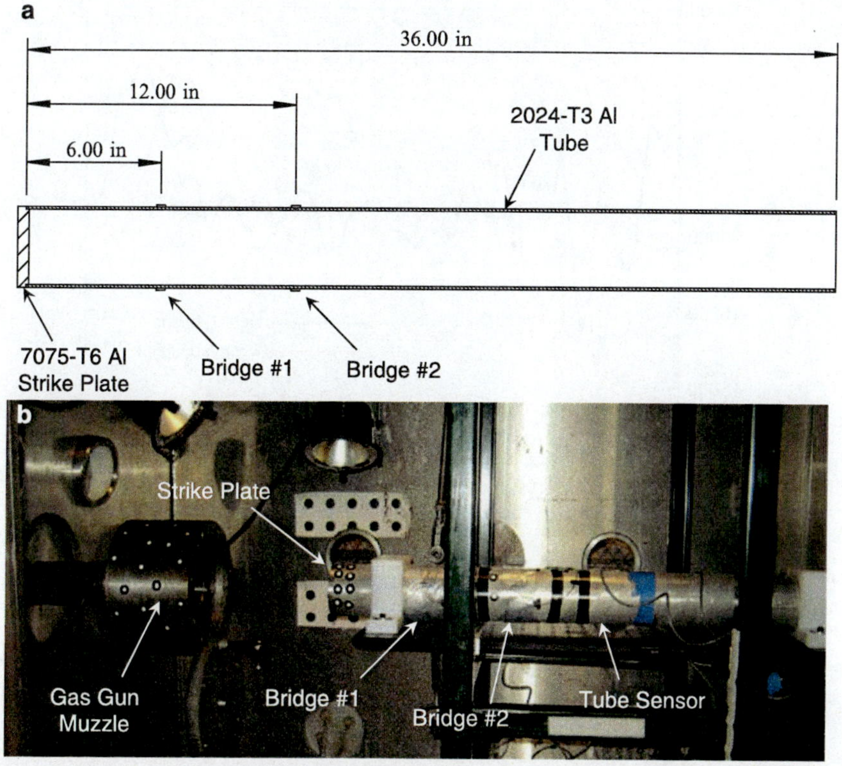


**Fig. 31.2** Tube sensor: (**a**) section schematic of the tube, (**b**) photo of the sensor and experimental setup

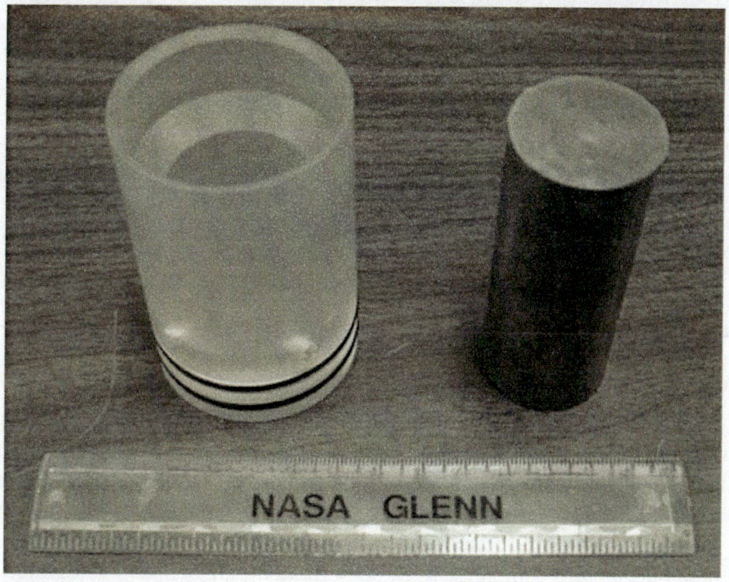


**Fig. 31.3** Polycarbonate sabot and gelatin projectile

micro-spheres needed to attain the desired specific gravity of the projectile. The ratio of gelatin to water is defined on a mass basis; the percent gelatin is defined by the ratio of gelatin mass to the total mass of gelatin and water. The desired specific gravity of the projectiles was 0.95. A detailed projectile manufacturing procedure was followed to ensure reproducible projectile characteristics.

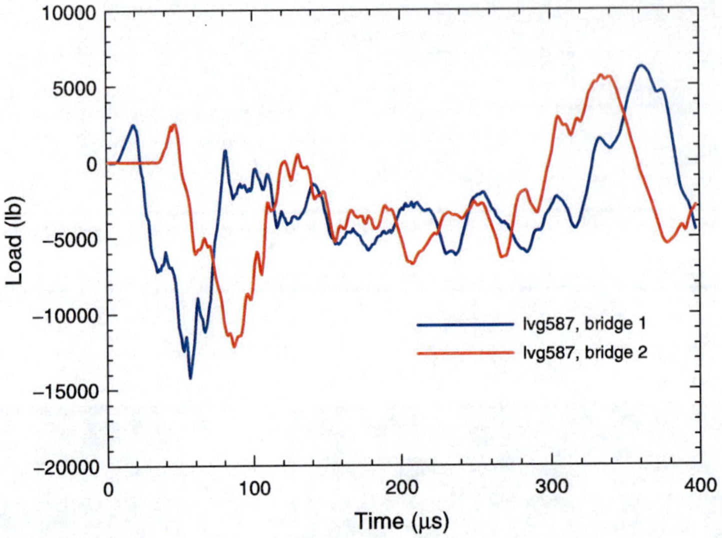


**Fig. 31.4** Tube sensor data from a gelatin impact test, projectile velocity: 504 ft/s

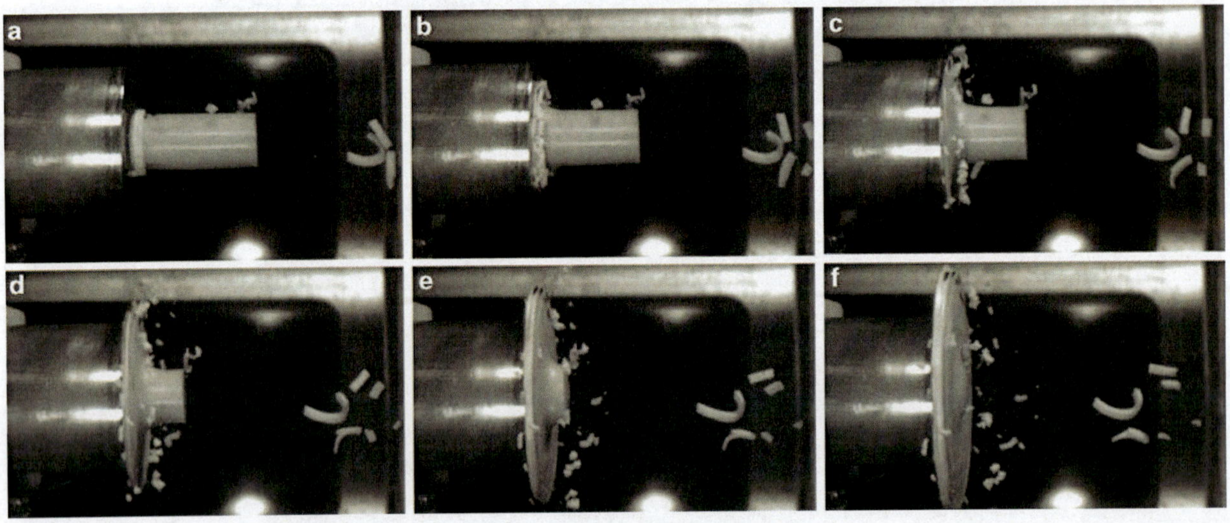


**Fig. 31.5** Sequence of a gelatin projectile impacting the strike plate of the tube sensor at 504 ft/s

The gelatin projectiles were accelerated by a single stage light gas gun using helium as the driving gas. The gun has a bore diameter of 2 in and a 12 ft long barrel. Instrumentation systems were placed inside a vacuum chamber mated to the gun barrel. The projectiles were carried down the barrel in polycarbonate sabots (Fig. 31.3) that were arrested by a sabot stopper at the muzzle. When arrested, the sabot remains in contact with the stopper providing a seal which prevents the high pressure propellant gas in the barrel from affecting the vacuum chamber pressure.

Tube sensor data was recorded using a Spectral Dynamics 2805 data acquisition module at a rate of 1.25E6 Hz. The projectile orientation and speed were computed using a calibrated pair of Vision Research Phantom V7.3 cameras and GOM's PONTOS digital image correlation point tracking software. The cameras recorded images at a rate of 10,000 fps and a resolution of 704 (horizontal) × 424 (vertical) pixels. The software tracked the three dimensional positions of points marked on the projectile as a function of time. A third Phantom V7.3 camera served as a backup system for measuring projectile speed. This camera was oriented normal to the projectile path and acquired images at 20,000 fps.

Figures 31.4 and 31.5 present results from an experiment in which the projectile impacted the strike plate at 504 ft/s. The output history of the tube sensor is shown in Fig. 31.4. The blue trace is the output from bridge #1 and the red trace is the output from bridge #2. Both bridges show an initial tensile load of roughly 2,500 lb followed by a compressive peak of −13,000 lb and then a 240 μs long region where the load oscillates about −4,000 lb. The initial tensile pulse is discussed in the next section. The compressive peak measured by bridge #1 (−14,500 lb) has a larger magnitude than that measured by bridge #1 (−12,500 lb). There are also other minor differences between the measurements from bridge #1 and #2. This is evidence of dispersion. The three regimes of Wilbeck's model for soft projectiles [11] can be seen in the data. The initial compressive peak corresponds to the shock development regime. The release wave regime coincides with the drop in compressive force. The steady flow regime coincides with the region of relatively constant force. Figure 31.5 shows the impact sequence from the high speed camera oriented normal to the projectile path. Figure 31.5a shows the projectile just prior to impact. Figure 31.5b shows the projectile just after impact. The impact sequence develops in Fig. 31.5c, d, e and f. 100 μs elapse between each image, so the sequence represents a 500 μs duration. The white debris visible in the images is light foam that was used to support the projectile in the sabot.

## 31.3 Numerical Simulations

LS-DYNA [12] simulations were used to evaluate the performance of the tube sensor. Three-dimensional, eight-noded, constant stress solid elements were used to mesh the strike plate and the tube. The mesh was constructed of elements with approximate aspect ratios of 1.0 and lengths of 1/32 in. The strike plate had 112 elements across its diameter and 16 elements through the thickness. The tube had four elements through the thickness and 1,152 elements along its length. The simulation contained roughly 1.6 million total elements. Both the strike plate and the tube were modeled as elastic materials with $E$, $\nu$ and $\rho$ as 73.1 GPa, 0.33 and 2.78 g/cc, respectively. An assumed, evenly distributed pressure history was applied to the center of the strike plate. It was applied over a 1.25 in diameter circle, the same diameter of the gelatin projectile. This pressure multiplied by the area gives a dynamic load history to compare to the tube sensor output.

Simulated dynamic forces are compared to experimental results in Fig. 31.6. The solid black trace is the input load history applied to the center of the strike plate. The input load has a rise time of 5 μs and a peak compressive load of −9,000 lb. After peak load, it decays to a force of −3,800 lb and remains there for 400 μs. Simulated forces, calculated from the strain histories of elements at locations that correspond with bridge #1 and #2 on the actual sensor are represented by dashed blue and red traces, respectively. Solid blue and red traces are actual experimental data measured by the tube sensor. The simulated dynamic loads agree quite well with those measured experimentally. The loads measured by the tube, however, are substantially different than the input load history. The peak compressive load measured the tube sensor is roughly 40% higher than the peak input force. The steady flow region (constant load), however, is reasonably well captured by the sensor until the reflected pulse arrives at the bridge. Another obvious discrepancy between the input and output load is the presence of the initial tensile pulse that was both measured experimentally and calculated in the simulation. The source of the tensile

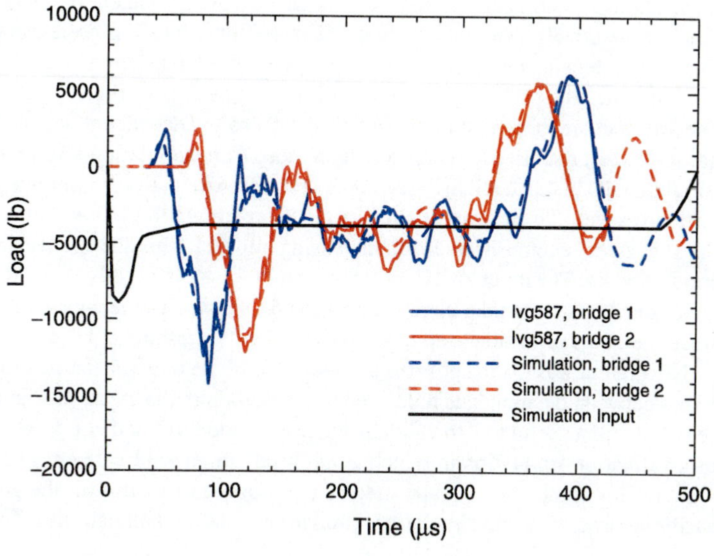


**Fig. 31.6** Comparison of experimental and simulated dynamic forces

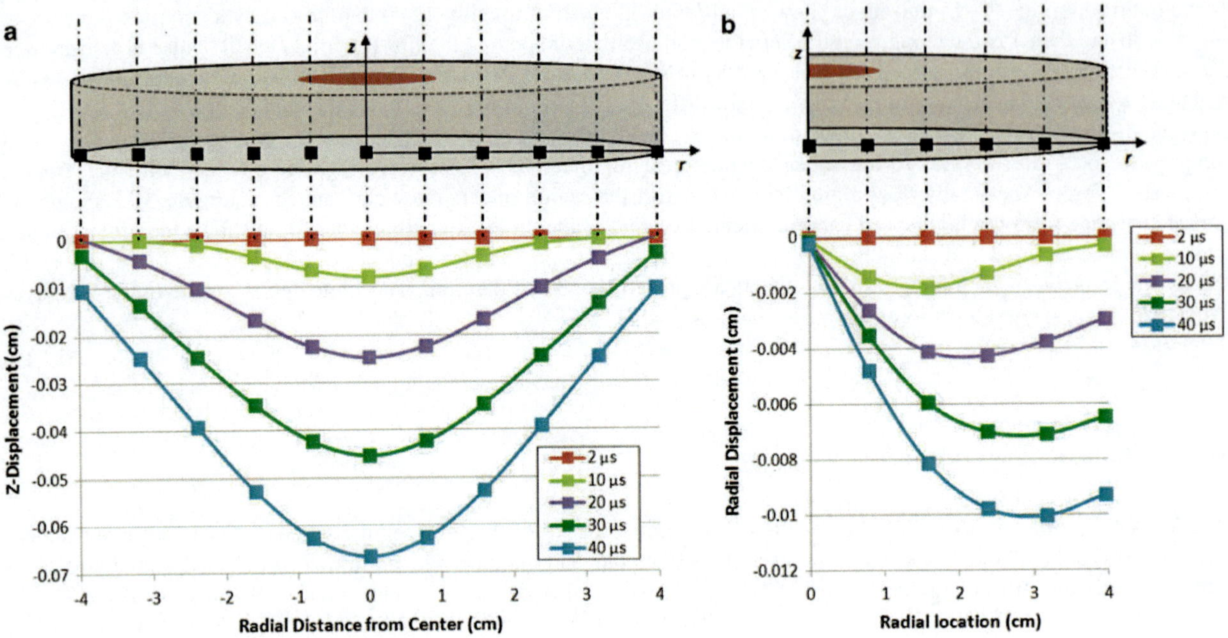


**Fig. 31.7** Strike plate deformation at different times after load: (**a**) axial bending, (**b**) radial contraction

pulse is deformation in the strike plate. Figure 31.7 shows the simulated strike plate deformation at various times after load initiation. As the load increases, the strike plate bends elastically into the tube. Figure 31.7a is a plot of the axial deflection of points distributed along the diameter of the strike plate. The bending draws the material of the strike plate radially inward. Figure 31.7b shows the radial displacement of points distributed along the radius of the plate. As the load accumulates negative radial displacements arise. The tube is attached to the strike plate, so it also is drawn inward. This uniform radial contraction creates the tensile pulse evident in both the experimental and simulated results.

The results in Fig. 31.6 show that the tube sensor output significantly overestimates the peak compressive input load. Additional simulations were run to determine how the pressure load input characteristics affect the measurement error. These characteristics include the rise time and shape of the load history. Results from this study are shown in Fig. 31.8. Results are presented from four simulations where the tube sensor was subjected to different load histories. The load histories are represented by dashed traces. The input load history for simulation #1, represented by the blue trace, is identical to the one shown in Fig. 31.6. It has a rise time of 5 μs, a peak load of −9,000 lb followed by an unloading tail to a constant load of −3,800 lb. The input history for simulation #2 (red trace) is identical to that of simulation #1 except it has a 25 μs rise time. Simulations #3 (green trace) and #4 (black trace) are subjected to a −9,000 lb step load history with 25 μs and 100 μs rise times, respectively. The tube output, measured at a location corresponding to bridge #1, is represented by solid traces. The initial tensile pulse amplitude decays with increasing rise time. In all four simulations, the tube sensor output overestimates the peak compressive load input. The data in Fig. 31.8 illustrate that this overshoot is dependent on both the rise time and the shape of the load pulse. It is not solely dependent on rise time, since the overshoot from simulation #1, which has the fastest rise time (5 μs), is less than those from simulation #2 and #3 (both have 25 μs rise times). The overshoot from simulation #3 is substantially higher than that from simulation #2, showing that this phenomenon is dependent on the shape of the load as well. The impulse of the input load for simulation #3 is significantly larger than that from simulation #2. Therefore, the overshoot becomes worse with increasing impulse. This also explains why simulation #2's overshoot is larger than that of simulation #1. The input load impulse of simulation #2 is slightly larger than that from simulation #1, since there is more area under the 25 μs ramp than there is under the 5 μs ramp. The results of simulation #4 show that the rise time also plays a role since the 100 μs rise time step load gives the least overshoot.

No attempt was made optimize the geometry of the tube sensor. It is likely that the geometry can be modified to minimize errors between the input and measured loads. Although the measured output is not always identical to the input load, the tube sensor is still a useful tool to validate numerical models of ballistic gelatin. The experimental data can be compared to results of a validation simulation in which a proposed numerical treatment of a gelatin projectile is fired into a model of the tube sensor. This comparison can be used as a performance metric for the gelatin model. The simple geometry of the sensor is easily incorporated into simulations, making validation simulations relatively straightforward.

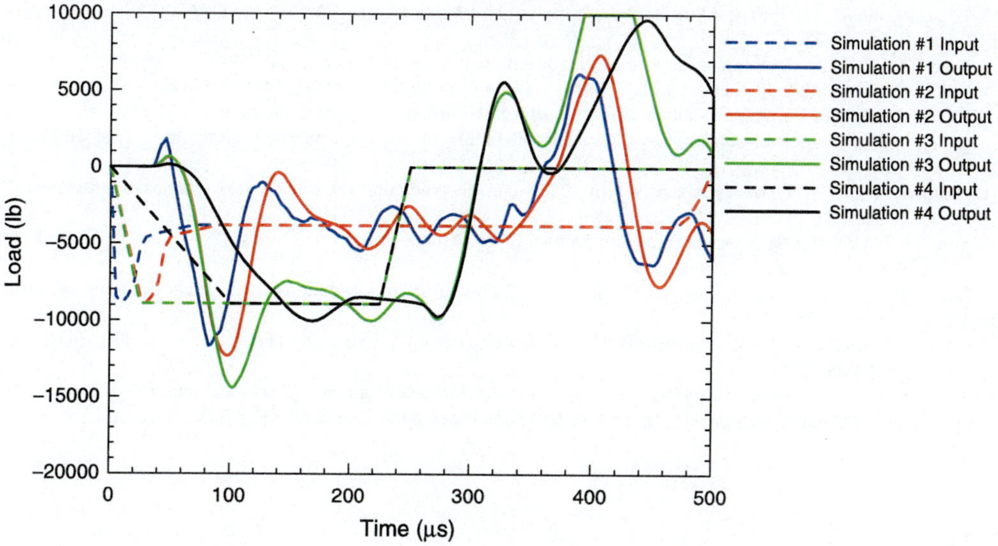


**Fig. 31.8** Results from simulations of the sensor tube subjected to four different load histories

## 31.4 Summary and Conclusions

A tube sensor was used to measure dynamic impact forces from ballistic gelatin projectiles. The sensor consists of a 0.5 in thick 7075-T6 aluminum strike plate attached to 36 in long 2024-T3 aluminum tube with a 3.5 in outside diameter and 0.125 in thick wall. The tube was instrumented with electrical resistance strain gages at two locations: 6 in and 12 in from the edge of the tube. 1.25 in diameter, 3.0 in long gelatin projectiles were fired normal to the tube's strike plate at velocities of roughly 500 ft/s using a single stage light gas gun. Data from the tube sensor's strain gages were recorded and high speed cameras were used to determine the projectile's velocity and orientation prior to impact.

LS-DYNA simulations were used to investigate the performance of the tube sensor. Simulated output from the tube matched the experimental data well. The deformation of the tube's strike plate was identified as the source of an initial tensile pulse found in both experimental data and the simulation results. A numerical investigation of the sensor's response to various input load histories was conducted. The results show that the tube sensor overestimates the peak compressive load for all four input loads considered. The magnitude of this overshoot is dependent on both the rise time and shape of the load input. Fast rise-time, high impulse load histories lead to large overshoots. Reducing the impulse and increasing the rise time of the load history reduce the overshoot. Measurement errors can likely be reduced by optimizing the sensor's geometry.

The tube sensor is a useful tool to validate numerical models of ballistic gelatin projectiles. It can easily be incorporated into validation simulations allowing the simulated and experimental sensor response to be directly compared.

**Acknowledgements** Support was received by the National Aeronautics and Space Administration Aviation Safety Project. The authors would like to acknowledge the useful advice and assistance of Duane Revilock, Charles Ruggeri and Jeff Hammel of the NASA Glenn Research Center Ballistic Impact Facility. This work was supported in part by an allocation of computing time from the Ohio Supercomputer Center (OSC).

## References

1. Tiffany CF, Gallagher JP, Babish CA (2010) Threats to aircraft structural safety, including a compendium of selected structural accidents/incidents. Aeronautical Systems Center Technical Report #ASC-TR-2010-5002
2. Loss of thrust in both engines after encountering a flock of birds and subsequent ditching on the Hudson River, US Airways Flight 1549, Airbus A320-214, N106US, Weehawken, New Jersey, January 15,2009, NTSB Accident Report #NTSB/AAR-10/3, PB2010-910403, 4th May 2010

3. McCallum SC, Constantinou C (2005) The influence of bird-shape in bird-strike analysis. In: Proceedings of the 5th European LS-DYNA users conference, Birmingham, 2005
4. Wilbeck JS, Rand JL (1981) The development of a substitute bird model. J Eng Power 103:725–730
5. McCarthy MA, Xiao JR, McCarthy CT, Kamoulakos A, Ramos J, Gallard JP, Melito V (2004) Modelling of bird strike on aircraft wing leading edge from fibre metal laminates – part 2: modelling of impact with SPH bird model. Appl Compos Mater 11:317–340
6. Wang FS, Yue ZF (2010) Numerical simulation of damage and failure in aircraft windshield structure against bird strike. Mater Des 31:687–695
7. Ivancevic D, Smojver I (2011) Hybrid approach in bird strike damage prediction on aeronautical composite structures. Compos Struct 94:15–23
8. Mau RH, Meguid SA, Ng TY (2008) Transient three dimensional finite element analysis of a bird striking a fan blade. Int J Mech Mater Des 4:79–96
9. Lavoie MA, Gakwaya A, Nejad Ensan M, Zimcik DG, Nadlall D (2009) Bird's substitute tests results and evaluation of available numerical methods. Int J Impact Eng 36:1276–1287
10. Barber JP, Taylor HR, Wilbeck JS (1975) Characterization of bird impacts on a rigid plate: part 1, Air Force Flight Dynamics Laboratory Technical report #AFFDL-TR-75-5
11. Wilbeck J (1978)Impact behavior of low strength projectiles, Air Force Materials Laboratory technical report #AFML-TR-77-134
12. LS-DYNA keyword user's manual, Version 971, Livermore Software Technology Corporation (LSTC), May, 2007

# Chapter 32
# Material Models Sensitivity of Tungsten-Based Penetrators Impacting on Confined Boron-Carbide

**Costas G. Fountzoulas and J.C. LaSalvia**

**Abstract** LaSalvia et al. (LaSalvia presentation (2010) Ballistic impact damage in hot- pressed boron carbide 34 ICACC, Daytona Beach, 25–29 Jan 2010) studied experimentally the interaction of confined hot-pressed boron carbide ($B_4C$) targets impacted by laboratory-scale tungsten-based long-rod penetrators. To better understand the physics involved, Fountzoulas et al. (Fountzoulas CG, LaSalvia JC (2011) Simulation of the ballistic impact of tungsten-based penetrators on confined hot-pressed boron carbide targets. In: Proceedings of 35th ICACC, Advances in Ceramic Armor VII, p 261) studied by modeling and simulation the ballistic behavior of these targets. To satisfactorily replicate the experimental damage of the targets during impact, the material strength and failure models were iteratively modified. Despite numerous iterations, the damage replication of the target was only partially successful. The fracture of $B_4C$ was able to be replicated to some extent but without being able to stop its penetration by the projectile, a disagreement with the experimental observations. The current effort reports on the sensitivity and modification of the existing strength and failure $B_4C$ material models of the ANSYS/AUTODYN library to predict the tensile failure to accurately simulate the ballistic response of ceramics.

**Keywords** Tungsten-based penetrators • Johnson-Holmquist (JH) material models • Model parameter sensitivity • Confined targets • Boron carbide • Impact • Penetration

## 32.1 Introduction

Ceramics are materials that possess characteristics such as low density, high hardness and high compressive strength that make them ideal for use in light armor; however, ceramics are also brittle and have a low tensile strength, which complicates the design of such systems [1]. The pioneering experimental and numerical work of Wilkins and co-workers gave a great deal of insight into the penetration process and failure phenomenology of ceramics utilized in composite armor [2]. Their observations of ceramic failure were further expounded upon by the subsequent work of Shockey et al. [3] who performed a series of experiments with confined ceramic targets impacted by tungsten-nickel-iron rods at velocities from 0.8 to 1.4 km/s and from the recovered targets were able to detail the failure sequence in the ceramic. It was determined that if the projectile velocity is sufficient or if the ceramic is not adequately confined, the projectile loading begins to introduce tensile cracks near the impactor periphery. Ceramics possess low tensile strength and when large tensile stress fields are introduced in the radial direction, they may form ring cracks, concentric around the impact point. Initially shallow, upon continued loading these ring cracks propagate in the direction of maximum principal tensile stress, typically on angles from 25° to 75° normal to the surface. They progress to the back surface of the ceramic and form large Hertzian cone cracks. As the loading increases further, the compressive strength beneath the penetrator is exceeded and microcracking begins. After the cone cracks form, the principal stresses redistribute to a circumferential direction, which result in tensile radial cracks forming like spokes on a hub from the impact site. Lateral cracks also form below the impact surface and intersect the radial and cone cracks forming fragments, which can break away from the impact site leaving a crater [3].

C.G. Fountzoulas (✉)
U.S. Army Research Laboratory, RDRL-WMM-B, MD 21005-5069 Aberdeen Proving Ground, USA
e-mail: Constantine.Fountzoulas.civ@mail.mil

J.C. LaSalvia
U.S. Army Research Laboratory, RDRL-WMM-E, APG, MD 21005-5069 Aberdeen Proving Ground, USA

V. Chalivendra et al. (eds.), *Dynamic Behavior of Materials, Volume 1: Proceedings of the 2012 Annual Conference on Experimental and Applied Mechanics*, Conference Proceedings of the Society for Experimental Mechanics Series,
DOI 10.1007/978-1-4614-4238-7_32, 

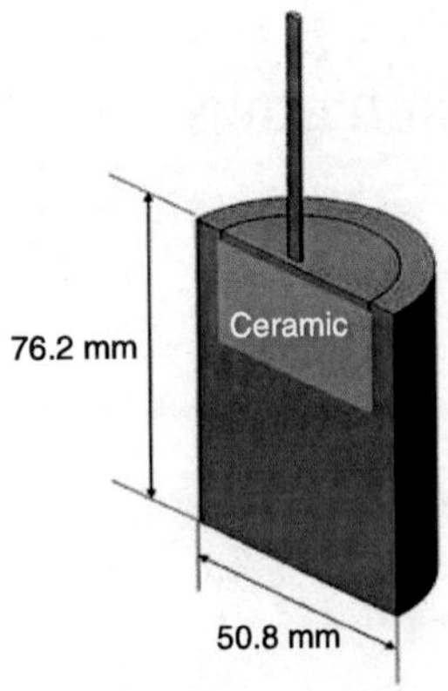


- PAD B4C (hot-pressed $B_4C$), BAE Advanced Ceramics
- 38.1 mm diameter by 38.1 mm thick cylinders
- Ti6Al4V "cups" (50.8 x 76.2 mm)
- Slip-fit (<+/-0.05 mm)
- 3.175 mm thick Ti6Al4V cover (welded)
- 93% Walloy (WHA), L/D 20 , 3.175 x 63.5 mm, 8.94 g ($\rho$= 17.9 g/cm$^3$, Y $\approx$ 1.4 GPa)
- Impact velocity 819, 1052, 1063, 1198 and 1205 m/s

| $\rho$ (g/cm$^3$) | Grain Size (μm) | E, ν (GPa) | HK4 (GPa) | $K_{IC}$ (MPa*m$^{1/2}$) | $\sigma_b$ (MPa) | Phases |
|---|---|---|---|---|---|---|
| 2.52 | 15 | 450, 0.17 | 18.9 | 2.9 | 400 | B4C, C |

**Fig. 32.1** Target configuration and $B_4C$ measured properties

Although fractured and fragmented, the ceramic still resists penetration. If confined, the penetrator can only proceed by moving the ceramic out of the way. To do so, the projectile must pulverize it and force the pulverized material to flow laterally and then rearward [3]. This zone of pulverized ceramic ahead of the projectile is known as the comminuted zone or the Mescall zone, after the work of Mescall et al. [4]. Ceramic penetration resistance is a function of its compressive strength, hardness, pulverization, ceramic powder flow properties and abrasiveness, which serves to erode and consume the lead surface of the projectile [3].

Subsequent studies by Bless et al. [5] and Hauver et al. [6] also observed that if the ceramic is sufficiently confined, the projectile will flow laterally on the surface of the ceramic and not penetrate. These phenomena have become known as interface defeat or infinite dwell [4]. Recent x-ray experiments by Lundberg et al. [7] have been able to capture the transition from the projectile dwelling on the ceramic surface to penetration.

However, perhaps the most widely utilized and well known of the material models developed for ceramics are the ones developed by Johnson and Holmquist [8, 9], which have become known as JH-1 and JH-2. Johnson, Holmquist and Beissel have also extended the ceramic model to account for phase transformation and this version is known as the JHB model [10]. One significant difference between the models is that the JH-2 allows for a gradual degradation of the ceramic strength from the intact surface as the damage accumulates, while the ceramic strength in the JH-1 and JHB models does not degrade from the intact strength curve until the material is fully damage (D = 1.0).

This work capitalizes on the partially successful results of the previous study by Fountzoulas et al. [11] to duplicate the damage of the confined $B_4C$ and further investigates the ability of the existing JH-2 ceramic material models to predict the observed damage, such as the lack of comminuted region, and short and long cracking introduced in confined cylinders of $B_4C$ struck by cylinder of tungsten carbide at high (819–1,205 m/s) velocities [9]. The effect of specific strength and failure model modified parameters on the damage development is described and the contribution and sensitivity of these parameters on the accuracy of the solution when compared to the experimentally determined damage patterns are discussed.

## 32.2 Experimental

Details of the impact experiments have been reported by LaSalvia et al. [12] and the experimental setup is shown in Figure 32.1. The targets were center impacted at velocities ranging from 819 to 1,205 m/s. A cross section showing the damage of the $B_4C$ at impact velocity at 1,205 m/s is shown in Fig, 32.2. The target was confined at bottom of the Ti-6Al-4 V "cup" (V = 0 m/s).

## 32.3 Numerical Simulations

The ballistic behavior of all the targets (Fig. 32.1) was studied 3D models and simulated using the non-linear ANSYS/AUTODYN commercial package [13]. The material models used were obtained from the AUTODYN library [13]. All materials were discretized using smooth particle hydrodynamic (SPH) solver with a particle size of 0.25 mm. The

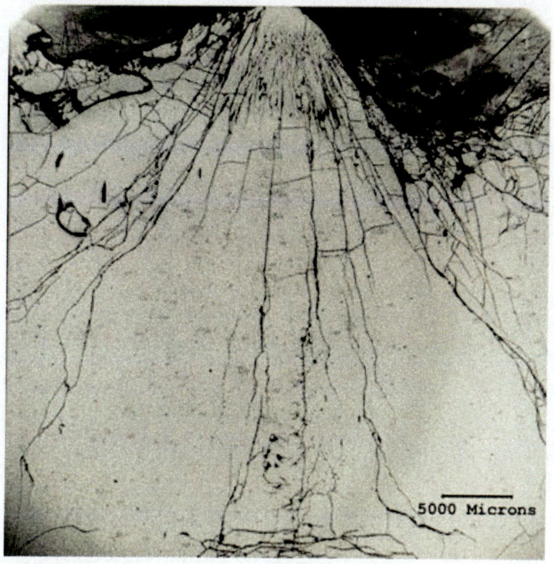


**Fig. 32.2** SEM photomicrograph of B4C failure pattern at 1,205 m/s impact velocity

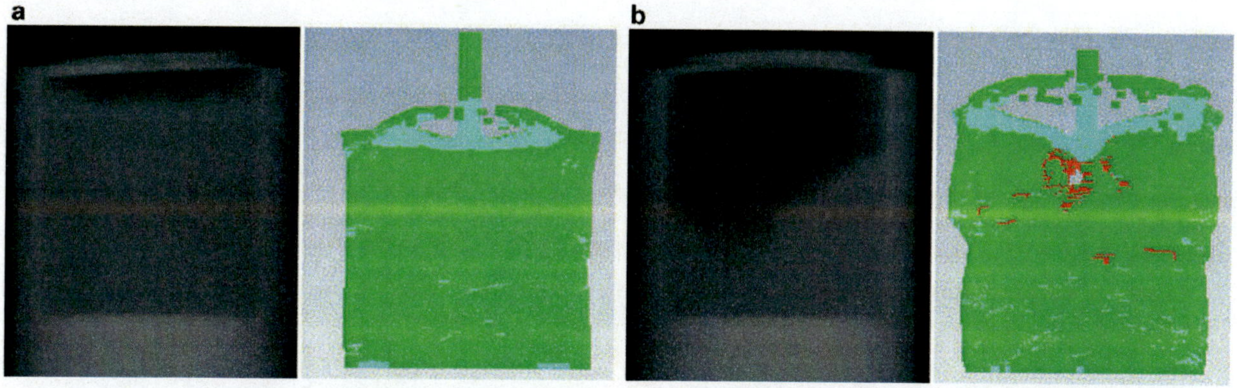


**Fig. 32.3** Static x-ray radiographs of $B_4C$ and simulated damaged status (40 μs) using hydro-tensile failure model. (**a**) 819 m/s: (i) radiograph (ii) simulation, (**b**) 1,205 m/s: (i) radiograph (ii) simulation

Ti-6Al-4 V sleeve and cover were modeled using a Puff equation of state (EOS), Von Mises strength and a Grady spall failure models taken from the Autodyn Material Library. The WHA projectile was modeled using a shock EOS, and a Johnson-Cook strength model [3]. The $B_4C$ was modeled using a polynomial EOS, JH-2 strength and failure models. In addition, the tensile failure criterion for the $B_4C$ was set to either the hydro-tensile or minimum pressure, or principal stress with or without crack softening. In our effort to capture the damage of the $B_4C$ various criteria were used. Figures 32.3 and 32.4 show the inability of the EUTODYN library models to capture the failure pattern at impact velocity of 1,250 m/s.

### 32.3.1 *Capturing the Overall Failure Pattern*

Table 32.1 shows the modifications of shear modulus, bulk modulus and hydro tensile limit of the Autodyn library values, which reproduced most successfully the failure pattern of the $B_4C$ and WHA projectile. Figure 32.5 shows the simulated failure pattern of the encapsulated $B_4C$ at 1,205 m/s impact velocity obtained with unmodified Autodyn library strength and failure models. Although the projectile was stopped the simulated failure pattern of the ceramic is poor compared to the experimental results. Figures 32.6 to 32.10 show the simulation results obtained with the modified material models. For

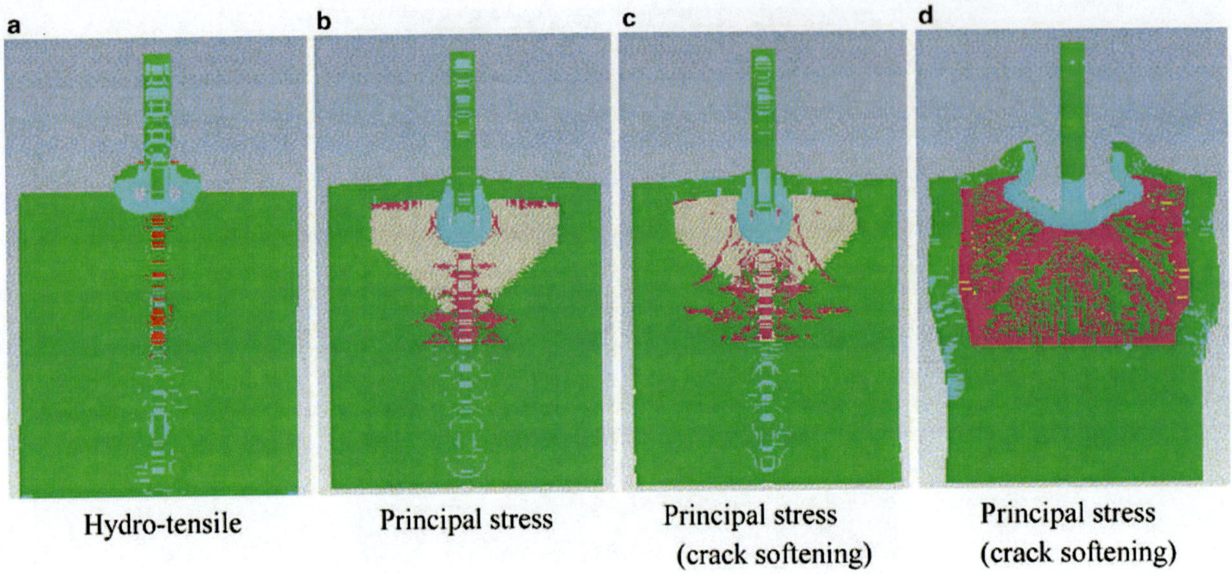


**Fig. 32.4** Simulated damage after 25 μs at 1,205 m/s impact velocity. (**a–c**) 2D axisymmetric, (**d**) planar symmetry

**Table 32.1** Most successful material models parameter modifications

| Parameter | AUTODYN library | Modifications (most successful) |
|---|---|---|
| G (Shear Modulus) [***Strength***] (GPa) | 199 | 100 |
| K (Bulk Modulus) [***EOS***] (GPa) | 230 | 117 |
| HTL (Hydro Tensile Limit) [***Failure***] (GPa) | −1.9 | −1.0 |

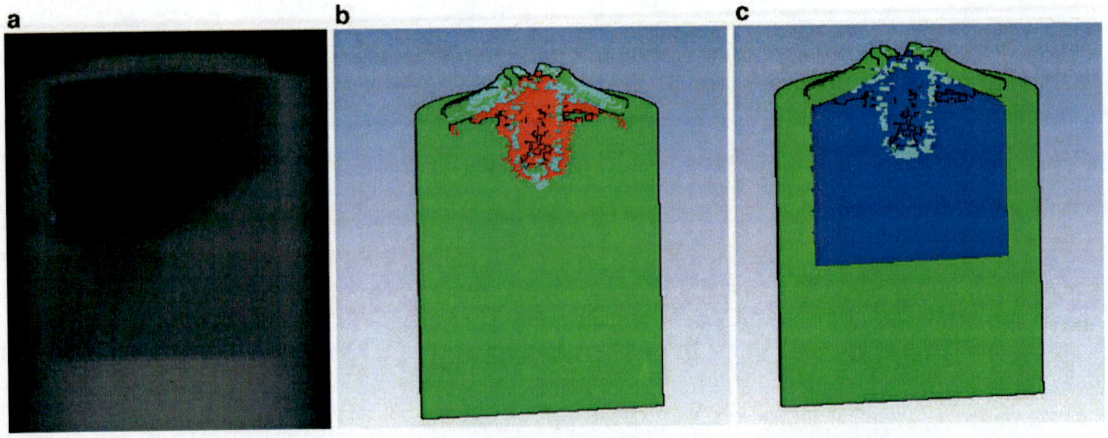


**Fig. 32.5** Simulation using Autodyn library parameters: impact velocity 1,205 m/s. (**a**) x-ray radiograph, (**b**) simulated damage, (**c**) simulated impactor position

impact velocities 1,052, 1,063, 1,198 and 1,205 m/s the simulations duplicated successfully the extent of damage under the cover plate (Figs. 32.6, 32.7, 32.8, and 32.9). However, for impact velocity 819 m/s the damage extent under the cover was smaller than observed experimentally (Fig. 32.10). The projectile was stopped for all impact velocities. Although the overall simulated damage appears to be like a mirror image of the actual one, the extent of the simulated damage is remarkably similar (Figs. 32.6 and 32.7). The simulated damage extent at 1,063 m/s appears larger and similar to the simulated damage at higher observed at 1,205 and 1,198 m/s (Figs. 32.6 and 32.7). The simulated damage at impact velocity 1,052 m/s was almost identical to the experimental damage.

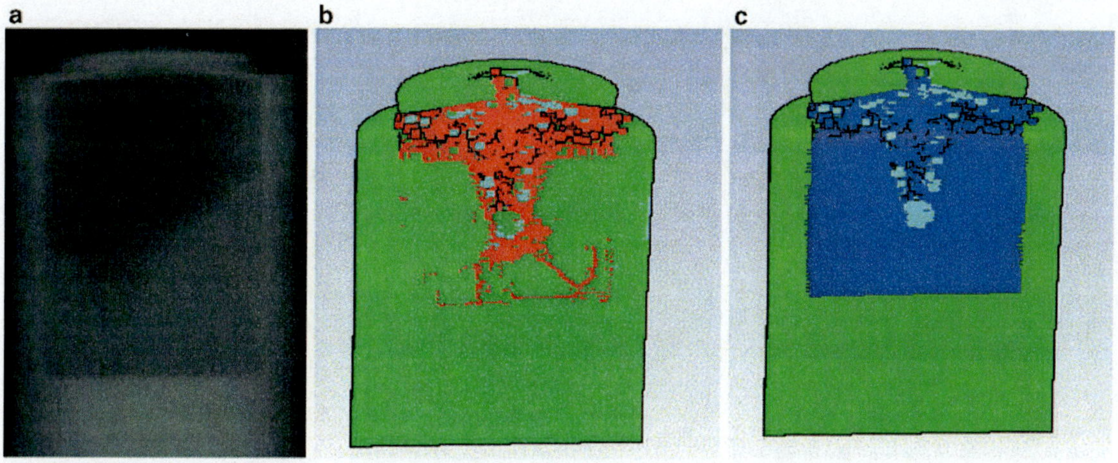


**Fig. 32.6** Simulation using modified parameters: impact velocity 1,205 m/s. (**a**) x-ray radiograph, (**b**) simulated damage, (**c**) simulated impactor position

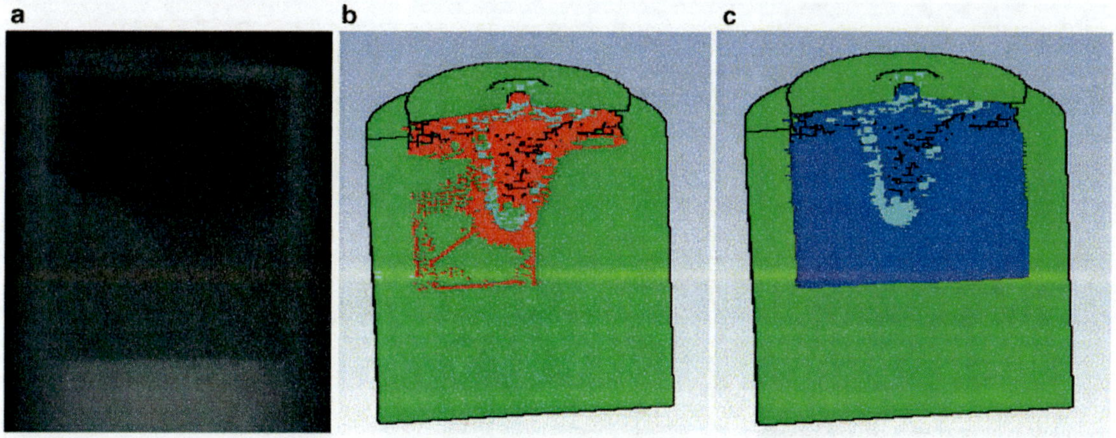


**Fig. 32.7** Simulation using modified parameters: impact velocity 1,198 m/s. (**a**) x-ray radiograph, (**b**) simulated damage, (**c**) simulated impactor position

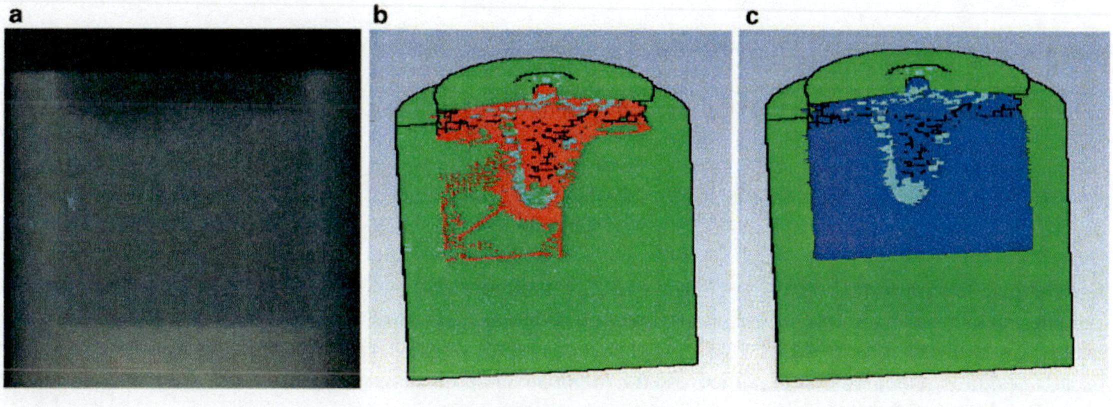


**Fig. 32.8** Simulation using modified parameters: impact velocity 1,063 m/s. (**a**) x-ray radiograph, (**b**) simulated damage, (**c**) simulated impactor position

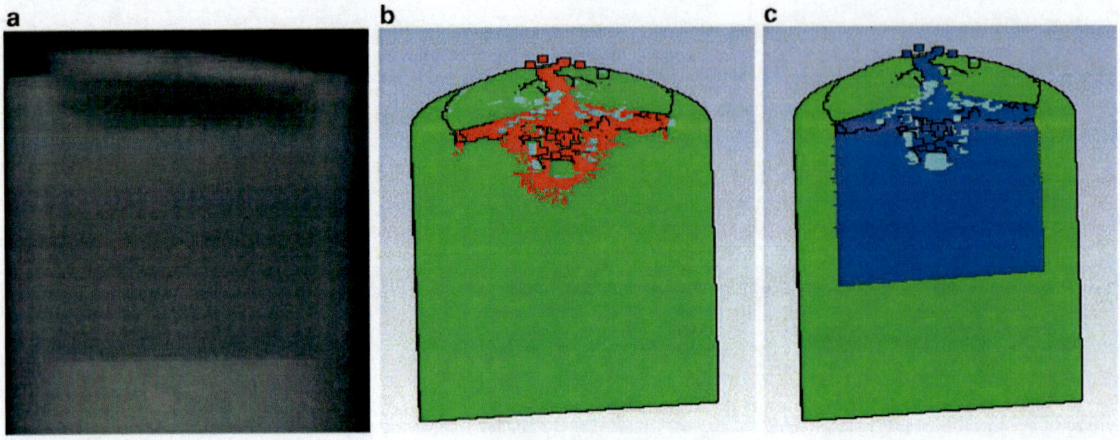

**Fig. 32.9** Simulation using modified parameters: impact velocity 1,052 m/s. (**a**) x-ray radiograph, (**b**) simulated damage, (**c**) simulated impactor position

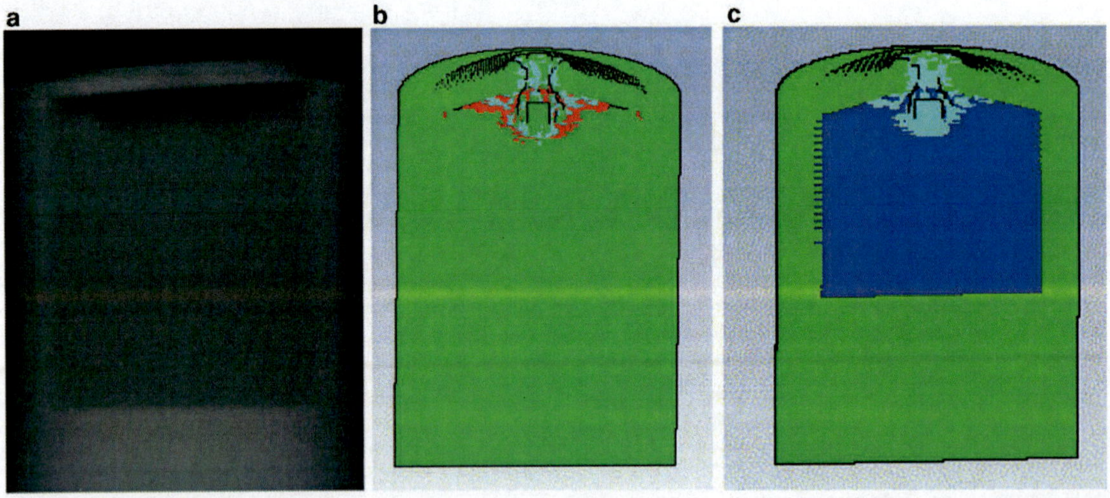

**Fig. 32.10** Simulation using modified parameters: impact velocity 819 m/s. (**a**) x-ray radiograph, (**b**) simulated damage, (**c**) simulated impactor position

However, the duplication of the only experimental existing detailed cracking pattern at 1,205 m/s (Fig. 32.2) is very poor (Fig. 32.6).

### *32.3.2 Duplication of Cracking Pattern (Fig. 32.2)*

To capture the detailed cracking pattern shown in Fig. 32.2 the options of principal tensile failure stress, fracture energy, crack softening and stochastic failure of the failure material model of $B_4C$ were used. The only parameter varied was the fracture energy, which was set at 23.7 J/m$^2$ (AUTODYN library value), 47.4 J/m$^2$ and 71.1 J/m$^2$ respectively. As it is shown in Fig. 32.11 this option, modified or unmodified, did not duplicate either the cracking pattern of the ceramic or the stopping of the impactor to any satisfactory degree.

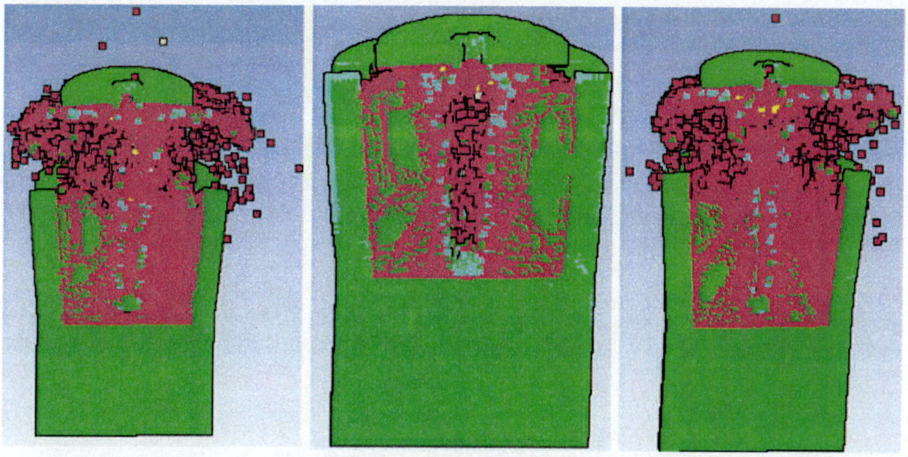

**Fig. 32.11** Principal tensile failure with fracture energy at 1,205 m/s velocity. (**a**) 23.7 J/m$^2$, (**b**) 47.4 J/m$^2$, (**c**) 71.1 J/m$^2$

## 32.4 Discussion

The current modifications of the shear modulus, bulk modulus and hydro tensile limit successfully duplicated the extent of the damage under the cover plate at impact velocities 1,052, 1,198 and 1,205 m/s Although the simulation at 1,063–819 m/s produced larger and smaller damage extent under the cover respectively, the simulated damage was comparable to the experimental damage. The projectile was stopped for all impact velocities and the cover plate was broken, in agreement with the experimental results.

As it is shown in Table 32.1, to successfully reproduce the failure pattern of the $B_4C$, we had to reduce the shear and bulk modulus, and the hydro tensile to about 50 % of the Autodyn library corresponding values. This may be partially attributed to the smaller values of these properties of the $B_4C$ due to its recently observed experimentally amorphization [14, 15].

The increase of the fracture energy associated with the principal tensile failure criterion produced excessive cracking of the B4C ceramic not observed experimentally (Fig. 32.2) even for fracture energy three times the published one. Moreover, the projectile stopped at almost the end of the ceramic, not also observed experimentally.

The failure of the existing strength and failure material models of all the parts of the target architecture to capture the observed experimentally cracking pattern of the ceramic may also be attributed to the inability of these models to account for the continuously developing compressive stresses caused by the ceramic confinement during the impact.

The simulations provide a significant amount of insight into the problem and could provide another measure that could be quantified through experiments. The simulations also showed that the higher the target cracking the more the retardation of the penetration process. The authors believe that due to the ceramic confinement, the strength of the failed ceramic is higher than the predicted by the $B_4C$ existing failure model, thus resulting in more effective defeat of the impacting WHA cylindrical impactor.

## 32.5 Conclusions

The Johnson-Holmquist series of ceramic material models have been developed for the high velocity impact and penetration of ceramics; their applicability to the confined hot-pressed boron carbide target hit by cylindrical WHA impactor is of interest. The ability of the phenomenological Johnson-Holmquist ceramic model to predict the observed damage patterns induced by the WHA impactor striking confined cylinders of hot-pressed boron carbide at velocities from 819 to 1,205 m/s was investigated. The simulation results of this investigation using modified shear and bulk modulus, hydro tensile limit and principal tensile stress criterion with increased fracture energy can be summarized as follows:

- The modified JH2 material models reproduced the overall $B_4C$ damage from 1,052 to 1,204 m/s impact velocity rather accurately when compared to the x-ray radiographs
  - However, the overall reproduction for 819 m/s impact velocity was not as good as the rest of them
  - The 1,198–1,205 m/s simulations also reproduced the asymmetrical damage as shown in the radiographs

- High impact velocity simulation did not reproduce showed the experimental cracking pattern of the $B_4C$
- The projectiles were stopped for all impact velocities
- The principal tensile stress criterion with increased up to three times the fracture energy value produced excessive ceramic cracking and decreased the resistance of it in penetration by the WHA projectile.

Our efforts to duplicate the cracking pattern of the $B_4C$ by modifying further the existing JH-2 material models continue for more accurate prediction of the failure pattern. Moreover, our efforts are anticipated to be improved by any measurement of the properties of the amorphous phase of the $B_4C$.

## References

1. Fountzoulas CG, Normandia MJ, LaSalvia JC, Cheeseman BA (2005) Numerical simulation of the cracking of silicon carbide tiles impacted by tungsten carbide spheres. In: Proceedings of 22nd IBS, Vancouver
2. Wilkins ML, Cline CF, Hondol CA (1969) Technical report UCRL-50694, Lawrence Radiation Laboratory
3. Shockey DA, Marchand AH, Skaggs SR, Cort GE, Burkett MW, Parker MW (1990) Failure phenomenology of confined ceramic targets and impacting rods. Int J Imp Eng 9(3):263–275
4. Mescall J, Weiss V (1984) Materials behavior under high stress and ultrahigh loading rates. Part II. In: Proceeding of the 29th Sagamore army conference, army materials and mechanics research center, Watertown
5. Bless SJ, Brar NS, Bless SJ, Brar NS, Kanel G, Rosenberg Z (1992) Failure waves in glass. J Am Cer Soc 75(4):1002–1004
6. Hauver GE, Netherwood PH, Benck RF, Kecskes LJ (1994) Enhanced ballistic performance of ceramic targets 1994. In: Proceedings of the19th Army Science Conference, Orlando
7. Lundberg P, Renström R, Holmberg L (2001) An experimental investigation of interface defeat at extended interaction times. In: Proceedings of 19th international symposium, vol 3. Ballistics, pp 1463–1469
8. Johnson GR, Holmquist TJ (1990) A computational constitutive model for brittle materials subjected to large strains, high strain rates and high pressure In: Proceedings of EXPLOMET conference, San Diego
9. Johnson GR, Holmquist TJ (1994) An improved computational constitutive model for brittle materials. In: Schmidt SC, Shaner JW, Samara GA, Ross M (eds) High pressure science and technology – 1993, vol 2. AIP Press, New York, pp 981–984
10. Holmquist TJ, Johnson GR, Beissel SR (2003) Response of Aluminum Nitride (Including a Phase Change) to Large Strains, High Strain Rates and High Pressure. J Appl Phys 94(3):1639–1646
11. Fountzoulas CG, LaSalvia JC (2011) Simulation of the ballistic impact of tungsten-based penetrators on confined hot-pressed boron carbide targets. In: Proceedings of 35th ICACC, Advances in Ceramic Armor VII Ceramic Engineering and Science Proceedings, Volume 32 (eds J. J. Swab, S. Widjaja and D. Singh), John Wiley & Sons, Inc., Hoboken, NJ, USA, p 261 doi:10.1002/9781118095256
12. LaSalvia presentation (2010) Ballistic impact damage in hot- pressed boron carbide. In: 34 ICACC, Daytona Beach, pp 25–29
13. ANSYS/AUTODYN Vol. 12.1, Manual, Century Dynamics Inc., Concord, CA
14. Daphalapurkar N, Hu G, Ramesh KT, Johns Hopkins University (2012) Specimen size effects on the dynamic failure strength of brittle materials. In: 36 ICACC presentation, Daytona Beach, 22–27 Jan 2012
15. LaSalvia JC, Shanholtz ER, U.S. Army Research Laboratory (2012) Further results on nanocrystallization features in ballistically-impacted boron carbide. In: 36 ICACC presentation, Daytona Beach, 22–27 Jan 2012

# Chapter 33
# Equation of State of Explosive Detonation Products

**Zhao Yanhong, Liu Haifeng, and Lu Guo**

**Abstract** The equation of state (EOS) of gaseous detonation products is investigated using Ross's modification of hard-sphere variation theory and the improved one-fluid van der Waals mixture model. The condensed phase of carbon is taken as a mixture of graphite, diamond, graphite-like liquid and diamond-like liquid. For a mixed system of detonation products, the free energy minimization principle is used to calculate the equilibrium compositions of detonation products by solving chemical equilibrium equations. The calculated detonation velocity and pressure of explosive are in good agreement with experimental ones. The equilibrium compositions of detonation products of PETN explosive with different densities at the Chapman-Jouquet (CJ) point are computed and the results are close to those of BKW and LJD EOS. The adiabatic expansion paths from the CJ point is obtained and compared with the results of JWL EOS, it is found that the calculated value of gamma is monotonically decreasing using the presented theory in this paper, while double peaks phenomenon appears using JWL EOS.

**Keywords** Equation of state • Detonation products • Statistical-physical theory • Chemical equilibrium

## 33.1 Introduction

The equation of state (EOS) of detonation products is essential for describing detonation properties of explosives, and it is the core parameter in numerical simulations of the detonation process of explosives. Due to the extremely complicated physical phenomenon existed in explosive detonation products, there are still more or less problems in the EOS used in practical engineering applications. In the very beginning stage of the detonation products, its state is near the CJ point, the temperature is as high as several thousand degrees and the pressure is up to tens of GPa, the density of the mixture is about $2 \sim 3$ g/cm$^3$. The interaction between the molecules of the stage is similar to that of solids or liquids. With the expansion of the product, the pressure drops to the ambient pressure, the molecular interaction shows the characteristics of gas [1].

Since the 1960s, scientists in United States, France and the former Soviet Union have done a lot of work in the study on the EOS of detonation products. According to semi-empirical theory and methodology, many mathematical and physical models are proposed [2], such as Jones-Wilkins-Lee (JWL) EOS, Becker-Kistiakowsky-Wilson (BKW) EOS, Lennard-Jones -Devonshire (LJD) EOS, Weeks-Chandler-Anderson (WCA) EOS. Numerous codes and computation methods are developed over the years, such as BKW, APPEGE, RUBY, LAMINEUR, TIGER and CHEQ. All of them have their successful usage for solving specific problems. The BKW EOS (BKW, APPEGE, RUBY, LAMINEUR and TIGER) is widely used in most of them. However, CHEQ is based on liquid perturbation theory which is advanced in 1970s and 1980s and the condensed phase of carbon is taken as a mixture of graphite, diamond, liquid carbon.

In this paper, we proposed a new theoretical model for explosive detonation products and a code based on it is developed. The EOS parameters of the code are only determined by the component parameters. Once the potential parameters of products are given, it can be used to in a variety of explosives without any other experimental data. Therefore, the strong predictive ability of it is of great help in the development of new explosives.

Z. Yanhong (✉) • L. Haifeng • L. Guo
Institute of Applied Physics and Computational Mathematics, 2 Fenghao East Rd, 100094 Beijing, China
e-mail: zhao_yanhong @iapcm.ac.cn; stallenlu@gmail.com

V. Chalivendra et al. (eds.), *Dynamic Behavior of Materials, Volume 1: Proceedings of the 2012 Annual Conference on Experimental and Applied Mechanics*, Conference Proceedings of the Society for Experimental Mechanics Series, DOI 10.1007/978-1-4614-4238-7_33, 

## 33.2 Theoretical Model

We assume that explosive produces t species including s gaseous species and p solid species. All species are composed of c types of different chemical elements. In a chemical reaction, the number of atoms of each element in the reactants is equal to that of the resultants by the mass conservation law and it can be formulated by

$$\sum_{i=1}^{s} a_{ij}n_i + \sum_{i=s+1}^{t} a_{ij}n_i = q_j \qquad j = 1,\cdots,c \tag{33.1}$$

Where $a_{ij}$ is the number of atoms of j element in i component, $n_i$ is mole concentration of the i component and $q_j$ is the total mole number of the j element. For chemical equilibrium, the chemical potentials of the reactants must equal to that of the resultants. According to the minimum free energy in chemical equilibrium state, we can obtain the chemical equilibrium equations [3].

The Helmholtz free energy A of detonation products is given by

$$A(n_c,\{n_i\}, V,T) = A_f(\{n_i\}, V,T) + A_c(n_c,V,T) \tag{33.2}$$

where $A_f$ and $A_c$ are the Helmholtz free energy of fluid and solid respectively, $\{n_i\}$ and $n_c$ are molar concentrations of fluid and solid respectively. At the same time, the Helmholtz free energy $A_f$ of fluid is given by

$$A_f(\{n_i\}, V, T) = A_{id}(\{n_i\}, V, T) + A_{int}(\{n_i\}, V, T) \tag{33.3}$$

where $A_{id}$ is the Helmholtz free energy of ideal gas and $A_{int}$ is excess free energy from molecular interactions.

The Helmholtz free energy $A_{id}$ of ideal gas can be evaluated with the theory of Rossini, Pennigton and the harmonic-oscillator rigid rotator theory [4]. In this paper, the calculated thermodynamic quantities are fitted by power series

$$S_{id} = a_0 + a_1T + a_2T^2 + a_3T^3 + a_4T^4 + a_5T^5 + a_6T^6 + a_7T^7 \tag{33.4}$$

Since $\frac{dH}{dT} = T\frac{dS}{dT}$, we have

$$H_{id} - H_{id}^0 = \frac{a_1T^2}{2} + \frac{2a_2T^3}{3} + \frac{3a_3T^4}{4} + \frac{4a_4T^5}{5} + \frac{5a_5T^6}{6} + \frac{6a_6T^7}{7} + \frac{7a_7T^8}{8} + C \tag{33.5}$$

Since $A = H - TS$, we have

$$\frac{A_{id} - H_{id}^0}{T} = -\left(a_0 + \frac{a_1T}{2} + \frac{a_2T^2}{3} + \frac{a_3T^3}{4} + \frac{a_4T^4}{5} + \frac{a_5T^5}{6} + \frac{a_6T^6}{7} + \frac{a_7T^7}{8}\right) + \frac{C}{T} \tag{33.6}$$

where $S_{id}$ and $H_{id}$ are entropy and enthalpy of ideal gas respectively, $H_{id}^0$ and C are the heat of formation and integration constant respectively. T is temperature and $a_i$ is fit parameter (it yields accurate results in the range 300 K $\le T \le$ 20,000 K.). The values of $a_i$, C and $H_{id}^0$ are listed in Table 33.1, the units of free energy and enthalpy are cal/mole and the temperature unit is K.

Assuming the intermolecular potentials of fluid species is

$$\phi_{ij}(r) = \varepsilon_{ij}\left\{\frac{6}{\alpha_{ij}-6}\exp\left[\alpha_{ij}\left(1-\frac{r}{r_{ij}^*}\right)\right] - \frac{\alpha_{ij}}{\alpha_{ij}-6}\left(\frac{r}{r_{ij}^*}\right)^{-6}\right\} \tag{33.7}$$

using the improved one-fluid van der Waals mixture model [5], the mixture can be taken as unary phase fluid. With the mole fractions $x_i$ $(= n_i/\sum n_j)$ and potential parameters $(\varepsilon_{ij}, \alpha_{ij}, r_{ij}^*)$, the $\varepsilon$, $r^*$ and $\alpha$ of the effective potential $\phi(r)$ can be evaluated by the following extended mixing rule

$$(r^*)^3 = \sum_{ij} x_ix_j\left(r_{ij}^*\right)^3,\ \varepsilon = \frac{\sum_{ij} x_ix_j\varepsilon_{ij}\left(r_{ij}^*\right)^3}{(r^*)^3},\ \alpha = \frac{\sum_{ij} x_ix_j\alpha_{ij}\varepsilon_{ij}\left(r_{ij}^*\right)^3}{\varepsilon(r^*)^3} \tag{33.8}$$

**Table 33.1** Fit parameters of equation-of-state for ideal gas

| Products | $H_2O$ | $CO_2$ | CO | $H_2$ | $NH_3$ | $CH_4$ | $N_2$ | $O_2$ |
|---|---|---|---|---|---|---|---|---|
| $a_0/10$ | 4.30877 | 4.81830 | 4.44249 | 2.88461 | 4.24498 | 3.91445 | 4.30266 | 4.60832 |
| $a_1/10^{-2}$ | 4.30877 | 2.30500 | 1.47415 | 1.36599 | 2.17124 | 2.63616 | 1.45588 | 1.53437 |
| $a_2/10^{-6}$ | −4.47360 | −6.39802 | −4.22175 | −3.89829 | −5.12185 | −5.74793 | −4.16261 | −4.39714 |
| $a_3/10^{-10}$ | 7.59656 | 11.0798 | 7.45936 | 6.98084 | 8.18083 | 8.61813 | 7.35787 | 7.83972 |
| $a_4/10^{-14}$ | −7.68565 | −11.2678 | −7.69300 | −7.27472 | −7.95793 | −7.99650 | −7.59235 | −8.14251 |
| $a_5/10^{-18}$ | 4.47512 | 6.56310 | 4.53302 | 4.31288 | 4.51285 | 4.37763 | 4.47312 | 4.81553 |
| $a_6/10^{-22}$ | −1.38115 | −2.02292 | −1.41040 | −1.34684 | −1.36706 | −1.29101 | −1.39093 | −1.50107 |
| $a_7/10^{-27}$ | 1.74707 | 2.55426 | 1.79416 | 1.71769 | 1.70568 | 1.57753 | 1.76804 | 1.91161 |
| $C/10^3$ | 2.73289 | 2.71734 | 2.55617 | 2.62875 | 2.41723 | 1.94193 | 2.54940 | 2.62514 |
| $H^0_{id}/10^4$ | −5.7107 | −9.3968 | −2.7201 | 0.00000 | −9.3680 | −1.6000 | 0.00000 | 0.00000 |

And $\varepsilon_{ij}, \alpha_{ij}, r^*_{ij}$ are evaluated from the combining rules [6]

$$\varepsilon_{ij} = l_{ij}(\varepsilon_{ii}\varepsilon_{jj})^{1/2}, \alpha_{ij} = m_{ij}(\alpha_{ii}\alpha_{jj})^{1/2}, r^*_{ij} = k_{ij}(r^*_{ii} + r^*_{jj})/2 \tag{33.9}$$

where $k_{ij} = l_{ij} = m_{ij} = 1$ for all species except $k_{H2O-CO2} = 0.96$, $k_{N2-CO2} = 1.035$ and $k_{N2-H2O} = 1.03$. Meanwhile, the potential parameters of $H_2O$, $N_2$, $CO_2$, CO, $O_2$, NO, $NO_2$, $H_2$, $NH_3$ and $CH_4$ (gas phase components) of detonation products refer to reference [6].

The excess free energy $A_{int}$ is calculated using Ross's modification of hard-sphere variation theory. It meets the following inequality

$$A_{int} \le A_{HS} + \frac{2\pi\rho}{T}\int_d^{\infty} \phi(r) g_{PY}\left(\frac{r}{d}, \eta\right) r^2 dr + F_{12}(\eta)NKT \tag{33.10}$$

where $A_{HS}$, $g_{PY}(\frac{r}{d}, \eta)$ and $F_{12}(\eta)$ are excess free energy of hard-sphere, Percus-Yevick hard-sphere radial distribution function and soft-sphere correction term respectively. $A_{HS}$ and $F_{12}(\eta)$ are functions of $\eta$

$$A_{HS} = NkT(4\eta - 3\eta^2)/(1 - \eta)^2$$

$$F_{12}(\eta) = (\eta^4/2 + \eta^2 + \eta/2)$$

where $\eta = \pi \rho d^3/6$, $\rho = N/V$, d is diameter of hard-sphere. The chosen value of d makes the right side of inequality (10) reach minimum value, and the minimum energy is current state Helmholtz free energy $A_{int}$ of current state.

In earlier studies, the carbon component is mostly taken as graphite, and later it's gradually viewed as a mixture of graphite and diamond. Fried [7] studied the four kinds of carbon phase (graphite, diamond, graphite-like liquid carbon, diamond-like liquid carbon) in detail, the given results are in good agreement with experimental and ab initio data. Fried proposed the formulas for calculating Gibbs free energy of four kinds of carbon phase (they are used in range: $0 \le P \le 600$ GPa,$300\text{ K} \le T \le 15{,}000$ K). Using the EOS and chemical potential presented by Fried, we obtain the Gibbs free energy of graphite, diamond, graphite-like liquid carbon, diamond-like liquid carbon in given (P, T). The carbon phase which most possibly exist in the detonation products is evaluated using Gibbs free energy minimization and other thermodynamic quantities are calculated using thermodynamic relationship as follow

$$p = -\frac{\partial A}{\partial V}$$

$$S = -\frac{\partial A}{\partial T}$$

$$E = -T^2 \frac{\partial}{\partial T}\left(\frac{A}{T}\right)$$

$$\mu_i = \left(\frac{\partial A}{\partial n_i}\right)_{T,V,n_j}$$

The specific expressions are referred to reference [8]

## 33.3 Results and Discussions

The shock-compression state of explosive detonation products satisfies the Hugoniot condition

$$E - E_0 = \frac{1}{2}(P + P_0)(V_0 - V)$$

where the subscript 0 refers to the initial undetonated state of explosive. The detonation velocity is given by

$$D = V_0\left(\frac{P - P_0}{V_0 - V}\right)^{1/2}$$

We can solve for $D_{CJ}$ and other thermodynamic quantities by locating the minimum D-point in the D-vs.-P Hugoniot plot. Table 33.2 shows the detonation velocity and pressure of eight explosives. Comparing with experimental data [4], the calculated values are in good agreement with experimental results.

The molar concentrations of detonation products of PETN explosive (1.77 g/cm$^3$) is calculated and then compared with the results of BKW and LJD [4] (Listed in Table 33.3). For study of the relationship between explosive detonation products and the initial density of explosives, PETN of different densities are calculated and the results are shown in Fig. 33.1. It can be seen in Fig. 33.1 that with the increasing of density, the variation of detonation products containing C is the largest.

**Table 33.2** Comparison on detonation parameters of explosives at CJ-point

| Explosives | Chemical formula | Density (g/cm$^3$) | $D_{Experiment}$ (km/s) | $D_{This\ work}$ (km/s) | $P_{Experiment}$ (GPa) | $P_{This\ work}$ (GPa) |
|---|---|---|---|---|---|---|
| HMX | $C_4H_8N_8O_8$ | 1.894 | 9.12 | 9.22 | 39.00 | 38.52 |
| RDX | $C_3H_6N_6O_6$ | 1.767 | 8.64 | 8.71 | 33.80 | 33.01 |
| LX-14 | $C_{1.521}H_{2.917}N_{2.588}O_{2.659}$ | 1.835 | 8.80 | 8.73 | 37.00 | 35.14 |
| HNB | $C_6N_6O_{12}$ | 2.020 | 9.34 | 9.33 | 43.00 | 40.45 |
| RX-23-AB | $H_{7.140}N_{2.495}O_{3.616}$ | 1.384 | 7.48 | 7.53 | 18.60 | 18.75 |
| RX-23-AC | $H_{10.31}N_{5.32}O_{0.95}$ | 1.135 | 7.88 | 7.97 | 16.90 | 16.23 |
| Comp_B3 | $C_{2.031}H_{2.641}N_{2.177}O_{2.653}$ | 1.720 | 7.89 | 8.04 | 29.50 | 27.82 |
| Cyclotol | $C_{1.75}H_{2.59}N_{2.38}O_{2.69}$ | 1.743 | 8.25 | 8.32 | 31.30 | 30.20 |

**Table 33.3** Detonation products of 1 mol PETN (1.77 g/cm$^3$) at CJ point

| Products | BKW | LJD | Calculated value |
|---|---|---|---|
| $H_2O$ | 4.00 | 3.93 | 3.869 |
| $H_2$ | | 0.048 | 0.027 |
| $O_2$ | | | |
| $CO_2$ | 3.89 | 3.78 | 3.843 |
| CO | 0.223 | 0.447 | 0.435 |
| $NH_3$ | | | 0.066 |
| $CH_4$ | | 0.012 | 0.002 |
| NO | | 0.062 | 0.008 |
| $N_2$ | 2.0 | 1.97 | 1.963 |
| C | 0.89 | 0.76 | 0.719 |

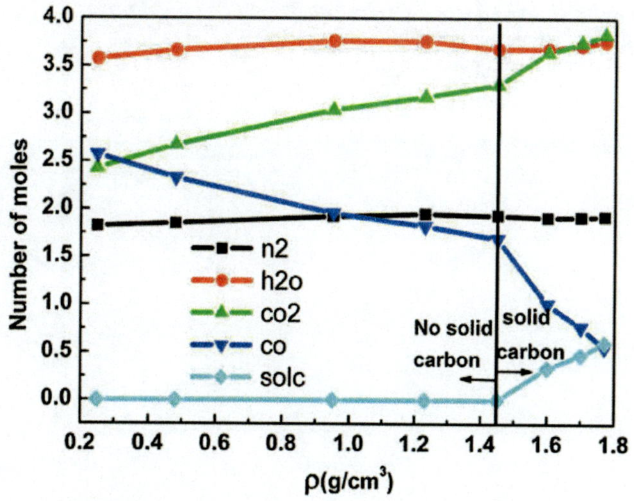


**Fig. 33.1** Molar concentrations of detonation products of 1 mole PETN at different initial densities at the CJ point

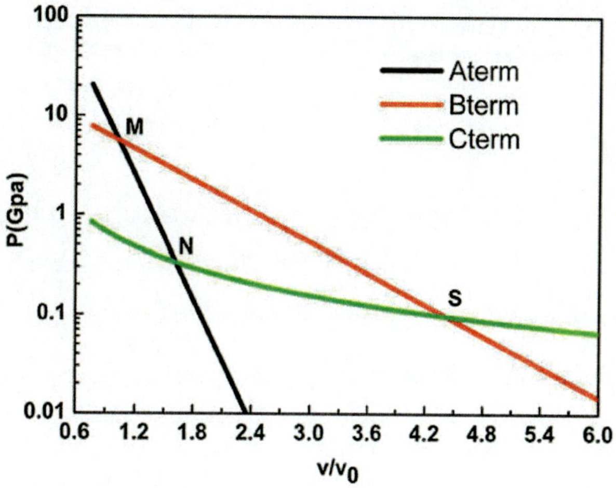


**Fig. 33.2** Comparison among three terms in JWL EOS

The molar concentration of CO is decreasing, but that of $CO_2$ and solid carbon are increasing. Due to the smaller explosive energy of low-density explosives, its chemical reaction is not complete, but for high-density explosives, the reaction is more complete due to the higher explosive energy and the detonation products are in more stable state.

The power of explosives is a more concerned ability in engineering application. The working process of detonation products is an adiabatic expansion from the CJ point. Nowadays, the JWL EOS is widely used in practical engineering application

$$P_s = A\exp(-R_1\bar{v}) + B\exp(-R_2\bar{v}) + C\bar{v}^{-(1+w)}$$

where A, B, C, $R_1$, $R_2$, ω are parameters depended on the cylinder test. $\bar{v} = V/V_0$ refers to relative volume. In JWL EOS, the pressure is expressed as the sum of two exponential function terms and a power function term of relative volume. The three terms respectively play a major role in high, medium and low pressure range. Figure 33.2 shows the relationship of them, where Aterm stands for$A\exp(-R_1\bar{v})$, Bterm represents $B\exp(-R_2\bar{v})$ and Cterm refers to$C\bar{v}^{-(1+w)}$.

In Fig. 33.3, our result of the adiabatic expansion paths from the CJ point is compared with that of JWL EOS. The two pressures are very close to each other near the CJ point. The pressure evaluated by JWL EOS is slightly higher than our result in the range from the CJ point to the point where relative volume is about 2.8. With further expansion of detonation products, the JWL-pressure becomes lower than our result. Due to the above characteristics, Fig. 33.3 shows that the area under the curve of this paper is approximatively equal to that of JWL curve. It indicates that the power of our EOS is same as JWL EOS. However, in later working process, the JWL curve is less smooth than our curve and it is also indicated by Fig. 33.4.

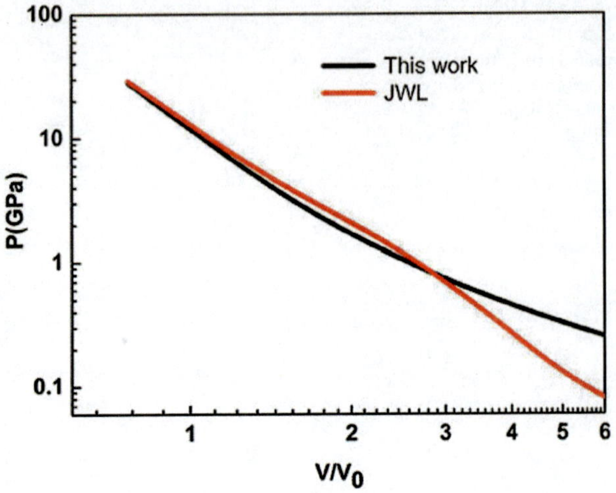


**Fig. 33.3** Comparison of the adiabatic expansion paths from the CJ point

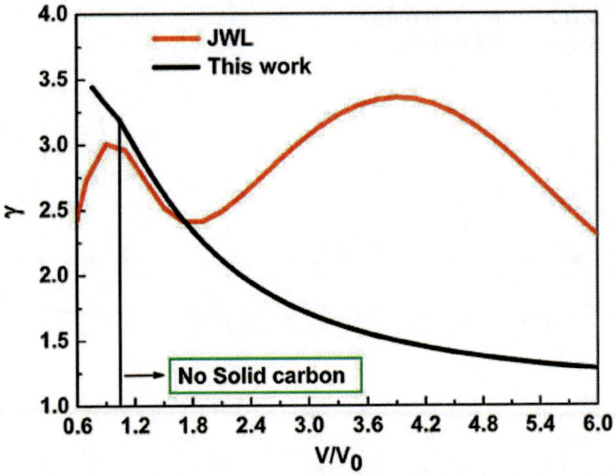


**Fig. 33.4** Comparison of the adiabatic gamma along the CJ adiabatic expansion paths

In Fig. 33.4, the adiabatic gamma $\gamma$ along the CJ adiabatic expansion paths computed by this paper is compared with JWL result. It indicates that with increasing of relative volume, the $\gamma$ of this paper is monotonically decreasing, but it has two peaks on JWL curve. The double peaks indicate that there are reversible reactions in adiabatic expansion process, which is contrary to the calculation results in Fig. 33.5. Figure 33.5 shows the variation of molar concentrations of detonation products along the CJ adiabatic expansion paths with relative volume. During adiabatic expansion process, Fig. 33.5 indicates that complex chemical reactions are still undergoing among detonation products and the molar concentration of each product is changing. However, no matter how complex the chemical reaction is, it proceeds in one direction monotonously with increasing of relative volume. Basically there is no reversible reaction, so the value of $\gamma$ should be monotonically decreasing. Therefore, the double peaks of JWL equation of state is not physical phenomenon. Comparing Figs. 33.2 and 33.4, it indicates that the first peak of $\gamma$ of JWL curve is an alternating process of Aterm line and Bterm line and the second peak is a corresponding process of Bterm line and Cterm line. It shows that the double peaks phenomenon is decided by the function form of JWL equation of state.

It can be seen in Fig. 33.4 that the value of $\gamma$ has a sudden change around the relative volume of 1.05. This is mainly because, during the expansion process from CJ point, free carbon no longer precipitates in detonation products near relative volume of 1.05, but they completely convert to gaseous compounds. This can also be seen from the variation of molar concentrations of products in Fig. 33.5. In the process from CJ point to the relative volume of 1.05, the molar concentrations of $CO_2$ and solid C are gradually decreasing, but the molar concentration of CO is increasing. Therefore, the main reaction is

$$CO_2 + C = 2CO$$

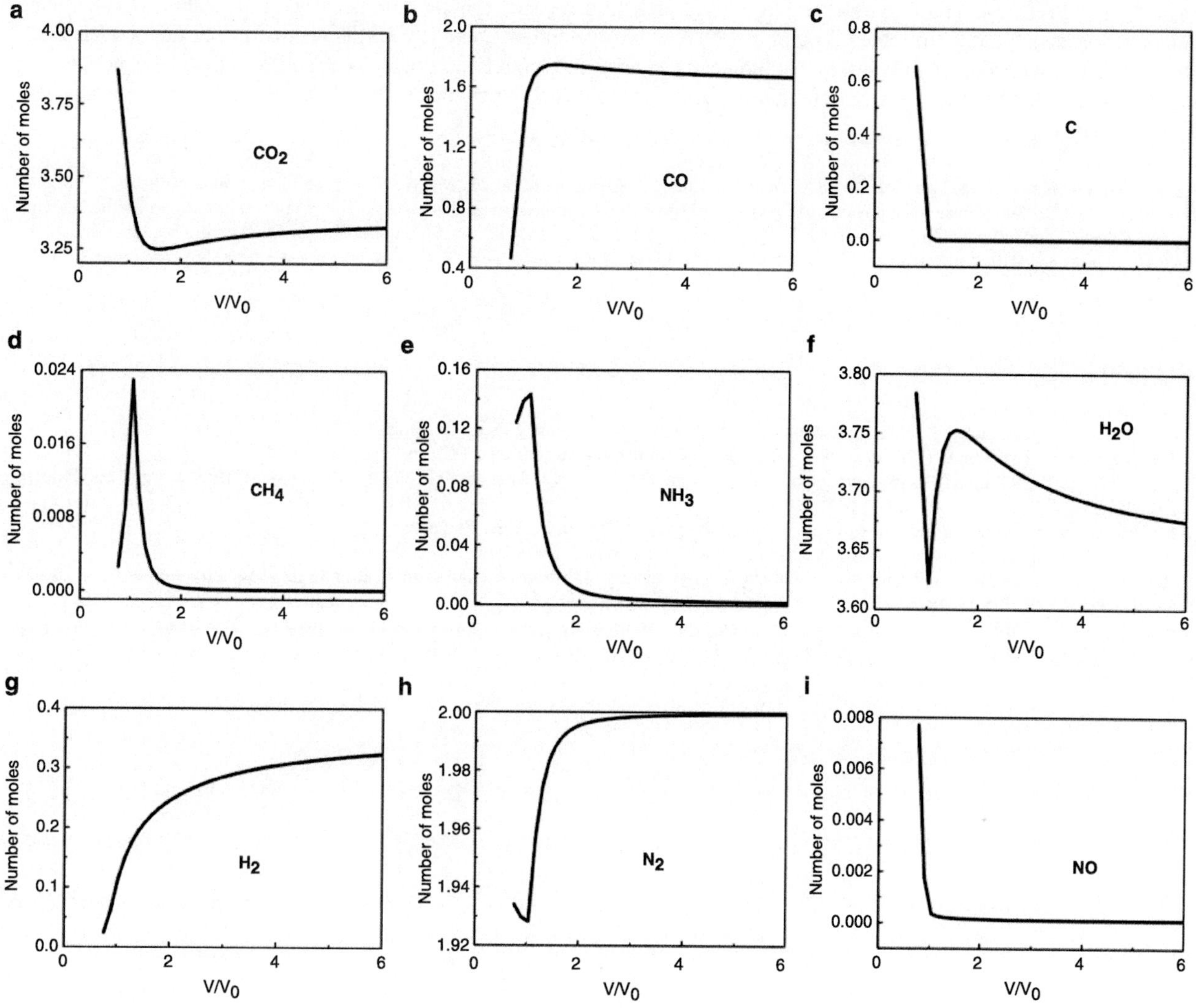


**Fig. 33.5** Variation of molar concentrations along the CJ adiabatic expansion paths with relative volume

Other products are also affected by the above reaction and the corresponding molar concentrations change suddenly. When detonation products expansion to a large volume (near relative volume of 6.0), Most of the C element of explosives exists in $CO_2$, a small amount of CO generated due to the lack of O. The majority of H element exists in $H_2O$ and a small quantity of $H_2$ created for the lack of O. The N element mainly exist $N_2$.

## 33.4 Conclusions

In this paper, detonation parameters of a variety of explosives are evaluated based on statistical-physical theory and the calculated detonation velocity and pressure are in good agreement with experimental results. We also calculate the equilibrium compositions of detonation products of PETN explosive at CJ point and the results are close to calculations of BKW and LJD EOS. The results indicate that with increasing of explosive density, the largest changes occur in products containing C element. The molar concentration of CO decreases, but that of $CO_2$ and solid carbon increase. With detailed analysis, it is pointed out that the chemical reaction of low-density explosives are not complete due to its lower energy capacity, however, reaction of high-density explosives are complete, such that their products are more stable. We obtain the adiabatic expansion paths from the CJ point of PETN explosive and compare it with the results of traditional JWL EOS. It indicates that the value of $\gamma$ decreases monotonically in our results, but it has double peaks in JWL EOS. It is pointed out that the double peaks is essentially determined by the specific function form of JWL EOS, therefore, it's not physical phenomenon.

**Acknowledgements** Our thanks are due to Profs. Guangcai Zhang, Haifeng Song, Guicun Ma, Hui Zheng, and Drs. Gongmu Zhang, Qili Zhang, Hongzhou Song, Bo Sun, Shuaichuang Wang, Mingfeng Tian, for helpful comments and discussions. This work was supported by the National Natural Science Foundation of China (Grant No. 11102026) and the Science Foundations of China Academy of Engineering Physics (Grant Nos. 2009A0101004 and 2010B0102016).

## References

1. Sun CW, Wei YZ, Zhou ZK (2000) Application of detonation. Defence Industry Press, Bei Jing
2. Chirat R, Pittion-Rossillon G (1981) A new equation of state for detonation products. J Chem Phys 74(8):4634
3. Liu HF, Chen DQ, Zhang SZ (1996) Equation of state of detonation products and the possible phase transition for CHBr3. J High Press Phys 10 (4):284
4. Mader CL (1979) Numerical modeling of detonation. University of California Press, Berkeley
5. Ree FH (1983) Simple mixing rule for mixtures with exp-6 interactions. J Chem Phys 78(1):409
6. Ree FH (1984) A statistical mechanical theory of chemically reacting multiphase mixtures: application to the detonation properties of PETN. J Chem Phys 81(3):1251
7. Fried LE, Howard WM (2000) Explicit Gibbs free energy equation of state applied to the carbon phase diagram. Phys Rev B 61(13):8734
8. Zhao YH, Liu HF, Zhang GM (2007) Equation of state of detonation products based on statistical mechanical theory. Acta Phys Sin 56(8):4791

# Chapter 34
# Incipient and Progressive Damage in Polyethylene Under Extreme Tensile Conditions

**Jevan Furmanski, E.N. Brown, C.P. Trujillo, D.T. Martinez, and G.T. Gray III**

**Abstract** The Dynamic-Tensile-Extrusion (Dyn-Ten-Ext) test was developed at LANL by Gray and coworkers to probe the tensile response of materials at large strains (>1) and high strain-rates (>1,000/s) by firing projectiles through a conical die at 300–700 m/s. This technique has recently been applied to various polymers, such as the fluoropolymers PTFE (Teflon) and the chemically similar PCTFE, which respectively exhibited catastrophic fragmentation and distributed dynamic necking. This work details investigations of the Dyn-Ten-Ext response of high density polyethylene, both to failure and sub-critical conditions. At large extrusion ratios (~7.4) and high velocities, such as those previously employed, HDPE catastrophically fragmented in a craze-like manner in the extruded jet. At more modest extrusion ratios and high velocities the specimen extruded a stable jet that ruptured cleanly, and at lower velocities was recovered intact after sustaining substantial internal damage. Thermomechanical finite element simulations showed that the damage corresponded to a locus of shear stress in the presence of hydrostatic tension. X-ray computed tomography corroborated the prediction of a shear damage mechanism by finding the region of partially damaged material to consist of macroscopic shear-mode cracks nearly aligned with the extrusion axis, originating from the location of damage inception.

**Keywords** Large-strain • High strain-rate • Damage

## 34.1 Introduction

The structural performance of polymers is highly sensitive to changes in their environment, with particularly strong dependence on temperature and deformation rate. The effects of these variables have long been studied in polymers, notably resulting in efforts to unify these effects though time-temperature equivalence relations. However, for the most part, investigations of the time-dependent material behavior of polymers have been restricted to the small-strain regime (such as Dynamic Mechanical Analysis (DMA) or Thermomechanical Analysis (TMA) where strains are limited to be well within the viscoelastic regime at 0.005 or less), which is insufficient for certain structural problems where large strains are attained at a high strain-rate (e.g., deformation near a rapidly propagating crack tip, or damage and failure due to an extreme insult). Few experimental approaches are available for unambiguously assessing the deformation response of materials under extreme mechanical loading conditions, i.e., simultaneous large strains ($\varepsilon > 1$) and high strain-rates ($d\varepsilon/dt > 10{,}000\ s^{-1}$) [1], see Fig. 34.1.

Servohydraulic load frames can achieve large uniform strains, but are generally limited to strain-rates less than 10 $s^{-1}$ in compression, and even less in tension. The high strain-rate response of polymers has been probed above $d\varepsilon/dt > 1{,}000\ s^{-1}$ using a Split Hopkinson Pressure Bar (SHPB) which imposes a one-dimensional stress wave on a specimen, but is limited in strain to approximately $0.05 < \varepsilon < 0.3$. While a wide range of researchers have employed the SHPB to study the high strain-rate response of polymers in compression [2–12], the high strain-rate polymer tensile response employing tensile-SHPB has been less widely studied. The data is typically limited to slightly lower strain rates to account for the longer gage length and typically does not extend to large enough strain to capture the failure response [3, 8, 13]. Results including SHPB testing of

J. Furmanski (✉) • C.P. Trujillo • D.T. Martinez • G.T. Gray III
MST-8, Mail Stop G755, Los Alamos National Laboratory, Los Alamos 87545, NM, USA
e-mail: jevan@lanl.gov

E.N. Brown
P-23, Mail Stop H803, Los Alamos National Laboratory, Los Alamos 87545, NM, USA

V. Chalivendra et al. (eds.), *Dynamic Behavior of Materials, Volume 1: Proceedings of the 2012 Annual Conference on Experimental and Applied Mechanics*, Conference Proceedings of the Society for Experimental Mechanics Series, DOI 10.1007/978-1-4614-4238-7_34, 

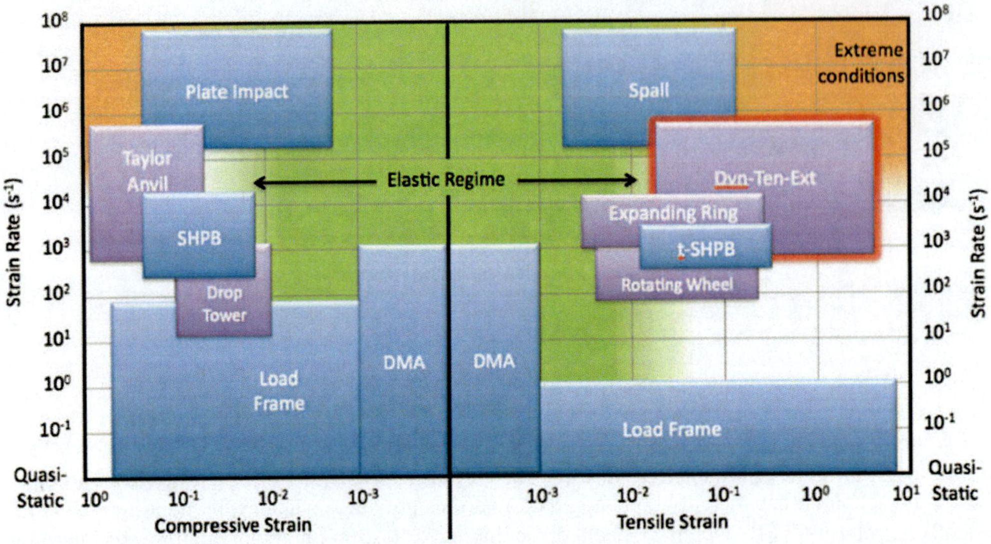


**Fig. 34.1** Comparison of the strain and strain-rate regime covered by Dyn-Ten-Ext against class mechanical characterization techniques. Techniques indicated in blue represent characterization methods with constant strain-rate and a steady homogeneous stress state at the continuum level prior to the onset of damage. Techniques indicated in purple represent integrated methods where the strain-rate evolves throughout the test and the stress state is complex. The regimes of strain and strain-rate shown are illustrative, with the exact conditions being dependent on the material being studied and extensive efforts in the literature are seeking to extend the applicability of many of these techniques

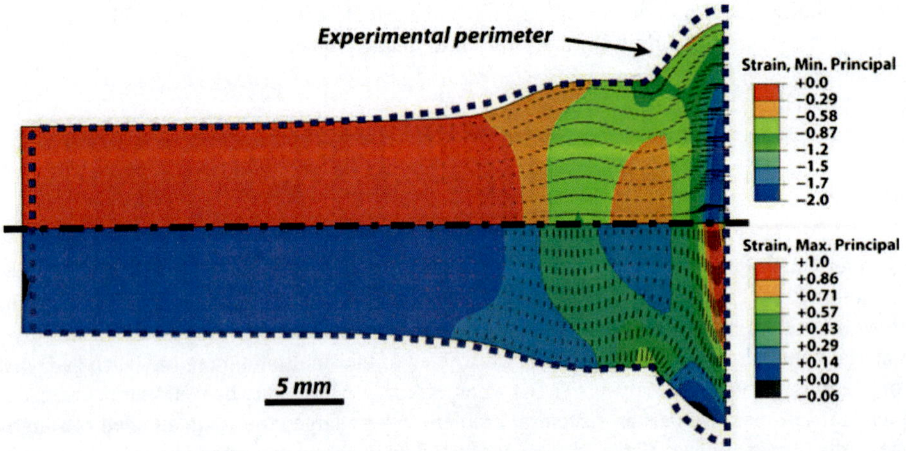


**Fig. 34.2** FEA and Taylor rod impact experimental results (outlined) for HDPE at 234 ms$^{-1}$. The instant of maximum compression approximately 80 μs after impact and just prior to rebound is shown (see [15]). Principal strain vectors are overlaid to show direction of flow. The maximum compressive logarithmic strain is approximately −2.0, while the maximum tensile strain approaches 1.0. The agreement between experiment and simulation was taken as validation of modeled material behavior, for use in modeling Dyn-Ten-Ext (Fig. 34.4)

polymers have extended time-temperature relations beyond small-strains up to approximately $\varepsilon = 20\%$ [5], providing a substantial improvement of our understanding of their viscoelastic and plastic flow response. Nevertheless, there is as yet no simple way to impose extreme mechanical conditions on a specimen in a uniform and easily interpretable manner.

There has been some success in imposing extreme mechanical conditions in a controlled manner, albeit by relaxing the requirement of uniform specimen deformation. The most prominent effort in this area was first conducted by G. I. Taylor for the study of the high-rate flow stress of steel [14]. His insight was that high-rate deformations could not be achieved through a sudden imposed acceleration of a loading apparatus, but could be achieved through sudden deceleration via impact of a rapidly moving specimen (velocity 100–800 m s$^{-1}$) against a massive anvil. During these Taylor cylinder impact experiments, also known as the Taylor anvil experiment, extreme plastic deformation occurs at strain-rates exceeding 10,000 s$^{-1}$ near the impact face, driven by the momentum of the trailing part of the projected specimen, much of which may not plastically deform at all (Figs. 34.2 and 34.3a).

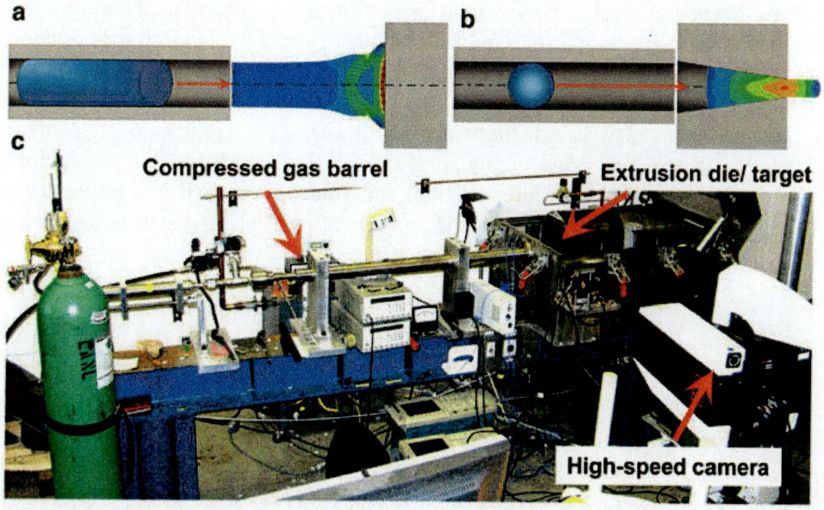


**Fig. 34.3** (**a**) Schematic of Taylor rod impact and (**b**) Dyn-Ten-Ext, with FEA simulation results showing maximum principal strain in HDPE. *Dark blue* denotes unstrained material, while red regions exhibit a strain of approximately 100 %. (**c**) Materials Structure /Property Relations team Taylor gun facility at Los Alamos National Laboratory

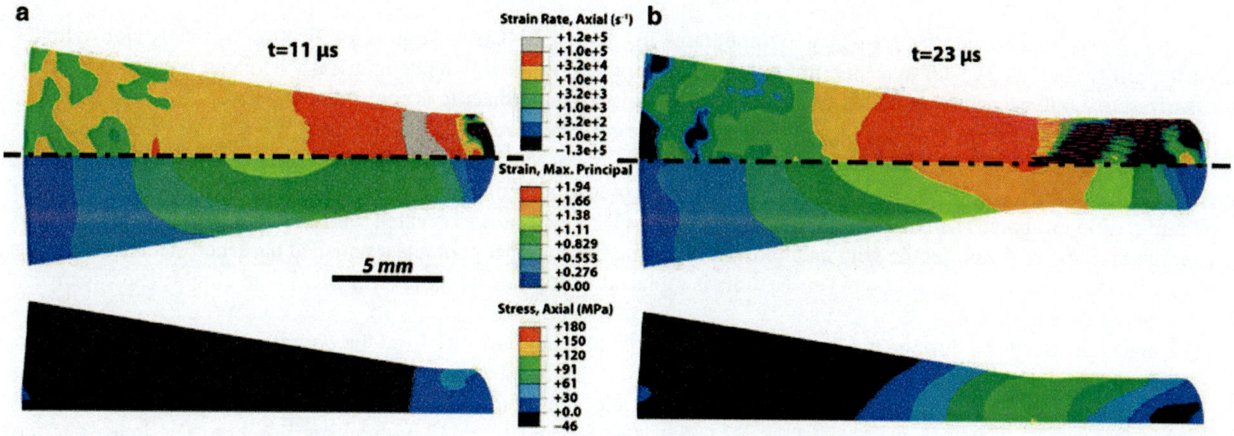


**Fig. 34.4** Dyn-Ten-Ext simulation for HDPE at an initial velocity of 447 ms$^{-1}$ and exit diameter 2.8 mm, showing axial logarithmic strain, strain rate (s$^{-1}$) and axial tensile stress (MPa) at two time points after specimen-die contact. The axial strain exceeds 1 in the critical section during extrusion, and the strain-rate likewise exceeds 10$^{4}$ s$^{-1}$ in this region. The critical section is under tension, while the tip and tail in the die remain in compression. The strain-rate diminishes rapidly a short distance past the die exit, after which point plastic deformation has effectively ceased

The Dynamic-Tensile-Extrusion (Dyn-Ten-Ext) test was developed by G.T. Gray and co-workers at Los Alamos National Laboratory (LANL) to examine extreme tensile conditions in materials [16, 17]. The apparatus utilizes the same hardware as a Taylor cylinder impact test, except that a conical extrusion die is fixed to the end of the gun barrel, forcing the specimen to extrude through it at a high velocity (Fig. 34.3c) and the anvil is not employed. The leading edge of the specimen is relatively unaffected by the extrusion process, but the trailing portion rapidly decelerates inside the die, and thus the extruded ligament between the two ends is pulled in high strain-rate tension, typically to large strains and, ultimately, to failure (Fig. 34.3b). As shown in Fig. 34.4, a finite element simulation of high-density polyethylene (HDPE) undergoing Dyn-Ten-Ext, through a die of exit diameter 3.60 mm with an initial velocity of 447 s$^{-1}$, predicts that the ligament achieves a strain of 1–1.9 at a strain-rate of 10$^{4}$–10$^{5}$ s$^{-1}$ in the critical tensile section during extrusion, thus achieving extreme tensile test conditions.

Thus, as an integrated test that generates extreme tensile deformation, Dyn-Ten-Ext provides a means to challenge the quantitative accuracy of constitutive models parameterized under less extreme conditions. Additionally, it provides a direct qualitative assessment of the mechanisms of deformation and damage under extreme conditions, which may or may not differ from those observed at lower strain-rates. For example, recent work has shown that the failure phenomena observed in Dyn-Ten-Ext qualitatively matched expectations derived from the microstructure and low strain-rate response in PTFE and PCTFE [18], and Polyurea [19].

The primary aim of this work is to identify the (potentially distinct) mechanisms of damage inception and progression in several formulations of polyethylene, and to include these mechanisms in 3-D dynamic finite element models for the analysis of extreme mechanical events in structures and components. In some cases, there may be multiple potential damage mechanisms acting in concert (such as shear failure, crazing, and void coalescence in a polymer), which would benefit from a sound theoretical approach to the deconvolution and analysis of parallel deformation mechanisms.

## 34.2 Materials and Methods

Industrial grade polyethylene extruded sheets of HDPE from Cope Plastics (Godfrey, IL) were studied. These materials were the same pedigree as previously investigated by Brown et al. [11]. The sheets were ~25 mm thick. The density was 969.8 ± 1.4 kg/m$^3$ as measured by He pycnometry. All compression samples were machined from the plates in the through thickness direction. As previously reported [4], HDPE exhibits a melt transition based on peak of the DSC melt endotherm of ~134°C. Below the melt the TMA trace of UHMWPE exhibit smooth increases with temperature that can be described by coefficients of thermal expansion of 124.5 × 10$^{-6}$/°C. DSC traces from HDPE exhibit no transitions in the form of endo- or exo-therms that might reflect a glass transition. Integrating the area under the DSC melt endotherm from 100°C to 150°C yields heat of melt of 236.9 J/g for. Taking the heat of melt for a perfect PE crystal to be 293 J/g from the percent crystallinity of the HDPE was 80.9%.

As previously mentioned, the Dyn-Ten-Ext apparatus in use at LANL is a slight modification to the Taylor cylinder impact Facility (Fig. 34.3). This system employs a helium driven gas gun with a pneumatically actuated breech that can be pressurized to 4,600 psi and released remotely. The exit velocity of the projectile is controlled by choice of breach pressure, with velocities from 100 to 600 ms$^{-1}$ easily achieved. The polymer projectile is loaded in though the breech, and the specimen is adhered weakly to a burst diaphragm made of thin metal foil or paper to even out the dynamic pressure during breech valve actuation and also to prevent the projectile from prematurely creeping down the barrel. The barrel has a smooth 7.62 mm (0.300 in.) bore. The barrel discharges into a target chamber that is evacuated with a roughing pump for testing to prevent interaction of the projectile with an atmosphere. While the chamber contains a polished hardened steel anvil during Taylor cylinder impact testing, for Dyn-Ten-Ext this is replaced with a catch tank loosely filled with cloth for soft recovery of any specimen fragments.

The main data collected during Dyn-Ten-Ext testing consist of initial velocity from the bore prior to interaction with the die and high-speed multiframe photography of the extruded material for obtaining deformed shape and speed, which is then compared to FEA simulations to understand the material behavior during the test. The specimen initial velocity is measured via two pressure transducers on the barrel near the target chamber and captured with an oscilloscope. The images of the extruded material are captured with an Imacon 200 framing camera (Hadland, Santa Cruz CA), typically with a 50 ns shutter time and 3–5 μs interframe time, while illumination is provided opposite the camera by a Powerlight 2599 flash lamp (Photogenic, Bartlett IL). The velocity of extruded material is collected from the captured images and is calibrated with a transparent scale near the die exit. The camera resolution gives a pixel-to-pixel spatial error of ~80 μm (~1% of the initial sphere diameter).

The Dyn-Ten-Ext specimen geometry can be a sphere or a hemisphere-ended cylinder, both of which use a diameter and length of 7.60 mm. The die uses a conical reducing section inclined 9° (half-angle) to the extrusion axisThe die exit diameter is specified depending on the desired draw ratio to be imposed prior to the onset of dynamic self-extrusion after the die exit. The nominal design of the die uses an exit diameter of 2.80 mm (0.110 in.), which imposes an axial true strain $\varepsilon \sim 2$ at the die exit.

Finite element analysis was conducted in ABAQUS CAE 6.10-2 (Simulia, Providence RI) using a 2-D axisymmetric model with linear quadrilateral elements. The material model employed J2 plasticity with isotropic hardening and adiabatic heating, piecewise fit to HDPE quasi-static and SHPB stress–strain curves at various temperatures from [4]. The resulting stress–strain curves were shifted in strain-rate and temperature via a well-calibrated time-temperature equivalence relation for the material [4]. The resulting time-temperature equivalent model was then validated against HDPE Taylor Impact images from [15] as shown in Fig. 34.2, with good agreement. The validated model was then applied to Dyn-Ten-Ext, with the projectile experiencing contact with an analytical surface representing the die, with the same contact and friction model used in the Taylor Anvil analysis.

## 34.3 Results

Dynamic-Tensile-Extrusion (DTE) was developed to study deformation of materials under what are best described as principally-tensile conditions, in contrast to the principally-compressive conditions in Taylor Impact. For large tensile strains at high strain-rates, tensile hydrostatic tension (or positive stress triaxiality) can drive damage mechanisms that would remain inactive under hydrostatic compression. For this reason, DTE is employed preferentially for studying extreme damage and failure in materials, and thus DTE is the focus of this work. To be certain, Taylor Rod Impact is well-established and does elicit large tensile (radial and hoop) strains, and is employed as a critical companion experiment for DTE.

An exemplary set of DTE experimental results for high density polyethylene (HDPE) appears in Fig. 34.5. A range of results is obtained as the input velocity is increased, progressing from limited damage to terminal damage progression, to catastrophic failure by a separate mechanism than seen at lower velocities. This is a useful set of results, because it captures the evolution of damage under extreme tensile conditions to failure, and also reveals that two separate mechanisms (or at least pathways) of damage are competing during the deformation.

The progression of damage to failure is evident upon cross-section of the recovered specimens (Fig. 34.6). It is clear that damage progresses along a pseudo-axial path that links an internal flaw to the exterior edge of the specimen, where rupture occurs at higher velocities at the die exit. Closer examination of the intact specimen (Fig. 34.7) reveals an internal failure at the specimen axis in the form of a macroscopic shear crack, which is followed by a shear-mode pull-out of the core behind the flaw in a strain concentrating process. X-ray tomography reveals the damage mechanism to be a shear-cracking process (Fig. 34.7b).

The apparent mechanism of damage is a pressure-mediated shear mode, as there is significant shear and tensile deformation during deformation inside the die which does not result in flaws. Numerical analysis reinforces this inference, by associating the damage with stress triaxiality and pressure-shear failure. Figure 34.8 shows the logarithmic planar shear strain field (LE12) with a chevron flaw in HDPE, where there is no apparent localization of strain due to the constitutive deformation behavior in the model. However, there is a locus of concentrated planar shear stress (S12) along the observed crack path. This provides a powerful impetus for a shear-stress driven damage mechanism to be added to the constitutive model, which would tend to relax the stress and convert it to a surface of strain localization.

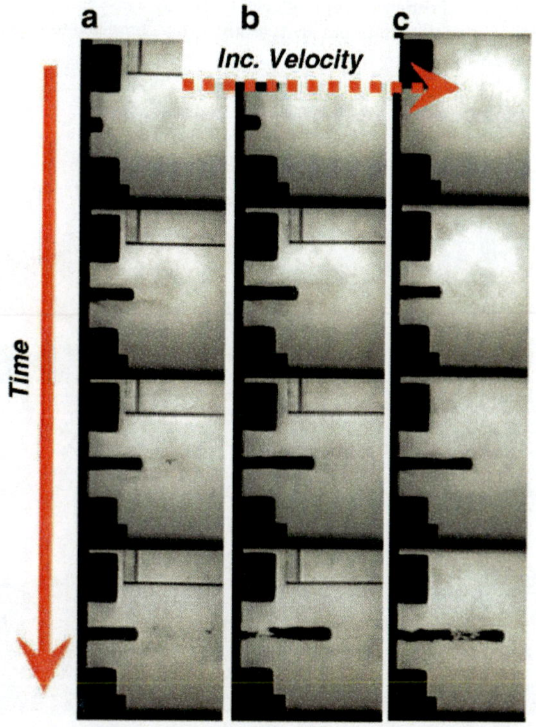


**Fig. 34.5** Dynamic-Tensile-Extrusion of high density polyethylene (HDPE), as captured at the die exit by a high speed framing camera. Three conditions are shown: (**a**) moderate velocity test, arresting with an intact but damaged specimen, (**b**) a higher velocity test, resulting in delayed rupture following internal damage progression, and (**c**) a high velocity test, where catastrophic fragmentation failure occurs near the leading end of the specimen

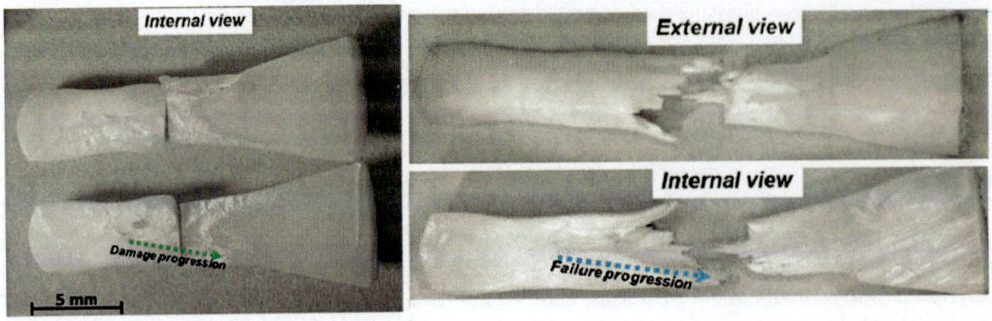


**Fig. 34.6** Section views of specimens (**a**) and (**b**) in Fig. 34.3. *Left*: Specimen (**a**); Local failure initiation at the specimen axis and damage progression rearward are evident in the section. *Right*: Specimen (**b**); similar failure initiation to that in specimen (**a**), but damage has progressed to failure along same path. Internal failure appears to substantially precede rupture of the external shell

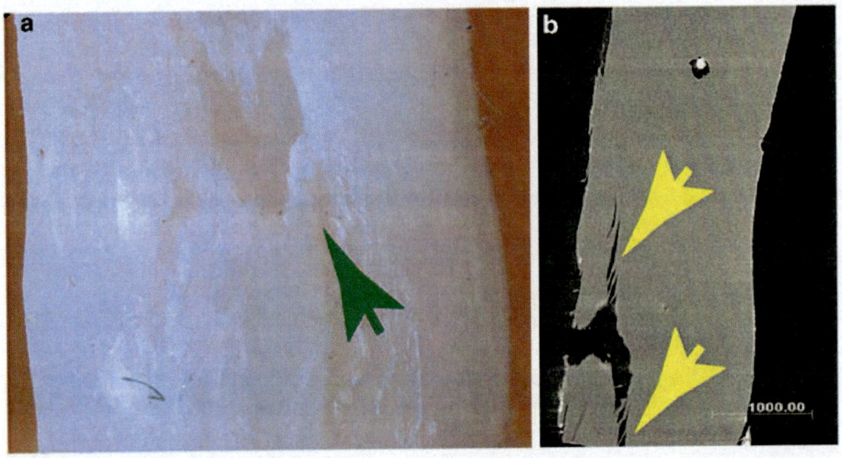


**Fig. 34.7** Postmortem of low-velocity HDPE Dyn-Ten-Ext with stable damage. (**a**) Cross-section of recovered specimen, showing chevron tensile failure and shear pull-out damage. (**b**) X-ray computed tomography showing shear-mode cracks in damage region ahead of and behind chevron failure

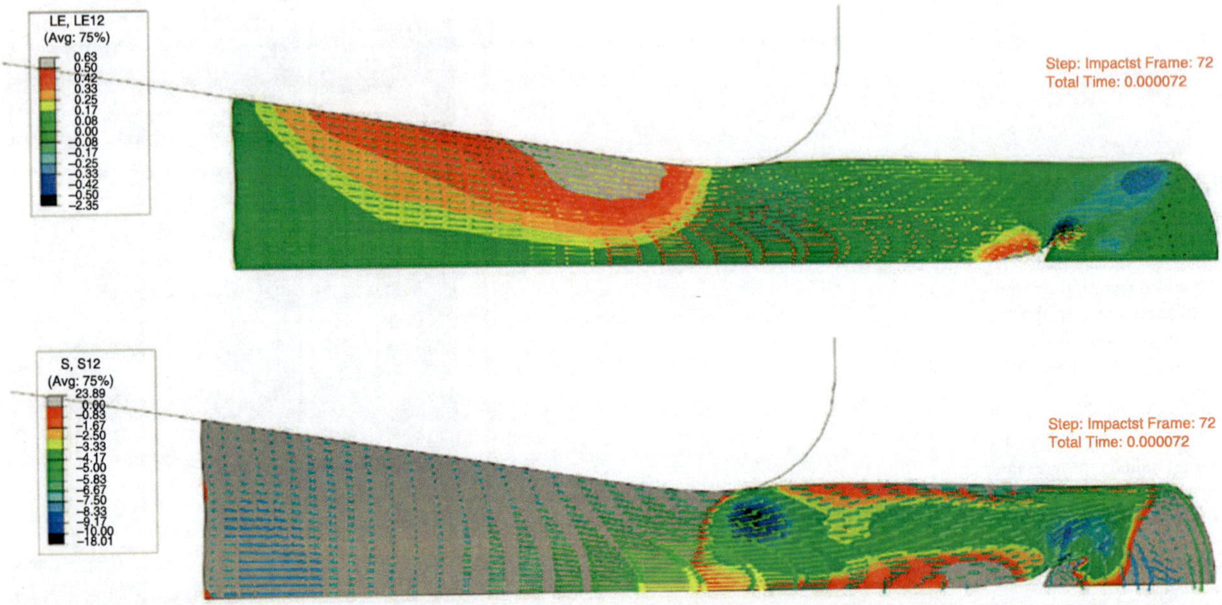


**Fig. 34.8** Planar logarithmic shear strain and shear stress fields during DTE of HDPE after the formation of an axial chevron flaw. Maximum principal components are shown as vectors. There is no apparent strain localization predicted from the continuum deformation, but there is a concentration of shear stress along the damage path. This motivates the inclusion of a shear-stress driven damage mechanism

**Acknowledgements** Los Alamos National Laboratory is operated by LANS, LLC, for the NNSA of the US Department of Energy under contract DE-AC52-06NA25396. This research was supported by Campaign 2: Dynamic Behavior of Materials and the Joint DoD/DOE Munitions Program.

## References

1. Funk DJ et al (2009) A summary report on the 21st century needs and challenges of compression science workshop. Los Alamos National Laboratory Report LA-UR-09-07771
2. Brown EN, Rae PJ, Gray GT (2006) The influence of temperature and strain rate on the tensile and compressive constitutive response of four fluoropolymers. J De Physique Iv 134:935–940
3. Brown EN, Rae PJ, Orler EB (2006) The influence of temperature and strain rate on the constitutive and damage responses of polychlorotrifluoroethylene (PCTFE, Kel-F 81). Polymer 47(21):7506–7518
4. Brown EN et al (2007) Influence of molecular conformation on the constitutive response of polyethylene: a comparison of HDPE, UHMWPE, and PEX. Exp Mech 47(3):381–393
5. Gray III GT, Blumenthal WR (2000) Split-Hopkinson pressure bar testing of soft material. ASM Handbook Volume 8: Mechanical Testing and Evaluation, 2000
6. Jordan JL, Foley JR, Siviour CR (2008) Mechanical properties of Epon 826/DEA epoxy. Mech Time-Depend Mater 12(3):249–272
7. Jordan JL et al (2007) Compressive properties of extruded polytetrafluoroethylene. Polymer 48(14):4184–4195
8. Rae PJ, Brown EN, Orler EB (2007) The mechanical properties of poly(ether-ether-ketone) (PEEK) with emphasis on the large compressive strain response. Polymer 48(2):598–615
9. Rae PJ, Dattelbaum DM (2004) The properties of poly (tetrafluoroethylene) (PTFE) in compression. Polymer 45(22):7615–7625
10. Richeton J et al (2006) Influence of temperature and strain rate on the mechanical behavior of three amorphous polymers: characterization and modeling of the compressive yield stress. Int J Solids Struct 43(7–8):2318–2335
11. Sarva SS, Boyce MC (2007) Mechanics of polycarbonate during high-rate tension. J Mech Mater Struct 2(10):1853–1880
12. Siviour CR et al (2005) The high strain rate compressive behaviour of polycarbonate and polyvinylidene difluoride. Polymer 46(26):12546–12555
13. Rae PJ, Brown EN (2005) The properties of poly(tetrafluoroethylene) (PTFE) in tension. Polymer 46(19):8128–8140
14. Taylor G (1948) The use of flat-ended projectiles for determining dynamic yield stress .1. theoretical considerations. Proceed Royal Soc London Series -Math Phys Sci 194(1038)
15. Brown EN, Trujillo CP, Gray GT (2007) Influence of polyethylene molecular conformation on Taylor impact measurements: a comparison of HDPE, UHMWPE, and PEX. Shock Compression of Condensed Matter – 2007, Pts 1 and 2 955: 691–694
16. Cao F et al (2008) Dynamic tensile extrusion response of tantalum. Acta Mater 56(19):5804–5817
17. Gray GT et al (2006) Influence of shock prestraining and grain size on the dynamic-tensile-extrusion response of copper: Experiments and simulation. Shock Compression of Condensed Matter – 2005, Pts 1 and 2, 845: 725–728
18. Brown EN, Trujillo CP, Gray G.T (2009) Dynamic-tensile-extrusion response of Fluoropolymers. Shock Compression of Condensed Matter – 2009, Pts 1 and 2, 1195: 1233–1236
19. Furmanski J et al (2011) Dynamic-tensile-extrusion of polyurea. In: Shock compression of condensed matter

# Chapter 35
# Experimental Analysis of Dynamic Deformation and Damage in Composite Sandwich Structures Subjected to Underwater Impulsive Loads

**Siddharth Avachat and Min Zhou**

**Abstract** The analysis of dynamic response of sandwich structures is complicated due to material heterogeneity, complex loading conditions and competing failure mechanisms. This investigation focuses on overall structural response, deformation, damage, and energy-absorption in air-backed and water-backed/submerged composite sandwich structures. The damage and failure characteristics of individual components of the sandwich structures are studied using laser-based in-situ diagnostics and postmortem analysis. Simply-supported composite sandwich plates with varying core-densities are subjected to a range of underwater impulsive loads using a novel projectile-impact based impulsive loading facility called the Underwater Shock Loading Simulator (USLS). In-situ high-speed digital imaging and postmortem analysis are used to study the dynamic deformation and failure characteristics of individual components; focusing on the effects of loading rate, core-characteristics and material heterogeneity on structural response. In finite-element simulations, the underwater blast loading intensity is considered using the Mie-Gruneisen equation-of-state of a linear Hugoniot form and core crushing is accounted for through a modified Drucker-Prager model. Results indicate that the core-density has a significant influence on dynamic deformations and failure modes. On a per-weight basis, low-density foam cores consistently outperform high-density foam cores, undergoing lesser deflections and transmitting smaller impulses. Polymeric foams experience considerable rate-effects and exhibit extensive shear cracking and collapse under high-magnitude multi-axial underwater impulsive loads. Calculations reveal a significant difference between the response of air-backed and water-backed/submerged structures. The experiments and computations offer approaches for improving the blast mitigation capabilities of submerged composite sandwich structures in the critical parts of a ship structure like keel, turbine-blades and rudders.

**Keywords** Composite materials • Blast resistance • Fluid structure interactions • Material-structure-property • Relations • Experimental mechanics

## 35.1 Introduction

Marine structures are designed to operate in hostile environments consisting of corrosive sea-water, hot and cold temperature extremes, transient dynamic loads like hull-slamming and complex three-dimensional hydrodynamic loads. Additionally, naval structures are required to withstand weapons impacts and blast loads resulting from surface and underwater explosions. Recent assessments of marine structures have demonstrated that sandwich composites can provide good blast mitigation due to their high strength-to-weight ratios and high shear-and-bending resistances. Characterization of the behavior of composite materials and polymeric foams under impulsive loading is a prerequisite for the analysis and design of effective, blast-resistant sandwich composites. Experimental and numerical studies have established that sandwich structures with thin, stiff faces and thick, low-density cores out-perform monolithic plates when deformation is dominated by bending. It has been demonstrated that overall deflection experienced by sandwich plates is significantly lesser than monolithic plates of equivalent mass [1–9]. The forces and impulses transmitted by sandwich structures are also smaller than monolithic structures [1, 4, 5]. Core design greatly influences dynamic response of sandwich structures. The dynamic strength of the core is an important factor in overall structural response. Stiff cores perform poorly while light cores exhibit higher blast mitigation.

S. Avachat (✉) • M. Zhou
The George W. Woodruff School of Mechanical Engineering, Georgia Institute of Technology, Atlanta, GA 30332, USA
e-mail: sidavachat@gatech.edu

V. Chalivendra et al. (eds.), *Dynamic Behavior of Materials, Volume 1: Proceedings of the 2012 Annual Conference on Experimental and Applied Mechanics*, Conference Proceedings of the Society for Experimental Mechanics Series, DOI 10.1007/978-1-4614-4238-7_35, 

Recent assessments of blast-loaded structures show that FSI effects need can be exploited to improve blast mitigation in marine structures[5, 10–13]. Experiments focusing on different core topologies and specimen sizes have been carried out by Espinosa et al. [14] and McShane et al. [15], using gas-gun based impact loading to generate underwater pressure impulses and by Dharmasena et al. [16] using explosive sheets to generate planar pressure impulses. Shukla and co-workers [17–21] examined the dynamic response of woven E-glass composite facesheets and stitched core sandwich structures to air-based shock loading concluding that stitched cores exhibit superior mechanical performance. They also showed that sandwich structures with graded cores provide good blast mitigation.

One of the most important factors influencing the behavior of sandwich composites under dynamic loads is the dynamic strength of the core. In this paper, we report the dynamic response of sandwich composites with different core densities but identical total masses subjected to high-intensity underwater impulsive loads. The focus is on the deformations and failure modes of sandwich composites under impulsive loads, with the aim to design more blast-resistant structures. Sandwich composites are subjected to a range of underwater impulsive loads and panel responses are compared to those of monolithic fiberglass plates of equivalent mass. Planar underwater impulses are generated using a novel experimental setup called the Underwater Shock Loading Simulator (USLS). A simply-supported beam configuration is chosen because the location of the failure modes in such a setting allows accurate time-resolved measurements using high-speed digital imaging. Specifically, high-speed digital imaging enables the study of overall deflection, face-wrinkling, core-face debonding, core-compression, core shear-cracking and rupture and their dependence on load intensity and core characteristics. The impulses are measured using high-dynamic range piezoelectric pressure transducers and a high-frequency data acquisition system.

## 35.2 Materials

Fiber-matrix composites are composed of two distinct phases: (1) reinforcements like glass-fibers, carbon-fibers, and (2) matrix materials like epoxy, polyester, etc. The strength and stiffness of the finished composite is determined by the volume fraction of glass fibers and directionality of fibers with respect to external loads. The manufacturing technique used here is called Vacuum Assisted Resin Transfer Molding (VARTM). Glass-reinforced composites are stacked in a pattern to create a "quasi-isotropic" layup – (0/45/–45/90)s. This process is repeated to achieve the desired thickness of 3 mm for sandwich structures faces and 6 mm for monolithic structures. Average fiber-volume is ~60%. Material properties of the composite materials are given in

The core is made of Divinycell H-100 PVC foam [22] for which stress–strain response is described by a volumetric hardening model in which the evolution of the yield surface is driven by the volumetric plastic strain. The response consists of three distinct regimes: (1) initial nearly elastic deformation, (2) plateau region in which deformation occurs at relatively constant stress; and (3) lock-up/densification stage beyond which the material becomes fully compacted as shown in Fig. 35.1. Low yield stresses and long stress-saturated plateaus indicate that these materials can absorb significant amounts of energy. High strain-rate studies on PVC foams show a weak dependence on strain-rate [23]. Sandwich structures are constructed by joining the facesheets to the PVC foam cores using epoxy adhesive. The total areal mass of the structures is kept constant to facilitate comparison of blast resistance; this leads to 30 mm thick HP60, 20 mm thick HP100 and 10 mm thick HP200 cores.

## 35.3 Experiments

### *35.3.1 Experimental Apparatus*

The Underwater Shock Loading Simulator (USLS) has been developed as part of a research effort to study the dynamic response of marine structures to underwater blast loading. A gas-gun based impact loading is used to create underwater pressure impulses. A projectile is accelerated down the length of a gas-gun barrel; after exiting the barrel it strikes a flyer plate fitted in a sealed water-chamber. The stress-wave generated inside the flyer plate is transmitted through the flyer-water interface into the water-chamber. By varying the projectile velocity and mass, pressure waves of varying magnitudes can be generated in the water-chamber. By varying the flyer-plate thickness, the decay time for the pressure wave can be changed. The pressure wave generated in the water-chamber travels down the length of the chamber (~700 mm) and impinges on the target. Figure 35.2 shows a schematic of the sectional view of the experimental setup and the placement of high-speed camera and illumination. Figure 35.3 shows the photograph of the USLS with different components. Diagnostics for the USLS consist of a high-speed camera (Imacon 200D) capable of capturing 200 million frames per second and a resolution of 1,368 × 1,368. The enclosure is fitted with transparent PMMA sheets to enable in-situ high-speed digital imaging of sandwich structures subjected to impulsive loads.

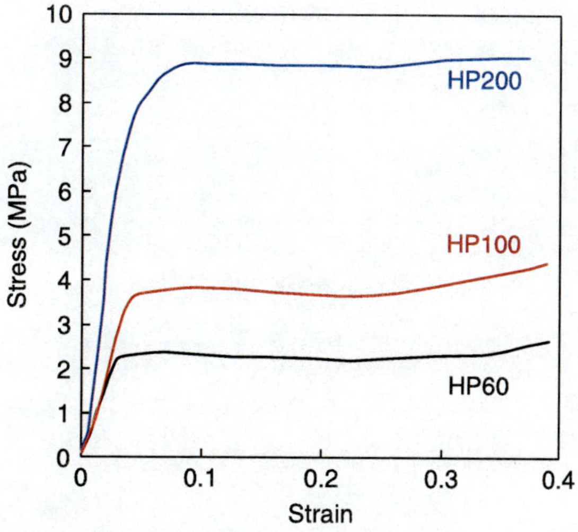


**Fig. 35.1** Stress–strain curve of HP60, HP100 and HP200 at a strain-rate of ~1,000 $s^{-1}$

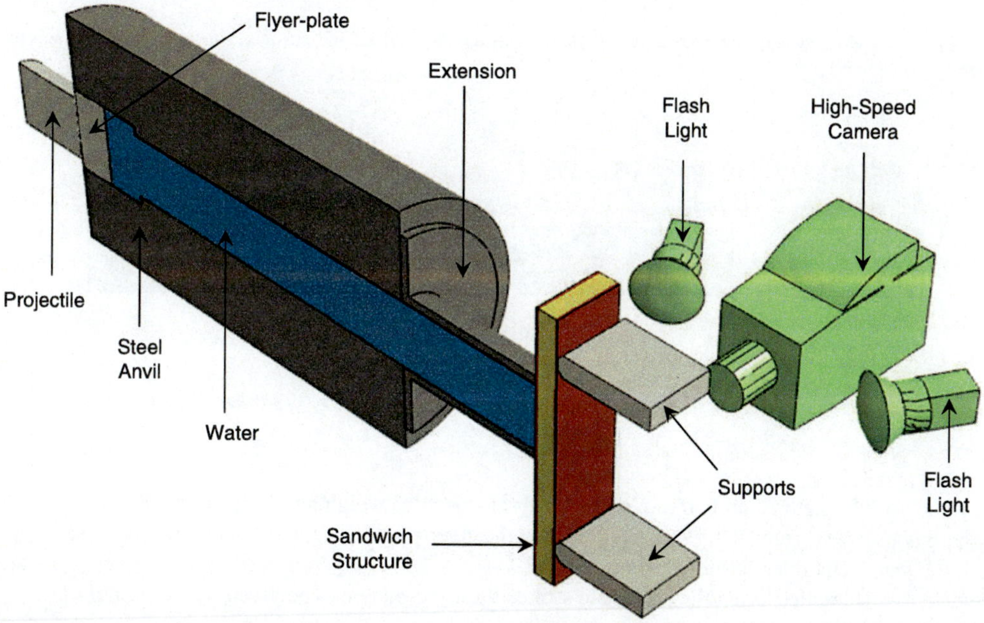


**Fig. 35.2** Sectional view of Underwater Shock Loading Simulator (USLS) and simply-supported sandwich structure showing the placement of high-speed camera and flash-lights. A high-velocity projectile hits the flyer-plate and creates a stress-wave which travels through the flyer-plate and into the water, generating an impulse identical to one produced by an underwater explosion

### *35.3.2 Loading Configurations and Impulsive Loads*

Simply-supported composite sandwich structures (shown in Fig. 35.2) are subjected to underwater impulsive loads of varying magnitudes. A simply-supported loading configuration emulates the behavior of a hull with stiffeners along the side. Sandwich structures analyzed here are in the form of beam specimens of length 300 mm and width 80 mm. Simply-supported behavior and thickness to length ratio of less than 0.1 ensures beam bending behavior. The facesheet thickness is 3 mm and the core-thicknesses are adjusted to keep total mass constant. Projectiles and flyer-plates are machined from

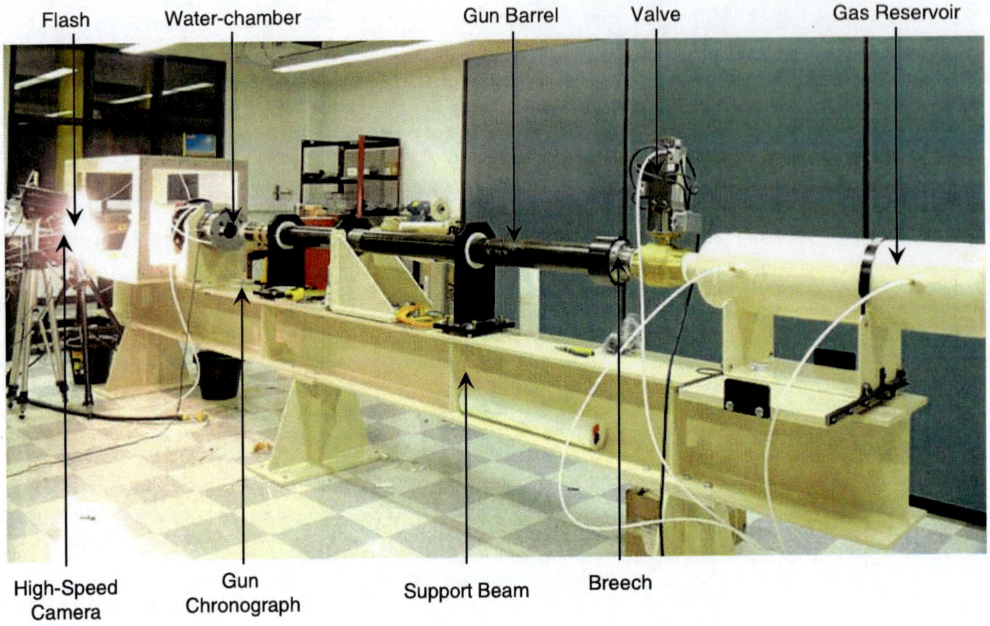


**Fig. 35.3** Photograph of the Underwater Shock Loading Simulator (USLS). Pictured are the gas-reservoir, gun-barrel, water-chamber and the Imacon 200D high-speed camera

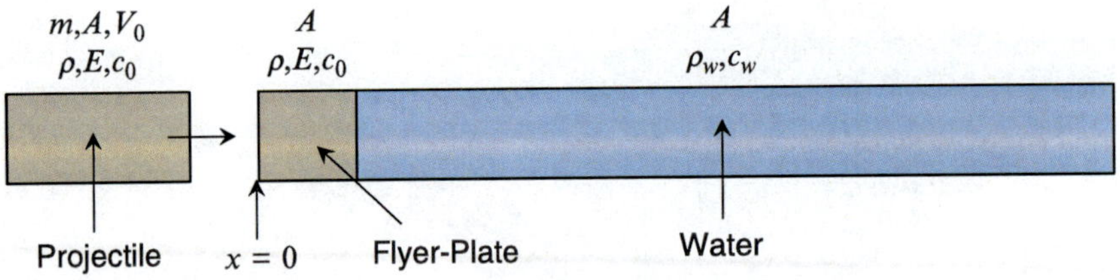


**Fig. 35.4** Schematic of the plate-impact and transmission-reflection problem at two interfaces – (1) projectile-flyer plate and (2) flyer-plate-water

aluminum alloy and are of length 80 and 50 mm respectively. Projectiles weigh approximately 0.8 kg. When the projectile, propelled by the gas escaping from the gun-barrel, strikes the flyer-plate, it creates a compressive stress-pulse which is transmitted into the water-chamber as shown in Fig. 35.4.

The stress-wave $\sigma(x,t)$ created when the projectile strikes the flyer-plate can be given by the equation

$$\sigma(0,t) = \frac{1}{A}F(0,t) = \frac{E}{c_0}V_0\left[\exp\left(-\frac{EA}{mc_0}t\right)\right], \tag{35.1}$$

where $E$ is the elastic modulus of material, $c_0$ is the speed of sound in the material, $m$ is the mass of the projectile, $V_0$ is the velocity of the projectile, $A$ is the cross-sectional area of the projectile and flyer-plate and $t$ is the time. The wave transmitted in to the water-chamber $\sigma_T$ is given by

$$\sigma_T = \frac{2\rho_w c_w}{\rho_0 c_0 + \rho_w c_w}\sigma_I, \tag{35.2}$$

where $\sigma_I = \sigma(x,t)$, $\rho_0$ is the density of flyer-plate, $c_0$ is the speed of sound in the flyer-plate material, $\rho_w$ is the density of water and $c_w$ is the speed of sound in water. Figure 35.5 shows the stress-pulse due to projectile impact and the impulse transmitted in to the water-chamber. Decay time is on the order of ~200 μs.

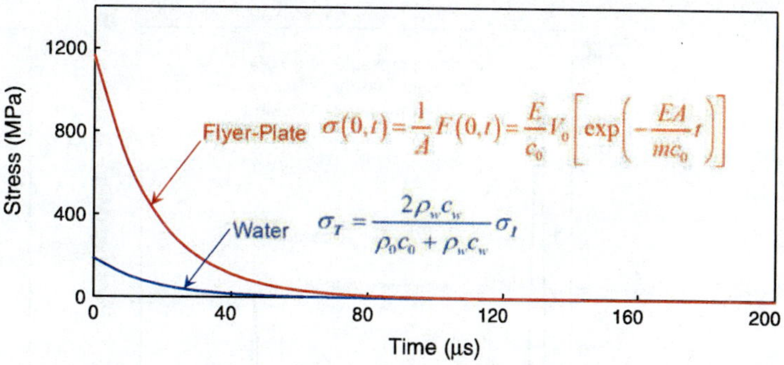


**Fig. 35.5** The stress-wave generated when the projectile strikes the flyer-plate and the pressure-wave transmitted into the water-chamber

In the USLS, depending on the projectile velocity and mass, peak-pressures ranging from 10 to 300 MPa can be generated in the shock tube. The pressure at any point in the water-chamber is given by the equation

$$p(t) = p_0 \exp\left(-\frac{t}{t_0}\right), \tag{35.3}$$

where $p_0$ is the peak pressure, $t$ is time at which measurement is carried out and $t_0$ is the decay time. The area under the curve of the pressure vs. time plot is the impulse imparted by the wave, measured in $Pa \cdot s$. Impulse $I$ is given by

$$I = \int_0^t p(t)\,dt. \tag{35.4}$$

The non-dimensional impulse $\bar{I}$ is given by

$$\bar{I} = \frac{I}{\rho_w c_w \sqrt{A}}, \tag{35.5}$$

where $I$ is impulse, calculated as the area under the curve of the pressure versus time plot, $\rho_w$ is the density of water, $c_w$ is the speed of sound water and $A$ is the area under impulsive loading.

Figure 35.6 shows the pressure histories corresponding to four different projectile velocities. Pressures are measured using dynamic pressure transducers capable of maximum pressure measurement up to 500 MPa. The rise time of the pressure-pulses is on the order of 25 μs and the decay time is on the order of 800 μs. The solid lines show experimentally measured pressure histories while the dotted lines show the numerically calculated pressure histories. The cylindrical shape of the shock tube allows a uniform pressure to be applied to the target over the area of contact. The impulse magnitudes are $\bar{I} = 0.015,\ 0.035,\ 0.055$ and $0.065$ .

## 35.4 Results

The USLS provides a unique means to generate high-intensity underwater impulsive loads under well-controlled conditions. Average strain rates achieved by the USLS are $10^4\ s^{-1}$. Dynamic deformations and failure modes are captured by in-situ high-speed digital imaging, with the aim to characterize the fundamental mechanisms of damage and failure. Attention is focused on the earliest stages of deformation because load-carrying capacity is highest at the earliest stages.

Figure 35.7 shows a sequence of high-speed photographs for a monolithic composite plate subjected to underwater impulsive loading. The projectile velocity is 75 m/s and the peak-pressure is 95 MPa – the impulse corresponds to $\bar{I} = 0.035$ shown in Fig. 35.6c. A thin film used for sealing the water-chamber can be seen in addition to the water coming out of the water-chamber. The deformation can be divided into two regimes – (1) flexural wave propagation and (2) overall deflection. Typically, flexural waves originate at the fluid–structure interface and travel towards the supports within ~50 μs. Overall structural deflection initiates when the back-face (in this case the entire structure) starts moving. The displacement of the back-face is tracked at the mid-plane and compared with that of other structures.

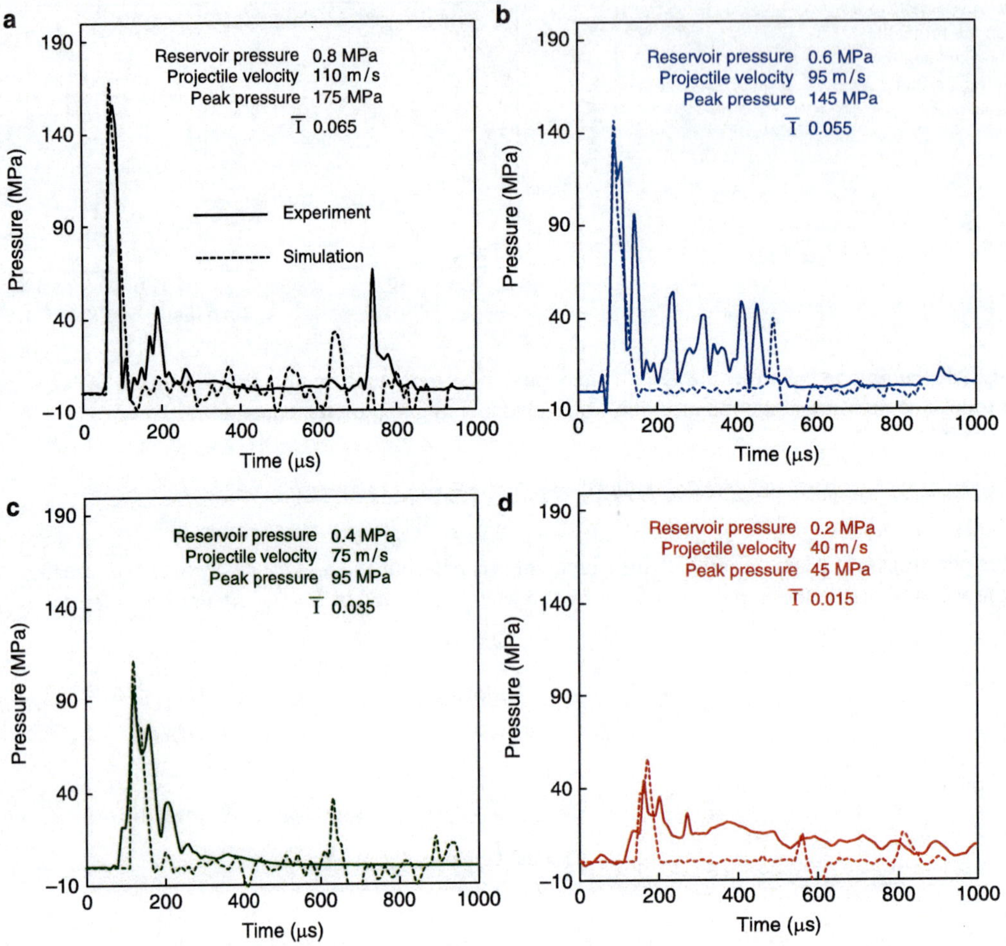


**Fig. 35.6** Comparison of numerical and experimental pressure histories in the water-chamber for four different projectile velocities and impulse magnitudes $\bar{I} = 0.015,\ 0.035,\ 0.055$ and $0.065$

While the deformation and damage in sandwich structures can be tracked using high-speed digital imaging, the monolithic composite plate is quite thin and damage mechanisms are hard to discern. These damage mechanisms are revealed in post-mortem photographs of the monolithic composite plate – shown in Fig. 35.8a–c. Figure 35.8a shows the deformed monolithic plate with easily identifiable "hinges" formed near the loading area; Fig. 35.8b, c show delamination between successive layers in the laminate, matrix-cracking, fiber-pullout and rupture. While the deflection is relatively uniform over the length of the composite plate, damage mechanisms are predominantly observed near the circumference of the shock-tube and near the supports. This indicates a significant role of shear stresses in damage creation.

Figure 35.9 shows a sequence of high-speed photographs of a composite sandwich structure with a Divinycell HP200 core subjected to underwater impulsive loading. Immediately after the onset of deformation, flexural waves travel through the front-face, severing the core-facesheet bond. If the core-facesheet bond is very strong, a layer of core material is torn away by the facesheet due to the low tensile strength of PVC core material. Core-face debonding and core-failure, due to cracking and fragmentation, is observed at $t = 50\ \mu s$. Core-crushing, a mechanical property that makes composite sandwich structures very attractive for marine applications commences at $t = 150\ \mu s$ and is restricted to the region close to the loading area. The back-face ruptures at $t = 900\ \mu s$.

Figure 35.10 shows a sequence of high-speed photographs of a composite sandwich structure with a Divinycell HP100 core subjected to underwater impulsive loading. Core-facesheet debonding and composite face wrinkling failure and core-indentation can be seen at $t = 150\ \mu s$. In this type of failure mechanism, the core-material fails in a highly localized region and causes compressive loading and buckling in the front-face. Shear-dominated cracks originate near the supports at $t = 300\ \mu s$ and lead to core fracture. Back-face ruptures at $t = 900\ \mu s$.

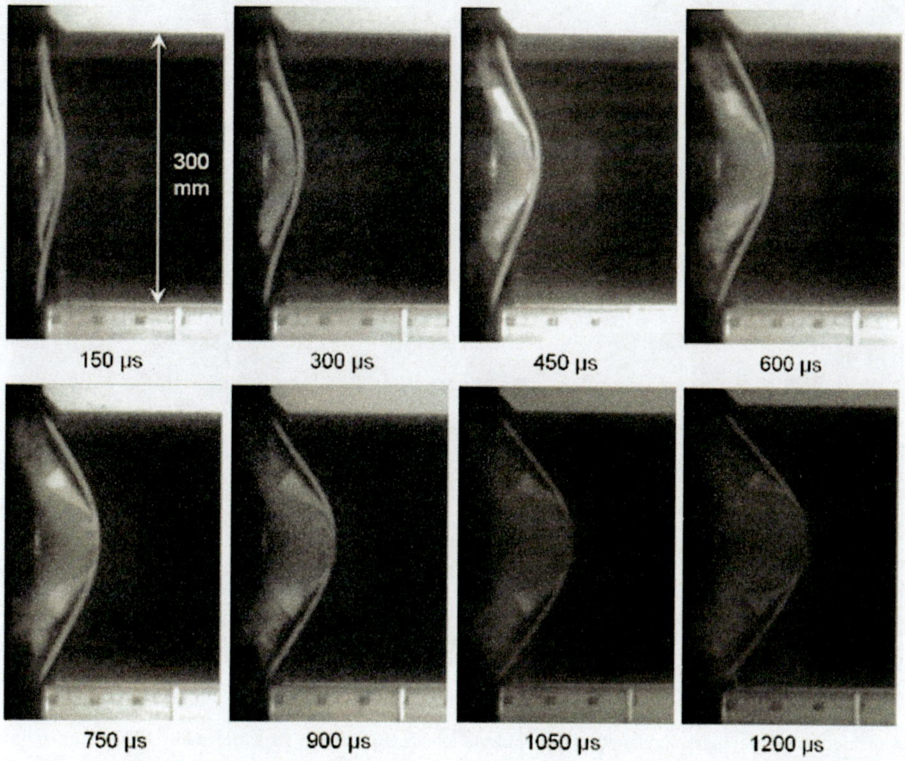


**Fig. 35.7** Sequence of high-speed photographs showing the deformation in a monolithic composite plate subjected to underwater impulsive loading with $\bar{I} = 0.035$

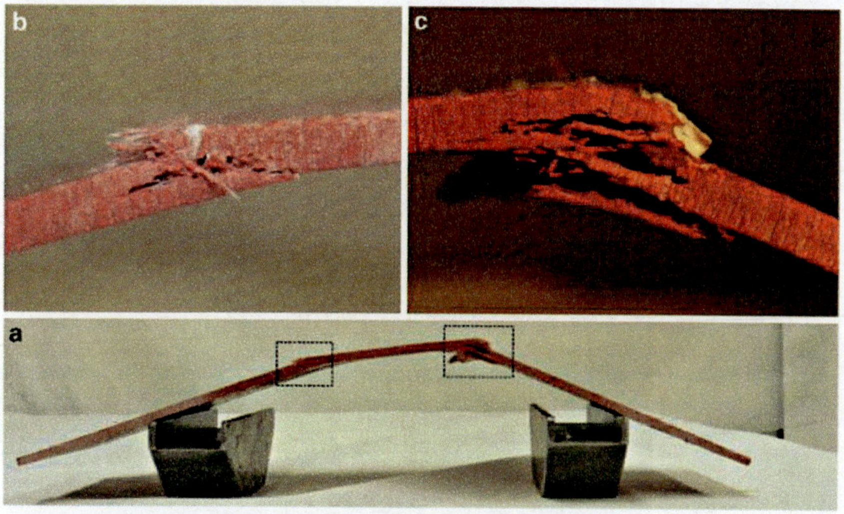


**Fig. 35.8** Post-mortem photographs of impulsively loaded composite plates with cross-sections showing inter-laminar delamination, matrix-cracking, fiber-matrix debonding, fiber-pullout and intra-laminar cracking

Figure 35.11 shows a sequence of high-speed photographs of a composite sandwich structure with a Divinycell HP60 core subjected to underwater impulsive loading. The dynamic response of the sandwich structure with a HP60 core is quite different than those with HP100 or HP200 cores – in that there is no core-shear cracking, front-face-wrinkling and core-face debonding. Core-compression commences immediately after the onset of loading at $t = 150$ $\mu s$ and inclined cracks originate near the loading circumference area. These cracks propagate from the front-face to the back-face and branch into three

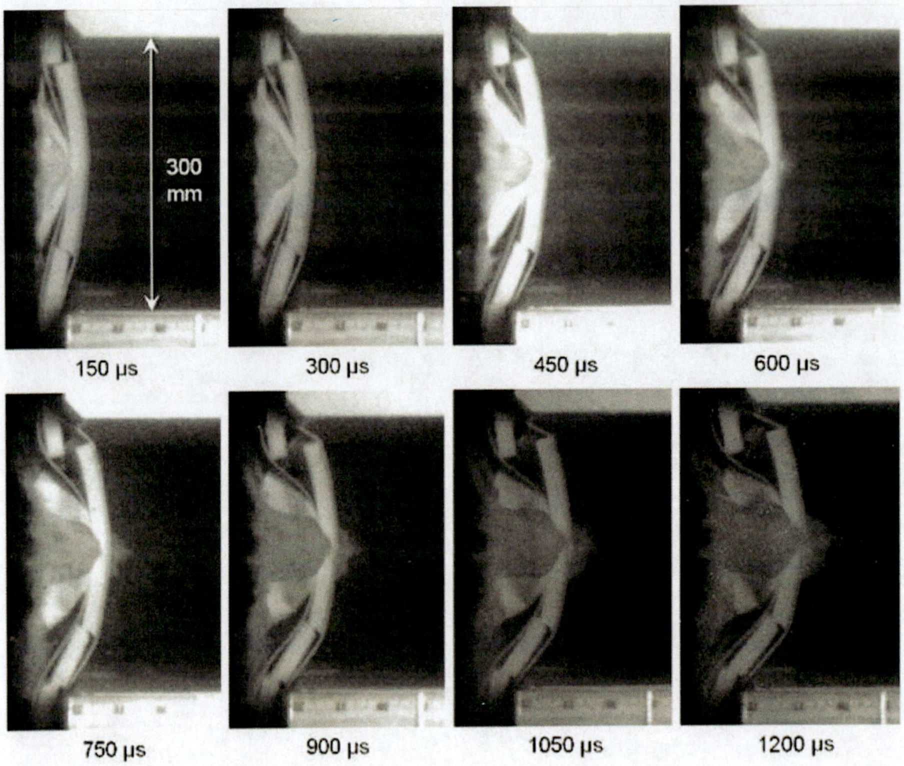


**Fig. 35.9** Sequence of high-speed photographs showing the deformation in a sandwich structure with HP200 core subjected to underwater impulsive loading with $\bar{I} = 0.035$ . Large-scale core-front-face debonding and core-fragmentation can be observed. The core fractures prior to core-compression and rupture occurs at $t = 900$ $\mu s$

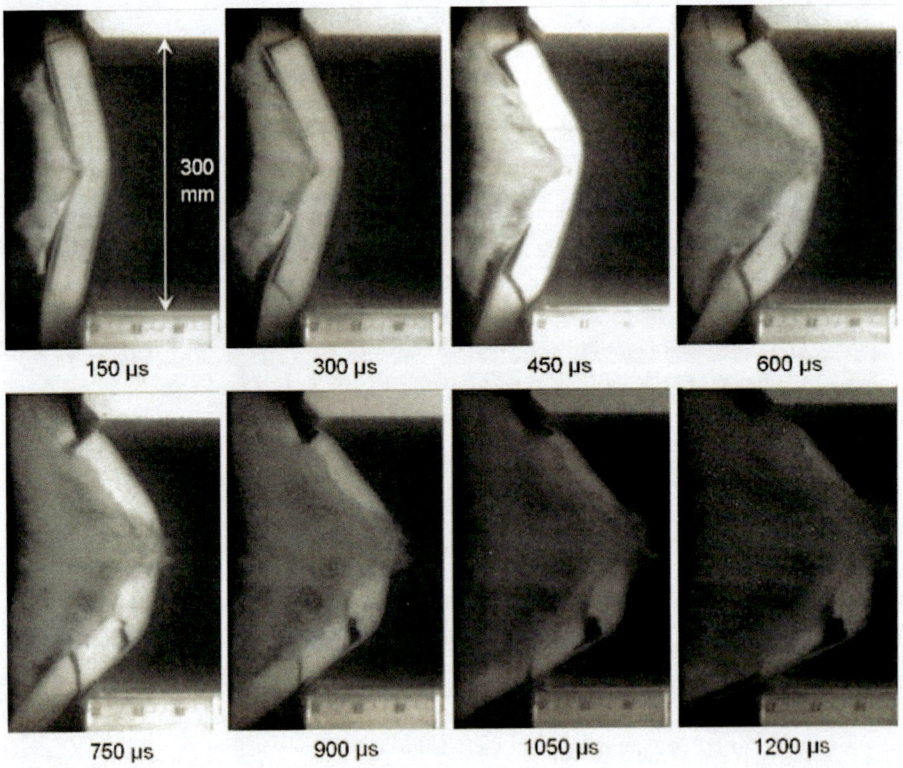


**Fig. 35.10** Sequence of high-speed photographs showing the deformation in a sandwich composite with HP100 core subjected to underwater impulsive loading with $\bar{I} = 0.035$. Front-face wrinkling and core-indentation occurs at $t = 300$ $\mu s$. Inclined cracks initiated at $t = 600$ $\mu s$ followed by rupture at $t = 900$ $\mu s$

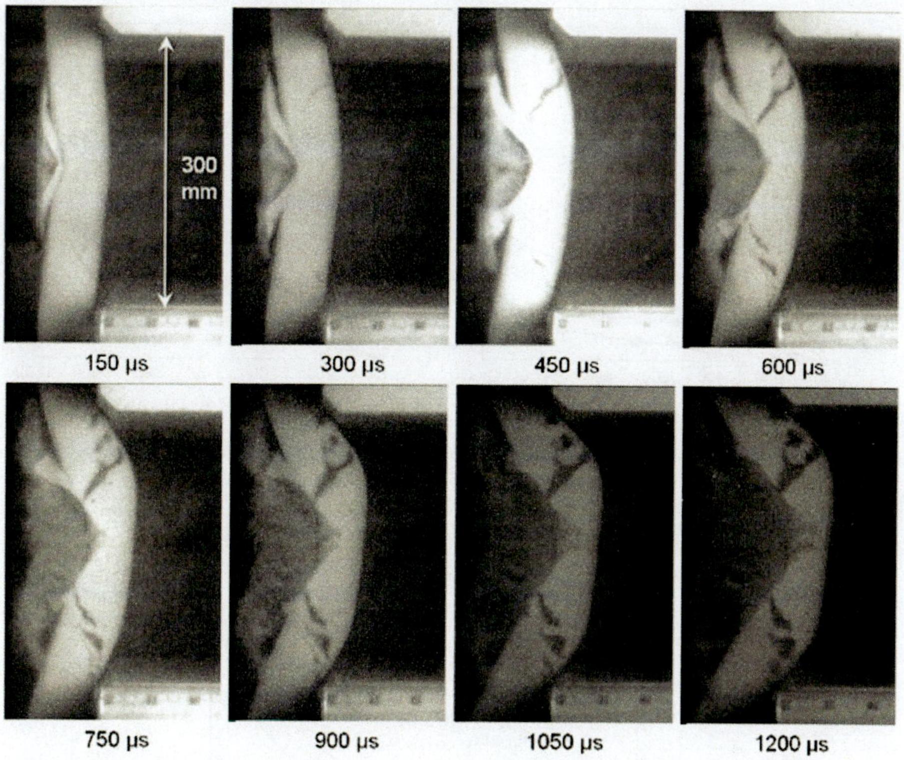


**Fig. 35.11** Sequence of high-speed photographs showing the deformation in a sandwich composite with HP60 core subjected to underwater impulsive loading with $\bar{I} = 0.035$. Deformation in the core is quite uniform and bending deformation occurs prior to core-cracking. Core-face debonding is relatively less widespread and facesheet wrinkling does not occur. Core-crushing occurs simultaneously with core-tensile cracking. Rupture is not observed

cracks ( at $t = 450$ $\mu s$) near the back-face – causing core-back-face debonding. Core-compression and core-cracking occur simultaneously with crack propagation through the core. This structure does *not* experience large-scale core-face debonding or rupture of the back-face.

Comparison of the high-speed photographs of different sandwich structures subjected to identical impulsive loads reveals that core-density plays an important role in dynamic response. High-density cores undergo fracture and fragmentation prior to core-compression. These cores also experience rupture due to failure in the front and back faces. For high-density cores, failure in the front face is primarily in the form of face-wrinkling and matrix-cracking. Compared to high-density cores, low-density cores exhibit considerably different dynamic deformations. For these structures, the front-face is less susceptible to wrinkling and rupture. Core-compression occurs simultaneously with core-cracking; core-cracking is primarily due to tensile loads as a result of bending-deformation.

The time-histories of midpoint deflections for each composite structure (obtained using high-speed photographs of the side-view) are shown in Fig. 35.12. The monolithic composite structure is used as a benchmark for comparison with other structures. The lesser the deflection is compared to the monolithic composite plate; the better is the blast-resistance. Figure 35.12a shows the normalized deflection, $\Delta/L$ (where $\Delta$ is deflection and $L$ is the span of the sandwich beam), of composite structures to an underwater impulsive load corresponding to $\bar{I} = 0.035$ shown in Fig. 35.6c. Monolithic composite experiences the greatest deflection at the highest rate in comparison to the sandwich structures. The sandwich structure with HP100 core initially exhibits a rate of deformation that is identical to the sandwich structure with HP200 core but the dynamic response of HP100 core diverges at $t = 400$ $\mu s$ and the rate of deflection reduces. For the sandwich structure with HP60 core, the rate of deflection is the lowest of all three sandwich structures. In terms of overall deflection at $t = 1000$ $\mu s$, the sandwich structure with HP200 core deflects ~20% less than the monolithic composite, sandwich structure with HP100 core deflects ~60% less than the monolithic composite and the sandwich structure with HP60 core deflects ~70% less than the monolithic composite. Figure 35.12b shows the time histories of midpoint displacements of composite structures subjected to an underwater impulsive load corresponding to $\bar{I} = 0.065$ shown in Fig. 35.6a. The deflections follow a similar trend as described for Fig. 35.12a. Due the higher load intensity, the sandwich structure with HP200 core deflects

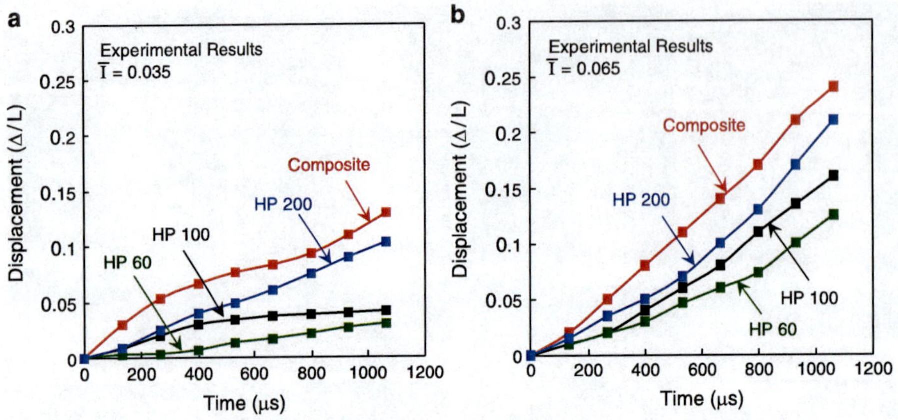


**Fig. 35.12** Experimentally measured mid-point displacements as a function of time for different structural configurations for (**a**) $\bar{I} = 0.035$ and (**b**) $\bar{I} = 0.035$

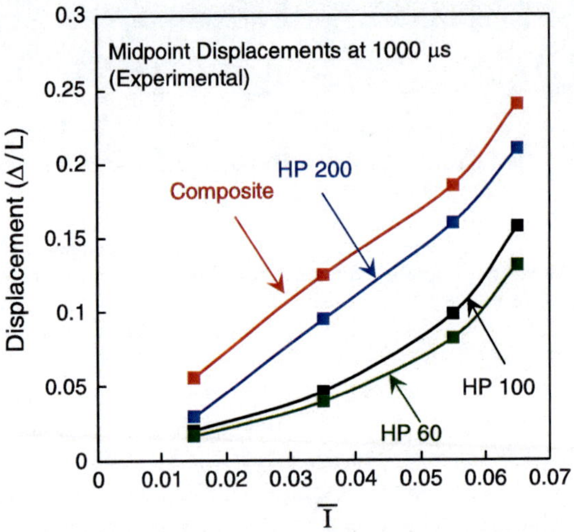


**Fig. 35.13** Displacement as a function of peak pressure for composite structures with different cores. On a per-weight basis, low-density cores consistently outperform high-density cores. Sandwich structures are superior to monolithic composite plates

~10% less than the monolithic composite, sandwich structure with HP100 core deflects ~30% less than the monolithic composite and the sandwich structure with HP60 core deflects ~45% less than the monolithic composite.

Figure 35.13 shows the normalized deflection, $\Delta/L$ (where $\Delta$ is deflection and $L$ is the span of the sandwich beam) at $t = 1000\ \mu s$ as a function of peak pressure for all the composite structures tested here. The monolithic composite shows highest deflection for all load intensities followed by the sandwich structure with HP200 core. Compared to the monolithic sandwich structure (if the $\Delta/L$ for monolithic composite is 100%), the deflection in the sandwich structure with HP200 core is ~20% smaller, that with HP100 core is ~60% smaller and that with HP60 core is ~66% smaller. The deflections experienced by sandwich structures with HP60 and HP100 cores are comparable, with the HP60 core outperforming the HP100 core by a slender margin – particularly at higher load intensities.

To obtain material-property-performance relationships, dimensional analysis is carried out. The impulse is normalized using equation (5), the core-properties are normalized by dividing the core-density by the density of the parent material (in this case PVC). The measured quantities – deflection and force are plotted as functions of normalized impulse, $\bar{I}$, and normalized density, $\rho_{core}/\rho_{PVC}$, in Fig. 35.14a, b respectively. The contour plots demonstrate that at low values of $\bar{I}$, deflection and reaction-force are both relatively insensitive to $\rho_{core}/\rho_{PVC}$. However, as $\bar{I}$ increases, high-density cores i.e. higher values of $\rho_{core}/\rho_{PVC}$ show a reduction in blast resistance. In terms of deflection and reaction-force, the performance of

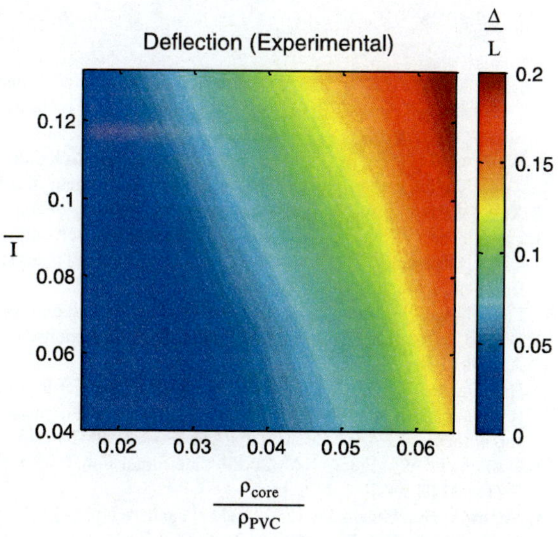


**Fig. 35.14** Material-property-performance map showing deflection ($\Delta/L$) as a function of material properties ($\rho_{core}/\rho_{PVC}$) and the load intensity ($\bar{I}$). The contours show that low-density cores undergo lesser deflection and consistently outperform high-density cores on a per-weight basis

HP60 and HP100 is closer, but HP60 slightly outperforms HP100 cores. Clearly, provided the dimensional requirements are satisfied, thick cores of low-density foams consistently provide the highest blast-mitigation to underwater impulsive loads.

## 35.5 Conclusions

An experimental characterization of the dynamic response of sandwich composites subjected to high-intensity underwater impulsive loads is presented. A new gas-gun based experimental facility is used to generate a range of impulsive loads. The deformation and failure modes in sandwich structures are studied using high-speed digital imaging. Experiments show that the core plays an important role in the dynamic response of sandwich composites. The density and stiffness of the core determines the failure mode experienced by the sandwich structure. Monolithic composites subjected to impulsive loads undergo matrix-cracking, delamination and fiber-pullout and form characteristic hinges near the loading circumference. HP200 cores undergo face-wrinkling, large-scale shear cracking and fragmentation, followed by core-crushing and ultimately rupture. HP100 cores experience core-indentation failure due in which the core experiences collapses in a small region and the front-face ruptures. This is followed by cracking and overall bending response. HP60 cores exhibit simultaneous core-compression and tensile cracking. Face-wrinkling is negligible and rupture is not observed.

Time-resolved deflections measured by high-speed photography indicate that sandwich structures are superior to monolithic structures. Low-density cores consistently outperform high-density cores, undergoing lesser deflections at all impulse magnitudes. For high magnitude impulsive loads, the sandwich structure with HP200 core deflects ~10% less than the monolithic composite, sandwich structure with HP100 core deflects ~30% less than the monolithic composite and the sandwich structure with HP60 core deflects ~45% less than the monolithic composite.

Uniaxial testing has shown that the dynamic response of polymeric foam cores exhibit mild strain-rate dependence. However, under complex multi-axial loads, the strain-rate effects are significantly higher. Strain-rate effects are much more pronounced for HP200 cores and are less dominant for HP100 and HP60 cores. The differing trends in structural response reflect the effects of different deformation and failure mechanisms in foams and fiber-reinforced composites under impulsive loads. The USLS provides a useful means with which the comparative load-carrying capacity in sandwich composites under dynamic loads can be obtained experimentally. The underlying trends in dynamic response are revealed by design maps which incorporate material properties, impulse magnitudes and deflections.

## References


1. Liang YM et al (2007) The response of metallic sandwich panels to water blast. J Appl Mech Trans ASME 74(1):81–99
2. Dharmasena KP et al (2010) Dynamic compression of metallic sandwich structures during planar impulsive loading in water. European J Mech Solids 29(1):56–67
3. Wei Z et al (2008) The resistance of metallic plates to localized impulse. J Mech Phys Solids 56(5):2074–2091
4. Xue ZY, Hutchinson JW (2003) Preliminary assessment of sandwich plates subject to blast loads. Int J Mech Sci 45(4):687–705
5. Xue ZY, Hutchinson JW (2004) A comparative study of impulse-resistant metal sandwich plates. Int J Impact Eng 30(10):1283–1305
6. Hutchinson JW, Xue ZY (2005) Metal sandwich plates optimized for pressure impulses. Int J Mech Sci 47(4–5):545–569
7. Qiu X, Deshpande VS, Fleck NA (2003) Finite element analysis of the dynamic response of clamped sandwich beams subject to shock loading. European J Mech Solids 22(6):801–814
8. Fleck NA, Deshpande VS (2004) The resistance of clamped sandwich beams to shock loading. J Appl Mech Trans ASME 71(3):386–401
9. Qiu X, Deshpande VS, Fleck NA (2004) Dynamic response of a clamped circular sandwich plate subject to shock loading. J Appl Mech Trans ASME 71(5):637–645
10. Hutchinson JW (2009) Energy and momentum transfer in air shocks. J Appl Mech Trans ASME 76(5):1–7
11. Kambouchev N, Radovitzky R, Noels L (2007) Fluid–structure interaction effects in the dynamic response of free-standing plates to uniform shock loading. J Appl Mech Trans ASME 74(5):1042–1045
12. Deshpande VS, Fleck NA (2005) One-dimensional response of sandwich plates to underwater shock loading. J Mech Phys Solids 53(11):2347–2383
13. Vaziri A, Hutchinson JW (2007) Metal sandwich plates subject to intense air shocks. Int J Solids Struct 44(6):2021–2035
14. Espinosa HD, Lee S, Moldovan N (2006) A novel fluid structure interaction experiment to investigate deformation of structural elements subjected to impulsive loading. Exp Mech 46(6):805–824
15. McShane GJ et al (2008) Dynamic rupture of polymer-metal bilayer plates. Int J Solids Struct 45(16):4407–4426
16. Dharmasena KP et al (2008) Mechanical response of metallic honeycomb sandwich panel structures to high-intensity dynamic loading. Int J Impact Eng 35(9):1063–1074
17. Tekalur SA, Bogdanovich AE, Shukla A (2009) Shock loading response of sandwich panels with 3-D woven E-glass composite skins and stitched foam core. Compos Sci Technol 69(6):736–753
18. Tekalur SA, Shukla A, Shivakumar K (2008) Blast resistance of polyurea based layered composite materials. Compos Struct 84(3):271–281
19. LeBlanc J et al (2007) Shock loading of three-dimensional woven composite materials. Compos Struct 79(3):344–355
20. Grogan J et al (2007) Ballistic resistance of 2D and 3D woven sandwich composites. J Sandwich Struct Mater 9(3):283–302
21. Wang EH, Gardner N, Shukla A (2009) The blast resistance of sandwich composites with stepwise graded cores. Int J Solids Struct 46 (18–19):3492–3502
22. DIAB Inc., S.D., DeSoto, Texas 75115, USA. http://www.diabgroup.com/europe/literature/e_pdf_files/man_pdf/H_man.pdf. Accessed 5 May 2011
23. Tagarielli VL, Deshpande VS, Fleck NA (2008) The high strain rate response of PVC foams and end-grain balsa wood. Compos Part B Eng 39(1):83–91

# Chapter 36
# Failure Characterization of AA 2024 T3 Panel Subjected to Close-In Blast

**Tomoko Sano, C. Fountzoulas, C.F. Yen, C. Chen, and M. Nansteel**

**Abstract** Aluminum alloys are the prime metallic materials used for light weight industrial applications. AA 2024 T3 series alloys have complex microstructure, fatigue resistance, and exhibit good strength to weight ratio. To increase safety of the passenger and cargo transportation systems, characterization of the damage in AA 2024 T3 alloys subjected to close-in blast is being conducted. In this research, two panels, one that failed and the other that did not, under the same close-in blast conditions are compared. The effect of the explosion on the microstructure and the microhardness of AA 2024 as a function of the distance away from the center of explosion are studied by scanning electron microscopy, x-ray diffraction, energy dispersive spectroscopy, and Knoop microhardness. It was found that no new phases were created, but the damage wave subjected the panel at the point of failure to tensile stresses. The results of this research will be discussed in detail.

**Keywords** AA 2024 • Explosive Loading • Failure

## 36.1 Introduction

An air blast from an explosive charge can cause significant damage to buildings and other structures. The damage is caused by the sudden release of energy, which creates a shock wave that applies a blast load on the structure [1].

The objective of the analysis is to study possible microstructural and stoichiometric variation of the AA 2024 T3 alloy with an emphasis at the center of the explosion to determine critical aspects of panel failure. The topographical observation of the surface of the aluminum alloy was conducted for detection of immediately recognizable microstructural and chemical differences. Detailed characterization was conducted with the use of scanning electron microscopy (SEM), energy dispersive spectroscopy (EDS), X-ray diffraction (XRD) and Knoop hardness testing.

## 36.2 Experimental

Two panels were characterized for this study after the close-in blast tests. Both panels were subjected to one unit charge of C4 explosive. The panels were 0.18 cm thick, 121.9 cm square, and clamped over a 7.62 cm wide area at the boundary. The closest approach distance from the nearest edge of the charge to the center of the panel surface was 7.12 cm

T. Sano (✉) • C. Fountzoulas • C.F. Yen
U.S. Army Research Laboratory, RDRL-WMM-B, 4600 Deer Creek Loop, Aberdeen Proving Ground, MD 21005, USA
e-mail: tomoko.sano.civ@mail.mil

C. Chen
U.S. Department of Homeland Security, Science and Technology Directorate, Transportation Security Laboratory, William J. Hughes Technical Center, Atlantic City International Airport, NJ 08405, USA

M. Nansteel
Battelle Memorial Institute, Transportation Security Laboratory, William J. Hughes Technical Center, Atlantic City International Airport, NJ 08405, USA

V. Chalivendra et al. (eds.), *Dynamic Behavior of Materials, Volume 1: Proceedings of the 2012 Annual Conference on Experimental and Applied Mechanics*, Conference Proceedings of the Society for Experimental Mechanics Series, DOI 10.1007/978-1-4614-4238-7_36, 

**Fig. 36.1** Sample from the ruptured panel. The *black* edge is where the rupture initiated

**Fig. 36.2** Triangular sample examined from the not-ruptured panel, in relation to the blast area

for both panels. The first panel failed (blast wave penetrated through) but the second panel did not. The first panel will be referred to as the "ruptured" panel and the second panel as the "not-ruptured" panel.

From the ruptured panel, a section of the center of the blast was obtained. Figure 36.1 shows an image of this sample, with the edge where the rupture initiated highlighted in black. For the not-ruptured panel, a triangular sample (Fig. 36.2) was removed from the center of the panel, similar in location to the ruptured panel sample, for SEM, EDS, and XRD analysis. SEM and EDS analyses were conducted by using the FEI Nova NanoSEM600 microscope and EDAX Genesis, respectively. XRD analysis was conducted by using a Rigaku MiniFlex II Desktop X-ray Diffractometer. The x-ray was a CuK$\alpha$1 radiation of wavelength $\lambda = 0.15406$ nm, at 2.0°/min scan speed and 0.02° sampling width. The hardness of the alloy was measured by using a Wilson Rockwell, Wilson Instruments, hardness tester.

## 36.3 Results

### 36.3.1 SEM and EDS Analysis

Evaluation by SEM of the surface of the ruptured sample showed no significant change in surface deformation away from the center of the blast. Figure 36.3a shows the heavily damaged region in ruptured panel sample. Microcracks were observed in all regions of the ruptured sample. For the sample from the not-ruptured panel, a gradual decrease in surface deformation was observed away from the explosion. As with the ruptured panel, the not-ruptured panel showed similar surface features close to the blast center, and cracks were observed throughout the sample (see Fig. 36.3b).

EDS analysis of sample surfaces from the ruptured panel and not-ruptured panel were conducted. No significant difference was observed between the two panels. The observed sample chemistries, though not always consistently, were Al, Cu, Mg, Zn, Mn, Si, Fe, Ti, and Cr, as shown in Fig. 36.4. This is in agreement with the typical AA 2024 composition [2]. But the current analysis also showed traces of S, Na, K, N, O, Ag, and C on the surface. It is very possible that these are residual elements that became situated on the sample surfaces from the blast.

### 36.3.2 X-ray Diffraction Analysis

The effect of the blast on the microstructure of ruptured AA 2024 T3 panels was studied by x-ray diffraction (XRD). A summary of the XRD results, comparing the reference spectrum to that of the ruptured panel and the not ruptured panel is shown in Fig. 36.5. The XRD spectra of the ruptured and not-ruptured panels were taken at areas close to the center of the blast. Both spectra displayed the typical spectrum of aluminum and showed that the explosion did not create any new phases. However, shifts in the XRD peaks with respect to the reference XRD peak were observed, indicating local deformation.

Table 36.1 shows the interplanar spacing, the variation of the lattice parameter of the ruptured sample compared to those of the reference sample, and the strain. Table 36.2 shows the same sets of data for the not–ruptured sample. The strain across the panel was computed as follows:

$$\bar{\varepsilon} = \frac{\sum \varepsilon_{hkl}}{5} \tag{36.1}$$

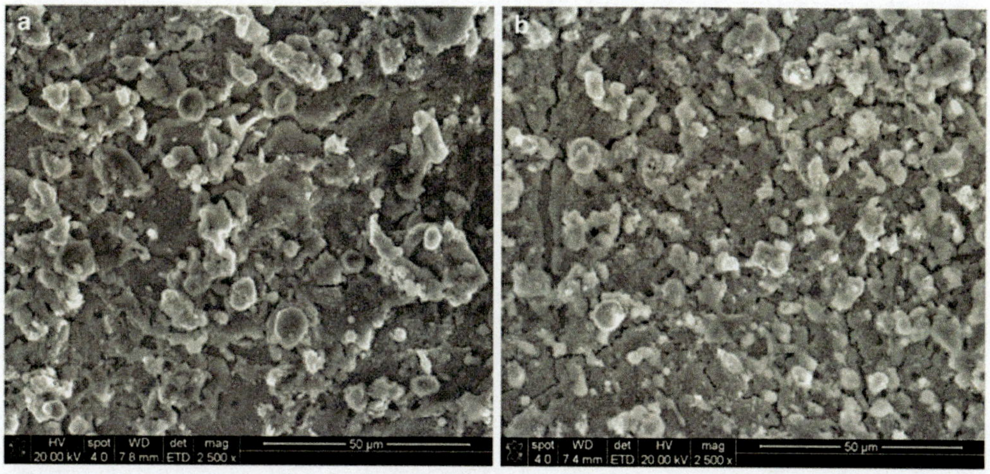


**Fig. 36.3** Damaged surfaces from (a) the blast center from the ruptured panel, (b) area by the blast center from the not-ruptured panel. Both images were captured at the same SEM conditions and magnifications

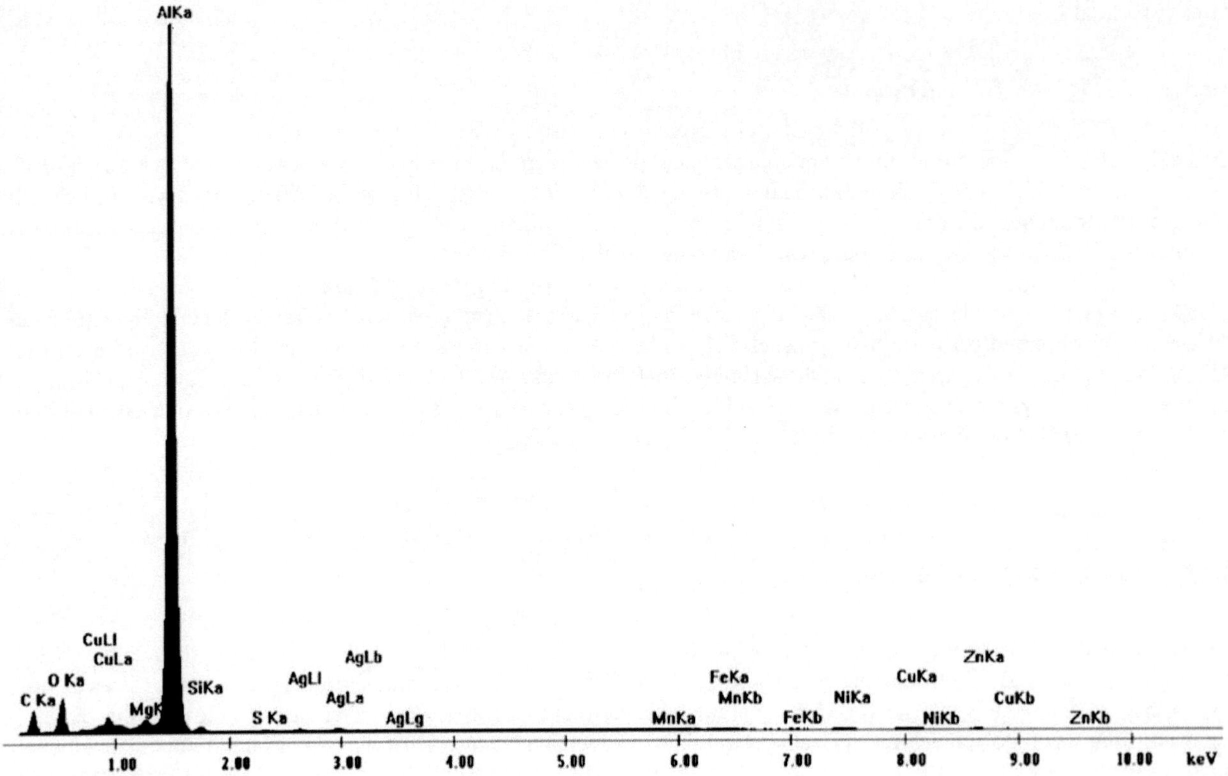


**Fig. 36.4** An example EDS spectra

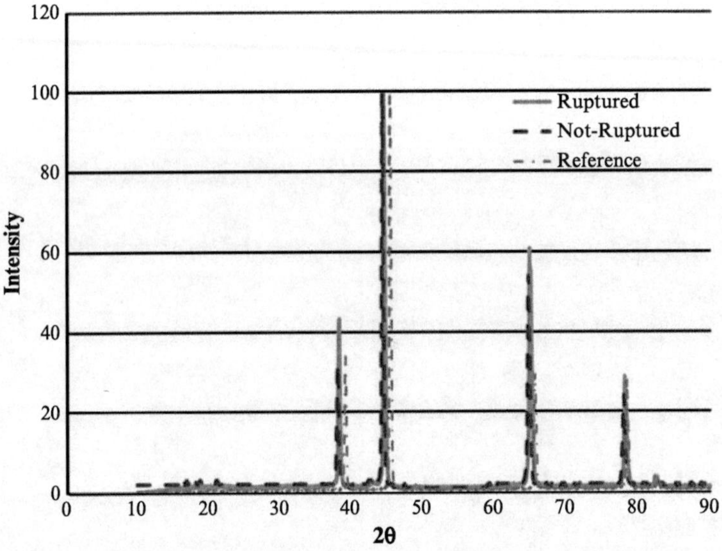


**Fig. 36.5** Comparison of the XRD spectra between the ruptured sample, not-ruptured sample, and the reference

**Table 36.1** Interplanar spacing variation across the ruptured panel

| Plane (hkl) | Interplanar spacing $d_{hkl}$ (Å) | Reference interplanar spacing $d_{hkl}$ (Å) | Lattice parameter (Å) | Strain (% $\varepsilon_{hkl}$) |
|---|---|---|---|---|
| 111 | 2.34 | 2.29 | 4.06 | 2.18 |
| 200 | 2.03 | 1.99 | 4.06 | 2.01 |
| 220 | 1.43 | 1.42 | 4.06 | 0.99 |
| 311 | 1.22 | 1.22 | 4.06 | 0.00 |
| 222 | 1.17 | 1.17 | 4.06 | 0.00 |
| | | | Ave = 4.06 | Ave =1.04 |

**Table 36.2** Interplanar spacing variation across the not-ruptured panel

| Plane (hkl) | Interplanar spacing $d_{hkl}$ (Å) | Reference interplanar spacing $d_{hkl}$ (Å) | Lattice parameter (Å) | Strain (% $\varepsilon_{hkl}$) |
|---|---|---|---|---|
| 111 | 2.35 | 2.29 | 4.08 | 2.62 |
| 200 | 2.04 | 1.99 | 4.08 | 2.51 |
| 220 | 1.44 | 1.42 | 4.06 | 1.41 |
| 311 | 1.22 | 1.22 | 4.06 | 0.00 |
| 222 | 1.14 | 1.17 | 3.94 | −2.56 |
| | | | Ave = 4.04 | Ave = 0.80 |

$$\varepsilon_{hkl} = \frac{d_{hkl} - d_{ref\_hkl}}{d_{ref\_hkl}} x100 \tag{36.2}$$

where:
$\bar{\varepsilon}$ is the average strain
$\varepsilon_{hkl}$ is the stain in the (hkl) interplanar spacing
$d_{ref\_hkl}$ is the (hkl) interplanar spacing of the reference spectrum and
$d_{hkl}$ is the is the (hkl) interplanar spacing

As indicated in Tables 36.1 and 36.2, the average lattice parameter for the ruptured sample was 4.06 Å and for the not-ruptured sample, 4.04 Å. The average strain around the blast center, computed using Eq. 36.1 with the reference XRD interplanar spacing, was 1.04% for the ruptured sample and was 0.80% for the not-ruptured sample. The strain of the not-ruptured sample was 23% less than the corresponding strain of 1.04% of the ruptured sample, indicating that on average, smaller tensile stresses were experienced in the not-ruptured panel. In a prior investigation [3] with different Al 2024 panels, the ruptured panel showed a strain variation in which the tensile stress changed to a compressive stress at the point of rupture, then back to tensile stress away from the center of the blast, while the not-ruptured sample from the prior report did not show a similar change in the stress state. Since only one XRD pattern was obtained for each sample, the stress variations could not bre observed for the panels reported in this paper.

### *36.3.3 Ultimate Tensile Strength (UTS) Measurement*

The UTS of both ruptured and not-ruptured AA 2024 T3 panels was determined by converting the measured Rockwell Hardness B values across the panels into strength. Figure 36.6a shows a summary of the average UTS of the ruptured panel, split into 4 areas away from the blast failure edge, and Fig. 36.6b shows that of the not-ruptured panel, including the semicircular areas outside the triangle. (No data points were taken for Area 2 for the not-ruptured panel.) Higher UTS values were obtained closer to the edge where the ruptured initiated. Both samples saw reduction in the UTS away from the blast center.

Compared to the average UTS of 415 MPa of the non deformed reference sample, the UTS of both ruptured and not-ruptured panels increased significantly. These values agree well with the published values of the AA 2024 T3 alloy [4]. The 19% UTS increase for the ruptured panel and 13% increase for the not-ruptured panel, compared to the not deformed

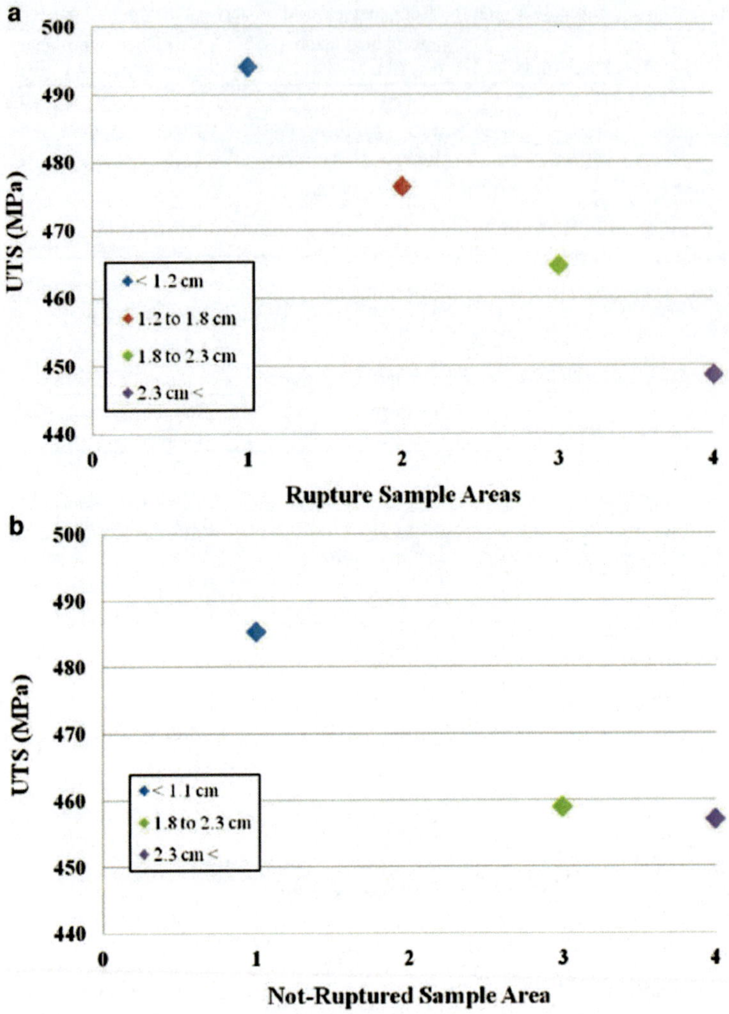


**Fig. 36.6** (**a**). The average UTS across the rupture sample away from the ruptured edge and (**b**) the average UTS across the not-ruptured sample

reference sample, is attributed to the strain hardening of the alloy due to the explosion. The variation in the UTS values and the reason one panel ruptured while the other did not, could be due to the location of the convergence of the compression then reflected tension waves in the panels created by the blast, and slight differences due to experimental variability. A definitive reason for the performance differences was not determined at this time.

## 36.4 Conclusions

Two AA 2024 panels subjected to close-in blast loading, one ruptured and the other not-ruptured, were characterized for possible microstructural and strain variation as a function of the distance from the center of the explosion. Neither significant microstructural nor chemical differences were detected. XRD results showed a tensile strain variation outwards from the center of the blast. At this stage of the study, the difference in the performance of the plates appear to be due to experimental variation causing strain hardening from the blast, but additional investigation by numerical analysis of the blast is needed to map the developed stresses on the panel.

## References

1. Kinney GF (1962) Explosive shocks in air. The Macmillan Company, New York, NY p 117
2. The Aluminum Association (2009), International alloy designations and chemical composition limits for wright aluminum and wrought aluminum alloys. www.aluminum.org
3. Sano T, Fountzoulas C, Yen C–F, Chen C, Nansteel M,(2011) Proceedings of the materials science & technology 2011 conference & exhibition, 2011 p 902
4. Cullity WD (2000) Materials science and engineering. An introduction, 5th edn. Wiley, Reading, MA p 798

## References

# Chapter 37
# Shock Waves in a Model Particulate Composite

**Michael Rauls and Guruswami Ravichandran**

**Abstract** Propagation of shock waves in heterogeneous media is examined. A model particulate composite is proposed, consisting of silica glass spheres (100–200 μm diameter) embedded in a poly methyl methacrylate (PMMA) matrix. Various volume fractions and particle sizes will be examined to determine and optimum configuration for dissipation of energy as the shock passes through the sample thickness. Composite fabrication techniques are discussed and an experimental and modeling approach is proposed.

**Keywords** Shock waves • Particulate composite • Wave scattering

Understanding the propagation of shock waves in heterogeneous materials is of considerable interest in many applications such as impact and blast [1]. The structure of shock waves in heterogeneous solids is related to the solid's microstructure and material properties. Results from experiments conducted on a model particulate composite consisting of glass spheres in a Polymethyl Methacrylate (PMMA) matrix are presented. The composite is fabricated by a compression molding process in which PMMA powder is plasticized to flow and envelop the distributed glass micro spheres. The specimens are subjected to normal plate impact using a powder gun at velocities in the range of 1–2 km/s. The particle velocity at the rear surface is measured using a VISAR system. The measured velocity profiles are examined in relation to the microstructure of the particulate composite. The effect of material parameters on the propagation of shock waves in heterogeneous materials is discussed.

Repeatable fabrication of sample specimens is critical to the investigation of heterogeneous material under shock loading. The large density mismatch between silica glass and PMMA presents issues for entrainment methods that rely on polymerizing Methyl Methacrylate (MMA). Silica glass rapidly sinks out of the mixture while it is fluid and accumulates on the bottom of the mold during curing. A new method is proposed where medium molecular weight PMMA powder is mixed with the appropriate weight of glass micro-spheres to achieve the desired volume fraction on particulate in the composite. The mixed powders are then transferred to a mold where heat and pressure are applied for 10–12 h.

A composite sample fabricated using this method is shown in Fig. 37.1 with 100–200 μm silica glass beads distributed in a 22 % volume fraction. It is clear from the above that the beads remain uniformly distributed after the compression molding process. The horizontal striations are gouges from a diamond saw, while the other pits are from 1,200 grit sand paper. Higher volume fraction mixtures may exhibit clumping of glass beads or limited flow of the matrix plastic. Further processing and sample preparation is required to achieve higher volume fractions.

The plate impact experiments performed on the previously described samples (nominal thickness, 6 mm) will be used to conduct shock wave studies on the sample. The composite samples will be highly polished or have a reflective coating applied to the rear surface at the center in order to obtain the free surface velocity, $u_p$, with a velocity interferometer set up as described below.

An impactor of a previously characterized homogeneous material, such as PMMA or aluminum, impacts a target specimen with a known initial microstructure (Fig. 37.2 a). The impact time on the target is recorded with shorting pins, while the shock speed in and particle velocity profile of the specimen is measured with a VISAR system. When a steady shock front is assumed, the Lagrangian shock speed can be calculated as $U_s = h/\Delta t$.

M. Rauls • G. Ravichandran (✉)
Division of Engineering and Applied Science, California Institute of Technology, Pasadena, CA 91125, USA
e-mail: ravi@caltech.edu

V. Chalivendra et al. (eds.), *Dynamic Behavior of Materials, Volume 1: Proceedings of the 2012 Annual Conference on Experimental and Applied Mechanics*, Conference Proceedings of the Society for Experimental Mechanics Series,
DOI 10.1007/978-1-4614-4238-7_37, 

**Fig. 37.1** Microstructure of a particulate composite of micro glass spheres in a PMMA Matrix, 5× magnification

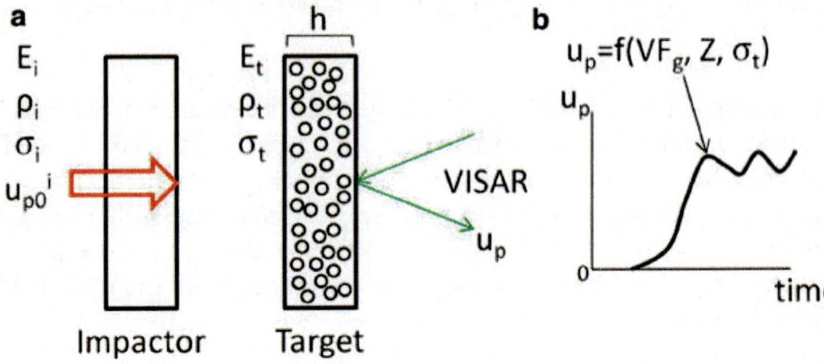


**Fig. 37.2** (**a**) Plate impact experiment arrangement, (**b**) Surface velocity profile

The rear surface particle velocity is however expected to be more complex. The impedance mismatch between PMMA and glass $R = Z_g/Z_p = 5.3$ creates scattering within the composite [1]. It was found in Ref. [2] that the shape of the measured shock wave depends mostly on changes to the binder material, indicating that impedance ratio is as influential a parameter as specimen geometry. The interaction of the various shock fronts and wave trains creates a velocity profile with a finite rise time related to the impedance mismatch, stress in the shocked state, and volume fraction of glass. Higher impedance mismatch and mid-range volume fractions of glass contribute to increased rise times and lower particle acceleration [3] in a layered composite. It is the goal of this work to characterize and model the previously mentioned parameters effects on the rise time, as well as average stress in the shocked state. The assumptions of interpreting the shock wave response of materials using the steady state Hugoniot will be critically examined.

**Acknowledgment** The support of the Air Force Office of Scientific Research through the University Center of Excellence: High-Rate Deformation Physics of Heterogeneous Materials at the California Institute of Technology is gratefully acknowledged.

## References

1. Chen X, Chandra N (2004) The effect of heterogeneity on plane wave propagation through layered composites. Compos Sci Technol 64:1477
2. Jordan JL, Herbold EB, Sutherland G, Fraser A, Borg J, Richards DW (2011) Shock equation of state of multi-constituent epoxy-metal particulate composites. J Appl Phys 109:013531
3. Zhuang S, Ravichandran G, Grady DE (2003) An experimental investigation of shock wave propagation in periodically layered composites. J Mech Phys Solids 51:245

# Chapter 38
# Performance Evaluation of a Composite Armor Concept by Finite Element Simulations of Ballistic Tests

**Advait Bhat, Balaji Jayakumar, and Jay C. Hanan**

**Abstract** A Hybrid-Composite-Armor (HCA) insert concept for a level III NIJ 0101.06 ballistic vest standard was successfully tested using 7.62 FMJ Lead core projectiles. Improvement in ballistic performance was observed for the HCA in terms of 24% reduction of Back Face Signature (BFS) by incorporating a cellular honeycomb layer between UHMWPE laminates. This improvement is in comparison to monolith UHMWPE laminates of the same areal density as the HCA. Observations made through BFS measurements and damage zone estimation confirms the superior performance of the HCA by inclusion of the honeycomb layer. Numerical simulations in finite element code ABAQUS were carried out to verify the improvement in ballistic performance by inclusion of a honeycomb layer. Constitutive material models for strength and damage for the projectile, face-plates, and honeycomb layer were implemented and ballistic impact simulations resulting change in velocity plots were used to compare performance. Volumetric stress distributions plots were used to estimate stress transfer and damage characteristics in the backing layer. Comparisons were made for with-honeycomb, spaced plates, and non-spaced plate configurations. Simulations were capable of predicting the ballistic limit for these cases and at least showing the relative value of design changes for HCA concepts that can be used for further improvements.

**Keywords** AA 2024 • Explosive Loading • Failure

## 38.1 Introduction

Personnel body armor continues to improve in terms of performance and endurance through the use of newer designs and materials. Multilayered composite armor systems (Hybrid Composites) are at the forefront of this technology with ceramics and high strength fiber composites being their integral components [1]. The choice of materials suitable for armor relies predominantly on factors like weight reduction, multi-hit capability, impact velocity, and maximum allowable rear face bulge deformation or Back Face Signature (BFS) for the expected projectile threat. Layering order of materials and their chosen areal densities were selected to maximize the energy absorption capacity of the armor system during each stage of the projectile impact with due consideration of wave propagation effects [2, 3]. Collapsible energy absorbers like honeycombs and foams have also been incorporated in armor systems to improve energy dissipation and attenuation of stress wave propagation [3–5]. The use of such porous materials can hence be particularly effective to reduce the kinetic energy transfer to the armor wearer and protect from Behind Armor Blunt Trauma (BABT). BABT is a clinical term for non-penetrating injuries to the torso due to BFS post projectile impact [6].

A multilayered Hybrid Composite Armor (HCA) insert was designed using UHMWPE laminates and amorphous metal based honeycombs to minimize BFS against deformable lead core 7.62 mm FMJ projectile according to the National Institute of Justice (NIJ) 0101.06 standard [7–9]. Preliminary experimental evaluation through ballistic testing indicated the HCA met the minimum BFS and projectile velocity criteria set for a NIJ level III insert. Post impact CT scanning, the tested inserts revealed a 9% reduction in damage cross section area per shot for the HCA in comparison to the UHMWPE monolith baseline. A 24% reduction in BFS compared to the baseline was also reported for the HCA. Honeycomb inclusion in the armor insert indicated an improved multi-hit capability and BABT protection. A focus of this investigation is revalidation of

A. Bhat • B. Jayakumar • J.C. Hanan (✉)
Mechanical and Aerospace Engineering, Oklahoma State University, Stillwater, OK 74078, USA
e-mail: jay.hanan@okstate.edu

V. Chalivendra et al. (eds.), *Dynamic Behavior of Materials, Volume 1: Proceedings of the 2012 Annual Conference on Experimental and Applied Mechanics*, Conference Proceedings of the Society for Experimental Mechanics Series, DOI 10.1007/978-1-4614-4238-7_38, 

the results observed from the HCA for a new prototype. Effectiveness of the honeycomb layer was investigated through 3-D finite element models developed in ABAQUS® 6.10 software code. Contribution of the honeycomb layer during ballistic impact was evaluated by comparing change in velocity plots and ballistic limits for with-honeycomb, non-honeycomb and spaced plate configurations.

## 38.2 Finite Element Simulations

### 38.2.1 *Explicit 3-D Finite Element Analysis (FEA) for Dynamic Impact Simulation*

The explicit code in ABAQUS works by direct computation of the dependent variables in terms of known quantities or provided inputs. An explicit dynamic analysis was also selected for the reason that it was computationally efficient and consistent for large models with relatively short dynamic response times, comparatively larger deformations and large number of small time increments. This analysis system was also quite flexible and allowed for fixed or automatic time incrementation depending upon the requirement. To formulate the contact between the impacting projectile and the armor plates the General Contact algorithm was used. This algorithm accounts for contact through element based surfaces that allows for material degradation. The classic erosion model in ABAQUS was used to simulate the material damage response by individual element deletion as the damage parameter $D$ reached unity. The software also allowed selection of post-damage-initiation material response and how failure evolved. Dynamic impact models for high velocity (>400 m/s) are prone to issues of excessive distortions, convergence problems, node penetrations, and warping of elements during the dynamic contact. Section controls, mass scaling, and material modeling options were used in the present study to solve these problems.

### 38.2.2 *Model Development*

The primary components of the finite element model were the constituents of the target plate and the impacting projectile. The geometry of the impacting projectiles was modeled to exactly match the volumetric attributes of the .308 WIN 7.62 mm FMJ round. 3D dynamic FEA is computationally extensive due to the large number of elements and node points involved. For faster execution of the analysis procedure and solution computation time reduction, only a quadrant of the actual volume was considered for all FEA owing to the axisymmetric nature of the problem. As the total time duration of the dynamic impact is in the order of $2 \times 10^{-4}$ s, the choice of reduced volume does not affect the solution with wave propagation issues. The time duration also allows for neglecting friction effects on contact. Solid part meshes were then created using hexagonal or hexagonal dominated elements (C3D4) and then structured and sweep mesh construction was done as required. For the cellular honeycomb, conventional shell elements (S4R) were used with wall thickness assignment through section properties. Section controls were assigned to all elements in the models to control excessive distortion. The mesh was generated in discrete biased zones (denser in the contact region, coarser at the far edge to improve result quality without severely increasing computation time). The element sizes varied across the length of the plate model. Meshed parts were created to enable mesh surface generation required for the ease of the erosion model. Necessary boundary conditions had to be evaluated and assigned to accurately model the field experiments. As only a quadrant of the entire volume was considered for analysis, the inside edge surfaces of the model were subjected to symmetry conditions along the X and Y axes. For the outer edge surfaces, their motion being completely restricted due to the test setup, a Pinned boundary condition was chosen. Initial velocity assignment to the projectile was done by regulating the predefined field values. Values for velocity change with time were procured as an output through a node set near the center of mass of the projectile. Steps for model build up are shown in Fig. 38.1.

Contribution of the honeycomb layer to the overall performance during ballistic impact was evaluated by comparing change in velocity plots and ballistic limits for with-honeycomb, spaced plate, and non-spaced plate configurations. These plots were used to calculate the ballistic limit of each configuration by comparing the impact velocities and the corresponding residual velocities post impact. The influence of spacing distance on the residual velocities was also explored. Volumetric stress distribution plots were used to identify damage zone sizes for each configuration.

For each configuration, OHFC copper bullets were considered. Lead was not chosen in the present study due to the excessive deformation problems associated with its use in high velocity impact FEA. Description of the issues faced is

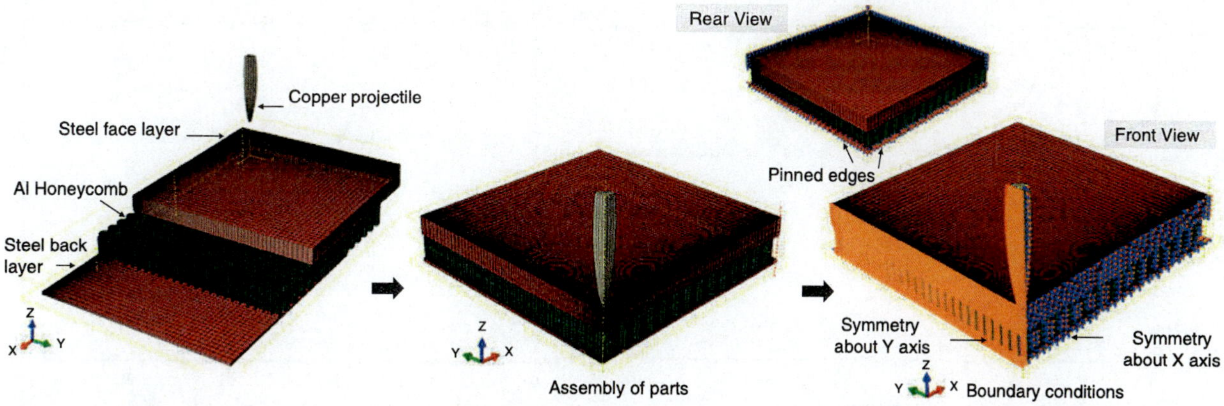


**Fig. 38.1** Steps involved in development of the FEA model are shown. Assembly of parts is followed by application of boundary conditions described in Sect. 2.2

available elsewhere [8]. Due to the relative complexity of modeling UHMWPE, steel was considered as the face sheet material. Since UHMWPE is planar isotropic, choosing steel as face sheet will not alter the dynamic response of the honeycomb core as the load will be applied uniformly along the leading edges of the honeycomb cross section. Properties for both OHFC copper and steel were taken from the ABAQUS user's manual [10]. The extended Shear and Ductile damage model was used for this steel while the Johnson-Cook plasticity and damage criterion was used for copper. Strain and strain energy based damage evolution was used for both these materials. Aluminum 2024 was considered as the base material for the corrugated honeycomb. Dynamic properties for 5052 aluminum are not available in published literature. Johnson-Cook plasticity and damage was also considered for aluminum. Material properties for Al 2024 were adopted [11].

## 38.3 Results and Discussion

General observation and the CT scan imaging had shown the fracture and damage modes observed in HCA to be identical to those reported by the scientific community so far for UHMWPE based armors [8]. After the initiation of contact, the first few UHMWPE layers failed in shear dominated failure. The projectile undergoes lateral expansion in process resulting in an increased cross section area that comes in contact with later layers of UHMWPE. This caused a change in failure mode from shear dominated failure to in-plane tensile failure and delamination due to resulting inter-laminar shear forces.

Numerical modeling to verify the contribution of the honeycomb layer during ballistic impact was performed using ABAQUS FEA. Simulations were used to compute the change in velocity during the period of impact, residual velocity after perforation of all impacted layers. A comparison of the residual velocities for with-honeycomb, spaced plate and non-spaced plate configurations was obtained through change in velocity plots received from the node set created on the projectile. Observed results are for a 0.5 mm node size along the thickness of the target plates. A change in velocity plot for the with-honeycomb configuration is shown in Fig. 38.2a. The time period for this plot corresponds to the contact time of the projectile with the components of the armor plate.

The plot indicates that the ballistic limit of the with-honeycomb configuration is 650 m/s. The plot also provides residual velocity values for each impact velocity. Similar results were computed for the spaced and the non-spaced configurations. The summary of the results is shown in Fig. 38.3. The residual velocity gives an indication of the amount of kinetic energy absorbed during the deformation process. Results show that the non-spaced configuration absorbs more energy close to the ballistic limit velocity. It has 7.6% lower residual velocity compared to the with-honeycomb and spaced configurations. However, for higher impact velocities (>750 m/s), the performance of non-spaced configuration appears to decrease by 2.3% in terms of the residual velocity.

Inclusion of spacing appears to improve the ballistic performance in this high velocity regime. To verify this hypothesis, numerical simulations with varying spacing distance were performed for the spaced plate configurations. Results are presented in Fig. 38.2b. An 11% decrease in residual velocity was observed for a 4X increase in the spacing. The trends observed in all the change in velocity plots shown in Fig. 38.2 shows no specific discretization in slope of the curve based on resistance to perforation. This is due to the high magnitude of impact velocities considered in this study (>640 m/s) and

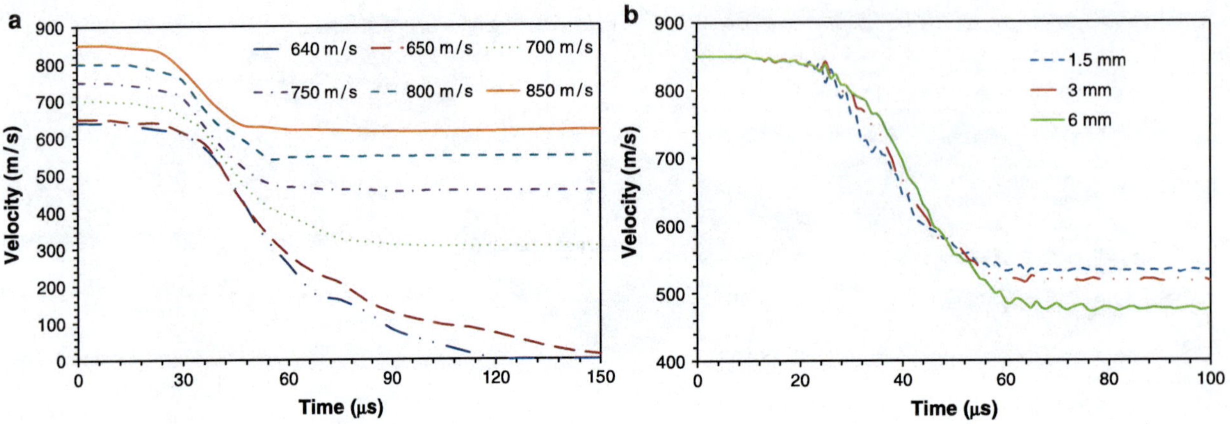


**Fig. 38.2** (**a**) Change in velocity plot numerically obtained for the with-honeycomb configuration; (**b**) Change in velocity plot for increasing plate spacing

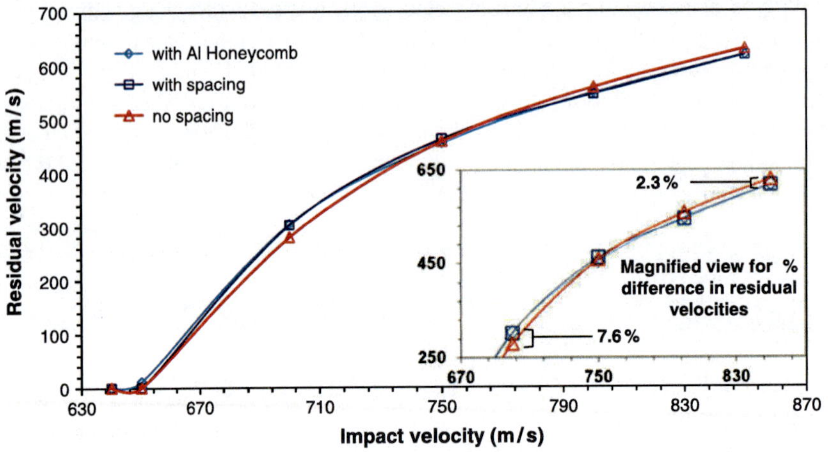


**Fig. 38.3** Ballistic limit predictions from the change in velocity plots for all the three configurations

smaller plate thicknesses; resulting in minimized interaction time between plate and projectile. Typically, impact at low velocities for thin face sheet high thickness honeycombs sandwich plate would result in three discrete zones in the change in velocity plots.

Experimental testing provides limited information concerning damage evolution and the process of projectile penetration during the impact. The current finite element simulations are able to provide information on these aspects through 3-D visualization of volumetric stress field distributions. This proves an effective way to compare the size of the damaged area for all individual components of the target plate.

The stress distribution versus time history for the backing layer is an indicator of the communication of stresses through the thickness of the target plate. Delayed stress transfer and minimized damage zones enable minimizing the BFS as well as the corresponding BABT. Figure 38.4 shows the stress distribution observed in the backing steel layer for the with-honeycomb and non-spaced configurations at time interval of 100 μs at 850 m/s impact velocity. The size of the damaged zone observed in the with-honeycomb configuration is smaller than the non-spaced one, similar to what was observed in experimental testing of the HCA. This was also indicated by the through-thickness damage evolution observed from volumetric stress distribution plots at increasing time increments for each configuration. Sample volumetric plots for the with-honeycomb plate are shown in Fig. 38.5. Results from the present FEA are in agreement to those by Ivañez et al. [12]. Observations made from change in velocity plots, ballistic limits, and 3-D volumetric stress distribution plots portray similar trends for the impact characteristics of the aluminum based collapsible absorber core they presented. Visual results from the current modeling procedure do not clearly match the expected damage zone sizes seen experimentally on the HCA.

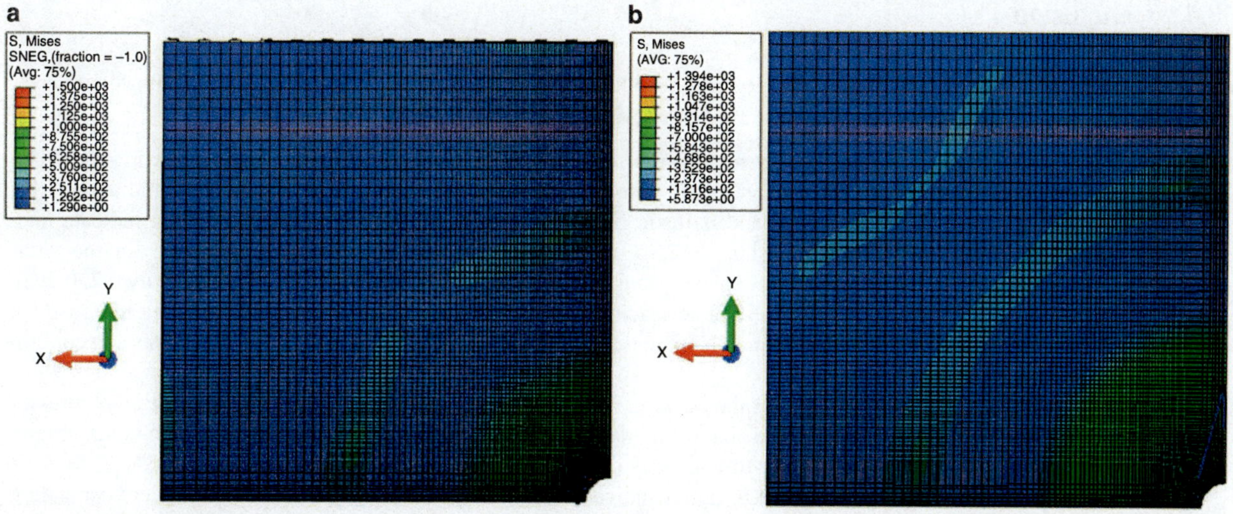


**Fig. 38.4** Stress distribution in the back layer at 850 m/s and 100 μs after impact; (**a**) with-honeycomb configuration, (**b**) non-spaced configuration

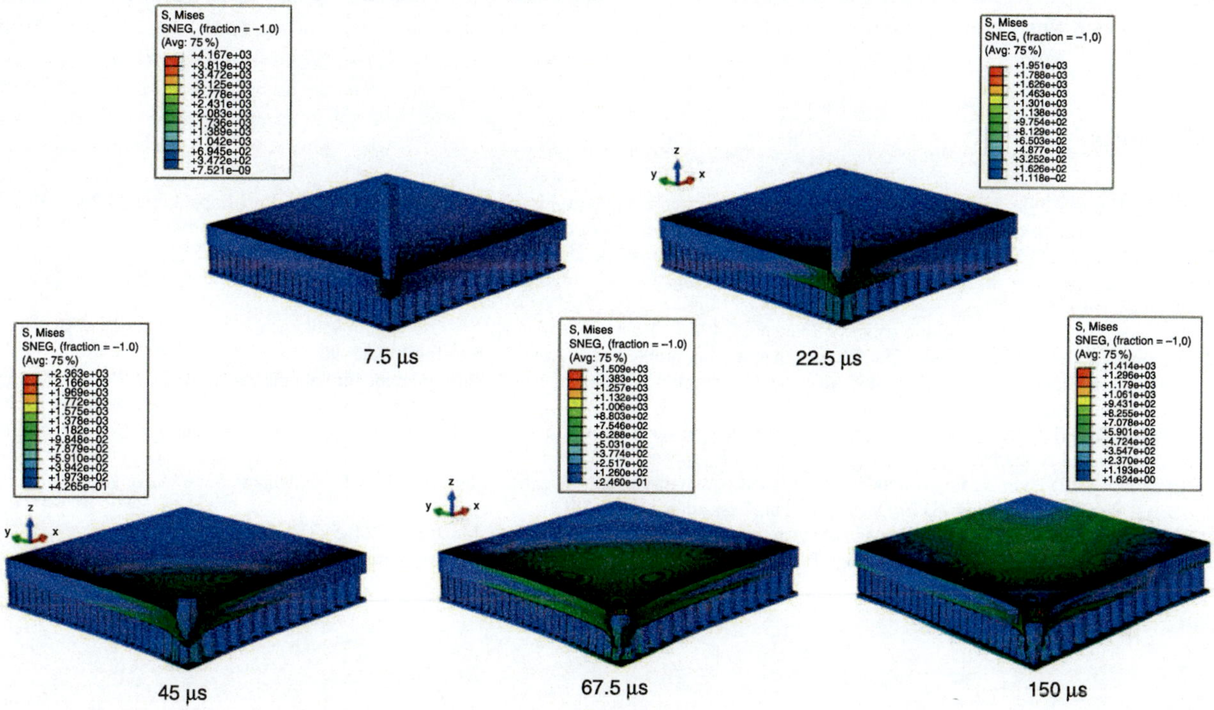


**Fig. 38.5** Volumetric stress distribution plots for the with-honeycomb configuration at varying time increments (impact velocity 640 m/s)

This difference can be attributed to the use of steel face and backing layers in the current model instead of UHMWPE. Since steel is isotropic, a high through thickness modulus results in localization of damage before the projectile interacts with the honeycomb layer. Current FEA simulations indicated that not more than three honeycomb cells interact directly with projectile. In the case of HCA, delamination occurred for the middle laminates of UHMWPE causing an increased area of interaction with the honeycomb. The reduced number of interacting honeycomb cells in the current FEA has led to minimal kinetic energy absorption by the honeycomb layer. This is evident through only 2.3% reduction of residual velocity in comparison to non-spaced plates. Repeating the FEA with an actual UHMWPE composite model or a higher cross-section area projectile can provide more insight.

## 38.4 Conclusion

A preliminary FEA was developed using dynamic explicit code in ABAQUS to evaluate the influence of a honeycomb layer on the ballistic performance of the armor plate.

1. Verification of performance improvement by inclusion of honeycomb layer was also done through FEA in ABAQUS. Three configurations (with-honeycomb, spaced-plate, non-spaced) were subjected to ballistic impact simulations by a copper projectile. Change in velocity plots and ballistic limit predictions from obtained results did not indicate high variation as expected by honeycomb inclusion. A 2.3% reduction in residual velocity was calculated for the with-honeycomb configuration in comparison to the spaced plate one for the high velocity regime (impact velocity >750 m/s). Verification of residual velocity as a function of plate spacing was also carried out for this high velocity regime. An 11% decrease in residual velocity was observed for a 4X increase in the spacing. Further numerical investigations can confirm and elaborate on this dependence.
2. The stress distribution versus time history for the backing layer was used an indicator of the communication of stresses through the thickness of the target plate. The size of the damaged zone observed in the with-honeycomb configuration was smaller than the non-spaced one similar to what was observed in the experimental testing of the HCA [7, 8]. This delayed stress transfer and minimized damage zones indicated minimizing of the BFS as well as the corresponding BABT.
3. Discrepancies in the present modeling scheme were identified. Localization of damage by inclusion of steel as well as high velocity and interaction time kinetics appear to have a significant influence on the observed results. As a part of future work, an actual UHMWPE composite model or a higher cross-section area projectile will be used to compare performance with experimental results of the HCA.

## References

1. Mahfuz H et al (2000) Investigation of high-velocity impact on integral armor using finite element method. Int J Impact Eng 24:203–217
2. Ben-Dor G, Dubinsky A, Elperin T (2000) The optimum arrangement of the plates in a multi-layered shield. Int J Impact Eng 37:687–696
3. Gama BA et al (2001) Aluminum foam integral armor: a new dimension in armor design. Compos Struct 52:381–395
4. Senf H, Strassburger E, Rothenhausler H (1997) Investigation of bulging during impact in composite armour. J Phys IV France 7:C3-301–C3-306
5. Dunn ES (1994) Impact absorbing armor. US Patent 5,349,893 1994
6. Cannon L (2001) Behind armour blunt trauma: an emerging problem. J R Army Med Corps 147:87–96
7. Jayakumar B, Bhat AR, Hanan JC (2009) Mechanical properties of amorphous metal honeycombs for ballistic applications. ASME Int Mech Eng Congress Exposit Proceed 2:73–74
8. Bhat AR (2009) Finite element modeling and dynamic impact response evaluation for ballistic applications. MS thesis, Department of Mechanical and Aerospace Engineering, Oklahoma State University
9. Hanan JC (2012) Hybrid composite body armor insert patent disclosure, Invention Number 2012–012, Oklahoma State University
10. ABAQUS – Version 6.10 EF, User Documentation, Dassault Systems, 2011
11. Wierzbicki T et al (2005) Calibration and evaluation of seven fracture models. Int J Mech Sci 47(4–5):719–743
12. Ivañez I et al (2011) Numerical modelling of foam-cored sandwich plates under high-velocity impact. Compos Struct 93:2392–2399

# Chapter 39
# Ballistic Impact on Woven Glass/Epoxy Composites at High and Low Temperatures

**Yougashwar Budhoo, Feridun Delale, and Benjamin Liaw**

**Abstract** This paper investigates the effect of temperature on the ballistic limit of woven glass/epoxy composite panels (6 × 4 × 0.25 in.) at five different temperatures: −60°C, −20°C, R.T, 75°C and 125°C. The studies were conducted by combining experimental and 3-D dynamic finite element approaches. The specimens tested were made of plain-weave S-2 glass fibers/toughened epoxy. The composite panels were impacted using an in-house gas gun. The time-histories of impact-induced dynamic strains was recorded. The damaged specimens were inspected visually and using the ultrasonic C scan method. A 3-D dynamic finite element (FE) software, with Chang-Chang composite damage model, was then used to simulate the experimental results of the ballistic impact tests. Good agreement between experimental and FE results has been achieved.

**Keywords** Finite element (FE) • Woven composite • Ballistic impact

## 39.1 Introduction

The main idea of composite material is to combine different materials to produce a new material with performance unattainable by the individual constituents. It gives flexibility to the designer to tailor the new material with properties to obtain peak performance for a particular application. The essence of the concept of composites is this: the bulk phase accepts the load over a large surface area, and transfers it to the reinforcement, which being stiffer, increases the strength of the composite. The significance here lies in the fact that there are numerous matrix materials and as many fiber types, which can be combined in countless ways to produce just the desired properties.

As the demand for materials with higher stiffness-to-weight and strength-to-weight ratios, tailoring flexibility, more resistant to harsh environment and damage tolerance arises; researchers and engineers seek to have more profound understanding of the behavior of different composites under various conditions. Investigations into woven composites have been extensive over recent years mainly because of their excellent impact resistance [1–5], dimensional stability, ease of handling etc. Some of these researches also focused on the effect of different environments, such as temperature [4–8] on the behavioral response of composite materials.

Y. Budhoo (✉)
Department of Engineering and Technology, Vaughn College of Aeronautics and Technology, 86-01, 23 Ave, East Elmhurst, Queens, NY 11369, USA
e-mail: yougashwar.budhoo@vaughn.edu

F. Delale • B. Liaw
Department of Mechanical Engineering, The City College of New York, 140th and Convent Ave, New York, NY 10031, USA

V. Chalivendra et al. (eds.), *Dynamic Behavior of Materials, Volume 1: Proceedings of the 2012 Annual Conference on Experimental and Applied Mechanics*, Conference Proceedings of the Society for Experimental Mechanics Series, DOI 10.1007/978-1-4614-4238-7_39, 

## 39.2 Experimental Procedure

### 39.2.1 Materials

The individual constituent materials combined to form the composite material used in this research are, S-2 glass woven fabrics placed in SC-79 toughened epoxy resin matrix. IM-7 graphite woven fabric and SC-79 epoxy matrix form the non-hybrid laminate called GL composite. The GL composite panel has a dimension of 100 mm × 100 mm × 25 mm. The panel consists of 39 layers of woven graphite fabric.

### 39.2.2 Experimental Setup

The ballistic impact tests were performed using an in house designed gas-gun setup equipped with an environmental chamber. Inside the chamber is a fixture which holds the specimen to be impacted. A 22 caliber copper bullet fitted in to a plastic sabot was used as the projectile. The sabot is very important and provides a tight fit of the projectile in the gun barrel. The gun consists of helium tank, temporary gas storage vessel, solenoid valve and stainless-steel barrel (Figs. 39.1, 39.2, 39.3, and 39.4).

**Fig. 39.1** Ballistic gas gun

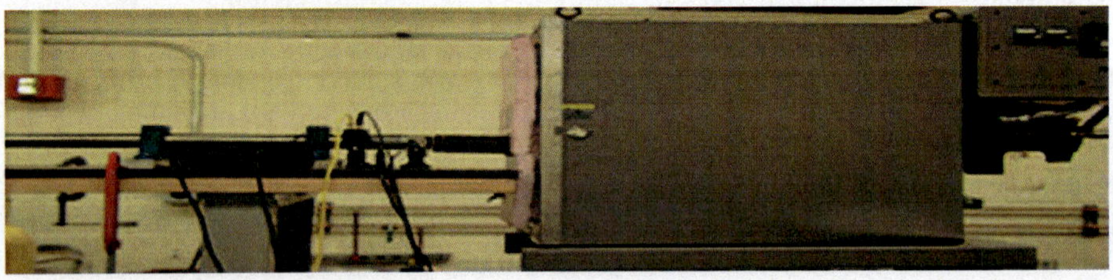

**Fig. 39.2** Ballistic gas gun and environmental chamber

**Fig. 39.3** Fixture

**Fig. 39.4** Sabot and bullet

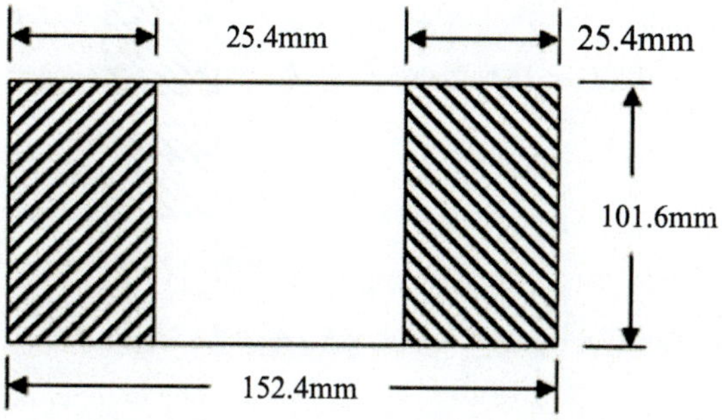


**Fig. 39.5** Ballistic impact test specimen, hatched region represents the clamped portion (All pictures in this section, 'experimental set up' is taken from my previous work [3])

The figures above show the gas gun with the environmental chamber fixture and bullet with sabot. Figure 39.5 shows the dimensions and clamped areas of the specimen to be impacted. The boundary conditions of the clamped sides are assumed to be fixed-fixed.

Compressed helium was used to launch the projectile. The velocity of the projectile can be varied by changing the pressure used to launch the projectile or by changing the location of the bullet in the gun barrel. Placing the bullet closer to the exit of the gun barrel lowers the velocity of the bullet since the pressure used to launch the bullet will be smaller as compared to when the bullet is placed further away from the exit. The velocity of the bullet is measured as follows: two holes are drilled in the barrel where two laser beams continuously feed two optical sensors. When the projectile passes through the holes, it blocks the lasers, the sensors sense the change in intensity of light and an oscilloscope captures the voltage change. The velocity of the projectile is then calculated from the distance between the two sensors divided by the time that projectile spends traveling between the two holes. As the bullet exits the gun barrel it strikes the specimen that is in the environmental chamber at the specific test temperature. The specimens were impacted at three different temperatures: room temperature (RT), 75°C and 125°C. The specimens were first soaked at the desired test temperature for thirty minutes before impact. This is done to eliminate any temperature gradient in the test specimens (Fig. 39.6).

The force-time histories and the deflection-time histories cannot be measured with our set-up. However, two strain gages placed one inch away from the impact point, one in the radial direction and the other in the hoop direction, measure the dynamic strain histories of the specimen during impact.

Two strain gauges were mounted on the impact side (i.e., the top surface) of each specimen. These strain gages were mounted at a distance of 25.4 mm from the center of the composite panel, where the impactor impacted the specimens. Figure 39.7 shows a schematic of the strain gage orientations and locations on the specimen. SG-1 measured the radial strain while SG-2 measured the hoop strain.

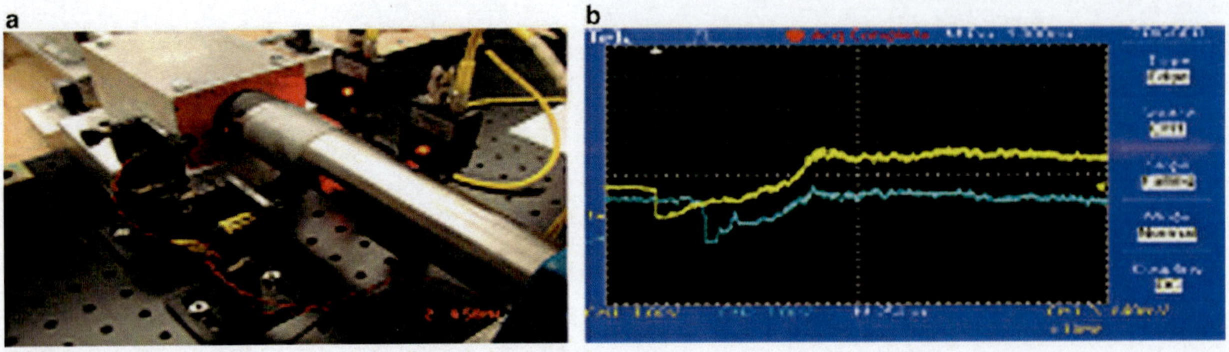


**Fig. 39.6** Setup to measurement velocity during ballistic impact. (**a**) Lasers & optical sensors, (**b**) voltage-time histories

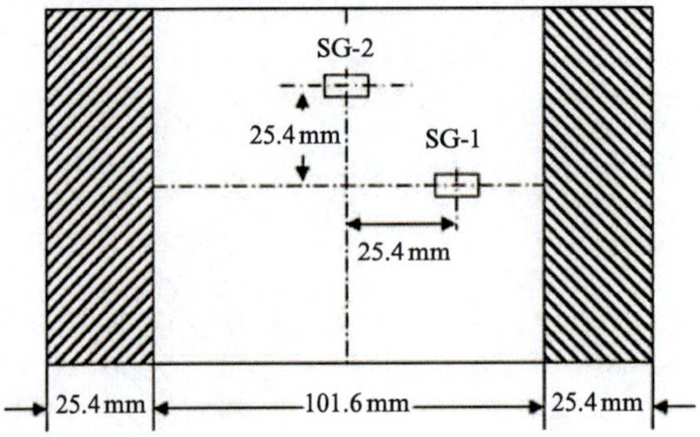


**Fig. 39.7** A typical composite panel specimen mounted with two strain-gages on the ballistic impact side

### *39.2.3 Post Damage Evaluation*

A modular and expandable ultrasonic system was used to scan and conduct the damage evaluation for the impacted specimens. Scanning was performed using the through-transmission technique. Flat and focused 5 MHz transducers (6.35 mm in diameter) were used to scan these post-impacted composite specimens.

## 39.3 Experimental Results

During the ballistic impact tests at various test temperatures, the main objective was to obtain an estimate of the ballistic limit of the four layups of composite. There are many definitions of ballistic limit, however, in this research the definition of ballistic limit was taken from [3] where ballistic limit is looked at in terms of not only velocity but also energy. During ballistic impacts, one of three outcomes is possible [3]:

1. The projectile impacts the front surface of the composite with a certain velocity, perforates it and exits with a certain residual velocity. In this case the projectile's initial energy is larger than the energy than the composite can absorb.
2. The projectile partially penetrates the composite i.e. the projectile gets stuck in the composite or rebounds. In this case the projectile's initial energy is less than what the composite can absorb.
3. The projectile perforates the target completely with a zero exit/residual velocity. In this case the initial velocity of the projectile is called the **ballistic limit** or $V_{50}$. Here, the entire energy of the projectile is absorbed by the target.

In the experiments conducted, the procedure used to obtain the ballistic limit was by finding an impact velocity which causes perforation in the target by the projectile and one that causes only partial penetration of the projectile in the target. More tests were conducted within this range of these two extremes velocities in order to find the maximum impact velocity

**Table 39.1** Parameters of ballistic impact tests on woven GL composite panels

| Materials | Temperature | Impact velocity | Penetration type | Impact energy |
|---|---|---|---|---|
| GLASS | 125°C | 282 m/s | Partial (stuck) | 119 J |
| | 125°C | 299 m/s | Partial (stuck) | 134 J |
| | 125°C | 325 m/s | Full | 158 J |
| | 75°C | 322 m/s | Partial | 156 J |
| | 75°C | 346 m/s | Partial | 180 J |
| | 75°C | 365 m/s | Full | 200 J |
| | R.T | 360 m/s | Partial | 194 J |
| | R.T | 387 m/s | Partial (stuck) | 225 J |
| | R.T | 400 m/s | Partial[a] | 240 J |
| | R.T | 403 m/s | Full | 244 J |
| | −20°C | 382 m/s | Partial (stuck) | 219 J |
| | −20°C | 404 m/s | Full | 245 J |
| | −60°C | 397 m/s | Partial[a] | 236 J |
| | −60°C | 404 m/s | Full | 245 J |

[a] Severe splitting on the back surface of specimen

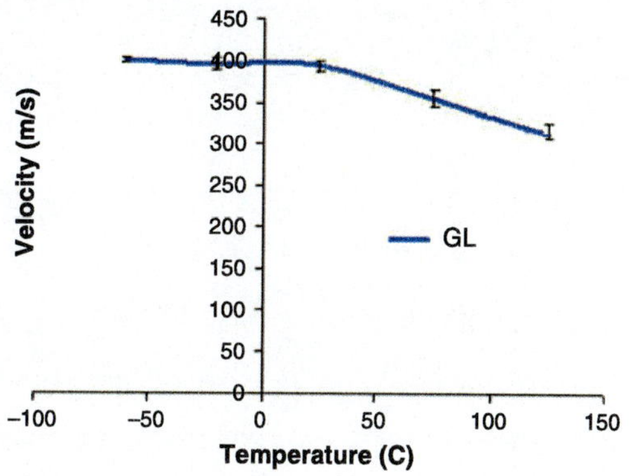


**Fig. 39.8** Temperature vs. ballistic limit for all layups of composite

which causes a zero exit/residual velocity. Due to the limited number of specimens available for testing, the exact ballistic limit could not be obtained experimentally. Hence, minimal tests were conducted as needed to obtain a reasonable range in velocities i.e. minimum velocity that causes perforation and maximum velocity that causes partial penetration.

In Table 39.1, the impact velocities which cause full and partial penetration in the composite specimens are given.

The interval of $V_{50}$ velocity for all panels was roughly estimated using experimental data. It can be seen that the impact resistance and ballistic limit tends to decrease slightly with an increase of temperature.

The impact resistance and ballistic limit as a function of temperature can be seen but to have a clearer understanding of the behavior, the impact velocity is been plotted against temperature in Fig. 39.8. The ballistic limit in Fig. 39.8 is obtained by taking the average of the minimum velocity that causes perforation and the maximum velocity that causes partial penetration at that particular test temperature. The error bar is added to each curve which represents the difference in these two velocities. It can be seen from the figure that as temperature decreases, the ballistic limit for GL increases.

### 39.3.1 *Dynamic Strains*

The ballistic impact event takes place in an extremely short time. Accordingly the frequency of the recording device should be high enough to capture the strain gauge readings in such a short time. The oscilloscope used in this study to capture the strain output from the strain gages has a maximum frequency of 500 KHz. In other words, it was only able to record a strain data every 2 μs.

Assuming that ballistic impact event takes 100 μs, the dynamic strain curve from experiment would have captured only 50 data points, resulting in significant under sampling. Thus, the strain curve constructed using these data points will not represent the actual response of the composite during the impact. The strain measurements obtained are mainly for vibration

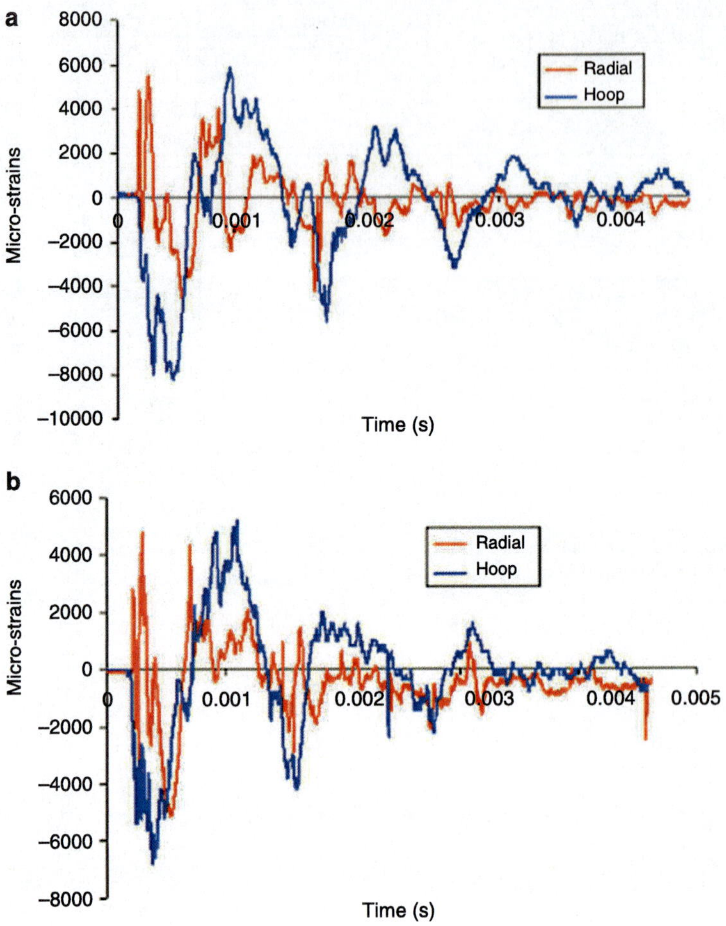


**Fig. 39.9** Hoop and radial experimental strain histories for GL at (**a**) 125°C (**b**) −60°C

of impacted composite panel and not the transient response of composite during ballistic impact. Experiments were conducted at five temperatures but only the strain results at extreme temperatures are presented here, i.e.125°C and−60°C.

Figure 39.9 shows the dynamic strain histories of composites for GL at 125°C and −60°C.

In order to better investigate the damage mechanisms and damage patterns of the composites and how they change with temperature, optical pictures were taken of specimen, then they were C-scanned and finally cut in two halves to analyze the cross-sectional view.

### *39.3.2 Delamination*

Figure 39.10 show ultrasonic C-scans of the impacted composites. C-Scan images exhibited increased delamination up to the ballistic limit and gradual decrease was observed after this critical velocity.

## 39.4 Finite Element Simulation

Finite element analysis (FEA) is a very useful tool in solving complex problems and also predicts the behavior of structures when actual testing may not be cost effective or practical. In this study, ballistic impact problems at various temperatures are modeled and analyzed using a 3-D nonlinear, dynamic finite element code. To validate the reliability of the model, the finite element results were compared to those obtained experimentally.

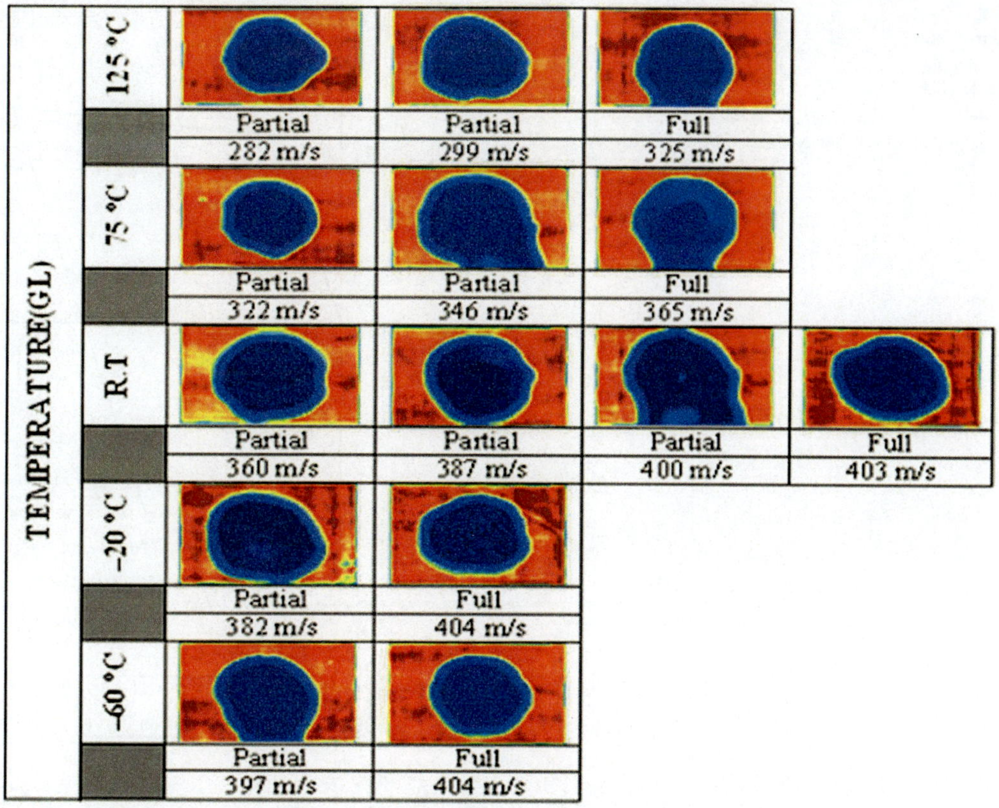


**Fig. 39.10** Ultrasonic C-scan images of impacted GL composites at various test temperatures velocities

Since the geometries of the composite plate and bullet are symmetric, it is customary that only a quarter of these parts be modeled to save computational time. All the parts were modeled using 8-node solid elements. All modeling and meshing was done using the same commercial finite element software.

## 39.4.1 *Material Models*

### 39.4.1.1 Mat_Plastic_Kinematic (Mat_003)

This model is used for isotropic and kinematic hardening plasticity with the option of including rate effects. $E_t$ is the slope of the second arm of the bilinear stress strain curve (Fig. 39.11) and $l_o$ and $l$ are undeformed and deformed lengths of uniaxial tension specimen respectively. This material was used for the 22 caliber copper bullet.

### 39.4.1.2 Mat_Composite_Damage(Mat_022)

In modeling the composite plate, MAT_022 also called the Chang-Chang composite damage model was used. This model is an orthotropic material model with optional brittle failure for composites. It requires nine material parameters to represent the composite material. From tensile testing at high temperatures, the material behavior of GR and GL specimens was found to be non linear. In order to use MAT_22 (best suited to describe a linear elastic behavior) to describe our materials' behavior, we consider the strain energy. Here we calculate the strain energy from the non linear stress strain behavior, and then knowing the strain energy and the initial Young's modulus, a new linear stress–strain curve is redrawn where the strain energy and Young's modulus is maintained but the failure strain and ultimate stress are changed. Figure 39.12 gives an example of the approximation of the true stress–strain curve for a material. The approximate curve was inputted in the finite

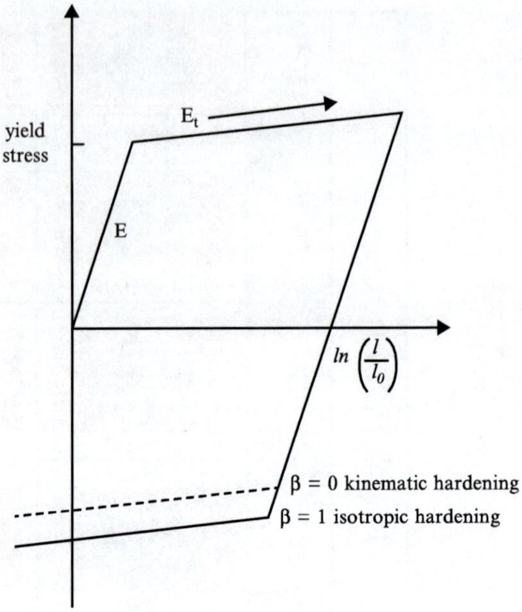


**Fig. 39.11** Elastic–plastic behavior with kinematic and isotropic hardening. (Taken from LS-DYNA theoretical manual)

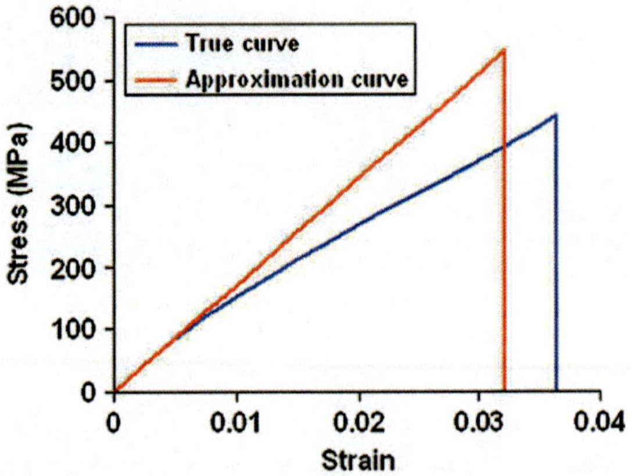


**Fig. 39.12** Linearization of non linear stress–strain curve obtained from tensile testing while maintaining strain energy

element software instead of the true curve. A linear fit of the true curve could be used, but wasn't because this curve only represent material behavior at quasi-static loading. In the approximate curve, the failure strain is smaller and ultimate strength is larger which the expected material behavior at higher strain rate.

In the Chang-Chang model five material parameters are used in three possible failure criteria which are as follows:

$S_1$ = Longitudinal tensile strength
$S_2$ = Transverse tensile strength
$S_{12}$ = Shear strength
$C_2$ = Transverse compressive strength
$\alpha_{=}$ Non linear shear stress parameter

Here, $S_1$, $S_2$, $S_{12}$, and $C_2$ are obtained from material strength measurements and $\alpha$ is defined by a material shear stress–strain measurements.

The three failure critera are matrix cracking, fiber breakage and compressive failure.

1. Matrix cracking:

$$F_{matrix} = \left(\frac{\sigma_2}{S_2}\right)^2 + \bar{\tau}$$

where

$$\bar{\tau} = \frac{\dfrac{\tau_{12}^2}{2G_{12}} + \dfrac{3}{4}\alpha\tau_{12}^4}{\dfrac{S_{12}^2}{2G_{12}} + \dfrac{3}{4}\alpha S_{12}^4}$$

where $\sigma_2$ is the tensile stress in the transverse direction whereas $\tau_{12}$ is the in-plane shear stress between fibers and matrix. Failure is assumed whenever $F_{matrix}>1$. Once this type of failure occurs, the material constants $E_2$ (Young's modulus in the transverse direction), $G_{12}$ (in-plane shear modulus in the 1-2 plane), $\nu_{12}$ and $\nu_{21}$ (generalized Poisson's ratios in the 1–2 plane) are set to zero.

2. Compressive failure:

$$F_{comp} = \left(\frac{\sigma_{12}}{2S_{12}}\right)^2 + \left[\left(\frac{C_2}{2S_{12}}\right)^2 - 1\right]\frac{\sigma_2}{S_2} + \bar{\tau}$$

Failure is assumed when $F_{comp}>1$. When this type of failure occurs, the material constants $E_2$, $\nu_{12}$ and $\nu_{21}$ are all set to zero.

3. Fiber breakage:

$$F_{fiber} = \left(\frac{\sigma_1}{S_1}\right)^2 + \bar{\tau}$$

where $\sigma_1$ is the tensile stress in the longitudinal direction. Failure is assumed when $F_{fiber}>1$. After fiber breakage, $E_1$ (Young's modulus in the longitudinal direction), $E_2$, $G_{12}$, $\nu_{12}$ and $\nu_{21}$ are all set to zero.

#### 39.4.1.3 Mat_Add_Erosion (Mat_00)

This failure criterion was added to all the materials used in our modeling although it is used mostly for material models which do not have a failure or erosion criterion.

It is basically an additional criterion which when satisfied; the element is deleted from the calculation. It is a good tool to use when comparison is needed between animation and real specimen, because it prevents animation from having excessively distorted elements. In this model, erosion is set to occur when the tensile failure strain of the material is reached.

### *39.4.2 Contact and Delamination Model*

The contact between the composite layers was defined using the option, CONTACT–AUTOMATIC–SURFACE–TO–SURFACE–TIEBREAK.

By using this contact definition, delamination between the composite layers is governed by the criterion

$$\left[\frac{\max(0, \sigma_n)}{NFLS}\right]^2 + \left(\frac{\sigma_s}{SFLS}\right)^2 \geq 1$$

where $\sigma_n$ and $\sigma_s$ are normal and shear stresses acting on the layer interface, respectively, while *NFLS* and *SFLS* are normal and shear failure strengths of the layer interface, respectively. It is assumed that delamination occurs mainly by shear failure; hence, NFLS is set to a very large number so delamination in the model is controlled by SFLS. Since SFLS is not known experimentally, it was varied in order to fit the experimental results. Additionally, **ERODING_SURFACE_TO_ SURFACE** contact model was used between the impactor and the composite. This model allows elements to be eroded when certain failure criteria are met. In this study strain-based failure criterion was used for element erosion; that is, when $\varepsilon \geq \varepsilon_{erosion}$, the element

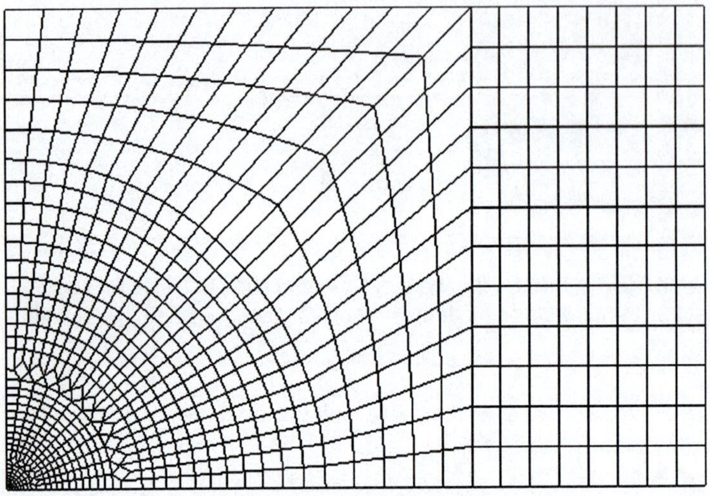

**Fig. 39.13** Quarter model of meshed plate used in ballistic impact simulation

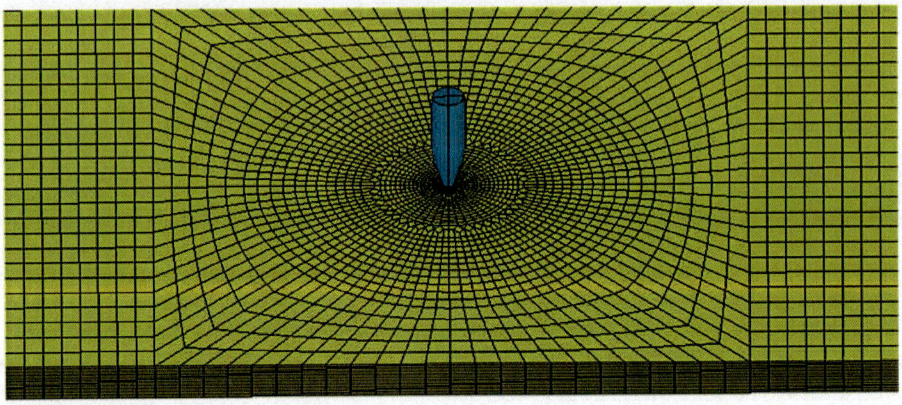

**Fig. 39.14** Mesh of entire composite plate and projectile for ballistic impact

was eroded and removed from calculation. The contact and delamination models as well as the erosion option were used for all composites.

### 39.4.3 Boundary Conditions

To model the impact event properly, the correct boundary conditions should be chosen. Preliminary simulations were carried out, and it was seen that a fixed-fixed boundary condition was the best suited to model the low velocity impact. A fixed-fixed boundary condition was therefore chosen for all the low velocity impact simulations.

Figure 39.15 shows the mesh of a quarter model of the plate used in simulating the drop impact test. Fine mesh around the center and course mesh away from center was used (Fig. 39.13).

Figure 39.14 shows the entire drop weight model of both the composite panel and the impactor for a non hybrid specimen for the low velocity impact. The material properties used to define the composite were obtained from tensile testing and are presented in Table 39.2.

These properties were taken from my previous work [9]. The entire table of material properties for high and low temperatures is presented here, but only the high temperature properties were used in the modeling.

Most of the finite element procedures were followed from my previous work [10].

**Table 39.2** Key material properties used in modeling

| | | $E_1 = E_2$ ($GP_a$) | $E_3$ ($GP_a$) | $G_{12}$ ($GP_a$) | $G_{13} = G_{23}$ ($GP_a$) | $V_{12}$ | $V_{13} = V_{23}$ | P (kg/m$^3$) |
|---|---|---|---|---|---|---|---|---|
| Glass | −60°C | 19.39 | 7.26 | 2.98 | 2.65 | 0.152 | 0.370 | 1,766 |
| | −20°C | 18.92 | 6.78 | 2.78 | 2.49 | 0.145 | 0.631 | 1,761 |
| | R.T | 17.04 | 6.50 | 2.74 | 2.30 | 0.130 | 0.410 | 1,755 |
| | 75°C | 16.13 | 5.85 | 1.89 | 2.03 | 0.128 | 0.436 | 1,750 |
| | 125°C | 15.03 | 5.46 | 1.44 | 1.87 | 0.121 | 0.453 | 1,744 |
| Graphite | −60°C | 43.82 | 6.43 | 3.97 | 2.32 | 0.140 | 0.382 | 1,392 |
| | −20°C | 43.71 | 5.75 | 3.93 | 1.96 | 0.129 | 0.462 | 1,387 |
| | R.T | 42.55 | 5.20 | 3.08 | 1.81 | 0.121 | 0.427 | 1,381 |
| | 75°C | 41.49 | 3.43 | 1.99 | 1.05 | 0.116 | 0.627 | 1,375 |
| | 125°C | 40.39 | 2.53 | 1.67 | 0.76 | 0.099 | 0.649 | 1,370 |

### *39.4.4 Determination of Damping Constant*

In analyzing the dynamic strain histories recorded during the ballistic impact tests, it can be seen that the response of the composite was damped. Slipping between layers of laminated structures such as composites introduces high internal damping to the system; therefore damping was required to be incorporated into the FE model. However, the damping constant is not easy to determine. In order to successfully incorporate a reasonable damping constant into the model, two options were used. The first option is to incorporate damping between layers in the form of a viscous damping coefficient in terms of percentage of critical damping. This helps to eliminate undesirable oscillation in contact. Between layers of similar materials, 10% damping was used and between layers of dissimilar materials, 20% damping was used. The second option where damping was introduced was through DAMPING_PART_STIFFNESS. In this option a Raleigh damping coefficient is specified which is uniformly applied to each element of the part. LS-DYNA suggests a value of 0.01–0.25 or in other words, 1–25% of the critical damping coefficient. Different values were used in the model and the strain outputs from the model were compared to those obtained experimentally. Best results were obtained when 10% damping was used.

### *39.4.5 Comparison of Strain-Time Histories*

In this study FE predictions were compared with experimental results mainly for dynamic strain histories and post-impact damage patterns, since it was not possible to obtain force and energy-time histories experimentally. As shown in Fig. 39.15, the dynamic strains obtained from experiments and the FE results are in good agreement. The strains obtained using the Chang-Chang model exhibited some magnitude and frequency mismatch in comparison with experimental data. This may be because the nonlinear material behavior could not be represented perfectly by the linear Chang-Chang model. It should be noted that the strain comparisons made here are mainly for vibration of impacted composite beams. It would be better to compare the transient response of composite during ballistic impact. However ballistic impact takes place in an extremely short time. In order to capture strains in such a short time, the frequency of the strain amplifier should be very high.

### *39.4.6 Comparison of Post-impact Damage Patterns*

The finite element models successfully predicted the damage observed during the experiments. The broken fibers appeared at the tension side of the experimental pictures for all lay-up sequences, but were not seen in the FE prediction. This is because of the strain based element erosion criterion used in the FE model. The area where broken fibers are seen represents failed material. On the other hand the failed elements are eroded and removed from calculation in the FE model, so they aren't seen in the comparisons (Fig. 39.16).

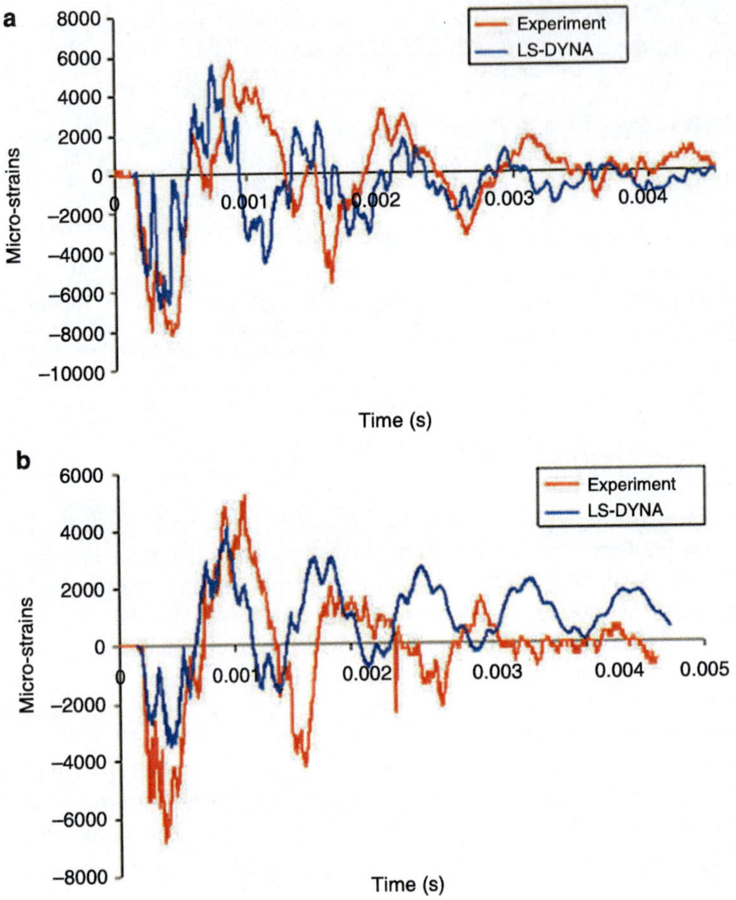


**Fig. 39.15** Comparison of LS-DYNA and experimental strain histories fo GL at (**a**) 125°C (**b**) −60°C

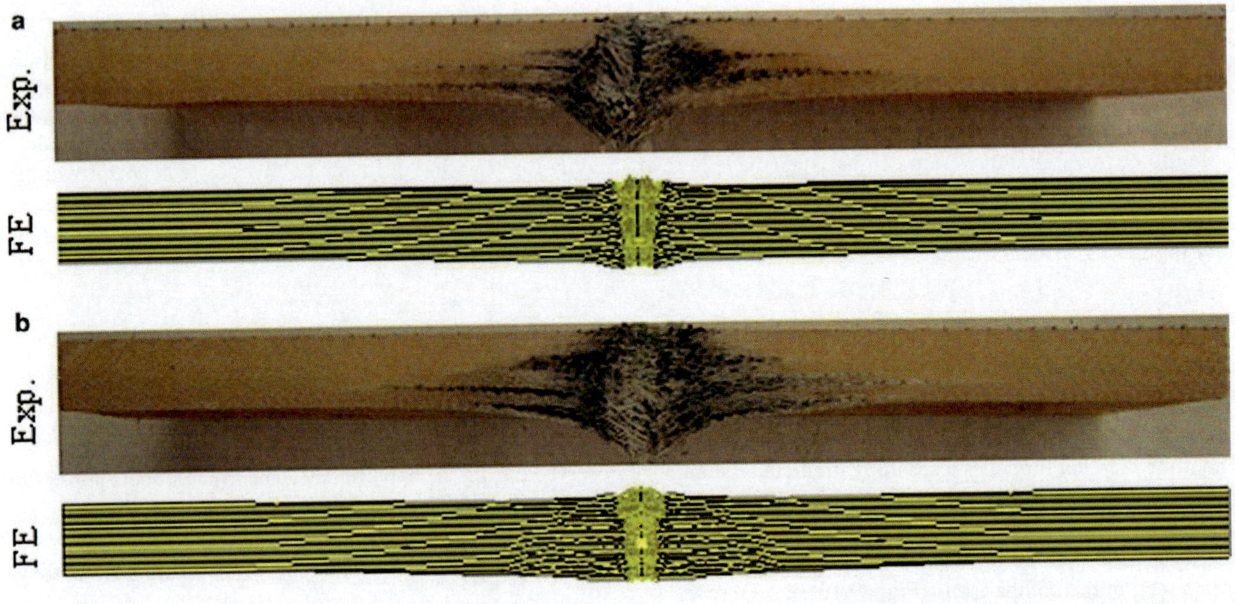


**Fig. 39.16** Comparisons of post impact damage patterns for non-hybrid GL impacted at (**a**) 125°C and 365 m/s (**b**) –60°C and 404 m/s

## 39.5 Conclusions

(a) The FE model incorporated with MAT_22 (Chang-Chang material model) was successfully used to simulate ballistic impact tests.
(b) The validity of the FE model for ballistic impacts was confirmed when compared to experimental results such as; strain histories and post-impact damage patterns.
(c) Delamination increased with increasing velocity for partial penetration cases. It then decreased with an increasing velocity for the full penetration cases.
(d) The interval of the ballistic limit velocity can be estimated using ballistic tests and Ultrasonic C-Scans.

## References

1. Budhoo Y, Liaw B, Delale F (2010) Effect of temperature on the impact damage of composite materials. In: Proceedings of the ASME 2010 international mechanical engineering congress & exposition IMECE2010, Vancouver, Canada
2. Budhoo Y, Delale F, Liaw B (2010) Temperature effect on drop-weight impact of woven composites. In: Proceedings of SEM 2011 annual conference & exposition on experimental and applied mechanics, Uncasville, CT, USA
3. Budhoo Y, Liaw B, Delale F (2010) Temperature effect on the ballistic impact of hybrid woven composites. In: Proceedings of the ASME 2011 international mechanical engineering congress & exposition IMECE2011, Denver, CO, USA
4. Hirai Y, Hamada H, Kim JK (1998) Impact response of woven glass-fabric composites- II. Effect of temperature. Compos Sci Technol 58(1):119–128
5. Im KH, Cha CS, Kim KS, Yang IY (2001) Effect of temperature on impacts damages in CFRP composite laminates. Compos Part B 32:669–682
6. Parlevliet PP, Bersee HEN, Beukers A (2006) Residual stresses in thermoplastic composites- A study of the literature-Part 1: formation of residual stresses. Appl Sci Manufact 37(11):1847–1857
7. Dlouhy I, Chlup Z, Boccaccini DN, Atiq S, Boccaccini AR (2003) Fracture behavior of hybrid glass matrix composites: thermal ageing effects. Compos Part A: Appl Sci Manufact 34(12):1177–1185
8. Naik NK, Shrirao P, Reddy BCK (2006) Ballistic impact behavior of woven fabric composites: formulation. Int J Impact Eng 32:1521–1552
9. Budhoo Y, Delale F, Liaw B (2010) Effect of temperature on tensile behavior of woven composites. In: Proceedings of SEM 2011 annual conference & exposition on experimental and applied mechanics, Uncasville, CT, USA
10. Budhoo Y, Liaw B, Delale F (2010) Effect of temperature on ballistic impact of graphite/epoxy composite. In: Proceedings of the IMAC20, annual conference & exposition on experimental and applied mechanics, Los Angeles, CA, USA

# Chapter 40
# Comparison of Mechanical Properties of Polymer-Based Multi-phase Particulate Composites

**Jennifer L. Jordan and Jonathan E. Spowart**

**Abstract** Multi-constituent particulate composites consist of individual particles of more than one material dispersed throughout and held together by a polymer binder. The mechanical and physical properties of the composite depend on the mechanical and physical properties of the individual components, particularly the binder; their loading density; the shape and size of the particles; the interfacial adhesion; residual stresses; and matrix porosity. Multi-constituent composites with cast-cure epoxy binder have been presented recently. In this study, the microstructure is varied by injection molding PMMA-based composites. The dynamic mechanical properties of PMMA-based and epoxy-based composites are measured using a split Hopkinson pressure bar. The mechanical properties of these composites are compared.

**Keywords** Particulate composite • PMMA • High strain rate

## 40.1 Introduction

Polymer composites comprised of metallic particles distributed throughout a contiguous polymer matrix can often be modified to produce advanced composites that exhibit multifunctional characteristics. For example, epoxy with Ni and Al particles [1, 2] to produce high strength materials with exothermic reactive properties, or Teflon® (PTFE) can be reinforced with Al and W particles [3]. The properties of the particulate composites often depend on varying particle size, loading fractions, particle type, and the adhesion between the particulate and the matrix [2–6].

Several studies on epoxy-based composites with similar microstructures have been reported. These studies have shown that particle size [3, 7–9], shape [10], and concentration [11] and properties of the constituents can affect mechanical properties. In $Al_2O_3$ particle-filled epoxy (Epon 828/Z), increasing the particle concentration and decreasing the particle size is found to increase the stress corresponding to 4% plastic strain [12]. A study of aluminum particle filled epoxy (DGEBA/MTHPA) composites has found that a small amount of filler (~ 5 vol.%) increases the compressive yield stress, but additional amounts of filler decrease the compressive yield stress [13]. However, tests on glass-bead-filled epoxy (DOW DER 331/bisphenol-A) found that increasing the volume fraction increased both the yield stress and fracture toughness of the material [14, 15]. In another study on a similar material, decreasing the aluminum particle size from micro to nano resulted in increased epoxy crosslink density and subsequently increased both static and dynamic strength [2].

This paper will present the experimental results comparing aluminum and nickel particles in PMMA prepared by injection molding with the same particles in epoxy prepared in a cast-cure process.

J.L. Jordan (✉)
Air Force Research Laboratory, AFRL/RWME 2306 Perimeter Road, Eglin AFB, FL 32542
e-mail: jennifer.jordan@eglin.af.mil

J.E. Spowart
AFRL/RXBC, Wright-Patterson AFB, OH 45433, USA

V. Chalivendra et al. (eds.), *Dynamic Behavior of Materials, Volume 1: Proceedings of the 2012 Annual Conference on Experimental and Applied Mechanics*, Conference Proceedings of the Society for Experimental Mechanics Series,
DOI 10.1007/978-1-4614-4238-7_40, © The Society for Experimental Mechanics, Inc. 2013

## 40.2 Experimental Approach

The samples for this study were prepared using a factorial design of experiments approach in order to maximize the number of variables tested with the minimum number of test specimens. The variables tested were aluminum particle size (5, 30 or 50 μm), volume percent of aluminum (20, 30, 40 vol.%), and the volume percent nickel (0, 5, 10 vol.%). Two levels and a centerpoint were chosen for the samples. The binder was a thermoplastic, PMMA. These samples compare directly with the epoxy-based samples discussed in previous papers [1, 16], with the exception that the PMMA based samples added a centerpoint to test for curvature. Table 40.1 shows the details of the factorial design for the PMMA-based samples; the epoxy-based samples are numbered the same starting with MNML and excluding the centerpoint.

Dynamic compression experiments were conducted using a split Hopkinson pressure bar (SHPB) [17] system at a strain rate of approximately 5,000/s. The bar system is comprised of 1,524 mm long, 12.7 mm diameter incident and transmitted bars of 6061-T6 aluminum. The striker is 610 mm long and made of the same material as the other bars. The samples, which were nominally 5 mm diameter by 2.5 mm thick, are positioned between the incident and transmitted bars. The bar faces were lightly lubricated with grease to reduce friction. A complete description of this testing system can be found in Reference 17.

## 40.3 Results and Discussion

Both sets of samples, PMMA-based and Epoxy-based, were tested in compression at 5,000/s. Figure 40.1 shows representative curves from the both materials containing 20 vol.% of 5 μm Al and 10 vol.% Ni. The compressive response of these materials is very different. The epoxy-based composite shows a rise to a peak stress, followed by a small decrease, a region of perfect plasticity and then work hardening, similar to other epoxy-based composites [18]. This is consistent with the deformation of the epoxy binder [17] with the strain softening after the peak decreased potentially due to particle-particle interaction. In contrast, the PMMA-based material shows a rapid rise to a peak stress, followed by a rapid strain softening, and then perfect plasticity at a very low stress. This is consistent with behavior of particulate composites that fail at the peak stress, where the low residual stress is the loading of sample fragments [19].

The peak stress from both the PMMA-based and epoxy-based samples are plotted versus the total volume fraction of particles in Fig. 40.2. The open symbols are the small aluminum samples and the closed symbols are the large aluminum samples. Additional data from the epoxy-based samples at the same strain rate and a slightly lower strain rate are included for comparsion [1]. The trend lines are included to guide the eye rather than as linear fits to the data. There is good agreement between the previously acquired epoxy-based data [1] and that measured in this study. The strengths of the binder materials are included for comparison. For both materials, the small aluminum samples generally have a higher strength than the large aluminum samples. For the epoxy-based, the strength of the composites increases with increasing volume fraction of particles. For the PMMA-based composites, the strength of the composites decreases with increase in volume fraction of particles. This decrease is believed to be due to the failure of the samples at the peak stress, which could be caused at lower stresses with a higher fraction of particles. These samples were loaded to 50 vol.% total particles. It is interesting that at this point, both the PMMA-based materials and epoxy-based materials seem to be converging to a common peak stress. This stress may be similar to that for a collection of particles as the particle-particle interaction begins to dominate the stress–strain response of the composites, irrespective of the particular polymeric binder used.

**Table 40.1** Material configurations from a two-level, three factor design of experiments including a centerpoint

| Material | Al particle size (μm) | Al volume fraction (%) | Ni volume fraction (%) |
|---|---|---|---|
| RXRW-1 | 50 | 40 | 10 |
| RXRW-2 | 5 | 40 | 10 |
| RXRW-3 | 50 | 20 | 10 |
| RXRW-4 | 5 | 20 | 10 |
| RXRW-5 | 50 | 40 | 0 |
| RXRW-6 | 5 | 40 | 0 |
| RXRW-7 | 50 | 20 | 0 |
| RXRW-8 | 5 | 20 | 0 |
| RXRW-9 | 30 | 30 | 5 |

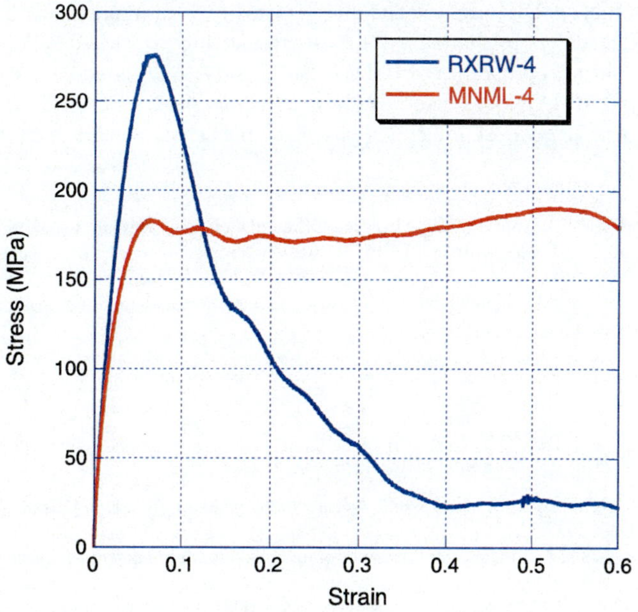


**Fig. 40.1** Stress–strain response of RXRW-4, PMMA-Al-Ni, and MNML-4, Epoxy-Al-Ni, at a strain rate of 5,000/s

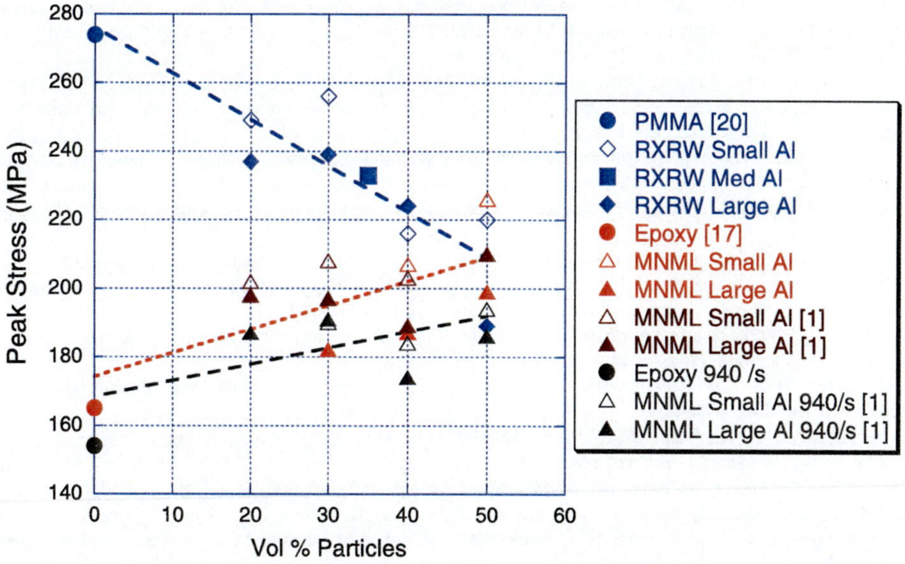


**Fig. 40.2** Volume fraction of particles versus peak stress measured using a split Hopkinson pressure bar for RXRW samples (PMMA-Al-Ni) and MNML samples (Epoxy-Al-Ni) where the experimental strain rate is 5,000/s. Open symbols represent samples that contain 5 μm aluminum and closed symbols represent samples with 50 μm aluminum [20]

## 40.4 Summary

Multi-constituent particulate composites consist of individual particles of more than one material dispersed throughout and held together by a polymer binder. The mechanical and physical properties of the composite depend on the mechanical and physical properties of the individual components, particularly the binder; their loading density; the shape and size of the particles; the interfacial adhesion; residual stresses; and matrix porosity. Multi-constituent composites with cast-cure epoxy binder have been presented recently. In this study, the microstructure is varied by injection molding PMMA-based composites. The dynamic mechanical properties of PMMA-based and epoxy-based composites are measured using a split Hopkinson pressure bar. The PMMA-based materials show a rise to peak stress followed by a sharp strain softening, indicating failure of the samples. In contrast, the epoxy-based samples show a nearly perfectly-plastic stress after a rise to peak stress.

**Acknowledgements** This research was sponsored by the Air Force Office of Scientific Research (AFOSR), Drs. David Stargel and Joycelyn Harrison, Program Managers. The authors would like to thank Brad White, who created the epoxy-based samples and performed and analyzed the previous characterization of these materials.

Opinions, interpretations, conclusions and recommendations are those of the authors and are not necessarily endorsed by the United States Air Force.

## References

1. Jordan JL, Spowart JE (2012) Microstructure design and optimization of highly-filled polymer-based composites, Air force research laboratory, AFRL-RW-EG-TR-2012-002
2. Martin M, Hanagud S, Thadhani NN (2007) Mechanical behavior of nickel + aluminum powder-reinforced epoxy composites. Mat Sci Eng A 443:209–218
3. Cai J, Walley SM, Hunt RJA, Proud WG, Nesterenko VF, Meyers MA (2008) High-strain, high-strain-rate flow and failure in PTFE/Al/W granular composites. Mat Sci Eng 472A:308–315
4. Patel NR (2004) Intermediate Strain Rate Behavior of Two Structural Energetic Materials. Master's Thesis, Georgia Institute of Technology
5. Fu S-Y, Feng X-Q, Lauke B, Mai Y-W (2008) Effects of particle size, particle/matrix interface adhesion and particle loading on mechanical properties of particulate-polymer composites. Composites: Part B 39:933–961
6. Clyne TW, Withers PJ (1993) An introduction to metal matrix composites, 1st edn. Cambridge University Press, England
7. Ferranti L, Thadhani NN (2007) Dynamic mechanical behavior characterization of epoxy-cast Al + $Fe_2O_3$ thermite mixture composites. Metal Matls Trans A 38A:2697–2715
8. Cai J, Nesterenko VF, Vecchio KS, Jiang F, Herbold EB, Benson DJ, Addiss JW, Walley SM, Proud WG (2008) The influence of metallic particle size on the mechanical properties of polytetrafluoroethylene-Al-W powder composites. Appl Phys Let 92:031903
9. Herbold EB, Nesterenko VF, Benson DJ, Cai J, Vecchio KS, Jiang F, Addiss JW, Walley SM, Proud WG (2008) Particle size effect on strength, failure, and shock behavior in polytetrafluoroethylene-Al-W granular composite materials. J Appl Phys 104:103903
10. Ramsteiner F, Theysohn R (1984) On the tensile behavior of filled composites. Composites 15:121–128
11. Ferranti JL, Thadhani NN, House JW (2006) Dynamic mechanical behavior characterization of epoxy-cast Al + $Fe_2O_3$ mixtures. AIP Conf Proc 845:805–808
12. Oline LW, Johnson R (1971) Strain rate effects in particulate-filled epoxy. ASCE J Eng Mech Div 97:1159–1172
13. Goyanes S, Rubiolo G, Marzocca A, Salgueiro W, Somoza A, Consolati G, Mondragon I (2003) Yield and internal stresses in aluminum filled epoxy resin. A compression test and positron annihilation analysis. Polymer 44:3193–3199
14. Kawaguchi T, Pearson RA (2003) The effect of particle-matrix adhesion on the mechanical behavior of glass filled epoxies. Part 1. A study on yield behavior and cohesive strength. Polymer 44:4229–4238
15. Kawaguchi T, Pearson RA (2003) The effect of particle-matrix adhesion on the mechanical behavior of glass filled epoxies. Part 2. A study on fracture toughness. Polymer 44:4239–4247
16. Jordan JL, Spowart JE, White B, Thadhani NN, Richards DW (2009) Investigation of dynamic mechanical properties of multi-constituent particulate composites based on factorial design of experiments. In: Society for Experimental Mechanics Annual Conference on Experimental and Applied Mechanics 2009, Society for Experimental Mechanics, pp 256–265, Orlando, FL
17. Jordan JL, Siviour CR, Foley JR (2008) Mechanical properties of Epon 826/DEA epoxy. Mech Time Depend Mater 12(3):249–272
18. Song B, Chen W, Montgomery ST, Forrestal MJ (2010) Mechanical response of an alumina-filled epoxy at various strain rates. J Compos Mater 43(14):1519–1536
19. Rumchik CG, Jordan JL (2007) Effect of aluminum particle size on the high strain rate properties of pressed aluminized explosives. In: Elert M, Furnish MD, Chau R, Holmes N, Nguyen J (eds) Shock Compression of Condensed Matter – 2007, pp 795–798, American Institute of Physics, Melville, NY
20. Jordan JL, Siviour CR (2010) unpublished data

# Chapter 41
# Energy Absorption of Sandwich Tubes Under Lateral Loading

**Zhihua Fan, Jianhu Shen, Guoxing Lu, and Dong Ruan**

**Abstract** The energy absorption of sandwich tubes impacted between two rigid plates was investigated experimentally and numerically in the paper. Based on the collapse patterns observed in quasi-static experiment, several typical sandwich specimens were made and tested. The specimens were placed on the bottom platen of an Instron machine and gained a constant upwards velocity, followed by the impact with the top rigid platen. Their deformation history and load-compression curves were recorded. The energy absorption of sandwich tubes were then calculated and analyzed. Three different crushing patterns have been identified from previous experiments. The dynamic enhancement of energy absorption of the sandwich tubes were only observed in collapse pattern III under the tested velocity up to 10 m/s. Finite element (FE) models using ABAQUS were developed and validated against experimental results and the strain rate effect of metallic foam was considered. They were used to explore the detailed energy-absorption characteristics beyond the experimental range for impact velocities up to 100 m/s. The dynamic enhancement occurred for each configuration of sandwich tubes when the impact velocities were greater than 20 m/s. It was found that increasing the compression velocity leads to an increase in total plastic energy dissipation. Sandwich tubes with a thicker foam core are proved to be the optimum design.

**Keywords** Empty tubes • Sandwich tubes • Collapse pattern • Energy absorption • Quasi-static • Dynamic response

## 41.1 Introduction

Recently, much attention on energy-absorbing devices (e.g. tubular structures) in crashworthiness applications has been attracted. As a commonly used energy absorber, structural responses of thin-walled metallic tubes have been widely investigated due to their higher energy absorption capacity [1]. Metallic foams have recently emerged as a novel material in energy-absorbing devices and hence the filling of metallic foams is a general method which may enormously improve the energy absorption capacity of tubes [1]. Research has identified the remarkable improvement in the energy absorption capacity of foam-filled single and double sandwich tubes in axial crushing [2–4], though not necessarily with better specific energy absorption (SEA). Meanwhile, the crashworthiness of foam-filled single tubes was enhanced under lateral crushing [5]. Thus, similar energy absorption enhancement might be expected for new topological sandwich tubes. The behavior of double tubes with metallic foams sandwiched under dynamic lateral compression has been less reported in the literature. Previous research on quasi-static responses has already identified such improvement [6].

In the current paper, energy absorption of empty and sandwich tubes under dynamic lateral compression is reported. A series of tests on empty tubes and sandwich tubes were performed by using an Instron VHS8800 machine, in which the compression speed was kept constant. The specimens were tied on the bottom platen and then they moved together, followed by an impact with the top platen. The dynamic force and deformation histories were recorded at the same time by

Z. Fan • G. Lu (✉)
School of Mechanical and Aerospace Engineering, Nanyang Technological University,
50 Nanyang Avenue, Singapore 639798, Singapore
e-mail: fanz0001@e.ntu.edu.sg; gxlu@ntu.edu.sg

J. Shen • D. Ruan
Faculty of Engineering and Industrial Sciences, Swinburne University of Technology, John Street, Hawthorn, VIC 3122, Australia

V. Chalivendra et al. (eds.), *Dynamic Behavior of Materials, Volume 1: Proceedings of the 2012 Annual Conference on Experimental and Applied Mechanics*, Conference Proceedings of the Society for Experimental Mechanics Series,
DOI 10.1007/978-1-4614-4238-7_41, 

**Fig. 41.1** Photograph of sandwich tubes specimens subjected to dynamic impact

**Table 41.1** Specimens for empty tubes and glued sandwich tubes (stroke: 75% of diameter)

| Test | Diameter $D_o$ (mm) | Thickness $t_o$ (mm) | Diameter $D_i$ (mm) | Thickness $t_i$ (mm) | Mass $m$ (g) | Velocity $v$ (m/s) | Collapse pattern | E.A. (J) | SEA (J/g) |
|---|---|---|---|---|---|---|---|---|---|
| ET01, ET02 | 99.8 | 1.92 | – | – | 83.1 | 10.0 | – | 62.3 | 0.75 |
| ET03 | 79.6 | 2.00 | – | – | 69.1 | 10.0 | – | 95.5 | 1.38 |
| ET04 | 49.9 | 2.96 | – | – | 64.0 | 10.0 | – | 232 | 3.63 |
| SWT01 | 99.9 | 1.87 | 50.3 | 1.90 | 180.0 | 10.0 | II | 431 | 2.39 |
| SWT02 | 99.7 | 1.87 | 49.9 | 2.94 | 204.0 | 1.0 | III | 588 | 2.88 |
| SWT03 | 99.6 | 1.89 | 79.8 | 1.97 | 172.0 | 10.0 | I | 260 | 1.51 |
| SWT04 | 99.6 | 1.89 | 59.8 | 1.86 | 177.0 | 10.0 | II | 346 | 1.70 |
| SWT05, SWT06 | 99.6 | 1.91 | 50.0 | 2.95 | 201.0 | 10.0 | III | 796 | 3.96 |
| SWT07 | 99.9 | 1.93 | 50.0 | 1.90 | 182.0 | 1.0 | II | 434 | 2.12 |

a high-speed data acquisition package. FE code ABAQUS/Explicit was then employed to analyze the quasi-static and dynamic response of the sandwich tubes. The strain-rate effect of metallic foam was also considered [7]. A comparative study of the energy absorption between quasi-static and impact tests on sandwich tubes is described, in terms of energy partition and SEA, to further explore the role of each component in dissipating energy and to optimize structural design subjected to quasi-static and dynamic crushing.

## 41.2 Experimental Details

The sections of some specimens are shown in Fig. 41.1, illustrating the aluminium alloy tubes with an annular region filled with an aluminum foam core or the sandwich tubes. The material of the metallic tubes was AA6060-T5 and the total length was fixed at 50 mm for each specimen. The yield stress $\sigma_y$ of thin-walled tube from tensile tests was 150 MPa. The density of tube wall material $\rho$ is 2,760 kg/m$^3$. Cylindrical tubes of foam were cut from an initial aluminium foam block with the nominal relative density of 9%, provided by Alporas®, Germany. The material properties of foam core from uniaxial compressive tests were the same as those in [7]. The thixotropic epoxy liquid adhesive was used to paste, separately, the two empty tubes with a foam core. Empty tubes and sandwich tubes were tested separately on the VHS8800 machine and their details are listed in Table 41.1.

A photograph of the experimental setup is shown in Fig. 41.2a. The top platen is clamped and the ending position of the bottom platen is controllable. In our tests, two constant compressive rates were applied, namely 1 m/s and 10 m/s. During the initial tests, the empty tubular specimen was placed either on the bottom platen or top platen, in order to explore the effect of these two methods. Subsequently, for all the rest of tests, the specimen was only placed on the bottom platen and then moved upwards together with the bottom platen. A high-speed camera was used to track the deformation profile during crushing. The specimens in Fig. 41.2b subjected to quasi-static loading were conducted in [6] and executed for the further energy absorption analysis, including energy partitions as well as specific energy efficiency to determine the optimum design.

## 41.3 Experimental Results

Due to the collision between the tubular specimen and compression platen, the contact force between the top platen and the tube, and the bottom platen and the tube could be different because of the inertia of the tube. To obtain a reliable compression force using the current experiment set-up in Fig. 41.2a, two impact cases were conducted for empty tubes with the same dimension (ET01 and ET02). Method I was to attach the tube to the bottom platen and move upwards with the bottom platen

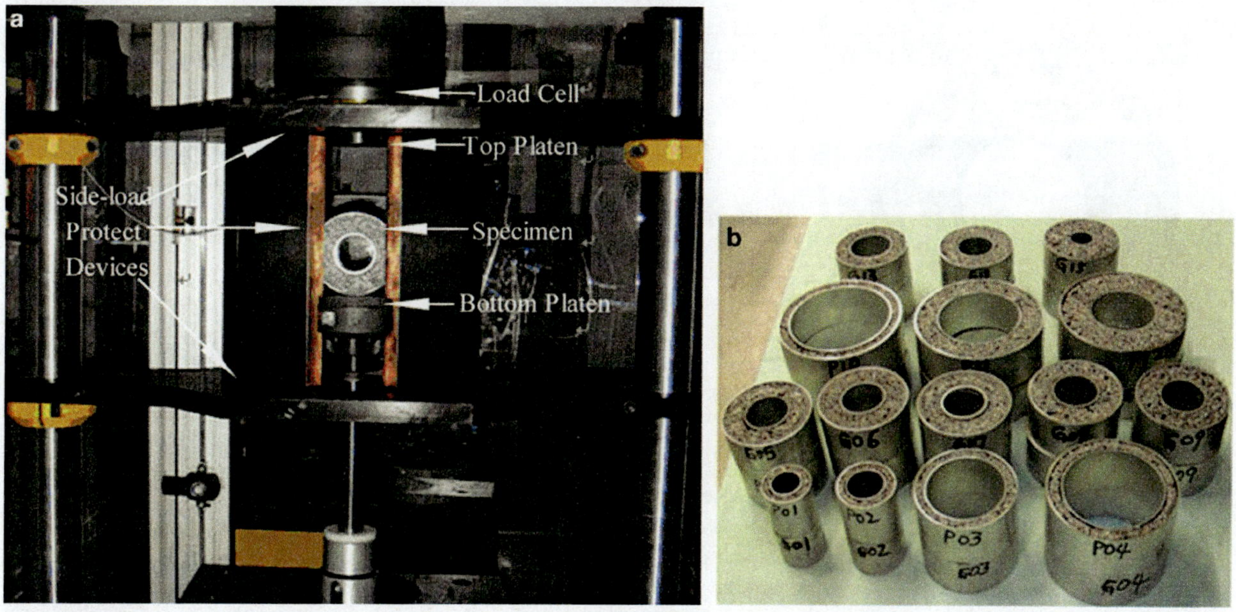


**Fig. 41.2** (**a**) Experimental setup for dynamic tests; (**b**) specimens for quasi-static tests [6]

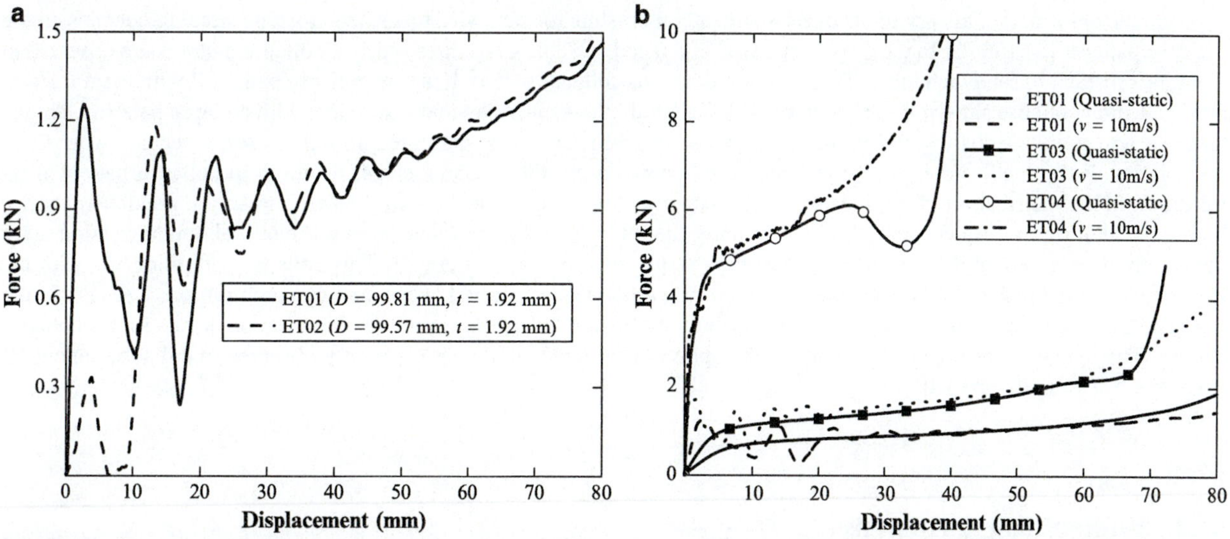


**Fig. 41.3** Comparison of force-displacement curves of (**a**) specimens ET01 and ET02 tested using method I and II, respectively; (**b**) empty tubes in the quasi-static and dynamic crushing ($v = 10$ m/s)

until it impacts the top platen. Method II was to attach a tube to the top platen and keep still until the bottom platen impacted it. Figure 41.3a shows the force-displacement histories for these two cases. The two methods resulted in diverse forces in the initial stage, but later the two forces were almost identical. For method II, the initial contact force was very small compare to that in the former case. This phenomenon was caused by the collision between tube and top platen which resulted in the cell wall of the tube bouncing away from the top platen at the beginning of the compression process. This elastic bouncing-back behaviour was observed from the high speed video. The contact force in method I was more reasonable to represent the dynamic compression force of the tubes between two plates. Thus this impact method was selected for all other tubular specimens.

Figure 41.3b shows the comparison of load-deformation results of several empty tubes between quasi-static and dynamic tests with an impact velocity of $v = 10$ m/s. Apart from the initial difference in the early deformation phase, the crushing

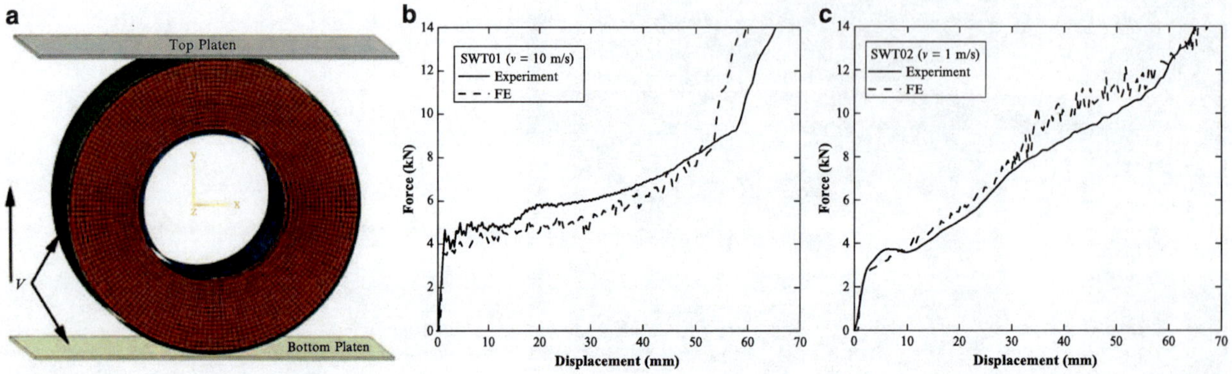


**Fig. 41.4** (**a**) Schematic of an FE model under impact; (**b**) and (**c**) validation of FE models against experimental results

strength are nearly the same under each loading case. From the deformation profile, the progressive collapse demonstrates three phases, i.e., initial tube wall collision, steady dynamic collapse and unloading phase. In the first phase, the contact force with large fluctuations is greatly influenced by the collision of the tube. The fluctuations are damped by the plastic deformation of the tube, in a mode similar to that in quasi-static compression [6]. Following this initial phase, a slight increase in the compressive strength of the plastic collapse phase is found, which comes up with a series of peaks and troughs, because of the hardening effect in the material and associated geometric change [2]. To avoid possible damage to the apparatus, the crushing process was stopped before self-contact of inner surface of the tube and this caused an unloading phase in the load-deformation curve.

From the deformation history of the quasi-static lateral crushing of sandwich tubes, the crushing strength depends on the collapse patterns [6]. Similar to the quasi-static case and listed in Table 41.1, three types of collapse patterns are observed in dynamic crushing of sandwich tubes. There are simultaneous collapse pattern I, plastic collapse pattern II with foam fracture and sequential collapse pattern III, respectively [8]. Generally, for sandwich tubular structure with collapse pattern I, double tubes deform simultaneously in the plastic collapse and the foam core is less deformed. In some cases, especially when thick foam is involved, ductile fracture of the foam core occurs during the crushing stage, resulting in collapse pattern II. A transition in the distribution of deformation might occur as the foam becomes even thicker. Unlike the previous patterns, both tubes deform at different time. In the beginning the inner tube does not deform at all. Noticeable deformation takes place only after a certain displacement, indicating a sequential collapse pattern III. This collapse pattern occurs with the specimens of the thickest foam (specimen SWT06), in both the quasi-static loading and dynamic impact. Because of the inherent large variation in the size, shape and distribution of foam cells, non-symmetric deformation occurs in this collapse pattern. The dynamic enhancement of energy absorption of the sandwich tubes were only observed in collapse pattern III under the test velocity up to 10 m/s.

## 41.4 Finite Element Analysis

Experimentally the maximum velocity of dynamic compression is 10 m/s limited by the capacity of the machine. In order to investigate the behaviour of tubes at higher velocities and to further understand the mechanics and mechanisms of plastic deformation, a finite element (FE) analysis was conducted. Fracture was not introduced in the simulation.

In FE analysis shown in Fig. 41.4a, element distortion control of foam was introduced since large plastic deformation was observed in the tests. The inner surface of the outer tube and the outer surface of the inner tube were tied to the foam core to model perfect bonding without failure. Self contact was applied to the inner surface of the inner tube and the friction coefficient was set at 0.1. Both platens were modeled as rigid bodies. The top platen, which was at the proximal surface, was fixed during the whole crushing process. To replicate the impact method I adopted in previous experiments [8], an initial velocity was applied to the sandwich tube and the bottom platen. Good agreements between the numerical and experimental results are shown in Fig. 41.4b, c. Thus, FE modeling was validated and can be used to explore the energy absorption of sandwich tubes under higher velocities in the next section.

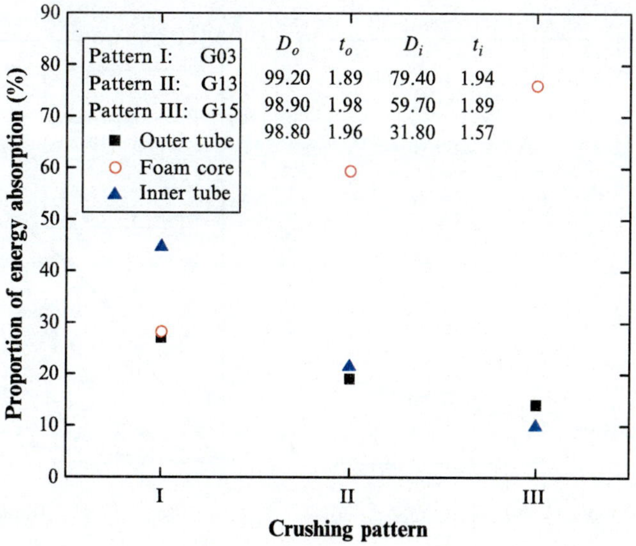


**Fig. 41.5** Energy partition for each crushing pattern in quasi-static

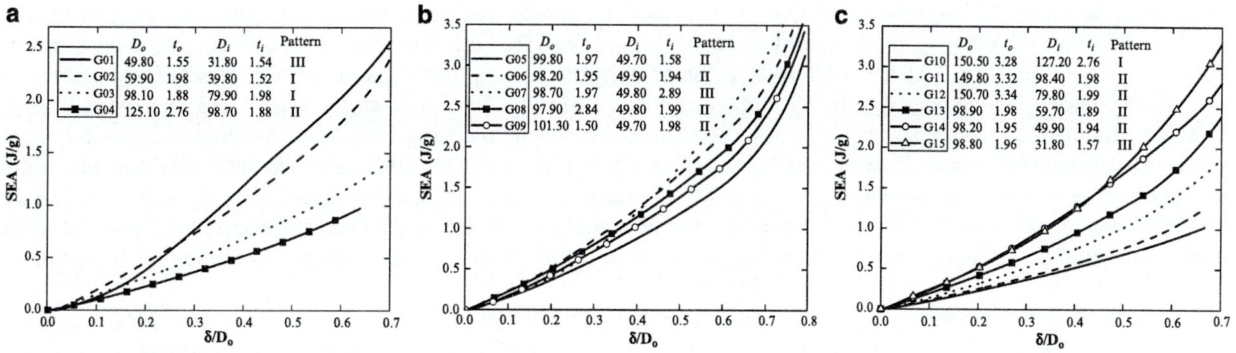


**Fig. 41.6** Plots of specific energy absorption (SEA) versus non-dimensional deformation for sandwich tubes under quasi-static loading: (**a**) group 1; (**b**) group2; (**c**) group 3

## 41.5 Energy Absorption Characteristics

To better understand energy dissipation mechanism in dynamic crushing, the energy absorption characteristics subjected to quasi-static loading should be examined in further detail. The deformation model and collapse force for each specimen were discussed in Ref. [6]. To illustrate the role of each component during the crushing, the corresponding plastic energy absorbed by sandwich tubes were partitioned, in terms of each component's plastic strain energy, which cannot be measured in the experiments and was only available from FE simulation. Figure 41.5 shows the proportion of strain energy shared by each component in those typical crushing patterns during crushing tests. Herein, typical specimens G03, G13 and G15 were selected due to their almost identical outer tube dimensions. The dimensions of the specimens are listed in Fig. 41.5 for reference. Specimen G15 has the largest value of diameter ratio, $D_o/D_i$, the foam core is the thickest. It shows that the transition in crushing pattern leads to a decrease in the proportions of plastic energy for both tubes. However, the foam core dominates the plastic energy during plastic collapse, which consumes up to 75% of the total energy dissipation at a compression strain around 80%, particularly in a more efficient collapse pattern III, due to the plastic deformation of foam as well as interaction between the tube and the foam filler.

In order to investigate the superiority of tubular structure with foam sandwiched subjected to lateral compression, all the experimental results in glued cases are studied in terms of SEA, which was defined as the energy absorbed per unit weight of a crushed structure, as plotted in Fig. 41.6. The specimens were divided into three groups according to their geometrical

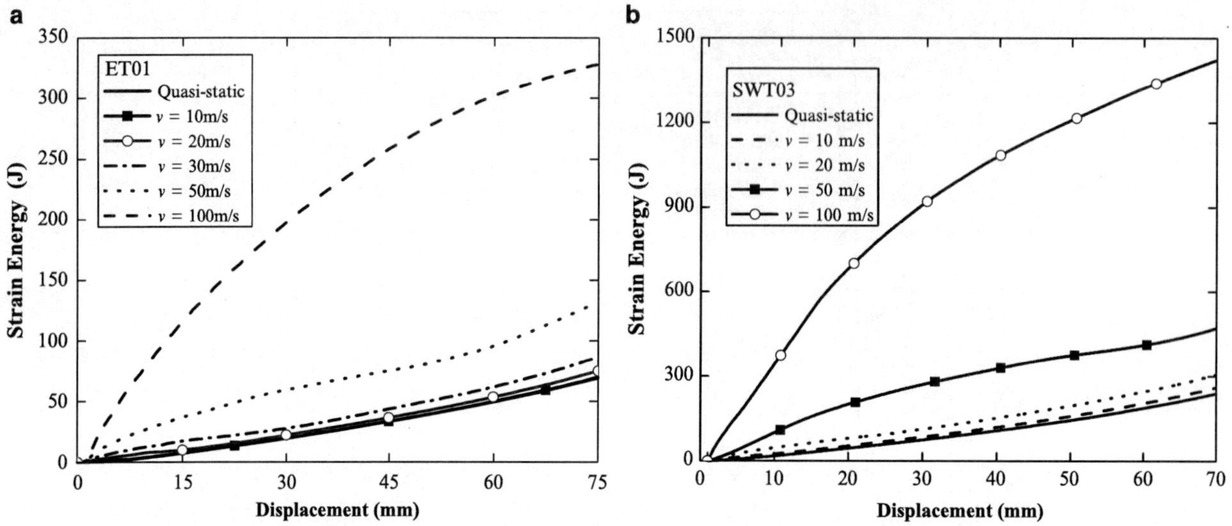


**Fig. 41.7** Energy absorption for (**a**) an empty tube, and (**b**) a sandwich tube under different impact velocities

features. For group 1 in Fig. 41.6a, the foam cores of all the specimens have the same thicknesses but different diameters of inner and outer tubes. Although specimen G01 has the smallest diameter and lightest weight, it is the most weight efficient structure with the SEA value up to 2.6 J/g at 70% strain. The specimen G04 is less efficient even with the largest thickness of outer tube. For specimens G02 and G03 with the same crushing pattern, the value of SEA is significantly greater in magnitude compared to the ones with larger average radius of curvature. Overall, it suggests that specimens deform in the crushing pattern III are favorable to energy absorbers. Likewise, for the second group in Fig. 41.6b, tubes have the same geometrical parameters except different thickness of inner tubes in specimens G05, G06 and G07. The specimen G07 dissipates energy much more than those with smaller thicknesses of inner tubes. Specimens G08 and G09, in which the only different geometrical parameter is the thickness of outer tube, the one with larger thickness of outer tube is much better in energy dissipation. This finding demonstrates that sandwich structures having thicker inner or outer tubes are the optimum design for quasi-static loading for a fixed crushing pattern.

Figure 41.6c plots the specific energy absorption of specimens in the last group with different foam core thicknesses, but similar diameters and thicknesses of outer tubes. As expected, specimen G15 with the most foam core filling is the one that absorbs the most energy. It is concluded that depending upon the level of plateau stress, the foam core may change the original collapse mode into a more efficient mode with higher energy efficiency. Besides, the interaction between the foam core and tube walls provides additional enhancement in energy dissipation. When foams are compressed, densification occurs after the plateau stage and the stress increases rapidly with the strain, which might lead to a higher stoke efficiency and a larger average force as well as specific energy absorption. A large amount of material in such sandwich tube sustains plastic deformation, which dissipates plastic energy remarkably.

The values of energy absorption (E.A) and SEA for specimens under dynamic experiments are also given in Table 41.1. It is clearly shown that sandwich tubes with collapse pattern III is the most weight efficient, with the maximum SEA value of 3.96 J/g. In order to understand the dynamic influence, particularly the velocity effect, the sandwich tubes were numerically compressed at different velocities up to 100 m/s. Empty tube (ET01), sandwich tubes with the same outer tube diameter but a thin foam core (SWT03) and a thick foam core (SWT06) were analyzed. From the FE simulation on specimens ET01 and SWT03, the variation of energy dissipation in each specimen for different impact velocities up to 100 m/s is plotted in Fig. 41.7, corresponding to an end displacement of 70 mm. It is evident that when the compression speed is lower than a certain value, saying 20 m/s as shown, no dynamic enhancement on energy absorption is observed. Small discrepancy appears only at the initial stage where localized plastic deformation takes place at the upper part of the tubes. Noticeable difference in the curves exists when the loading velocity is beyond that value (20 m/s). It is clearly shown that the energy dissipation capacity is enhanced sharply as the increase in the compression velocity, for both empty and sandwich tubes. Although the mass of specimen SWT03 (172 g) is about double of the mass of specimen ET01 (83.1 g), the value of energy dissipation of the former at $v = 100$ m/s is 1,413 J, which is more than four times than the latter. It shows potential specific energy (SEA) improvement of current sandwich tubes compared to empty tubes.

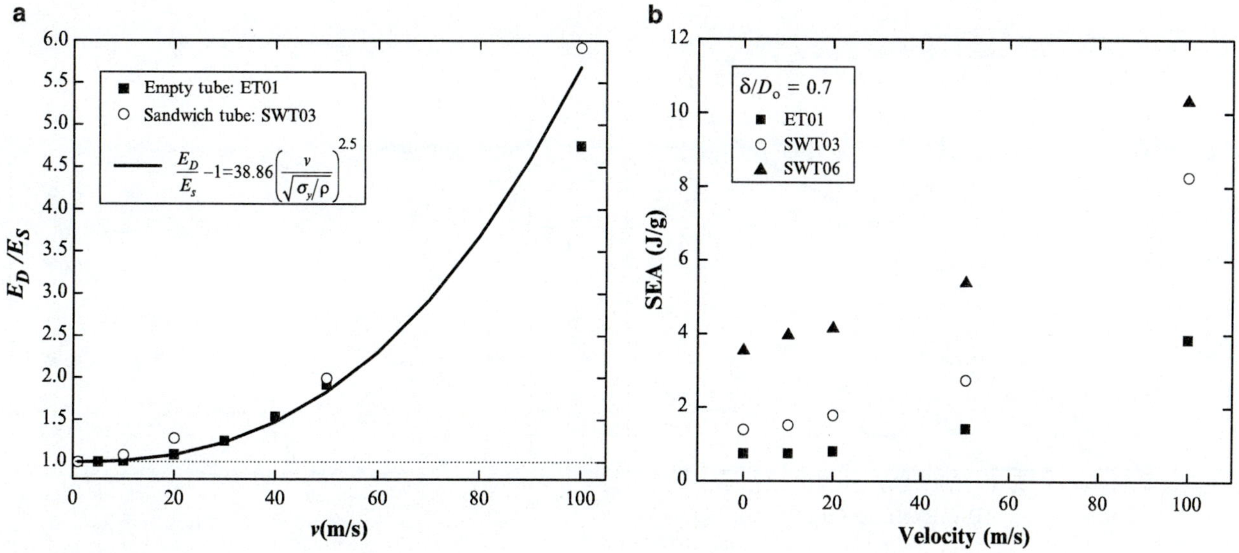


**Fig. 41.8** (**a**) Plot of energy absorption enhancement for both empty and sandwich specimens over its quasi-static counterpart; (**b**) plot of SEA versus compressive velocity for ET01, SWT03 and SWT06

Furthermore, the ratio of energy dissipation for each specimen under dynamic impact and quasi-static compression is plotted in Fig. 41.8a. It is interesting to note that the points seem to line into a single curve, no matter whether it is for empty or sandwich tubes. When the velocity is less than a certain value, e.g. 20 m/s, the ratio of dynamic energy to quasi-static energy is around 1.1. Afterwards, the energy absorption capability under the dynamic crushing increases significantly with the compression speed, and especially it is over five times when the compression velocity reaches 100 m/s. For the empty tubes studied, the dynamic energy enhancement over its quasi-static counterpart is obtained, empirically, which is also plotted in Fig. 41.8a,

$$\frac{E_D}{E_S} = 1 + 38.86\left(\frac{v}{\sqrt{\sigma_y/\rho}}\right)^{2.5} \tag{41.1}$$

To evaluate the efficiency of novel structures and new materials, SEA was also introduced and extensively studied. The values of SEA against compression velocity for those specimens are plotted in Fig. 41.8b. Generally, the dynamic effect (velocity) would lead to an increase in the SEA value for all the specimens. It is illustrated that for specimen SWT06 with the thickest foam and smallest inner tube, the value of SEA is almost four times than that of the empty tube under the same loading profile. This result demonstrates that sandwich tubes having a thicker foam core are structures with better weight efficiency.

To evaluate the function of foam cores, the total energy absorbed is partitioned for each component. Figure 41.9a shows the partition of a sandwich tube (SWT06) with a thick foam core under an impact velocity of 10 m/s. When the dynamic displacement is less than 20 mm, the inner tube does not contribute to energy dissipation until the crushing layer becomes severely deformed and stiffened. The impact energy is almost equally dissipated by the outer tube and the foam core. Afterwards, all the three components start to dissipate energy. It is noted that at the strain of 0.7, the foam core dissipates 56.2% of the total strain energy. For the outer tube and inner tube, it is 28.5% and 15.3%, respectively. To further investigate the effect of velocity on energy absorption, a graph of energy dissipation for the tube under different compression speeds is plotted in Fig. 41.9b. The foam core absorbs more than half of the whole internal strain energy, and this percentage (approximately 60%) does not vary significantly with the compression speed. However, for compression velocity less than 50 m/s, the outer tube absorbs more energy than the inner one. On the contrary, when the compression velocity becomes large (above 50 m/s), the inner tube deformed severely with more strain energy dissipation. Overall, for the sandwich tubes with collapse pattern III, the foam core absorbs the largest portion of energy up to 60%, which again confirms the efficient enhancement of foam-filling in structural design.

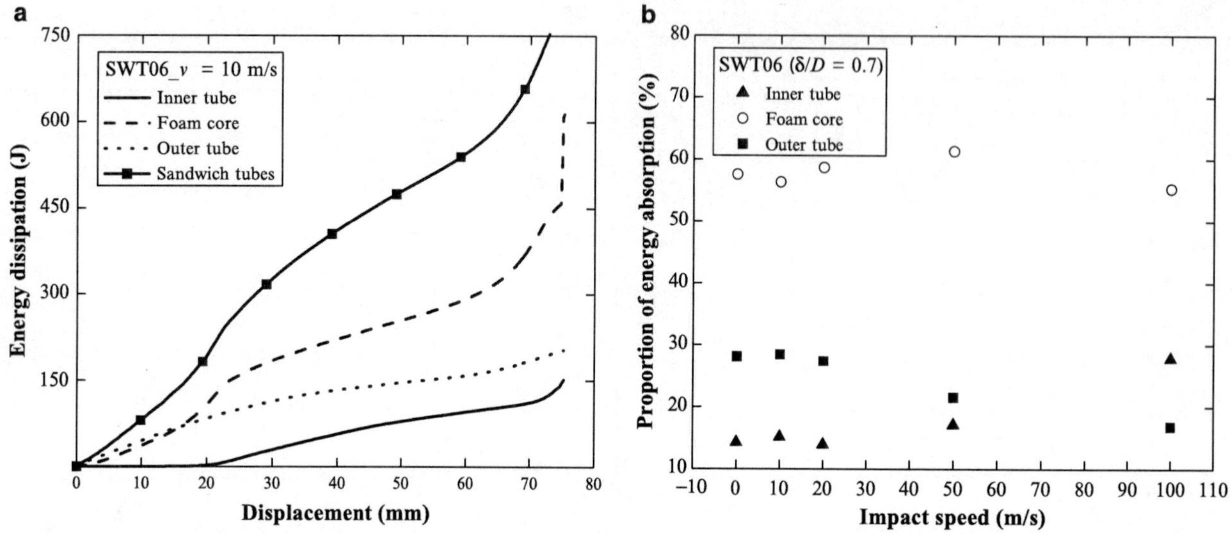


**Fig. 41.9** Energy absorption characteristics for each component of SWT06: (**a**) energy dissipation versus displacement; (**b**) partition of energy versus speed

## 41.6 Conclusion

This paper presents both experimental and finite element results on the energy absorption of empty tubes and sandwich tubes compressed by two flat plates. The dynamic lateral crushing tests of several empty tubes and sandwich tubes with different geometrical parameters have been carried out using an INSTRON machine. Through the comparison of load-deflection curves for both impact methods, a suitable and desirable impact method was chosen for the rest of impact tests on sandwich tubes to avoid rebound phenomenon. Based on experimental results and detailed FE analysis on empty and sandwich tubes subjected to quasi-static and dynamic loading, energy absorption characteristic were clarified, in terms of total energy absorbed, specific energy absorption as well as energy partition. For both empty and sandwich tubes, significant enhancement has been shown for the energy absorption in dynamic loadings when the impact velocity is greater than 20 m/s, for which the inertia effect would be responsible. It is evident that sandwich tubes with a thicker metallic foam core are more energy efficient, not only in quasi-static situation, but also in dynamic loading cases.

## References

1. Lu G, Yu TX (2003) Energy absorption of structures and materials. Woodhead, Cambridge
2. Reid SR, Reddy TY, Gray MD (1986) Static and dynamic axial crushing of foam-filled sheet metal tubes. Int J Mech Sci 28:295–322
3. Güden M, Kavi H (2006) Quasi-static axial compression behavior of constraint hexagonal and square-packed empty and aluminum foam-filled aluminum multi-tubes. Thin-Walled Struct 44:739–750
4. Reddy TY (1988) Wall R J. Axial compression of foam-filled thin-walled circular tubes. Int J Impact Eng 7:151–166
5. Hall IW, Güden M, Claar TD (2002) Transverse and longitudinal crushing of aluminum-foam filled tubes. Scr Mater 46:513–518
6. Fan Z, Shen J, Lu G (2011) Investigation of lateral crushing of sandwich tubes. Procedia Eng 14:442–449
7. Shen J, Lu G, Ruan D (2010) Compressive behaviour of closed-cell aluminium foams at high strain rates. Compos Part B: Eng 41:678–685
8. Fan Z, Shen J, Lu G and Ruan D (2011) Investigation of dynamic behavior of sandwich tubes with metallic foam cores compressed by two platens. In: Proceedings of the 3rd international symposium on plasticity and impact (ISPI'2011), Hong Kong, 8–12 Dec 2011

# Chapter 42
# Improvement of Dynamic Properties of PLLA/PBAT Polymer Alloys Using Dialkyl Peroxide

**Masahiro Nishida, Hiroki Ichihara, Hiroki Watanabe, Norio Fukuda, and Hiroaki Ito**

**Abstract** Polymer alloys of poly(lactic acid) (PLA) and poly(butylene adipate/terephthalate) (PBAT) were used, and the effects of the mixing ratio of PLA and PBAT and the addition of dialkyl peroxide (compatibilizing agent) were examined. The stress–strain curves of PLA/PBAT and PLA/PBAT/dialkyl peroxide specimens were measured using the tensile split Hopkinson pressure bar (Kolsky Bar) method. The mixing ratio of PLA and PBAT and the dialkyl peroxide addition affected the stress–strain curve behavior. The dialkyl peroxide addition increased the elongation at break and changed the fracture surface to a whitened form that was markedly uneven. The elongation at break increased with increasing PBAT content.

**Keywords** Bioplastics • Tensile split hopkinson bar • Polymer alloys • Compatibilizing agent

## 42.1 Introduction

The increasing use of plastic products worldwide is causing considerable damage to the environment; therefore, biodegradable plastics (plastics that can decompose in the natural environment) and bio-based plastics (plant-derived or recyclable—resource-based plastics) are being extensively investigated, and new biodegradable and bio-based plastics are continuously being developed.

Poly(lactic acid) (PLA) is a typical biodegradable bio-based polymer (plant-derived polymer). In Japan, PLA is already being used to manufacture many industrial products such as the interior parts of cars, parts of computer cases, and cell-phone cases. Many studies have been conducted to determine other industrial products that can be manufactured using PLA. However, such applications are limited to machine parts that are subject to low loading. Polymer blends/alloys or natural fiber reinforcing have been required to overcome the low impact resistance and the brittleness of PLA. Because poly (butylene adipate/terephthalate) (PBAT) is a ductile and biodegradable polymer, polymer alloys of PLA and PBAT have great potential for high impact strength. For example, Fukuda, one of the authors, showed that the Izod impact strength of PLA and PBAT polymer alloys increases when dialkyl peroxide is used, as shown in Fig. 42.1 [1]. In particular, when the mixing ratio of PLA/PBAT/dialkyl peroxide was 60/40/1, the Izod impact strength was 60 kJ/m$^2$, which is comparable to that of polycarbonate (PC). Coltelli et al. examined the addition of acetyl tributyl citrate (ATBC) and PBAT to PLAT in order to improve tensile properties at low strain rate [2]. Signori et al. studied the effect of processing at high temperature on the molecular weight distribution, the morphology and the thermo-mechanical properties of PLA and PBAT polymer alloys [3]. However, less work has been reported on the basic mechanical properties of such plastics at high strain rates. The authors' group has been studying the mechanical properties of PLA/PBAT polymer alloys at high strain rates [4].

In the present study, the stress–strain curves of PLA and PBAT polymer alloys were measured at high strain rates using a tensile split Hopkinson pressure bar (Kolsky bar). The effects of the addition of dialkylperoxide (compatibilizing agent)

M. Nishida (✉) • H. Ichihara • H. Watanabe
Department of Mechanical Engineering, Nagoya Institute of Technology, Gokiso-cho, Showa-ku, Nagoya, Aichi 466-8555, Japan
e-mail: nishida.masahiro@nitech.ac.jp

N. Fukuda
Headquarters, Aichi Center for Industry and Science Technology, 1267-1 Akiai, Yakusa-cho, Toyota-shi, Aichi 470-0356, Japan

H. Ito
Industrial Research Center, Aichi Center for Industry and Science Technology, 1-157-1, Onda-cho, Kariya-shi, Aichi 448-0013, Japan

V. Chalivendra et al. (eds.), *Dynamic Behavior of Materials, Volume 1: Proceedings of the 2012 Annual Conference on Experimental and Applied Mechanics*, Conference Proceedings of the Society for Experimental Mechanics Series,
DOI 10.1007/978-1-4614-4238-7_42, 

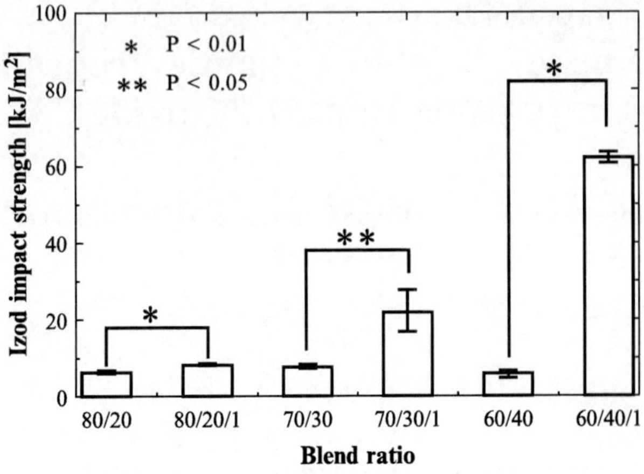


**Fig. 42.1** Impact resistance of different PLA alloys [1]

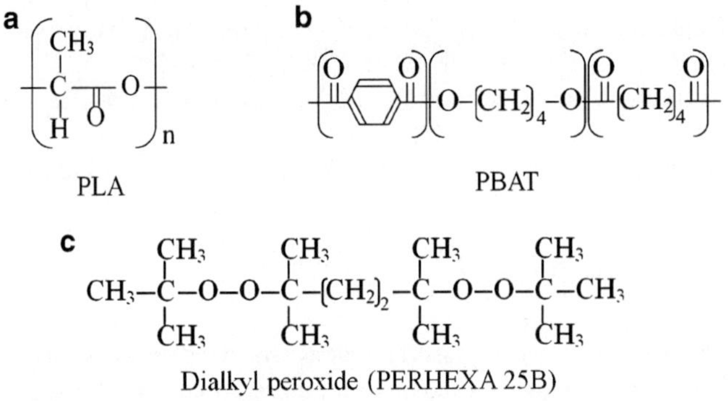


**Fig. 42.2** Chemical structural formula of PLA, PBAT and dialkyl peroxide

and the mixing ratio of PLA and PBAT on the maximum stress, the elongation at break, and fracture morphology were also discussed.

## 42.2 Experimental Methods

### *42.2.1 Materials*

We used PLA and PBAT alloys prepared using PLA from Toyota Motor Corporation (Toyota Eco-Plastic S-17) and PBAT from BASF (Ecoflex). In order to examine the effect of a compatibilizing agent, dialkyl peroxide (NOF Corporation, PERHEXA 25B) was used. The mixing ratios of PLA and PBAT were 60/40, 50/50, and 30/70. The mixing ratios of PLA/PBAT/dialkyl peroxide were 60/40/1, 50/50/1, and 30/70/1. The chemical structural formulas of PLA, PBSL and dialkyl peroxide are shown in Fig. 42.2. We prepared the polymer alloys using a twin-screw extruder (Technovel Co., Osaka Japan) at 180°C. The screw speed was 400 rpm, and the feed rate was 100 g/min. After melt mixing, the strands prepared by the twin-screw extruder were cooled rapidly, pelletized, and then dried. Next, 10-mm-thick plates were prepared using a conventional hot press at 190°C and 5 MPa for 30 min. Figure 42.3 shows photographs of the cryo-fractured surfaces of the specimens before dynamic tensile tests. These were taken using a scanning electron microscope (SEM). In the case of the PLA and PBAT specimens, a sea–island structure consisting of the PLA matrix and PBAT particles of the same size (domain) (1–6 μm) can be observed. When the compatibilizing agent was used, particles and phase separation were not clearly observed.

### *42.2.2 Dynamic Tensile Tests*

Dynamic tensile test specimens with a gage mark area of approximately 2 × 2 mm and a gage length of 4 mm as shown in Fig. 42.4 were used. Specimens for dynamic tensile tests were produced from plates of 5 mm thickness using a milling

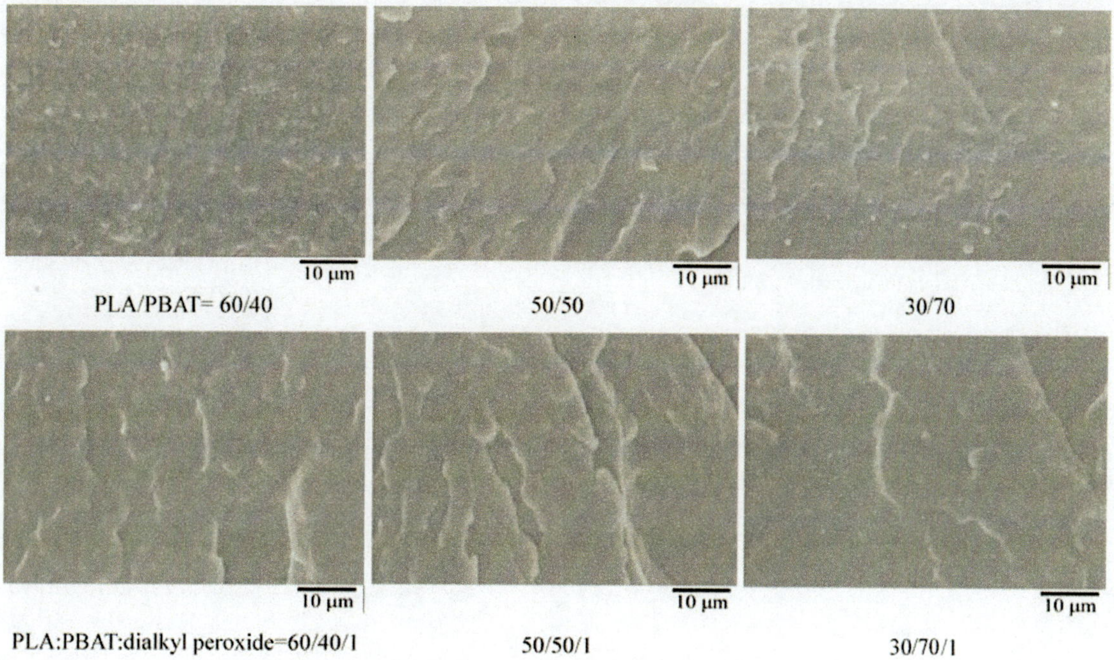


**Fig. 42.3** Photographs of cryo-fractured surfaces taken using a scanning electron microscope

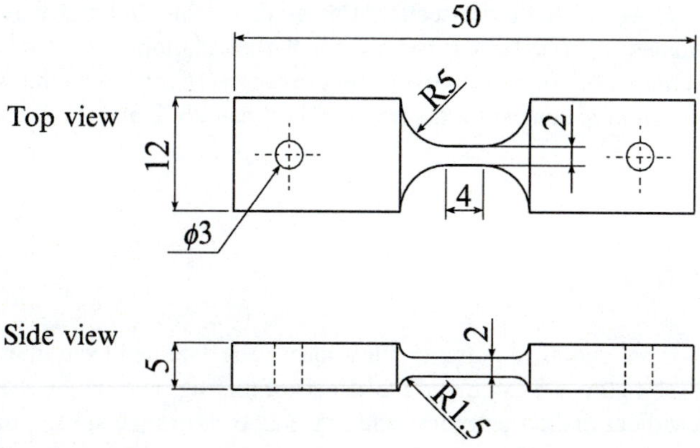


**Fig. 42.4** Specimen shape of dynamic tensile tests

machine. At high strain rates of 500–1,000 $s^{-1}$, the tensile properties of the specimens were examined by the tensile split Hopkinson (Kolsky) bar test [5], as shown in Fig. 42.5. The input and output bars were made of stainless steel (SUU304), and their diameters and lengths were 12 mm and 2,000 mm respectively. Strain gages were applied to both sides of the input and output bars at distances of 1,750 mm and 350 mm from the specimen, respectively. Because the stress histories were almost the same on both ends of the specimens, the strain and stress of the specimens were calculated from the strain of the bars as measured by the strain gages, using the following equations [6–8]:

$$\sigma_{av}(t) = \frac{AE}{A_s}\varepsilon_T(t) \tag{42.1}$$

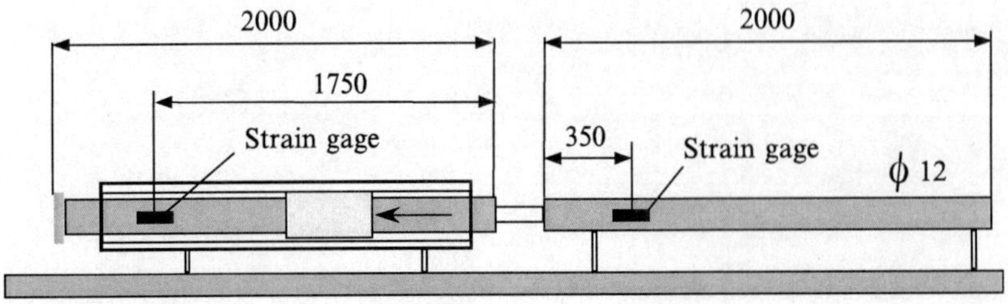


**Fig. 42.5** Experimental set-up for dynamic tensile tests

**Table 42.1** Material constants of input and output bars

| | |
|---|---|
| Density | $8.0 \times 10^3$ kg/m$^3$ |
| Velocity of bar wave, $c_3$ | 4970 m/s |
| Young's modulus, $E$ | 200 GPa |

$$\varepsilon_{av}(t) = \frac{2c_3}{L}\int_0^t [\varepsilon_I(t) - \varepsilon_T(t)]dt \tag{42.2}$$

Here $\varepsilon_I$ and $\varepsilon_T$ are the axial strains induced in the input bar by the incident wave, and in the output bar by the transmitted wave, respectively. $E$ and $c_3$ are Young's modulus and elastic wave velocity, respectively, of both the input and the output bars. $L$ is the gage length. $A$ and $A_S$ are the cross-sectional areas of the input/output bars and specimens, respectively. The material constants of the stainless steel (SUU304) bars used in the calculations are listed in Table 42.1. We preserved the specimens in a desiccator at a humidity of 30–40 % until just before use in order to avoid the effect of moisture absorption on the specimens. The specimens were maintained at a temperature between 23°C and 25°C.

## 42.3 Results and Discussion

Figure 42.6a, b show the stress–strain curves of polymer alloy specimens obtained from the strain history and stress history using Eqs. 42.1 and 42.2. For each figure, we examined the effects of mixing ratio on the stress–strain curves. Figure 42.6a shows the stress–strain curves without dialkyl peroxide, and Fig. 42.6b shows the stress–strain curves of dialkyl peroxide addition. Because the strain rate changed slightly during compression, it was determined using the averaged value of the strain rate–strain curve [9]. The results of PLA/PBAT = 80/20 and 70/30 [5] were added to Fig. 42.6a. The results of PLA/PBAT/dialkyl peroxide = 80/20/1 and 70/30/1 [5] were added to Fig. 42.6b.

Regardless of dialkyl peroxide addition, the maximum stress decreased and the elongation at break increased with increasing PBAT content. The specimens of PLA/PBAT/dialkyl peroxide = 50/50/1 and 30/70/1 did not break at a strain of 0.34 under this experimental condition. Comparing Fig. 42.6a, b, it was found that the dialkyl peroxide addition clearly increased the elongation at break. In particular, when PLA/PBAT = 60/40, 50/50 and 30/70, the dialkyl peroxide addition dramatically increased the elongation at break. When PLA/PBAT = 80/20, the dialkyl peroxide addition only slightly

**Fig. 42.6** Effect of PBAT content on stress–strain curve

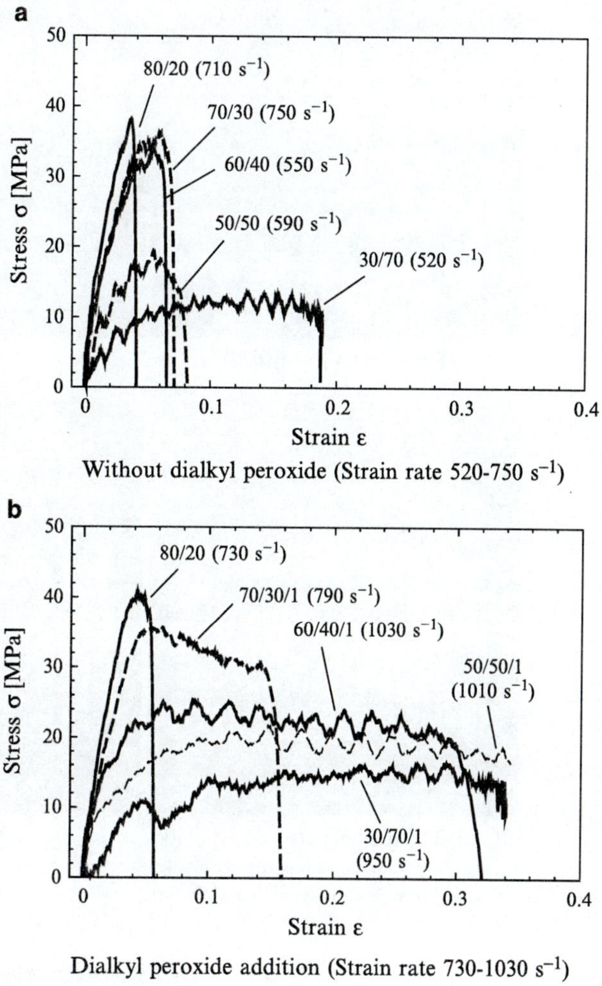


increased the elongation at break. When the mixing ratio of PBAT was minor, the effect of dialkyl peroxide addition was small. Compared to the elongation at break, the dialkyl peroxide addition had less or no effect on the maximum stress.

Figure 42.7 shows photographs of the fracture surface after the dynamic tensile tests. For comparison, we also observed the specimens after dynamic tensile tests which were carried out by Ichihara et al. in 2011 [5]. Figure 42.8 shows the fracture surfaces for PLA/PBAT = 80/20 and 70/30 and PLA/PBAT/dialkyl peroxide = 80/20/1 and 70/30/1.

When PLA/PBAT = 80/20 and 70/30 and PLA/PBAT/dialkyl peroxide = 80/20/1, the fracture surfaces were flat and smooth. When PLA/PBAT = 60/40, 50/50 and 30/70, several spherical and semi-spherical holes were observed. It seems that fractures at boundaries between the PLA rich phase and the PBAT rich phase occurred. When PLA/PBAT/dialkyl peroxide = 70/30/1, 60/40/1, 50/50/1 and 30/70/1, the fracture surfaces were not flat, and stress whitening was observed near the fracture surfaces.

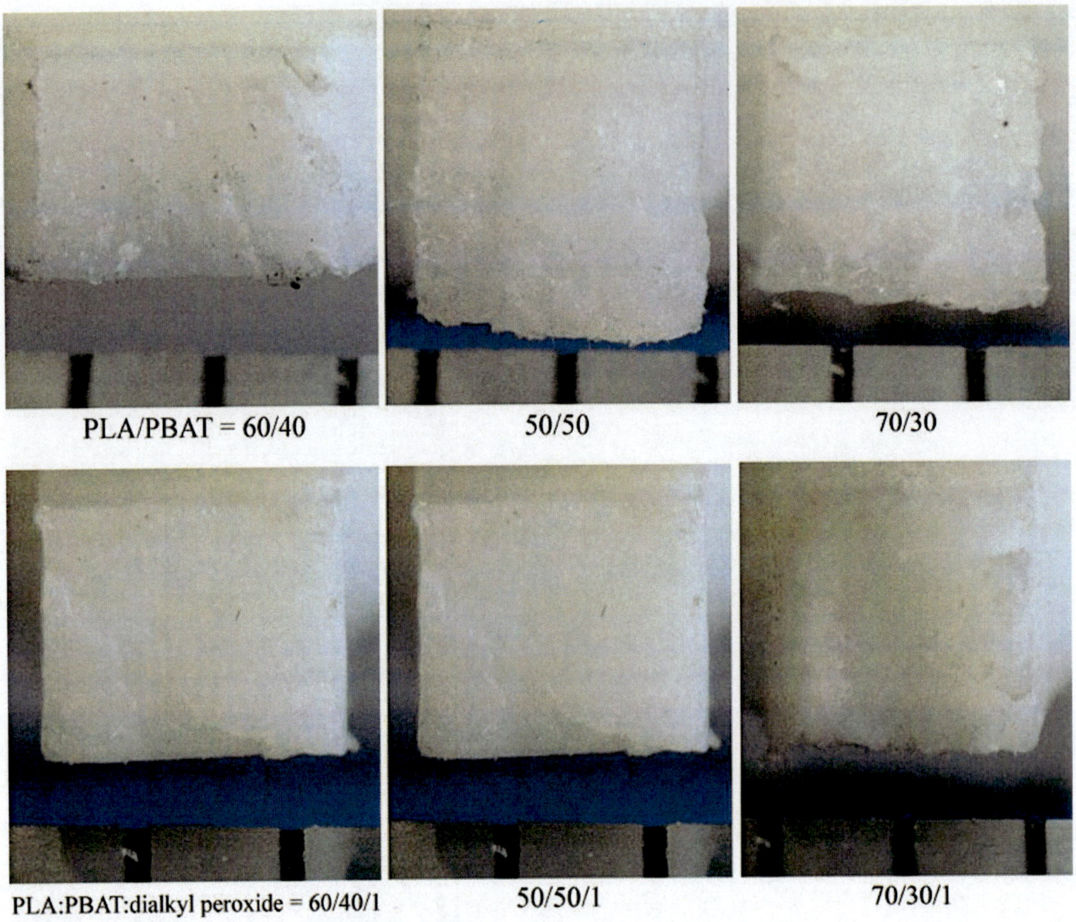


**Fig. 42.7** Fracture surface after dynamic tensile test

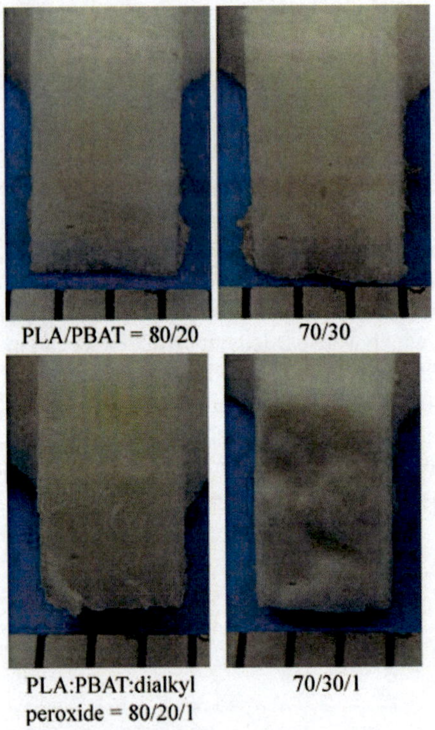


**Fig. 42.8** Observation of dynamic tensile test specimen in previous study [5]

## 42.4 Conclusions

The stress–strain curves of PLA/PBAT and PLA/PBAT/dialkyl peroxide specimens were measured using the tensile split Hopkinson pressure bar (Kolsky Bar) method. The dialkyl peroxide addition increased the elongation at break and changed the fracture surface to a whitened form that was markedly uneven. When dialkyl peroxide was added, the elongation at break increased with increasing PBAT content. When PLA/PBAT/dialkyl peroxide = 70/30/1, 60/40/1, 50/50/1, 30/70/1, the area underneath the stress–strain curve before the breaking point, increased in size.

**Acknowledgment** The authors are greatly indebted to Dr. K. Ogawa of the Space Dynamics Laboratory in Japan for his valuable advice with respect to the tensile split Hopkinson pressure bar.

## References

1. Fukuda N, Matsubara H, Kitagawa R (2007) The crystallization behaviors and mechanical properties of poly(lactic acid)/poly(butylenes adipate terephthalate) alloys. In: Proceedings of 59th symposium on macromolecules, in Nagoya pp 5657–5658 2007
2. Coltelli M-B, Maggiore ID, Bertoldo M, Signori F, Bronco S, Ciardelli F (2008) Poly(lactic acid) properties as a consequence of poly(butylene adipate-co-terephthalate) blending and acetyl tributyl citrate plasticization. J Appl Polym Sci 110:1250–1262
3. Signori F, Coltelli M-B, Bronco S (2009) Thermal degradation of poly(lactic acid) (PLA) and poly(butylenes adipate-co-terephthalate) (PBAT) and their blends upon melt processing. Polym Degrad Stab 94:74–82
4. Nishida M, Ichihara H, Fukuda N (2011) Evaluation of dynamic compressive properties of PLA/PBAT polymer alloys using split Hopkinson pressure bar. Eng Trans 59(1):23–30
5. Nishida M, Ichihara H, Fukuda N (2011) Dynamic properties of PLA/PBAT polymer alloys. In: Proceedings of annual international workshop 2011, dynamic behavior of structures and materials, interaction and friction, in Metz 2011
6. Gray GT III (2000) Classic split Hopkinson pressure bar testing. In: Mechanical testing and evaluation, vol 8, ASM Handbook. ASM, Materials Park, pp 462–476
7. Chen W, Song B (2010) Split Hopkinson (Kolsky) bar: design. Testing and Applications, Springer
8. Ogawa K (2001) Impact tensile characteristics of 6061-T6 aluminum alloy. J Jpn Inst Light Met 51:175–181, in Japanese
9. Nakai K, Yokoyama T (2008) Strain rate dependence of compressive stress–strain loops of several polymers. J Solid Mech Mater Eng 2(4):557–566

# Chapter 43
# Operational Equations of State for Modeling High-Rate Phenomena

**Michael Grinfeld**

**Abstract** The concept of operational equations of state (EOS) is discussed and illustrated with several examples. We discuss several related concepts: (1) the complete and incomplete equations of state, (2) the skeleton of the equation of state, (3) the thermodynamic degrees of freedom of the EOS. Special emphasis is given to the operational equations of state associated with the data related to isentropic and high-rate (shock-like) loading. We present several novel explicit formulas for the recovery of the complete and incomplete operational equations of state for hydrocode. It is assumed that for the process under study that the substances can be treated as possessing only two thermodynamic degrees of freedom.

## 43.1 Introduction

Thermodynamics issues associated with equations of state (EOS) are the focus of this paper. The famous physicist and thermodynamics expert Percy W. Bridgman made many valuable contributions [1] while developing thermodynamics foundations of the EOS for high pressures. He once noticed [2]: "There are as many formulations of the Second Law as there have been discussions of it." This author's half-century of experience with thermodynamics showed that the situation is the same with interpretations of almost all key concepts of thermodynamics, not only with the Second Law. In fact, all researchers, even from the same community, have different visions about the most fundamental concepts of thermodynamics. This is a quite natural and universal phenomenon. The same happens with any discipline having a wide set of applications. This is why many authors using thermodynamics begin with discussion of their own understanding of thermodynamics. The main purpose here is just to make sure that the author is on the same page with the readers and that he uses a close or a far-distant vocabulary of the basic thermodynamics terms. The terminology, accepted in the contemporary EOS community and in this paper, in particular, is quite different from the terminology used in this author's thermodynamics monograph [3].

The Bridgman's main focus was on the high-pressure static experiments. The discussions of the EOS aspects associated with shock-wave loading can be found in the valuable reviews [4–9].

The citations [1–9] show that the EOS are the mandatory ingredients, which should be added to the universal conservation laws in order to make possible the analytical and numerical modeling. Formulation of the adequate EOS generally includes both theoretical schemes and experimental data. Theoretical schemes and experimental data are not independent from each other but mutually related (see, for instance, [1]). Historically, the EOS first appeared in continuum mechanics and thermodynamics. Some researchers are even inclined to say that the EOS is the heart of thermodynamics. The current understanding of the EOS concepts is the result of a long-term evolution. We shall briefly discuss this below. Since that period until now, the fulfillment of the First and Second Laws of thermodynamics is treated as the most important demand for the reasonable EOS.

Bridgman insisted that thermodynamics is a purely macroscopic topic which cannot be substantiated by any microscopic principles [2]. Indeed, the purely thermodynamic, i.e., macroscopic, approach to the EOS is possible. Moreover, it is the approach which historically appeared first in the theory of EOS. It is completely based on the macroscopic measurements and macroscopic concepts. Basically, thermodynamics formulates quite strong a priori limitations on the potentially possible EOS.

M. Grinfeld (✉)
U.S. Army Research Laboratory, ATTN: AMSRD-ARL-WM-TD, Bldg. 390, Aberdeen Proving Ground, MD 21005, USA
e-mail: michael.greenfield4.civ@mail.mil

V. Chalivendra et al. (eds.), *Dynamic Behavior of Materials, Volume 1: Proceedings of the 2012 Annual Conference on Experimental and Applied Mechanics*, Conference Proceedings of the Society for Experimental Mechanics Series, DOI 10.1007/978-1-4614-4238-7_43, © The Society for Experimental Mechanics, Inc. 2013

Those limitations formally express the fact that neither the *perpetuum mobile* of the first kind nor the perpetuum mobile of the second kind can be made with the help of the substance under study. No knowledge whatsoever of the underlying microstructure is required for the applicability of the purely thermodynamic approach. On the other hand, many interesting and rather universal features of real substances' thermodynamics are not able to be explained, in principle. In this very sense, thermodynamic approach is always mentioned as the phenomenological approach. Thermodynamics principles, taken alone, are not sufficient for establishing the ultimate EOS. When using purely thermodynamic approach, all necessary information, other than the information, demanded by the two fundamental laws, the phenomenological thermodynamics should take directly from the macroscopic measurements. The philosophy of the purely thermodynamics approach is quite simple as compared with, say, quantum and statistical mechanics. The additional advantage of phenomenological thermodynamics, as compared with other branches of physics, consists in using only the quite simple mathematics and extremely limited number of assumptions.

In this paper, we are dealing exclusively with the purely thermodynamic (phenomenological) approach. However, it seems reasonable to add a couple of words about the approaches to the EOS based on the methods of quantum and statistical mechanics. Many macroscopic features of substances, which cannot be interpreted in terms of phenomenological thermodynamics, can be understood based on the hypotheses about the microscopic features of substances. Quantum mechanics and statistical mechanics are the main contemporary tools that allow converting the microscopic hypothesis into the macroscopic EOS. For applicability of the procedures, dictated by quantum and statistical mechanics, various experimental data are still needed. But those data are not of macro- but microscopic nature. Basically, the required experimental data appear to be the energy levels of the substances under study. When those levels are known, then, the EOS can be immediately expressed in the form of the canonical functional, dictated by the principles of quantum mechanics or statistical physics. Those instruments and known as statistical sums, functional densities, and likewise. When applying those universal techniques, nobody explicitly cares about the validity of the Laws of thermodynamics. This does not mean though that thermodynamic Laws became unimportant. Quite on the contrary, historically, the Laws of thermodynamics played an exceptional role as the major heuristic principles in the conceptual development of microscopic theories. When the basic principles of quantum mechanics and, especially of statistical mechanics were developed, those principles were explicitly implemented in the theories. Therefore, when using the procedures of statistical mechanics correctly, the Laws of thermodynamics are satisfied automatically. It should be mentioned also that the philosophy and mathematics used in quantum and statistical mechanics are much more complex and sophisticated than in phenomenological thermodynamics.

As previously mentioned, the purely phenomenological thermodynamic approach is nothing else but the thermodynamically consistent combination of the two main Laws of thermodynamics, from one hand, and experimental data, from the other hand. Thermodynamics dictates the vigilance when choosing the required experimental data. If not a sufficient data were used, then, the model is incomplete or indeterminate. If too much data were used, then, typically, it becomes impossible to include all the data into the EOS in a thermodynamically consistent way. Thermodynamics analysis is nothing else but the development of the procedures allowing choosing the consistent data for the thermodynamically correct EOS.

In conclusion of this general discussion, it is worthy of mentioning that different combinations of experimental data can be consistently converted into the EOS. Therefore, one and the same substance can be characterized by different thermodynamically consistent EOS. The most relevant choice of experimental data is dictated by the available measuring instruments, available experimental set-ups, and the range of parameters of particular interest for practitioners. For instance, the data, relating to the isentrope at $0\,°K$, and the analogous data, relating to the Hugoniot adiabat at high temperatures, can be equivalent from the mathematical point of view. But obviously, the first data is much more relevant to use when dealing with the low temperature physics problems, whereas the second data is much more relevant when dealing with the analysis of high-rate phenomena. Therefore, we will focus more on the EOS based on the second sort of data.

## 43.2 The Complete and Incomplete EOS

We assume that for the processes under study the substance's state can be described by fixing two thermodynamic parameters $(V,T)$, $(V,P)$, $(V,E)$, etc, where $V$is the specific volume, $P$- pressure, and $E$ – the specific internal energy, and $T$– the absolute temperature. All these pairs are thermodynamically equivalent in the sense that each of them uniquely determines the thermodynamic state of the two-parameter substance. However, for different applications or for the available experimental data some of the possible pairs appear to be more convenient than others. For example, the hydro-code-based analysis explicitly employs the pair$(V,E)$.

The EOS can be sub-divided into two big classes – the incomplete and complete ones. Historically, the incomplete EOS appeared first, even before postulating the First and Second Laws of thermodynamics and accepting the concept of the internal energy $E$. The Boyle-Mariotte and the Gay-Lussac relationships, which can be combined in the formula

$$PV = RT \tag{43.1}$$

where $R$ is the universal gas constant. It is historically one of the first examples of the incomplete EOS. After the discovery of the First Law they were coined as the thermal EOS as opposite to the simplest caloric EOS

$$E = C_V T \tag{43.2}$$

where $C_V$ is the specific heat capacity at constant volume.

The caloric EOS is also incomplete. The two incomplete EOS (43.1) and (43.2) allow the natural generalizations like

$$P = P(V,T), \ E = E(V,T) \tag{43.3}$$

where $P = P(V,T)$ and $E = E(V,T)$ are mathematically "reasonable" functions of their arguments.

Before understanding the key role of the Second Law of thermodynamics, the thermal and caloric EOS were treated as completely independent from each other. Also, the temperature at that time $T$ was not the absolute temperature in its contemporary understanding but the temperature of the so-called gaseous thermometer.

The very concept of the complete EOS appeared only after the formulation of the Second Law of thermodynamics. It appeared together with the concepts of the absolute temperature and the specific entropy $S$. The Second Law of thermodynamics implies that the thermal and caloric EOS are not independent from each other. The choice of one of them implies certain constraints on the choice of the other, and vice versa. At the same time, one of those EOS does not completely determine the other. In other words, both EOS contain a certain amount of equivalent information and, at the same time, each of them contains the information which the other does not.

One might conclude that the Second Law of thermodynamics emphasizes the necessity of having two incomplete EOS for the full description of the substance (43.3): the incomplete thermal EOS $P = P(V,T)$ and the incomplete caloric EOS $E = E(V,T)$. Fortunately, the situation is much better than that. In fact, the Second Law claims that there is a single thermodynamic function which contains all the thermodynamic information. One of such functions is the function defining the specific internal energy as function of the volume $V$and entropy $S$:

$$E = E(V,S) \tag{43.4}$$

Contrary to the function $E = E(V,T)$, when dealing with the function (43.4) – $E = E(V,S)$, we have all the information contained in two incomplete EOS (43.3). At the same time, when choosing any function $E = E(V,S)$ there is no risk of facing any thermodynamic contradictions similar to those associated with choosing independently two or several incomplete EOS. One can see that, when dealing with the internal energy $E$, the independent variables $(V,S)$are much more convenient than other pairs like $(V,T)$, $(V,P)$ and others. That is why the function $E = E(V,S)$ was coined as the canonically associated with the thermodynamic pair$(V,S)$. It can also be called the complete EOS. For the perfect gas, the complete EOS (43.4) differs from both (43.1) and (43.2), and it reads

$$E(V,S) = C_V T^* \left[ e^{\frac{S-S^*}{C_V}} \left( \frac{V}{V^*} \right)^{-\frac{R}{C_V}} - 1 \right] + E^* \tag{43.5}$$

where $(V^*, S^*)$ is an arbitrary but fixed state in the $(V,S)$ plane, and $E^*$ and $T^*$ are the values of the internal energy and the absolute temperature in the fixed state.

It is important that the complete EOS for one and the same substance can be written with the help of other independent variables. In particular, when using the independent variables $(V,T)$, the canonically associated thermodynamic potential is the free energy $F$, and the complete EOS should be presented as

$$F = F(V,T) \tag{43.6}$$

When using the independent variables $(V,E)$, the canonically associated thermodynamic potential is the entropy $S$, and the complete EOS should be presented as

$$S = S(V,E) \tag{43.7}$$

We call the function $S = S(V,E)$ the complete EOS for hydro-code. It should be clearly distinguished from the widely used incomplete EOS for hydro-code, which has the form $P = P(V,E)$. The pressure is not canonically associated with the pair and some essential thermodynamic parameters cannot be found with the help of the pressure-function.

An example of the complete EOS for hydro-code from the publication [10] reads

$$S(V,E) = R\ln\frac{V-b}{V^*-b} + C_V\ln\frac{E-E^*+C_VT^*}{C_VT^*} \tag{43.8}$$

where $b$ is a constants.

## 43.3 The Operational EOS

Another example of the complete EOS for hydrocode is also taken from [10]

$$S(V,E) = S^* + C_V\ln\frac{E-E^*+C_VT^*}{E_S^*(V)-E^*+C_VT^*} \tag{43.9}$$

where $E_S^*(V)$ is a function of one variable. This function has the physical meaning of the internal energy distribution on the isentrope passing through the state $(V^*,E^*)$.

We call (43.9) the operational EOS. Why operational? There are two different reasons for that.

First, and foremost, the Eq. 43.9 explicitly contains information about the experimental operations which are required for the recovery of this EOS from experimental data. It says, that in order to make this EOS useful for engineering applications one has to make the experiments allowing to determine the function $E_S^*(V)$. These measurements can be done by measuring the pressure function $P_S^*(V)$ in the adiabatic environment. With the help of this function the operational EOS (43.9) can be rewritten as

$$S(V,E) = S^* + C_V\ln\frac{E-E^*+C_VT^*}{C_VT^* - \int\limits_{V^*}^{V} \varpi P_S^*(\varpi)} \tag{43.10}$$

The operational EOS (43.9), (43.10) are mathematically equivalent but they explicitly emphasize the role of different experiments.

There is another reason for using the adjective "operational" when talking about the EOS similar to (43.9) and (43.10). This reason is purely mathematical. These EOS for hydro-code are functions of the hydro-code thermodynamic variables$(V,E)$. At the same time, they are *operators* mapping the experimentally measured functions $E_S^*(V)$or $P_S^*(V)$ into the set of the EOS for hydro-code $S(V,E)$.

The classical example of the incomplete EOS for hydro-code is known as the Mie-Gruneisen EOS. For more than century, the Mie-Gruneisen EOS keeps playing an exceptional role in the high-rate ballistics and geophysics. For the Mie-Gruneisen EOS the function $P(V,E)$ reads

$$P(V,E) = P_{ref}(V) + \frac{\Gamma(V)}{V}(E - E_{ref}(V)) \tag{43.11}$$

The three functions $P_{ref}(V)$, $E_{ref}(V)$, $\Gamma(V)$ in the Eq. 43.11 should be determined from experimental measurements or from different theoretical conjectures. The Mie-Gruneisen EOS is indeed incomplete. For instance, it does not allow calculating the specific heat capacity $C_V$of the substance which the complete EOS must do.

## 43.4 The Degrees of Freedom of the EOS

Formally, there are six constants in Eqs. 43.5 or 43.8. Some of them reflect the arbitrariness in the choice of the fixed state, other reflect the arbitrariness in the choice of the standard energy and entropy levels. Changing those constants does not affect the physics of the models under study. At the same time, by changing the constants $C_V$ and $R$ or $b$ we actually change the physical properties of the substance. In this very sense, the EOS (43.5) or (43.8) have two physically sound degrees of freedom. These degrees of freedom are physical constants.

The complete EOS (43.9) or (43.10) have one scalar physical degree of freedom $C_V$ and one functional degree of freedom $E_S^*(V)$or $P_S^*(V)$.

The Mie-Gruneisen EOS is instructive from the standpoint of its thermodynamic degrees of freedom. As a matter of fact, this EOS has three functional degrees of freedom: $P_{ref}(V),\ E_{ref}(V),\ \Gamma(V)$. The Mie-Gruneisen EOS is not only a function of two variables hydro-code variables $(V,E)$ but it is also an operator mapping the functions $P_{ref}(V),\ E_{ref}(V),\ \Gamma(V)$ into the EOS (43.11). So, it can be called an incomplete operational EOS with three functional degrees of freedom.

## 43.5 The EOS Skeletons

None of the presented EOS are universal, but based on some theoretical assumptions. Some assumptions include physical conjectures and hypotheses regarding the microstructure or the phenomenological or mathematical assumptions, etc. Some assumptions are based on formal aesthetics-driven mathematical generalizations. For example, Courant and Friedrichs [4] mathematically analyzed a class of model EOS of the form

$$E(V,S) = E_1(V) + E_2(S) \tag{43.12}$$

where $E_1(V)$ and $E_2(S)$ are rather general functions. Courant and Friedrichs coined it as the EOS with separable internal energy. The Eq. 43.12 defines a complete EOS with two functional degrees of freedom. However, it is not yet an operational EOS because it does not mentioned explicitly any required physical experiments.

We call the forms like (43.12) the skeletons of the EOS. The skeletons should be simple and should allow rather general analytical consideration. The skeleton should be converted into the operational EOS. For instance, in the publication [11] we considered a generalized Courant-Friedrichs skeleton

$$E(V,S) = X(S)\Omega(V) + \Delta(S) \tag{43.13}$$

with three arbitrary functions $X(S),\ \Omega(V),\ \Delta(S)$.

In the case of constant heat capacity $C_V$, the skeleton (43.13) can be converted in the following operational EOS:

$$E(V,S) = C_V T_S^*(V) e^{\frac{S-S^*}{C_V}} - \int_{V^*}^{V} d\bar{\omega} P_S^*(\bar{\omega}) + E^* \tag{43.14}$$

where $T_S^*(V)$ and $P_S^*(V)$ are the values of the temperature and pressure on the isentrope passing through the state $(V^*,S^*)$.

Obviously, the operational EOS (43.14) has the skeleton (43.13) with the following functions $X(S),\ \Omega(V),\ \Delta(S)$:

$$\mathrm{X}(S) = \int_{S^*}^{S} d\xi e^{\int_{S^*}^{\xi} d\eta C_V^{-1}(\eta)},\quad \Omega(V) = T_S^*(V),\ \Delta(V) = \int_{V^*}^{V} d\bar{\omega} P_S^*(\bar{\omega})\,. \tag{43.15}$$

The analogy of the Courant-Friedrich skeleton (43.12) when dealing with the EOS for hydro-code would be the skeleton with two functional degrees of freedom of the form

$$S(V,E) = S_1(V) + S_2(E) \tag{43.16}$$

which can be called the EOS with separable entropy. Obviously, the EOS (43.8) and (43.9) both have skeletons of the form (43.16). The skeleton (43.16) implies that the heat capacity $C_V$ of the substance must be a function of the internal energy $E$, only

$$C_V = C_V(E) \tag{43.17}$$

and the skeleton (43.17) can be rewritten as follows:

$$S(V,E) - S^* = S_1(V) + \int_{E^*}^{E} \frac{d\xi}{H(\xi)}; \quad H(\xi) \equiv A + \int_{E_*}^{\xi} \frac{d\eta}{C_V(\eta)} \tag{43.18}$$

where $A$ is a constant. The function $H$ has physical meaning of the absolute temperature. In order to find the function $C_V(E)$ one has to make the experiment on isochoric heating. That would be the first step to converting the skeleton (43.16) into the operational EOS. The second one should contain the information about the experiment for determination of the function $S_1(V)$. For the case of isentropic loading and constant heat capacity the conversion is described by the complete operational EOS (43.9) and (43.10).

## 43.6 The Operational EOS Based on the Hugoniot Measurements

So far, the operational EOS were based on the measurements related to isentropic loading. For the applications in high-rate mechanics and geophysics the experimental data is often taken from the non-adiabatic shock-wave loading, especially from the measurements related to the Hugoniot adiabat. Such measurements are hard to get and their amount is quite limited. When experimenting with shock waves, the experimenters basically measure (1) the velocity of the shock-front and (2) the particles velocity behind the shock front. These measurements allow to calculate the internal energy density as function of the specific volume behind right behind front: $E_H^*(V)$.

Consider the Hugoniot adiabat starting at the state $(V^*, E^*)$. If the only available measurement on the Hugoniot adiabat is the measurement of the internal energy density $E_H^*(V)$, it does not make much sense to consider EOS with skeletons allowing too many degrees of freedom. Instead, it makes sense to use the simplest skeleton (43.18). The corresponding operational EOS, then, reads

$$S(V,E) = S^* + \int_{E^*}^{E} d\varepsilon \frac{1}{H(\varepsilon)} + \int_{V^*}^{V} d\varpi \frac{2E_H(\varpi) - 2E^* - P^*(V^* - \varpi)}{(V_* - \varpi)H(E_H(\varpi))} \tag{43.19}$$

Thus, the complete operational EOS appears to be the integral operator with respect to the experimental data.

If we assume, in addition, that the heat capacity $C_V$ is constant, then the operational EOS (43.19) can be simplified to read

$$S(V,E) = S^* + C_V \ln \frac{E - E_* + C_V T^*}{C_V T^*} + C_V \int_{V^*}^{V} d\varpi \frac{2E_H(\varpi) - 2E_* - P^*(V^* - \varpi)}{(V_* - \varpi)[E_H(\omega) - E_* + C_V T^*]} \tag{43.20}$$

When more experimental data is available it makes sense to consider EOS with skeletons having bigger number of thermodynamic degrees of freedom. If the heat capacity $C_V$ is constant, then, the most general skeleton for the hydro-code can be presented in the form

$$S(V,E) = \mathrm{A}(V) + C_V \int_{E^*}^{E} \frac{d\varepsilon}{\varepsilon - E^* + C_V \Lambda(V)} \tag{43.21}$$

where $\mathrm{A}(V)$ and $\Lambda(V)$ are two functions to be determined from measurements. Thus, we have a skeleton for the operational EOS with one scalar $C_V$ and two functional degrees of freedom.

In order to recover the complete EOS with the skeleton (43.21), we have to express the functions $\mathrm{A}(V)$ and $\Lambda(V)$ in terms of experimental data. Assume that the experimental data allowed us to find two functions – $E = E_S^{\otimes\otimes}(V)$ and $E = E_H^{**}(V)$. The former describes the internal energy on the isentrope, passing through the state $(V^{\otimes}, E^{\otimes})$, and the latter describes the Hugoniot adiabat, passing through the state $(V^*, E^* = C_V T^*)$ The recovery of the operational EOS is actually equivalent to expressing the functions $\mathrm{A}(V)$ and $\Lambda(V)$ in terms of the functions $E = E_S^{\otimes\otimes}(V)$, and $E = E_H^{**}(V)$.

The explicit formula of the corresponding operational EOS reads [12]:

$$S(V,E) - S^{\otimes} = C_V \ln \frac{E - CT^*\varepsilon(V,V^*)}{E_S^{\otimes\otimes}(V) - CT^*\varepsilon(V,V^*)} \tag{43.22}$$

Now, we have to describe the function $\varepsilon(V,V^*)$in terms of experimental measurements. The description requires several steps. First, having known the functions $E = E_S^{\otimes\otimes}(V)$, and $E = E_H^{**}(V)$, we can calculate the pressure functions $P = P_S^{\otimes\otimes}(V)$ and $P = P_H^{**}(V)$ for the isentrope and the Hugoniot adiabat, respectively, using the relationships

$$P_S^{\otimes\otimes}(V) = -\frac{dE_S^{\otimes\otimes}(V)}{dV} \tag{43.23}$$

and the Rankine-Hugoniot relationship

$$P_H^{**}(V) = -\frac{2E_H^{**}(V) - 2E^* + P^*(V - V^*)}{V - V^*} \tag{43.24}$$

Before calculating the function $\varepsilon(V,V^*)$, one has to calculate first the functions $(V)$ and $\pi(V)$:

$$(V) \equiv V^* \frac{P_S^{\otimes\otimes}(V) - P_H^{**}(V)}{E_S^{\otimes\otimes}(V) - E_H^{**}(V)}, \quad \pi(V) \equiv \frac{V^*}{CT^*} \frac{E_S^{\otimes\otimes}P_H^{**} - E_H^{**}P_S^{\otimes\otimes}}{E_H^{**} - E_S^{\otimes\otimes}} \tag{43.25}$$

Now, the functions $\tau(V,\theta)$ and $\varepsilon(V,\theta)$ can be calculated with the help of the formula

$$\tau(V,\theta) \equiv e^{-\frac{1}{V^*}\int_\theta^V d\xi(\xi)}, \quad \varepsilon(V,\theta) \equiv \frac{1}{V^*}\int_\theta^V d\eta\pi(\eta)\tau(V,\eta) \tag{43.26}$$

At last, temperature $T(V,E)$ and the pressure $P(V,E)$ functions can be presented in the forms

$$T(V,E) = T^*\left[\frac{E}{C_V T^*} - \varepsilon(V,V^*)\right] \tag{43.27}$$

and

$$P(V,E) = \frac{P_H^{**} - P_S^{\otimes\otimes}}{E_H^{**} - E_S^{\otimes\otimes}}E - \frac{E_S^{\otimes\otimes}P_H^{**} - E_H^{**}P_S^{\otimes\otimes}}{E_H^{**} - E_S^{\otimes\otimes}}, \tag{43.28}$$

respectively.

## 43.7 Conclusion

We discussed the concept of the operational EOS for two-parameter substances. The operational EOS is an ordinary function with respect to the chosen thermodynamic variables. At the same time, it is an operator with respect to the chosen experimental data.

The operational EOS can be complete or incomplete. The complete EOS is the thermodynamic potential canonically associated with the chosen independent thermodynamic variables.

We discussed the concept of a skeleton of the EOS. The skeleton is something close to the concept of a general solution of ordinary differential equations or differential equations with partial derivatives. We discussed the concept of the degrees of freedom of the EOS.

We gave several explicit formulas of the operational EOS related to the experimental measurements associated with isentropic and shock loadings.

## References

1. Bridgman PW (1931) The physics of high pressure. Bell and Sons, London
2. Bridgman PW (1941) The nature of thermodynamics. Harvard University Press, Cambridge
3. Grinfeld MA (1991) Thermodynamic methods in the theory of heterogeneous systems. Longman, New York
4. Courant R, Friedrichs KO (1948) Supersonic flow and shock waves. Interscience, New York
5. MacDonald JR (1969) Review of some experimental and analytical equations of state. Rev Mod Phys 41:316–349
6. Zharkov VN, Kalinin VA (1971) Equations of state for solids at high pressures and temperatures. Consultants Bureau, New York
7. Zel'dovich B Ya, Raizer P Yu (2002) Physics of shock waves and high-temperature hydrodynamic phenomena. Dover, Mineola
8. Eliezer S, Ghatak A, Hora H (2002) Fundamentals of equations of state. World Scientific, Singapore
9. Menikoff R (2007) Empirical EOS for solids. In: Horie I (ed) Shock wave science and technology reference library, 2, solids 1. Springer, Berlin
10. Grinfeld MA (2011) A novel equation of state for hydrocode. In: Elert ML et al (eds) Proceedings of the conference, Shock compression of condensed matter, Chicago, 2011
11. Grinfeld MA (2011) The generalized courant-friedrichs equation of state for hydrocode. In: Elert ML et al (eds) Proceedings of the conference, Shock compression of condensed matter, Chicago, 2011
12. Grinfeld MA (2012) Operational equations of state. 3. Recovery of the EOS for hydrocode from the measured heat capacity, isentrope and hugoniot adiabat, ARL technical report, 2012

# Chapter 44
# Influence of Loading Kinetics on the Shock Response of Polycrystalline Copper

**Juan P. Escobedo, Carl P. Trujillo, Darcie Dennis-Koller, Ellen K. Cerreta, and Curt A. Bronkhorst**

**Abstract** This work examined the effect of shock loading kinetics on the early stages of damage evolution in Cu samples. The tensile loading characteristics were tailored by controlling the geometry and material used for the flyer plates in the plate impact experiments. All the Cu specimens used in this study had similar microstructures with an average grain size of 100μm. This work has shown that the interplay of the energy supplied by the shock impulse and the time spent in its dissipation seem to be responsible for the characteristics of the resultant damage fields. For an increasing value in the coupled parameter of energy*time, the damage fields appear to evolve from the early stages of damage, in the form of void nucleation and initial void growth, to the void coalescence stage. In this regard, a damage field consisting of well-coalesced voids, which created a nearly complete fracture surface, was observed for the highest value of energy*time. Additionally, a framework is postulated in which damage development can be predicted from the real-time, continuum level response measurements, i.e. free surface velocity. Under this framework, good agreement between the characteristics of the free surface velocities and the predicted, as well as the experimentally observed, damage fields is found.

**Keywords** Shock loading • Dynamics failure • Damage evolution • EBSD

## 44.1 Introduction

Extensive work to date has shown that dynamic damage in ductile metals is a process of void nucleation, growth, and coalescence [1–13]. These works have also established that dynamic damage is a complex process strongly influenced by the microstructure but also by the dynamic loading profile or shock wave shape imparted to the specimen [6–8, 11]. This intertwined interplay of parameters and mechanisms (i.e. shock hardening, tensile stress rates) precludes the assessment of individual contributions from specific material or loading characteristics. Thus, to address this problem, a comprehensive study has been undertaken that aims at the determination of the individual contributions of microstructure and loading profile to the mechanisms responsible for damage development. To this end, an experimental, two-fold approach has been followed with emphasis on tracking the microstructural evolution at the early stages of damage: (1) to investigate the effect of microstructural features, impact experiments was repeated on high purity copper specimens with grain sizes in the 30–200 μm range and (2) to investigate the effect of kinetics, the grain boundary density was held constant while the loading condition was altered. For all cases, the plate impact experiments were conducted at low peak compressive stresses, on the order of ~1.5–2.5 GPa, such that the tensile stresses developed within the target material during the experiment were low enough to cause only incipient or early stages of damage. The damage state was characterized by means of optical and electron-backscatter diffraction (EBSD) microscopy, as well as x-ray micro-tomography. The combination of these experimental techniques led to a more comprehensive understanding of the damage evolution that could be correlated with likely deformation mechanisms.

The results corresponding to the effects of microstructure on the mechanisms for damage evolution have recently been reported [13–15]. The main findings can be summarized as follows: (1) damage initiated preferentially at grain boundaries with misorientation angles in the ~10–57° range, while at low angle and special Σ3 boundaries dynamic failure seems to

J.P. Escobedo (✉) • C.P. Trujillo • D. Dennis-Koller • E.K. Cerreta • C.A. Bronkhorst
Los Alamos National Laboratory, Los Alamos, NM 87545, USA
e-mail: jpescobedo@lanl.gov

V. Chalivendra et al. (eds.), *Dynamic Behavior of Materials, Volume 1: Proceedings of the 2012 Annual Conference on Experimental and Applied Mechanics*, Conference Proceedings of the Society for Experimental Mechanics Series, DOI 10.1007/978-1-4614-4238-7_44, 

be mitigated; (2) the mechanisms for damage evolution were dictated by the grain size, which inherently determined the proximity of the nucleated voids; and (3) the previous meso-scopic length scale mechanisms were reflected in the characteristics of the continuum level free surface velocity (FSV) histories. It was found that the re-acceleration slope after the minima provided a direct indicator of the damage kinetics. In this regard, a steep slope was associated with a fast damage growth rate, dominated predominately by coalescence.

Building on the previous observations, the second part of the study was initiated to probe the effect of loading kinetics on the fundamental mechanisms of void nucleation, growth, and coalescence. The results presented in this paper correspond to this effort. As previously stated, the specimens used in this study possessed similar microstructures. Additionally, because the characteristics of the tensile loading profile cannot be experimentally accessed in these experiments, the decompression times were used as indicators to determine the kinetic aspects of the tensile pulse. These times are directly measured from FSV histories to determine average decompression rates in shock experiments. It should be noted that, although these rates have been generally used as indicators of the loading kinetics in other studies [5, 6, 11, 12], such a parameter is not an entirely accurate representation of the characteristics of the tensile pulse. Additional experiments designed by using one-dimensional wave propagation calculations are currently being performed. The details of the experimental design to achieve a given tensile profile as well as the corresponding results will be presented in due time. Those results in conjunction with the results and their interpretation presented herein will collectively allow for a comprehensive evaluation of the influence of tensile loading profile on the mechanisms responsible for damage evolution.

## 44.2 Experimental Methods

The plate impact experiment consists of a single monolithic flyer launched against a sample target (Fig. 44.1a). Upon impact, an initial shock compressive state is generated in the sample for durations on the order of less than a microsecond. Then, as the sample decompresses from the shocked state, release waves travel from the free surfaces towards the interior of the target. Their interaction results in the creation of tensile stresses in the interior of the specimen (Fig. 44.1b). If these tensile stresses are high enough, damage in ductile materials is created by the nucleation, growth, and coalescence of voids. This

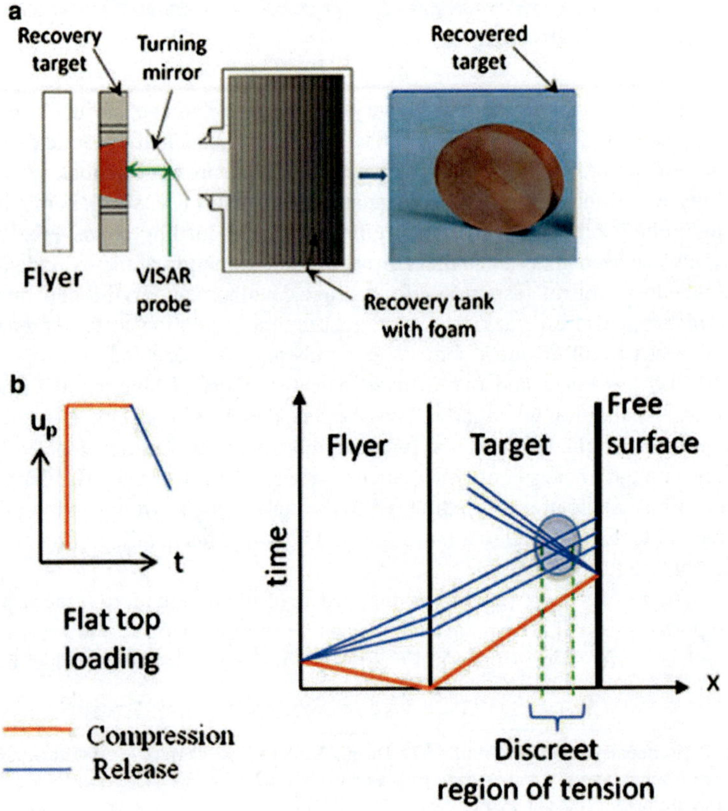


**Fig. 44.1** (**a**) Plate impact experiment, (**b**) compressive and release wave interactions

ultimately forms a fracture zone within the target. This zone is termed the spall region [1, 7]. The real-time, sub-microsecond response is assessed by measurement of the free surface velocity of the targets. This task can be performed by means of velocimetry interferometer for any surface reflector (VISAR) [16] or photon Doppler velocimetry (PDV) systems [17, 18]. Then by use of the Hugoniot equation of state of the specific material being studied [1], the mechanical state, i.e. stress, can be extracted.

To conduct these experiments, copper targets were prepared as circular disks that were 15 or 20 mm in diameter and 4 mm in thickness. The copper samples were press-fit into target assemblies formed by three concentric momentum trapping rings to mitigate perturbations from edge release waves. The rings were made of copper to ensure similar mass impedance across the target assembly. Diagnostic pins were positioned around the perimeter of the target assembly to measure the flyer velocity, tilt, and to trigger diagnostic equipment. Velocity profiles were obtained using VISAR or PDV probes mounted 90° off of the shock direction axis to prevent any subsequent impact of the target with the probe. Further experimental details can be found in [7, 8, 13]. All targets were prepared from fully annealed 99.999% pure oxygen-free high-conductivity (OFHC) copper. The samples used for this study were annealed under vacuum at 850°C for 1 h to produce microstructures with average grain sizes of 100 μm. The results of the complete metallographic characterization have been reported elsewhere [13]. The fraction of boundaries with misorientations of 60°, that fulfill the special Σ3 boundary criteria, were held to a value of 0.65% for all samples utilized in this study.

Following impact, each recovered specimen was diametrically sectioned and mounted such that the cross section of the spalled sample was exposed for post-impact characterization. The deformed specimens were prepared following standard metallographic techniques up to a 0.05 μm colloidal silica final polish, performed on a vibratory polisher. Finally, they were electrochemically polished using a solution of two parts of phosphoric acid to one part of water at ~1.9 V for ~10 s. Optical microscopy was performed on a Zeiss microscope equipped with an automated stage. Image processing was done using Image J software. EBSD analysis was performed on a Phillips XL30 FEG SEM using a voltage of 20 kV, a spot size of 4 and with 1.5 μm step size. Data was acquired and analyzed using orientation imaging microscopy (OIM) software by TexSEM Laboratories (TSL) of EDAX. The OIM software was used to calculate properties obtained from EBSD measurements, such as misorientation maps.

## 44.3 Results

A group of representative experiments conducted under the different loading conditions pursued in this study is listed in Table 44.1. The experiments were conducted at peak compressive stresses of 1.5 and 2.5 GPa. For each compressive stress, copper or z-cut quartz flyers were utilized to impose different decompression rates. In this regard, the rate at which the sample decompresses from the peak state is a function of the interaction of the release waves in the target and the flyer. This interaction is, in turn, a function of intrinsic material properties such as acoustic speeds and density. Hence, using flyers of two different materials resulted in two different loading tensile profiles imposed on the targets. In addition, to further control the shock loading characteristics and isolate the effect of the decompression rate parameter, the pulse duration (PD), which represents the time that the samples were under shock compression, was held similar for all experiments. This task was experimentally achieved by tailoring the flyer thickness.

The free surface velocity (FSV) histories for the experiments reported herein are shown in Fig. 44.2. Values of about 85 m/s were measured for the peak free surface velocity in Exps. 1–2; and about 143 m/s in Exps. 3–4. The pulse durations were also measured in these profiles. Their values are listed in Table 44.1.

Moreover, Table 44.1 can be read as follows. The first three columns list the flyer parameters. As such, they represent the imposed loading conditions. Columns 4 (FSV) and 5 (PD) represent the measurable material response, directly extracted from the FSV profiles. The last three columns correspond to stress, density and energy. These are calculated material properties at the peak compressive state. They were calculated by considering the measured FSV and the Hugoniot parameters for copper ($C_o = 3.94$ and $s = 1.489$) as input for the Hugoniot jump conditions [1]:

**Table 44.1** Shot parameters for the plate impact experiments

| Exp. ID | Flyer Material | Flyer Thickness (mm) | Flyer Velocity (m/s) | Peak particle velocity (m/s) | Pulse duration (μs) | Compressive stress (GPa) | Shocked density (gr/cm$^3$) | ΔEnergy (J/gr) |
|---|---|---|---|---|---|---|---|---|
| 1 | Quartz | 2.0 | 131 | 85 | 0.38 | 1.5 | 9.092 | 2.488 |
| 2 | Copper | 1.4 | 88 | 86 | 0.36 | 1.5 | 9.092 | 2.488 |
| 3 | Quartz | 2.0 | 215 | 143 | 0.37 | 2.5 | 9.002 | 0.675 |
| 4 | Copper | 1.4 | 140 | 144 | 0.36 | 2.5 | 9.002 | 0.675 |

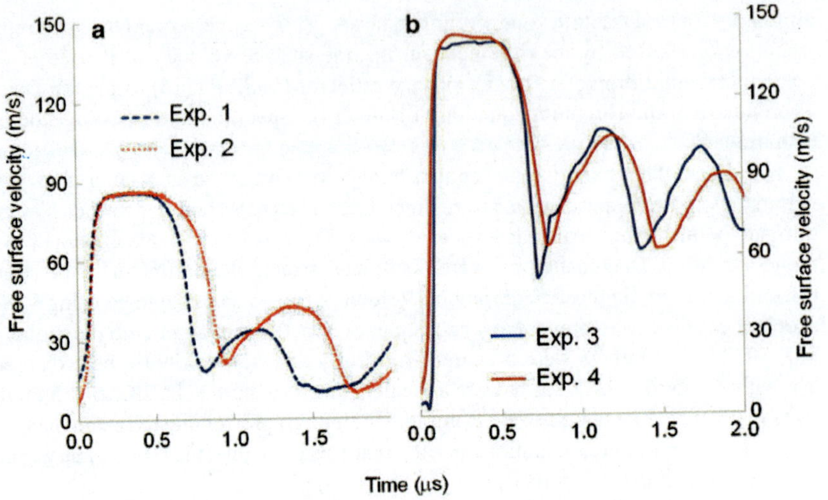


**Fig. 44.2** Free surface velocities (FSV) of experiments at peak stresses of (**a**) 1.5 GPa (**b**) 2.5 GPa

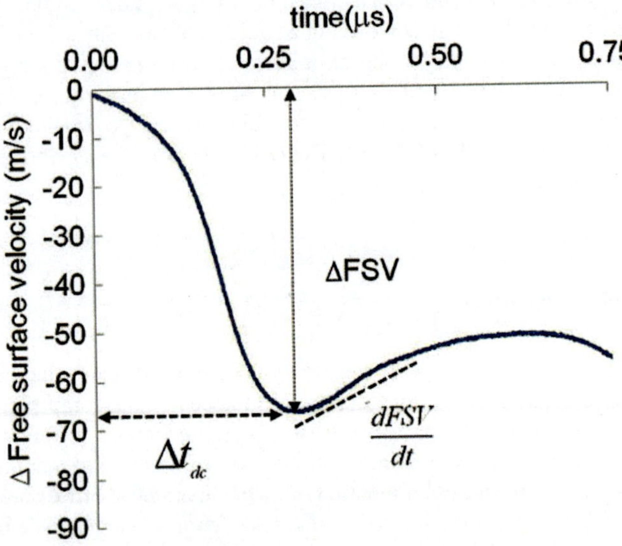


**Fig. 44.3** Parameters of interest in the pullback wave

$$\rho_0 D = \rho(D - u_p) \quad (44.1)$$

$$\sigma - \sigma_0 = \rho_0 D u_p \quad (44.2)$$

$$\Delta E = \frac{(\sigma - \sigma_0)(\frac{1}{\rho_0} - \frac{1}{\rho})}{2} \quad (44.3)$$

Where $u_p$ = FSV/2 (mm/μs) is the particle velocity, ρ is the density, σ is the uniaxial strain stress (GPa), E is the specific internal energy (J/gr) and D is the shock velocity (mm/μs). The last parameter was calculated as $D = 3.94 + 1.489\, u_p$. The zero sub index indicates initial values for the density and stress: $\rho_o = 8.93$ gr/cm$^3$ and $\sigma_0 = 0.0001$ GPa (atmospheric pressure).

Of interest in the FSV profile are the selected areas indicated in Fig. 44.2. The characteristics of the FSV in this area have been typically associated with the tensile stage in plate impact experiments. These areas encompass the sections in the FSV profile that start at the decay from peak state to the minima and the first rebound. This part of the FSV is generally referred to as the pullback wave. As an example, the pullback wave from Exp. 1 is shown in detail in Fig. 44.3. Parameters of interest are the magnitude of the drop in the FSV profile (ΔFSV), as well as the decompression and recompression rates.

**Table 44.2** Calculated values from FSV data

| | Pull-back characteristics | | | | | |
|---|---|---|---|---|---|---|
| Exp. No. | ΔFSV (m/s) | $\Delta t_{dc}$ (ms) | dFSV/dt (m/s/μs) | Spall strength (GPa) | Density at minima (gr/cm$^3$) | Estimated $\Delta V_{void}$ (cm$^3$/gr) |
| 1 | 66.4 | 0.278 | 82 | 1.28 | 8.950 | 0.00025 |
| 2 | 63 | 0.386 | 95 | 1.21 | 8.954 | 0.00030 |
| 3 | 88 | 0.203 | 130 | 1.70 | 8.993 | 0.00078 |
| 4 | 75.5 | 0.263 | 163 | 1.45 | 9.007 | 0.00096 |

The measurements of the different parameters are reported in Table 44.2. As before, the first 3 columns indicate measured values and the rest are calculated properties. The ΔFSV values were directly measured from the FSV profiles. The round shape in the pullback wave at the onset of the decay precluded the calculation of constant decompression rates. This rounding has been associated with the non-linearity of the plastic process occurring in the sample [5]. Consequently, rather than reporting decompression rates, the $\Delta t_{dc}$ times are reported in Table 44.2. They represent the time it took the sample to decay from the peak state to the minima, as indicated in Fig. 44.3. For a given stress level, longer times were measured in the experiments using copper flyers.

The next parameter of interest is the initial re-acceleration rate of the FSV after the minima (dFSV/dt), as shown in Fig. 44.3. This rate indicates the rate at which the sample is shock recompressed. This recompression rate, in turn, has been shown to correlate with the rate of damage growth occurring in the sample for a variety of materials [13, 19]. The measured values suggest a faster damage development in the experiments using Cu flyers (Exps. 2 and 4) as compared to those performed using quartz, for a given peak stress.

The first calculated property is the spall strength. This is a property generally associated with the resistance to failure in dynamic experiments [1–13]. The spall strength, was calculated using the following relationship for a material that exhibits elasto-plastic behavior [10]:

$$\sigma_{spall} \equiv \rho_0 C_L \Delta FSV \left(1 + \frac{C_L}{C_o}\right)^{-1} \tag{44.4}$$

where $C_o = 3.91$ mm/μs, $C_L = 4.77$ mm/μs, are the bulk and longitudinal sound speeds, respectively. The calculated values are listed in Table 44.2. For a given peak stress, smaller spall strengths were calculated in experiments conducted using copper flyers (Exps. 2 and 4). In addition, higher spall strengths are observed at larger peak stresses. The last two columns of Table 44.2 will be addressed in the Discussion section.

To complement the inferences made from FSV profiles, optical microscopy and EBSD measurements were performed on the recovered samples. Optical, bright field, images of the cross sections are shown in the Fig. 44.4a–d. The damage content, in the form of void area fraction ($V_A$), was measured. The values are listed in Table 44.3. For a given stress, more damage was measured in the samples conducted at lower decompression rates (Exp. 2 and 4). In addition, more damage was measured at higher peak compressive stresses. To further examine the damage fields, EBSD misorientation maps are shown in Fig. 44.4e–h. These areas are from the center of the recovered samples as indicated in the optical micrographs in Fig. 44.4a–d. In maps of this type, lattice misorientation within a grain is correlated with plastic dissipation processes, such as dislocation slip.

For Exp. 1 mostly incipient damage, in the form of scattered voids, with a moderate amount of plastic dissipation is observed in the misorientation map (Fig. 44.4e). Then, more damage and a larger amount of plastic dissipation are observed in Fig. 44.4f for Exp. 2. As the peak stress increased to 2.5 GPa, void coalescence started to dominate the damage field as observed in Exp.3 (Fig. 44.4g). The larger coalescence observed in Exp. 4 caused the creation of a nearly complete fracture surface in this sample (Fig. 44.4h). The characteristics of the damage fields seem to correlate with the respective recompression rates listed in Table 44.2. More well-developed damage, in the form of void growth and coalescence, can be associated with larger recompression rates in the FSV. It should be noted that this correlation is valid irrespective of the compressive stress.

## 44.4 Discussion

For a given compressive stress, agreement was found between the damage fields and the details of the FSVs for each sample. For instance, the conventional interpretation for the spall strength value suggests that the lower spall strength displayed in the experiments using copper flyers should correlate with more damage developed in these samples. This has been shown to

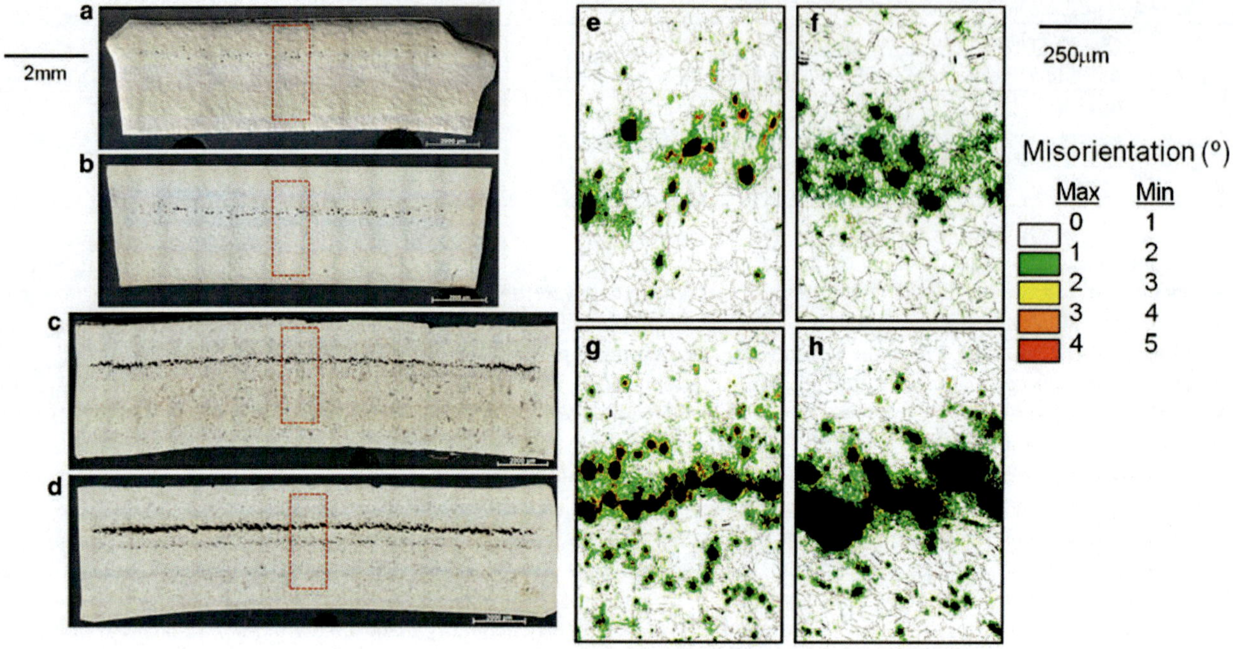


**Fig. 44.4** Optical micrographs of cross sections from (**a**) Exp.1, (**b**) Exp2, (**c**) Exp 3, (**d**) Exp 4. (**e–h**) Misorientation maps from the corresponding selected areas

**Table 44.3** Predicted and measured damage

| Exp. ID | Energy*time (J*μs) | Estimated damage dFSV/dt*$\Delta V_{void}$ | Measured damage ($V_A$) |
|---|---|---|---|
| | | k(μm$^3$) | |
| 1 | 0.19 | 0.041 | 0.27 |
| 2 | 0.26 | 0.057 | 0.59 |
| 3 | 0.50 | 0.204 | 1.75 |
| 4 | 0.65 | 0.312 | 3.56 |

be accurate. However, the same interpretation would suggest that for Exp. 3 and 4 that displayed larger spall strengths, less damage should exist. However, this trend was not observed. The major limitation of the spall strength value is that it does not provide an accurate insight into the details for the damage kinetics occurring in the sample. Thus, with the objective to establish better correlations between FSVs and expected damage, in what follows, we postulate a more comprehensive interpretation of the different parameters measured in the FSVs to lend more insight into the kinetics of damage.

The density changes during decompression for each of the experiments are shown in Fig. 44.5. The initial densities (at $t = 0$) are listed in Table 44.1. They correspond to the calculated densities for the shocked state. Then, as shown in Fig. 44.5, the density for each sample decreases as the sample decompresses. The final densities shown in the paths are associated with the minima in the FSV upon release (pullback wave). The densities for each sample at this state are listed in Table 44.2. Finally, the slopes of these paths were calculated considering the measured $\Delta t_{dc}$, which as previously discussed, is the time for the sample to decompress from the peak state to the minima.

In this regard, a typical interpretation of the minima of the pullback wave is that it corresponds to a state in which the mass impedance, in essence density, has reached a value low enough to cause a wave reflection towards the free surface. This value cannot be experimentally determined. Hence, for the current discussion, the density of copper at room conditions (8.93 gr/cm$^3$) was taken as the density needed to create the mentioned wave reflections (dotted line). It is observed in Fig. 44.5 that the change in density achieved solely due to shock release was not enough to lower the density to the reference value. Thus, the difference in density had to be compensated by cavity formation occurring in the interior of the sample that aided to lower the density to the reference value. These cavities likely resulted from void nucleation and initial growth. The associated change in density due to void formation is referred to as $\Delta\rho_{void}$ in Fig. 44.5. From each $\Delta\rho_{void}$ the specific void volume ($\Delta V_{void}$) was calculated for each experiment. The calculated values are listed in Table 44.2. This interpretation of the

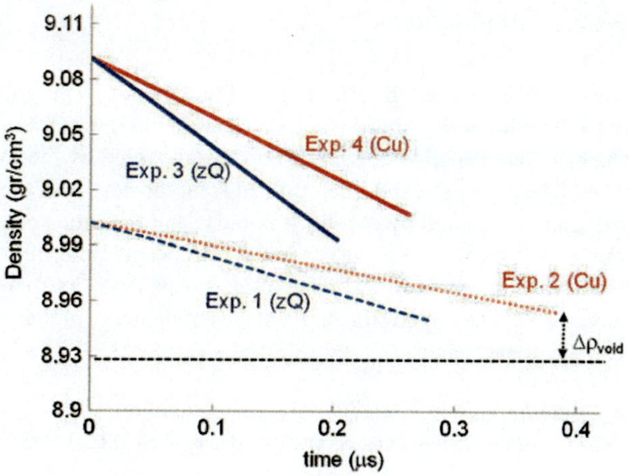


**Fig. 44.5** Changes in density due to shock release for each condition

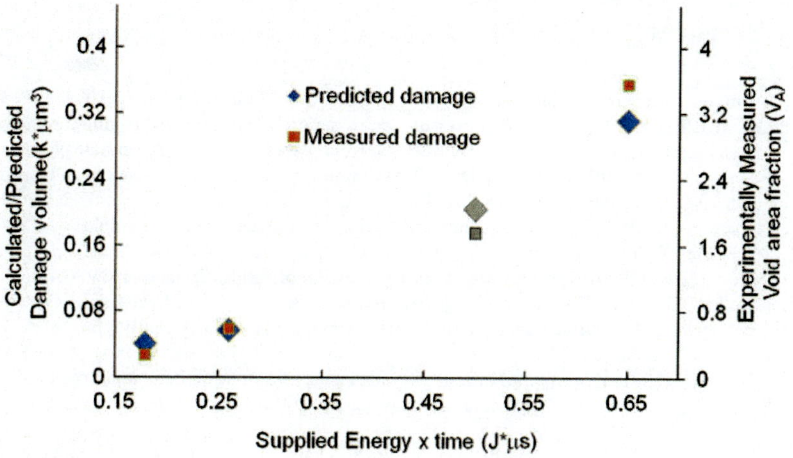


**Fig. 44.6** Predicted and measured damage as a function of imparted energy*time

details of the FSVs is in better agreement with the damage fields observed in each sample. For instance, upon release from higher stresses (Exps. 3 and 4), a larger difference is observed in Fig. 44.5 between the reference density and density attained due to shock release. Thus, larger $\Delta V_{void}$ were needed in these cases as compared with Exps. 1 and 2.

These results suggest that more damage, in the form of voids, was created at higher compressive stresses. This inference agrees with the experimental measurements. Additionally, the decompression rate determined how close the density approached the reference value. As such, a faster rate caused a larger decrease in density and, hence, a smaller $\Delta\rho_{void}$ was needed. In this case also, the calculated values for $\Delta V_{void}$ seem to correlate with the experimental observations. In experiments using quartz flyers, which displayed larger decompression rates, less damage was observed in the specimens as compared with the experiments using copper flyers.

Furthermore, if $\Delta V_{void}$ is taken as the nucleated voids with initial growth at the early stages of damage, an estimate of the total damage can be calculated by considering the damage growth rate. As previously stated, the damage growth rates were indirectly assessed by considering the reacceleration rates (dFSV/dt) extracted from the FSVs (Table 44.2). The results from the ($\Delta V_{void}$) x (dFSV/dt) calculations, as well as the experimentally measured void content are shown in Fig. 44.6 and listed in Table 44.3.

The abscissa axis shows a parameter that couples the energy supplied to the specimen (peak compressive stress) and the decompression times. It physically represents the time ($\Delta t_{dc}$) it took the specimen to dissipate the imparted energy ($\Delta E$) upon release from a given compressive stress. These values are also listed in Table 44.3. Similar qualitative trends are observed in

The proportionality constant ***k*** should be adapted to account for additional phenomenological processes that have been unaccounted in this framework. The caveat with this prediction is that more damage may have developed after the first pullback wave, due to the subsequent tensile-compressive stresses that the specimen experienced at later times. However, the arrival of release waves that originated from the lateral faces of the sample, named edge waves, should cause a decrease in the stresses amplitudes. Hence, it can be assumed that most of the damage developed during the first tension-compression cycle (pullback).

## 44.5 Conclusions

This work examined the effect of loading kinetics in the early stages of damage evolution in shock loaded Cu samples. The experimental observations and their interpretation presented herein suggest that to predict damage from FSV measurements, the calculated spall strength must be accompanied by the measurement and analysis the decompression and recompression rates. These values can be determined from the details of the pullback waves. Additionally, the combination and interplay of the energy supplied by the shock impulse and the time spent in its dissipation seem to be responsible for the damage fields. For an increasing coupled parameter of energy*time, more damage was predicted, as well as experimentally observed. Finally, as more energy (higher peak stress) was dissipated for longer periods of time (lower decompression rate), the damage fields evolved from early stages of damage, in the form of void nucleation and initial void growth, to the later stages of void coalescence.

**Acknowledgement** Los Alamos National Laboratory is operated by LANS, LLC, for the NNSA of the US Department of Energy under contract DE-AC52-06NA25396. Funding was provided by the LDRD-DR Grant 20100026.

## References

1. Antoun T, Seaman L, Curran D, Kanel G, Razorenov S, Utkin A (2002) Spall fracture. Springer, New york
2. Murr LE (1981) Effects of peak pressure, pulse duration, and repeated loading on the residual structure and properties of shock defromed metals and alloys. In: Meyers M, Murr LE (eds) Shock waves and high strain -rate phenomena in metals. Plenum Press, New York, pp 753–777
3. Seaman L, Curran DR, Shockey DA (1976) Computational Models for Ductile and Brittle-Fracture. J Appl Phys 47(11):4814
4. Curran DR, Seaman L, Shockey DA (1977) Dynamic Failure in Solids. Phys Today 30(1):46
5. Johnson JN (1981) Dynamic Fracture and Spallation in Ductile Solids. J Appl Phys 52(4):2812
6. Johnson JN, Gray GT, Bourne NK (1999) Effect of pulse duration and strain rate on incipient spall fracture in copper. J Appl Phys 86(9):4892
7. Gray GT III (1993) High pressure shock compression of solids. Springer, New York
8. Koller DD, Hixson RS, Gray GT, Rigg PA, Addessio LB, Cerreta EK, Maestas JD, Yablinsky CA (2005) Influence of shock-wave profile shape on dynamically induced damage in high-purity copper. J Appl Phys 98(10):103518
9. Meyers MA, Aimone CT (1983) Dynamic Fracture (Spalling) of Metals. Prog Mater Sci 28(1):1
10. Kanel GI (2001) Distortion of the wave profiles in an elastoplastic body upon spalling. J Appl Mech Tech Phys 42:358
11. Minich RW, Cazamias JU, Kumar M, Schwartz AJ (2004) Effect of microstructural length scales on spall behavior of copper. Metall Mater Trans A 35A(9):2663
12. Peralta P, DiGiacomo S, Hashemian S, Luo SN, Paisley D, Dickerson R, Loomis E, Byler D, McClellan KJ (2009) Characterization of Incipient Spall Damage in Shocked Copper Multicrystals. Int J Damage Mech 18(4):393
13. Escobedo JP, Dennis-Koller D, Cerreta EK, Patterson BM, Bronkhorst CA, Hansen BL, Tonks D, Lebensohn RA (2011) Effects of grain size and boundary structure on the dynamic tensile response of copper. J Appl Phys 110(3):033513
14. Patterson BM, Escobedo-Diaz JP, Dennis-Koller D, Cerreta EK (2012) Dimensional Quantification of Embedded Voids or Objects in Three Dimensions Using X-Ray Tomography. Microscopy and microanalysis, doi:10.1017/S1431927611012554 Published online by Cambridge University press, 3 Feb
15. Cerreta EK, Escobedo JP, Perez-Bergquist A, Koller DD, Trujillo CP, Gray Iii GT, Brandl C, Germann TC (2012) Early stage dynamic damage and the role of grain boundary type. Scripta Mater, 66(9):638 doi:10.1016/j.scriptamat.2012.01.051
16. Barker LM, Hollenback RE (1972) Laser Interferometer for Measuring High Velocities of Any Reflecting Surface. J Appl Phys 43(11):4669
17. Hemsing WF (1979) Velocity Sensing Interferometer (Visar) Modification. Rev Sci Instrum 50(1):73
18. Strand O, Goosman DR, Martinez C, Whitworth TL, Kuhlow WW (2006) Compact system for high-speed velocimetry using heterodyne techniques. Rev Sci Instrum 77(8):083108
19. Escobedo JP, Gupta YM (2010) Dynamic tensile response of Zr-based bulk amorphous alloys: Fracture morphologies and mechanisms. J Appl Phys 107(12):123502

# Chapter 45
# Experimental Characterization and Modeling of the 2024-T3 Aluminum Alloy Under Close-in Blast Loading

**Jian Yu, Chian-Fong Yen, Charles Chih-Tsai Chen, and Mark Nansteel**

**Abstract** Thin 2024-T3 aluminum alloy panels were subjected to close-in blast loading. Panels of various thicknesses were tested against different blast loadings and approach distances. A finite elemental close-in blast simulation model (LSDYNA ALE method) was first calibrated with the experimental data, and then it was used to predict damage behaviors at different blast loadings and approach distances. Simulation results were comparable to the experimental data.

**Keywords** Aluminum alloys • Blast simulation

## 45.1 Introduction

Aluminum alloys are the prime metallic materials used for light weight industrial applications. Experimental characterization of the damage in AA 2024 T3 alloy subjected to close-in blast loading is one of the critical steps to assess and improve the safety of passenger and cargo transportations. To evaluate the survivability of thin structures, shock-hole blasts on Aluminum Alloy 2024-T3 panels were tested at different explosive loadings and approach distances. The tests established the critical approach distance, the minimum distance at which the blast would breach the panel, as a function of the explosive loading. Finite element simulations of the shock-hole tests were carried out using Johnson–Cook (J-C) material models, which include strain and strain rate hardening, thermal softening effects and failure. LS-DYNA ALE method was used to implement the near-field blast simulation. The spall strength parameter of the JC model was calibrated with one set of experimental data at one particular explosive loading. The calibrated J-C model was able to predict failure behaviors at other explosive loadings and approach distances. Numerical predictions were comparable to the experimental observations.

## 45.2 Airblast Shock-Hole Blast Testing

A series of close-in shock-hole blast tests were conducted to characterize the critical breach behavior of 2024-T3 aluminum alloy panels at different explosive loadings and approach distances. Each panel was subjected to a selected size of charge of C4 explosive. Two thicknesses were tested, 0.10 cm and 0.18 cm, respectively. The panel was bolted along its square edges into a test fixture with a circular opening for the testing. The fixture was considered rigid for the current analysis study.

J. Yu (✉) • C.-F. Yen
U.S. Army Research Laboratory, 4600 Deer Creek Loop, Building 4600, RDRL-WMM-B, Aberdeen Proving Ground, MD 21005, USA
e-mail: jian.h.yu.ctr@mail.mil

C.C.-T. Chen
U.S. Department of Homeland Security, Transportation Security Laboratory, TSL-100, Building 315, Atlantic City International Airport, NJ 08405, USA

M. Nansteel
Battelle Memorial Institute, Transportation Security Laboratory, William J. Hughes Technical Center, Atlantic City International Airport, NJ 08405, USA

V. Chalivendra et al. (eds.), *Dynamic Behavior of Materials, Volume 1: Proceedings of the 2012 Annual Conference on Experimental and Applied Mechanics*, Conference Proceedings of the Society for Experimental Mechanics Series, DOI 10.1007/978-1-4614-4238-7_45, © The Society for Experimental Mechanics, Inc. 2013

**Fig. 45.1** A typical failure pattern of a blast penetrated panel. This failure pattern has six ruptured petals

A given amount of spherical explosive charge was mounted directly above the center of the circular fixture. The distance between the edge of the spherical charge and the surface of the panel is called the approach distance. The detonation of the explosive introduced a shockwave overpressure on the panel. Figure 45.1 shows the photographs of a typical breached panel (blast penetrated through).

### 45.2.1 *Test Results: Effect of Explosive Loading and Approach Distance on Panel Fracture*

The shockwave formed by the detonation of the explosive would initiate a localized failure on the surface of the panel that was directly adjacent to the charge. This localized failure usually develops during the first 100 microseconds after the close-in blast detonation. The failure is mostly attributed to the initial shock impulse from the blast. If the panel is breached during the initial shock, the subsequent overpressure from the blast causes cracks to propagate from the breached point. The final failure pattern is characterized by the multiple cracks that form ruptured petals. The number of ruptured petals formed after the blast is dependent on the explosive loading and approach distance. Generally, the number of petals increases as explosive loading increases and approach distance decreases.

### 45.2.2 *Damage Evaluation*

A ruptured petal was cut from the panel for Secondary Electron Microscopy (SEM) analysis. The surface of the petal was examined from the center of the blasted area to the unaffected area at the edge of the panel. SEM analysis revealed a gradual change in surface deformation on the petal. Figure 45.2 shows the difference between the highly damaged surface at blast center and the undamaged surface. The damage to the blasted surface is a characteristic of a ductile failure by void coalescence. This kind of the damage is usually observed in a high energy impact failure event.

## 45.3 Modeling and Simulation

Utilizing LS-DYNA, a finite element analysis of the shock-hole blast testing system was developed and used to model the deformation and failure of the Aluminum Alloy 2024-T3 panels subjected to explosive loading. An Arbitrary Lagrangian Eulerian (ALE) solver adopted within LS-DYNA was employed to allow full tracking of the interaction between the explosive and air fluids and the solid panel due to the capability of performing adaptive re-meshing to ensure the continuing presence of a high quality mesh. A finite element model for the shock-hole blast testing system developed for performing simulations of explosive blasting is shown in Fig. 45.3. The analysis considered the entire testing system. Both the test panel and the fixture were modeled using 4-noded shell elements. An air region was modeled utilizing 3D ALE elements, which over-lapped with the panel and fixture. Figure 45.3 also shows an ALE region of explosive which was generated by using an

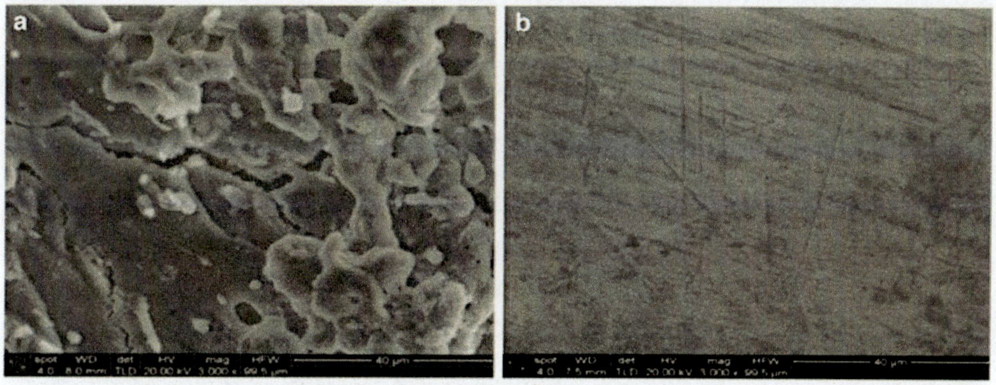


**Fig. 45.2** (**a**) surface at the center of the blast, (**b**) surface at the unaffected area

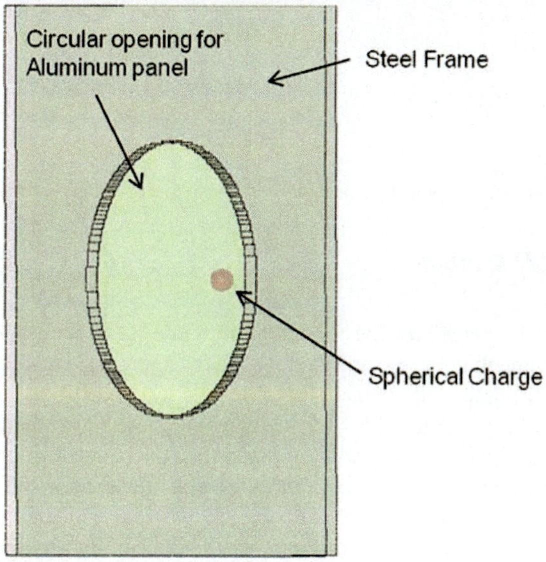


**Fig. 45.3** Blast simulation setup

**Table 45.1** J-C parameters for Aluminum 2024 T-3

| J-C parameters | A (MPa) | B (MPa) | n | C | m | $D_1$ | $D_2$ | $D_3$ | $D_4$ | $D_5$ |
|---|---|---|---|---|---|---|---|---|---|---|
| 2024-T3 | 369 | 684 | 0.73 | 0.0083 | 1.7 | 0.13 | 0.13 | −1.5 | 0.011 | 0.0 |

LS-DYNA input statement option of *Initial_ALE_Volume to establish the proper charge size and location. During the analysis, a fluid–structure interaction option of *Constrained_Largrange_in_Solid provided within LS-DYNA is used to efficiently model the explosive loading applied to the panel and fixture. A rate-dependent Johnson and Cook (J-C) elastoplastic model and an EOS provided within LS-DYNA were used to model the Aluminum Alloy 2024-T3 [1]. The fixture was modeled as a rigid steel body. The required material parameters for the J-C model of 2024-T3 are listed in Table 45.1. The EOS for Al 2024-T4 was used in place of Al 2024-T3 [2]. The Jones-Wilkins-Lee (JWL) equation of state was used to simulate the pressure distribution resulting from the detonation of C4 explosive [3]. The spall strength parameter, or the maximum principal stress spall model, of Johnson-Cook's material model for aluminum 2024-T3 was first calibrated using one set of experimental data at one particular explosive loading. This spall strength parameter was also used to simulate other loadings at different critical approach distances.

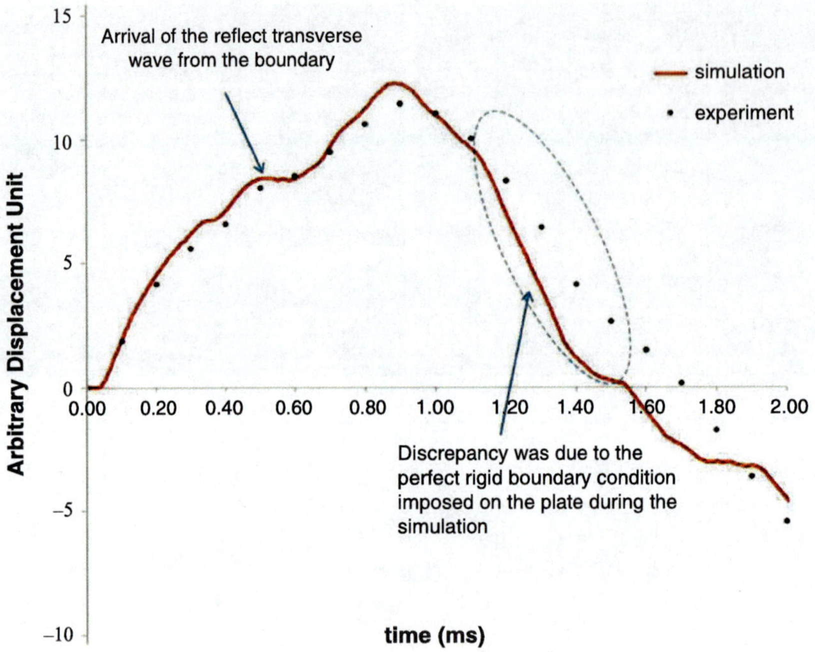


**Fig. 45.4** Comparison between simulation and experiment

### *45.3.1 Validation of the ALE Method*

A simulation of shock-hole testing on a panel was performed to correlate with one particular experiment where the panel does not fail. In that particular experiment, the explosive loading and the approach distance were setup in a way such that the blast would not breach nor plastically deform the panel. The objective was to calibrate the arbitrary Lagrangian–Eulerian (ALE) formulation for C4 blast simulation. Figure 45.4 shows the comparison of the center displacement of panel between the experiment and simulation. Since the boundary condition imposed on the panel during the simulation was perfectly rigid, the panel behaved more rigidly when comparing to the experiment. Unlike the experiment where the test fixture did not have compliance toward the blast loading, the simulated test fixture was treated as a rigid body. So the simulated tensile wave was not damping out as quickly as in the experiment. The simulated blast loading of the C-4 on the panel is very accurate and the ALE formulation for blast is valid. The simulated blast loading result was in good agreement with the experimental data. It even matched the reflected transverse wave from the boundary before the deflection of the plate reached its maximum. The discrepancy occurred only in the unloading of the plate. The simulated rigid boundary, which was imposed on the simulated plate, contributed to this discrepancy. The simulated blast data of the panel almost duplicated perfectly the experiment's results. The ALE formulation is adequate for blast simulation.

### *45.3.2 J-C Spall Strength Parameter Calibration*

From the experiment's data, the critical approach as function of explosive charge loading was known. The spall strength parameter was adjusted for the simulation model until the result showed a breach at an approach distance that matched with the experiment. For example, the spall strength parameter was calibrated for a panel at a thickness of 0.10 cm at an explosive loading of 1.71X unit of C4 and a critical approach distance of 1. Once the spall strength parameter was calibrated, simulations were performed to predict the failure pattern at different approach distances. Table 45.2 shows the comparison between the predicted number of ruptured petals and the experimental observation. The same calibrated spall strength parameter was used to simulate the panel at a thickness of 0.18 cm. The results are presented in Table 45.3. The simulation was able to predict the critical approach distance within 7 % error. It also was able to predict the ruptured pattern.

**Table 45.2** Ruptured petals prediction and experimental observation for the 0.10 cm thick panel. The highlighted boxes indicate the calibration of spall strength parameter at the critical approach distance

| Normalized approach distance | | | | | | | | | | | | | | | | | | | | |
|---|---|---|---|---|---|---|---|---|---|---|---|---|---|---|---|---|---|---|---|---|
| Loading 1.71X | 0.53 | 0.57 | 0.62 | 0.66 | 0.7 | 0.74 | 0.79 | 0.83 | 0.87 | 0.91 | 0.96 | 1 | 1.04 | 1.09 | 1.13 | 1.17 | 1.21 | 1.26 | 1.3 | 1.34 |
| Loading 1.71X | | | | | | | | | | | | | | | | | | | | |
| Simulated results: number of petals | 10 | 9 | 8 | 8 | 8 | 8 | 6 | 6 | 6 | 6 | 5 | 5 | no breach | no breach | no breach | no breach | no breach | no breach | no breach | no breach |
| Experimental results: average number of petals | 8 | 8 | 8 | 7 | 5 | 6 | 5 | 6 | 6 | 5 | 5 | 4 | no | no | no | no | no | no | no | no |

**Table 45.3** Ruptured petals prediction and experimental observation for the 0.18 cm thick panel

| Normalized approach distance | | | | | | | | | |
|---|---|---|---|---|---|---|---|---|---|
| Loading 1.71X | 1 | 1.1 | 1.2 | 1.3 | 1.4 | 1.5 | 1.6 | 1.7 | 1.8 |
| Simulated results: number of blasted petals | 6 | 6 | 6 | 5 | 5 | 4 | no breached | no breached | no breached |
| Experimental results: average number of blasted petals | 6 | 6 | 6 | 5 | 3 | no breached | no breached | no breached | no breached |

## 45.4 Conclusion

Simulations of different explosive loadings and approach distances of a close-in blast on aluminum (2024 T-3) panels were performed. The blast simulation of C4 was validated with experimental data. The spall strength in the J-C model was calibrated a particular approached distance at which the blast loading was just enough to breach the panel. The simulation model was able to predict the critical approach distance for panels with different thickness.

## References

1. Katayama M, Kibe S, Toda S (1995) A numerical simulation method and its validation for debris impact against the whipple bumper shield. Int J Impact Eng 17:465–476
2. Lesuer D (2000) Experimental investigations of material models for Ti-6Al-4V titanium and 2024-T3 aluminum, DOT/FAA/AR-00/25. Available to the U.S. public through the National Technical Information Service (NTIS), Springfield, Virginia 22161
3. Dobratz B, Crawford P (1985) LLNL explosive handbook, UCRL-52997 Rev.2, Lawrence Livermore National Laboratory

# Chapter 46
# Interaction of Oxygen and Carbon Dioxide at High Pressure and Temperature

**Liu Haifeng, Zhao Yanhong, and Lu Guo**

**Abstract** The interaction potential parameters of original molecules and those of atoms and molecules of dissociation products are determined by shock Hugoniot data of molecular fluid in this article. At low pressure, Ross's modification of hard-sphere perturbation theory is used to evaluate interaction potential parameters of original molecules. At high pressure, potential parameters of dissociation products are calculated using chemical equilibrium method originated from statistical mechanical theory. Comparing to those from the corresponding states scaling relations, the potential parameters obtained in this article using the above method have a better description for impact experiments in a wide pressure range.

**Keywords** Interaction potential parameters • Oxygen • Carbon dioxide • Dissociation

## 46.1 Introduction

The majority of the basic components of condensed explosives are C, H, N and O atoms. A number of theories show that the Chapman-Jouguet (CJ) detonation products of such explosives are mainly liquid products involving $N_2$, $CO_2$, $H_2O$, CO, $NH_3$, $H_2$, $CH_4$, NO, $O_2$ and free carbon. The extended model of hard-sphere perturbation theory [1] can well describe the Hugonoit properties of these liquid products. In this model, the key issue is the accuracy of molecular interactions. Although the potentials of $H_2$, $N_2$ and $H_2O$ evaluated by ab initio calculations [2–4] are effective, the calculations are very cumbersome and difficult to accurately describe molecular interactions in a wide range. In shock compression state, the final density of simple liquid is three times as large as the normal density and its temperature can be as high as thousands of degrees. Therefore, the thermodynamic properties of liquid are determined mainly by short-range repulsive interaction forces. There are two commonly used potentials: Lennard-Jones (L-J) potential and Exponential-6 (Exp-6) potential. In L-J potential, it has only two parameters and they can be easily determined by experiments. The quantum mechanical calculations show that in a very short distance, the repulsive part of LJ potential is not very stiff and it can not describe many-body interactions. While the three parameters Exp-6 potential is more flexible. It can more accurately describe the interaction between molecules and its repulsive part is consistent with the results of molecular beam scattering experiments [5]. To some extent, the Exp-6 potential involves an effective description of many-body interactions of liquid molecules. Therefore, it is used in this article.

The most commonly used method to ascertain parameters of Exp-6 potential is based on corresponding states scaling relations [1, 5]. In the method, the parameters of Ar are taken as reference values and then the parameters of new molecules are determined using the critical temperature and critical volume (or critical pressure) of itself and Ar. In the case of lack of enough data of shock experiments, it is used to determine the potential parameters. In our earlier work [6, 7], we calculated the properties of detonation products of PETN using potential parameters obtained by this method.

With the rapid development of experimental techniques in recent years, the shock data of lots of detonation products have been measured, thus these data can be used to obtain more accurate potential parameters. Meanwhile, in ultra-high pressure detonation, molecules dissociate into atoms and ionize into ions [8, 9]. For lack of direct experimental data which can be

L. Haifeng (✉) • Z. Yanhong • L. Guo
Institute of Applied Physics and Computational Mathematics, 2 Fenghao East Rd, 100094 Beijing, China
e-mail: liu_haifeng@iapcm.ac.cn

V. Chalivendra et al. (eds.), *Dynamic Behavior of Materials, Volume 1: Proceedings of the 2012 Annual Conference on Experimental and Applied Mechanics*, Conference Proceedings of the Society for Experimental Mechanics Series,
DOI 10.1007/978-1-4614-4238-7_46, 

used to collect data for ascertain the potential parameter, how to determine the interactions between atomic and ionic products has long been a research priority in the detonation field.

Taking $O_2$ and $CO_2$ for example, we introduce a new method to determine the parameters of potential function, in which the shock Hugoniot data is used. The extended hard-sphere perturbation theory of Ross is used to calculate interaction potential parameters of original molecules at low pressure and they are evaluated using chemical equilibrium method based on statistical physics at high pressure. In the range that the temperature is under 20,000 K and the pressure is under 100 GPa, the effect of ionization can be neglected. For our calculation is in that range, only the dissociation is taken into account and we get a good result.

## 46.2 Theoretical Model and Method

### *46.2.1 Extended Hard-Sphere Perturbation Theory for One-Component Fluid*

The EOS of one-component gas is described by the extended hard-sphere perturbation theory of Ross. Assuming the Helmholtz free energy $A_f$ of liquid can be divided into the sum of two parts:

$$A_f(V,T) = A_{id}(V,T) + A_{\text{int}}(V,T) \tag{46.1}$$

where $A_{id}$ is the free energy of ideal gas and $A_{\text{int}}$ is the excess free energy from molecular interactions [10–13]. The Helmholtz free energy for ideal gas term is calculated with the theory of Rossini (monatomic system), Pennigton and Kobe (diatomic system) and the harmonic-oscillator rigid rotator theory (polyatomic system) [14]. The calculated thermodynamic quantities are fitted by power series as follow

$$S^0 = a_0 + a_1T + a_2T^2 + a_3T^3 + a_4T^4 + a_5T^5 + a_6T^6 + a_7T^7$$

$$H^0 - H_0^0 = \frac{a_1T^2}{2} + \frac{2a_2T^3}{3} + \frac{3a_3T^4}{4} + \frac{4a_4T^5}{5} + \frac{5a_5T^6}{6} + \frac{6a_6T^7}{7} + \frac{7a_7T^8}{8} + C$$

$$\frac{F^0 - H_0^0}{T} = -\left(a_0 + \frac{a_1T}{2} + \frac{a_2T^2}{3} + \frac{a_3T^3}{4} + \frac{a_4T^4}{5} + \frac{a_5T^5}{6} + \frac{a_6T^6}{7} + \frac{a_7T^7}{8}\right) + \frac{C}{T}$$

where $S^0$, $H^0$, $F^0$ are entropy, enthalpy and free energy of ideal gas respectively, $H_0^0$ and $C$ are the heat of formation and integration constant in the integration of $\frac{dH}{dT} = T\frac{dS}{dT}$ respectively. $T$ is temperature and $a_i$ is fitting parameter (used in the range 300 $K \leq T \leq$ 20,000 $K$.). The values of $a_i$, $C$ and$H_0^0$are listed in Table 46.1, the units of free energy and enthalpy are cal/mole and the temperature unit is K.

The excess free energy $A_{\text{int}}$ [15] is calculated by following inequality

$$A_{\text{int}} \leq A_{HS} + \frac{2\pi\rho}{T}\int_d^\infty \varphi(r)g_{PY}\left(\frac{r}{d},\eta\right)r^2dr + F_{12}(\eta)NkT \tag{46.2}$$

**Table 46.1** Fitting parameters for EOS of ideal gas

| Products | $O_2$ | O | $CO_2$ | CO |
|---|---|---|---|---|
| $a_0/10$ | 4.60832 | 3.68020 | 4.81830 | 4.44249 |
| $a_1/10^{-2}$ | 1.53437 | 1.00695 | 2.30500 | 1.47415 |
| $a_2/10^{-6}$ | −4.39714 | −3.13573 | −6.39802 | −4.22175 |
| $a_3/10^{-10}$ | 7.83972 | 5.77935 | 11.0798 | 7.45936 |
| $a_4/10^{-14}$ | −8.14251 | −6.09086 | −11.2678 | −7.69300 |
| $a_5/10^{-18}$ | 4.81553 | 3.62710 | 6.56310 | 4.53302 |
| $a_6/10^{-22}$ | −1.50107 | −1.13458 | −2.02292 | −1.41040 |
| $a_7/10^{-27}$ | 1.91161 | 1.44761 | 2.55426 | 1.79416 |
| $C/10^3$ | 2.62514 | 2.10351 | 2.71734 | 2.55617 |
| $H_0^0/10^4$ | 0.00000 | 5.90000 | −9.3968 | −2.7201 |

where $A_{HS}$ is excess free energy of hard-sphere, $\varphi$(r) is interaction potential, $g_{PY}(\frac{r}{d},\eta)$ is Percus-Yevick hard-sphere radial distribution function and $F_{12}(\eta)$ is the soft-sphere correction term.

$$A_{HS} = NkT(4\eta - 3\eta^2)/(1-\eta)^2$$

$$F_{12}(\eta) = (\eta^4/2 + \eta^2 + \eta/2)$$

where $\eta = \pi\rho d^3/6$ and $\rho = N/V$. d is diameter of hard-sphere and its value makes the right side of inequality (46.2) reach minimum value. The minimum energy is the Helmholtz free energy $A_{int}$ of current state is the minimum energy. $\Phi$(r) is the Exp-6 potential

$$\varphi(r) = \varepsilon\left\{\frac{6}{\alpha-6}\exp\left[\alpha\left(1-\frac{r}{r^*}\right)\right] - \frac{\alpha}{\alpha-6}\left(\frac{r}{r^*}\right)^{-6}\right\}$$

Where $\varepsilon$, $\alpha$ and $r^*$ are potential parameters. Ross proved that the calculated pressure and energy of soft-sphere model of hard sphere perturbation theory are agreed well with Monte Carlo results [16].

### 46.2.2 Equivalent One-Component Fluid Model for Mixture System

For detonation products system, new molecules and atoms will generate by chemical reaction of molecular products at high pressure, therefore, it's a mixture system. The vdw1f model is used to describe gaseous product of the mixture, that is, the mixture can be taken as an effective one-component fluid. The parameters $\varepsilon$, $\alpha$ and $r^*$ of potential $\varphi$(r) are evaluated by the extended mixing rule [9]

$$\varepsilon = \frac{\sum_{ij} x_i x_j \varepsilon_{ij}(r^*_{ij})^3}{(r^*)^3} \quad \alpha = \frac{\sum_{ij} x_i x_j \alpha_{ij}\varepsilon_{ij}(r^*_{ij})^3}{\varepsilon(r^*)^3} \quad (r^*)^3 = \sum_{ij} x_i x_j (r^*_{ij})^3 \tag{46.3}$$

And $\varepsilon_{ij}$, $\alpha_{ij}$, $r_{ij}^*$ are obtained by extended Lorent-Berthelor combination rule [1]

$$\varepsilon_{ij} = l_{ij}(\varepsilon_{ii}\varepsilon_{jj})^{\frac{1}{2}} \alpha_{ij} = m_{ij}(\alpha_{ii}\alpha_{jj})^{\frac{1}{2}} r^*_{ij} = \frac{k_{ij}(r^*_{ii} + r^*_{jj})}{2} \tag{46.4}$$

where $l_{ij}$, $m_{ij}$ and $k_{ij}$ are correction coefficients. The values of them are set to 1. $x_i = n_i/\sum_j n_j$ is component molar concentration and calculated by solving chemical equilibrium equations. For a system involving t types of components, *s* types are gaseous component and *t-s* types are solid. They are composed of *c* kinds of chemical elements. The mass conservation law indicates that the number of atoms of each element in reactants is equal to that of resultants in chemical reactions and it is formulated by

$$\sum_{i=1}^{s} a_{ij}n_i + \sum_{i=s+1}^{t} a_{ij}n_i = q_j \, j = 1, c \tag{46.5}$$

where $a_{ij}$ is the number of atoms of *j* element in *i* component, $n_i$ is mole concentration of the *i* component and $q_j$ is the total mole number of the *j* element. In chemical equilibrium state, the chemical potentials of reactants must equal to that of resultant. According to the minimum free energy law, the chemical equilibrium equations [6, 14, 17] can be derived.

### 46.2.3 Optimization of Potential Parameters

The shock compression state of fluid is determined by the Hugonoit relationship as follow

$$E_H - E_0 = \frac{1}{2}(P_H + P_0)(V_0 - V_H)$$

where $V_0, E_0, P_0$ and $V_H, E_H, P_H$ are parameters of the initial state and final state respectively. The initial state of each kind of fluid is set as the initial experimental value.

The potential parameters of $O_2$-$O_2$, $CO_2$-$CO_2$ and CO-CO interactions are determined using extended hard-sphere perturbation theory with shock Hugoniot data of molecular products at low pressure. At high pressure, molecules will dissociate into atoms, such as $O_2$ can generate O atoms and $CO_2$ will yield a mixture of gaseous products (CO, $O_2$, O) and solid carbon. In the mixture, the interaction potential parameters of gas molecules are set the value ascertained at low pressure. For O-O interaction potential parameters, we firstly assume a set of initial values and then calculate parameters of $O_2$-O interaction in accordance with formula (46.4). The shock Hugoniot curve is obtained using the effective one-component fluid model and then compared with experimental shock Hugoniot curve. If the given accuracy is not satisfied, the parameters of O-O interaction have to be adjusted to meet accuracy requirement. We use the modified Murchan equation as the equation of state for solid products [6].

## 46.3 Results and Discussions

Using the new model and method proposed in this paper, the Exp-6 potential fitting parameters of detonation products are obtained and listed in Table 46.2. Shock Hugoniot curves of $O_2$ are shown in Fig. 46.1. The discrete points refer to experimental results [18, 19] and other curves are calculated results of theoretical model. The parameters for black curve are from reference [1]. They are evaluated using corresponding states scaling relations and experimental data of Ar; it is found that the calculated value is smaller than experimental value in low pressure range (less than 40 Gpa). Without consideration of oxygen dissociation, our potential parameters are better for describing the EOS of low pressure range. For a system taking into account of oxygen dissociation, with new evaluated parameters of $O_2$ and O in this paper, we get the green curve. In Fig. 46.1, it shows that the dissociation curve (green) nearly overlaps with the non-dissociation curve (red) below 40 Gpa and the two curves gradually separate from each other above 40 Gpa. Due to the reproduction of experimental data of the

**Table 46.2** Fitting parameters for Exp-6 potential

| Products | $\varepsilon^*$/K | $r^*_{ii}$/nm | $\alpha_{ii}$ | Reference |
|---|---|---|---|---|
| $O_2$ | 125.0 | 0.384 | 13.0 | 1 |
| $O_2$ | 125.0 | 0.385 | 13.2 | This work |
| O | 708.0 | 2.40 | 10.9 | This work |
| $CO_2$ | 245.6 | 0.417 | 13.0 | 1 |
| $CO_2$ | 245.6 | 0.428 | 13.1 | This work |
| CO | 108.3 | 0.412 | 13.0 | 1 |

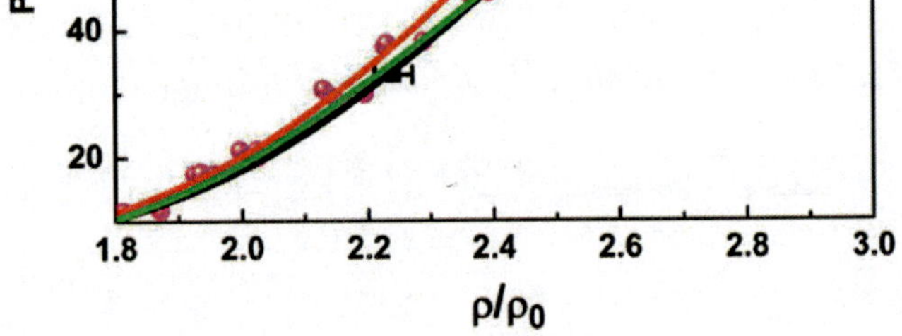


**Fig. 46.1** Hugoniot curves of liquid $O_2$

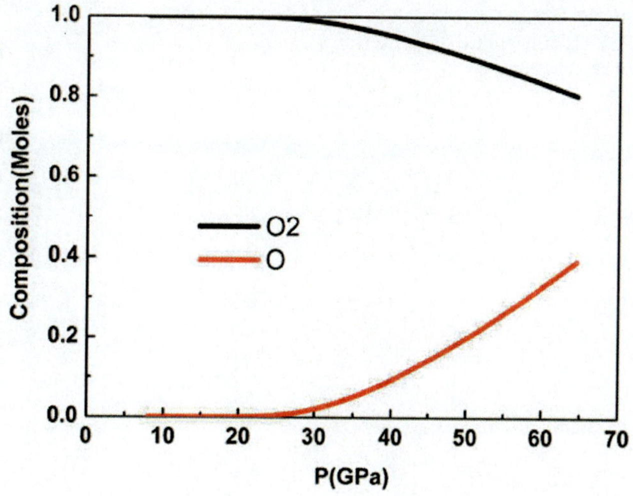


**Fig. 46.2** Variation of products of liquid $O_2$ under shock compression

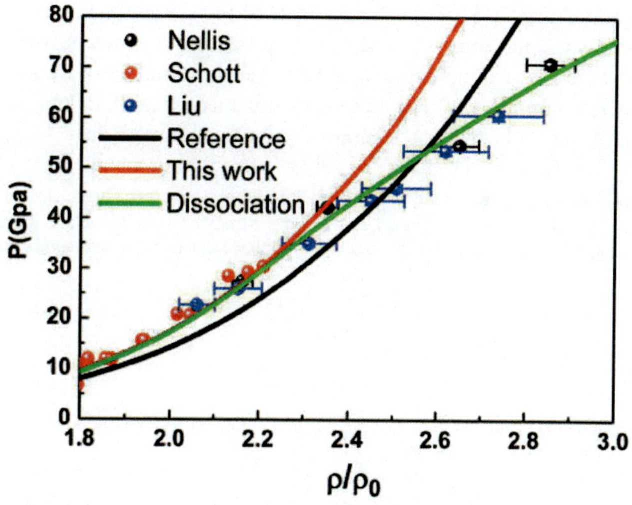


**Fig. 46.3** Hugoniot curves of liquid $CO_2$

dissociation model, we inter that the molecular dissociation is more important at high pressure and it can be also seen in Fig. 46.2, which shows the relationship between shock pressure and components of detonation products.

Due to difficulties for sample preparation of liquid carbon dioxide, there is no Hugoniot experimental data of liquid carbon dioxide published in literatures before early 1990s. Nellis [20] measured four shock Hugoniot points of liquid $CO_2$ using two-stage light gas gun technique in 1991 and later in 1998, Liu [11] measured seven new points of liquid $CO_2$ with the same technique. All these points are valuable data for the determination of the equation of state of liquid carbon dioxide. At high pressure, carbon dioxide will partly dissociate into CO, $O_2$, O and solid carbon. The potential parameter of $O_2$-O interaction is chosen as the previously determined value and those of $CO_2$-CO interaction is ascertained using their experimental data at low pressure. The EOS of solid carbon is Murnaghan equation. Figure 46.3 is comparison chart for shock Hugoniot curves of $CO_2$ and the relationship between shock pressure and components of dissociation products is shown in Fig. 46.4.

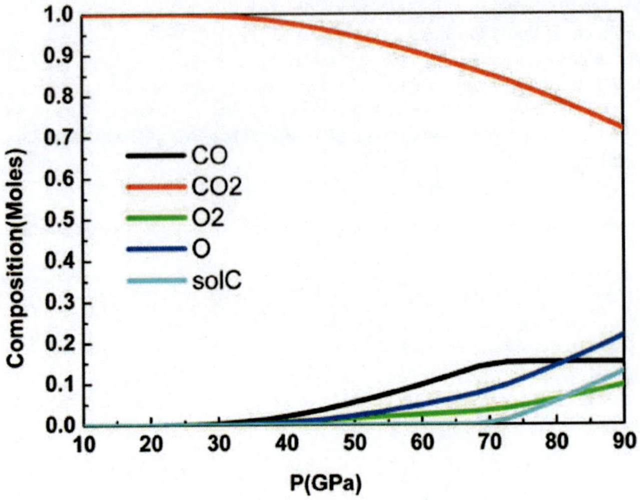


**Fig. 46.4** Variation of products of liquid $CO_2$ under shock compression

## 46.4 Conclusions

In this paper, we propose a new method to evaluate the interaction potential parameters of original molecules and those of atoms and molecules of dissociation products, in which only the shock Hugoniot data of molecular fluid is used. Taking $O_2$ and $CO_2$ for example, the obtained potential parameters are investigated in detail. It shows that the data of shock Hugoniot experiment are well reproduced by our theoretical data and the relationship between variation of components and law of dissociation of fluid substances are quantitatively given.

**Acknowledgements** Our thanks are due to Profs. Guangcai Zhang, Haifeng Song, Guicun Ma, Hui Zheng, and Drs. Gongmu Zhang, Qili Zhang, Hongzhou Song, Bo Sun, Shuaichuang Wang, Mingfeng Tian, for helpful comments and discussions. This work was supported by the National Natural Science Foundation of China (Grant No. 11102026) and the Science Foundation of China Academy of Engineering Physics (Grant Nos. 2009A0101004).

## References

1. Ree FH (1984) A statistical mechanical theory of chemically reacting multiphase mixtures: application to the detonation properties of PETN. J Chem Phys 81(3):1251
2. Ree FH, Bender CF (1974) Nonadditive interaction in molecular hydrogen at high pressure. Phys Rev Lett 32(3):85
3. Ree FH, Winter NW (1980) Ab initio and Gordon–Kim intermolecular potentials for two nitrogen molecules. J Chem Phys 73(1):322
4. Matsuoka O, Clementi E, Yoshimine M (1976) CI study of the water dimer potential surface. J Chem Phys 64(4):1351
5. Ross M, Ree FH (1980) Repulsive forces of simple molecules and mixtures at high density and temperature. J Chem Phys 73(12):6146
6. Zhao YH, Liu HF, Zhang GM (2007) Equation of state of detonation products based on statistical mechanical theory. Acta Phys Sin 56(8):4791
7. Zhao YH, Liu HF, Zhang GM (2009) Equation of state of detonation products for PETN explosive. Chin J High Pres Phys 23(2):143
8. Gu YJ, Zheng J, Chen ZY, Chen QF, Cai LC (2010) The equation of state of H2 + He fluid mixture in the region of partial dissociation. Acta Phys Sin 59(7):4508
9. Chen QF, Cai LC, Chen DQ, Jing FQ (2005) Pressure dissociation of dense hydrogen. Chin Phys 14(10):2077
10. Yang XD, Xie W, Wu BJ, Hu D, Jing FQ (1998) Theoretical calculation for the Hugniot curves of liquid nitrogen. Chin J HighPres Phys 12(1):1
11. Liu FS, Chen XM, Chen PS, Sun Y, Jing FQ (1998) Equation of state of liquid $CO_2$ at high temperature and high densities. Chin J HighPres Phys 12(1):28
12. Yang XD, Hu D, Jing FQ (1999) Studies of EOS for detonation products: liquid nitrogen, liquid helium and water. Chin J HighPres Phys 13(2):93
13. Chen QF, Cai LC, Jing FQ, Chen DQ (1999) Theoretical calculation of the Hugonoit curves for liquid deuterium and hydrogen. Acta Phys Sin 48(3):485
14. Mader CL (1979) Numerical modeling of detonation. University of California Press, Berkeley
15. Ross M (1979) A high-density fluid perturbation theory based on an inverse 12th power hard-sphere reference system. J Chem Phys 71(4):1567
16. Ree FH (1983) Simple mixing rule for mixtures with exp-6 interactions. J Chem Phys 78(1):409
17. Liu HF, Chen DQ, Zhang SZ (1996) Equation of state of detonation products and the possible phase transition for CHBr3. Chin J HighPres Phys 10(4):284
18. Nellis WJ, Mitchell AC (1980) Shock compression of liquid argon, nitrogen, and oxygen to 90 GPa (900 kbar). J Chem Phys 73(12):6137
19. Marsh SP (1980) LASL shock Hugoniot data. University of California Press, Berkeley
20. Nellis WJ, Mitchell AC, Ree FH, Ross M, Holmes NC, Trainor RJ, Erskine DJ (1991) Equation of state of shock-compressed liquids: carbon dioxide and air. J Chem Phys 95(7):5268

# Chapter 47
# Response of Curved Carbon Composite Panels to Shock Loading

**Puneet Kumar, David S. Stargel, and Arun Shukla**

**Abstract** Dynamic experiments were conducted using a shock tube facility coupled with high speed digital image correlation system to understand the effect of plate curvature on blast response of 32 layered carbon fiber-epoxy panels. Three different geometries of carbon fiber panels were evaluated. These panels had a radii of curvatures of infinity (panel A), 304.8 mm (panel B), and 111.76 mm (panel C). Panels with dimensions of 203.2 mm × 203.2 mm × 2 mm were held with clamped boundary conditions during the shock loading. The input and reflected pressure profiles from the shock loading were carefully analyzed to obtain the impulse and energy during the loading process. As a result of shock loading, Panel A had a mid-point deflection of 14 mm before failure initiated, whereas it was 18 mm for the panel B and 10 mm in panel C. Macroscopic postmortem analysis was also performed to comprehend and compare the different mechanisms of failure observed in the three panels, which showed types of failure mechanism in all the three panels: fiber breakage and inter-layer delamination. The fiber breakage was induced on the face exposed to shock loading and continued inside. The delamination was visible on the side of the specimen as well as on the face exposed to the shock loading. The study showed that panel C can mitigate the highest intensity (pressure) shock waves without the initiation of catastrophic damage in the panel. In regards to the other two panels investigated, panel B could sustain the least shock wave intensity before failure. Panel A could mitigate the blast pressure having intensity in between the intensities impinging on the other two panels.

**Keywords** Carbon composite • Blast loading • Shock tube • Digital image correlation

## 47.1 Introduction

A controlled experimental study has been conducted to understand the effect of plate curvature on the blast response of carbon composite panels. Accidental explosions or bomb blasts cause extreme loading on structures, which have both flat and curved geometries (as in the USS Cole [1] and Oklahoma State [2] bombing). Therefore it is important to understand the effect of curvature on the blast response. Also, a better understanding of the effect of curvature will help in manufacturing new structures with better blast resistant property. Carbon composite panels having three different radii of curvature were subjected to shock loading using a shock tube in order to study their dynamic response. The choice of these panels was motivated by the study on the ballistic response of these panels by Stargel [3]. Real-time and post-mortem analysis was conducted on the panels to evaluate the effects of plate curvature on blast mitigation. In particular, the midpoint transient deflection, velocity, and macroscopic post-mortem analysis of the panels has been used to characterize the response of curved panels when subjected to a controlled blast loading.

Experimentally shock loading on structure can be imparted using two different methods: one is by using explosives and the other is by using shock tubes. The use of real explosives is dangerous and has spherical wave fronts and pressure

P. Kumar • A. Shukla (✉)
Dynamic Photomechanics Laboratory, Department of Mechanical, Industrial and Systems Engineering, University of Rhode Island, Kingston, RI 02881, USA
e-mail: shuklaa@egr.uri.edu

D.S. Stargel
Air Force Office of Scientific Research, Aerospace, Chemical, and Material Sciences, 875 N. Randolph Street, Suite 325, Room 3112, Arlington, VA 22203-1768, USA

V. Chalivendra et al. (eds.), *Dynamic Behavior of Materials, Volume 1: Proceedings of the 2012 Annual Conference on Experimental and Applied Mechanics*, Conference Proceedings of the Society for Experimental Mechanics Series, DOI 10.1007/978-1-4614-4238-7_47, 

signatures which are spatially complex and difficult to measure. On the contrary, shock tubes offer the advantage of planar wave fronts and wave parameters that can be easily controlled. Above all, these parameters are easy to replicate when using a shock tube and as such shock tube was the preferred choice of applying blast loading in our experiments.

There is a large volume of literature dealing with the blast loading of structures [4–13]. For brevity of space, only a few studies are mentioned here. Franz et al. [14] studied the response of glassfiber chopped-strand mat laminates to air pressure blast. They found matrix cracking, delamination/debonding, and penetration as final damage in monolithic and layered laminates with varying areal density. Ochola et al. [15] concentrated on strain rate sensitivity of both carbon fiber reinforced polymer and glass fiber reinforced polymer by testing a single laminate configuration with strain rate varying from $10^{-3}$ and $450\ s^{-1}$. Results showed that the dynamic material strength for GFRP increases with increasing strain rate and the strain to failure for both CFRP and GFRP decreased with increasing strain rate. LeBlanc and Shukla [16] studied the underwater shock loading response of E-glass/Vinyl ester curved composite panels. They used the 3D-DIC system for measuring the transient response during the experiments. They also compared the experimental results to simulation results obtained from the commercially available Ls-Dyna finite element code, which showed a high level of correlation using the Russell error measure. Shen et al. [17] experimentally investigated the response of sandwich panels with aluminum face sheets and aluminum foam core. Panels with varying curvatures (two different curvatures), and different core/face sheet configurations were tested at three different blast intensities. They found that the initial curvature of the sandwich panel changes the deformation mode and improved the performance of the structure when compared to equivalent flat plate.

The present study was performed using carbon fiber composite panels having three different curvatures. The results from this study show that the plate curvature affects the blast mitigation property. As the radius of curvature decreases (becomes more sharply curved), the response of panels to sustain shock loading changes.

## 47.2 Experimental Procedure

### *47.2.1 Specimen Geometry and Material Details*

Panels with three different radii of curvature were utilized in the experiments: infinite radius of curvature (Panel A), 304.8 mm radius of curvature (Panel B) and 111.8 mm radius of curvature (Panel C). The specimens are shown in Fig. 47.1. The specimens were fabricated using ready-to-cure sheets of unidirectional AS4/3501-6 material manufactured by the Hercules corporation of Magna, Utah. These sheets were manufactured by pulling a row of uniformly spaced fibers through resin bath. The spacing between the fibers determines the ratio of fiber volume to total volume of the material. After stacking 32 plies of theses unidirectional material in the desired orientation, the panels were vacuum bagged and then placed in autoclave for curing. The curved panels were manufactured in the same way as described above with a difference that cylindrically curved pre-forms were used for laminate lay-up and curing.

The orientation of the individual unidirectional plies that comprised the composite laminate was selected to simulate quasi-isotropic properties ($[0/90/+45/-45]_{4s}$). Each experiment was repeated three times to assure consistent results. The specimens were 203 mm long × 203 mm wide × 2 mm thick, made out of 32 layers of carbon fibers. For the case of curved panels, the arc length of the curved edges corresponds to the plate length. The material properties of the laminates are listed below in Table 47.1 [18].

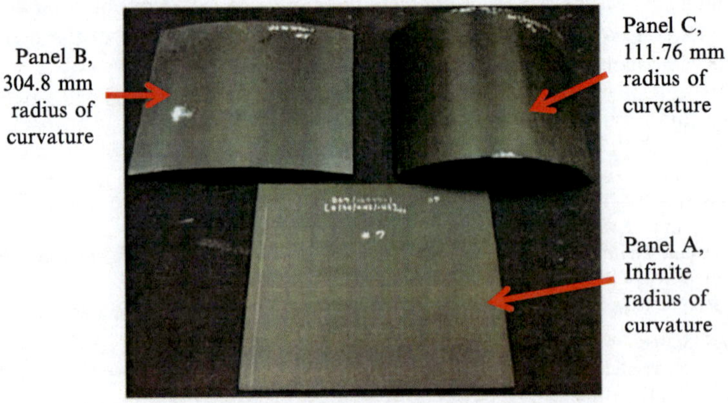


**Fig. 47.1** Specimen geometries

Table 47.1 Material properties

| Elastic modulus | $E_x = 147$ GPa, $E_y = 11$ GPa, $E_{xy} = 5$ Gpa |
|---|---|
| Tensile strength in fiber direction | 2,004 MPa |
| Compressive strength in fiber direction | 1,197 MPa |
| Tensile strength in transverse direction | 53 MPa |
| Compressive strength in transverse direction | 200 MPa |
| Shear strength | 137 MPa |

Fig. 47.2 URI shock tube facility

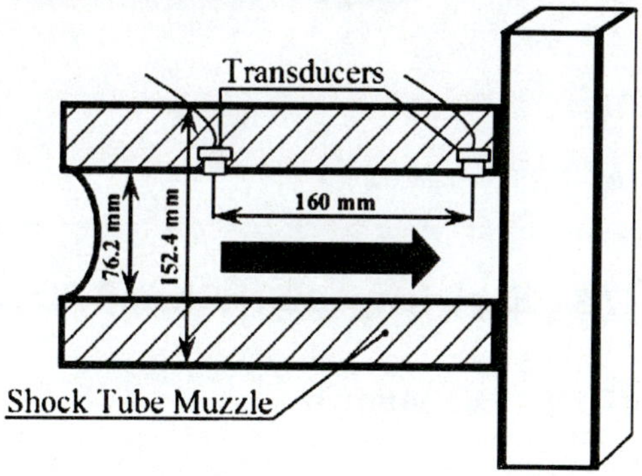


Fig. 47.3 Shock tube muzzle

### 47.2.2 Shock Loading Apparatus and Loading Conditions

The shock tube apparatus used in this study to obtain the controlled dynamic loading is shown in Fig. 47.2. A complete description of the shock tube and its calibration can be found in [19]. The shock tube consists of a long rigid cylinder, divided into a high-pressure driver section and a low pressure driven section, which are separated by a diaphragm. By pressurizing the high-pressure section a pressure difference across the diaphragm is created. When this pressure differential reaches a critical value, the diaphragm ruptures. The subsequent rapid release of gas creates a shock wave, which travels down the shock tube to impart shock loading on the specimen at the muzzle end.

The shock tube utilized in the present study has an overall length of 8 m, consisting of driver, driven, converging and muzzle sections. The diameter of the driver and driven section is 0.15 m. The final muzzle diameter is 0.07 m. Two pressure transducers (Fig. 47.3), mounted at the end of the muzzle section measure the incident shock pressure and the reflected shock pressure during the experiment. The incident shock wave pressure was kept constant for all of the experiments. A typical pressure profile obtained at the transducer location closer to the specimen is shown in Fig. 47.4. The specimens were shock loaded at three different pressures varying from 3 to 8 MPa. At each of the three pressures, experiments were repeated three times to validate the consistency. The specimens were held with clamped boundary conditions on all the four edges. Appropriate fixtures for holding each of the plate geometries are shown in Fig. 47.5.

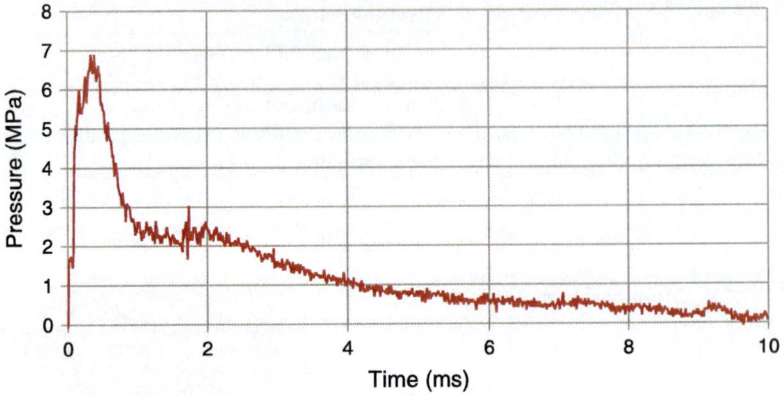


**Fig. 47.4** Typical pressure profile

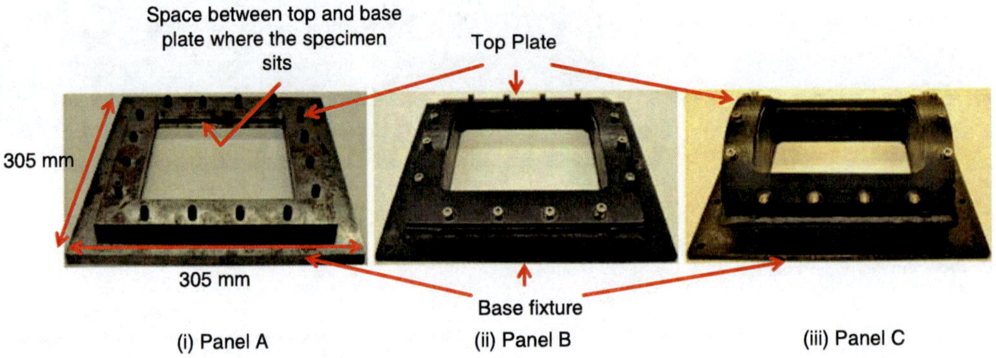


**Fig. 47.5** Loading fixture

## 47.3 Experimental Results and Discussion

### *47.3.1 DIC Analysis*

The DIC technique is used to obtain the out-of-plane deflections and velocities as well as the in-plane strains on the back surface for all the three geometries. The shock tube used in this study provides a uniform pressure pulse over a circular area of 4,562 mm$^2$ (muzzle area). This is verified by the DIC image of the out of plane displacement on flat plate during shock impingement as shown in Fig. 47.6. The image taken at 50 μs shows that the flat panel had a uniform deflection of 3 mm ± 0.2 mm within a central region of diameter 72 mm. As the muzzle diameter of the shock tube is 76 mm, the uniform deflection contour constitutes about 92 % of the total loaded area. This out of plane deflection decays to 0 mm from 3 mm in an area of thickness 3 mm beyond this central region.

The full-field deflection at the failure loading for the panels is shown in Fig. 47.7. The total deflection in these panels subjected to blast loading comprises of two distinct regions, namely, the indentation region followed by the flexural deflection (Fig. 47.8). During indentation, localized deflection superpose onto the overall deflection. In case of flexure, the overall deflection starts to overpass the localized deflection. The indentation deflection is localized around the loading area whereas a full field deflection over the area of the specimen being loaded is predominant in flexural mode. Also, the boundary conditions affect the deflection in case of flexural mode. Panels B and C had elliptical deflection contour at 200 μs. Since both these panels are curved, the shock wave impinges on the projected area and creates localized elliptical deflection contour. Panel B had fluid–structure interaction for first 200 μs at which time the boundary condition starts affecting the deflection development. Around 200 μs, the elliptical deflection contour in panel B starts changing to rectangular shape because of the boundary conditions. Also, the deflection mode in panel B changes from indentation to flexural mode at this time. The deflection in panel C continues to develop further with elliptical contour. The deflection is primarily because of indentation and boundary conditions do not affect the deflection contours. As such, it retains its elliptical shape throughout the loading process.

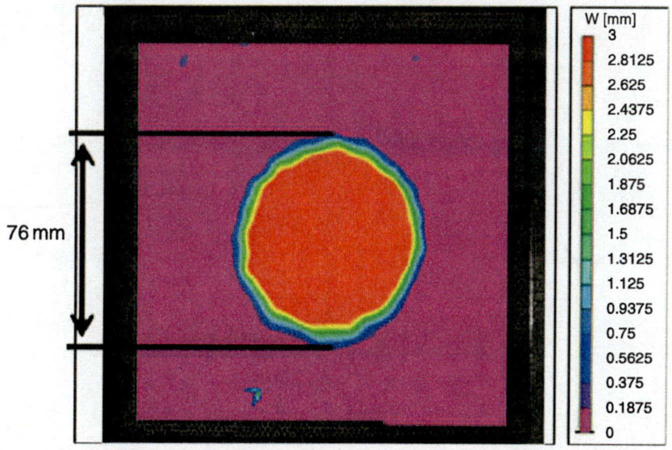


**Fig. 47.6** Out of plane deflection at t = 50 μs

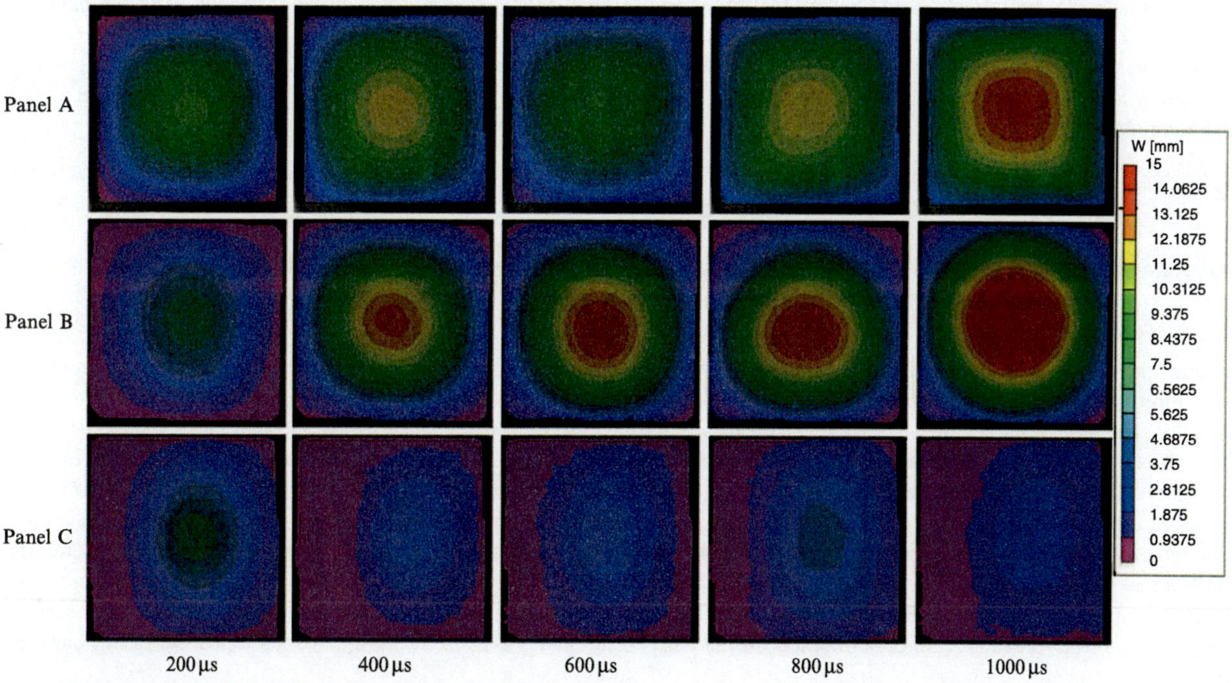


**Fig. 47.7** Full field deflection of panels from 3D-DIC analysis

**Fig. 47.8** Deformation mechanism observed in panels

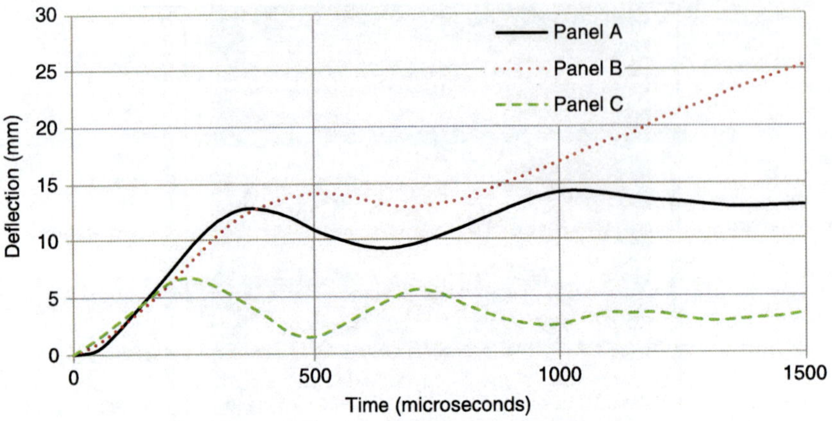


**Fig. 47.9** Out of plane deflection at the center point

From the full-field DIC analysis, the out-of-plane deflection and velocities and the in-plane strain data were extracted at the center point of the three panels (Fig. 47.9). Due to brevity of space out-of-plane velocities and in-plane strain data will be presented in the conference. From the time-deflection history (Fig. 47.9) at the center point of the panels, it is seen that the deflection rate (35 m/s) is almost same in all the three panels, but panels A & B attained a higher deflection as compared to panel C. This can be explained based on the fact that because of the curvature, the loading area decreases in Panel C. This also shows that the Panel C is stiffer than the other two panels as it can sustain higher pressures and have a lower deflection as compared to the other two panels. Panel A and B had a similar deflection trend till 1,000 μs. At this time, panel B failed catastrophically and this explains the change in deflection development in the panel.

### 47.3.2 *Macroscopic Post-Mortem Analysis*

The post-mortem image of the shock loaded flat carbon composite panel A (at failure loading) is shown in Fig. 47.10. There is inter-layer delamination and fiber breakage evident in the post-mortem analysis. The fiber breakage initiated from the clamping edges. To better understand the failure, the panel was divided into three regions and a close postmortem analysis was done on each of the regions. The close-up postmortem images of region 1 (Fig. 47.10) shows that there is fiber breakage in multiple layers. The fiber breakage continues up to 7 layers from the side which was subjected to shock loading. Carbon composite panels are brittle. When they are subjected to shock loading, there is deflection in the panel, but the clamping tries to restrain this deflection. This restrain from the clamping causes the fiber breakage initiation. There is fiber breakage and delamination which extends to the third layer, is visible in region 2. As seen in the macroscopic postmortem image of region 2, there is transverse fiber breakage. The delamination and fiber breakage in region 2 also started from the clamping boundary as in the region 1. There is fiber breakage and delamintaion in region 3 around the clamping boundary. At the same time there is fiber breakage and delamination on the edge of the panel as seen in the macroscopic post-mortem image of region 3.

Again, to brevity of space the post-mortem analysis for panel B and panel C will be presented during the conference.

## 47.4 Conclusions

Three types of panels with varying curvature have been subjected to a controlled shock loading using a shock tube. 3D DIC technique coupled with high speed photography is used to obtain the out-of-plane deformation/velocity and in-plane strain on the back face of all the three panels.

1. The macroscopic post-mortem analysis and DIC deflection, velocity and in-plane strain analysis shows that panel C (112 mm radius of curvature) is capable of sustaining the highest threshold failure load.

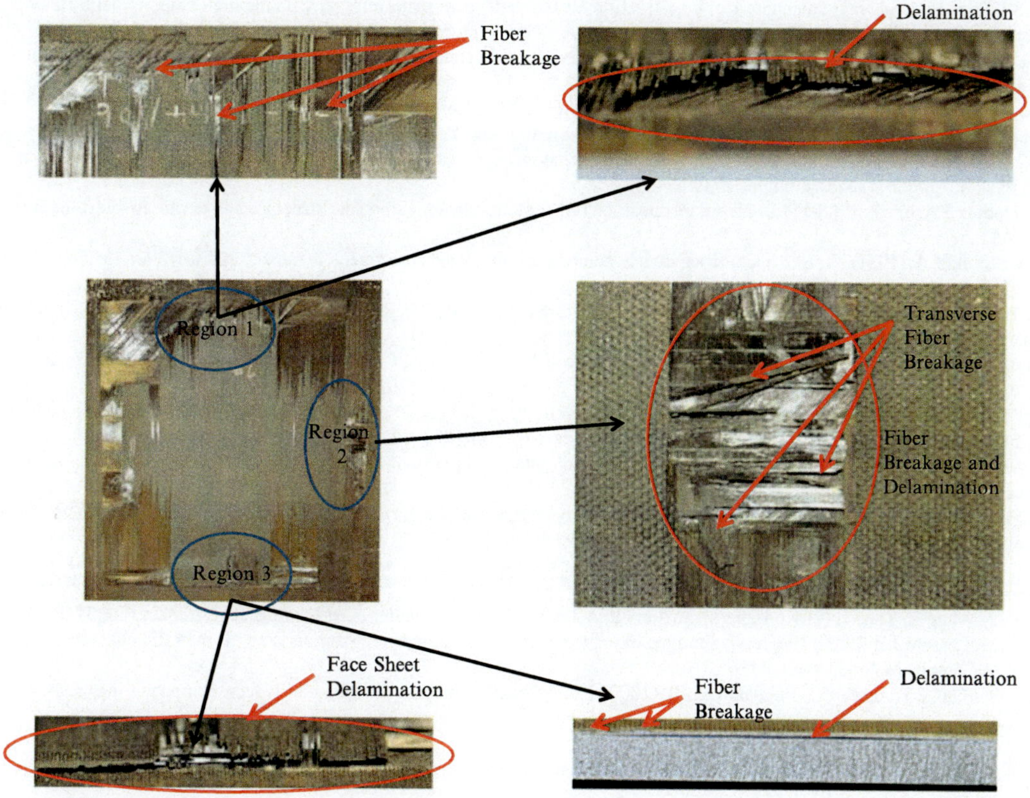


**Fig. 47.10** Post-mortem image of Panel A

**Table 47.2** Threshold loading and damage observed in each panel

| | Threshold blast pressure | Damage observed |
|---|---|---|
| Panel A | 3.65 MPa | No visible damage |
| | | At P = 4.68 MPa, there is fiber breakage, and surface delamination extending up to 7 layers |
| Panel B | 3 MPa | No visible damage |
| | | At P = 3.58 MPa, there is catastrophic failure with fiber breakage and delamination |
| Panel C | 7.78 MPa | No visible failure except burning of carbon fiber on the surface exposed to blast loading |
| | | At P = 8 MPa, there is catastrophic failure with fiber breakage |

2. The flexural deformation decreases and indentation deformation increases as the radius of curvature decreases. There is a limit to which the radius of curvature can be decreased. As the radius of curvature reduces to a limiting value, the shock wave will glide over the surface.

Overall, the threshold blast pressure, which is defined as the pressure just above which the panel fails and the damage observed (at threshold loading) in each of the panels are summarized in Table 47.2.

**Acknowledgement** The authors acknowledge the financial support provided by the Department of Homeland Security (DHS) under Cooperative Agreement No. 2008-ST-061-ED0002.

## References

1. Perl R, O'Rourke R (2001) Terrorist attack on USS Cole: background and issues for congress, CRS report for congress
2. The Oklahoma City bombing: improving building performance through multi-hazard mitigation, Federal Emergency Management Agency Mitigation Directorate, FEMA 277, Aug 1996

3. Stargel DS (2005) Experimental and numerical investigation into the effects of panel curvature on the high velocity ballastic impact response of aluminum and composite panels, Doctor of Philosophy Dissertation, University of Maryland, College Park
4. Chun L, Lam KY (1995) Dynamic analysis of clamped laminated curved panels. Compos Struct 30:389–398
5. Rajendran R, Lee JM (2009) Blast loaded plates. Mar Struct 22:99–127
6. Biggs JM (1964) Introduction to structural dynamics. McGraw-Hill, New York
7. Clough RW, Penzien J (1975) Dynamics of structures. McGraw-Hill, New York
8. Tekalur SA, Shivakumar K, Shukla A (2008) Mechanical behavior and damage evolution in E-glass vinyl ester and carbon composites subjected to static and blast loads. Compos B 39:57–65
9. Arora H, Hooper PA, Dear JP (2011) Dynamic response of full-scale sandwich composite structure subject to air-blast loading. Compos A 42:1651–1662
10. Jackson M, Shukla A (2011) Performance of sandwich composites subjected to sequential impact and air blast loading. Compos B Eng 42:155–166
11. Andrews EW, Moussa NA (2009) Failure mode maps for composite sandwich panels subjected to air blast loading. Int J Impact Eng 36:418–425
12. Langdon GS, Lemanski SL, Nurick GN, Simmons MC, Cantwell WJ, Schleyer GK (2007) Behaviour of fibre–metal laminates subjected to localised blast loading: Part I – Experimental observations. Int J Impact Eng 34:1202–1222
13. Lemanski SL, Nurick GN, Langdon GS, Simmons MC, Cantwell WJ, Schleyer GK (2007) Behaviour of fibre metal laminates subjected to localised blast loading – Part II: quantitative analysis. Int J Impact Eng 34:1223–1245
14. Franz T, Nurick GN, Perry MJ (2002) Experimental investigation into the reponse of chopped strand mat glassfibre laminates to blast loading. Int J Impact Eng 27:639–667
15. Ochola RO, Marcus K, Nurick GN, Franz T (2004) Mechanical behavior of glass and carbon fiber reinforced composites at varying strain rates. Compos Struct 63:455–467
16. LeBlanc J, Shukla A (2011) Dynamic response of curved composite panels to underwater explosive loading: experimental and computational comparisions. Compos Struct 93:3072–3081
17. Shen J, Lu G, Wang Z, Zhao L (2010) Experiments on curved sandwich panels under blast loading. Int J Impact Eng 37:960–970
18. Shokrieh MM, Lessard LB (2000) Progressive fatigue damage modeling of composite materials, Part II: material characterization and model verification. J Compos Mater 34:1081–1116
19. LeBlanc J, Shukla A, Rousseau C, Bogdanovich A (2007) Shock loading of three-dimensional woven composite materials. Compos Struct 79:344–355

# Chapter 48
# Perforation of Aluminum Plates with 7.62 APM2 Bullets

**Michael J. Forrestal, T.L. Warren, and T. Børvik**

**Abstract** We conducted an experimental analytical study to understand the mechanisms and dominant parameters for 7.62 mm APM2 bullets that perforate 6082-T651 aluminum plates at oblique impacts. Tests were conducted with the full bullet and the hard core only to show that the hard core dominates the perforation process. Models show good agreement with measured residual and ballistic-limit velocities.

**Keywords** Perforation • Aluminum plates • APM2 Bullets

## 48.1 Introduction

For this paper, we study the perforation of 6082-T651 aluminum armor plates with 7.62 mm APM2 bullets at oblique impacts. We conducted perforation experiments with the full 10.7 g, APM2 bullets and only the 5.25 g, hard steel cores from the full bullet. Our data shows the perforation process is dominated by the hard steel core. Other papers concerned with 7.62 APM2 bullets are presented in [1–5].

In the next sections, we describe the projectiles and targets, and present residual and ballistic-limit velocity data.

## 48.2 Projectiles

Figure 48.1 shows the dimensions and the parts that make up the APM2 bullet. The 7.62 mm-diameter, 10.7 g, APM2 bullet consists of a brass jacket, an end cap, lead filler, and a 5.25 g, ogive-nose, hard steel core. The steel core has density $\rho_p = 7{,}850$ kg/m$^3$, hardness $R_c$ 63, and $\psi = \text{CRH} = 3.0$ (caliber-radius-head).

M.J. Forrestal (✉)
1805 Newton Pl, NE Albuquerque, NM 87106, USA
e-mail: mjforrestal@gmail.com

T.L. Warren
3804 Shenandoah Pl, NE Albuquerque, NM 87111, USA
e-mail: tlwarre@msn.com

T. Børvik
Structural Impact Laboratory (SIMLab), Centre for Research-based Innovation (CRI) Department of Structural Engineering, Norwegian University of Science and Technology, NO-7491 Trondheim, Norway
e-mail: tore.borvik@ntnu.no

V. Chalivendra et al. (eds.), *Dynamic Behavior of Materials, Volume 1: Proceedings of the 2012 Annual Conference on Experimental and Applied Mechanics*, Conference Proceedings of the Society for Experimental Mechanics Series, DOI 10.1007/978-1-4614-4238-7_48, © The Society for Experimental Mechanics, Inc. 2013

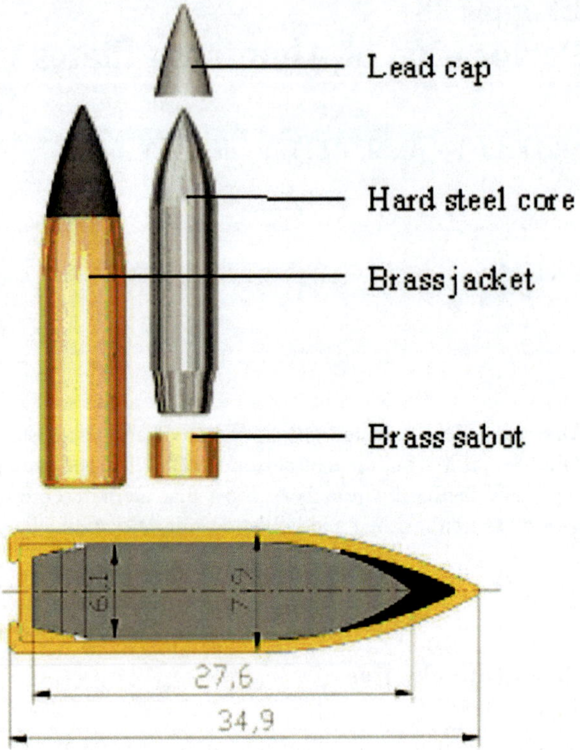


**Fig. 48.1** Geometry and dimensions of the 7.62 mm APM2 bullet (in mm)

## 48.3 6082-T651 Aluminum Target Plate

We obtained 20-mm-thick plate material from Alcoa Europe. The supplier provided inspection certification sheets with tensile data from four tests in the rolling direction. The ultimate tensile strengths were between $\sigma_u = 298$–$302$ MPa, and yield strengths at the 0.2 % offset were between $\sigma_o = 256$–$261$ MPa. As a check, we also conducted three tensile tests in the rolling direction. We obtained $\sigma_u = 290$–$291$ MPa and $\sigma_o = 264$–$266$ MPa. Therefore, tensile data from Alcoa Europe are in close agreement with our tensile data.

## 48.4 Ballistic Results

We conducted tests at normal impact, $\beta = 0°$, and at oblique impacts of $\beta = 15$, 30, and 45°. Figures 48.2, 48.3, 48.4, 48.5, and 48.6 show residual velocity versus striking velocity data. In the figures, the full bullet data are curve-fit with the Lambert-Jonas empirical model [6, 7] given by

$$V_r = \left(V_s^p - V_{bl}^p\right)^{\frac{1}{p}} \tag{48.1}$$

In Eq. 48.1, $V_s$ and $V_r$ are the striking and residual velocities, $p$ is the empirical constant used to best-fit the data with the least squares method, and $V_{bl}$ is the ballistic-limit velocity.

For the parameters investigated in this study, we show that the hard steel core dominates the perforation process.

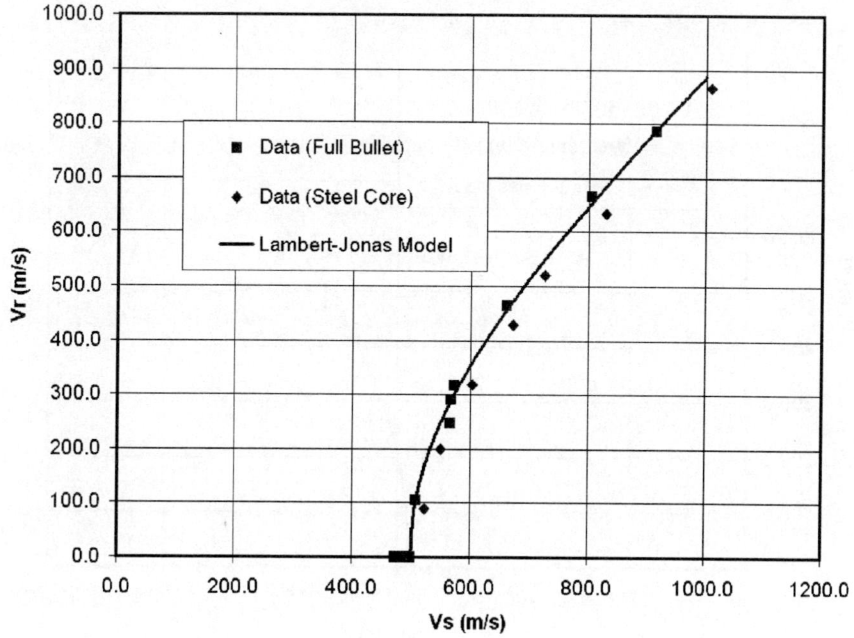


**Fig. 48.2** Residual velocity versus striking velocity ($\beta = 0.0^0$)

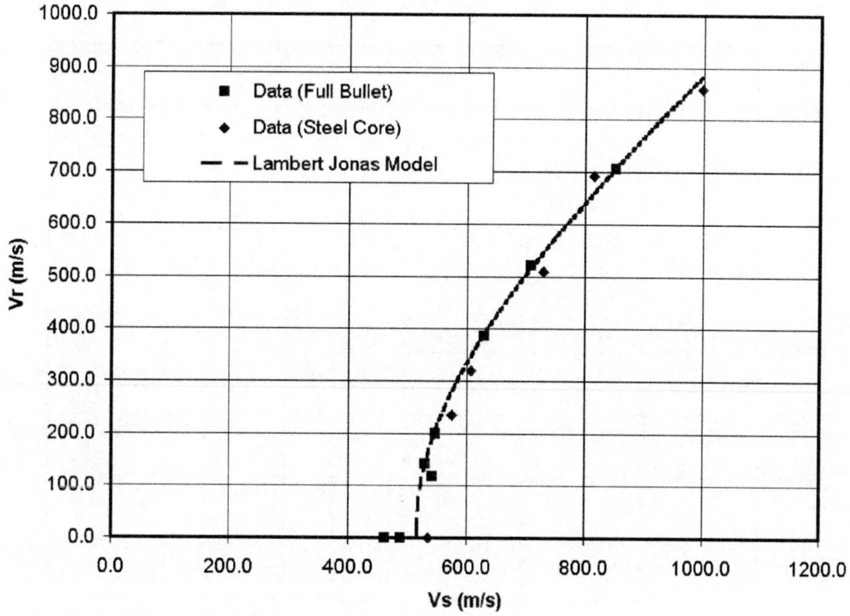


**Fig. 48.3** Residual velocity versus striking velocity ($\beta = 15.0^0$)

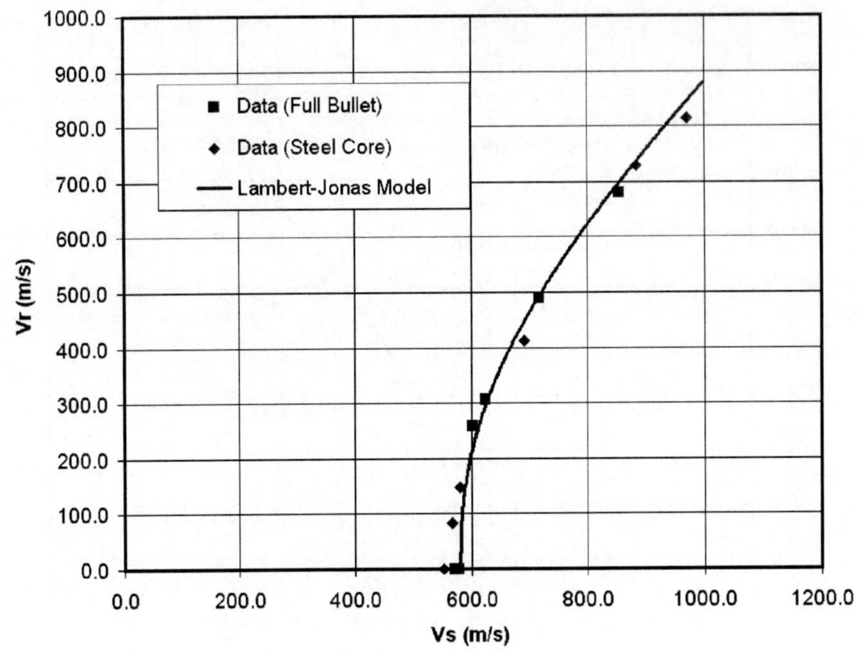


**Fig. 48.4** Residual velocity versus striking velocity ($\beta = 30.0^0$)

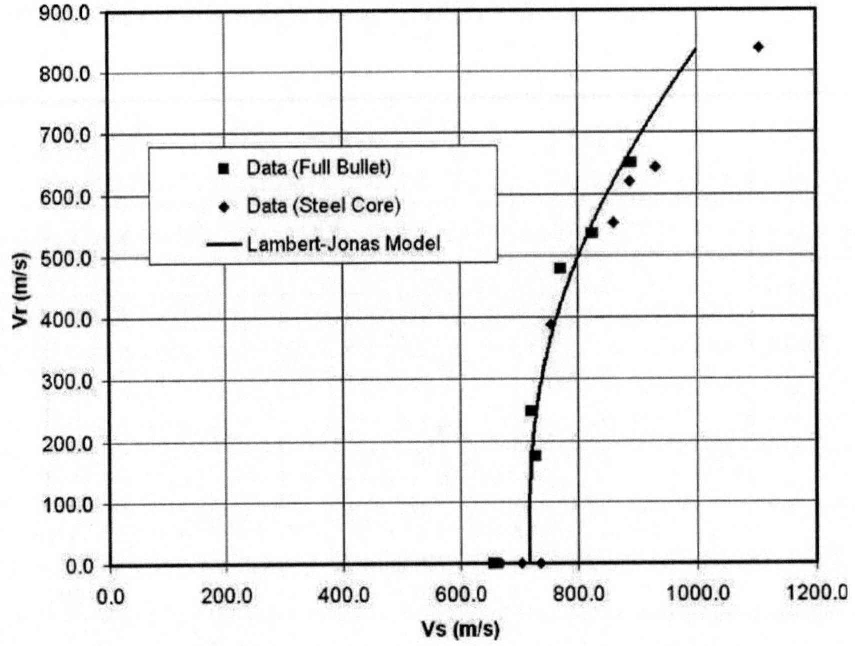


**Fig. 48.5** Residual velocity versus striking velocity ($\beta = 45.0^0$)

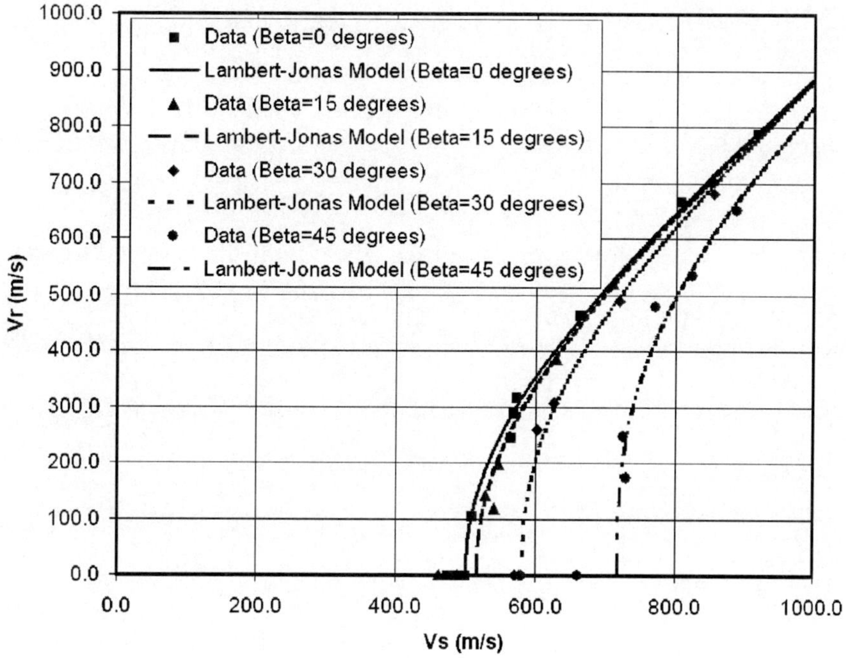


**Fig. 48.6** Residual velocity versus striking velocity for the full bullet

**Acknowledgement** The financial support of this work from the Structural Impact Laboratory (SIMLab), Centre for Research-based Innovation (CRI) at the Norwegian University of Science and Technology (NTNU), is gratefully acknowledged.

## References

1. Gooch WA, Burkins MS, Squillacioti RJ (2007) Ballistic testing of commercial aluminum alloys and alternative processing techniques to increase the availability of aluminum armor. In: Proceedings of the 23rd international symposium on Ballistics, Spain, pp 981–988
2. Børvik T, Forrestal MJ, Warren TL (2010) Perforation of 5083-H116 aluminum armor plates with ogive-nose rods and 7.62 mm APM2 bullets. Exp Mech 50:969–978
3. Forrestal MJ, Børvik T, Warren TL (2010) Perforation of 7075-T651 aluminum armor plates with 7.62 mm APM2 bullets. Exp Mech 50:1245–1251
4. Rosenberg Z, Surujon Z, Yeshurun Y, Ashuach Y, Dekel E (2005) Ricochet of 0.3in AP projectile from inclined polymeric plates. Int J Impact Eng 31:221–233
5. Dorogoy A, Bittel D, Brill A (2011) Experementation and modeling of inclined ballistic impact in thick polycarbonate plates. Int J Impact Eng 38:804–814
6. Lambert JP, Jonas GH (1976) Towards standardization in terminal ballistic testing: velocity representation. Ballistic Research Laboratories, Report BRL-R-1852, USA
7. Zukas JA (1990) High velocity impact dynamics. Wiley, New York, pp 424–427

# Chapter 49
# *A Priori* Pulse Shaper Design for Constant-Strain-Rate Tests of Elastic-Brittle Materials

**John T. Foster and Erik E. Nishida**

**Abstract** Pulse shaping techniques have been used for many years now in Kolsky bar testing of brittle materials. The use of pulse shapers allow the experimentalist to conduct high-strain-rate tests on brittle materials while ensuring that the sample will achieve a state of dynamic stress equilibrium before it fails, as well as to achieve a constant-strain-rate loading state for a large portion of the test. The process of choosing the appropriate pulse shaper system has typically been one of trail-and-error, sometimes requiring many experimental trails to achieve optimal results. Advances in analytic modeling of Kolsky bar tests now make it possible, in an a priori fashion, to design a pulse shaper system to produce a known constant-strain-rate experiment. This paper describes the approach of coupling these analytic models to an optimization technique to quickly find a pulse shaper system that will produce an experiment at a known constant-strain-rate. Experiments were conducted and the model predictions compared to resulting strain-rate histories for a G-10 material.

## 49.1 Introduction

Kolsky bar[1] experimental methods [1–3] are well known techniques for performing high-rate material testing, usually to produce a family of stress-strain relationships at strain-rates from $10^2$/s to $10^4$/s. A minimal Kolsky apparatus consists of an incident, transmission, and striker bar, all cylindrical in geometry, and typically with length-to-diameter ratios of greater than 20 to ensure one-dimensional wave propagation. Figure 49.1 shows a schematic of a Kolsky bar for reference.

By definition, elastic-brittle materials experience no post-yield or viscous behavior and are therefore only tested at high-strain-rates to determine the maximum stress (or strain) in which they fail. Experimental observations have been made in brittle materials that show an increase in failure strength at high-strain-rates [4, 5]; therefore, it is necessary to conduct these types of experiments. Care must be taken when testing brittle materials with a Kolsky bar because the failure strains are typically very small, and the sample can fail before a dynamic stress equilibrium state has been achieved, thereby invalidating the standard data reduction techniques. Pulse shaping techniques [6, 7] have been used for many years now to lengthen the rise-time of the incident stress pulse allowing several wave propagations in the sample before it fails. Analysis has shown that several wave propagations is sufficient for the sample to achieve a near homogenous stress-state under reasonable assumptions of equalibrium [8]. An additional motivation for pulse shaping is to ensure that the experiment is conducted at a near constant-strain-rate for accurate reporting of the data. Recent analysis [9] has shown that, again under reasonable assumptions, an elastic-brittle sample will always achieve a dynamic equilibrium state *before* it achieves a constant strain rate; therefore, the experimentalist can focus their pulse-shaping techniques on achieving a constant-strain-rate and automatically ensure valid and accurate reporting of data.

Analytic models have shown utility in the design-of-experiments for the selection of pulse-shapers which greatly reduce the number of experimental trails required in the laboratory for optimal test conditions. In this paper, we combine these

[1] Also known as a split-Hopkinson bar.

J.T. Foster (✉)
Mechanical Engineering Department, The University of Texas, One UTSA Circle, San Antonio, TX 78249, USA
e-mail: john.foster@utsa.edu

E.E. Nishida
Terminal Ballistics Technology Department, Sandia National Laboratories, Albuquerque, NM, USA

V. Chalivendra et al. (eds.), *Dynamic Behavior of Materials, Volume 1: Proceedings of the 2012 Annual Conference on Experimental and Applied Mechanics*, Conference Proceedings of the Society for Experimental Mechanics Series, DOI 10.1007/978-1-4614-4238-7_49, © The Society for Experimental Mechanics, Inc. 2013

**Fig. 49.1** Schematic of Kolsky bar

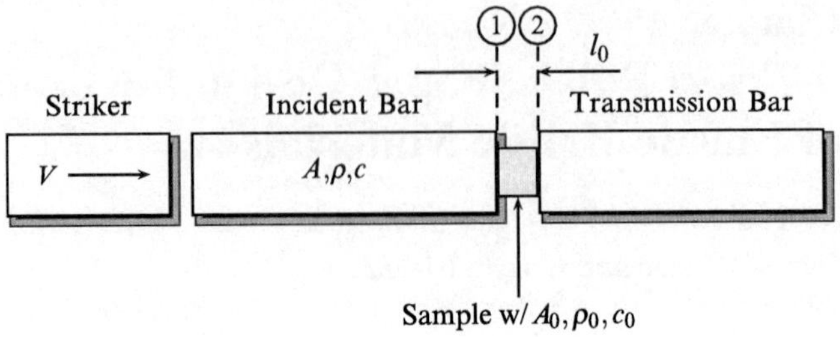


analytic models with a unique extension of a widely used non-linear optimization technique to show that pulse-shaper systems can be optimized a priori to achieve constant-strain-rate experiments at any achievable strain-rate[2] for a given elastic-brittle material. A review of the important equations is presented along with a description of the optimization technique, results from a typical optimization trial, and comparisons from model predictions and actual experiments conducted on an elastic-brittle G-10 material are shown in the following sections.

## 49.2 Analytic Models for Pulse-Shaper Design and Sample Strain History

From the derivations in Frew et al. [4], the time-rate-of-change of strain, $\varepsilon_0(t)$, in an elastic sample is governed by the differential equation

$$\frac{d\varepsilon_0(t)}{dt} = \frac{2A_0E_0}{\rho c A l_0}\varepsilon_0(t) = \frac{2\varepsilon_i(t)}{\rho c l_0 E}, \tag{49.1}$$

where, ρ and $c$ are the density and one-dimensional compressive wave speed, respectively, $E$ is the the one-dimensional elastic modulus defined by $E = \rho c^2$. The subscript notation convention adopted uses the 0 subscript for sample parameters and unsubscripted for bar parameters unless otherwise noted. The $\varepsilon_i(t)$ term in Eq. (49.1) represents the incident strain, governed by a complex piecewise-continuous set of differential equations that depend of the choice of pulse-shaper system (which includes striker length, material choice, and impact speed). These equations presented briefly are

$$\frac{d\varepsilon_p(t)}{dt} = \frac{V_{st}}{h_p}\left\{1 - K\left(\frac{1}{\rho c} + \frac{1}{\rho_{st}c_{st}}\right)\frac{g(\varepsilon_p(t))}{1-\varepsilon_p(t)} - \frac{2K}{\rho_{st}c_{st}}\sum_{i=1}^{n}\left[\frac{g(\varepsilon_p(t-i\tau))}{1-\varepsilon_p(t-i\tau)}\right]\right\}, \tag{49.2a}$$

for $n \leq t < (n+1)\tau$ and $n = 0, 1, 2, \ldots$, using the summation convention $\sum_{i=1}^{0}[\cdot] = 0$ and introducing the new subscripts $st$ and $p$ to indicate striker and pulse-shaper parameters, respectively. $h_p$ is the pulse shaper thickness. Equation (49.2) requires the definition

$$g(x) = \frac{x^j}{1-x^k}, \tag{49.2b}$$

which is the one-dimensional response function of the pulse shaper with $j$ and $k$ being material specific fitting parameters. The additional parameters $K$ and $\tau$ are defined as

$$K = \frac{\sigma_{p_0}a_p}{AV_{st}} \quad \text{and} \quad \tau = \frac{2L_{st}}{c_{st}}. \tag{49.2c}$$

where $\sigma_{p_0}$ is another material response parameter, $a_p$ is the cross-sectional area of the pulse-shaper, and $L_{st}$ is the length of the striker bar. Finally, the pulse-shaper strain, $\varepsilon_p(t)$, can be related to the incident bar strain with

[2] There are limits on achievable strain-rates for brittle materials [8–10].

$$\varepsilon_i(t) = \frac{\sigma_{p_0} a_p}{AE} \frac{g(\varepsilon_p(t))}{1 - \varepsilon_p(t)} \tag{49.3}$$

Now Eqs. (49.1)–(49.3) can be solved simultaneously through the implementation in a numerical ordinary differential equation solver which will solve the correct equations for each additional wave reflection in the striker bar (i.e. value of $n$). In the interest of brevity, not presented here were a set of modified equations that incorporated the elastic unloading effects in the pulse shaper. A complete discussion of these equations can be found in [6] and it is noted that these effects were included in the numerical simulator developed for this study.

## 49.3 Optimization Technique

The optimization technique used in this study was the Levenburg-Marquardt [11] algorithm which is a commonly used and robust method often used in non-linear least squares curve fitting. The algorithm ultimately requires the solution of the linear matrix equation

$$(\mathbf{J}^{\mathbf{T}}\mathbf{J} + \lambda \, \mathrm{diag}(\mathbf{J}^{\mathbf{T}}\mathbf{J}))\Delta\vec{\beta} = \mathbf{J}^{\mathbf{T}}\left[\vec{y} - f\left(\vec{\beta}\right)\right], \tag{49.4}$$

for $\Delta\vec{\beta}$ at multiple iterations, each one adjusting the parameters $\vec{\beta}$ by $\Delta\vec{\beta}$ until some desired convergence is achieved. The $\vec{\beta}$ is representative of a vector of optimization parameters (i.e. the pulse-shaper system parameters in our case) and the term $\left[\vec{y} - f\left(\vec{\beta}\right)\right]$ is the objective function we are seeking to minimize. For the problem of interest in this study, $\vec{y}$ is just a vector of equal constants representing the target strain-rate which we wish to conduct an experiment and $f\left(\vec{\beta}\right)$ is the time derivative of the sample strain found through the solution of Eqs. (49.1)–(49.3), i.e. the sample strain-rate time history. The $\lambda$ term is a "damping" parameter that is adjusted at each iteration, effectively interpolating the algorithm between a standard Gauss-Newton algorithm and the Steepest Descent algorithm. This achieves a good balance between algorithm robustness and convergence rate.

The only term yet undefined in Eq. (49.4) is the Jacobian matrix $\mathbf{J}$. This represents the gradient of $f$ with respect to $\vec{\beta}$, or in indicial notation

$$J_{mn} = \frac{\partial f(t_m, \beta_n)}{\partial \beta_n} \tag{49.5}$$

where, for the problem of interest in this study, the $t_m$ are the discrete solution times of strain-rate history and the $\beta_n$ are the individual pulse-shaper system parameters. Notice that the term $\mathbf{J}^{\mathbf{T}}\mathbf{J}$ is an approximation to the Hessian of the system. Ideally one would prefer to use an analytic Hessian for algorithm robustness and speed of convergence, but when analytic solutions are not available we need to ensure the Jacobian is calculated with as much accuracy as possible. The calculation of this Jacobian is where a unique solution is offered in this work. Typically, when no analytic solutions are available the Jacobian term is calculated through finite-difference approximation. Finite-differences would only be accurate to $O\left(\Delta\vec{\beta}\right)$ or $O\left(\Delta\vec{\beta}^2\right)$ depending on the choice of method (i.e. backward/forward difference or central difference). Restricting the allowable size of $\Delta\vec{\beta}$ to ensure an accurate Jacobian would defeat the advantage of the Levenburg-Marquardt algorithm which allows for large steps to be taken in the direction of smaller gradients avoiding slow convergence. On the other hand, if a finite-difference Jacobian is calculated with large values of $\Delta\vec{\beta}$ it may not be accurate enough for the method to converge at all. Additionally, finite-difference Jacobians require additional functional evaluations and require storing information from previous iterations that could otherwise be discarded. A simple,[3] accurate solution is offered for the calculation of the Jacobian term in the sequel.

[3] It is noted that one could use Automatic Differentiation to calculate accurate derivates, but this typically requires a modest amount of programming skill and/or implementation in a programming language that supports classes and overloaded operators.

It is possible to approximate derivatives quite accurately with a technique based on a complex variable Taylor series expansion of a function. This method was first described by Lyness and Moler [12, 13] and has more recently been rediscovered [14, 15] for use in engineering analysis. The basic idea is that a model parameter can be made to be complex and expanded in a Taylor series about a small perturbation in the imaginary plane by some arbitrary value $h$ as follows

$$f(\beta_n + ih) = f(\beta_n) + \frac{\partial f(\beta_n)}{\partial \beta_n}\frac{ih}{1!} + \frac{\partial f^2(\beta_n)}{\partial \beta_n^2}\frac{(ih)^2}{2!} + \frac{\partial f^2(\beta_n)}{\partial \beta_n^3}\frac{(ih)^3}{3!} + \cdots \tag{49.6}$$

Now we can take the imaginary part of both sides of Eq. (49.6) and solve for the first derivative ignoring terms of $O(h^2)$ to yield an the estimate

$$\frac{\partial f(\beta_n)}{\partial \beta_n} \approx \frac{Im(f(\beta_n + ih))}{h}. \tag{49.7}$$

This shows that an estimate of the first derivative of a function can be made by only utilizing one functional evaluation of a perturbed model parameter up the imaginary axis. The step-size $h$ is arbitrary and can be made as small as practical (even to machine precision) to yield an accurate estimate of this derivative. The only disadvantage of this technique is the necessity of requiring your functions to accept complex numbers as arguments; however, in a dynamically-typed engineering language like Matlab, *Mathematica*, or Python this is a non-issue. Because the subtle simplicity of this approach maybe hard to recognize immediately, we will quickly illustrate with a example:

*Example.* Using $f(x) = \cos(x)$, we have

$$f(x + ih) = \cos(x + ih),$$

or through trigonometric identities we have

$$f(x + ih) = \cos(x)\cosh(h) - i\sin(x)\sinh(h),$$

Applying Eq. (49.7)

$$\frac{\partial f(x)}{\partial x} = -\frac{\sin(x)\sinh(h)}{h},$$

Since $h$ is arbitrary we desire it to be as small as possible or

$$\frac{\partial f(x)}{\partial x} = \lim_{h\to 0} -\frac{\sin(x)\sinh(h)}{h},$$

using L'Hôpital's rule

$$\frac{\partial f(x)}{\partial x} = \lim_{h\to 0} -\frac{\sin(x)\cosh(h)}{1},$$

of course, as $h \to 0$ then $\cosh(h) \to 1$ thereby recovering the exact derivative

$$\frac{\partial f(x)}{\partial x} = -\sin(x).$$

The complex Taylor series expansion technique described above was the method utilized for determining the Jacobian entries in the Levenburg-Marquardt optimization method for finding the pulse-shaper design parameters for testing an elastic-brittle material at a given strain-rate. The results of an optimization trial are shown in the next section with applications to Kolksy bar experiments performed on a G-10 sample material.

## 49.4 Optimization of Pulse-Shaper Parameters for the Constant-Strain-Rate Testing of G-10

It was our desire to test the homogenous response of the composite material G-10 to determine any rate-dependence upon the failure strain. The optimization technique described above was used to assist in the selection of pulse-shaper design parameters. While technically possible to allow the optimization process to determine all non-material property parameters such as sample length and diameter as well as striker length, these are parameters that are not readily changed in the course of conducting experiments in the laboratory.

For this optimization study it was assumed that the sample sizes were fixed and there where only striker bars of fixed length to choose from. The parameters that where allowed to be optimized where the pulse-shaper thickness and diameter and the striker impact speed, all other parameters required to solve Eqs. (49.1)–(49.3) were given and are summarized as follows: the incident and transmission bars for the Kolsky apparatus were 19.05mm in diameter and made from 7075-T6 aluminum with $\rho = 2,700\,\mathrm{kgm}^{-3}$ and $c = 5,051\,\mathrm{ms}^{-1}$, the striker bar was also 19.05mm in diameter at a length of $L_{st} = 3,\ 048$ mm made of C350 maraging steel with $\rho_{st} = 8,170\,\mathrm{kgm}^{-3}$ and $c_{st} = 4,950\,\mathrm{ms}^{-1}$, the pulse-shaper material was annealed copper with one-dimensional response parameters, $\sigma_{p0} = 625$MPa, $j = 0.32$, $k = 4.25$, and elastic modulus (used when the pulse-shaper unloading not presented here) of $E_p = 110$GHzPa. The sample being testing was composite material commonly called G-10 or FR-4. The material consists of a continuous filament glass cloth material with an epoxy resin binder. The samples tested and reported in this work were prepared into cylinders with a diameter of 9.525mm and length $l_0 = 5.08$mm. The samples where prepared with the direction of laminate glass cloth parallel to the length. From quasi-static tests conducted in a standard test frame it was known that the elastic modulus of the material is $E_0 = 120$GHzPa (in the orientation tested).

While the Levenburg-Marquardt algorithm is quite robust, ultimately like all non-linear solvers it requires a set of initial conditions that are close to minimum you are seeking. With visual tools developed to aid us in our pulse-shaping we where able to come up with an initial "guess" to feed the optimization routine. These initial parameters where as follows: $V_{st} = 19\mathrm{ms}^{-1}$, a pulse shaper diameter of 5.08mm which corresponds to $a_p = 20.27\mathrm{mm}^2$ and a pulse shaper thickness $h_p = 1.27$mm. The solution with these initial conditions in indicated by "Iteration 1" in Fig. 49.2. The initial guess shows a result that varies in strain-rate by approximately 200/s during the first 100μs of the simulation. The optimization algorithm was allowed to run while targeting a strain-rate of 500/s. Figure 49.2 shows a sequential selection of the output from the optimization with the final result indicated by "Iteration 13." Notice this final result has a much smaller variance and is very near a constant 500/s for the first 100μs. The final results for the optimized pulse-shaper parameters where $V_{st} = 19\mathrm{ms}^{-1}$, a pulse-shaper diameter of 6.07mm (0.239 in) and thickness $h_p = 1.55$mm (0.061 in). While it would be very easy to machine a pulse-shaper to these exact dimensions, it would become prohibitively costly over time. Therefore, in practice in the laboratory, pulse-shapers are simply punched out of existing stock material using a selection of fixed punch sizes. A punch of 0.25 in diameter was selected as was stock copper material at a thickness of 0.062 in which where the closest matches to the measurements provided by the optimization code.

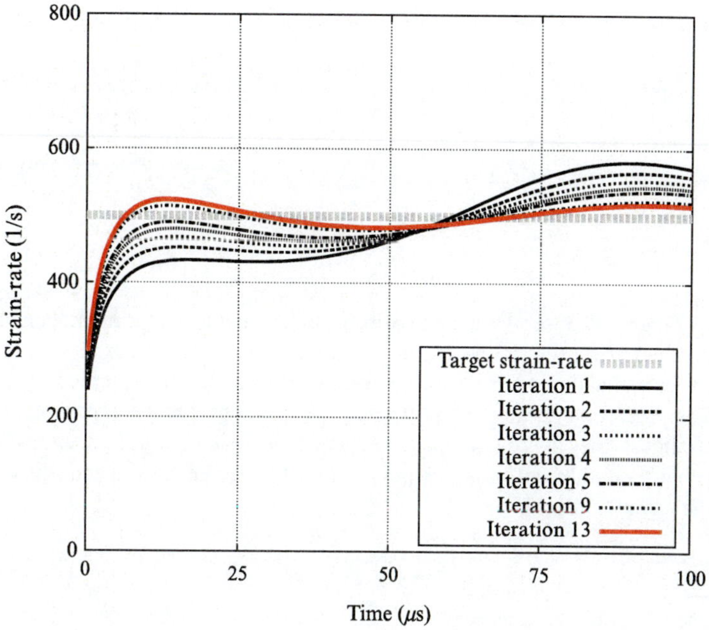


**Fig. 49.2** Sequential iteration of the solution for strain-rate using the Levenburg-Marquardt optimization algorithm targeting a strain rate of 500/s

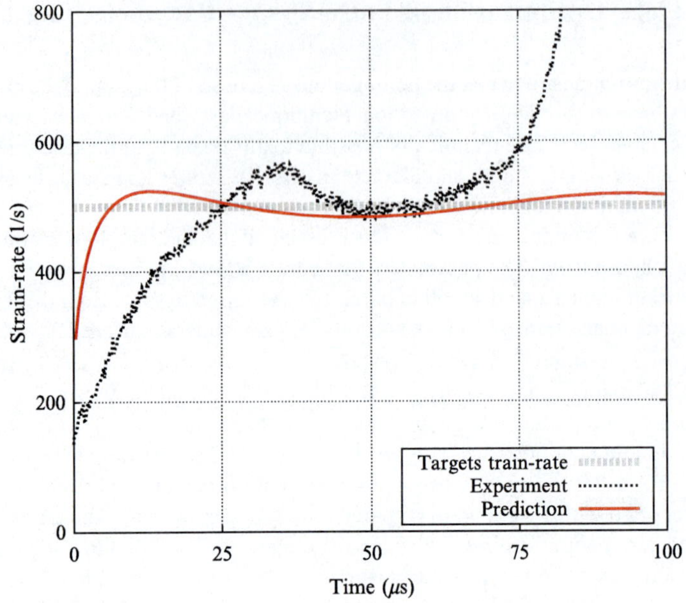


**Fig. 49.3** Comparison of strain-rate time histories for the optimized pulse shaper prediction and actual experiment conducted at a nominal strain rate of 500/s

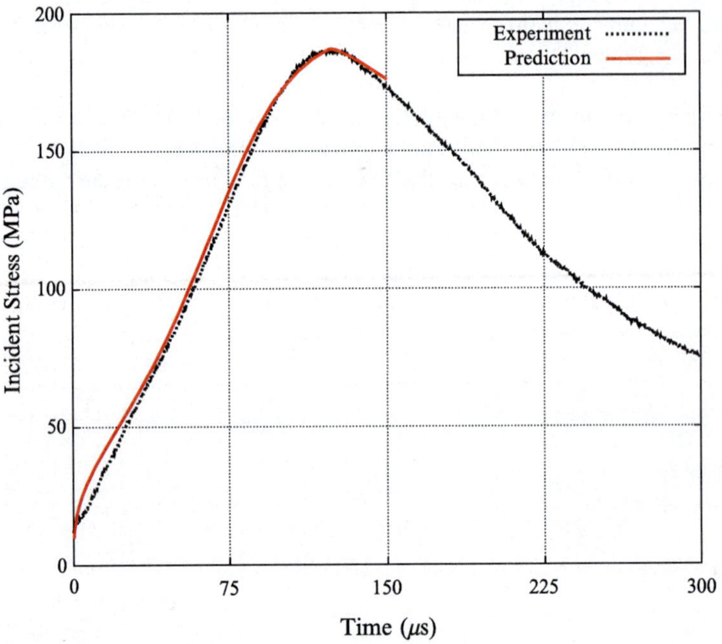


**Fig. 49.4** Comparison incident stress waves for the optimized pulse shaper prediction and actual experiment conducted at a nominal strain rate of 500/s

Figure 49.3 shows a comparisons of an actual experiment conducted and the predictions utilizing the optimized parameters for the strain-rate time histories. Notice that the experimental result for strain-rate history approaches the design strain-rate and remains near there for 20–30 μs before the sample fails and the strain-rate runs away at approximately 75μs. The prediction does not closely match the initial part of the experimental curve because Eq. (49.1) being solved for sample strain assumes that the sample is instantaneously in equilibrium. In reality, it takes a finite amount of time for the sample to achieve equilibrium[4] at which point Eq. (49.1) becomes valid and is in close agreement with the experimental data (ignoring the sample failure).

[4] See [9] for further discussion.

**Fig. 49.5** True stress as a function of true strain curves for two samples of G-10 conducted at a nominal strain-rate at 500/s

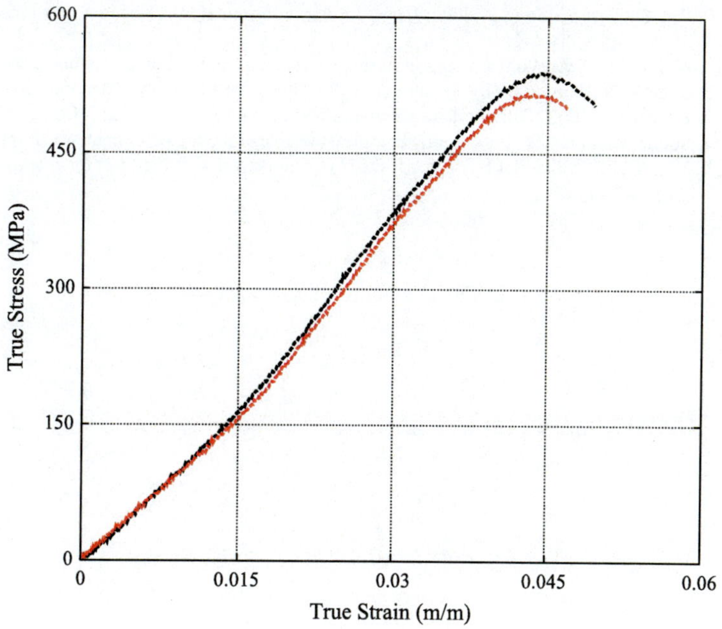


Figure 49.4 shows a comparison of predicted incident stress pulse using the optimized pulse-shaper parameters to the actual experiment and they are in very close agreement.

For the sake of completeness a sample of the stress-strain curves produced for G-10 at a nominal strain-rate of 500/s are shown in Fig. 49.5. As anticipated, the material behaves in a linearly-elastic brittle manner with a failure strain slightly over 4.5%

## 49.5 Summary

A technique was presented for optimizing a pulse-shaper system design to allow for the *a priori* selection of parameters that produce a near constant-strain-rate for the Kolsky bar testing of elastic-brittle materials. If utilized, this process will greatly reduce the need for trial-and-error pulse shaper design in the laboratory. The predictions from the method are shown to be in good agreement with an experimental trial conducted on G-10 at a nominal strain-rate of 500/s.

**Acknowledgements** This work was partially supported by Sandia National Laboratories. Sandia is a multiprogram laboratory operated by Sandia Corporation, a Lockheed Martin Company, for the United States Department of Energy under Contract DE-AC04-94AL85000.

## References

1. Kolsky H (2003) Stress waves in solids. Dover, Mineola
2. Graff KF (1991) Wave motion in elastic solids. Dover, New York
3. Chen W, Song B (2010) Split hopkinson (Kolsky) bar: design, testing and applications. Springer, New York
4. Frew DJ, Forrestal MJ, Chen W (2001) A split Hopkinson pressure bar technique to determine compressive stress-strain data for rock materials. Exp Mech 41:40–46
5. Kimberly J, Ramesh KT (2011) The dynamic strength of an ordinary chondrite. Meteorit Planet Sci 46(11):1653–1669
6. Frew DJ, Forrestal MJ, Chen W (2002) Pulse shaping techniques for testing brittle materials with a split Hopkinson pressure bar. Exp Mech 42 (1):93–106
7. Frew DJ, Forrestal MJ, Chen W (2005) Pulse shaping techniques for testing elastic-plastic materials with a split Hopkinson pressure bar. Exp Mech 45(2):186–195
8. Ravichandran G, Subhash G (1994) Critical appraisal of limiting strain rates for compression testing of ceramics in a split hopkinson pressure bar. J Am Ceram Soc 77(1):263–267
9. Foster JT (2012) Comments on the validity of test conditions in Kolsky bar experiments of elastic-brittle materials. Exp Mech. Accepted

10. Pan Y, Chen W, Song B (2005) Upper limit of constant strain rates in a split hopkinson pressure bar experiment with elastic specimens. Exp Mech 45(5):440–446
11. Marquardt DW (1963) An algorithm for least-squares estimation of nonlinear parameters. J Soc Ind Appl Math 11(2):431–441
12. Lyness JN, Moler CB (1967) Numerical differentiation of analytic functions. SIAM J Numer Anal 4(2):202–210
13. Lyness JN (1968) Differentiation formulas for analytic functions. Math Comput 22(102):352–362
14. Squire W, Trapp G (1998) Using complex variables to estimate derivatives of real functions. SIAM Rev 40:110–112
15. Voorhees A, Millwater H, Bagley R (2011) Complex variable methods for shape sensitivity of finite element models. Finite Elem Anal Des 47 (10):1146–1156

# Chapter 50
# Time Correlation Between Acoustic, Electromagnetic and Neutron Emissions in Rocks Under Compression

**A. Carpinteri, G. Lacidogna, A. Manuello, G. Niccolini, and O. Borla**

**Abstract** Acoustic (AE), electromagnetic (EME), and neutron (NE) emissions were measured during laboratory compression tests on rock specimens loaded up to failure.

Tests were performed on magnetite and basalt specimens at constant displacement rate. AE signals were detected by applying to the specimen surface a piezoelectric (PZT) transducer with resonance frequency of 78 kHz. EME signals were revealed by the induced current in a closed circuit due to change in the magnetic flux during specimen compression. The specimens were also monitored by means of $He^3$ proportional counter and neutron bubble detectors.

The observed EME signals were strictly correlated in time with AE signals for all the tested specimens. Bursts of AE and EME activity are observed when significant stress drops occur. The neutron emission signals were mainly detected during stress-drops in the load vs. time diagrams, or at the end of the tests when the specimens showed catastrophic behaviors characterized by snap-back instabilities. Applications of all these monitoring techniques to earthquake forecasting seem to be possible.

**Keywords** Fracture precursors • Acoustic emission • Electromagnetic emission • Neutron emissions • Piezonuclear reactions

## 50.1 Introduction

It is possible to demonstrate experimentally that brittle fracture in solid materials is accompanied by the release of different forms of energy.

By subjecting quasi-brittle materials such as granitic rocks to compression tests, for the first time bursts of neutron emission during the failure process were observed [1–5], necessarily involving nuclear reactions, besides the well-known acoustic emission (AE) [6–13], and the phenomenon of electromagnetic radiation (EM) [14–19], which is highly suggestive of charge redistribution during material failure and at present under investigation.

The phenomenon of EME is regarded as an important precursor of critical phenomena in Geophysics, such as rocks fracture, volcanic eruptions and earthquakes [19,20]. For example, anomalous radiation of geo-electromagnetic waves were observed before major earthquakes. At the laboratory scale, rocks and concrete under compression generate AE and EM emissions nearly simultaneously. This evidence suggests that also EM emissions are generated during crack growth, reinforcing the idea that the EME phenomenon can be applied as a forecasting tool for earthquakes.

While the mechanism of AE is fully understood, being provided by transient elastic waves due to stress redistribution following fracture propagation [6–13], the origin of EME from fracture is not completely clear and different attempts have been made to explain it.

An explanation of the EME origin was related to dislocation phenomena [16], which however are not able to explain EME from fracture in brittle materials where the motion of dislocations can be neglected [17]. Frid et al. [17] and

A. Carpinteri (✉) • G. Lacidogna • A. Manuello • O. Borla
Politecnico di Torino, Department of Structural Engineering & Geotechnics, C.so Duca degli Abruzzi 24, 10129 Torino, Italy
e-mail: g.niccolini@inrim.it

G. Niccolini
INRIM - National Research Institute of Metrology, Strada delle Cacce 91, 10135 Torino, Italy

V. Chalivendra et al. (eds.), *Dynamic Behavior of Materials, Volume 1: Proceedings of the 2012 Annual Conference on Experimental and Applied Mechanics*, Conference Proceedings of the Society for Experimental Mechanics Series,
DOI 10.1007/978-1-4614-4238-7_50, 

Rabinovitch et al. [11] recently proposed a model of the EME origin where, following the rupture of bonds during the crack growth, mechanical and electrical equilibrium are broken at the fracture surfaces with creation of ions moving collectively as a surface wave on both faces. Lines of positive ions on both the newly created faces (which maintain their charge neutrality unlike the capacitor model) oscillate collectively around their equilibrium positions in opposite phase to the negative ones. The resulting oscillating dipoles created on both faces of the propagating fracture act as the source of EME [21].

As regards the neutron emissions, in this work we present experimental tests performed on brittle rock test specimens (Magnetite and Basalt), using $He^3$ neutron device and bubble type BD thermodynamic neutron detectors.

For brittle specimens of larger dimensions, neutron emissions, detected by $He^3$, were found to be of about three orders of magnitude higher than the ordinary natural background level at the time of the catastrophic failure. These emissions fully confirm the previous tests [1–5] and are due to piezonuclear reactions, which depend on the different modalities of energy release during the tests. For specimens with sufficiently large size and slenderness, a relatively high energy release is expected, and hence a higher probability of neutron emissions at the time of failure.

The experimental analysis carried out by the authors may open a new possible scenario, in which the stress state of the elements firstly involves the generation of micro-cracks accompanied by mechanical energy release in the field of ultrasonic vibrations that can be measured using suitable AE equipments. Hence, the formation of coherent EM fields occurs over a wide range of frequencies, from few Hz to MHz. This excited state of the matter could be a precursor of subsequent resonance phenomena of nuclei able to split and produce neutron bursts in the presence of stress-drops or sudden catastrophic fractures.

## 50.2 Experimental Set Up

Experimental compression tests were performed on brittle rock test specimens under monotonic displacement control. The materials used for the tests are non-radioactive Magnetite and Basalt. In these tests, 27 cylindrical specimens with different size and slenderness are used. The compression tests were performed at the Fracture Mechanics Laboratory of the Politecnico di Torino. In particular, in this work the experimental results of four specimens (3 of Magnetite and 1 of Basalt) were examined. They were subjected to uniaxial compression using a MTS servo-controlled hydraulic testing machine with a maximum capacity of 1,000 kN. Each test was performed in piston travel displacement control by setting constant piston velocity. The test specimens were arranged in contact with the press platens without any coupling materials, according to the testing modalities known as "test by means of rigid platens with friction".

The tested materials, shapes and sizes of the specimens, the employed piston velocities and the peak load values are listed in Table 50.1.

The AE activity emerging from the compressed specimens was detected by attaching to the specimen surface a piezoelectric (PZT) transducer, resonant at 78 kHz, which is able to convert the high-frequency surface movements due to the acoustic wave into an electric signal (the AE signal). Sensitivity of the transducer in the low-frequency range were measured by placing it on a shaker excited with all frequencies in the range 0–10 kHz (white noise). The result of this calibration at low frequencies was 1.2 $\mu V/(mms^{-2})$. Resonant sensors are more sensitive than broadband sensors, which are characterized by a flat frequency response in their working range, and then they can be successfully used in monitoring large-sized structures.

The EME detecting device, realized at the National Research Institute of Metrology (INRIM), is constituted by three pickup coils with different number of turns, made of a 0.2 mm copper wire, that are positioned around the monitored specimen. This instrumentation, which acquires data in the frequency range from few Hz up to 4 MHz, exploits the induction Faraday's law: the induced voltage in a closed circuit (loop) is proportional to the change in the magnetic flux throughout the circuit. The first coil, constituted by 5 turns, works in a frequency range from 300 KHz to 4 MHz. The other two coils

**Table 50.1** Tested materials and features

| Specimen | Material | Geometry of the specimen | | | Displacement velocity (mm/s) | Peak load (kN) |
|---|---|---|---|---|---|---|
| | | D (mm) | H (mm) | $\lambda = H/D$ | | |
| P1 | Magnetite | 40 | 40 | 1 | 0.001 | 197.97 |
| P5 | Magnetite | 90 | 90 | 1 | 0.002 | 932.38 |
| P8 | Magnetite | 20 | 20 | 1 | 0.0005 | 45.05 |
| P9 | Basalt | 50 | 100 | 2 | 0.001 | 177.64 |

constituted by 125 turns and 500 turns, respectively work in the frequency range from few kHz to 20 kHz, and from few Hz to 1 kHz.

Due to the difficulties in neutron measurements in the presence of electromagnetic disturbances, EME measurement carries out both the validation of NE signals and the monitoring of EME from fracturing. The simultaneous presence of sharply peaked EME signals (well characterized and far from the continuous magnetic noise) and NE signals can be regarded as signature of ongoing damage process. As a further check on NE signals, a set of passive neutron detectors, based on the superheated bubble detection technique and insensitive to electromagnetic noise, were employed. A detailed description of the used neutron detectors is given in the following.

### 50.2.1 $He^3$ Neutron Proportional Counter

The $He^3$ detector used in the compression tests under monotonic displacement control, and by ultrasonic vibration, is a $He^3$ type (Xeram, France) with electronics of preamplification, amplification, and discrimination directly connected to the detector tube. The detector is powered with 1.3 kV, supplied via a high voltage NIM (Nuclear Instrument Module). The logic output producing the TTL (transistor-transistor logic) pulses is connected to a NIM counter. The device was calibrated for the measurement of thermal neutrons; its sensitivity is 65 cps/$n_{thermal}$ ($\pm$ 10% declared by the factory) i.e., a thermal neutron flux of 1 thermal neutron/s $cm^2$ corresponds to a count rate of 65 cps.

Considering that the fracture of dielectric materials, such as rocks, can lead to the emission of charged and neutral particles (electrons, photons, hard X-rays), in order to avoid possible false neutron measurements the output of the detector is enabled for detecting signals only exceeding a fixed amplitude. This threshold value was determined by measuring the analog signal of the detector by means of a Co-60 gamma source (half-life: 5,271 years, type decay: $beta^-$, beta maximum energy: 317.8 keV, gammas: 1173.2 and 1332.5 keV). The presence of an interfering capacity on the charge preamplifier input increases the electronic noise and consequently the probability of spurious counts. For this reason, the coaxial cable used for connecting detector and charge preamplifier presented a low capacity (36 pF/m) and a short length (about 50 cm). Moreover, during the experimental measurements, the front-end electronics was screened with aluminum foils, and the $He^3$ tube was immersed in a sound-absorbing substance such as polystyrene in order to avoid possible accidental impacts and vibrations.

### 50.2.2 Neutron Bubble Detectors

A set of passive neutron detectors insensitive to electromagnetic noise and with zero gamma sensitivity was used in compression tests under cyclic loading. The dosimeters, based on superheated bubble detectors (BTI, Ontario, Canada) (Bubble Technology Industries (1992)) [22], are calibrated at the factory against an Am-Be source in terms of NCRP38 (NCRP report 38 (1971)) [23]. Bubble detectors provide instant visible detection and measurement of neutron dose. Each detector is composed of a polycarbonate vial filled with elastic tissue-equivalent polymer, in which droplets of a superheated gas (Freon) are dispersed. When a neutron strikes a droplet, the latter immediately vaporizes, forming a visible gas bubble trapped in the gel. The number of droplets provides a direct measurement of the equivalent neutron dose. These detectors are suitable for neutron integral dose measurements, in the energy ranges of thermal neutrons ($E = 0.025$ eV) and fast neutrons ($E > 100$ keV).

All the signals (AE, EME and NE) were acquired by a National Instruments Digitizer with eight channels simultaneously sampling at 1 MSa/s. The trigger was set to the AE channel with a detection threshold fixed at 20 mV to filter out the background noise. For all the specimens, the recorded AE, EME and NE time series were related to the time history of the applied load.

## 50.3 Test Results

### 50.3.1 AE and EME Measurements

All specimens were tested in compression up to failure, showing a brittle response with a rapid decrease in load-carrying capacity when deformed beyond the peak load (Figs. 50.1 and 50.2). Experimental evidence indicates the presence of AE,

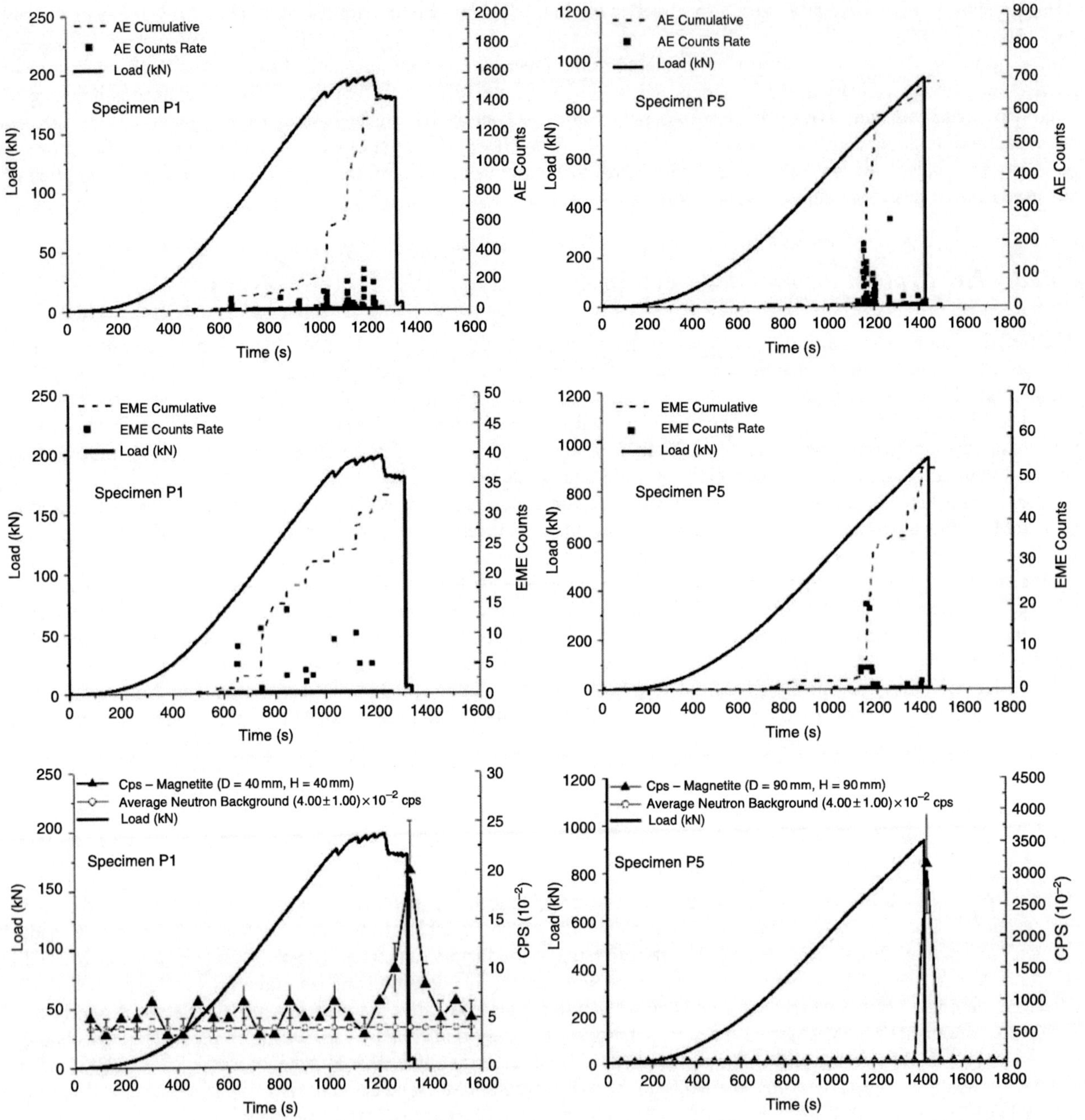


**Fig. 50.1** Load versus time curve of the Magnetite specimen P1 (*left*); accumulated number and rate of AE (*upper left*); accumulated number and rate of EME (*middle left*); NE count rate (*lower left*). Load versus time curve of the Magnetite specimen P5 (*right*); accumulated number and rate of AE (*upper right*); accumulated number and rate of EME (*middle right*); NE count rate (*lower right*)

EME and NE activity. It is interesting to note that in this experimental campaign – unlike those carried out by the authors on specimens of other materials, such as concrete, Luserna stone, Carrara marble and Syracuse limestone [14,15] – the EME activity is much more widespread during the loading process, and not just concentrated at the moment of the final collapse.

As a matter of fact, in the Magnetite specimen P1, whose behaviour is described by the load vs. time curve of Fig. 50.1 (left), the observed bursts of AE and EME activity can be clearly correlated with the stress drops occurred before the moment of the collapse, Fig. 50.1 (upper and middle left). In the Magnetite specimen P5, characterized by a perfectly brittle behaviour without evident stress drops before the final collapse (note the linearity till failure in the load vs. time curve of Fig. 50.1 (right)), the specimen failure was preceded by two closely correlated bursts of AE and EME activity at nearly 80%

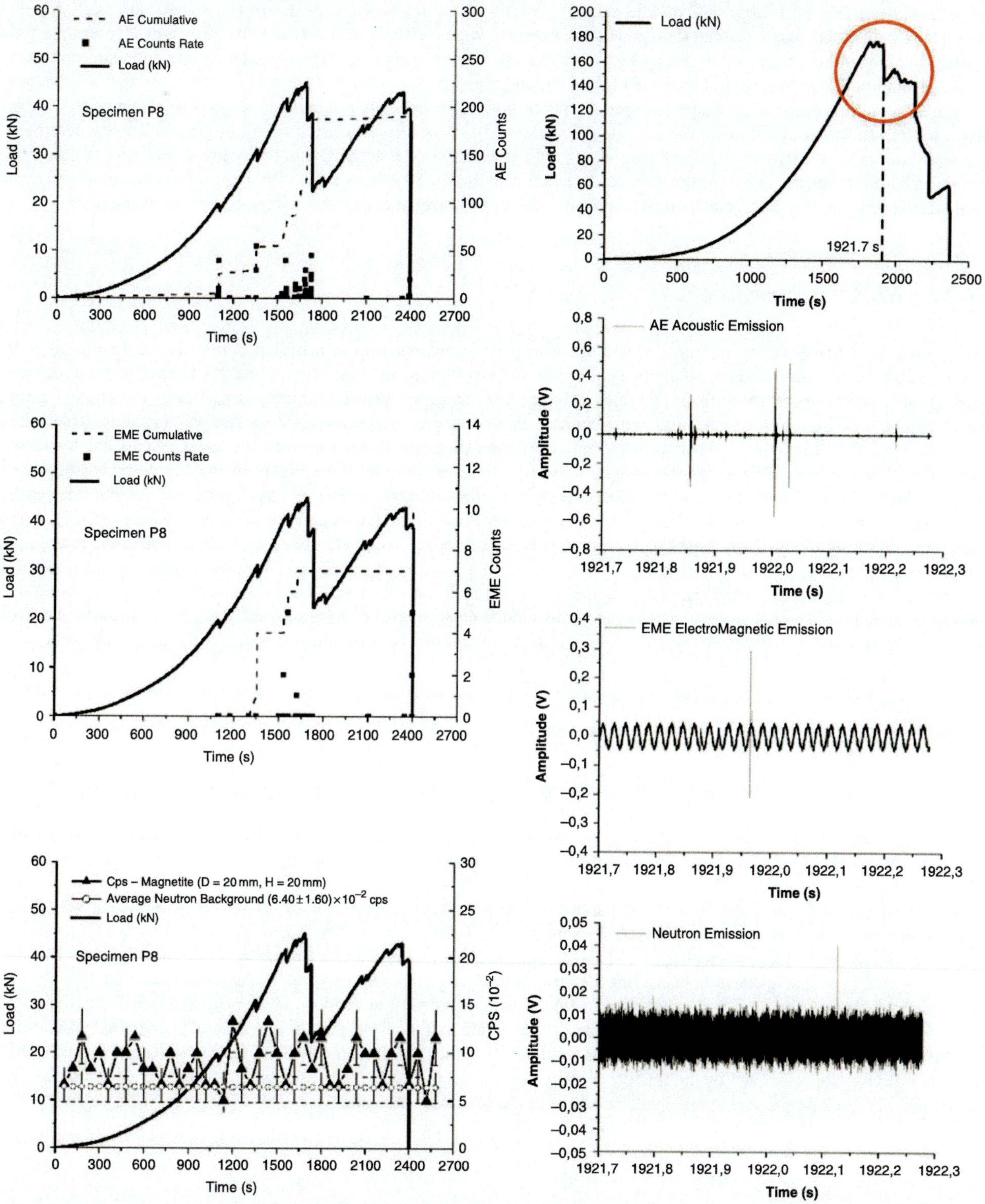


**Fig. 50.2** Load versus time curve of the Magnetite specimen P8 (*left*); accumulated number and rate of AE (*upper left*); accumulated number and rate of EME (*middle left*); NE count rate (*lower left*). Load vs. time curve of the Basalt specimen (*upper right*); AE, EME (*middle right*) and NE signals (*lower right*) detected in a time window of 0.6 s, starting at 1921.7 s from the beginning of the test

of the peak load, Fig. 50.1 (upper and middle right). This activity, particularly as regards electromagnetic emission, can be due to the behaviour under loading of Magnetite rocks that – being rich in iron, about 65% in weight – determine the formation of magnetic charges generated by friction during the loading process, and their spontaneous release independently of the formation of macro-cracks at the time of final collapse.

The load vs. time curve of the Magnetite specimen P8 is double-peaked with a significant stress drop at about 60% of the test duration, followed by a drop in the AE rate, Fig. 50.2 (left). This momentary relaxation in the AE activity describes the well-known Kaiser effect [24], which consists in, after stress drops, AE activity being very low during the reloading of the material until the stress exceeds the previous reached values, Fig. 50.2 (upper left). This relaxation was observed also in the EME activity, Fig. 50.2 (middle left), confirming the close correlation degree between these two phenomena.

### *50.3.2 NE Measurements*

As regards the NE measurements, the $He^3$ neutron detector was switched on at least 1 h before the beginning of each compression test, in order to reach the thermal equilibrium of electronics, and to make sure that the behaviour of the device was stable with respect to intrinsic thermal effects. The average measured background level is ranging from $(4.00 \pm 1.00)\cdot 10^{-2}$ to $(6.40 \pm 1.60)\cdot 10^{-2}$ cps. Neutron measurements of specimen P8 yielded values comparable with the ordinary natural background, whereas in specimens P1 the experimental data exceeded the background value by about five times. For specimen P5 the neutron emissions achieved values of about three orders of magnitude higher than the ordinary background. In Figs. 50.1 and 50.2 (left), the load vs. time diagram and the neutron count rate evolution for each specimen are shown. Moreover, bursts of NE activity were observed at the failure time of specimens P1 and P5, Fig. 50.1 (lower), confirming the need of catastrophic ruptures, i.e., characterized by sudden release of the stored strain energy, to obtain such anomalous neutron emissions.

Furthermore, during the compression tests a rise in the thermal equivalent neutron dose was observed, if analysed by neutron bubble detectors, consistently with the increment in the neutron level measured by the $He^3$ device. In particular, for specimen P5, a value of more than 1,000 times higher than the ordinary background was found at the end of the test.

### *50.3.3 AE, EME and NE Time Correlation*

Test results from the Basalt specimen (P9) provide a high degree of correlation among the three emission time series and the loading time history. As an example, considering a time window of 0.6 s starting at 1921.7 s from the beginning of the test, AE bursts followed by an EME pulse in the kH frequency gap are shown in Fig. 50.2 (right). Similar simultaneous EM pulses were observed in the Hz to MHz range that, for reasons of space, are not included in the figure. The time window is related to the evident stress drop indicated by a circle in the load vs. time diagram, Fig. 50.2 (right).

As already discussed in [14], AE and EME signals, from a growing fracture, follow a time delay consistent with their propagation velocities. Being $d$ the distance fracture source – AE transducer, $v_{AE}$ and $v_{EME}$ the average propagation velocities of AE and EME waves with $v_{AE} << v_{EME}$, the time delay can be estimated by $\Delta t = d/v_{AE}$. Inserting $d = 10^{-1}$ m and $v_{AE} = 10^3$ m s$^{-1}$, an estimation of the time delay for the considered event is $\Delta t = 10^{-4}$ s $= 100$ μs.

If then we consider that the main crack propagation, in the specified time-window, takes place (begins its first motion) when the first AE peak of great amplitude is recorded, the generation of the EME pulse follows the AE burst also considering the different average propagation velocities of AE and EME signals. Therefore, the EME signal seems to spread during the mechanical vibration generated by the fracture. Finally, the NE event is observed at the time of catastrophic failure of the specimen, Fig. 50.2 (upper and lower right), clearly after the first two emissions.

## 50.4 Conclusions

The experimental evidence presented in this paper confirms AE and EME signals as failure precursors in natural materials like rocks. The observed EME were strictly correlated in time with AE signals in all the tested specimens. Bursts of AE and EME activity were observed when significant stress drops occur. This suggests the use of electromagnetic measurements to enhance monitoring systems based on the AE technique. Bursts of NE activity were observed only when the specimen fails in a sudden, catastrophic way. In particular, for specimens with sufficiently large size, the neutron flux was found to be of about three orders of magnitude higher than the background level at the time of catastrophic failure. Therefore, the observed acoustic, electromagnetic, and neutron activities from laboratory experiments look promising for effective applications also at the geophysical scale.

A further study on the generating times of AE and EME signals from propagating fracture has been performed on a basalt specimen under compression test. The results obtained from this analysis show how the crack formation is accompanied by mechanical energy release in the field of ultrasonic vibrations detected by AE sensors. Then, the generation of coherent EM fields occurs over a wide range of frequencies, from few Hz to MHz, during the fracture propagation. This excited state of the matter could be a precursor of subsequent resonance phenomena of nuclei able to produce fission and neutron bursts in the presence of stress-drops or sudden catastrophic fractures.

## References

1. Carpinteri A, Cardone F, Lacidogna G (2009) Piezonuclear neutrons from brittle fracture: early results of mechanical compression tests. Strain 45:332–339
2. Cardone F, Carpinteri A, Lacidogna G (2009) Piezonuclear neutrons from fracturing of inert solids. Phys Lett A 373:4158–4163
3. Carpinteri A, Cardone F, Lacidogna G (2010) Energy emissions from failure phenomena: mechanical, electromagnetic, nuclear. Exp Mech 50:1235–1243
4. Carpinteri A, Borla O, Lacidogna G, Manuello A (2010) Neutron emissions in brittle rocks during compression tests: monotonic vs cyclic loading. Phys Mesomech 13:268–274
5. Carpinteri A, Lacidogna G, Manuello A, Borla O (2011) Energy emissions from brittle fracture: neutron measurements and geological evidences of piezonuclear reactions. Strenght Fract Complex 7:13–31
6. Mogi K (1962) Study of elastic shocks caused by the fracture of heterogeneous materials and its relation to earthquake phenomena. Bull Earthq Res Inst 40:125–173
7. Lockner D (1993) The role of acoustic emissions in the study of rock fracture. Int J Rock Mech Min Sci Geomech 7:883–889
8. Ohtsu M (1996) The history and development of acoustic emission in concrete engineering. Mag Concr Res 48:321–330
9. Rundle JB, Turcotte DL, Shcherbakov R, Klein W, Sammis C (2003) Statistical physics approach to understanding the multiscale dynamics of earthquake fault systems. Rev Geophys 41:1–30
10. Niccolini G, Schiavi A, Tarizzo P, Carpinteri A, Lacidogna G, Manuello A (2010) Scaling in temporal occurrence of quasi-rigid body vibration pulses due to macrofractures. Phys Rev E 82(046115):1–5
11. Carpinteri A, Lacidogna G (2006) Damage monitoring of an historical masonry building by the Acoustic Emission technique. Mater Struct 39:161–167
12. Carpinteri A, Lacidogna G (2006) Structural monitoring and integrity assessment of medieval towers. J Struct Eng (ASCE) 132:1681–1690
13. Carpinteri A, Lacidogna G (2007) Damage evaluation of three masonry towers by Acoustic Emission. Eng Struct 29:1569–1579
14. Lacidogna G, Carpinteri A, Manuello A, Durin G, Niccolini G, Agosto A (2010) Acoustic and electromagnetic emissions as precursor phenomena in failure processes. Strain 47(Suppl 2):144–152
15. Carpinteri A, Lacidogna G, Manuello A, Niccolini G, Schiavi A, Agosto A (2010) Mechanical and electromagnetic emissions related to stress-induced cracks. Exp Tech, 1–12. doi: 201010.1111/j.1747-1567.2011.00709.x
16. Misra A (1977) Theoretical study of the fracture-induced magnetic effect in ferromagnetic materials. Phys Lett A 62:234–236
17. Frid V, Rabinovitch A, Bahat D (2003) Fracture induced electromagnetic radiation. J Phys D 36:1620–1628
18. Hadjicontis V, Mavromatou C, Nonos D (2004) Stress induced polarization currents and electromagnetic emission from rocks and ionic crystals, accompanying their deformations. Nat Hazard Earth Syst Sci 4:633–639
19. Warwick JW, Stoker C, Meyer TR (1982) Radio emission associated with rock fracture: possible application to the great Chilean earthquake of May 22, 1960. J Geophys Res 87:2851–2859
20. Nagao T, Enomoto Y, Fujinawa Y et al (2002) Electromagnetic anomalies associated with 1995 Kobe earthquake. J Geodyn 33:401–411
21. Rabinovitch A, Frid V, Bahat D (2007) Surface oscillations. A possible source of fracture induced electromagnetic oscillations. Tectonophysics 431:15–21
22. Bubble Technology Industries (1992) Instruction manual for the Bubble Detector Spectrometer (BDS), Chalk River, Ontario, Canada
23. National Council on Radiation Protection and Measurements. Protection against neutron radiation, NCRP report 38, 1971
24. Kaiser J (1950) An investigation into the occurrence of noises in tensile tests, or a study of acoustic phenomena in tensile tests. Ph.D. dissertation, Munich (FRG), Technische Hochschule München

# Chapter 51
# Fluid Structure Interaction During Shock Loading in a Compressible Fluid

**Erheng Wang, Jefferson Wright, and Arun Shukla**

**Abstract** The compressibility of a fluid highly affects the fluid structure interaction during a shock loading. The currently available fluid structure interaction models all ignore the compressibility of a fluid; this is only valid for an underwater shock loading or a very weak air shock loading. In this paper, a new 1-D fluid structure interaction model based on the gas-dynamic theory (Rankine-Hugoniot relation) is proposed. This model considers the fluid compressibility and therefore is more accurate for a shock loading process in a compressible fluid, such as air. A series of well-designed 1-D shock wave loading experiments on free-standing monolithic metal plates using a shock tube apparatus are performed to verify the proposed model. The reflected pressure profile and the impulse imparted upon the specimens are all predicted. The results show that the current model can give a more accurate prediction on the reflected pressure profile and the imparted impulse on the specimen than that from the classic Taylor's model, which does not consider fluid compressibility.

**Keywords** Blast resistance • Blast loading • Fluid structure interaction

## 51.1 Introduction

The compressibility of a fluid is one of the most important properties of a fluid, especially for highly compressible fluids, such as air. During a shock wave loading, the fluid structure interaction is also highly affected by the compressibility of the fluid. However, the classic Taylor's model [1–4] does not incorporate the compressibility and considers the medium uncompressible. Kambouchev's model [5], an extension of Taylor's model, considers the compressibility of a fluid by imposing the momentum conservation equation. Due to the mathematical difficulties, they conduct their analysis under the acoustic limit condition, which means that the compressibility of the fluid is ignored. Therefore, the acoustic limit condition reduces the Kambouchev's model back to the Taylor's model. Recent experimental results [6] have indicated that these two models are not suitable for the fluid structure interaction during a shock wave loading in a compressible fluid. For example, during an air blast event there is a noticeable difference between the incident and reflected shock wave velocities [6], which has been assumed to have no difference in the above two models. Other researchers [7–9] give a semi-theoretical peak pressure of the reflected shock wave with normal incidence. Unfortunately, they cannot evaluate the time histories of the reflected pressure as well as the reflected impulse.

In this paper, a new fluid–structure-interaction model that considers the high compressibility of a fluid is proposed based on the one-dimensional gas-dynamic theory. The main purpose is to predict the reflected pressure profile and imparted impulse upon the plate given the incident pressure profiles. A series of well-designed 1-D shock wave loading experiments on free-standing monolithic Aluminum and Steel plates are performed to illuminate the fluid structure interaction process in air. The validation of the new model and the difference between the experimental and theoretical results induced by the inaccuracy of the experiments are discussed.

E. Wang (✉)
Department of Aerospace Engineering, University of Illinois at Urbana-Champaign, 104 S Wright St, Urbana, IL 61801, USA
e-mail: wange@egr.uri.edu

J. Wright • A. Shukla
Dynamic Photomechanics Lab, Department of Mechanical, Industrial and Systems Engineering, The University of Rhode Island, 92 Upper College Road, Kingston, RI 02881, USA

V. Chalivendra et al. (eds.), *Dynamic Behavior of Materials, Volume 1: Proceedings of the 2012 Annual Conference on Experimental and Applied Mechanics*, Conference Proceedings of the Society for Experimental Mechanics Series,
DOI 10.1007/978-1-4614-4238-7_51, 

## 51.2 Theoretical Consideration

Consider a planar shock wave impinging normally and uniformly upon an infinite free-standing monolithic plate with an areal density $\rho_s$. Figure 51.1 shows a typical shock wave loading process. A planar incident shock wave front is traveling from the left side to the right side (incident process, Fig. 51.1a). The right side of the incident wave front is undisturbed gas, while the left side of incident wave front is the high pressure gas driven by the incident shock wave. After the incident shock wave impinges on a static free-standing flat plate, a reflected planar shock wave is generated and travels from the right side to the left side (reflected process, Fig. 51.1b). The left side of the reflected wave front is still the high pressure gas driven by the incident shock wave, while the right side of the reflected wave front is now the gas that is disturbed by the reflected shock wave. Moreover, at this time the plate will move with a velocity that is equal to the particle velocity of the gas behind the reflected shock wave front. $U_+$ and $U_-$ are the velocities of the incident and the reflected shock wave fronts, respectively. The state of the gas can be defined using the following physical parameters: the pressure $p$, the particle velocity $u$, the sound velocity $c$ and the density $\rho$. The subscript 0 on the parameters denotes the initial state of the gas. Subscript 1 represents the state of the gas located behind the incident shock wave front and this is defined as the incident state. Subscript 2 represents the state of the gas located behind the reflected shock wave front and this will be defined as the reflected state.

The incident shock wave pressure profiles is assumed to have a typical exponential decay shape with respect to time (Fig. 51.2), which can be expressed as,

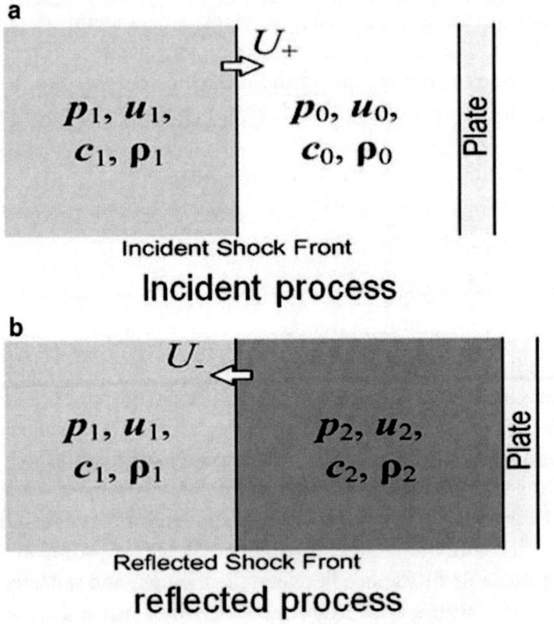


**Fig. 51.1** Sketch of the shock processes

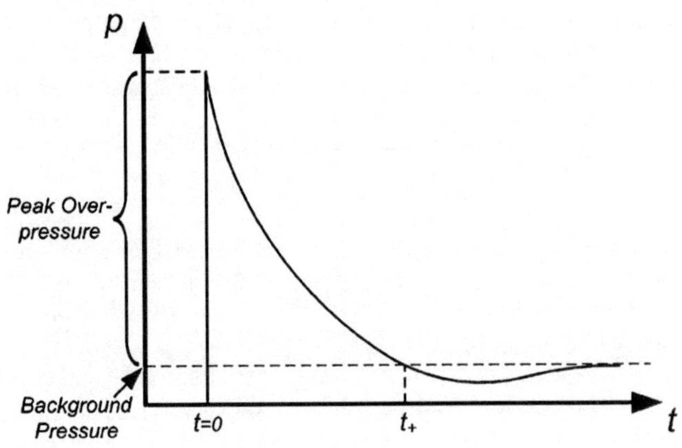


**Fig. 51.2** Typical shock pressure profile

$$p_1(t) = p_{1\_peak}e^{-\frac{t}{\theta_1}}, \quad 0 \leq t \leq \infty \tag{51.1}$$

where, $p_{1\_peak}$, $\theta_1$, and $t_{1+}$ are the peak pressure, time constant and positive time period of the incident shock wave, respectively. Generally, the physical parameters of the undisturbed gas ($p_0$, $u_0$, $c_0$, and $\rho_0$) and the incident pressure profile ($p_{1\_peak}$, $\theta_1$ and $t_{1+}$) can be measured in the real event. The question is how to evaluate the reflected pressure profile $p_2$ based on the known parameters.

### 51.2.1 *Taylor's Model*

In Taylor's model, the pressure $p_2$ at the certain location $x$ and time $t$ can be separated into three parts: the pressure of the incoming wave $p_I$, the pressure of the reflected wave from a rigid wall $p_{r1}$ and the pressure of the rarefaction wave $p_{r2}$, due to the acceleration process of the plate, [1–3]

$$p_2(x,t) = p_I + p_{r1} + p_{r2} = p_{1\_peak}e^{-\left(t-\frac{x}{U_+}\right)/\theta_1} + p_{1\_peak}e^{-\left(t+\frac{x}{U_+}\right)/\theta_1} - \rho_f U_+ \dot{w}\left(t + \frac{x}{U_+}\right) \tag{51.2}$$

Where, $U_+$ is the shock wave front velocity. $x$ represents the distance from the surface of the plate. $\theta_1$ is the decay time constant of the incident pressure profile. $\rho_f$ is the density of the fluid and $w$ is the displacement of the plate. Therefore, the pressure profile applied on the plate (at $x = 0$) is,

$$p_2(t) = 2p_{1\_peak}e^{-\frac{t}{\theta_1}} - \rho_f U_+ \dot{w}(t) \tag{51.3}$$

Application of Newton's second law to the plate gives,

$$\rho_s \ddot{w}(t) = 2p_{peak}e^{-\frac{t}{\theta}} - \rho_f U \dot{w}(t) \tag{51.4}$$

Where, $\rho_s$ is the areal density of the plate. Considering the initial conditions $w(t) = \dot{w}(t) = 0$ and defining a non-dimensional parameter $\psi = \left(\rho_f U_+ \theta_1\right)/\rho_s$, the solution of Eq. 51.4 is,

$$w(t) = \frac{2p_{peak}\theta_1^2}{\rho_s \psi(\psi - 1)}\left[(\psi - 1) + e^{-\frac{\psi t}{\theta_1}} - \psi e^{-\frac{t}{\theta_1}}\right] \tag{51.5}$$

Then the pressure in Eq. 51.3 can be expressed as,

$$p_2(t) = 2p_{1\_peak}e^{-\frac{t}{\theta_1}} - \frac{2p_{1\_peak}\psi}{\psi - 1}\left[e^{-\frac{t}{\theta_1}} - e^{-\frac{\psi t}{\theta_1}}\right] \tag{51.6}$$

Since the pressure should approach 0 at the time $t_{I+}$, $t_{I+}$ can be determined by substituting this condition into Eq. 51.6,

$$t_{1+} = \frac{\theta_1 \ln(\psi)}{\psi - 1} \tag{51.7}$$

Therefore, all of other parameters can be evaluated by the calculated pressure profile $p_2$.

### 51.2.2 *Present Model*

In the present model, we first assume that the reflected pressure profile also has a typical exponential decay shape.

$$p_2(t) = p_{2\_peak}e^{-\frac{t}{\theta_2}}, \quad 0 \leq t \leq \infty \tag{51.8}$$

Where, the unknown parameters $p_{2\_peak}$, $\theta_2$, and $t_{2+}$ are the peak pressure, time constant and positive time period of the reflected shock wave, respectively. Note that since the reflected pressure profile will decay to $p_0$ at the time $t_{2+}$, there are only two independent parameters in three unknown parameters $p_{2\_peak}$, $\theta_2$ and $t_{2+}$.

To simplify the fluid structure interaction analysis, two hypotheses are introduced: (1) the velocity of the reflected shock wave front does not change during the fluid structure interaction process. (2) The physical properties of the gas, such as pressure and particle velocity, located on each side of the reflected shock wave front are uniform. The feasibility and applicability of these two hypotheses have been discussed by Wang and Shukla [6]. The first hypothesis makes $U_-$ as a constant. The second hypothesis indicates that all of the physical parameters of the gas are only related to the initial peak value and the time (not related to the location). Therefore, the analysis process can be separated into two steps. The first step is to determine the peak value of all of the physical parameters. The second step is to determine their time history.

Considering the shock loading process described in Fig. 51.1, there are fourteen parameters $p_0$, $u_0$, $c_0$, $\rho_0$, $p_1$, $u_1$, $c_1$, $\rho_1$, $p_2$, $u_2$, $c_2$, $\rho_2$, $U_+$ and $U_-$ including five known parameters $p_0$, $u_0$, $c_0$, $\rho_0$ and $p_1$. They can be correlated by one-dimensional gas-dynamic theory for a shock wave in an ideal gas, which considers the compressibility of a fluid. The jump conditions for a shock wave front are as follows (using the incident shock wave process in Fig. 51.1a as an example) [6, 10]:

Conservation of mass:

$$\rho_0 v_0 = \rho_1 v_1 \tag{51.9}$$

Conservation of momentum:

$$\rho_0 v_0^2 + p_0 = \rho_1 v_1^2 + p_1 \tag{51.10}$$

Conservation of energy:

$$\frac{1}{2} v_0^2 + e_0 + p_0 \tau_0 = \frac{1}{2} v_1^2 + e_1 + p_1 \tau_1 \tag{51.11}$$

where $v$ is the relative particle velocity with respect to the shock wave front. $v_0 = u_0 - U_+$ and $v_1 = u_1 - U_+$. The Rankine-Hugoniot relations with two modified forms can be derived from these jump conditions,

$$\frac{p_1}{p_0} = (1 + \mu^2) M_0^2 - \mu^2 \tag{51.12a}$$

$$\frac{p_0}{p_1} = (1 + \mu^2) M_1^2 - \mu^2 \tag{51.12b}$$

$$(1 - \mu^2)(U_+ - u_0)^2 - (u_1 - u_0)(U_+ - u_0) = (1 - \mu^2) c_0^2 \tag{51.13a}$$

$$(1 - \mu^2)(U_+ - u_1)^2 - (u_0 - u_1)(U_+ - u_1) = (1 - \mu^2) c_1^2 \tag{51.13b}$$

where, $\mu^2 = (\gamma - 1)/(\gamma + 1)$; $\gamma$ is the adiabatic coefficient and $M$ is the Mach number, $M_1 = |u_1 - U_+|/c_1$ and $M_0 = |u_0 - U_+|/c_0$.

The jump conditions for the shock wave (Eqs. 51.9 ~ 51.11) and the modified Hugoniot relations (Eqs. 12 and 13) compose the system of governing equations. The same procedure can be applied to the reflected process (Fig. 51.1b) by changing the subscript 0–1, subscript 1–2 and subscript + to −. Then the peak values of the parameters of the gas behind the incident and reflected shock wave fronts can be calculated through above equations as follows,

$$U_+ = u_0 + c_0 \sqrt{\frac{(p_{1\_peak}/p_0) + \mu^2}{1 + \mu^2}} \tag{51.14}$$

$$u_{1\_peak} = u_0 + \frac{(1 - \mu^2)\left[(U_+ - u_0)^2 - c_0^2\right]}{U_+ - u_0} \tag{51.15}$$

$$c_{1\_peak} = \sqrt{\left(U_+ - u_{1\_peak}\right)^2 + \frac{\left(u_0 - u_{1\_peak}\right)\left(U_+ - u_{1\_peak}\right)}{1-\mu^2}} \tag{51.16}$$

$$\rho_{1\_peak} = \left(\frac{u_0 - U_+}{u_{1\_peak} - U_+}\right)\rho_0 \tag{51.17}$$

$$U_- = u_{1\_peak} - c_{1\_peak}\sqrt{\frac{\left(p_{2\_peak}/p_{1\_peak}\right) + \mu^2}{1+\mu^2}} \tag{51.18}$$

$$u_{2\_peak} = u_{1\_peak} + \frac{(1-\mu^2)\left[\left(U_- - u_{1\_peak}\right)^2 - c^2_{1\_peak}\right]}{U_- - u_{1\_peak}} \tag{51.19}$$

$$c_{2\_peak} = \sqrt{\left(U_- - u_{2\_peak}\right)^2 - \frac{\left(u_{1\_peak} - u_{2\_peak}\right)\left(U_- - u_{2\_peak}\right)}{1-\mu^2}} \tag{51.20}$$

$$\rho_{2\_peak} = \left(\frac{u_{1\_peak} - U_-}{u_{2\_peak} - U_-}\right)\rho_{1\_peak} \tag{51.21}$$

Equations 51.14, 51.15, 51.16 and 51.17 show that the parameter peak value of the gas behind the incident shock wave can be explicitly calculated by the known parameters. However, to determine the parameter peak value in Eqs. 51.18, 51.19, 51.20 and 51.21, we have to determine the reflected pressure $p_2$, which cannot be derived from above equations directly.

Since a shock wave front has a thickness of $10^{-8}$ m and a mach number larger than 1 [11, 12], the time that the shock wave front passes the face plate has a value between $10^{-10}$ and $10^{-11}$ s. A simple calculation shows that even for a peak reflected pressure as high as 1,000 MPa (nuclear weapon [7]) and an areal density of a plate as low as 1 kg/m$^2$ (about 0.4 mm aluminum foil), the velocity of the plate only has a value between $10^{-2}$ and $10^{-3}$ m/s, which can be thought of as negligible (zero). Notice the fact that the particle velocity $u_2$ of the gas behind the reflected shock wave front must be the same as the particle velocity of the plate in Fig. 51.1 [1, 10]. Therefore, the peak reflected pressure can be calculated by assuming the particle velocity $u_2$ of the gas behind the reflected shock wave front is zero at the time when the shock wave just impinges upon the plate. From Rankine-Hugoniot relations, the peak value of $u_2$ can be explicitly calculated as,

$$p_{2\_peak} = p_{1\_peak}\left\{\left[\frac{u_{1\_peak}}{(1-\mu^2)c_{1\_peak}} + \sqrt{\frac{u^2_{1\_peak}}{(1-\mu^2)^2 c^2_{1\_peak}} + 4}\,\right]^2 \frac{1+\mu^2}{4} - \mu^2\right\} \tag{51.22}$$

Now the peak reflected pressure can be calculated. The problem changes to how to determine the time history constant $\theta_2$ or $t_{2+}$ of $p_2$. First, though the shock wave is not a reversible process, the change of the parameters in the gas in front of and behind the shock wave front is a reversible, adiabatic process. This allows two more equations to correlate the pressure, density and sound velocity of the gas,

$$\rho c^2 = \gamma p \tag{51.23}$$

$$p = A\rho^{\gamma} \tag{51.24}$$

where, $A$ is a constant for a reversible adiabatic process. Then from Eqs. 51.1, 51.8, 51.14 and 51.15, the time history of the density, sound velocity and particle velocity of the gas can be correlated to the time history of the incident and reflected pressure.

$$\rho_1(t) = \left[\frac{p_1(t)}{p_{1\_peak}}\right]^{\frac{1}{\gamma}} \rho_{1\_peak} = \rho_{1\_peak}\exp\left(-\frac{t}{\gamma\theta_1}\right) \tag{51.25}$$

$$\rho_2(t) = \left[\frac{p_2(t)}{p_{2\_peak}}\right]^{\frac{1}{\gamma}} \rho_{2\_peak} = \rho_{3\_peak} \exp\left(-\frac{t}{\gamma\theta_2}\right) \tag{51.26}$$

$$c_1(t) = \left[\frac{\rho_1(t)}{\rho_{1\_peak}}\right]^{\frac{\gamma-1}{2}} c_{1\_peak} = c_{1\_peak} \exp\left[-\frac{(\gamma-1)t}{2\gamma\theta_1}\right] \tag{51.27}$$

$$c_2(t) = \left[\frac{\rho_2(t)}{\rho_{2\_peak}}\right]^{\frac{\gamma-1}{2}} c_{2\_peak} = c_{2\_peak} \exp\left[-\frac{(\gamma-1)t}{2\gamma\theta_2}\right] \tag{51.28}$$

$$u_1(t) = U_- + c_1\sqrt{\frac{[p_2(t)/p_1(t)] + \mu^2}{1+\mu^2}} \tag{51.29}$$

$$u_2(t) = \frac{(1-\mu^2)\left\{[U_- - u_1(t)]^2 - c_1^2\right\}}{U_- - u_1(t)} + u_1(t) \tag{51.30}$$

Note the signs of the particle velocities and shock wave velocities in Eqs. 51.14, 51.15, 51.18, 51.19, 51.29 and 51.30 indicate the direction of motion. In Fig. 51.1, a positive sign indicates the direction is the same as the propagating direction of the incident shock wave front and a negative sign indicates the direction is opposite that of the propagating direction of the incident shock wave front.

Second, If we consider the momentum conservation of the plate, all of the impulse from the reflected pressure $p_2$ has been transmitted onto the plate at the end of the fluid structure interaction process and the final velocity of the plate should be equal to the gas particle velocity $u_2$ at time $t = t_{2+}$. Therefore, the following equation can be constructed,

$$\int_0^{t_{2+}} [p_2(t) - p_0]dt = \rho_s u_2 \tag{51.31}$$

Where, $p_0$ and $\rho_s$ are known parameter. Since $\theta_2$ and $t_{2+}$ can be related by,

$$\theta_2 = \frac{t_{2+}}{\ln(p_{2\_peak}) - \ln(p_0)} \tag{51.32}$$

all of the other parameters in Eq. 51.31 are only related to $t_{2+}$. $t_{2+}$ can be calculated from this equation and consequently the reflected pressure is $p_2$ determined by Eqs. 51.8, 51.22 and 51.31. Therefore, all of the parameters of gas behind the incident and reflected shock wave fronts can be calculated.

## 51.3 One-Dimensional Shock Wave Loading Experiments on Free-Standing Monolithic Plates

### *51.3.1 Experimental Setup*

Shock tubes have been widely used for simulating blast loading as they can generate a repeatable tailored shock wave with a planar wave front [6, 10, 13–15]. The process in which this planar shock wave normally impinges upon a flat plate can be considered a one-dimensional process. Figure 51.3 shows the experimental setup. A monolithic metal plate is placed right in front of the muzzle of the shock tube and can move freely. A high speed digital camera, Photron SA1, was used to capture the real time side-view position of the plate. The axis of the camera lens is perpendicular to the axis of the shock tube, i.e. the moving direction of the plate. The distance between the camera and the plate was chosen to be approximately 2 m, which is more than 20 times that of the plate's dimension (~ 0.1 m) and 10 times that of the plate displacement in the images

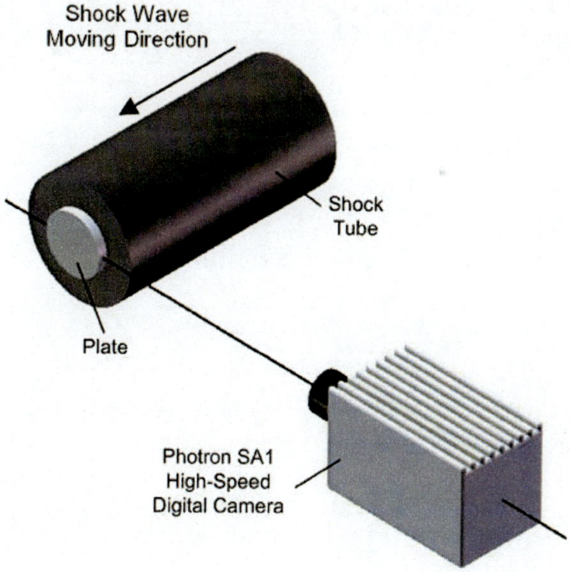


**Fig. 51.3** Experimental setup

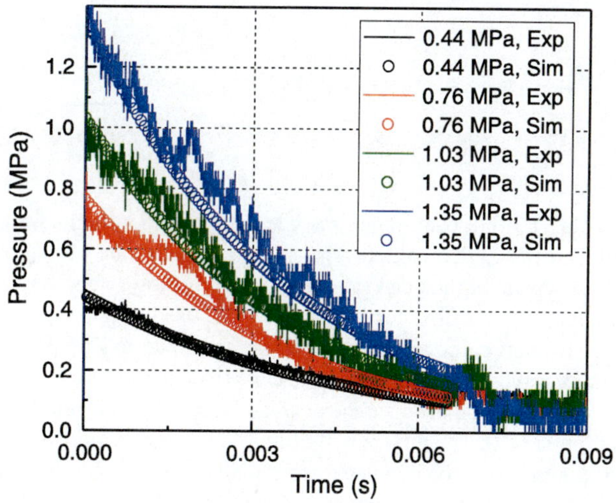


**Fig. 51.4** Incident pressure profiles with different peak pressure

(~ 0.2 m), to avoid image distortion during plate propagation. Two transducers (PCB CA102B) are mounted at the muzzle to measure the pressure profiles. The detail of the transducer location and data acquisition can be found in Ref. [6, 13].

The circular plate specimens are fabricated using 6061 T6 Aluminum and1,018 steel. The diameter of the flat circular plate specimen is 0.0777 m, which was slightly bigger than the inner diameter of the shock tube (0.0762 m) in order to ensure that the plate can cover the whole muzzle. The thickness of the specimens was 0.0064 m and the areal densities are 17.2 and 48.8 kg/m$^2$ for the Aluminum and Steel respectively.

### 51.3.2 Experimental Procedue and Data Analysis

During the experiments, the circular specimens were placed right in front of the muzzle end of the shock tube (as shown in Figs. 51.3 and 51.4). The specimens were then subjected to shock wave loadings with four different incident peak pressures: 0.44, 0.76, 1.03 and 1.35 MPa, respectively. The experimental and the simulated (fitted) pressure profiles for different peak pressures are shown in Fig. 51.5. The time constants $\theta_1$ and positive time periods $t_{1+}$ for each pressure profile are listed in

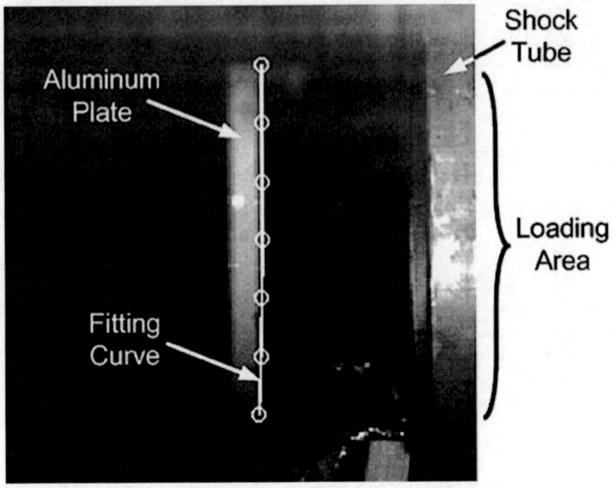


**Fig. 51.5** A typical side-view image and curve fitting of the front face

**Table 51.1** Parameters used for fitting the incident pressure profiles

| Peak incident pressure (MPa) | Positive time period $t_{1+}$ (s) | Time constant $\theta_1$ (s) |
|---|---|---|
| 0.44 | 0.0065 | 0.0044 |
| 0.76 | 0.0070 | 0.0035 |
| 1.03 | 0.0080 | 0.0034 |
| 1.35 | 0.0090 | 0.0034 |

Table 51.1. For each shock wave level, two experiments were conducted for the purpose of verifying data repeatability. The high-speed digital camera was set to a framing rate of 40,000 fps and an image resolution of 512 × 256 pixels.

A typical high-speed side-view image during a shock wave loading process is shown in Fig. 51.5. Curve fitting methods, such as cubic spline curve fitting, can be used to pick up the position of the front face of the specimen. An example of the seven-point cubic spline curve fitting is shown in Fig. 51.5. Since the specimen did not show any compression during the shock wave loading process, this curve can be used to represent the position of the specimen. Then, the displacement profile of each point on the specimen can be obtained by correlating the position of the specimen in each image to that in the image at time $t = 0$. The differential of this displacement profile with respect to the time gives the velocity profile. The momentum of the plate can be evaluated from the velocity profile and the areal density of the plate. It can be expressed as:

Horizontal momentum:

$$I_x^{plate} = \oint_S \rho_s u_x ds \tag{51.33}$$

where, $u_x$ is the $x$ direction (horizontal) velocity and $ds$ is the areal element of the plate.

## 51.4 Results and Discussion

### *51.4.1 Experimental Results*

The reflected pressure profiles measured by the pressure transducer are shown in Fig. 51.6. The high speed side-view images of the circular plates at different incident shock levels are shown in Figs. 51.7 and 51.8.

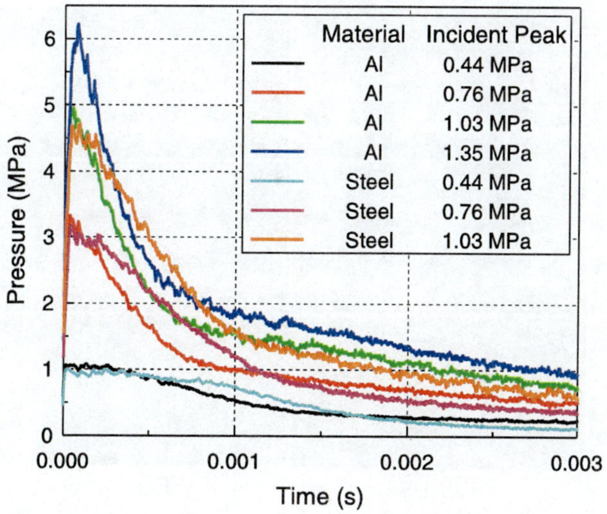


**Fig. 51.6** Experimental reflected pressure profiles

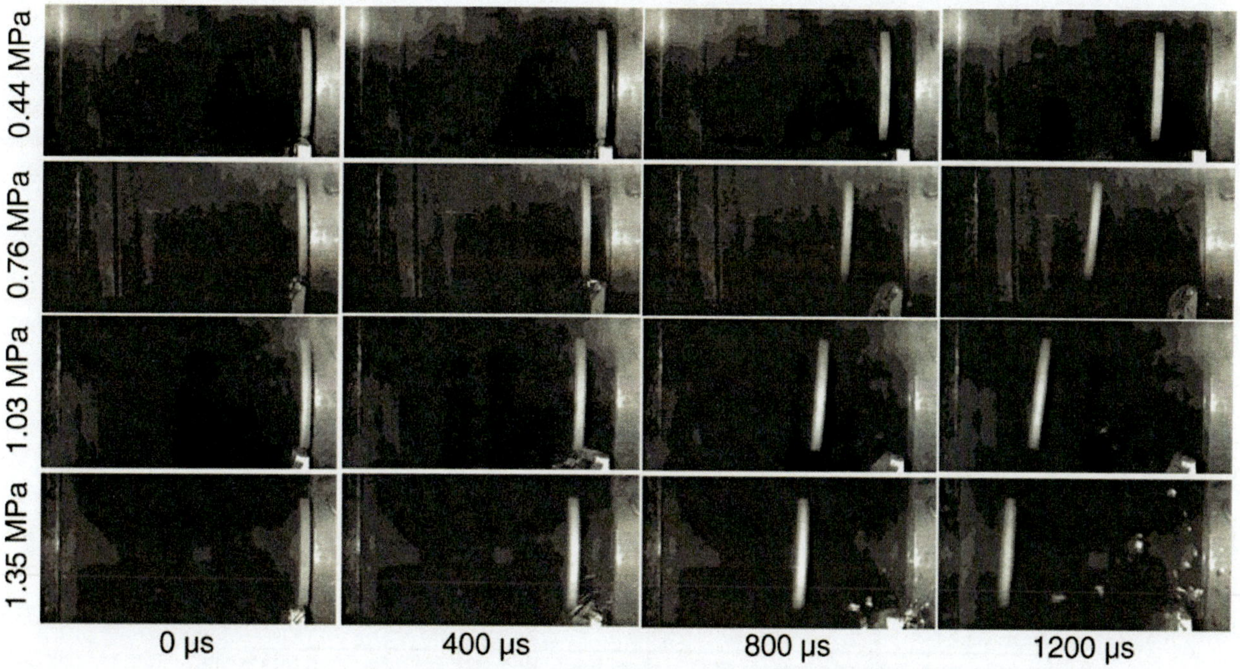


**Fig. 51.7** High-speed side-view images of the circular aluminum plate under different incident shock level

### *51.4.2 Impulse of the Reflected Pressure Pulse and Momentum of the Plate*

The impulse of the experimental reflected pressure pulse calculated by the left side of Eq. 51.31 and the momentum of the plate calculated by the right side of Eq. 51.31 are shown in Fig. 51.9 (only the results of the Aluminum plate are shown). It can be seen that the momentum of the plates agrees with the impulse of the reflected pressure very well. This verifies that momentum is conserved during the loading process as described by Eq. 51.31. Since the high speed side-view images can only be recorded for a short time period (generally less than 3 ms), the impulse of the reflected pressure can be used to determine the momentum of the plates for longer time periods.

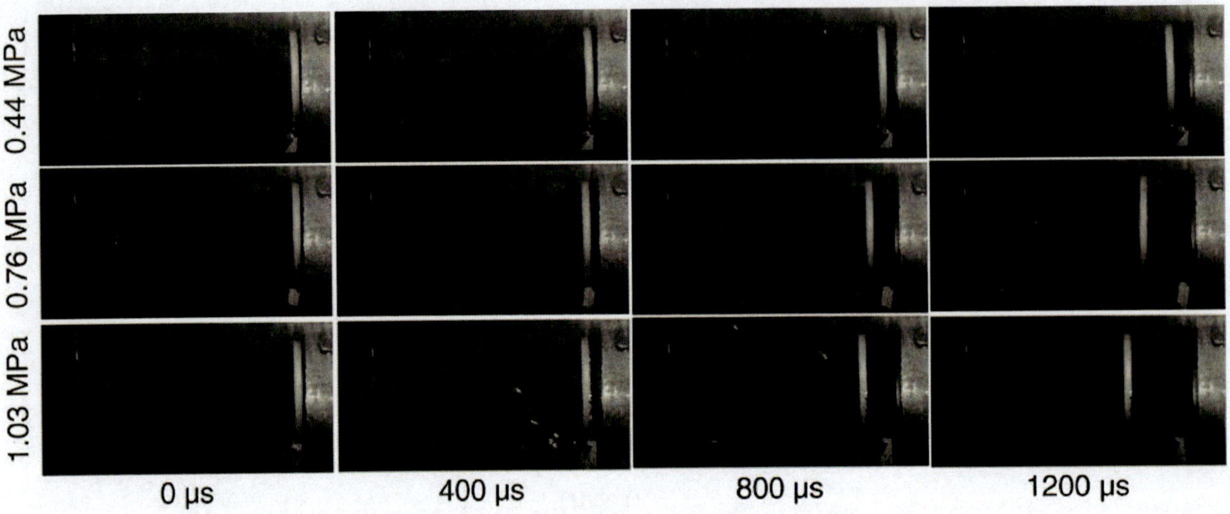


**Fig. 51.8** High-speed side-view images of the steel circular plate under different incident shock level

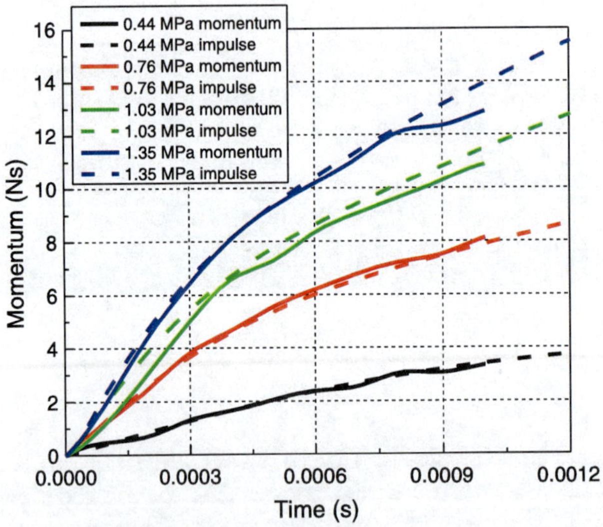


**Fig. 51.9** Comparison of the momentum of the plates and the impulse applied on the plates for different peak incident shock pressures

**Table 51.2** The input experimental parameters

| Parameters | Value |
|---|---|
| Pressure $p_0$ of the undisturbed gas | 0.101325 MPa |
| Density $\rho_0$ of the undisturbed gas | 1.21 kg/m$^3$ |
| Sound velocity $c_0$ of the undisturbed gas | 343 m/s |
| Particle velocity $u_0$ of the undisturbed gas | 0 m/s |
| Adiabatic coefficient $\gamma$ | 1.4 |

### *51.4.3 Evaluated Peak Reflected Pressure*

The parameters needed in the analytical model, such as the pressure, density, sound velocity and particle velocity of the undisturbed gas in front of the incident shock wave front, are listed in Table 51.2.

The experimental results of the peak reflected pressure as a function of the incident pressures are shown in Fig. 51.10. The peak reflected pressures determined by Eq. 51.6 from Taylor's model and by Eq. 51.22 from the current model are also

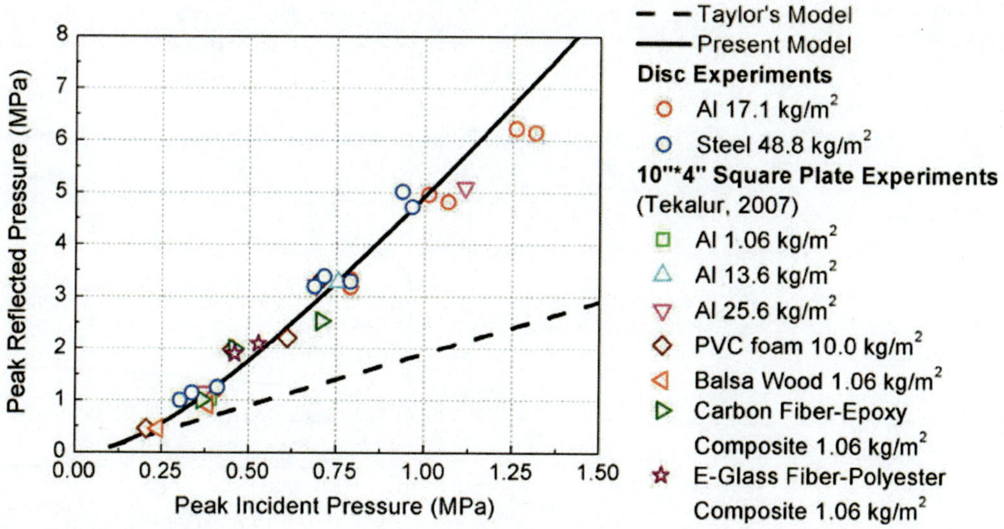


**Fig. 51.10** The comparison of the experimental and analytical relation between peak incident and peak reflected pressures

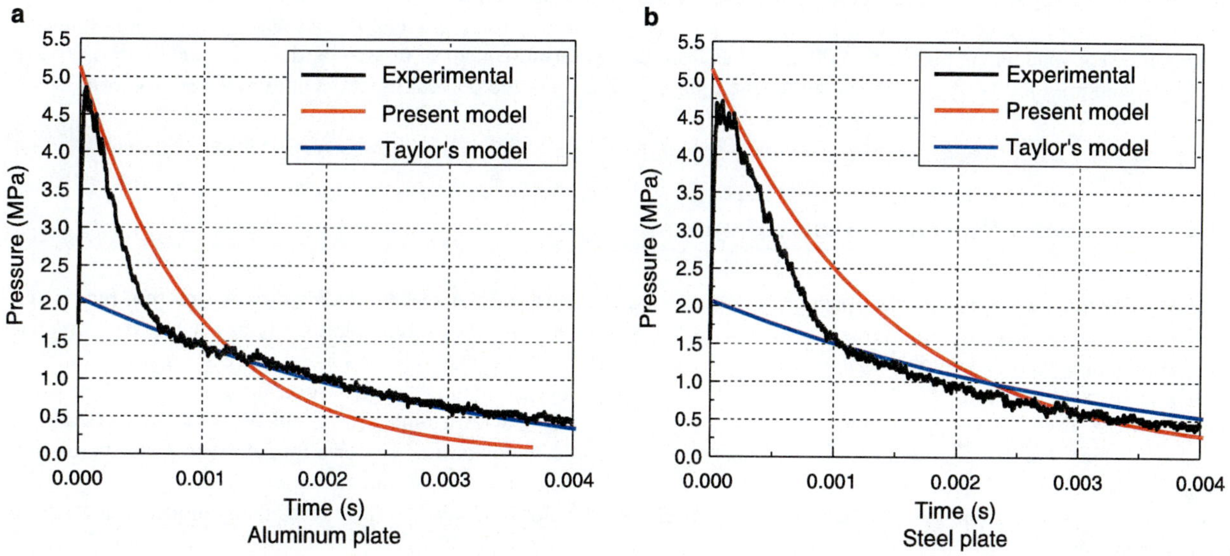


**Fig. 51.11** Experimental and predicted reflected pressure profiles

shown in this figure as well as the experimental data for various other materials tested in our lab [15]. It can be seen that the experimental data follows the predicted trend from the current model very well. Taylor's model gives much lower prediction results. It is evident that the reflected peak pressure only depends on the incident peak pressure, and the compressibility of the gas highly affects the results.

## *51.4.4 Evaluated Reflected Pressure Profiles*

The reflected pressure profile obtained from the experiment, Taylor's model (Eq. 51.6) and the present model for Aluminum and Steel plates are shown in Fig. 51.11. For simplification, only the results of 1.03 MPa peak incident shock pressure are provided. It can be seen that at the beginning (before 700 μs), the prediction from Taylor's model is much lower than the experimental data, while the present model gives a more accurate prediction especially at the very beginning (before 250 μs).

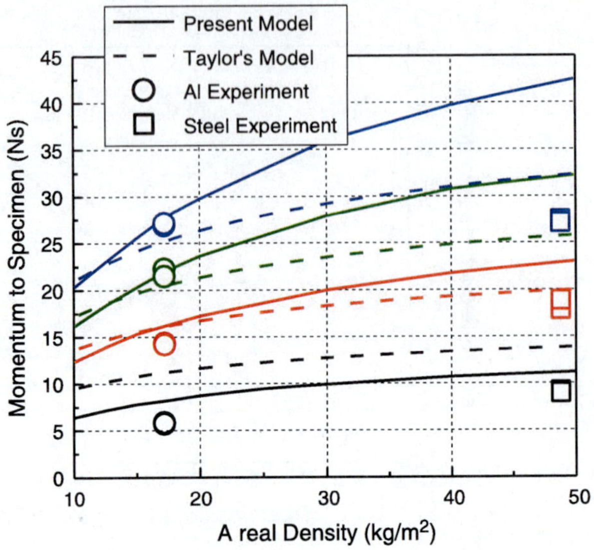


**Fig. 51.12** Experimental and predicted reflected pressure profiles. *Black*: 0.44 MPa; *Red*: 0.76 MPa; *Green*: 1.03 MPa; *Blue*: 1.35 MPa

However, after 700 μs, Taylor's model gives a better prediction. Actually, from Figs. 51.7 and 51.8, the plates are far from the muzzle and leave a considerable gap after 700 μs. The gas leaking from this gap in fact destroys the one-dimension assumption. The load on the plate after 700 μs comes from the wind (movement of the gas particles) behind the shock wave front. The compressibility of the gas can almost be ignored. Therefore, the present model is not proper in this region.

### *51.4.5 Evaluated Reflected Pressure Profiles*

The momentum transmitted into the specimens from the experiments and the results predicted by the present model and Taylor's model are shown in Fig. 51.12. For the plate with low areal density (Aluminum plate), Taylor's model over predicts the momentum transmitted onto the plates at low peak incident pressure (for example 0.44 MPa) while at high peak incident pressure (for example 1.35 MPa), it under predicts this momentum. The present model gives more accurate results than those obtained from Taylor's model especially under a shock wave loading with a high peak incident pressure (for example 1.03–1.35 MPa). For the plate with high areal density (Steel plate), it looks like Taylor's model gives better prediction especially at higher incident pressure. Considering the heavier specimen need more time to achieve the same velocity as the gas particle. Therefore, more gas will leak from the gap between the plate and the muzzle, consequently induce large error in the experiments.

## 51.5 Summary

A new fluid structure interaction model, which considers the compressibility of a gas, is proposed and implemented in this study. One-dimensional gas-dynamic theory (the Rankine-Hugoniot relation) for an ideal gas was utilized to consider the compressibility of the gas and evaluate its physical parameters such as density, particle velocity, and sound velocity in the gas. The conservation between the pressure impulse applied on the plate and the momentum of the plate itself was utilized to determine the reflected pressure profile. A series of shock wave loading experiments on free-standing monolithic Aluminum and Steel plates was conducted using a shock tube apparatus. The measured reflected pressure profiles and the momentum transmitted onto the plates were compared with the results predicted by the present model and previous models [1–4]. The present model predicts more accurate results for the peak reflected pressure, the reflected pressure profile and the momentum transmitted onto the plate. Further analysis shows that the peak reflected pressure is only related to the peak incident pressure which is confirmed by the experimental results.

**Acknowledgement** The authors kindly acknowledge the financial support of the Office of Naval Research (Dr. Y.D.S. Rajapakse) under grant no. N000140410268, and the Department of Homeland Security under Cooperative Agreement No. 2008-ST-061-ED0002.

## References

1. Taylor GI (1963) In: Batchelor G (ed) The scientific papers of Sir Geoffrey Ingram Taylor, volume III: aerodynamics and the mechanics of projectiles and explosions. Cambridge University Press, Cambridge, pp 287–303
2. Xue Z, Hutchinson JW (2004) A comparative study of impulse-resistant metal sandwich plates. Int J Impact Eng 30:1283–1305
3. Deshpande VS, Fleck NA (2005) One-dimensional response of sandwich plates to underwater shock loading. J Mech Phys Solids 53:2347–2383
4. Tan PJ, Reid SR, Harrigan JJ (2005) Discussion: the resistance of clamped sandwich beams to shock loading. (Fleck NA, Deshpande VS (2004) ASME J Appl Mech 71:386–401) J Appl Mech 72:978–979
5. Kambouchev N, Noels L, Radovitzky R (2006) Nonlinear compressibility effects in fluid-structure interaction and their implications on the air-blast loading of structures. J Appl Phys 100(6):063519
6. Wang E, Shukla A (2010) Analytical and experimental evaluation of energies during shock wave loading. Int J Impact Eng 37:1188–1196
7. Glasstone S (1964) The effects of nuclear weapons, United States Atomic Energy Commission
8. Baker WE, Cox PA, Westine PS, Kulesz JJ, Strehlow RA (1983) Explosion hazards and evaluation. Elsevier, Amsterdam
9. Smith PD, Hetherington JG (1994) Blast and ballistic loading of structures. Butterworth-Heinemann Elsevier Science Ltd, Oxford
10. Courant R, Friedrichs KO (1948) Supersonic flow and shock waves. Interscience, New York
11. Granger RA (1985) Fluid mechanics. Thomson Learning, Citrus Springs, FL USA,ISBN-13: 978-0030629518, 1985
12. http://aerorocket.com/Nozzle/Validate/Tshock/Shock.html, 2010
13. Shukla A, Ravichandran G, Rajapakse YDS (2009) Dynamic failure of materials and structures. Springer, New York. ISBN 144190445X
14. Stoffel M, Schmidt R, Weichert D (2001) Shock wave-loaded plates. Int J Solid Struct 38:7659–7680
15. Tekalur SA (2007) Failure of marine composite materials due to blast loading, Ph.D. Dissertation, University of Rhode Island

# Chapter 52
# Moderate Velocity Ball Impact of a Mock High-Explosive

**J. Furmanski, P. Rae, and B. Clements**

**Abstract** Modeling of thermal and mechanical events in high-explosive materials is complicated by the composite nature of the material, which experiences viscoelastic and plastic deformations and sustains damage in the form of microcracks that can dominate its overall behavior. A mechanical event of interest is projectile interaction with the material, which leads to extreme local deformation and adiabatic heating, which can potentially lead to adverse outcomes in an energetic material. Simulations of such an event predicted large local temperature rises near the path of a spherical projectile, but these were experimentally unconfirmed and hence potentially non-physical. This work concerns the experimental verification of local temperatures both at the surface and in the wake of a spherical projectile penetrating a mock (unreactive) high-explosive at ~700 m/s. Fast response thermocouples were embedded radially in a mid-plane of a cylindrical target, which was bonded around the thermocouples with epoxy and recorded by an oscilloscope through a low-pass filter with a bandwidth of 500 Hz. A peak temperature rise of 70 K was measured both at the equator of the projectile and in its wake, in good agreement with the temperature predicted in the minimally distorted elements at those locations by a finite element model in ABAQUS employing the ViscoSCRAM constitutive model. Further work is needed to elucidate the extreme temperature rises in material undergoing crushing or fragmentation, which is difficult to predict with meshed finite element methods due to element distortion, and also challenging to quantify experimentally.

**Keywords** Mock high-explosive • Shear heating • Thermal initiation

## 52.1 Introduction

A problem of interest to various communities in the weapons establishment is the mechanical insult of high-explosives and the resulting probability of unplanned detonation. Shock-driven detonation is relatively well understood [1], but a more complicated problem is a slow thermal runaway process which is driven by increasing reactivity of the material with increasing temperature but initiated by a sub-critical deformation event [2]. Recent work at Los Alamos National Laboratory has focused on coupling physically realistic material deformation models for high-explosives with computational models of the thermal activation of the various chemical reactions involved in the decomposition of the energetic component of the material [3]. The reaction model is highly sensitive to the initial temperature perturbation from a mechanical event that is passed from the mechanical analysis, and so confidence in the prediction of deformation-related temperature changes are critical to the assessment of explosives experiencing mechanical insult.

The constitutive model ViscoSCRAM was developed at LANL for high-explosive deformation over a large range of strains and strain-rates [4]. This model incorporates linear viscoelasticity, plasticity, equation-of-state with rate-dependent compressibility, and a statistical fracture mechanics model of distributed microcracking. Temperature rises from

J. Furmanski (✉)
Los Alamos National Laboratory, MST-8, Mail Stop G755, Los Alamos, NM 87545, USA
e-mail: jevan@lanl.gov

P. Rae
Los Alamos National Laboratory, WX-6, Mail Stop J564, Los Alamos, NM 87545, USA

B. Clements
Los Alamos National Laboratory, T-1, Mail Stop B221, Los Alamos, NM 87545, USA

V. Chalivendra et al. (eds.), *Dynamic Behavior of Materials, Volume 1: Proceedings of the 2012 Annual Conference on Experimental and Applied Mechanics*, Conference Proceedings of the Society for Experimental Mechanics Series, DOI 10.1007/978-1-4614-4238-7_52, 

deformation are therefore realized from accumulated viscoelastic work, plastic dissipation (shear), compressive heating during volume change, and from plasticity at the deforming tips of microcracks. It also includes frictional heating from evolving shear cracks that are under positive hydrostatic loads. While some components of ViscoSCRAM can be calibrated directly from laboratory experiments, others are largely empirical and potentially confounded in the thermal response, complicating the application of the model for thermal problems. Thus, an experiment which both mimics a physical problem of interest and directly measures the temperature of the deforming material in a modeled material can directly assess the reliability of ViscoSCRAM for dynamic thermo-mechanical problems.

This work focused on the validation of predictions of a moderate velocity (~700 m/s) sphere impact on a mock high-explosive using a previously calibrated ViscoSCRAM model of that material. Embedded fast-response thermocouples in the mock HE target recorded the local temperature near the projectile and in its wake, which was then directly compared to predictions of the same from ViscoSCRAM.

## 52.2 Materials and Methods

### *52.2.1 Material*

The material employed in this work was a mock high-explosive 900-21, which consists of ~95 % Barium Nitrate in a plastic binder material. This is intended as a mock material replacement of the DOE high-explosive PBX 9501, with much of the deformation mechanics shared between the two materials. The 900-21 was compression molded and machined into 6 in. diameter, 1 in. thick discs of material. These discs were then further machined and bonded into sandwich targets with thermocouples on an interior plane.

### *52.2.2 Experimental*

The two bare mock targets were formed by gluing two 6 in. diameter 900-21 discs together with 5 min Barco-Bond. At the glue location, 8 type K thermocouples comprised of 0.25 mm diameter wire (Omega, part number 5TC-TT-K-30-72) were located radially about the center impact location (Fig. 52.1). After come calibration of the experiment, the front surface of the targets were machined to achieve a desired arresting either on the thermocouple plane or just beyond it; and these two conditions were then tested and are reported here. The temperature rises were recorded on a Yokogawa multi-channel

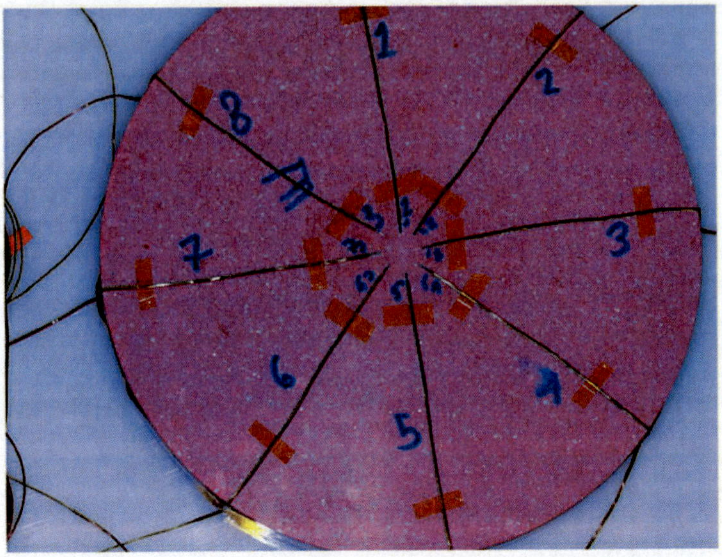


**Fig. 52.1** Mock target midplane with radial thermocouples, prior to bonding

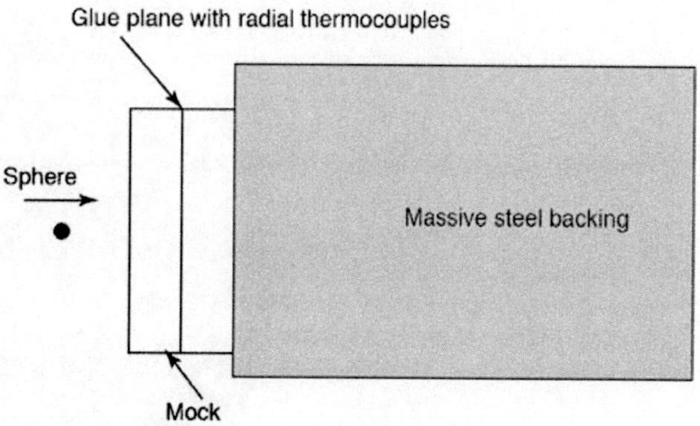


**Fig. 52.2** Experimental arrangement

oscilloscope using 8 thermocouple convertor channels at 200 k samples/s. The bandwidth of the thermocouple amplifiers was 500 Hz. In this way the temperature data in Celsius were recorded directly to a large ascii file.

Bare mock charges were used to reduce the problem of modeling the interaction of any metal cover plates including the tensile strength of the metal/mock glue line. However, to prevent extensive spall of the rear of the mock target, the whole assembly was held in direct contact with a massive at steel anvil at the rear by strapping it down to the anvil with tape under tension (Fig. 52.2). In this way, the back of the target had essentially infinite compressive stiffness but zero tensile strength at the target/anvil interface (only the target inertia need be considered). It was supposed that this provides a computationally tractable boundary condition.

Hardened steel spheres 7.60 mm in diameter were fired from the LANL Dynamic Materials Properties team 'Taylor' gun into 900-21 mock targets. To obtain the maximum velocity possible from the gun for this type of projectile, a small nylon obturator (a small plastic cup that forms a seal behind the projectile) was required together with two 2 mil stainless steel burst discs and a He gas pressure of 4,600 psi. The mass of the obturator was 0.340 g and in all shots it and the ball remained firmly lodged within the target. The projectile velocities were measured using the time between interruptions of two laser beams a known distance apart. The velocity for the first (mid-plane arrested) test was 709 m/s, and for the second increased penetration test, the velocity was 733 m/s. The sample chamber was pumped down to a roughing vacuum prior to firing to prevent the sealed chamber becoming pressurized upon gas release.

Photographs were taken prior to target assembly to locate the recording thermocouples relative to the target centerline. After shooting, the targets were disassembled in various ways and photographed to reveal the location of the stopped projectile.

### *52.2.3 Numerical*

ViscoSCRAM was employed as a user material subroutine (VUMAT) in the finite element code ABAQUS 6.9 (Simulia, Providence RI). The material had been previously calibrated to a range of mechanical stress–strain characterization, dilatometry, and shock-response tests conducted by LANL and its affiliates. The model employed C3D8R 3-dimensional hexahedral elements for the mock target with frictional contact with the steel projectile. The mesh is shown in Fig. 52.3. The finest elements had linear dimensions on the order of millimeters, and while various numbers of elements were invoked, a typical number was 50,000. Temperature was output from the model in the solid elements, though a number of highly distorted elements in the contact region returned non-physical temperatures and deformed shapes, and were effectively ignored in the analysis.

## 52.3 Results

The desired terminal embedded position of the steel projectile was achieved in both cases (Fig. 52.4), with the mid-plane of the projectile arresting in the thermocouple plane (Fig. 52.3a) in one, and in the other arresting approximately one diameter past the thermocouple plane (Fig. 52.3b). In the mid-plane case one thermocouple was destroyed by the projectile, but two others were in the highly deformed boundary region. The bond was improved in the second case, and the thermocouple plane could not be revealed in postmortem analysis. In both tests, no spall failure was observed on the back surface, as desired.

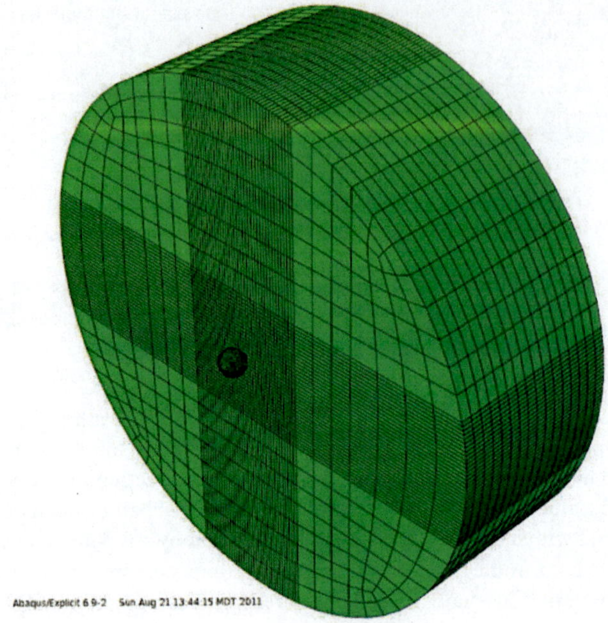


**Fig. 52.3** The finite element mesh used for the ABAQUS simulations. Impact speeds between 500 and 2,000 m/s were investigated

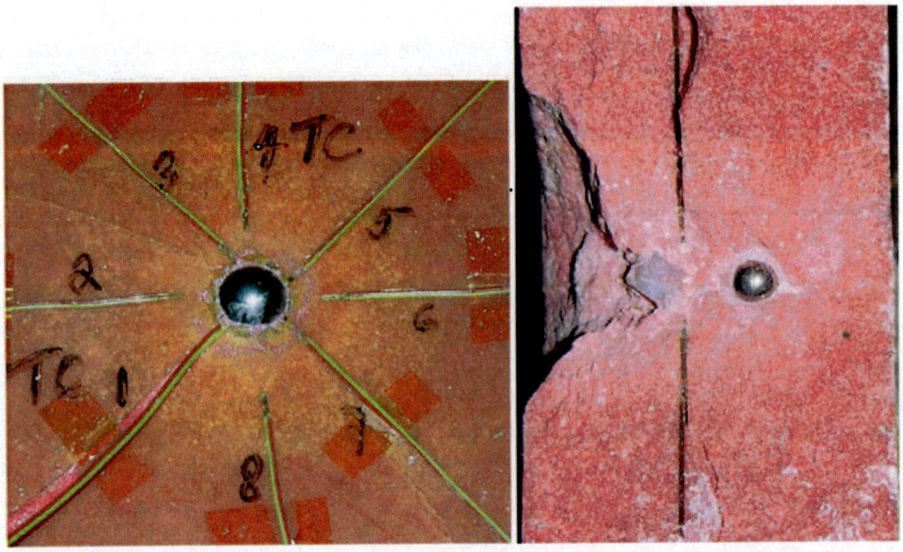


**Fig. 52.4** Arrested positions of projectiles from each conducted test, with relative position to thermocouple plane

The peak temperature rise at the mid-plane of the projectile is shown in Fig. 52.5. The maximum obtained temperature rise, 57.6°C, occurred in thermocouple channel 5, which Fig. 52.3 shows is quite close to the surface of the projectile. Channel 3 showed a significant rise of 40.2°C. Other channels show either more modest temperature rise, or a delayed temperature rise from thermal diffusion, settling at a temperature rise of 20–30°C. Channel 1 was destroyed by the projectile.

The maximum temperature rise in the second shot, where the thermocouples interacted with the wake behind the projectile, is shown in Fig. 52.6. The maximum temperature rise was 61.7°C, and occurred in channel 1. Other channels showed an approximate temperature rise of 20–30°C.

Figure 52.7 shows the results of one of our simulations done on the ball impact geometry. The viscoelastic behavior of the mock HE is well captured in that our model successfully accounts for the material flowing over the trapped ball (Fig. 52.5, compared to Fig. 52.4b). Damage was also conical in shape but far more diffuse than observed in the experiment.

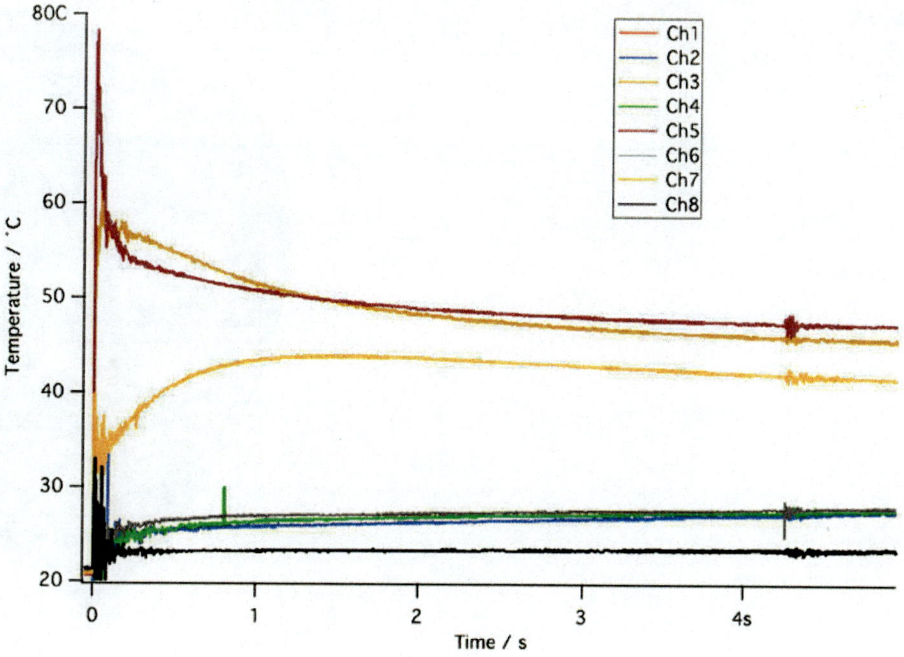


**Fig. 52.5** Thermocouple temperature measurements from the mid-plane arrested test

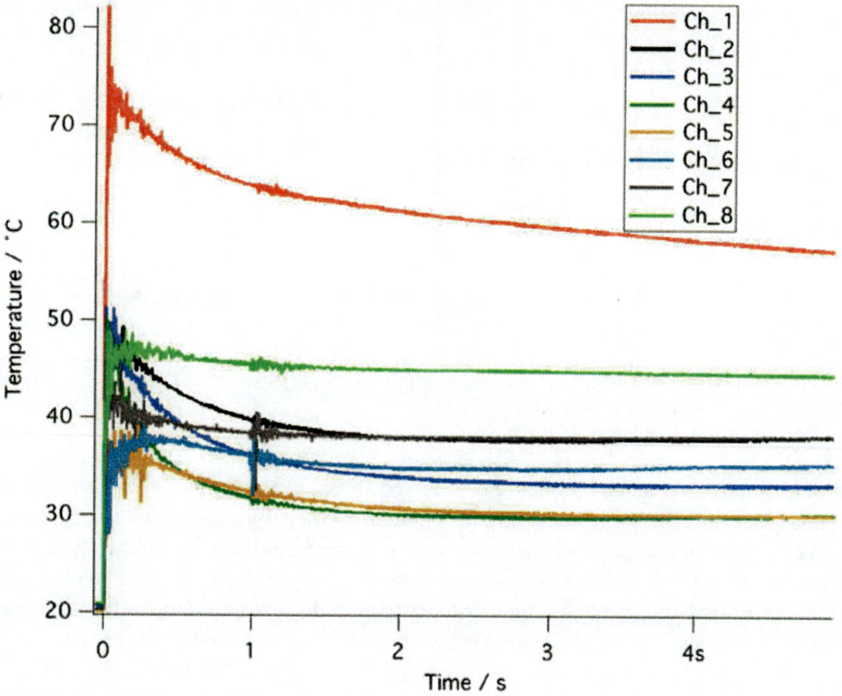


**Fig. 52.6** Thermocouple temperature measurements measured in the wake of the projectile in the second test

Perhaps the biggest discrepancy was that at high velocities our simulation showed significant bulging of the bottom of the mock HE when no solid back surface was applied. In the field of penetration, the bulge is referred to as "plug formation", which is common in materials that have a weak shear strength but a strong compressive strength. It is clear that our HE model overestimates this effect and must be accordingly modified.

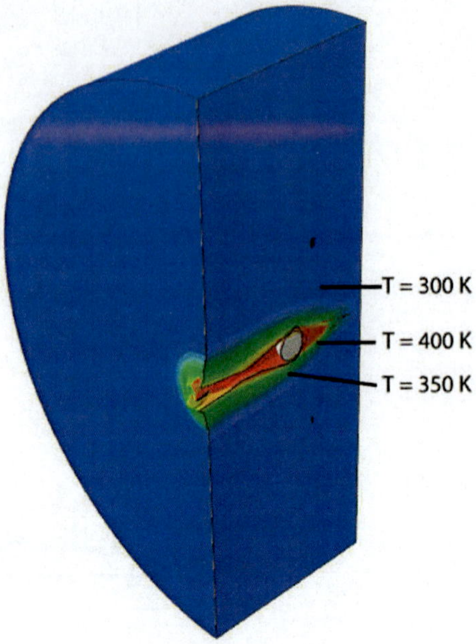


**Fig. 52.7** FEA predicted temperature field from ViscoSCRAM/ABAQUS simulation, excluding highly distorted elements

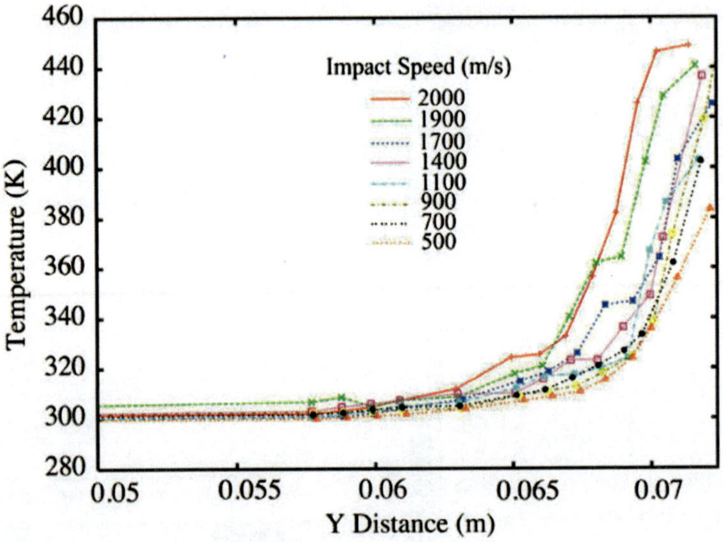


**Fig. 52.8** FEA predicted temperature distribution from projectile path (at *right*)

Temperatures are in qualitative agreement with those observed in the experiments as long as highly distorted elements were omitted (about one element thickness away from the penetration hole). By varying the impact speed, a measure of the temperature increase as a function of the speed could be examined. Figure 52.8 shows the temperature profile in the vicinity of the impact hole as a function of projectile velocity. The temperature of the closest element, as mention, was unphysical and thus omitted. The peak temperature values agree well with the experimental values. Additionally, the temperature gradient is approximately linear from the projectile path, decaying approximately 20–30°C above room temperature within ~5 mm.

## 52.4 Discussion

The measured peak temperature in both tests, ~60°C, agrees well with the simulation results, when highly distorted elements are ignored as non-physical. This agreement is encouraging for current application of the model, but more work is warranted to separately calibrate the temperature contribution of each of the thermally relevant deformation mechanisms of the material. Future work will also be applied to similar integrated impact tests of reactive energetic materials to test predictions of the numerical reactive codes under these conditions.

Element distortion was a common problem in the simulations and various means of controlling distortion was investigated. None were found to be overly satisfactory. For example, the method of element deletion could be used to remove highly distorted elements but as a result the penetration depth was typically significantly overestimated. Future calculations may attempt to use particle methods to overcome this problem.

There was little qualitative difference in the thermocouple response in the mid-plane arrested projectile, and the response recorded in the wake of the projectile. It was anticipated that the temperature field could continue to evolve in the highly deformed wake of material left in the path of penetration, but this appeared not to be an important factor.

It is clear that the temperature rises associated with deformation and flow of mock HE material during penetration at velocities ~700–750 m/s are relatively modest (~60°C). The polymer binder is a complex material, but it is unlikely that the temperature rise observed is strongly affected by the various endotherms and exotherms in the material that exist between room temperature and the bulk melting temperature of Estane rubber at ~105°C, particularly when the volume fraction of binder is so low. The other constituents of the mock, Barium Nitrate and Melamine have decomposition or transition temperatures above 350°C.

The low-pass filter bandwidth of the recording instrument, 500 Hz, limited the extent to which the temperature spike could be quantified. It is possible that temperature dynamics occurring over time scales less than 2 ms are outside the measurement bandwidth of the thermocouple amplifier but nevertheless present in the experiment. In the highest peak measurements, even from backward extrapolating the measured trace after the peak, a temperature rise of more than 80°C is unlikely. Insensible but real local temperature dynamics could have a significant effect on the predictions of the numerical reactivity code, and so further experiments with greater experimental bandwidth are planned to clarify this issue. Other advancements to the experiment for future work include a more sophisticated means of controlling excessive element deformation and removing highly distorted or damaged elements in a physically realistic and numerically stable manner.

**Acknowledgements** Los Alamos National Laboratory is operated by LANS, LLC, for the NNSA of the US Department of Energy under contract DE-AC52-06NA25396. This research was supported by Campaign 2: Dynamic Behavior of Materials and the Joint DoD/DOE Munitions Program.

## References

1. Cooper P (1996) Explosives engineering. Wiley-VCH, New York
2. Blaine Asay (2010) Shock wave science and technology reference library, vol 5, Non-shock initiation of explosives. Springer, New York
3. See Chapter 3 of Ref. 2
4. Bennett JG, Haberman KS, Johnson JN, Asay BW, Henson BF (1998) A constitutive model for the non-shock ignition and mechanical response of high explosives. J Mech Phys Solids 46:2303

# Chapter 53
# Blast Testing of Three Types of Panels with External Strengthening Systems

N.L. Carey and J.J. Myers

**Abstract** This research investigated the behavior of hybrid, plain, and steel fiber-reinforced concrete panels coated with various polyurea systems under blast loading. Four coatings were assessed including two plain polyureas and two discrete fiber-reinforced polyurea systems. Chopped E-glass fibers were discretely integrated in with the polyurea to develop a fiber-reinforced polyurea system. The addition of glass fiber to a polymer coating provides improved stiffness and strength to the composite system while the polyurea base material provides ductility. Seventeen panels were tested at the Missouri University of Science and Technology's experimental mine using a charge weight of 1.36 kg of C4 at a 305 mm standoff distance. Hybrid panels demonstrated higher blast mitigation performance compared to plain and steel fiber-reinforced concrete panels. Addition of plain polyurea or discrete fiber-reinforced polyurea system on the tension side improved panel performance by containing fragmentation during a blast event. Discrete fiber-reinforced polyurea systems exhibited less bulging compared to the plain polyureas due to higher stiffness.

**Keywords** Blast mitigation • Composite strengthening • Discrete fiber-reinforced polyurea • Steel fiber-reinforced concrete • Hybrid panel systems • Blast testing

## 53.1 Introduction

Over the past decade, the world has witnessed unprecedented levels of devastation from both natural disasters and terrorist attacks. The resulting fatalities and destruction have increased public awareness of the dire need for drastic improvements in infrastructure design and security. Thus, civil infrastructure may require structural strengthening or repair during service life in order to enhance the performance, which is a more economical solution compared to complete replacement.

Research initiatives have been advanced to investigate new materials that can be used for hazard mitigation (blast and/or seismic), strengthening/upgrade, or for general repair-retrofit applications. This research project focused on assessing the effectiveness of various blast mitigation materials and coating technologies to be used for enhancing the blast resistance of structural members. This study examined the behavior of hybrid, plain, and steel fiber-reinforced concrete (SFRC) panels coated with various polyurea systems under blast loading. Four coatings were evaluated including two plain polyureas and two discrete fiber-reinforced polyurea (DFRP) systems. Chopped E-glass fibers were discretely integrated in with the polyurea to develop a fiber-reinforced polyurea system. The addition of glass fiber to a polymer coating provides improved stiffness and strength to the composite system while the polyurea base material provides ductility [1]. Polyurea coating systems were previously evaluated for blast mitigation and impact resistance at Missouri University of Science and Technology (Missouri S&T), and were shown to be adequate in containing spalling and reducing fragmentation from blast and repeated impact [2, 3]. Previously, the authors and others have investigated the use of polyurea systems without discrete fiber for hazard mitigation and general strengthening [2–7]. Elastomeric polyurea coating is an elastic, ductile, and energy absorption material. It is corrosion resistant, durable, and lightweight. Several key characteristics include chemical and water resistance, excellent elongation, and quick curing. Because it is a low-viscosity liquid, it bonds very quickly and evenly during

N.L. Carey • J.J. Myers (✉)
Civil, Architectural, and Environmental Engineering, Missouri University of Science and Technology,
325 Butler Carlton Hall, 1401 North Pine Street, Rolla, MO 65409, USA
e-mail: jmyers@mst.edu

V. Chalivendra et al. (eds.), *Dynamic Behavior of Materials, Volume 1: Proceedings of the 2012 Annual Conference on Experimental and Applied Mechanics*, Conference Proceedings of the Society for Experimental Mechanics Series,
DOI 10.1007/978-1-4614-4238-7_53, © The Society for Experimental Mechanics, Inc. 2013

application. Polyurea coating is capable of withstanding regular thermal or dynamic movement as well. Despite requiring special equipment, experienced operators for mixing, and thorough surface preparation, application is easy and the material cures rapidly. Also, the material is not very sensitive to temperature and humidity during installation and service, which eliminates the need for additional arrangements for application procedures and curing environments. Polyurea also aids in the confinement of post-blast materials in compression loaded structures, which produces a residual load-bearing capability.

Furthermore, three panel types were investigated, including plain reinforced concrete (RC), SFRC, and hybrid system. Hybrid system was a bi-layered construction consisting of a SFRC base and a top layer of wood fiber-fly ash (WF-FA) material to serve as a sacrificial layer. WF-FA is a hybrid material recently developed at Missouri S&T. Hybrid panels were subjected to blast loading to evaluate the effectiveness of an additional sacrificial layer combined with SFRC base. Hybrid panel systems demonstrated enhanced performance compared to plain RC panels by reducing fragmentation and overall structural damage during impact and blast testing [2, 3]. The hybrid material's ability to absorb large amounts of energy limits the destruction caused by impact and/or blast loading; therefore, its use would be beneficial for infrastructure protection and construction applications. The development of energy absorbent construction materials is vital in providing enhanced protection from an explosive blast. This research also focused on evaluating the effectiveness of SFRC to resist damage during blast testing. Concrete is known to have low tensile strength and poor performance under dynamic tensile stresses, including repeated missile impacts, earthquake, and blast. Adding internal fibers to the concrete (i.e. fiber-reinforced concrete) has been shown to significantly increase energy absorption and ductility, improving performance, toughness, and impact resistance [8]. Moreover, fiber reinforcement enhances performance during both impact and blast testing [9, 10] and has been found to increase flexural and shear strength, stiffness, and reduce deflection and cracking [11–14].

## 53.2 Experimental Program

The main objective of this study was to examine the performance of hybrid, plain, and SFRC panels coated with various polyureas and DFRP systems under blast loading. Seventeen panels were fabricated and tested during this experimental activity. Table 53.1 provides detailed descriptions of the test specimens and investigated coating systems. The nominal dimensions of each panel were 1,180 × 1,180 mm. All panels included Grade 420 MPa 9.5 mm (ASTM No. 3) steel bars for longitudinal reinforcement. The first two sets consisted of 90 mm thick plain and SFRC panels with a 0.5% reinforcement ratio in each direction. The 3rd set consisted of hybrid panel systems composed of a 90 mm thick SFRC base with a 0.5% reinforcement ratio in both directions and a 51 mm sacrificial top layer of high-volume WF-FA material on the impact/blast face with a 0.16% reinforcement ratio in one direction only. Four coating configurations were evaluated consisting of two various elastomeric polyureas and two DFRP systems. DFRP systems consisted of chopped E-glass fibers discretely integrated in with a polyurea matrix.

**Table 53.1** Blast panel test matrix

| | |
|---|---|
| 1st set | Plain concrete (90 mm thick) |
| | 1 – Control specimen [1-1] |
| | 1 – Coated with polyurea A [1-2] |
| | 1 – Coated with discrete fiber-reinforced polyurea A system [1-3] |
| | 1 – Coated with polyurea B [1-4] |
| | 1 – Coated with discrete fiber-reinforced polyurea B system [1-5] |
| 2nd set | Concrete with Propex Novocon® 1050 FE steel fiber (90 mm thick) |
| | 1 – Control specimen [2-1] |
| | 1 – Coated with polyurea A [2-2] |
| | 1 – Coated with discrete fiber-reinforced polyurea A system [2-3] |
| | 1 – Coated with polyurea B [2-4] |
| | 1 – Coated with discrete fiber-reinforced polyurea B system [2-5] |
| | Plain Concrete (90 mm thick) |
| | 1 – Control specimen [2-6] |
| 3rd set | Hybrid (90 mm conc. with Propex Novocon® 1050 FE steel fiber & 51 mm WF-FA) |
| | 1 – Control specimen [3-1] |
| | 1 – Control specimen [3-2] |
| | 1 – Coated with polyurea A [3-3] |
| | 1 – Coated with discrete fiber-reinforced polyurea A system [3-4] |
| | 1 – Coated with polyurea B [3-5] |
| | 1 – Coated with discrete fiber-reinforced polyurea B system [3-6] |

Key: [set no. – specimen no.]

Table 53.2 Material mechanical properties at test age

| Panel type | | | Compressive strength, MPa | MOE, GPa | MOR, MPa |
|---|---|---|---|---|---|
| 1st set | Plain RC control panel [1-1] | | 23.6 | – | – |
| | Plain RC panels [1-2]–[1-5] | | 21.4 | 20.8 | 3.0 |
| 2nd set | Plain RC control panel [2-6] | | 29.2 | – | – |
| | SFRC panels [2-1]–[2-5] | | 26.5 | 21.2 | 3.5 |
| 3rd set | Hybrid control panel [3-1] | WF-FA | 11.6 | 7.0 | – |
| | | SFRC | 16.8 | 21.7 | 2.5 |
| | Hybrid control panel [3-2] | WF-FA | 10.7 | 4.9 | – |
| | | SFRC | 28.4 | 26.1 | – |
| | Hybrid (coating system: polyurea A [3-3], DFRP A system [3-4]) | WF-FA | 10.8 | 7.2 | – |
| | | SFRC | 29.8 | 26.7 | 3.2 |
| | Hybrid (coated with polyurea B [3-5]) | WF-FA | 11.8 | 7.0 | – |
| | | SFRC | 16.8 | 21.7 | 2.5 |
| | Hybrid (coated with DFRP B system [3-6]) | WF-FA | 10.8 | 7.0 | – |
| | | SFRC | 16.8 | 21.7 | 2.5 |

Key: [set no. – specimen no.]

Table 53.3 Steel fiber mechanical properties

| | |
|---|---|
| Fiber length, mm | 50 |
| Equivalent diameter, mm | 1.0 |
| Tensile strength, MPa | 1,050 |

Table 53.4 Mechanical properties of elastomeric polyureas

| Material | Tensile strength, MPa [15] | Elongation (%) [15] | Gel time (sec) | Tack free time (sec) |
|---|---|---|---|---|
| Polyurea A | 14.8 | 91 | 3–6 | 6–9 |
| Polyurea B | 19.3–20.7 | 430–445 | 11–13 | 78–85 |

### *53.2.1 Material Properties*

Seventeen panels were fabricated using standard ready-mix concrete and WF-FA material during hybrid panel system fabrication. WF-FA mix design was selected from a previous study conducted by Tinsley and Myers [3]. The average compressive strength, modulus of elasticity (MOE), and modulus of rupture (MOR) results at test age for each material are summarized in Table 53.2.

Propex Novocon® 1050 FE (flat end) steel fiber was used during SFRC batching. Novocon® 1050 FE is a cold drawn wire fiber with flat ends that provide an optimal anchorage within the concrete mix. Table 53.3 provides steel fiber mechanical properties as specified on product data sheet. During SFRC panel fabrication, steel fibers were added on site after concrete batching. Fibers were placed inside the drum of the concrete truck and mixed with fresh concrete for sufficient amount of time (min. 75 rotations at full mixing speed) to ensure uniform distribution of the fibers throughout the concrete. The dosage rate of 50 kg per cubic yard of concrete was used as suggested by the fiber manufacturer.

Four various coating configurations were investigated in this study consisting of two plain polyureas and two DFRP systems. Two types of elastomeric polyurea were examined (polyurea A and polyurea B) as prescribed in a previous fiber length optimization study undertaken by Carey and Myers [1]. Some panels were coated with plain polyurea A or polyurea B, and some with DFRP A or DFRP B systems. Both polyureas were two-component 100% solid reactive cures. Mechanical properties provided by the manufacturer for each elastomeric polyurea are displayed in Table 53.4. DFRP systems consisted of chopped E-glass fibers integrated into the polyurea matrix. Multi-end continuous low static E-glass fiber roving designed specifically for spray-up applications was used. Mechanical properties of the glass fiber provided by the manufacturer are listed in Table 53.5. Fiber length of 6.4 mm was used during fiber-reinforced polyurea coating application based on a fiber length optimization study conducted by Carey and Myers [1] concluding that samples with shorter fibers exhibited higher ductility due to the weaker bond between the fiber and matrix. During coating application, coupon specimens were fabricated to determine the mechanical properties of each coating system. In addition, ignition loss testing was conducted to determine the fiber volume fraction in DFRP specimens. Mechanical properties of DFRP systems are provided in Table 53.6 based upon experimental testing.

**Table 53.5** Mechanical properties of E-Glass fiber

| Mechanical properties | Dry range |
|---|---|
| Tensile strength [16], MPa | 59–98 |
| Tensile modulus [16], MPa | 7,542–14,893 |
| Flexural strength [17], MPa | 166–307 |
| Flexural modulus [17], MPa | 6,939–12,065 |

**Table 53.6** Mechanical properties of DFRP systems

| Coating designation | | Fiber volume fraction (%) | Modulus of elasticity, MPa | Max stress, MPa |
|---|---|---|---|---|
| 1st and 2nd set | DFRP A | 14.0 | 710.2 | 10.38 |
| | DFRP B | 7.0 | 723.9 | 7.94 |
| 3rd set | DFRP A | 9.5 | 475.7 | 5.94 |
| | DFRP B | 3.0 | 179.3 | 4.96 |

**Fig. 53.1** Coating application process. (**a**) Polyurea application process. (**b**) DFRP system application process

## 53.2.2 *Spray-on Polyurea Application*

Twelve panels were coated with 6.4 mm thick spray-on elastomeric polyurea systems by polyurea supplier/installer. Polyurea systems were applied on the tension side of each specimen, and in some cases along the sides to prevent coating debonding. In addition, a 0.05–0.08 mm thick primer was applied on the panel bottom surface prior to coating application to achieve an adequate and strong bond and to prevent delamination. During DFRP coating application, a mechanical chopper/spray gun was used to combine and simultaneously distribute a two-component polyurea mixture, chop the E-glass fiber roving, and disperse the glass fibers into the coating system creating a uniformly discretely integrated fiber-reinforced system. Coating application process is provided in Fig. 53.1. The coating systems were applied by experienced operator employing required safety procedures. The E-glass fiber roving used during specimen fabrication was specifically designed for chopping and spray-up applications. Fiber volume was adjusted during coating application by increasing the speed of fiber chopping and integration into the polyurea matrix.

## 53.2.3 *Blast Test Setup*

After specimen fabrication and coating with various polyurea systems, panels were transported to the Missouri S&T experimental mine for blast testing. Panels were tested using a charge weight of 1.36 kg of C4 at a 305 mm standoff distance as prescribed in a previous blast study by Lu [18] based on expected damage level. During testing, each panel was simply supported on the steel w-sections with a 64 mm bearing on each end. Charges were hung from the mine ceiling using a wire to a 305 mm standoff distance measured from the center of the explosive charge to the top face and centered over the test panel surface. Panel blast test setup is depicted in Fig. 53.2.

**Fig. 53.2** Blast test setup

## 53.3 Experimental Results

Plain RC panels (1st set) demonstrated surface cratering along the centerline on the top face of the panel after detonation due to crushing of the material by the compression from the shock wave and erosion by the gases from the blast. Most of the RC panels exhibited flexural failure during blast testing as a result of the initial pressure wave impact. Major surface spalling was observed in the center of the bottom face of the RC control panel [1-1] due to the shock wave propagation all the way through the material and the subsequent tensile reflections from the near free-face (see Fig. 53.3a). Tearing of the coating was observed only in one case, on the bottom/tension face of the RC panel coated with DFRP B system [1-5], due to internal voids formed within the system during the application when insufficient amount of polyurea matrix material is combined with a lot fiber or due to lower strength and higher ductility of the coating compared to DFRP A system (see Fig. 53.3b, c).

The second set consisted of five SFRC panels and one plain RC panel comparable strength. All panels in the second set exhibited minor surface cratering along the centerline on the top face of the panel after detonation due to crushing of the material by the compression from the shock wave and erosion by the gases from the blast. RC control panel [2-6], SFRC control panel [2-1], and SFRC panel coated with elastomeric polyurea A [2-2] exhibited flexural failure as a result of the initial pressure wave impact. Major surface spalling was observed in the center of the bottom face of the RC control panel [2-6] and SFRC control panel [2-1] due to the shock wave propagation all the way through the material and the subsequent tensile reflections from the near free-face (see Fig. 53.4a, b). Coating delamination and tearing were not observed on the bottom/tension face during the 2nd set panel testing (see Fig. 53.4c).

The third set consisted of six hybrid panels. Both hybrid control panels exhibited flexural failure after the charge detonation due to the initial pressure wave impact (see Fig. 53.5a, c). Hairline cracking was observed on the top face of WF-FA layer after detonation during each hybrid panel testing (see Fig. 53.5b). Hairline flexural cracks developed through the SFRC base along the midspan on the bottom/tension face of the control panels and panels coated with DFRP A and DFRP B systems. Coating bulging, delamination, and tearing were not observed on the bottom/tension face during the third set panel testing.

Fig. 53.3 The 1st set panel damage after detonation. (a) Panel 1-1 bottom face damage. (b) Panel 1-2 bottom face damage. (c) Panel 1-5 bottom face damage

Fig. 53.4 The 2nd set panel damage after detonation. (a) Panel 2-6 bottom face damage. (b) Panel 2-1 bottom face damage. (c) Panel 2-2 bottom face damage

Fig. 53.5 The 3rd set panel damage after detonation. (a) Panel 3-1 bottom face damage. (b) Panel 3-2 top face damage. (c) Panel 3-2 bottom face damage

## 53.4 Comparison of Test Results

The main objective of this research activity was to examine the effectiveness of hybrid material, SFRC, and coating systems for blast mitigation. Most of the panels exhibited flexural failure during blast testing. Residual midspan deflection and coating bulge size (in some cases) recorded for each panel are summarized in Table 53.7. Residual deflection was a permanent deflection following the blast event and was measured at the midspan from the top of the panel to the bottom of the crater (as applicable) using a straight edge level and a tape measure. Percent bulging was determined from the total

**Table 53.7** Residual deflection and bulge size summary

| Panel No., compressive strength, MPa | | Residual deflection, mm | Bulge size, mm | Total deformation, mm | Bulging % |
|---|---|---|---|---|---|
| 1st set | RC control [1-1], 23.6 MPa | 31 | – | 31 | – |
| | RC with polyurea A [1-2], 21.4 MPa | 25 | 32 | 56 | 56 |
| | RC with DFRP A [1-3], 21.4 MPa | 46 | 44 | 91 | 49 |
| | RC with polyurea B [1-4], 21.4 MPa | 64 | 76 | 140 | 55 |
| | RC with DFRP B [1-5], 21.4 MPa | 48 | 64 | 111 | 57 |
| 2nd set | RC control [2-6], 29.2 MPa | 32 | – | 32 | – |
| | SFRC control [2-1], 26.5 MPa | 38 | – | 38 | – |
| | SFRC with polyurea A [2-2], 26.5 MPa | 36 | 57 | 93 | 61 |
| | SFRC with DFRP A [2-3], 26.5 MPa | 12 | 14 | 26 | 54 |
| | SFRC with polyurea B [2-4], 26.5 MPa | 19 | 35 | 54 | 65 |
| | SFRC with DFRP B [2-5], 26.5 MPa | 13 | 21 | 33 | 62 |
| 3rd set | Hybrid control [3-1], SFRC = 16.8 MPa | 15 | – | 15 | – |
| | Hybrid control [3-2], SFRC = 28.4 MPa | 9 | – | 9 | – |
| | Hybrid with Poly. A [3-3], SFRC = 29.8 MPa | 5 | – | 5 | – |
| | Hybrid with DFRP A [3-4], SFRC = 29.8 MPa | 2 | – | 2 | – |
| | Hybrid with Poly. B [3-5], SFRC = 16.8 MPa | 2 | – | 2 | – |
| | Hybrid with DFRP B [3-6], SFRC = 16.8 MPa | 1 | – | 1 | – |

Key: [set no. – specimen no.]

deflection to evaluate coating behavior. During the analysis of the 1st set of panels, it was observed that plain concrete control specimen demonstrated less top surface cratering compared to the other four coated specimens due to slightly higher concrete strength. All six panels in the second set demonstrated similar top surface cratering, but SFRC panels exhibited less cracking on the top face compared to plain RC panels. The hybrid panels, third set, showed much higher performance compared to plain RC and SFRC panels. The hybrid control panel demonstrated lower deflection and lack of spalling of SFRC layer on the back face compared to plain concrete and SFRC control specimens. WF-FA combined with SFRC exhibited energy absorption and ductile characteristics based on low deflections and lack of bulging. The addition of plain polyurea and DFRP systems on the tension side improved panel performance by containing fragmentation during a blast event. DFRP systems exhibited less bulging compared to the plain polyurea coating due to higher stiffness, as shown in Table 53.7 by determining the ratio of the bulge size to the total deformation. DFRP A system demonstrated less bulging compared to DFRP B system due to higher fiber volume fraction. In the case of plain RC panel coated with DFRP B system, the tearing in the coating was a result of internal voids formed within the system during the application process or due to lower strength and higher ductility of the coating compared to DFRP A system. Based on the results, DFRP A system was determined to be the most advantageous for blast mitigation purposes.

## 53.5 Conclusions

In this research study, the behavior of hybrid, plain, and SFRC panels coated with various polyureas and DFRP systems was investigated for blast mitigation aspects. SFRC panels exhibited less cracking on the top face compared to plain RC panels. Hybrid panels demonstrated higher performance compared to plain RC and SFRC panels. WF-FA combined with SFRC exhibited energy absorption and ductile characteristics based on low deflections and the lack of bulging. The addition of plain polyurea or DFRP system on the tension side improved panel performance by containing fragmentation during a blast event. DFRP systems exhibited less bulging compared to plain polyurea coatings due to higher stiffness. DFRP A system demonstrated less bulging compared to DFRP B system due to higher fiber volume fraction.

## References

1. Carey NL, Myers JJ (2011) Discrete fiber reinforced polymer systems for repair of concrete structures: *polyurea-fiber characterization results*. American Concrete Institute Special Publication-275, March 2011, pp. 285–298
2. Carey NL, Myers JJ (2009) Blast and impact resistance of hybrid systems for barrier and wall panel applications. Center for Infrastructure Engineering Studies (CIES), Missouri University of Science and Technology, Rolla, 225 pp

3. Tinsley ME, Myers JJ (2007) Investigation of a high-volume fly ash-wood fiber material subjected to low-velocity impact and blast loads. Center for Infrastructure Engineering Studies (CIES), University of Missouri – Rolla, Rolla, 178 pp
4. Coughlin A (2008) Contact charge blast performance of fiber reinforced and polyurea coated concrete vehicle barriers. Master's Thesis, Pennsylvania State University, University Park, 123 pp
5. Hrynyk TD, Myers JJ (2007) Static evaluation of the out-of-plane behavior of URM infill walls utilizing modern blast retrofit systems. Center for Infrastructure Engineering Studies (CIES), University of Missouri – Rolla, Rolla, 202 pp
6. Tanizawa Y, Myers JJ (2009) In-plane behavior of strengthened unreinforced masonry infiill walls with alternative wood-fiber fly ash material. Center for Infrastructure Engineering Studies (CIES), Missouri University of Science and Technology, Rolla, 110 pp
7. Viswanath T (2007) Experimental study on the impact resistance of polyurea coated concrete. Master's thesis, Pennsylvania State University, University Park, 69 pp
8. ACI Committee 544 (1996) State-of-the-art report on fiber reinforced concrete (ACI 544.1R-96). American Concrete Institute, Farmington Hills, Michigan, 65 pp
9. Lok TS, Xiao JR (1999) Steel-fibre-reinforced concrete panels exposed to air blast loading. Proc Inst Civil Eng Struct Build 134(4):319–331
10. Luo X, Sun W, Chan SYN (2000) Characteristics of high-performance steel fiber-reinforced concrete subject to high velocity impact. Cem Concr Res 30(6):907–914
11. Campione G, Mangiavillano ML (2008) Fibrous reinforced concrete beams in flexure: experimental investigation, analytical modelling and design considerations. Eng Struct 30(11):2970–2980
12. Dinh HH, Parra-Montesinos GJ, Wight JK (2010) Shear behavior of steel fiber-reinforced concrete beams without stirrup reinforcement. ACI Struct J 107(5):597–606
13. Imam M, Vandewalle L, Mortelmans F (1994) Shear capacity of steel fiber high-strength concrete beams. ACI Spec Pub 149:227–242
14. Susetyo J, Gauvreau P, Vecchio FJ (2011) Effectiveness of steel fiber as minimum shear reinforcement. ACI Struct J 108(4):488–496
15. ASTM International (2006) Standard test methods for vulcanized rubber and thermoplastic elastomers—tension (ASTM D 412-06a). American Society for Testing and Materials, West Conshohocken, 14 pp
16. ASTM International (2010) Standard test method for tensile properties of plastics (ASTM D 638-10). American Society for Testing and Materials, West Conshohocken, 16 pp
17. ASTM International (2010) Standard test methods for flexural properties of unreinforced and reinforced plastics and electrical insulating materials (ASTM D 790-10). American Society for Testing and Materials, West Conshohocken, 11 pp
18. Lu B (2005) Application of displacement-based design method to blast-resistant reinforced concrete structures. Dissertation, Presented to University of Missouri – Rolla, Rolla, in partial fulfillment of the requirements for the degree of Doctor of Philosophy, 104 pp

# Chapter 54
# Ultra-High-Speed Imaging for Explosive-Driven Shocks in Transparent Media

**M.J. Murphy and S.A. Clarke**

**Abstract** An ultra-high-speed imaging system employing laser backlighting with spoiled coherence is developed to visualize shock waves driven into transparent media, such as air and polymethylmethacrylate (PMMA), through shock and detonation interaction. Resulting time-resolved image sequences yield two-dimensional visualizations of shock geometry as well as quantitative measurements of shock position history. One-dimensional measurements of detonator-driven shock position in both air and PMMA have recently been validated using simultaneous photonic Doppler velocimetry (PDV) measurements and predictive calculations. The imaging system is currently being utilized to visualize shock waves driven into air and PMMA samples through multiple interactions with detonating explosives. System development, implementation, and experimental data are discussed.

**Keywords** SWIFT • Explosive drive • Ultra-high-speed • Shock wave • Pseudo-streak • PMMA

## 54.1 Introduction

With the intent of experimentally characterizing the performance of initiators, detonators, and high explosives (HE), an early version of a laser-backlit schlieren system was introduced in [1] to outline key aspects of the implemented technique critical to successful visualization of explosive events. As presented, the system primarily provided qualitative data that allowed general comparisons to be made between flows generated by different explosive insults. The current shock wave image framing technique (SWIFT) developed in Detonator Technology at Los Alamos National Laboratory (LANL) combines ultra-high-speed image recording with spoiled-coherence laser backlighting to directly visualize explosive output directed into transparent media with multi-frame resolution and a much higher image quality than the original system. The qualitative information inherent in SWIFT images as simple flow visualizations provides direct observation of the complex flows under investigation, and data-reduction procedures have been developed to extract numerous quantitative data sets directly from the recorded images.

Implementation of the imaging system, a small range of experimental results, and examples of reduced quantitative data for explosive-driven shocks in both air and polymethylmethacrylate (PMMA) samples are presented in the ensuing discussion. Attention is given to experimental SWIFT data sets and the physical insights that stem from directly observing explosive-driven shock waves as they evolve in both space and time. The main goal of this report is to introduce SWIFT as a means of obtaining unique insights into explosive-drive phenomena, as well as any application involving strong shock waves propagating in transparent media.

M.J. Murphy (✉) • S.A. Clarke
W-6 Detonator Technology, Los Alamos National Laboratory, MS-P950, Los Alamos, NM 87545, USA
e-mail: mjmurphy@lanl.gov

V. Chalivendra et al. (eds.), *Dynamic Behavior of Materials, Volume 1: Proceedings of the 2012 Annual Conference on Experimental and Applied Mechanics*, Conference Proceedings of the Society for Experimental Mechanics Series,
DOI 10.1007/978-1-4614-4238-7_54, 

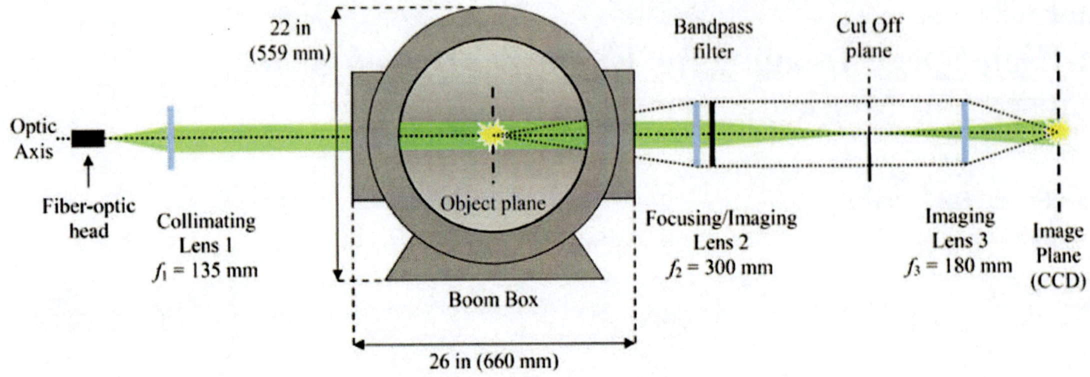


**Fig. 54.1** Schematic of SWIFT optical configuration for detonator/HE investigation

## 54.2 Experimental

The optical configuration for SWIFT measurements mirrors an inline lens-based schlieren system, as described in [2]. A schematic is included in Fig. 54.1 for reference, where the optics have been chosen to investigate length scales relevant to detonators (~ 10 mm field of view). High-quality Nikon lenses are employed to both avoid optical aberrations in the imaging train, as well as to take advantage of precise Nikon F-mount architecture. As a result, every aspect of the imaging system from light collimation to stand-off distance to magnification can be rapidly changed by appropriate lens selection without the need to re-align the inline optical train.

With regard to system magnification ($M_0$), the physical stand-off distance (primarily defined by the axial dimension of the "boom box" used to contain the explosive events) is chosen to match the infinite-focus focal length of the first imaging lens (Lens 2 in Fig. 54.1). As such, dual-lens imaging can be utilized in a manner analogous to infinity-corrected optical systems in microscopy, where the magnification is determined as the ratio of focal lengths of the second imaging lens to the first, i.e. $M_0 = f_3/f_2$. Precise matching of the stand-off distance to $f_2$ is not critical since the second imaging lens can be manually adjusted via the focusing ring to obtain sharp focus, with the corresponding magnification varying only slightly from $f_3/f_2$.

There are two critical system components in the SWIFT implementation that directly affect the quality of the recorded data. The first is a high-quality ultra-high-speed camera system employing multiple intensified image sensors that are well-aligned such that errors due to perspective, channel registration, and parallax are minimal. Such a camera system allows multi-frame image sequences to be recorded from a single event, which provides visualization of the temporal evolution of the flow. The employment of image intensifiers on the sensors allows ultra-short (5 ns) exposures to be utilized, which provide dual advantages for visualizing explosive events. First, nanosecond-scale exposures are required to freeze the motion of the supersonic shock waves in order to avoid blurring of the shock fronts during recording. Second, such short exposures are required to minimize light saturation of the sensors due to intense explosive luminance. In current SWIFT experiments, a SIMD camera system (Specialised Imaging) is employed that utilizes eight CCD sensors operating in dual-frame capacity to obtain 16-frame image sequences at frame rates up to 16 MHz, or 8-frame sequences at frame rates up to 200 MHz. Each image has nominally 1.3 Mpixel resolution, a 12-bit bit depth, and is recorded using 5 ns exposures. As depicted in Fig. 54.1, a precision optical band-pass filter (Andover 532FS02-50) with a bandwidth of 1.0 +/− 0.2 nm is used to block nearly all broadband explosive luminance from entering the camera.

The second critical component in the SWIFT setup is a customized 5 W continuous-wave laser (Spectra-Physics Reveal 5) having spoiled coherence and fiber-optic delivery for backlighting the experiments. The absence of coherence in the laser light averts major pitfalls that coherent light introduces into schlieren applications, as summarized in [2] and studied in [3]. The fiber-optic light delivery (SMA connector) provides diverging light from a finite-sized aperture (400 μm), which simplifies light collimation by eliminating the need for either focusing optics or a spatial filter (lack of coherence alleviates diffraction effects). As a result, for schlieren applications the end-user is left with an intense, monochromatic light source of finite-sized aperture that is suitable for backlighting explosive events having high self-luminance and supersonic shock motion.

It is important to note that in many SWIFT experiments, where detonators or HE are investigated by driving shock waves into transparent solids, a schlieren cutoff is typically not employed due to the high refractivity of the solid media and the high strength and curvature of the shock waves. In other words, the shock fronts refract the collimated light by such an extent that one visualizes a physical boundary of material density expanding outward, and a schlieren cutoff is not required to visualize this boundary since the refracted light fails to be captured by the aperture of the first imaging lens. Conversely, a conventional schlieren cutoff (typically a knife edge) is commonly utilized for initiators and detonators directed into gaseous media (low refractivity) when the shadowgraph effect due to the large spatial extent of the flow [2] is insufficient to clearly visualize the shock fronts. Recent work by [4] specifically focuses on extending schlieren and shadowgraph methodologies to shock waves in transparent, inert solids in order to extract shock Hugoniot parameters for the solids. Though the authors share a strong interest in this form of analysis for extracting Hugoniot information from SWIFT datasets, the challenges inherent in capturing shock-refracted light in solids resulting from detonation interaction have yet to be overcome.

## 54.3 Results and Discussion

As a typical detonator functions, a steady detonation wave developed in the detonating explosive interacts with the metal cup that contains the explosive and shock-compresses the metal to pressures exceeding the steady-state detonation condition, i.e. the Chapman-Jouguet (CJ) state [5]. By performing two distinct series of SWIFT experiments, where detonator output is directed into solid PMMA and ambient air, two drastically different flows resulting from the same functioning detonator can be investigated to facilitate understanding of how the detonator performs, as well as to compare with predictive calculations.

Figure 54.2 displays image sequences corresponding to the same detonator operated into PMMA and air using the same firing conditions. In PMMA, the images depict the outward expansion of a shock wave that forms at the interface between the metal housing and the PMMA sample due to a mismatch in the shock compressibility of each material. In air, the SWIFT images depict the outward expansion of the metal cup while it accelerates as an explosively-driven flyer, as well as the transient onset of a detached shock wave that forms ahead of the accelerating metal. Both image sequences suggest the primary output of the detonator has axially symmetric geometry, since no effort was made to key the output to the recording

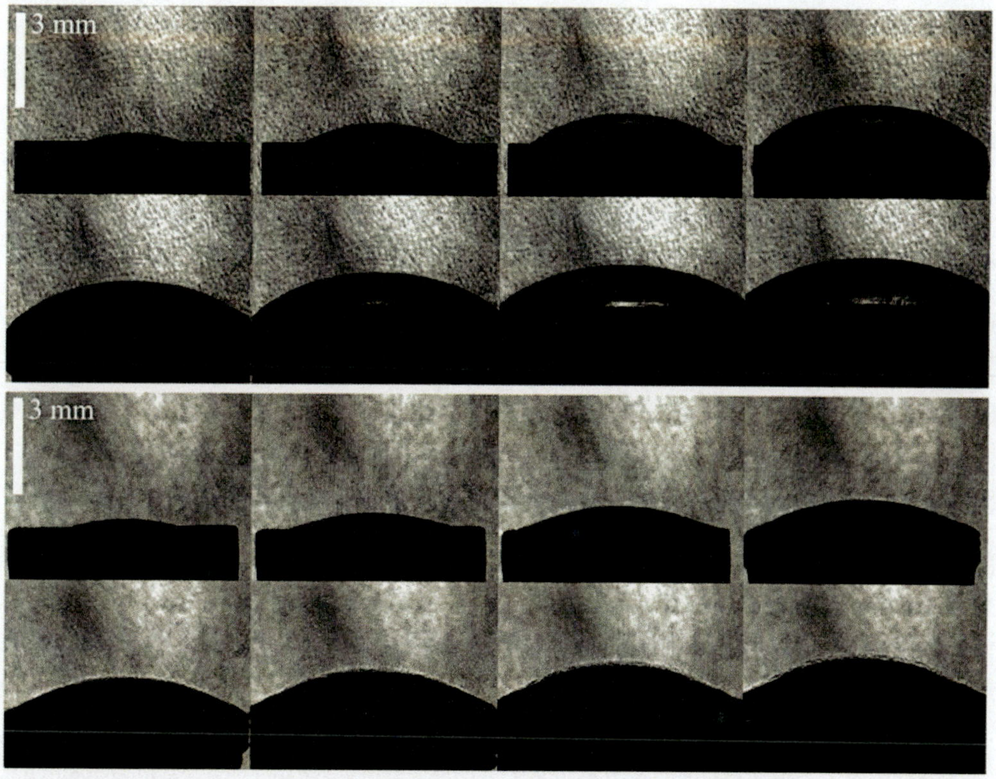


**Fig. 54.2** SWIFT images of detonator output into (*top*) PMMA and (*bottom*) air using 70 ns inter-framing

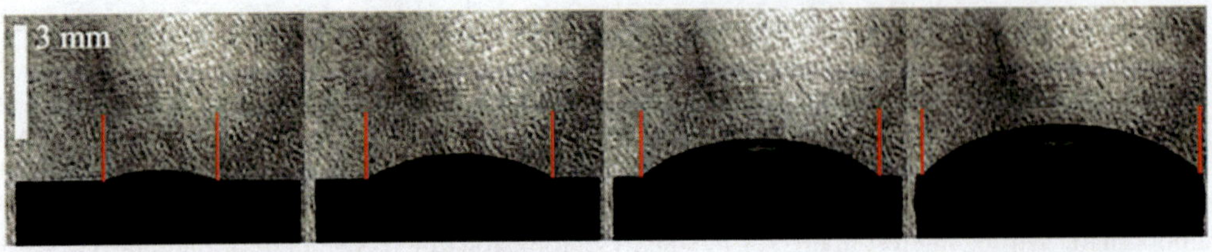


**Fig. 54.3** Subset of SWIFT image sequence for detonator output into PMMA with *vertical lines* denoting lateral shock position across the detonator surface

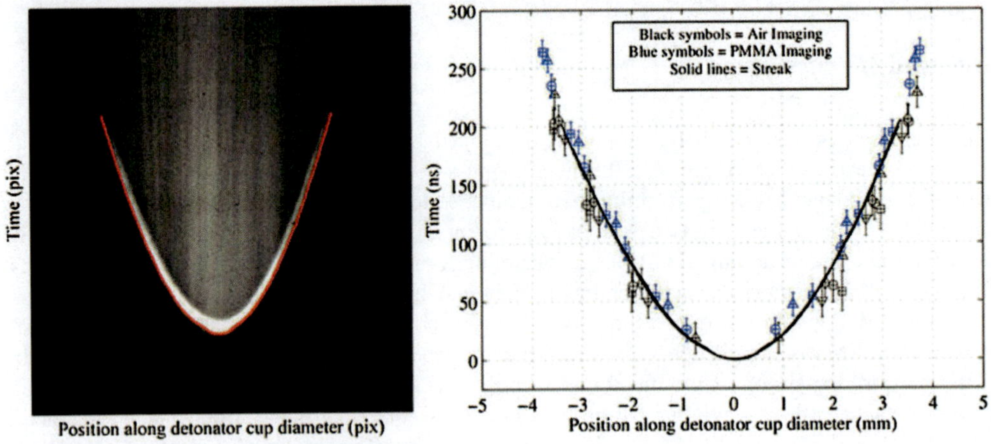


**Fig. 54.4** (*Left*) Experimental streak-camera data with an edge-detected curve overlay and (*right*) pseudo-streak data extracted from seven different SWIFT experiments plotted with the actual streak data

camera. Additionally, the results suggest that secondary shocks and/or wall effects inside the detonator have little impact on the behavior of the leading flow features.

In extracting precise quantitative data from SWIFT images, one-dimensional measurements of vertical shock position in PMMA and flyer position in air have been previously reduced to shock and free-surface velocities and compared to simultaneous photonic Doppler velocimetry (PDV) data with very good agreement [6]. In the same work, mass velocities in the detonator cup and shock velocities in the PMMA samples were converted to shock pressures using well-known velocity-Hugoniot parameters [7] and momentum conservation in order to compare to detonation- and shock-interaction calculations, as well as 1-D numerical simulations.

In this work, a different analysis is presented where 1-D detonator breakout across the surface of the metal cup is measured and compared to streak camera recordings taken across the detonator surface. Figure 54.3 includes the first four frames of the SWIFT data for the PMMA experiment displayed in Fig. 54.2, where red lines are included to track the lateral position of the shock wave driven into the metal cup from the supporting detonation wave. By defining the lateral-position origin as the geometric center of the shock wave and the reference time as the instant of first motion of the metal cup center (obtained from simultaneous PDV data), measurements of lateral shock position are paired with corresponding exposure times and plotted in position-time space as pseudo-streak data. Results from several different SWIFT experiments are included in Fig. 54.4 (right), and are plotted to overlay actual streak-camera data measured from the same detonator (Fig. 54.4 left). Very good agreement is demonstrated that suggests pseudo-streak data reduced from SWIFT images can be used to quickly describe detonator breakout, as well as to validate experimental streak records.

To investigate methods in which a detonator can be used to initiate HE through impact loading rather than detonation interaction, an experiment is designed based on the observation in Fig. 54.2 that a metal detonator cup becomes a thin, explosively-driven flyer plate when the detonator is operated into media with minimal or no impedance, e.g. gas or vacuum. The point of interest is to directly visualize detonator output into both air and PMMA after the cup has interacted with a metal cutter assembly. While this general application of detonators is not novel, and is similar to the small-scale floret test assembly [8], there is scarce literature available on the subject, and the authors have found no previous studies demonstrating direct visualization of the output flow.

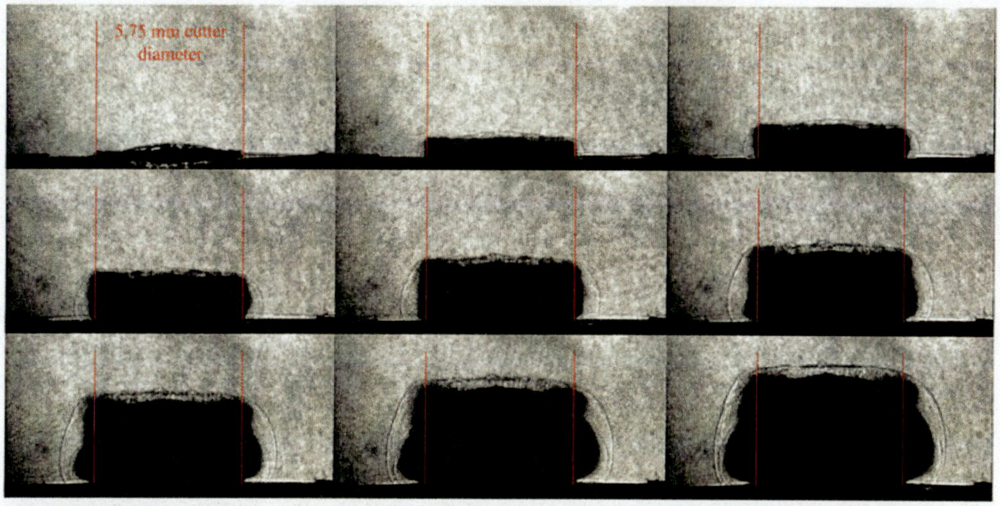


**Fig. 54.5** SWIFT images of the flow in air resulting from the interaction of a standard detonator with a stainless steel cutter assembly (*vertical lines* denote lateral locations of the cutter barrel)

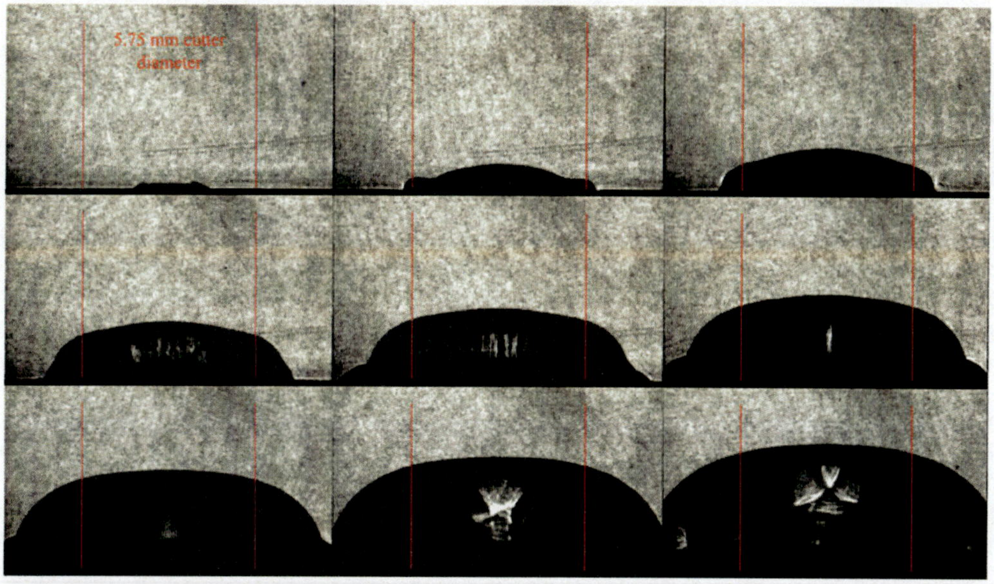


**Fig. 54.6** SWIFT images of the flow in PMMA resulting from the interaction of a standard detonator with a stainless steel cutter assembly (*vertical lines* denote lateral locations of the cutter barrel)

A simple cutter is machined out of stainless steel that accepts the detonator of interest on one end and has a 5.75 mm diameter through hole axially aligned with the detonator centerline. Since the aluminum cup of the detonator has a diameter of 7.6 mm, one expects a smaller-diameter flyer to be generated as the detonation wave shock loads the aluminum cup and causes it to shear at the cutter edges. The height of the through hole in the cutter is 2.5 mm, and the thickness of the aluminum cup is 127 μm. Two assemblies are experimentally investigated corresponding to detonator output into air and PMMA, respectively. The corresponding SWIFT images are displayed in Figs. 54.5 and 54.6, where the inter-frame time is 135 ns.

Notice in Fig. 54.5 that the central region of the flow is curved, as is observed in the bottom of Fig. 54.2, but it is difficult to directly observe a distinct flyer boundary early on. The central region is followed by a flat expansion front that resembles a piston having a lateral geometry consistent with the cutter diameter. As time progresses it becomes clear that the expanding detonation products from the detonator compose the expansion region and partially envelop the central portion of the flow containing the flyer. Since a schlieren cutoff is not utilized, the detached shock waves are visible due to shadowgraphy

**Fig. 54.7** SWIFT images of the flow in PMMA resulting from the detonation of an HE column (flow is from *top* to *bottom*)

through the large spatial extent of the flow, and the curvature of the shock fronts. The presence of the lateral detached shock waves supports the observation that the product gases are expanding supersonically, as one would expect. Results from simultaneous PDV verify that not only is a flyer generated through the cutter, but its velocity history is nearly identical to an unaltered detonator having no interaction with a cutter.

Considering Fig. 54.6, where the output is visualized within PMMA, frames 1 and 2 support the idea that a central flyer initially shocks the PMMA and is promptly followed by a flat expanding piston. Given that the PMMA sample is used here as a surrogate for HE, there is utility in knowing that within a 135 ns inter-frame delay a pellet of HE would incur a double insult that may affect its initiation and subsequent run-to-detonation. Further, frame 5 suggests the shock waves within the stainless steel cutter eventually shock-load the PMMA sample as well. Additional analysis is ongoing to relate the strength of the primary shock driven into PMMA to PDV measurements of the flyer velocity at the instant of impact. Nevertheless, for impact-loading applications, the SWIFT images in Figs. 54.5 and 54.6 present a consistent story on which design decisions can be based for an explosively-generated flyer plate using standard detonators.

As a final experimental example, the SWIFT system is used to visualize a small column of HE as it detonates. A standard detonator is used to initiate a column of XTX 8004 having nominal density, diameter, and height of 1.5 g/cm$^3$, 3 mm, and 20 mm, respectively. Both the detonator and HE are fully embedded in solid PMMA in order to facilitate visualization of the expected explosive-driven shock waves. Few detonation parameters are known for XTX 8004 apart from detonation velocity as a function of charge radius [9]. As a result, the SWIFT experiment has been performed as an initial feasibility test to evaluate the usefulness of the method for characterizing HE performance. Since an existing tube of XTX 8004 was utilized for the test, the detonator was made to initiate the HE at an angle to accommodate the end geometry of the tube. Future tests are being designed with an inline HE train, where the detonator is coaxially aligned to the HE column.

The resulting SWIFT image sequence is included in Fig. 54.7, where the effects of non-coaxial initiation are evident in the first few images. The shock waves expanding outward from the HE column are very clearly visualized, and the quality of the data allows precise measurements of shock position along the HE/PMMA interface to be made with minimal uncertainty. Resulting position measurements are plotted versus corresponding exposure times in Fig. 54.8, where independent plots are included that correspond to shock-position measurements made on each side of the HE column. Error bars denoting measurement uncertainty in the shock positions are not included in the plots, since their vertical extent is smaller than the data markers used to plot the measurements. Application of linear least-squares fitting to the position data yields slopes of 7.407 +/− 0.103 mm/μs and 7.536 +/− 0.077 mm/μs, corresponding to the left and right sides of the HE column, respectively.

The published value for the detonation velocity of 1.5 g/cm$^3$ XTX 8004 at a charge diameter of 3.13 mm is given as 7.30 mm/μs in [9]. Though the velocity values obtained from the SWIFT data fall on the high side of the published values, they deviate by less than 2–4 %. As initial results from a non-ideal feasibility study, where effects of HE initiation at an angle may play a critical role in the quantitative results, the authors find this agreement very good.

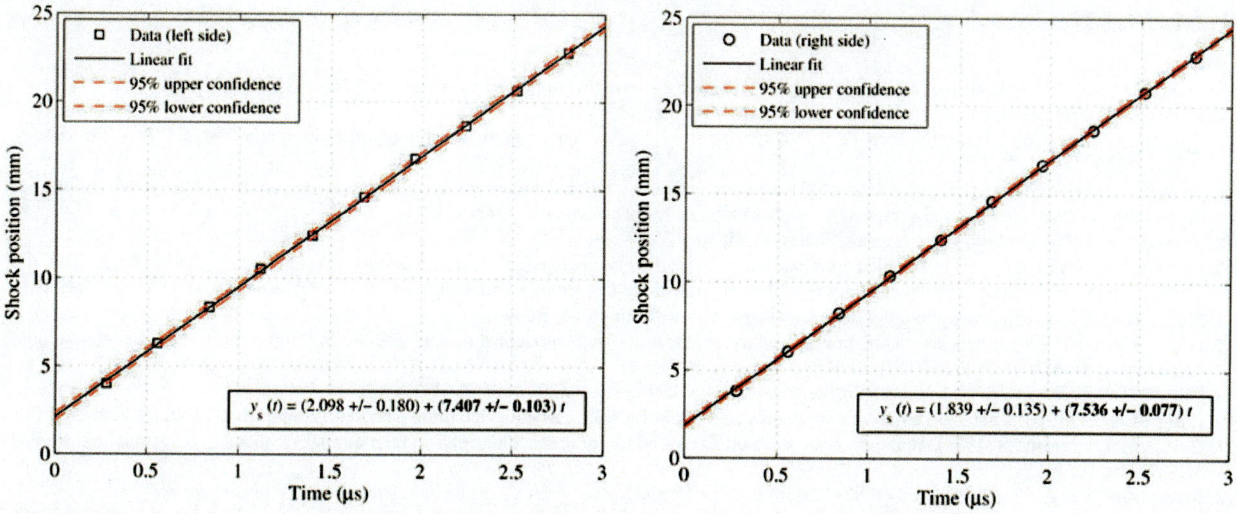


**Fig. 54.8** Extracted measurements of shock position along the *left* and *right* sides of the HE/PMMA interface with linear fits and confidence bands included

Additional investigation is underway to extract pressure at the HE/PMMA interface, and possibly equation-of-state parameters for the detonation products, directly from the SWIFT data. The analysis is based on proven data-reduction procedures for LANL's standard aquarium experiment [10–12], where PMMA replaces water as the immersion medium for the HE column.

## 54.4 Conclusions

Current SWIFT implementation is presented to demonstrate a higher quality of achievable data than was possible through its 5-year old predecessor. Improvements in optical configuration, coupled with the use of higher-quality lenses, a laser backlight with spoiled coherence, and a high-performance ultra-high-speed framing camera, combine to form an ensemble capable of investigating explosive-drive phenomena with high fidelity. As an example, visualizations of standard detonator output into both air and PMMA are presented as a means of studying output geometry, expected flow physics, and breakout time history. The latter results demonstrate good agreement with streak-camera data, and thus validate an additional data-reduction procedure for extracting quantitative results directly from SWIFT images.

The use of detonator output to generate explosively-driven flyer plates for impact-loading applications is presented, and the results provide unique design information based on the use of PMMA as an HE surrogate. The specific utility of the results emerges from accurate, time-resolved visualization of the complex shock-loading that occurs as a detonator/cutter assembly interacts with the PMMA sample.

Direct visualization of the flow resulting from HE detonation within PMMA provides data that is rich in information directly relevant to the understanding and characterization of how the HE performs. Measurements of shock position along the HE column are extracted from the SWIFT images, and detonation velocities are determined to within 4 % of published values in the worst case.

Overall, the SWIFT system offers the experimenter high-resolution images, ultra-high-speed frame rates, time-resolved visualization of a single event, a robust range of magnifications, and the ability to visualize shock waves in the presence of explosive phenomena and high self-luminance. The authors plan to broaden the use of the system to many more scientific flows involving complex shock interactions and initiator-scale events that span fractions of a millimeter.

**Acknowledgments** Funding was provided by the Joint DOD/DOE Munitions Program. Los Alamos National Lab is operated by Los Alamos National Security, LLC, under Contract No. DE-AC52-06NA25396 with the U.S. Department of Energy.

## References


1. Clarke SA et al (2007) High-speed multi-frame laser schlieren for visualization of explosives events. Proc SPIE 6638:6662
2. Settles GS (2001) Schlieren and shadowgraph techniques. Springer, New York
3. Oppenheim AK, Urtiew PA, Weinberg FJ (1966) On the use of laser light sources in schlieren-interferometer systems. Proc Roy Soc A 291:279–290
4. Svingala FR, Hargather MJ, Settles GS (2011) Modern optical methods for determining the shock Hugoniot of transparent solids. In: Proceedings of 28th international symposium on shock waves, Manchester, July 2011
5. Cooper PW (1996) Explosives engineering. Wiley-VCH, New York
6. Murphy MJ, Clarke SA (2011) Simultaneous photonic Doppler velocimetry and ultra-high speed imaging techniques to characterize the pressure output of detonators.In: Proceedings of 17th APS topical conference on shock compression of condensed matter, Chicago, June 2011
7. Marsh SP (1980) LASL shock hugoniot data. University of California Press, Berkeley
8. Lee K-Y et al (1998) Synthesis, detonation spreading and reaction rate modeling of fine TATB. In: Proceedings of 11th international detonation symposium, ONR 33300-5, 1998
9. Gibbs TR, Popolato A (1980) LASL explosive property data. Univesity of California, Berkeley
10. Craig BG et al (1978) Characterization of two commercial explosives, Los Alamos Scientific Laboratory report LA-7140, 1978
11. Goldstein S, Johnson JN (1981) Aquarium tests on aluminized ANFO. In: Proceedings of 7th international symposium on detonation, NSWC MP 82-334, 1981
12. Johnson JN, Mader CL, Goldstein S (1983) Performance properties of commercial explosives. Propellants Explos Pyrot 8:8–18

# Chapter 55
# Utilizing Knots to Reduce Dynamic Loads in Fire Service Rope Systems

**K. Boron, M. Obstalecki, P. Kurath, and G.P. Horn**

**Abstract** Rope systems are commonly used in applications including recreational sport climbing, industrial fall protection, and Fire Service rescue. These rope systems must be able to absorb energy in order to reduce the risk of impact injury to the individual experiencing the fall. However, there are several factors that go into the selection of the rope based on end use condition, whether that is weight, abrasion resistance or heat resistance. Thus some of the properties that are required for ropes to be effectively deployed for escape scenarios in the Fire Service, do not translate to optimal energy absorbing capabilities to prevent injuries to the individual attached to the system.

The present pilot study quantifies the energy absorbing capabilities of overhand and figure-eight knots tied in various firefighter escape ropes, including those constructed from Technora, nylon and composite ropes. Using a specially designed vertical drop apparatus, experiments are conducted to quantify the change in energy absorption characteristics that may result in a reduction in maximum arrest forces when knots are incorporated. The results show that the maximum arrest forces can be attenuated by up to 35% by simply incorporating a knot, but the time average force delivered is not changed by incorporating a knot and the rope system elongates significantly. The effect of these knots on reducing the strength of the system is quantified for comparison with maximum impact load measured during the dynamic loading, showing that the reserve load carrying capacity is significantly reduced by the incorporation of knots.

**Keywords** Dynamic load • Rope • Fire fighter

## 55.1 Introduction

Rope systems are simple structures that play critical life safety roles in a variety of industries and applications. Recently, the Fire Service has begun deploying personal escape systems in response to firefighter line of duty injuries and fatalities as a result of jumping from the second floor and above when fire conditions require a rapid exit from the interior of the structure. These tools are based largely on life safety rope systems in common use for recreational sport climbing and industrial fall protection. While a recreational sport climber may realistically expect to take a fall of several meters onto the ropes they pay out during the climb, these systems absorb the climber's potential energy lost during falls with "dynamic" ropes that absorb energy by elongating up to 100% prior to failure. In industrial fall protection, occupational risks are reduced by utilizing ropes and secondary energy absorption systems that use a combination of means to absorb energy through deformation, friction, or unrecoverable damage to structures such as tearing or ripping. Failure to absorb sufficient energy can result in

K. Boron • P. Kurath
Department of Mechanical Science and Engineering, University of Illinois at Urbana-Champaign, 1206 W. Green St, Urbana, IL 61801, USA

M. Obstalecki
Sibley School of Mechanical and Aerospace Engineering, Cornell University, 105 Upson Hall Ithaca, New York 14853, USA

Illinois Fire Service Institute, University of Illinois at Urbana-Champaign, 11 Gerty Drive, Champaign, IL 61820, USA

G.P. Horn (✉)
Department of Mechanical Science and Engineering, University of Illinois at Urbana-Champaign, 1206 W. Green St, Urbana, IL 61801, USA

Illinois Fire Service Institute, University of Illinois at Urbana-Champaign, 11 Gerty Drive, Champaign, IL 61820, USA
e-mail: ghorn@illinois.edu

V. Chalivendra et al. (eds.), *Dynamic Behavior of Materials, Volume 1: Proceedings of the 2012 Annual Conference on Experimental and Applied Mechanics*, Conference Proceedings of the Society for Experimental Mechanics Series, DOI 10.1007/978-1-4614-4238-7_55, 

large dynamic impact loads that can be damaging to the human body, the rope, and anchoring structure. According to Occupational Safety and Health Administration (OSHA) regulations, the ropes and fall arrest systems used in occupational settings should limit the maximum arrest force to 8 kN [1]. However, these limitations are imposed under the assumption that the worker is wearing a full body harness. When wearing a "seat harness" it has been suggested that this limit should be reduced to 4 kN [2] based on a study by Magdefrau who investigated the maximum arrest force and damage to a humanoid mannequin when wearing various harnesses [3]. At arrest forces as low as 4 kN, the mannequin absorbed enough energy to fracture its spine when using a seat harness, yet no visible damage was noted when dropped from the same distance using a full body harness despite higher maximum loads.

The systems employed in the Fire Service have important additional constraints that are not encountered in sport climbing or occupational settings including: significant weight and bulk restrictions, systems must be attached and deployed rapidly under adverse conditions, and the systems must be able to withstand elevated temperatures of up to 500°C. The leading common structural rope fibers that can satisfy these criteria are in the aramid family, which are light, easily packed and have a high temperature resistance. However, these fibers are orders of magnitude stiffer than the relatively compliant materials used in sport or industrial applications, resulting in larger force generation under dynamic loading. At the same time, weight and bulk restrictions limit the ability to incorporate secondary energy absorption systems such as those used in industrial fall protection.

During the on-going efforts to experimentally verify numerical simulations of firefighter escape scenarios [4], significant discrepancies were found between experiments and simulations when fire service knots were incorporated. While the model included dissipation effects of friction over a window sill, it did not account for frictional dissipation due to the knotted ends. When experiments were repeated with the ends fixed such that minimal slippage occurred, the disagreement between experiment and model was reduced, on average, from 37% to less than 7% for a large range of escape scenarios. It is apparent that a significant amount of energy can be absorbed by the inclusion of knots within the rope system, and it has been hypothesized that intelligent introduction of knots in Fire Service escape systems could provide an alternate means for energy absorption without the addition of the weight, bulk, and complexity of accessory energy absorption systems.

In reviewing the literature, the ability of knots to absorb energy in a dynamic loading scenario has received minimal treatment. Shad et al. introduced knot effects in their model by assuming a specific percentage of load was dissipated, but no experimental validation was provided [5]. Thus, no rigorous studies have been found despite significant research into dynamic loading of rope in the sport climbing industry [5–11], maritime industry [8, 10–14], worker safety [8, 10, 11, 14, 15] as well as in individual polymer strands [11, 16]. This paper presents new results that provide the first estimate of energy absorption due to frictional dissipation during knot tightening, specifically in ropes utilized in the Fire Service.

Introduction of knots into the system reduces the ultimate tensile strength of rope [6, 12]. In a study conducted on the strength of knots in dynamic climbing rope, Brown reported that the figure-eight loop end configuration reduced the strength of Beal Tiger (10 mm) dynamic climbing rope by 16% [6]. Milne and McLaren determined that the strength of polyester ropes used in sailing applications is reduced by 24% when a figure-eight loop end knot is utilized [12]. When introducing a knot into a rope structure, the load is no longer transferred evenly across the rope cross section. Furthermore, as a knot tightens it begins to pinch down on the rope, causing consistent failures at the entrance to the knot at much lower loads when compared to an unknotted rope system. While knots may introduce a compact and efficient means to absorb energy, it is important to ensure that they do not weaken the rope enough to cause failure prior to the complete absorption of the dynamic load.

This paper provides the first experimental measurement of the energy absorbing capability of common knots tied in the stiff rope systems utilized by the Fire Service. The energy absorption is characterized by the reduction in maximum arrest force felt while the rope system is subject to a relatively short fall on a short length of rope; a probable occurrence in Fire Service escape scenarios. Assuming ropes do not fail during the fall event, the entire change in potential energy from the drop must be absorbed by the mechanical extension of the rope, and frictional dissipation, particularly when knots are included. The maximum arrest force in each rope is compared to the minimum breaking strength of the knotted rope to determine the reserve strength in the escape system. All of these tests were conducted at room temperature, which does not account for the further strength reducing effects of elevated temperature deployment which has also been shown to significantly reduce rope strength [17–19]

## 55.2 Experimental Set Up

### 55.2.1 *Specimens*

Three different commercially available 7.5 mm diameter escape ropes are utilized for various aspects of this study. Technora – Technora rope (Technora sheath, Technora core) is constructed from 100% aramid fibers that allow for both high strength

**Table 55.1** Minimum breaking strength of the ropes utilized in this study (from rope manufacturers)

| Rope | MBS (kN) |
|---|---|
| Technora-technora | 25.6 |
| Technora-nylon | 19.2 |
| Nylon-nylon | 14.9 |

and heat-resistance [18, 19]. Previous studies have shown that Technora fibers enhance overall strength and abrasion resistance of rope while providing a measure of flame resistance [20]; however, they are also the stiffest of the ropes studied. Nylon – nylon rope (nylon sheath, nylon core) have relatively a low stiffness along with low heat-resistance and thus is no longer in common use by the Fire Service. The nylon-nylon rope is maintained in this study to compare with knot strength results in existing literature, but has not been dynamically tested with knots. A hybrid Technora (sheath) – nylon (core) rope provides a promising compromise that combines high heat and abrasion resistance of the Technora sheath with improved compliance for energy absorption from the nylon core. The minimum breaking strength (MBS, defined by NFPA 1983 standard as three standard deviations less than the mean breaking strength) for each of these ropes is provided in Table 55.1 [21].

The effect of knots on the maximum arrest force from dynamic loading of Technora-Technora and Technora-nylon firefighter escape rope is studied using four different configurations. The rope is firmly attached to the loadframe (anchoring point) in each configuration to emulate "sewn eye" terminations, which provide no slippage and minimal energy dissipation. The end attached to the drop weight is varied as follows: (1) rope end firmly attached, (2) figure-eight knot on a bight, and (3) overhand knot on a bight. A fourth scenario is tested where both ends were firmly clamped and a figure-eight knot is tied between the ends. To study the effect of knot pre-tightening, one series of Technora-Technora ropes is tested with a 130 N preload applied to the rope prior to dynamic loading and another with a 830 N preload. A 130 N (30 lb) pre-load simulates a firefighter physically tightening a knot by hand as would be the case in the escape scenario. The 830 N (187 lb) pre-load is utilized to determine the effect of pre-loading on the dynamic forces. A new rope sample is tied for each drop test conducted. Throughout the study, knot tying remained consistent by having a single person dress, pre-load, and compare the tied knots to a "base knot."

The quasistatic strength reduction from knotting each rope is quantified in two separate knot conditions: (1) figure-eight knot in the middle of the rope, and (2) a figure-eight on a bight at the securing ends of the rope. Fresh ropes were tested for each rope and knot combination.

### *55.2.2 Testing Apparatus*

The dynamic drop test frame utilized in this study is described in detail elsewhere [4] and is utilized here in the vertical drop configuration to minimize confounding effects of frictional dissipation at the sill. For this study, a fall factor (total drop height / total rope payout) of 0.25 was utilized, which coincides with the dynamic testing for firefighter escape rope that must be completed for the NFPA 1983 standard. The 600 mm rope length is subjected to a 150 mm vertical drop. While this fall factor may be smaller than one expected during sport climbing, it is considered to be a likely fall factor for Fire Service escape scenario. In order to ensure a consistent drop height and orientation, a rapid release clamp was employed. The current system utilizes a solid drop mass consisting of Olympic weightlifting plates as a substitute for the firefighter's mass. The solid mass ignores the human body's capability to absorb energy during a fall, focusing the data on rope properties rather than the dynamic effects of the human body. Eliminating the effects of the human body and friction over a simulated sill allows for more straightforward determination of the knot effects. The first round of testing was conducted using a mass of 84 kg, which is similar to the standard drop weight utilized in UIAA standards for dynamic rope testing. The experimental frame incorporates load cells located at the anchor point and falling mass (Omega, LCCD-10 K, Stamford, CT). A total of three trials were run for each configuration. Load values were sampled and recorded at 1,000 kHz using LabVIEW data acquisition software. The amount of rope that was pulled through each knot was recorded by tracking the displacement of a point marked on the rope immediately adjacent to the knot.

Quasi-static testing of rope strength with a knot was conducted using a Riehle screw-driven testing rig with a 44.5 kN load range. The crosshead speed of the test rig is a constant 5.4 mm/s. Data points were taken at 10 Hz using LabVIEW data acquisition software. Drum type fixtures [21] were used to secure the specimens with a figure eight knot in the middle to the Riehle machine. The ends of the rope were wrapped around each drum two times, the ends placed through rope clamps and an over hand knot was tied at the end. A total of five tests were conducted for each rope configuration.

## 55.3 Experimental Results and Discussion

### *55.3.1 Static Results*

Table 55.2 shows the quasistatic breaking strength of the different escape rope systems when figure-eight knots are tied in an end loop and middle configuration, including minimum breaking strength as calculated in Table 55.1. Comparing Tables 55.1 and 55.2 reveals a significant reductions in the rope strength due to the incorporation of knots. The relatively compliant Nylon-Nylon rope is the least affected by tying knots, with a reduction in minimum breaking strength of 23% and 36% for the end and middle configurations. The strength reduction for the end knot configuration is similar to that determined for polyester rope used in marine applications [12]. However, the same knots result in significantly larger strength reductions when tied in firefighter escape ropes constructed from Technora-Nylon (50% and 55% respectively) and Technora-Technora (55% and 59% respectively). When knots are tied in both of these ropes, the maximum load appears to be controlled by the failure of the outer sheath which carries the bulk of the load. While the nylon core in the composite rope appears to attenuate this strength loss slightly, the relative loss of strength in these fire resistant ropes is much larger than has been reported in the literature for other structural ropes [6, 12].

### *55.3.2 Dynamic Testing*

The maximum arrest force data from the each of the tested knot and rope combination is reported in Table 55.3 along with the calculated percentage of mean breaking strength from the unknotted ropes (for fixed-fixed sample) and knotted scenarios. Table 55.4 shows the total system elongation, which takes into account knot slip as well as permanent rope elongation that results from the large loads encountered. In all cases, the fixed terminations provide no measurable slip; the 14.7 mm elongation on the fixed-fixed sample was a result of mechanical extension of the rope from dynamic loading.

As Table 55.3 indicates, incorporating a knot within the escape system significantly reduces the impact force felt by the mass compared to systems that utilize only fixed terminations. The relatively small fall factor of 0.25 resulted in maximum forces of 10.5 kN in the fixed-fixed Technora-Technora samples, which is significantly larger than the 8 kN maximum arrest force allowable by OSHA regulations for industrial fall protection. When the tests were repeated after changing one termination to a figure-eight or overhand knot, the maximum arrest force decreases to a little over 8 kN, which is unaffected

**Table 55.2** Static strength testing for rope with a figure eight loop end configuration

| | Mean ± SD (kN) | MBS (kN) | % Reduction in MBS |
|---|---|---|---|
| **Figure eight loop end configuration** | | | |
| Technora-technora | 13.6 ± 0.8 | 11.3 | 55.9% |
| Technora-nylon | 11.4 ± 0.6 | 9.6 | 50.0% |
| Nylon-nylon | 11.5 ± 0.4 | 10.2 | 22.8% |
| **Figure eight middle configuration** | | | |
| Technora-technora | 12.6 ± 0.8 | 10.3 | 59.8% |
| Technora-nylon | 9.7 ± 0.4 | 8.6 | 55.2% |
| Nylon-nylon | 9.6 ± 0.3 | 8.6 | 35.6% |

**Table 55.3** Results from dynamic testing of different knot configurations

| Rope | Configuration | Maximum arrest force (kN) | % of MBS |
|---|---|---|---|
| Technora – technora (130 N preload) | Fixed-fixed | 10.5 | 41 |
| | Fixed – figure eight | 8.1 | 72 |
| | Fixed – overhand | 8.4 | 74 |
| | Fixed – figure eight – fixed | 7.8 | 76 |
| Technora – technora (830 N preload) | Fixed – fixed | 10.5 | 41 |
| | Fixed – figure eight | 8.1 | 72 |
| | Fixed – overhand | 8.2 | 73 |
| | Fixed – figure eight – fixed | 8.3 | 81 |
| Technora – nylon (830 N preload) | Fixed – fixed | 6.7 | 35 |
| | Fixed – figure eight | 4.0 | 42 |
| | Fixed – overhand | 4.3 | 45 |
| | Fixed – figure eight – fixed | 4.3 | 50 |

**Table 55.4** Total rope elongation for Technora-technora rope with 830 N preload

| Configuration | Elongation (mm) |
|---|---|
| Fixed-fixed | 14.7 |
| Fixed – figure eight | 49.8 |
| Fixed – overhand | 49.0 |
| Fixed – figure eight – fixed | 41.4 |

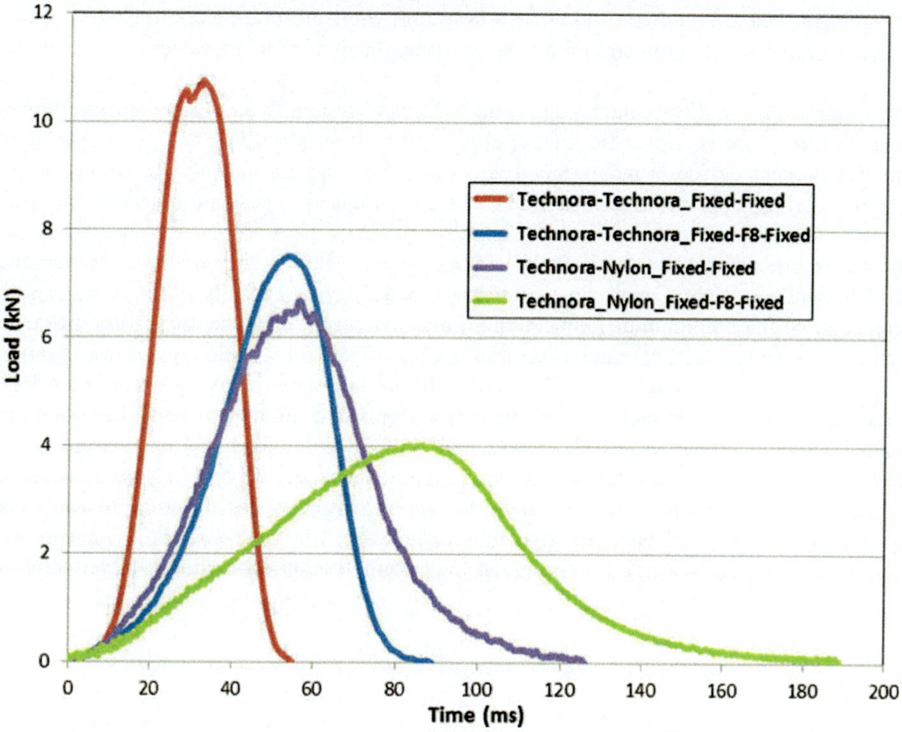


**Fig. 55.1** Load-time plots for the initial fall arrest from different rope and knot configurations under identical fall conditions (F8 indicates a figure-eight knot is incorporated into the escape rope system)

by the magnitude of the preload. Thus, the maximum arrest force is attenuated by approximately 2.4 kN or more than 20% with this simple modification. When a figure-eight knot is tied in–line with the rope, the maximum arrest force is reduced to 7.8 kN when a 130 N preload is applied; this value increases to 8.3 kN when the 830 N preload is applied. The addition of a single knot into a Technora-Technora rope brought the impact force closer to an acceptable level as defined by OSHA standards for an individual wearing a full body harness, yet the impact forces still remain twice as high as the 4 kN maximum arrest force recommended for the seat type harness commonly utilized for fire escape scenarios.

For the Technora-nylon system, the knots again reduced the impact force by approximately 2.5 kN, representing a larger percentage (37%) of the maximum arrest force experienced by the fixed-fixed sample. In all cases, this maximum arrest force is less than the 8 kN OSHA standard. With the different knots incorporated into the escape system, the maximum arrest force is near the 4 kN upper limit recommended by Sulowski et al. [2]. Thus, the additional energy absorption provided by knots may be critical to reducing the likelihood of injuries during Fire Service escape scenarios even with relatively compliant Technora-nylon rope bailout systems.

A typical load-time plot for the Technora-Technora and Technora-nylon rope systems with and without a knot are shown in Fig. 55.1. The amount of work done by the ropes in absorbing the momentum of the falling weight can be quantified by integrating the area under the load-time curve (sometimes called computing the "impulse" in biomechanics literature). The human body can survive higher forces for a shorter period of time, so it is important to determine if the knots expose the body to higher average loads over a longer period of time and can possibly be damaging. From this analysis, we can see that even though knots significantly attenuate the maximum arrest force delivered to the drop mass, the average work done by the rope is statistically unchanged regardless of the type of knot used or preload applied to the knot for both types of rope, remaining nearly constant at 250 kN-s for the Technora-Technora rope. In fact, though statistically insignificant, this value increases slightly when the knots are incorporated due to the additional extension of the rope as shown in Table 55.4.

Further, the Technora-Nylon ropes that reduce maximum arrest forces by 50–65% compared to that generated by the Technora-Technora ropes deliver a significantly larger impulse to the drop weight (310 kN-s), which is again unaffected by the presence of knots. The total deflection of this more compliant rope is significantly larger while absorbing the applied potential energy, thus the rope has to do more work to absorb the falling momentum. The total time for the load to be absorbed and fall completely arrested (and maximum force generated) by the Technora-Technora rope without knots is approximately 32 ms seconds, which is 20 ms shorter than Technora – Technora with the different knot configurations and approximately 25 ms shorter than the Technora – Nylon fixed-fixed configuration. However, it is not clear if these longer exposures to high forces present a more harmful scenario for a firefighter. In future work, the 84 kg solid drop mass may be replaced with an instrumented anthropomorphic dummy, weighted to that expected for a fully loaded firefighter (135 kg).

As expected, including a knot in the escape system causes the rope system to elongate significantly when compared to the fixed-fixed rope. This increase is due to the knot slipping and in turn, absorbing energy. The additional rope length may be a problem if a descent device is incorporated into the escape system, and this device is no longer reachable; however, the relatively short rope extension measured in this study should not cause a problem in the operation of descent devices.

Fortunately, the minimum breaking strength for all of the ropes tested is greater than the maximum arrest force encountered during the dynamic loading. However, the reserve load carrying capacity of the ropes can be determined by calculating the percentage of the minimum breaking strength reached during the dynamic loading from the same 0.25 fall factor drop for each of the rope and knot configurations. For the fixed-fixed rope (assuming no loss of strength by the fixed ends), the 0.25 drop resulted in loads that were 41% and 35% of the minimum breaking strength for Technora-Technora and Technora-Nylon respectively. This suggests that there is a significant amount of strength remaining in the rope to account for additional affects such as larger fall factors and strength reductions due to elevated temperatures and/or abrasion at the window sill. After the knots are tied in the Technora-Technora rope, the dynamic loads reached between 72% and 81% of the minimum breaking strength. Thus, the reserve capacity for this rope to carry additional load is significantly compromised. On the other hand, the dynamic loads reached in the Technora-Nylon rope are between 35% and 50% of the minimum breaking strength for the given knot configuration, providing a larger safety factor for these scenarios.

## 55.4 Summary and Future Work

Results from dynamic impact loading of Fire Service escape ropes show that securing the ropes using simple knots or including a figure-eight knot in-line would reduce the impact force felt by a falling mass (i.e. firefighter) by up to 37%. The significant reduction in maximum arrest forces can reduce the risk of injury to firefighters even in the relatively small falls expected when deploying escape systems. While this reduction in maximum arrest force is significant and important, these loads are still large enough to cause potential harm to the firefighter wearing a seat harness or truck belt. Despite the reduction in maximum arrest force, the total momentum of the falling mass must be absorbed by the rope system, thus incorporating knots will subject a firefighter to the same time averaged forces; lower in magnitude, but longer in duration. Ropes with knots have larger elongations as a result of the frictional dissipation of energy, but the additional extension measured here would not adversely impact the operation of descent control devices (if they are utilized in these systems).

At the same time, we have shown that knots significantly reduce the breaking strength of the aramid rope to a much larger extent than previously measure with traditional polymeric rope materials and designs. This reduction in strength is found to be relatively larger than the reduction in maximum arrest force such that the reserve strength of a Technora-Technora rope may be reduced from nearly 60% down to less than 20% for the relatively small fall factor studied here.

It is suggested that extending this study to include ropes that have multiple knots may provide additional energy absorption capabilities that may restore some of the reserve strength to the system and reduce dynamic loads below the 4 kN recommendation for the seat type harnesses in use by the Fire Service. Additional knots should be studied as well as dynamic loading from larger fall factors that may reach the dynamic breaking point of the rope. Finally, the utilization of an instrumented anthropomorphic dummy would allow more realistic determination of impact loads encountered by a firefighter due to the compliance of the human body, as well as quantification of injury risk for the different scenarios.

## References


1. OSHA 29 CFR Parts 1910 and 1926 safety standards for fall protection in the construction industry; 1926.502 Fall protection systems criteria and practices. – section (d) Personal fall arrest systems
2. Sulowski AC (2012) How good is the 8 kN maximum arrest force limit in industrial fall arrest systems? http://www.fallpro.com/fall-protection-info-center/online-articles/maximum-arrest-force-limit-in-fall-arrest-systems/. Accessed 29 Feb 2012
3. Magdefrau H (1989) Die Belastung des meschlichen Korpers beim Sturz ins Seil und deren Folgen. Ludwig-Maximilians-Universitat, Munich
4. Obstalecki M, Chaussidon J, Kurath P, Horn GP (2011) Prediction of dynamic forces in fire service escape scenarios. In: Proceedings of the 2011SEM annual conference and exposition, Uncasville
5. Schad R (2000) Analysis of climbing accidents. Accident Anal Prev 32(3):391–396
6. Brown A (2008) The strength of knots in dynamic climbing rope.http://personal.strath.ac.uk/andrew.mclaren/Alasdair_Brown_2008.pdf. Accessed 29 Feb 2012
7. Vogwell J, Minguez JM (2007) The safety of rock climbing protection devices under falling loads. Eng Fail Anal 14(1):114–1123
8. Leech CM (1987) Theory and numerical methods for the modelling of synthetic ropes. Commun Appl Math 3(1):407–413
9. Pavier M (1998) Experimental and theoretical simulations of climbing falls. Sports Eng 1(1):79–91
10. Hennessey CM, Pearson NJ, Plaut RH (2005) Experimental snap loading of synthetic ropes. Shock Vib 12(1):163–175
11. Wu H (1993) Frictional constraint of rope strands. J Textile Inst 84(2):199–213
12. Milne KA, McLaren AJ (2006) An assessment of the strength of knots and splices used as eye terminations in a sailing environment. Sports Eng 9(1):1–13
13. Utting WS, Jones W (1985) Tensile testing of a wire rope strand. J Strain Anal 20(3):151–164
14. Baszczynski K (2007) Dynamic strength tests for low elongation lanyards. Int J Occu Safety Ergon 13(1):39–48
15. Spierings AB, Stampfli R (2006) Methodology for the development of an energy absorber: application to worker security ropes. Int J Impact Eng 32(1):1370–1383
16. Saitta AM, Soper PD, Wasserman E, Klein ML (1999) Influence of a knot on the strength of a polymer. Macmillan Mag 399(1):46–48
17. Horn GP, Kurath P (2011) Failure of firefighter escape rope under dynamic loading and elevated temperatures. Conf Proc Soc Exp Mech Series 1:353–359
18. Fontenot N, Stenvers D, Gilmore J, Solomon A, van Berkel B, Grabandt O, Kong D (2009) Use of synthetic rope in high – temperature or fire environments. OCEANS 2009, MTS/IEEE Biloxi – Marine technology for our future: global and local challenges
19. Gilmore J, Stenvers D, Chou R (2008) Some recent developments of rope technologies – further enhancements of high performance ropes OCEANS
20. Su C, Luo G (2010) High performance flame-retardent composite yarns with oxidized fiber. Fibers Polym 11(7):1049–1054
21. NFPA (2001) Standard for life safety rope and system component. National Fire Protection Association, Quincy

# Chapter 56
# Tensile Behavior of Single PPTA Fibers Measured by the Kolsky Bar Using the Direct Fiber Clamping Method

**J.H. Kim, N.A. Heckert, S.D. Leigh, H. Kobayashi, W.G. McDonough, R.L. Rhorer, K.D. Rice, and G.A. Holmes**

**Abstract** The Kolsky bar test has been widely used in measuring material behavior under high strain rate conditions. In particular, this methodology has been used to characterize the high strain rate behavior of polymer and polymer composites during ballistic impact. Chen et al. measured the tensile properties of single poly(*p*-phenylene terephthalamide) (PPTA) fibers at high strain rates by gluing the fiber directly to the Kolsky bar. However, the application of this technique is somewhat limited because of the time-consuming nature of the gluing procedure [1]. Due to the highly stochastic nature of fiber tensile strength, large test-sample sizes are required to analyze distributions of fiber properties such as tensile strength, modulus, and strain-to-failure. The authors have investigated the direct gripping method to facilitate high throughput single fiber tensile testing at high strain rates. As an extension of these efforts, the polymethyl methacrylate (PMMA) platens used for directly gripping the fibers are replaced here by rubber platens to assess the impact of the platen material properties on the stress transfer from the Kolsky bar to the fiber at high strain rates. In this study, the system compliance for the PMMA grips is measured under the quasi-static condition to assess the stress transfer from the grip and to correct the modulus and failure strain of the fiber. In preliminary tests using a PMMA grip, the system compliance was obtained by the 5, 10 and 60 mm gauge lengths. These data will be compared with similar data using the rubber grips to investigate the influence of the gripping material properties.

**Keywords** Soft body armor • Kolsky bar • Fiber gripping • Tensile behavior

## 56.1 Introduction

High-strength polymer fibers such as poly(*p*-phenylene terephthalamide) (PPTA) are widely used in soft body armor applications, and the material properties of these fibers are often characterized under quasi-static test conditions, although fiber deformation during ballistic impact is much faster than under this test condition. Therefore, a measurement platform is needed to obtain accurate and reproducible material properties tested under a similar loading condition for ballistic events. To measure single fiber properties under such a high strain rate condition, Kolsky bar tests which require test specimens shorter than 10 mm have been utilized [1, 2]. Since fiber strengths are highly variable, a significant amount of data is often needed for statistical modeling. Therefore directly clamping a fiber rather than gluing a fiber onto the Kolsky bar, which limits the number of tests, has been proposed to increase throughput [3].

---



J.H. Kim • H. Kobayashi • W.G. McDonough • G.A. Holmes (✉)
Polymers Division (M/S 8541), National Institute of Standards and Technology, Gaithersburg, MD 20899, USA
e-mail: gale.holmes@nist.gov

N.A. Heckert • S.D. Leigh
Statistical Engineering Division (M/S 8980), National Institute of Standards and Technology, Gaithersburg, MD 20899, USA

R.L. Rhorer
Intelligent Systems Division (M/S 8220), National Institute of Standards and Technology, Gaithersburg, MD 20899, USA

K.D. Rice
Law Enforcement Standards Office (M/S 8102), National Institute of Standards and Technology, Gaithersburg, MD 20899, USA

V. Chalivendra et al. (eds.), *Dynamic Behavior of Materials, Volume 1: Proceedings of the 2012 Annual Conference on Experimental and Applied Mechanics*, Conference Proceedings of the Society for Experimental Mechanics Series, DOI 10.1007/978-1-4614-4238-7_56, © The Society for Experimental Mechanics, Inc. 2013

In our initial publication [4], quasi-static tests on 2 and 60 mm long gauge length samples were performed with gluing and direct clamp approaches to investigate gripping method effects on the tensile strengths. In the gluing approach, the fiber is glued to a template of known gauge length. With the direct clamp approach, the fiber is directly gripped between two polymethyl methacrylate (PMMA) blocks (Textechno Herbert Stein GmbH & Co. KG.[1]). The two gripping methods were found to yield different strength distributions. The direct grip results were observed to give the expected uniform increase in tensile strength as the gauge length decreased.

In the current direct grip approach, PMMA blocks were used to clamp the Kevlar 29 fibers with sufficient force to eliminate slippage of the fiber during the loading process. As an extension of the effort to develop a direct fiber gripping method, the effect of the platen material properties for the grip on the stress transfer is investigated under quasi-static test condition. To assess this effect, system compliance is measured from single fiber tensile tests. Obtaining the system compliance is important to correct modulus and failure strain, particularly for test results with short fibers. Since the system compliance is based on the load-displacement curve of the fiber tensile behavior, an effect caused by the gripping material properties may be present. In preliminary tests using a PMMA grip, the system compliance obtained by the 5, 10, and 60 mm gauge lengths is determined for the tensile tests. The system compliances using a rubber gripping materials will be compared with this result as an assessment of the effect of the platen material properties.

## 56.2 Experimental Procedure

### *56.2.1 Quasi-Static Loading with PMMA grip*

For the tensile test under quasi-static loading, PPTA fibers were directly clamped by PMMA fiber grips on the tensile testing machine. The fiber gripping technique used is shown in Fig. 56.1. To clamp a single PPTA fiber directly, the fiber is placed under a slight tensile load of 0.01 N by attaching a small mass to maintain the straightness and alignment with the grips, and a single PPTA fiber is clamped between two grips on both ends, so the fiber only contacts PMMA. The clamping force of the blocks is controlled by tightening a spring. The gauge lengths of the fiber specimens were 5, 10, and 60 mm. The tensile tests were performed under a constant strain rate 0.00056 $s^{-1}$ with the standard uncertainty in the load cell of 0.1%.

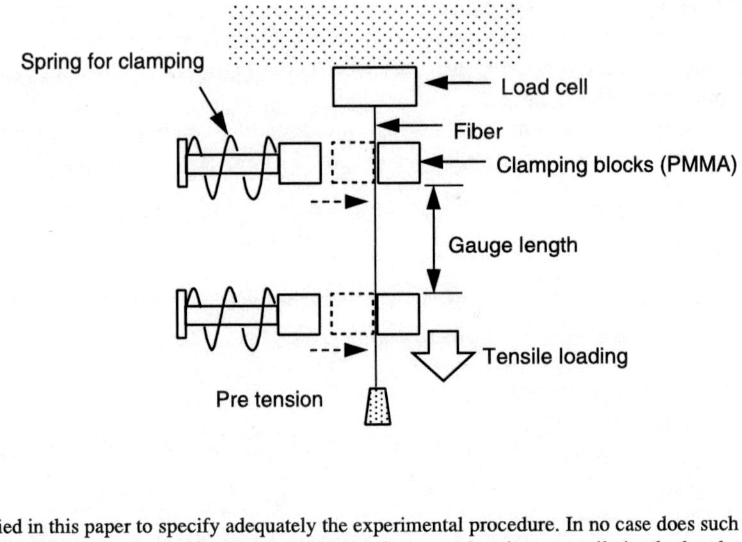


**Fig. 56.1** Schematic close up of the direct grip for quasi-static loading

[1] Certain commercial materials and equipment are identified in this paper to specify adequately the experimental procedure. In no case does such identification imply recommendation or endorsement by the National Institute of Standards and Technology, nor does it necessarily imply that the product is the best available for the purpose.

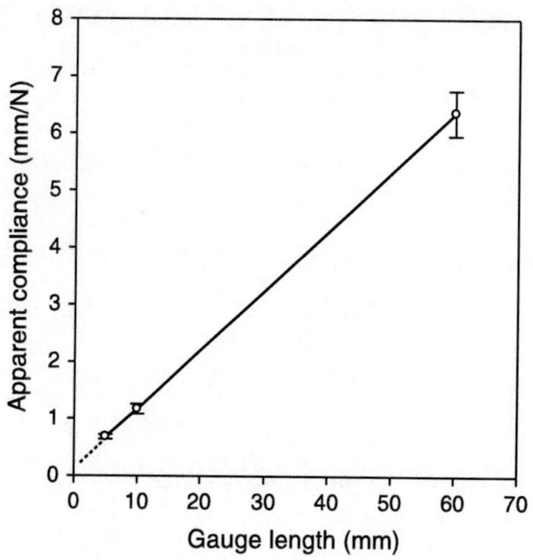


**Fig. 56.2** The apparent compliance ($C_a$) for the single PPTA fiber tensile test using the PMMA grip and the *error bars* refer the standard deviations of the $C_a$ values taken as an estimate of the standard uncertainty

## 56.3 Results and Discussion

During the tensile test, the slope of the initial straight line region is related to the axial stiffness of the material. Apparent compliance is the extension (ΔL) per unit force (ΔF) based on this slope (Eq. 56.1). If the extension of the specimen is not determined, the displacement of the crosshead from the testing machine is used as an estimate of the extension and the apparent compliance in this method depends on the test machine and grip system.

$$C_a = \left(\frac{\Delta L}{\Delta F}\right) \tag{56.1}$$

Figure 56.2 shows the apparent compliances for the tensile test with 5, 10, and 60 mm gauge lengths measured by the PMMA grip. The apparent compliance at the zero gauge length equals the system compliance, which is used for determining the corrected elastic modulus [5]. The system compliance obtained by the intercept of the apparent compliances plot for the 5, 10, and 60 mm gauge lengths is statistically distinguishable from zero ($t = 7.66$), so it is necessary to correct the apparent specimen modulus by the system compliance.

## 56.4 Conclusions

Preliminary tests using a PMMA fiber grip were carried out to investigate effects of gripping materials on single fiber tensile tests. For comparison with other types of gripping materials, the system compliance derived from a plot of apparent compliances with various fiber lengths was obtained by using the PMMA grip. The tensile test results show that the measured system compliance influences the initial load-displacement behavior of the fiber. This result supports requiring system compliance to be measured as a function of gripping materials properties. Moreover, the relation between the gripping material properties and system compliance will be useful for designing a clamp for the high strain rate test.

## References

1. Cheng M, Chen WN, Weerasooriya T (2005) Mechanical properties of Kevlar (R) KM2 single fiber. J Eng Mater Technol Trans ASME 127(2):197–203
2. Lim J, Zheng JQ, Masters K, Chen WNW (2010) Mechanical behavior of A265 single fibers. J Mater Sci 45(3):652–661
3. Kim JH, Heckert NA, Leigh SD, Kobayashi H, McDonough WG, Rice KD, Holmes GA (2011) Statistical analysis of fiber gripping effects on single fiber tensile test. In: Proceedings of the SAMPE 2011 technical conference, Long Beach
4. Kim JH, Rhorer RL, Kobayashi H, McDonough WG, Holmes GA (2010) Effects of fiber gripping methods on single fiber tensile test using Kolsky bar. Society of experimental mechanics, Indianapolis
5. Turek DE (1993) On the tensile testing of high modulus polymers and the compliance correction. Polym Eng Sci 33(6):328–333

# Chapter 57
# Effect of Strain Rate and Pre-twist on Strength of Single Kevlar® KM2 Fiber

**Brett Sanborn, C. Allan Gunnarsson, Paul Moy, James Q. Zheng, and Tusit Weerasooriya**

**Abstract** High performance fibers such as Kevlar KM2 used in protective vests are loaded primarily in tension at high loading rates. During the weaving and crimping process of fibers/yarns to make fabrics, individual fiber tensile strength may be reduced due to pre-straining of fibers by twisting. To systematically investigate the loading rate effects under different pre-twisted stress states, a miniaturized Kolsky tension bar is used to study single Kevlar KM2 fibers subject to varying levels of shear strain at high strain rates. Low and intermediate strain rate responses of the fibers are investigated using a Bose Electroforce loading system. In this paper, tensile strength as a function of pre-loaded shear strain on a single fiber is presented for different strain rates of loading. Fracture morphology of the fibers is also investigated using a high resolution scanning electron microscope to understand the mechanisms of failure.

**Keywords** Fracture • Kevlar • Shea strain • NEXAFS

## 57.1 Introduction

Studies of loading rate effects on tensile strength of single high performance fibers are crucial to expand and develop predictive models and finite element codes. Previous single fiber studies [1–5] have focused on determining longitudinal tensile strength at low and high strain rate and transverse compressive behavior at low rate. Aside from longitudinal tensile and transverse compressive behavior of single fibers, behavior at other complex stress states remains unexplored. During production of fabrics, the weaving and crimping of fibers and yarns, individual fiber tensile strength may be compromised due to damage accumulation from pre-straining of fibers from twisting, as shown by DeTeresa et al. [6] for Kevlar® 49 at low strain rate. Thus, the present work is concerned with studying the tensile strength of single Kevlar® KM2 fibers as a function of strain rate and level of pre-set shear strain.

## 57.2 Experiments

Experiments to study the strain rate effects on the deformation and failure behavior of Kevlar® KM2 at different pre-twisted stress states were conducted at low and intermediate strain rates using a Bose Electroforce Test system, and at high strain rate using a miniaturized Split-Hopkinson Tension bar (SHTB) modified to test single fibers. Details of the fiber SHTB are given in [1, 2]. Fibers glued to pre-cut cardboard frames were attached between a rotary-stage and load cell. The rotary-stage and the load cell were either connected to the grips of the Bose system or the end of the SHTB. The sides of the frame were cut and the rotary stage was rotated so that the fiber was twisted to the desired level of pre-shear strain. Experiments were conducted at 9 levels of pre-shear strains at low and intermediate strain rates and 10 levels of pre-shear strains at high strain rate. Ten experiments were conducted at each pre-shear strain to account for possible variations in strength due to the presence of internal microstructural defects and surface defects such as variations in fiber diameter.

B. Sanborn (✉) • C.A. Gunnarsson • P. Moy • J.Q. Zheng • T. Weerasooriya
US Army Research Laboratory, RDRL-WMP-B, Aberdeen Proving Ground, MD 21005, USA
e-mail: brett.sanborn2.ctr@mail.mil; carey.a.gunnarsson.civ@mail.mil; paul.moy.civ@mail.mil

V. Chalivendra et al. (eds.), *Dynamic Behavior of Materials, Volume 1: Proceedings of the 2012 Annual Conference on Experimental and Applied Mechanics*, Conference Proceedings of the Society for Experimental Mechanics Series, DOI 10.1007/978-1-4614-4238-7_57, © The Society for Experimental Mechanics, Inc. 2013

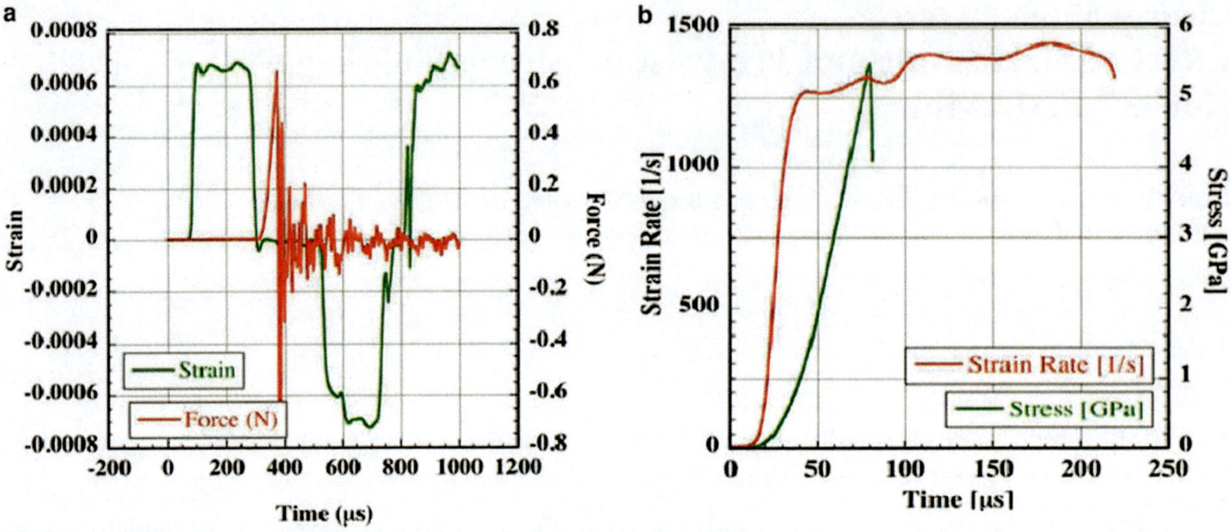


**Fig. 57.1** Results from a typical high strain rate experiment on single Kevlar® KM2 fiber. (**a**) Measured strain signals in the bar and force history from the piezoelectric load cell. (**b**) Strain rate and stress histories seen by the fiber

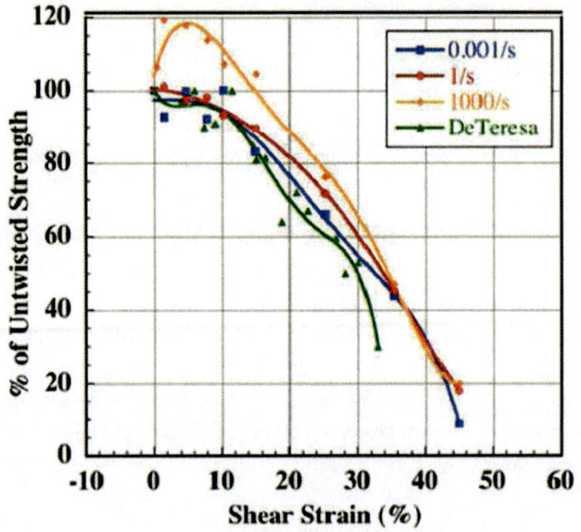


**Fig. 57.2** Summary of all experiments on pre-twisted Kevlar® KM2 and apparent rise in strength when pre-twisted fiber is pulled in high rate tension

## 57.3 Results

Figure 57.1a shows results from a typical high-rate SHTB experiment on a single Kevlar® KM2 fiber held in the untwisted state. A piezoelectric load cell was used instead of a transmission bar to measure the weak transmitted signal. Figure 57.1b shows the strain rate and stress history of the specimen. Figure 57.1b also shows that after an initial ramp loading time of 40 μs, the fiber deformed and failed at constant strain rate of about 1,250/s.

Results from pre-twist experiments at all strain rates are shown in Fig. 57.2. Along with the results of this study, the results obtained by DeTeresa et al. [6] have also been plotted for comparison. Each data point in Fig. 57.2 represents an average value of 10 experiments, and has been non-dimensionalized based on the untwisted average strength of the fiber at each strain rate. At low rate, the fiber retains at least 90% of the untwisted strength until 10% shear strain is reached, whereupon the strength begins to decrease as the pre-set shear strain is increased. The low rate data closely follows the results of DeTeresa et al., even though a different type of Kevlar® (Kevlar® 49) was used in that study. When the strain rate was increased three orders of magnitude to intermediate strain rate of 1/s, no notable change in strength was observed.

The fiber retained 90% of the untwisted strength and began to decrease past a shear strain of 8–10%. When the strain rate increased another three orders of magnitude to high strain rate of 1,000/s, however, the fiber increased in strength with increasing pre-shear strain up to a shear strain of about 1.6%. The fiber retained more than 100% of untwisted strength up to a shear strain of 15%, whereupon a similar drop off in strength was observed as the low and intermediate rate results. At a shear strain of about 1.6%, the fiber was about 20% stronger than the untwisted strength at the same strain rate. The base line value of 5.02 $\pm$ 0.47 GPa for untwisted fiber at high rate is consistent with results found by other researchers [1, 4]. The pre-twist level of 0.64% shear strain also shows an increase in strength consistent with the rise in strength seen at 1.6%, confirming that the increase in strength of fiber held at a pre-twisted strain is a rate effect.

## References

1. Cheng M, Chen W, Weerasooriya T (2005) Mechanical properties of Kevlar® KM2 single fiber. J Eng Mater Technol 127:197–203
2. Cheng M, Chen W, Weerasooriya T (2004) Experimental investigation of the transverse mechanical properties of a single Kevlar® KM2 fiber. Int J Solid Struct 41:6215–6232
3. Lim J, Zheng JQ, Masters K, Chen WW (2010) Mechanical behavior of A265 single fibers. J Mater Sci 45:652–661
4. Lim J, Chen WW, Zheng JQ (2010) Dynamic small strain measurements of Kevlar® 129 single fibers with a miniaturized tension Kolsky bar. Polym Test 29:701–705
5. Lim J, Zheng JQ, Masters K, Chen WW (2011) Effects of gage length loading rates, and damage on the strength of PPTA fibers. Int J Impact Eng 38:219–227
6. DeTeresa SJ, Allen SR, Farris RJ, Porter RS (1984) Compressive and torsional behavior of Kevlar® 49 fibre. J Mater Sci 19:57–72

# Chapter 58
# Effect of Low Temperatures on the Ballistic Limit of Hybrid Woven Composites

**Yougashwar Budhoo, Feridun Delale, and Benjamin Liaw**

**Abstract** This paper investigates the effect of low temperatures on the ballistic limit of hybrid woven composite panels (6 × 4 × 0.25 in.) at three different temperatures: R.T, −20°C and −60°C. Studies were also done on the stacking sequence and its effect on the ballistic limit at the above stated temperature. The studies were conducted by combining experimental and 3-D dynamic finite element approaches. The specimens tested were made of plain-weave hybrid S2 glass-IM7graphite fibers/toughened epoxy. The composite panels were impacted using an in-house gas gun. The time-histories of impact-induced dynamic strains was recorded. The damaged specimens were inspected visually and using the ultrasonic C-scan method. A 3-D dynamic finite element (FE) software, with Chang-Chang composite damage model, was then used to simulate the experimental results of the ballistic impact tests. Good agreement between experimental and FE results has been achieved.

**Keywords** Ballistic • Impact • Hybrid composite • Temperature • Ballistic limit • LS-DYNA

## 58.1 Introduction

Quite often composite materials are being exposed to various temperature environments. How the material respond to an impact under a change in environmental conditions therefore becomes necessary to be fully understood. Understanding the behavior a composite material gives researchers a greater opportunity to improve its performance. An important advantage of composite is the flexibility it gives the designer to tailor the new materials with properties to obtain peak performance for a particular application.

As the demand for materials with higher stiffness-to-weight and strength-to-weight ratios, tailoring flexibility, more resistant to harsh environment and damage tolerance arises; researchers and engineers seek to have more profound understanding of the behavior of different composites under various conditions. Investigations into woven composites have been extensive over recent years mainly because of their excellent impact resistance [1–3], dimensional stability, ease of handling etc. Some of these researches also focused on the effect of different environments, such as temperature [4–7] on the behavioral response of composite materials. This study focuses on the behavior of hybrid composites at room and low temperatures.

Y. Budhoo (✉)
Department of Engineering and Technology, Vaughn College of Aeronautics and Technology, 86-01, 23 Ave, East Elmhurst, Queens, NY 11369, USA
e-mail: yougashwar.budhoo@vaughn.edu

F. Delale • B. Liaw
Department of Mechanical Engineering, The City College of New York, 140th and Convent Ave, New York, NY 10031, USA

V. Chalivendra et al. (eds.), *Dynamic Behavior of Materials, Volume 1: Proceedings of the 2012 Annual Conference on Experimental and Applied Mechanics*, Conference Proceedings of the Society for Experimental Mechanics Series,
DOI 10.1007/978-1-4614-4238-7_58, 

## 58.2 Experimental Procedures

### *58.2.1 Materials*

The individual constituent materials combined to form the composite material studied in this research are: IM-7 graphite (IM7-GP 6000) and S2-glass (S2-4533 6000) woven fabrics in a SC-79 toughened epoxy matrix.

The IM-7 graphite woven fabric and SC-79 epoxy matrix form the non-hybrid composite called GR for short. The GR specimen contained 28 layers of graphite fabrics. For the case of the hybrid composite, it consists of a GR core, which is made up of 16 layers. This core is sandwiched between two outer laminates. Each outer laminate is made up of nine layers of S2-glass fabrics and SC-79 epoxy. The hybrid composite is called GL/GR/GL for short.

The composite was manufactured using the vacuum assisted resin transfer molding (VARTM) technique to stack the plain woven fabrics together. The specimens were cured at 177°C. Fiber volume fraction for all types was 55%. The final thickness of the specimens to be tested was 6.35 mm.

### *58.2.2 Experimental Setup*

The ballistic impact tests were performed using an in house designed gas-gun setup equipped with an environmental chamber. Inside the chamber is a fixture which holds the specimen to be impacted. A 22 caliber copper bullet fitted in to a plastic sabot was used as the projectile. The sabot is very important and provides a tight fit of the projectile in the gun barrel. The gun consists of helium tank, temporary gas storage vessel, solenoid valve and stainless-steel barrel (Figs. 58.1, 58.2, 58.3, and 58.4).

The figures above show the gas gun with the environmental chamber fixture and bullet with sabot. Figure 58.5 shows the dimensions and clamped areas of the specimen to be impacted. The boundary conditions of the clamped sides are assumed to be fixed-fixed.

Compressed helium was used to launch the projectile. The velocity of the projectile can be varied by changing the pressure used to launch the projectile or by changing the location of the bullet in the gun barrel. Placing the bullet closer to the exit of the gun barrel lowers the velocity of the bullet since the pressure used to launch the bullet will be smaller as compared to when the bullet is placed further away from the exit. The velocity of the bullet is measured as follows: two holes are drilled in the barrel where two laser beams continuously feed two optical sensors. When the projectile passes through the holes, it blocks the lasers, the sensors sense the change in intensity of light and an oscilloscope captures the voltage change. The velocity of the projectile is then calculated from the distance between the two sensors divided by the time that projectile

**Fig. 58.1** Ballistic gas gun

**Fig. 58.2** Ballistic gas gun and environmental chamber

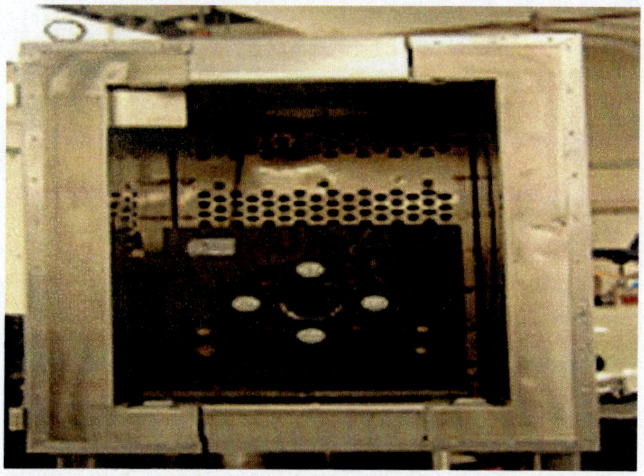

**Fig. 58.3** Fixture

**Fig. 58.4** Sabot and bullet

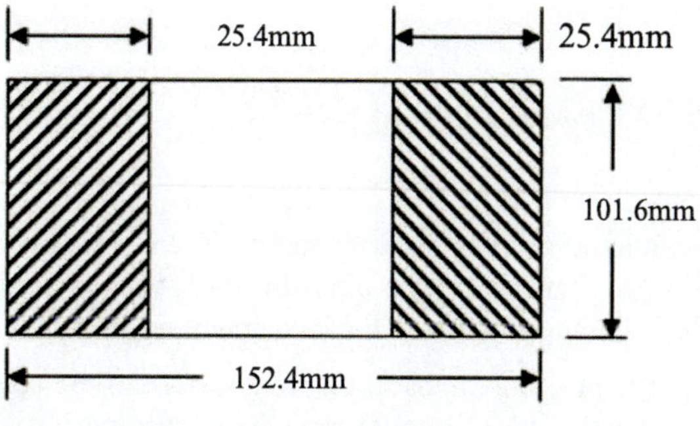


**Fig. 58.5** Ballistic impact test specimen, hatched region represents the clamped portion

spends traveling between the two holes. As the bullet exits the gun barrel it strikes the specimen that is in the environmental chamber at the specific test temperature. The specimens were impacted at three different temperatures: room temperature (RT), 75°C and 125°C. The specimens were first soaked at the desired test temperature for thirty minutes before impact. This is done to eliminate any temperature gradient in the test specimens (Fig. 58.6).

The force-time histories and the deflection-time histories cannot be measured with our set-up. However, two strain gages placed one inch away from the impact point, one in the radial direction and the other in the hoop direction, measure the dynamic strain histories of the specimen during impact.

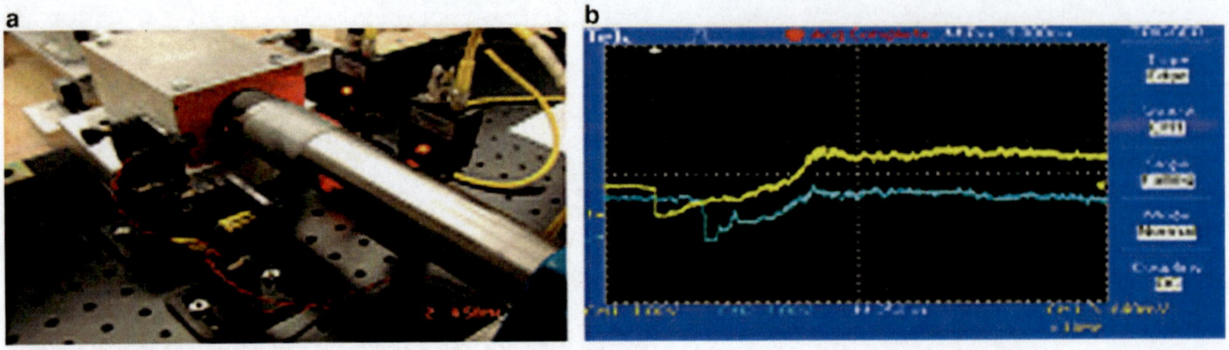


**Fig. 58.6** Setup to measurement velocity during ballistic impact. (**a**) Lasers and optical sensors. (**b**) Voltage-time histories

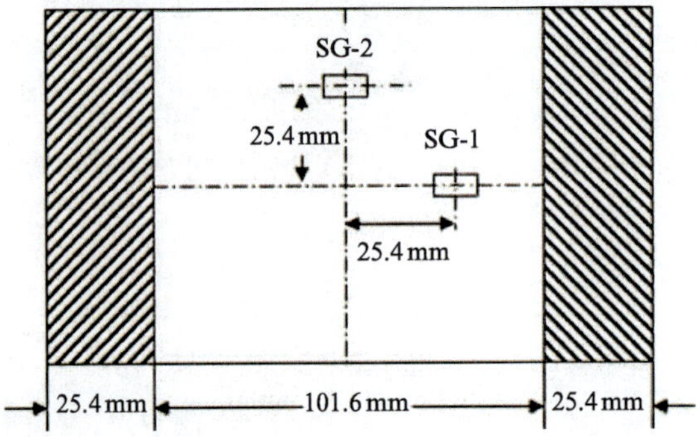


**Fig. 58.7** A typical composite panel specimen mounted with two strain-gages on the ballistic impact side

Two strain gauges were mounted on the impact side (i.e., the top surface) of each specimen. These strain gages were mounted at a distance of 25.4 mm from the center of the composite panel, where the impactor impacted the specimens. Figure 58.7 shows a schematic of the strain gage orientations and locations on the specimen. SG-1 measured the radial strain while SG-2 measured the hoop strain.

## 58.3 Experimental Results

During the ballistic impact tests at various test temperatures, the main objective was to obtain an estimate of the ballistic limit of the four layups of composite. There are many definitions of ballistic limit, however, in this research the definition of ballistic limit was taken from [8] where ballistic limit is looked at in terms of not only velocity but also energy. During ballistic impacts, one of three outcomes is possible [8]:

1. The projectile impacts the front surface of the composite with a certain velocity, perforates it and exits with a certain residual velocity. In this case the projectile's initial energy is larger than the energy than the composite can absorb.
2. The projectile partially penetrates the composite i.e. the projectile gets stuck in the composite or rebounds. In this case the projectile's initial energy is less than what the composite can absorb.
3. The projectile perforates the target completely with a zero exit/residual velocity. In this case the initial velocity of the projectile is called the **ballistic limit** or $V_{50}$. Here, the entire energy of the projectile is absorbed by the target.

In the experiments conducted, the procedure used to obtain the ballistic limit was by finding an impact velocity which causes perforation in the target by the projectile and one that causes only partial penetration of the projectile in the target. More tests were conducted within this range of these two extremes velocities in order to find the maximum impact velocity which causes a zero exit/residual velocity. Due to the limited number of specimens available for testing, the exact ballistic

**Table 58.1** Parameters of ballistic impact tests on woven GL/GR/GL composite panels

| Materials | Temperature | Impact velocity (m/s) | Penetration type | Impact energy (J) |
|---|---|---|---|---|
| GL/GR/GL | R.T | 306 | Partial | 140 |
| | R.T | 313 | Partial[a] | 147 |
| | R.T | 319 | Full | 153 |
| | −20°C | 301 | Partial | 136 |
| | −20°C | 320 | Full | 154 |
| | −60°C | 295 | Partial | 131 |
| | −60°C | 322 | Partial | 156 |
| | −60°C | 345 | Full | 179 |
| | −60°C | 368 | Full | 203 |

[a]Severe splitting on the back surface of specimen

**Table 58.2** Parameters of ballistic impact tests on woven GR/GL/GR composite panels

| Materials | Temperature | Impact velocity (m/s) | Penetration type | Impact energy (J) |
|---|---|---|---|---|
| GR/GL/GR | R.T | 327 | Partial | 160 |
| | R.T | 343 | Partial | 176 |
| | R.T | 362 | Full | 197 |
| | −20°C | 299 | Partial | 134 |
| | −20°C | 324 | Partial[a] | 158 |
| | −20°C | 338 | Full | 171 |
| | −60°C | 275 | Partial | 113 |
| | −60°C | 297 | Partial | 132 |
| | −60°C | 312 | Full | 146 |
| | −60°C | 358 | Full | 192 |
| | −60°C | 390 | Full | 228 |

[a]Severe splitting on the back surface of specimen

limit could not be obtained experimentally. Hence, minimal tests were conducted as needed to obtain a reasonable range in velocities i.e. minimum velocity that causes perforation and maximum velocity that causes partial penetration.

In Tables 58.1 and 58.2, the impact velocities which cause full and partial penetration in the composite specimens are given.

The interval of $V_{50}$ velocity for all panels was roughly estimated using experimental data.

The impact resistance and ballistic limit as a function of temperature can be seen, but to have a clearer understanding of the behavior, the impact velocity has been plotted against temperature in Fig. 58.8. The ballistic limit in Fig. 58.8 is obtained by taking the average of the minimum velocity that causes perforation and the maximum velocity that causes partial penetration at that particular test temperature. The error bar is added to each curve to represent the difference in these two velocities. It can be seen that at R.T the impact resistance of the hybrid composite with the GR as the outer layer performs better than the hybrid with GL as the outer layer. This behavior becomes reverse as the temperature drops to approximately below −50°C. R.T and 75°C was added to the curve just for comparison and was taken from [9]. The experimental procedure portion was also taken from previous work [9]

### *58.3.1 Dynamic Strains*

The ballistic impact event takes place in an extremely short time. Accordingly the frequency of the recording device should be high enough to capture the strain gauge readings in such a short time. The oscilloscope used in this study to capture the strain output from the strain gages has a maximum frequency of 500 KHz. In other words, it was only able to record a strain data every 2 μs.

Assuming that the ballistic impact event takes 100 μs, the dynamic strain curve from experiment would have captured only 50 data points, resulting in significant under sampling. Thus, the strain curve constructed using these data points will not represent the actual response of the composite during the impact. The strain measurements obtained are mainly for vibration of impacted composite panel and not necessarily the transient response of composite during ballistic impact.

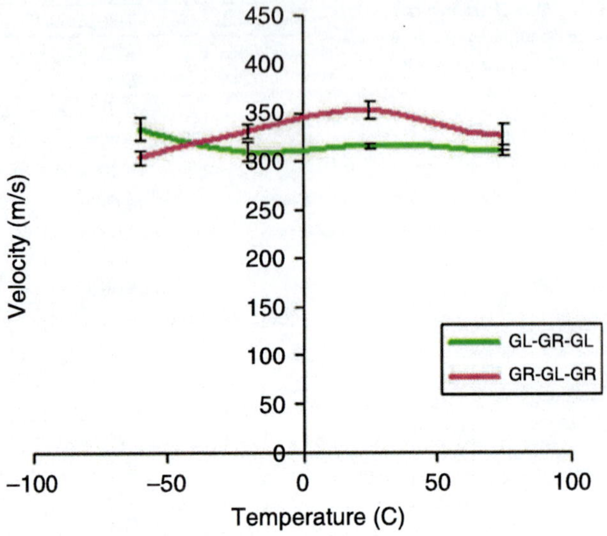


**Fig. 58.8** Temperature versus ballistic limit for all the layups

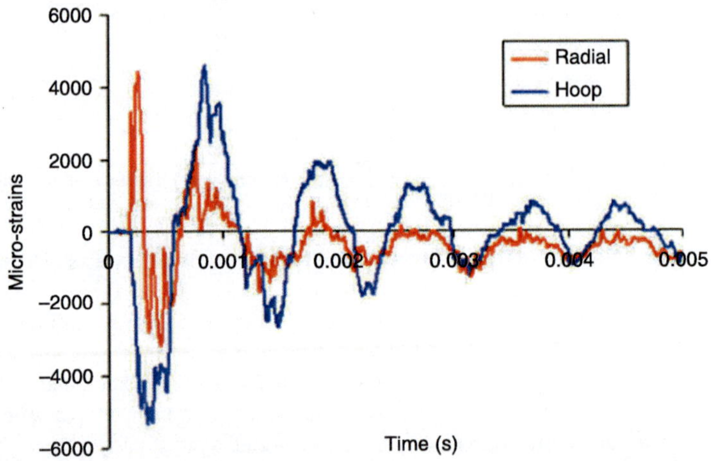


**Fig. 58.9** Hoop and radial experimental strain histories for GL/GR/GL at −60°C

Figures 58.9 and 58.10 show the dynamic strain histories for GL/GR/GL and GR/GL/GR composites at −60°C .It is observed that for the hybrid GL/GR/GL, the magnitude of the strains is larger than that of GR/GL/GR. One explanation for this is probably because GR is more brittle than GL, hence instead of large deformations, GR will fail. One important piece of information to keep in mind while comparing these strains is that they are measured on the surface of the impacted specimen. From [9] this behavior was also seen at high temperatures.

### 58.3.2 *Post Mortem Analysis of Specimens*

By looking at the specimens it is not possible to understand the behavior of the material during the impact event. Some information can however be obtained by inspecting the specimen, to do this, the impacted specimens were C-scanned and finally cut in two halves to analyze the cross-sectional view. Figures 58.11 and 58.12 show ultrasonic C-scans of the impacted composites. Delamination in hybrid composites where GL was used as an outer skin was more pronounced compared to that in the other hybrid composites. Looking that the C-Scan images the two hybrid seem to exhibit a slightly different behavior as the ballistic limit is approached, but the amount of temperature points is not sufficient to make a general conclusion. Beyond the ballistic limit, however, delamination was observed to decrease as the velocity increases.

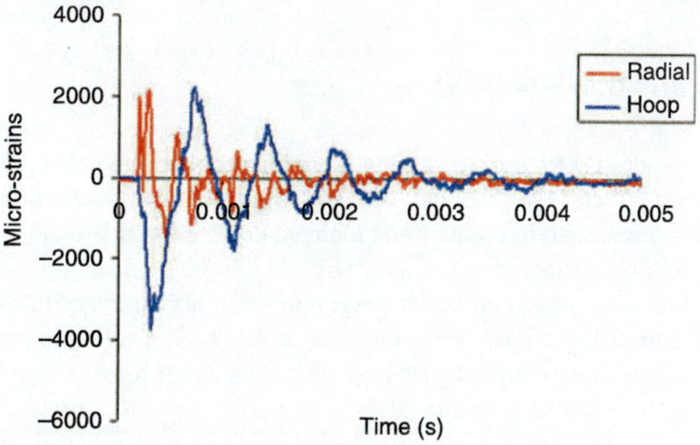


**Fig. 58.10** Hoop and radial experimental strain histories for GR/GL/GR at −60°C

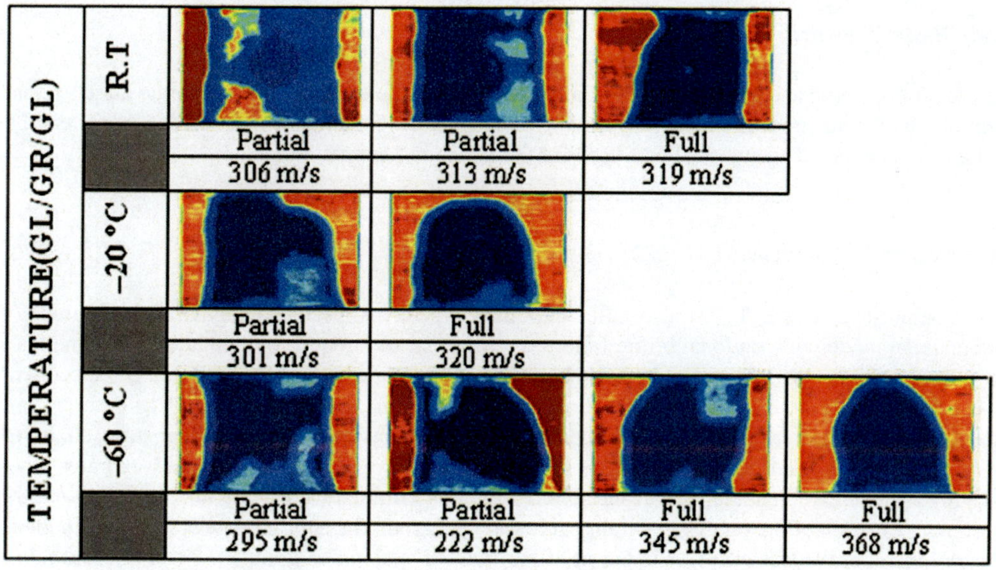


**Fig. 58.11** Ultrasonic C-scan images of impacted GL/GR/GL composites at various test temperature and velocities

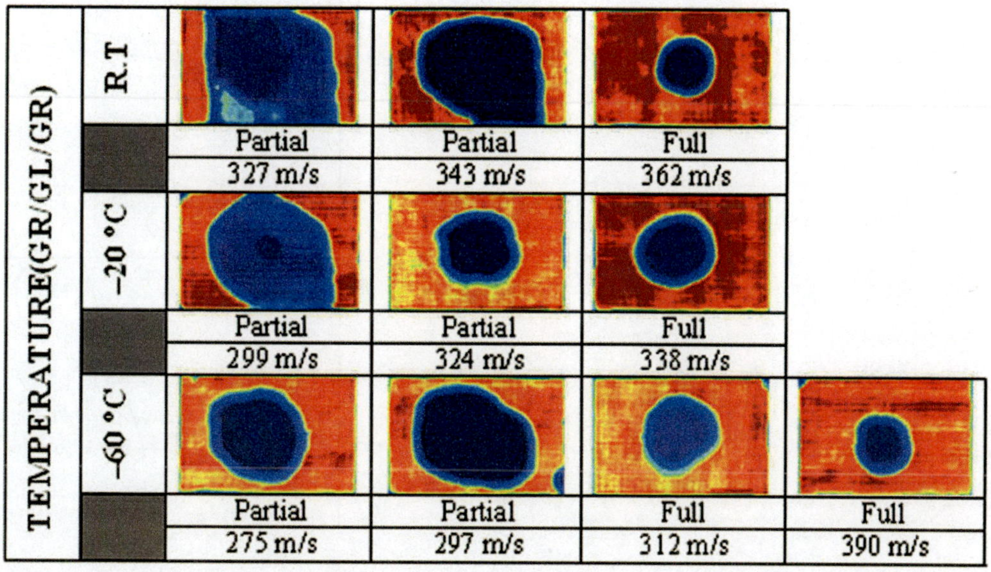


**Fig. 58.12** Ultrasonic C-scan images of impacted GR/GL/GR composites at various test temperature and velocities

## 58.4 Dynamic Finite Element Simulation

Finite element analysis (FEA) is a very useful tool in solving complex problems and also predicts the behavior of structures when actual testing may not be cost effective or practical. In this study, ballistic impact problems at various temperatures are modeled and analyzed using a 3-D nonlinear, dynamic finite element code. To validate the reliability of the model, the finite element results were compared to those obtained experimentally.

Since the geometries of the composite plate and bullet is symmetric, it is customary that only a quarter of these parts be modeled to save computational time. All the parts were modeled using 8-node solid elements. All modeling and meshing was done using the same commercial finite element software. The FEA setup for this research is the same as the author's previous work and is the same as the write up found in [9]

### *58.4.1 Material Models*

#### 58.4.1.1 Mat_Plastic_Kinematic (MAT_003)

This model is used for isotropic and kinematic hardening plasticity with the option of including rate effects. $E_t$ is the slope of the second arm of the bilinear stress strain curve (Fig. 58.13) and $l_o$ and $l$ are undeformed and deformed lengths of uniaxial tension specimen respectively. This material was used for the 22 caliber copper bullet.

#### 58.4.1.2 Mat_Composite_Damage(MAT_022)

In modeling the composite plate, MAT_022 also called the Chang-Chang composite damage model was used. This model is an orthotropic material model with optional brittle failure for composites. It requires nine material parameters to represent the composite material. From tensile testing at high temperatures, the material behavior of GR and GL specimens was found to be non linear. In order to use MAT_22 (best suited to describe a linear elastic behavior) to describe our materials' behavior, we consider the strain energy. Here we calculate the strain energy from the non linear stress strain behavior, and then knowing the strain energy and the initial Young's modulus, a new linear stress-strain curve is redrawn where the strain energy and Young's modulus is maintained but the failure strain and ultimate stress are changed. Figure 58.14 gives an example of the approximation of the true stress-strain curve for a material. The approximate curve was inputted in the finite element software instead of the true curve. A linear fit of the true curve could be used, but wasn't because this curve only

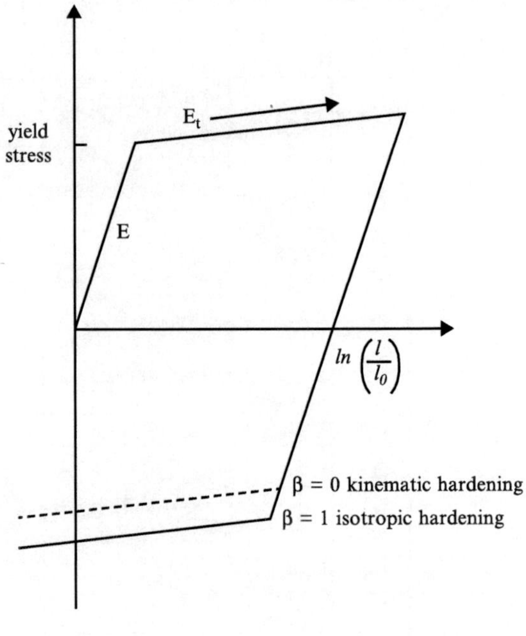


**Fig. 58.13** Elastic-plastic behavior with kinematic and isotropic hardening (Picture is taken from LS-DYNA theoretical manual)

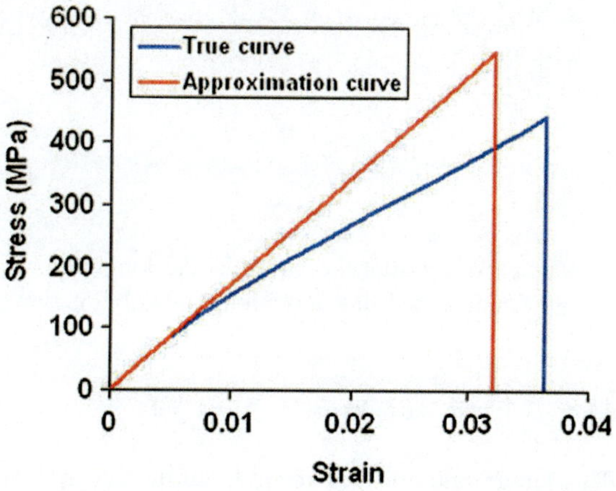


**Fig. 58.14** Linearization of non linear stress-strain curve obtained from tensile testing while maintaining strain energy

represent material behavior at quasi-static loading. In the approximate curve, the failure strain is smaller and ultimate strength is larger which the expected material behavior at higher strain rate.

In the Chang-Chang model five material parameters are used in three possible failure criteria which are as follows:

$S_1$ = Longitudinal tensile strength
$S_2$ = Transverse tensile strength
$S_{12}$ = Shear strength
$C_2$ = Transverse compressive strength
$\alpha$ = Non linear shear stress parameter

Here, $S_1$, $S_2$, $S_{12}$, and $C_2$ are obtained from material strength measurements and $\alpha$ is defined by a material shear stress-strain measurements.

The three failure criteria are matrix cracking, fiber breakage and compressive failure.

1. *Matrix cracking:*

$$F_{matrix} = \left(\frac{\sigma_2}{S_2}\right)^2 + \bar{\tau}$$

where

$$\bar{\tau} = \frac{\frac{\tau_{12}^2}{2G_{12}} + \frac{3}{4}\alpha\tau_{12}^4}{\frac{S_{12}^2}{2G_{12}} + \frac{3}{4}\alpha S_{12}^4}$$

here $\sigma_2$ is the tensile stress in the transverse direction whereas $\tau_{12}$ is the in-plane shear stress between fibers and matrix. Failure is assumed whenever $F_{matrix} > 1$. Once this type of failure occurs, the material constants $E_2$ (Young's modulus in the transverse direction), $G_{12}$ (in-plane shear modulus in the 1-2 plane), $\nu_{12}$ and $\nu_{21}$ (generalized Poisson's ratios in the 1-2 plane) are set to zero.

2. *Compressive failure:*

$$F_{comp} = \left(\frac{\sigma_{12}}{2S_{12}}\right)^2 + \left[\left(\frac{C_2}{2S_{12}}\right)^2 - 1\right]\frac{\sigma_2}{S_2} + \bar{\tau}$$

Failure is assumed when $F_{comp} > 1$. When this type of failure occurs, the material constants $E_2$, $\nu_{12}$ and $\nu_{21}$ are all set to zero.

3. *Fiber breakage:*

$$F_{fiber} = \left(\frac{\sigma_1}{S_1}\right)^2 + \bar{\tau}$$

where $\sigma_1$ is the tensile stress in the longitudinal direction. Failure is assumed when $F_{fiber}>1$. After fiber breakage, $E_1$ (Young's modulus in the longitudinal direction), $E_2$, $G_{12}$, $\nu_{12}$ and $\nu_{21}$ are all set to zero.

#### 58.4.1.3 Mat_Add_Erosion (MAT_00)

This failure criterion was added to all the materials used in our modeling although it is used mostly for material models which do not have a failure or erosion criterion.

It is basically an additional criterion which when satisfied; the element is deleted from the calculation. It is a good tool to use when comparison is needed between animation and real specimen, because it prevents animation from having excessively distorted elements. In this model, erosion is set to occur when the tensile failure strain of the material is reached.

### 58.4.2 *Contact and Delamination Model*

The contact between the composite layers was defined using the option, CONTACT_AUTOMATIC_SURFACE_TO_SURFACE_TIEBREAK.

By using this contact definition, delamination between the composite layers is governed by the criterion:

$$\left[\frac{\max(0, \sigma_n)}{NFLS}\right]^2 + \left(\frac{\sigma_s}{SFLS}\right)^2 \geq 1$$

where $\sigma_n$ and $\sigma_s$ are normal and shear stresses acting on the layer interface, respectively, while *NFLS* and *SFLS* are normal and shear failure strengths of the layer interface, respectively. It is assumed that delamination occurs mainly by shear failure; hence, NFLS is set to a very large number so delamination in the model is controlled by SFLS. Since SFLS is not known experimentally, it was varied in order to fit the experimental results. Additionally, ERODING_SURFACE_TO_ SURFACE contact model was used between the impactor and the composite. This model allows elements to be eroded when certain failure criteria are met. In this study strain-based failure criterion was used for element erosion; that is, when $\varepsilon \geq \varepsilon_{erosion}$, the element was eroded and removed from calculation. The contact and delamination models as well as the erosion option were used for all composites.

### 58.4.3 *Boundary Conditions*

To model the impact event properly, the correct boundary conditions should be chosen. Preliminary simulations were carried out, and it was seen that a fixed-fixed boundary condition was the best suited to model the low velocity impact. A fixed-fixed boundary condition was therefore chosen for all the low velocity impact simulations.

Figure 58.15 shows the mesh of a quarter model of the plate used in simulating the drop impact test. Fine mesh around the center and course mesh away from center was used.

Figure 58.16 shows the entire drop weight model of both the composite panel and the impactor for a non hybrid specimen for the low velocity impact. The material properties used to define the composite were obtained from tensile testing and are presented in Table 58.1.

In Table 58.3, the material properties used in the FEA simulation is presented. These properties were taken from previous work [10, 11]. The entire table of material properties for high and low temperatures is presented here, but only the high temperature properties were used in the modeling.

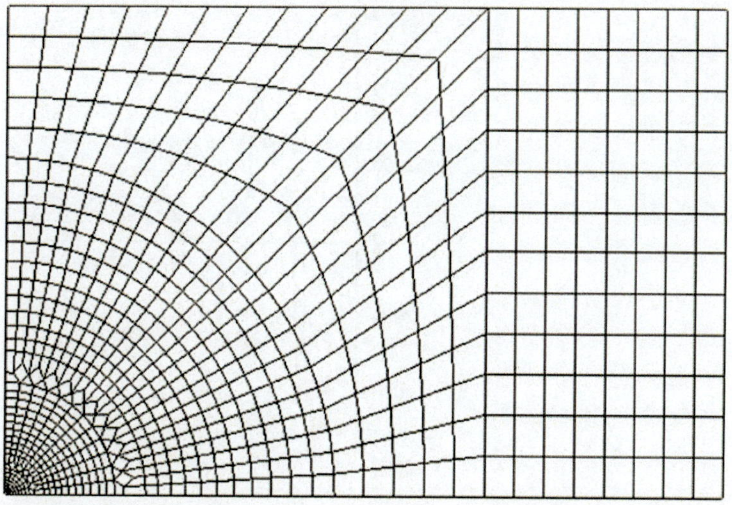

**Fig. 58.15** Quarter model of meshed plate used in ballistic impact simulation

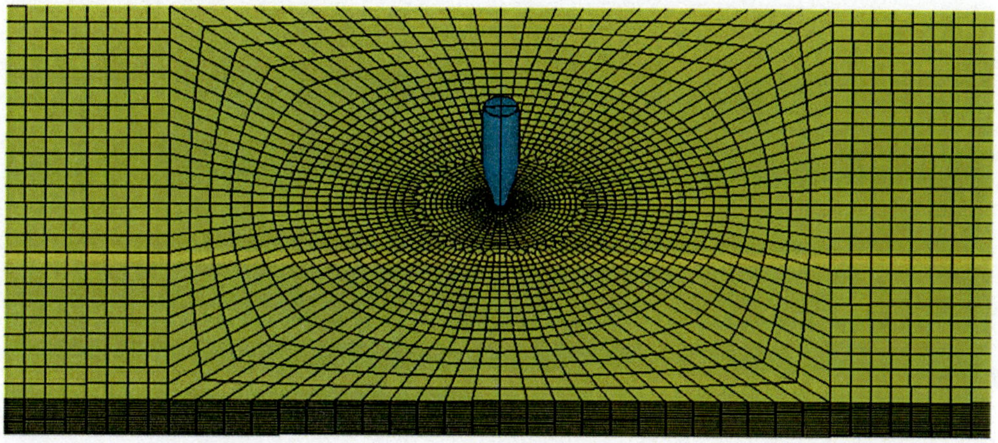

**Fig. 58.16** Mesh of entire composite plate and projectile for ballistic impact

**Table 58.3** Key material properties used in modeling

| | | $E_1 = E_2$ ($GP_a$) | $E_3$ ($GP_a$) | $G_{12}$ ($GP_a$) | $G_{13} = G_{23}$ ($GP_a$) | $V_{12}$ | $V_{13} = V_{23}$ | P (kg/m$^3$) |
|---|---|---|---|---|---|---|---|---|
| Glass | −60°C | 19.39 | 7.26 | 2.98 | 2.65 | 0.152 | 0.370 | 1766 |
| | −20°C | 18.92 | 6.78 | 2.78 | 2.49 | 0.145 | 0.631 | 1761 |
| | R.T | 17.04 | 6.50 | 2.74 | 2.30 | 0.130 | 0.410 | 1755 |
| | 75°C | 16.13 | 5.85 | 1.89 | 2.03 | 0.128 | 0.436 | 1750 |
| | 125°C | 15.03 | 5.46 | 1.44 | 1.87 | 0.121 | 0.453 | 1744 |
| Graphite | −60°C | 43.82 | 6.43 | 3.97 | 2.32 | 0.140 | 0.382 | 1392 |
| | −20°C | 43.71 | 5.75 | 3.93 | 1.96 | 0.129 | 0.462 | 1387 |
| | R.T | 42.55 | 5.20 | 3.08 | 1.81 | 0.121 | 0.427 | 1381 |
| | 75°C | 41.49 | 3.43 | 1.99 | 1.05 | 0.116 | 0.627 | 1375 |
| | 125°C | 40.39 | 2.53 | 1.67 | 0.76 | 0.099 | 0.649 | 1370 |

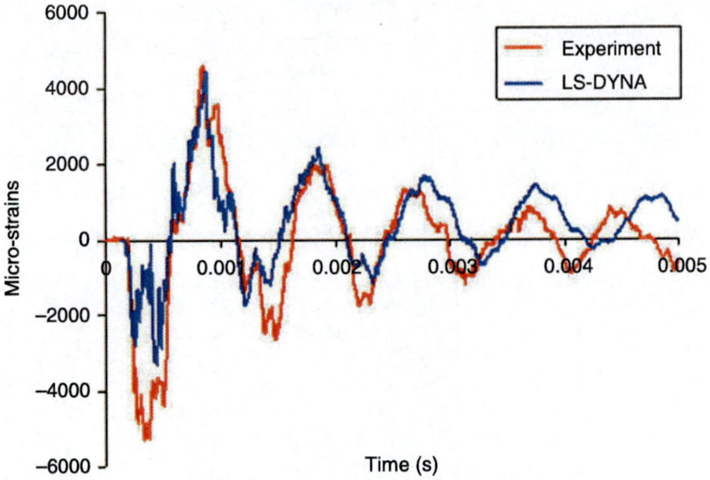


**Fig. 58.17** Comparison of LS-DYNA and experimental strain histories for GL/GR/GL at −60°C

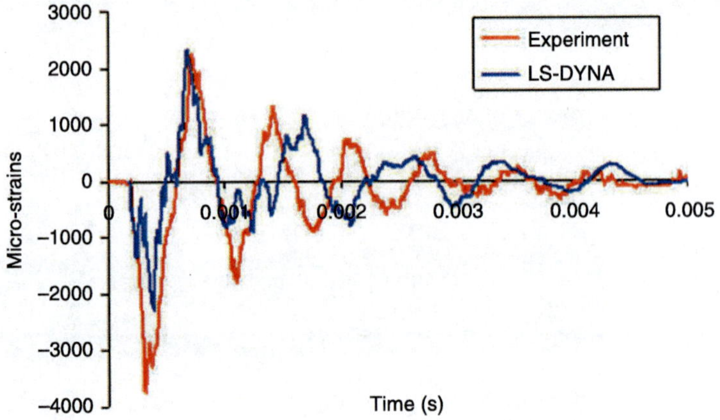


**Fig. 58.18** Comparison of LS-DYNA and experimental strain histories for GR/GL/GR at −60°C

### *58.4.4 Determination of Damping Constant*

In analyzing the dynamic strain histories recorded during the ballistic impact tests, it can be seen that the response of the composite was damped. Slipping between layers of laminated structures such as composites introduces high internal damping to the system; therefore damping was required to be incorporated into the FE model. However, the damping constant is not easy to determine. In order to successfully incorporate a reasonable damping constant into the model, two options were used. The first option is to incorporate damping between layers in the form of a viscous damping coefficient in terms of percentage of critical damping. This helps to eliminate undesirable oscillation in contact. Between layers of similar materials, 10% damping was used and between layers of dissimilar materials, 20% damping was used. The second option where damping was introduced was through DAMPING_PART_STIFFNESS. In this option a Raleigh damping coefficient is specified which is uniformly applied to each element of the part. LS-DYNA suggests a value of 0.01 to 0.25 or in other words, 1–25% of the critical damping coefficient. Different values were used in the model and the strain outputs from the model were compared to those obtained experimentally. Best results were obtained when 10% damping was used.

### *58.4.5 Comparison of Strain-Time Histories*

The dynamic strain histories and post-impact damage patterns were compared between experimental and numerical results, since it was not possible to obtain force and energy-time histories experimentally. In Figs. 58.17 and 58.18, the dynamic strains obtained from experiments were in good agreement to the numerical results with some magnitude and frequency mismatch. One possible reason for this may be due o the limitations of the material model used in representing the actual material behavior.

**Fig. 58.19** Comparisons of post impact damage patterns for non-hybrid GL/GR/GL impacted at –60°C and 345 m/s

**Fig. 58.20** Comparisons of post impact damage patterns for hybrid GR/GL/GR impacted at 125°C and 368 m/s

### *58.4.6 Comparison of Post-Impact Damage Patterns*

Figures 58.19 and 58.20 show the comparison of experimentally obtained and FE predicted post impact damage patterns of ballistic impact tests. Significant delamination between dissimilar layers during the ballistic impact tests of the hybrid composite was predicted by the finite element model. Looking at the actual specimen it can be seen that the finite element model gave a good prediction of this delamination. Looking at the cross-section of the specimens, the broken fibers are considered as failed. In the elements in the finite element model these fibers are not seen because of a strain based element erosion criterion that was used. This criterion is very important in the modeling to prevent extreme distortion of the element which can lead to numerical instability.

## 58.5 Conclusions

From the results discussed above, the following conclusions can be drawn:

1. The FE model incorporated with MAT_22 (Chang-Chang material model) was successfully used to simulate the ballistic impact tests.
2. The validity of the FE model for ballistic impact was confirmed after comparison with experimental results such as strain histories and post-impact damage patterns.
3. Experiments show that GR/GL/GR provides a better impact resistance than GL/GR/GL at temperatures above approximately 50°C.
4. After ballistic impact tests, specimens and projectile were examined and fiber shear out, shear plugging, delamination, matrix failure, fiber breakage and projectile deformation were observed.

## References

1. Khojin AS, Bashirzadeh R, Mahinfalah M, Jazar RN (2006) The role of temperature on impact properties of Kevlar/fiber glass composite laminates. Compos B Eng 37(7-8):593–602
2. Khojin AS, Mahinfalah M, Bashirzadeh R, Freeman B (2007) Temperature effects on Kevlar/hybrid and carbon fiber composite sandwiches under impact loading. Compos Struct 78(2):197–206
3. Wu E, Chang LC (1995) Woven glass/epoxy laminates subject to projectile impact. Int J Impact Eng 16(4):607–619

4. Hirai Y, Hamada H, Kim JK (1998) Impact response of woven glass-fabric composites- II. Effect of temperature. Compos Sci Technol 58 (1):119–128
5. Im KH, Cha CS, Kim KS, Yang IY (2001) Effects of temperature on impacts damages in CFRP composite laminates. Compos B 32:669–682
6. Parlevliet PP, Bersee HEN, Beukers A (2006) Residual stresses in thermoplastic composites: a study of the literature. Part 1: formation of residual stresses. Appl Sci Manuf 37(11):1847–1857
7. Dlouhy I, Chlup Z, Boccaccini DN, Atiq S, Boccaccini AR (2003) Fracture behavior of hybrid glass matrix composites: thermal ageing effects. Compos A Appl Sci Manuf 34(12):1177–1185
8. Naik NK, Shrirao P, Reddy BCK (2006) Ballistic impact behavior of woven fabric composites: formulation. Int J Impact Eng 32:1521–1552
9. Budhoo Y, Liaw B, Delale F (2010) Effect of temperature on ballistic impact of graphite/epoxy composite. In: Proceedings of the IMAC2011, annual conference and exposition on experimental and applied mechanics. Denver, CO, USA
10. Budhoo Y, Delale F, Liaw B (2011) Temperature effect on tensile testing of hybrid and non-hybrid composites. In: Proceedings of the SEM 2011 annual conference and exposition. Montville, CT, USA
11. Budhoo Y, Delale F, Liaw B (2011) Temperature effect on poisson's ratio of woven composites. In: Proceedings of the SEM 2011 annual conference and exposition. Montville, CT, USA

# Chapter 59
# Dynamics of Multilayered Structures of VACNTs with Metallic Inter-layers

**Ramathasan Thevamaran and Chiara Daraio**

**Abstract** We study the dynamic behavior of periodic multilayered structures composed of compliant freestanding vertically aligned carbon nanotube (VACNT) arrays alternating with rigid metallic inter-layers, subjected to small transient excitations. The presence of intrinsically nonlinear effects in VACNTs arrays enable us to create materials and structures with tunable properties in which, for example, the stiffness can be tuned over a large range. As a result, the dynamic stress propagation properties in these materials can be tailored to specific applications.

**Keywords** Carbon nanotubes • Wave propagation • Multilayered structures • Nonlinear dynamics • Localized deformation

## 59.1 Introduction

VACNT arrays synthesized by thermal chemical vapor deposition present a nonlinear foam-like behavior in quasistatic cyclic compressive loading and recoverable deformations up to 80% [1, 2]. In addition, they exhibit intrinsically graded functional properties that lead to strain localization and characteristic modes of deformation and buckling [1–3]. Further, they exhibit nonlinear viscoelastic behavior [4, 5] in response to applied stresses. These characteristics enable effective energy absorption when they are subjected to dynamic impacts [2, 6].

## 59.2 Synthesis of VACNTs and Experimental Setup

A floating catalyst thermal chemical vapor deposition (tCVD) process is used in our laboratory to synthesize vertically aligned carbon nanotubes [2]. The system consists a tube furnace with about 15 cm heating zone operating at 827°C temperature and at atmospheric pressure (Fig. 59.1). A feedstock consisting of ferrocene (Fe catalyst precursor) and toluene (carbon source) at the ratio of 1 g:50 ml is injected into the furnace tube at 1 ml $min^{-1}$, at a constant flow rate of 800 sccm, with argon as the carrier gas. The carbon nanotubes are grown on thermally oxidized silicon wafers. This growth process results in the growth of VACNT arrays of thickness in the order of ~1 mm and average bulk density of 0.229 $g/cm^3$. The synthesized samples are removed from substrate with a razor blade.

After removal from the growth substrate, the VACNT samples are cut into disks of diameter = 4.76 mm and length ~ 1 mm, and they are assembled in a multilayered structure alternating them with stainless steel cylinders (Fig. 59.2). The total length of the multilayer stack tested consists of 20 VACNT layers and 20 steel cylinders. The stainless steel cylinders have diameter and length = 4.76 mm. We test the dynamic response of these multilayer stacks subjected to small velocity impacts. For testing, we place the multilayer stacks in a support system composed of four horizontally aligned polycarbonate rods of 4.76 mm diameter and confine them with a rigid block mounted on an optical table.

R. Thevamaran (✉)
Graduate Aerospace Laboratories, California Institute of Technology, 1200 E California Blvd., MC 104-44, Pasadena, CA 91125, USA
e-mail: rthevama@caltech.edu

C. Daraio
Graduate Aerospace Laboratories, California Institute of Technology, 1200 E California Blvd., MC 105-50, Pasadena, CA 91125-3405, USA

V. Chalivendra et al. (eds.), *Dynamic Behavior of Materials, Volume 1: Proceedings of the 2012 Annual Conference on Experimental and Applied Mechanics*, Conference Proceedings of the Society for Experimental Mechanics Series,
DOI 10.1007/978-1-4614-4238-7_59, © The Society for Experimental Mechanics, Inc. 2013

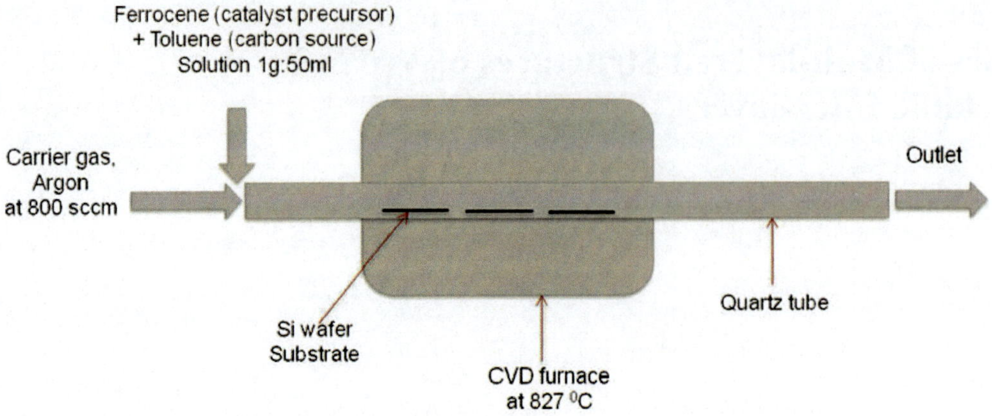


**Fig. 59.1** Schematic of the thermal chemical vapor deposition (tCVD) process for VACNT synthesis

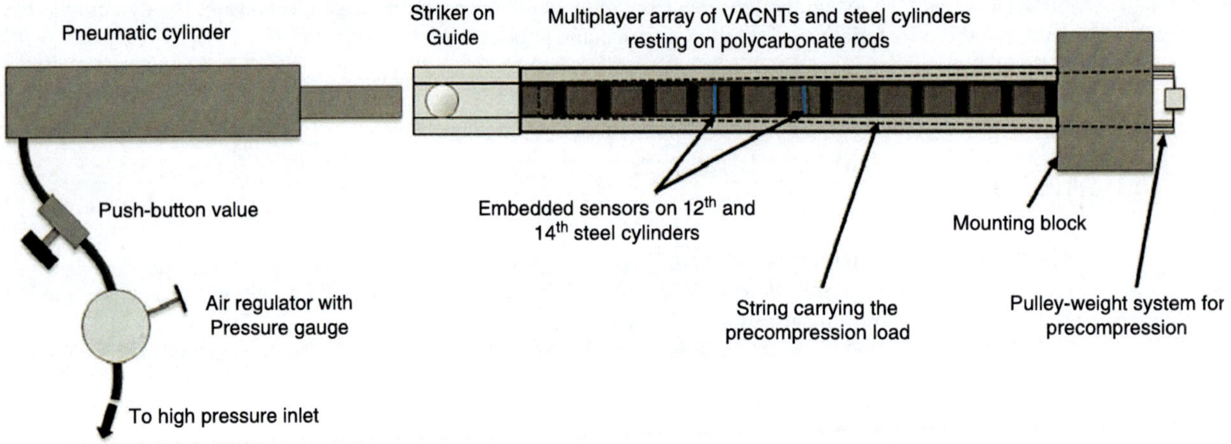


**Fig. 59.2** Schematic diagram showing a top view of the experimental set up used for the dynamic testing of the multilayered structures

Piezoelectric sensors, calibrated by the conservation of momentum, are embedded on the 12th and 14th steel cylinders (counting from the impacted end of the system) to measure the force-time profile of the propagating stress wave. Spherical strikers of 4.76 mm diameter were launched on a low friction PTFE (polytetrafluoroethylene) guiding rail to excite the multilayered structures. In our tests we used two different types of strikers: a PTFE sphere having a mass of 0.12 g and a stainless steel sphere having a mass of 0.45 g. Different static pre-compressions ($F_s$) were applied to the multilayer stacks using a pulley-weight system, to verify the tunability of the system's dynamic response.

## 59.3 Results and Discussion

We calculated the effective velocity of the propagating wave by dividing the measured distance between the sensors by the wave's travel time between them. The effective velocity of the propagating wave was observed to decrease with increasing dynamic forces normalized by the applied precompression ($F_d/F_s$), following a power-law decay (Fig. 59.3). Here the dynamic force, $F_d$ is taken to be the average of the peak forces measured by the sensors. The application of static pre-compression affects the response of the system: Higher precompression values result in higher wave speeds and in a decrease of the exponent of the power-law fits of Fig. 59.3. This is likely a result of the stiffening of the VACNT layers due to pre-stress.

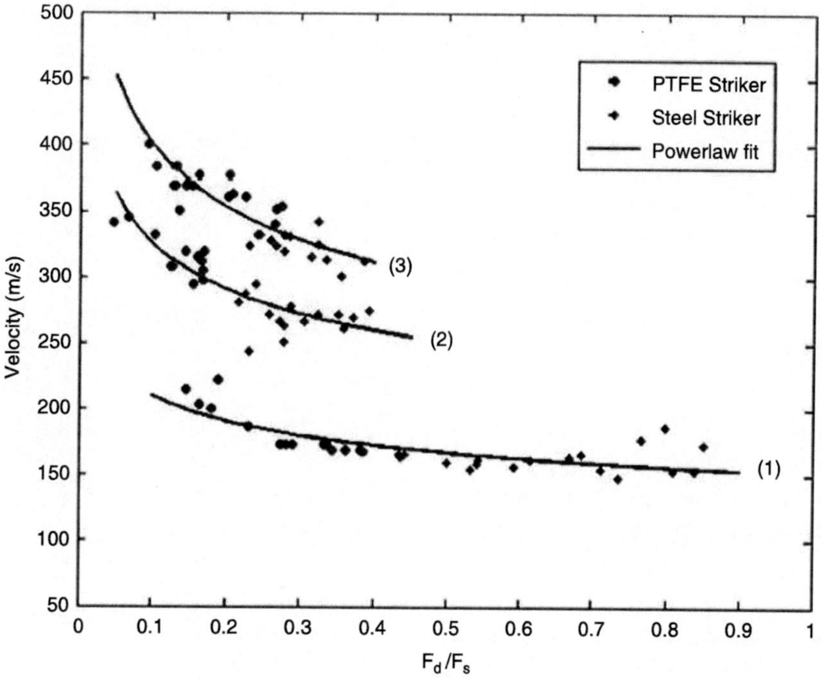


**Fig. 59.3** Experimental data showing the relation between the effective wave speed in the multilayer stacks as a function of the dynamic force amplitude ($F_d$) normalized by the static pre-compression ($F_s$). The data was obtained for three different values of static pre-compressions: (1) 5N, (2) 15N and (3) 20N. The *solid lines* show the power-law fits

Understanding the fundamental characteristics of these multilayered systems subjected to dynamic impact loading will enable the design of new, lightweight carbon nanotube-based materials with tunable properties. Applications for such materials can be found in vibration attenuation and small velocity impact absorption.

**Acknowledgement** This work is supported by the Institute for Collaborative Biotechnologies, under contract W911NF-09-D-0001 with the Army Research Office.

## References

1. Cao AY, Dickrell PL, Sawyer WG, Ghasemi-Nejhad MN, Ajayan PM (2005) Super-compressible foamlike carbon nanotube films. Science 310 (5752):1307–1310
2. Misra A, Greer JR, Daraio C (2008) Strain rate effects in the mechanical response of polymer anchored carbon nanotube foams. Adv Mater 20:1–5
3. Yaglioglu O (2007) Carbon nanotube based electromechanical probes. Massachusetts Institute of Technology, Ph.D. thesis, Cambridge, MA
4. Suhr J, Victor P, Ci L, Sreekala S, Zhang X, Nalamasu O, Ajayan PM (2007) Fatigue resistance of aligned carbon nanotube arrays under cyclic compression. Nat Nanotechnol 2(7):417–421
5. Deck CP, Flowers J, McKee GSB, Vecchio K (2007) Mechanical behavior of ultralong multiwalled carbon nanotube mats. J Appl Phys 101:023512-1-9
6. Daraio C, Nesterenco VF, Jin S (2004) Highly nonlinear contact interaction and dynamic energy dissipation by forest of carbon nanotubes. Appl Phys Lett 85(23):5724–5727

# Chapter 60
# Dynamic Contact Mechanics of Geologic Materials: Experiments, Modeling and Applications

**D.M. Cole, M.A. Hopkins, S.W. Broadfoot, and S.D. Sloan**

**Abstract** Although stress wave propagation in granular geologic materials is generally recognized to depend critically on the physical characteristics of the medium, there is a lack of understanding of the physical basis of this process. This knowledge gap inhibits the development of mechanistic models of wave propagation in granular materials and of soil-sensor interaction. To address this important issue, we are currently engaged in an effort to quantify and model the dynamic contact mechanics of granular materials using a combination of micro-scale experimentation, numerical modeling and field validation.

We have developed a dynamic contact experimental system that operates at sub-kHz frequencies. This laboratory based system employs piezoelectric crystals to monitor the force transmitted though contacts in a linear array of grains of geologic materials. The deformation over single or multiple contacts is measured, allowing the constitutive behavior of the contacts to be quantified. The paper describes this system in a configuration for normal loading, and presents initial results for a large-grained quartz sand. Our approach to developing detailed contact laws (that account for both elastic and inelastic deformation) is described.

The implementation of material-specific contact laws in a discrete element model is also described with a focus on small normal displacements. The DEM simulates wave propagation as a function of measurable grain-scale properties and includes the simulation of polyellipsoidal and polyhedral particle shapes and Hertzian and non-Hertzian laws for contact stiffness and frictional loss. In addition, the paper describes a number of research interests that are supported by this work.

**Keywords** Contact mechanics • Laboratory experiments • Geologic material • Dynamic Response

## 60.1 Introduction

We are currently engaged in an effort to improve our fundamental understanding of the dynamic contact mechanics of geologic materials using a combination of micro-scale experimentation and numerical modeling. There is increasing interest in the use of in-ground sensors for threat detection and in support of this work, a great deal of attention has focused on the acquisition and interpretation of seismic/acoustic signals on the macroscopic scale. This approach is proving to be highly effective, but the details of the underlying micro-physical processes associated with signal detection are not well understood and require further study. The present work pays particular attention to the grain-scale characteristics of the medium, and involves grain-grain and grain-sensor contact experiments using naturally occurring granular materials to arrive at a physically based understanding of the underlying contact mechanics.

The effort involves the development of dynamic contact laws (with and without damage) based on grain-scale experiments and the implementation of those laws in a discrete element model capable of simulating wave propagation through the soil and the subsequent interaction of the wave with buried sensors. Validation of the numerical simulations is planned and will involve physical test beds that correspond precisely to the virtual test bed constructed for the discrete element model. As of this writing, we have developed equipment for the dynamic contact experiments and conducted

D.M. Cole (✉) • M.A. Hopkins
ERDC-CRREL, 72 Lyme Rd, Hanover, NH 03755, USA
e-mail: David.M.Cole@usace.army.mil

S.W. Broadfoot • S.D. Sloan
ERDC-GSL, 3909 Halls Ferry Road, Vicksburg, MS 39180-6199, USA

V. Chalivendra et al. (eds.), *Dynamic Behavior of Materials, Volume 1: Proceedings of the 2012 Annual Conference on Experimental and Applied Mechanics*, Conference Proceedings of the Society for Experimental Mechanics Series, DOI 10.1007/978-1-4614-4238-7_60, © The Society for Experimental Mechanics, Inc. 2013

preliminary work on one of the granular materials that has been employed in our related work in the quasistatic regime. The present effort thus focuses on the equipment and our initial experimental results, but some comments on the modeling effort are provided for context.

## 60.2 Background

Geophones and more recently, common fiber optic cable are being used with success as buried sensors to provide in-ground seismic vibration-sensing capability. Both types of sensors are commercially available and while there are no theoretical constraints on the geologic media in which such sensors can be used, there is a lack of fundamental understanding of grain-to-sensor interaction and of the effects that variability in the granular media has on wave propagation and sensor coupling. The root of the problem is that sensors cannot be rigidly affixed to a granular medium, so even under the best of circumstances there are uncertainties with regard to the coupling between the ground and sensor.

Previous work on the coupling between geophones and soil [1–3] has improved our understanding of the mechanisms affecting geophone response and data quality. Similar work, however, is necessary to understand the relationship between buried fiber optic cable and the surrounding media, especially considering that sensor response is directly related to the physical interactions that produce strain along the cable. The ground coupling of these sensors is dependent on several factors (geologic media, compaction, type of back-fill material), and field tests indicate that performance depends significantly on the physical characteristics of the local geologic media. Despite these indications, a systematic, mechanistic evaluation has yet to be carried out for grain/sensor coupling interactions for buried fiber optic sensors.

There has been considerable progress in recent years regarding direct experimental observations of the contact behavior of geologic materials [4–6]. Unique equipment and methods have been developed for conducting quasistatic contact experiments on a wide range of materials and the associated contact laws are being developed and implemented in a discrete element model of a standard geotechnical triaxial cell [7]. Considerable progress has also been made in the development of complex virtual particle shapes that correspond well to the vagaries of naturally occurring materials [8].

We note that there is considerable interest in the matter of seismic propagation in granular media in general, and an associated body of literature exists that is too extensive to include here. It is useful to point out that there is interest in the underlying micromechanics of the propagation of waves in sands over the frequencies of interest in the present effort [9–11], and that the work discussed here will have a broad impact in the field of wave propagation in earthen materials.

As described in the following section, we employ piezoelectric crystals and strain gauge load cells to monitor the force transmitted through the grain-to-grain contacts. Although the grain arrangements and materials differ from the present case, the use of imbedded piezoelectric crystals has been applied to good effect in other recent work [12] that examined force transmission through linear arrays of grains in contact. Since the sub-kHz frequencies examined in the present effort are not optimal for the performance of the piezoelectric crystals, the experimental results presented below use the output of a high resolution load cell. The piezoelectric method is included, however, since it will be well suited for upcoming work involving grain-scale impact loading experiments.

## 60.3 Equipment and Methods

The experiment calls for applying tone bursts (e.g., typically a set of 10 cycles of a sinusoidally varying signal with a specified frequency and amplitude) to a linear array of grains as indicated in Fig. 60.1. Some of the elements in the array consist of sand grains bonded to piezoelectric crystals oriented to monitor normal force as indicated in Fig. 60.2. Because of

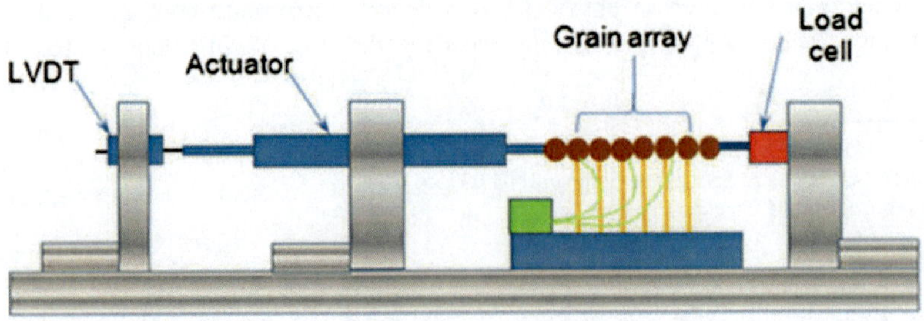


Fig. 60.1 Schematic of the text fixture showing a linear array of grains mounted in-line with the actuator

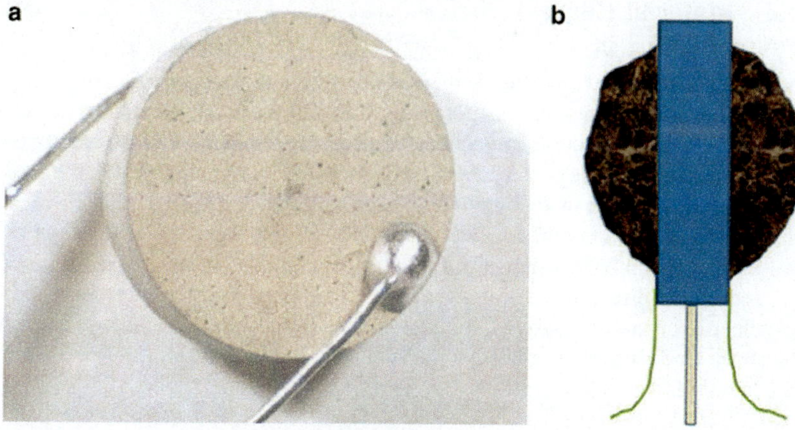


**Fig. 60.2** Photograph of the piezoelectric crystal employed to measure the normal force transmitted through grains (**a**) and a sketch showing how the crystal is bonded to two sand particles to form a single unit (**b**)

the inherent instability of the contacts in such an array, the grains are selected and positioned such that relatively low radius of curvature grains are in contact with high radius of curvature grains (virtually flat in some cases) to enhance the overall stability of the array.

As of this writing, we employ a tone burst generator (HP model 33120A) to generate the desired input signal to an analog amplifier which in turn powers the voice coil actuator as indicated in Fig. 60.1. We note that our earlier work cited above employed a similar system for quasistatic experiments because it does not generate the excessive electrical noise that characterizes pulse-width-modulated motion control systems. Such noise can interfere excessively with other components of the system and deteriorate performance – particularly sensitive deformation and force measurements. A computer-based system to generate the desired input signals in a more automated fashion is currently under development. The present system employs an eight-channel computer-based data acquisition system to monitor the piezoelectric crystals, axial deformation of the array and the input signal to the system.

A static normal force was maintained on the contacts by a spring acting between the actuator shaft and the LVDT mount. For the present experiments, this force was typically in the range of 1–2 N. Since a sinusoidal dynamic force is applied to the contacts, the static normal force must be at least equal to the zero-to-peak amplitude of the dynamic force to avoid loss of contact. Under a given amplitude of the dynamic force, both the elastic and inelastic components of the deformation become smaller (and increasing more difficult to measure) as the static normal force increases. As a consequence, we focused on relatively low values of static normal force to optimize the magnitude of the dynamic normal deformation to be measured. The sinusoidally varying dynamic force levels in these experiments were in the approximate range of $\pm 0.2$ to $\pm 1.0$ N.

The force and deformation readings were digitized at a rate that resulted in 100–400 readings per load cycle, which proved to be more than adequate to characterize the response while not overburdening the data collection system. Raw voltages are recorded and the appropriate conversion factors are applied in post-test analysis. The latter is performed with an approach that has been described previously [3, 4], in which the readings are smoothed using a piece-wise linear regression analysis that does not produce a phase shift. This analysis determines the average values of the peak-to-peak stiffness, hysteresis loop width and area and internal friction (the ratio of the hysteresis loop area to the maximum strain energy experienced during a cyclic of loading) for the second and subsequent loading cycles. The first loading cycle is typically affected by transients and is therefore omitted from the analysis. We estimate the precision of the force measurement system to be approximately $\pm 3 \times 10^{-3}$ N, and that of the deformation measurement system to be approximately $\pm 0.2$ μm.

Note that the configuration for these experiments consisted of one instrumented grain between two other grains, each of which is rigidly bonded to a mounting pin as described in our previous work cited above. One of these latter grains is bolted to the shaft of the actuator and the other to a rigid backing plate (see Fig. 60.1). In a deviation from the illustration in Fig. 60.1, the instrumented grain is not mounted on a wire as indicated, but is simply held in place by the action of the normal force acting on the grain array. Since the strain gauge load cell was employed in the present experiments, it was necessary to account for its elastic compliance in the deformation measurements obtained from the LVDT. This was done by configuring the system with all elements (mounting pins, piezoelectric crystal) in place with the exception of the sand grains (and their associated contacts). In place of the grains, a piezocrystal was bonded with epoxy to the specimen mounting pins to assure a solid connection with the testing system components. The result of this exercise was a value of 7.22 μm $N^{-1}$ for the system contribution to the measured deformation, which was in turn subtracted from all deformation measurements in the post-test analysis.

## 60.4 Material Characteristics

The experiments were conducted with grains of a smooth, large-grained river sand referred to as Vicksburg sand. Table 60.1 gives size and shape statistics of this material as determined from an analysis of digitized images of the indicated number of grains. We note that the discrete element modeling component of the effort generates virtual particles that replicate these population statistics. The particular grains of Vicksburg sand employed in the present experiments were selected on the basis of pairing contacts that mated relatively flat surfaces with curved surfaces with a radius of curvature in the 1–3 mm range. This was primarily done for the sake of stability of the grain array under the static normal force. The grain combination used in the experiments is shown in Fig. 60.3. The grains on the left and right of the figure were bonded to stainless steel pins and bolted direct to the test fixture. The assembly in the center consisted of two grains bonded to either face of a piezoelectric crystal measuring 1 mm in thickness and 4 mm in diameter (slightly larger than the average grain size of Vicksburg sand). The static normal force applied to the array was sufficient to hold them in position during the application of the dynamic force.

## 60.5 Quasistatic Contact Properties

As described elsewhere [4–6], we have conducted normal and sliding contact experiments on a number of terrestrial and lunar materials. Those experiments typically employ cyclic loading waveforms because of that approach provides comprehensive information on the elastic and inelastic components of deformation – both of which are needed to develop detailed contact laws. Figure 60.4 presents typical results for normal stiffness and hysteresis loop area vs. peak normal force from cyclic contact experiments on Vicksburg sand grains. The average radius of curvature of the grains referenced in Fig. 60.4 are 1.97 and 0.71 mm. Note that the normal stiffness increases with magnitude of the cyclic normal force, as does the area of the observed hysteresis loops. These experiments typically cover a range in loading frequency of $10^{-2}$ to 1 Hz to check for frequency effects on the contact properties. As is the case with the present dynamic experiments, the quasistatic experiments involve the application of sets of loading cycles (typically 5) with a given frequency and amplitude and the reported values are averages of the second and subsequent cycles. These results provide a reference point for single contact behavior for comparison with the dynamic results presented in the following section.

**Table 60.1** Average physical properties of Vicksburg sand

| Mean diameter[a] (mm) | Aspect ratio | Major axis (mm) | Minor axis (mm) | Roundness | Ra[b] (Pretest) (μm) |
|---|---|---|---|---|---|
| 2.946 ± 0.264 (n = 36) | 1.341 ± 0.203 | 3.441 ± 0.470 | 2.583 ± 0.259 | 1.310 ± 0.120 | 3 ± 1 (n = 34) |

[a] The n-values shown in this column apply to the aspect ratio, major and minor axes and roundness statistics. Two profiles were typically obtained per grain
[b] The n-values reported in this column indicate the number of grain profiles used to generate the Ra statistics

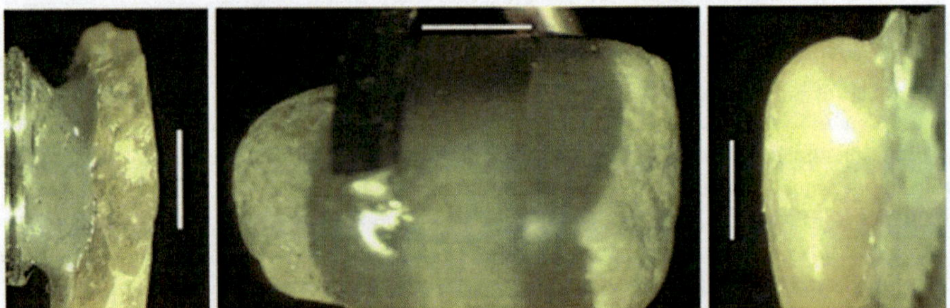

**Fig. 60.3** Photographs of the grain array used in the dynamic contact experiments (scale bars = 1 mm)

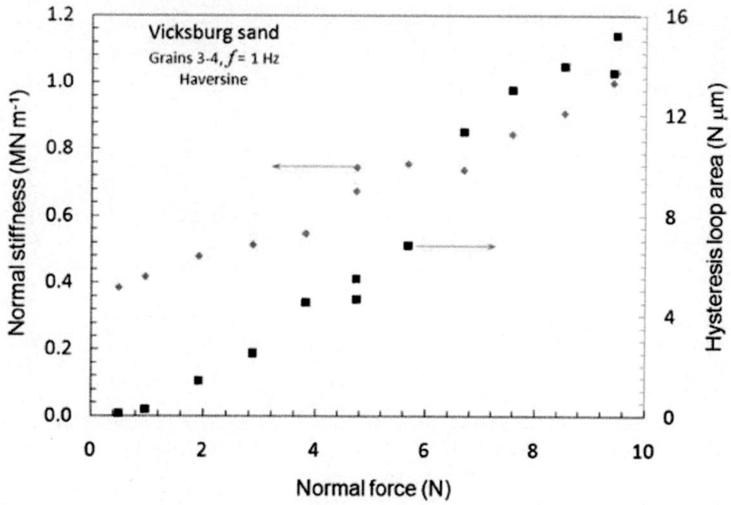


**Fig. 60.4** Representative values of normal stiffness and hysteresis loop area for Vicksburg sand grains

## 60.6 Experimental Results

Figure 60.5 presents typical force-displacement data for the indicated experimental conditions. It is immediately evident that the dynamic contact behavior involves substantial loss (as evidenced by the appreciable hysteresis loop areas) at the indicated force levels. Note that these results represent the response of three grains (one of which contains a piezoelectric crystal) and two contacts. Figure 60.6 plots the effective elastic stiffness of the grain array and internal friction as a function of frequency. It is evident that the stiffness observed in these experiments appears to be relatively independent of frequency, at least over the range examined in these experiments. It is somewhat puzzling that the internal friction increases with frequency while the stiffness remains relatively constant. We are examining this behavior to determine the extent to which experimental artifacts may be influencing the phase lag between the force and deformation measurements. This matter is considered further in Sect. 60.7. We note that although some materials, particularly those with high surface roughness and appreciable loss compliance, exhibit a mild frequency dependence at sub-1 Hz frequencies, such an effect has not been systematically present in the behavior of Vicksburg sand grains under normal loading. Figure 60.7 shows values of normal stiffness determined at lower frequencies plotted along with the present results. Note that the quasistatic data involve a grain pair with a single contact whereas the dynamic data involve three grains and two contacts. The overall lack of a frequency effect is suggested by these results. The lower stiffness values of the dynamic experiments are due primarily to the fact that there are two contacts contributing to the measured deformation.

## 60.7 Discussion

Regarding the scale of the present laboratory experiments and the transition from quasistatic to dynamic loading, it is evident that experiments involving a very small number of grains are effectively quasistatic despite the fact that the material would respond dynamically under similar excitation at large scales. Viewed another way, while a seismic wave with a frequency of 200 Hz, for example, is readily detectible as such at geophysical scales, its wavelength is so long that it appears as quasistatic loading on the scale of several grains. Despite this disconnect between laboratory and field scales, our approach is to determine the contact behavior and develop contact laws that apply to the appropriate frequency range and rely on the numerical modeling to capture the wave nature of the response at scale. Although the results reported here extend to a frequency of 200 Hz, our eventual goal is to achieve1 kHz in these experiments. The system we have developed is capable of this range, but it is evident that further refinements are needed to obtain consistent results at higher frequencies.

With respect to rate effects, which if present, are of critical importance in the modeling of contact behavior, we observed in our quasistatic experiments that rate effects typically emerged for materials that exhibited higher levels of inelastic deformation under cyclic loading. Moreover, the increased inelastic deformation invariably was associated with increased surface roughness of the particles. Although we do not as yet have a quantitative link between an objective measure of

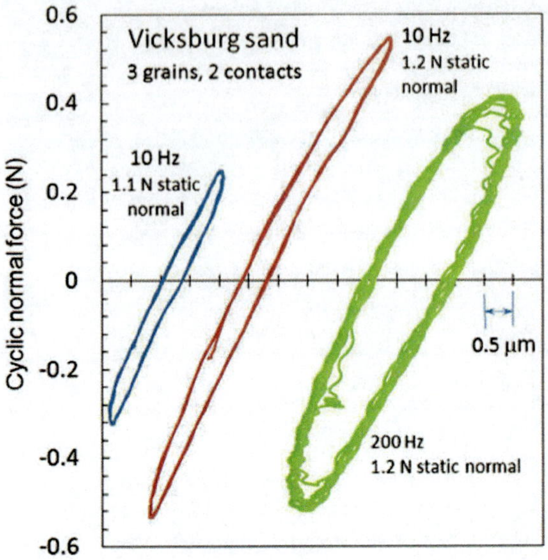


**Fig. 60.5** Typical hysteresis loops from the dynamic contact experiments involving three grains and two contacts

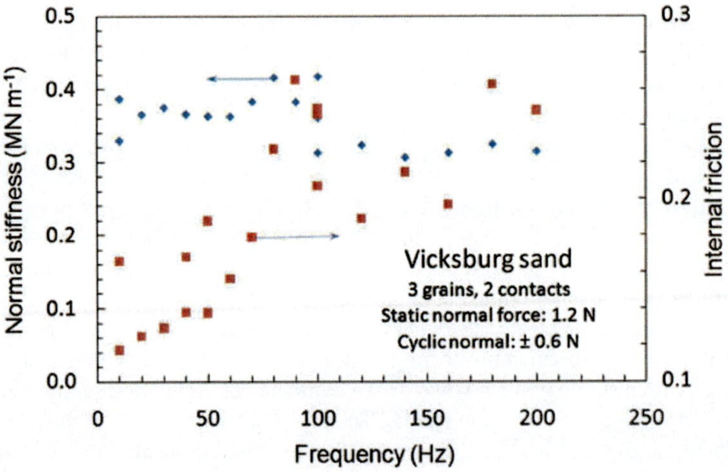


**Fig. 60.6** Normal stiffness and internal friction versus loading frequency from the dynamic contact experiments

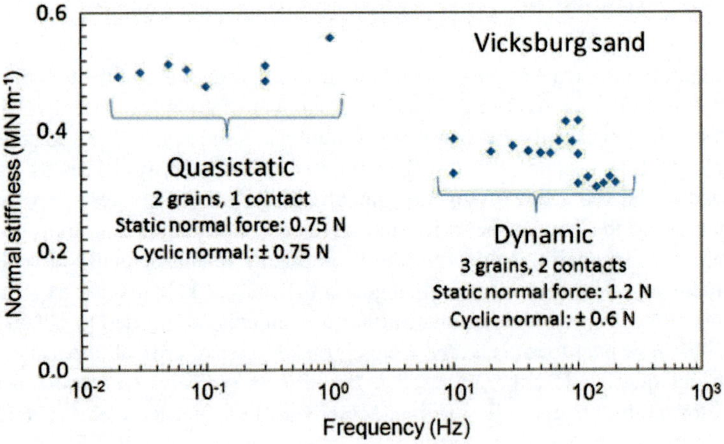


**Fig. 60.7** Normal stiffness versus frequency for quasistatic and dynamic contact experiments on Vicksburg sand grains

surface roughness and the associated contact properties (expressed through a loss compliance term as described in the references cited above), the net effect of roughness on the contact properties is implicitly captured in the values of the loss compliances that we determine for specific materials of interest.

We are encouraged that the dynamic measurements appear to be reasonably consistent with the quasistatic results obtained from completely separate experimental system, particularly with respect to the stiffness measurements. The hysteretic behavior observed in the dynamic experiments at relatively low frequencies is in reasonable agreement with values observed in quasistatic experiments. Specifically, our previous work indicates that internal friction values for the subject material averages approximately 0.1 for a variety of grain characteristics, and the dynamic experiments produce values of internal friction in the range of 0.1–0.3 for frequencies in the range of 10–30 Hz. We see, however, a trend for the internal friction to increase with frequency in the dynamic experiments. Given the above considerations of rate effects, it is puzzling why the internal friction values suggest a frequency effect while none is apparent in the stiffness values. To avoid spurious dynamic effects from the system, we have removed all nonessential components (such as the load cell protective covering, strain relief for the electrical leads, etc.) that could influence the results. Additionally, the circuitry for the LVDT has been carefully modified to operate with minimal phase lag over our frequency range, and possible contributions to phase lag from that device are of too small an order to explain the observed lag values. We will continue to explore this matter as the work progresses.

Although we did not present the measurements obtained from the piezoelectric crystals, some comments on their use are in order. Typical piezoelectric material has a modulus comparable to quartz (66 vs. 76 GPa, respectively) so its elastic response will nominally match that of the material to which it is bonded. What we did observe in these experiments, however, was that although the peak output voltage was in good agreement with the applied forces, their output decayed too quickly to provide an accurate measure of the actual force-time history that the grain experienced. Performance of course improved as the loading frequency increased, and we are in the process of determined the threshold frequency for acceptable response in this particular application. That being said, our main interest in using them relates to monitoring forces associated with impact loading of grains, an application to which their stiffness and high response rates make them particularly well suited.

Finally, with regard to the broader implications of the work, a physical and theoretical understanding of grain-scale properties is providing a deeper level of understanding soil-sensor interaction and continues to affect research interest in the areas of non-line-of-sight sensing, near-surface seismic sensing for perimeter detection and intrusion denial, ultra-light vehicle detection with seismic sensing, subterranean detection and infra-sound sensing. Additionally, an increased understanding of the underlying physics of soil-sensor interaction will shape our approach to improving sensor emplacement methods, locations and calibration algorithms to specific, near-surface soil properties.

## 60.8 Summary and Conclusions

The results of a preliminary set of dynamic contact experiments on a large-grained river sand (Vicksburg sand) have been presented. Loading conditions involved static background forces in the range of 1–2 N and superimposed dynamic forces in the range of $\pm 0.2$ to $\pm 1$ N. Dynamic loading was applied in tone bursts of 10 cycles with a given amplitude and frequency to a linear array of three grains, and the measured response included the applied force and the total deformation of the array in the direction of loading. Behavior over the range of 10–200 Hz was examined. The results showed a general lack of frequency effect over that range, and the observed normal stiffness values were on the order of values observed for Vicksburg sand under quasistatic loading conditions, given allowance for the number of individual contacts contributing to the measured deformation.

From our preliminary observations, and given the limitations described above, we conclude that the experimental approach provides grain-scale contact data that are in reasonable agreement with those obtained on the subject material under quasistatic loading conditions, primarily as indicated by normal stiffness values. Furthermore, the force and displacement measuring systems have adequate sensitivity to capture the relevant grain-scale contact behavior. These preliminary results indicate that further clarification is needed regarding the phase relationship between the applied force and the resulting deformation.

**Acknowledgments** Support for this work was provided by the ERDC Military Engineering Basic Research program. The authors express their appreciation to the following members of ERDC-CRREL's Engineering Support Branch: Douglas Punt, Christopher Williams and William Burch for their help and creative input in the design and fabrication of the equipment described in this paper.

## References


1. Hoover GM, O'Brien JT (1980) The influence of the planted geophone on seismic land data. Geophysics 45(8):1239–1253
2. Krohn CE (1980) Geophone ground coupling. Geophysics 49(6):722–731
3. Drijkoningen GG (2000) The usefulness of geophone ground-coupling experiments to seismic data. Geophysics 65(6):1780–1787
4. Cole DM, Taylor LA, Liu Y, Hopkins MA (2010) Grain-scale mechanical properties of lunar plagioclase and its simulant: initial experimental findings and modeling implications. In Proceedings of ASCE Earth and Space 2010, pp. 74–83
5. Cole DM, Peters JF (2008) Grain-scale mechanics of geologic materials and lunar simulants under normal loading. Gran Matter 10:171–185. doi:10.1007/s10035-007-0066-y
6. Cole DM, Peters JF (2007) A physically based approach to granular media mechanics: grain-scale experiments, initial results and implications to numerical modeling. Gran Matter 9(5):309–321. doi:10.1007/s10035-007-0046-2
7. Cole DM, Uthus L, Hopkins MA, Knapp BR (2010a) Normal and sliding contact experiments on gneiss. Gran Matter, pp 1434–5021 (Print) 1434–7636 (Online). doi: 0.1007/s10035-010-0165-z
8. Hopkins MA (2010) Polyhedra on the cheap. In: Discrete element methods simulation of discontinua: theory and applications. Queen Mary University of London, pp 46–50
9. Bachrach R, Dvorkin J, Nur AM (2000) Seismic velocities and Poisson's ratio of shallow unconsolidated sands. Geophysics 65(2):559–564
10. Buckingham MJ (1997) Theory of acoustic attenuation, dispersion, and pulse propagation in unconsolidated granular materials including marine sediments. J Acoust Soc Am 102(5):2579–2596
11. Buckingham MJ (2007) On pore-fluid viscosity and the wave properties of saturated granular materials including marine sediments. J Acoust Soc Am 122(3):1486–1501
12. Boechler N, Theocharis G, Job S, Kevrekidis PG, Porter MA, Daraio C (2010) Discrete breathers in one-dimensional diatomic granular crystals. Phys Rev Lett 104:2244301-1–2244301-4

# Chapter 61
# Experimental Testing of Micro-Particles Collision

**Wei-Hsun Lin and Chiara Daraio**

**Abstract** We performed experiments on the collisions of micrometer scale (200–300 μm) stainless steel (316 and 440 C) particles and studied the interaction properties and the coefficient of restitution. We used a novel experimental apparatus to enable non-contact measurements based on laser excitations/detection and high-speed photography. The colliding particles were aligned in v-shaped grooves on a silicon wafer, fabricated using anisotropic etching techniques. We used high-power lasers to excite one of the particles (striker), and to control the impact velocity with high repeatability. The motion of the striker particle was triggered by partial laser ablation of its surface. The displacements of the particles involved in the collision were recorded with a high-speed camera mounted on an optical microscope. The particle velocities were obtained from the recorded images using digital image correlation. We calibrated the setup by tracking the dynamic excitation of single particles, and tested collisions between two particles. This study introduces a new experimental approach to understand the fundamental dynamic response of micro-particle collisions, and to test the limits of validity of the Hertzian interaction law.

**Keywords** Collisions • Micro spheres

## 61.1 Introduction

In this paper, we investigate the dynamic contact mechanics between micro-particles of different materials. Nature presents a large number of micrometer-scale granular particles – such as in the foam of soils, dust, and sand. A full understanding of the collective behavior of these particles requires an understanding of the fundamental interactions between them. These interactions are not necessarily similar to particle interactions on the macroscale, which are dominated by elastic Hertzian contact interactions. At the microscale, other effects such as electrostatic, magnetic, and hydrodynamic forces may be of comparable strength to the elastic interactions. Knowledge of the interparticle interaction potentials at this small scale obtained by direct experimental investigations should benefit studies such as the flow of granular particles [1], packing of randomly jammed particles [2], non-equilibrium thermal processes in granular gasses [3], and the propagation of acoustic waves in granular media [4].

Although the collision of millimeter-scale spheres has being studied extensively [5, 6], it is difficult to apply existing techniques to the study of collision between micro-particles. As an example, Foerster et al. imaged the collisions between two spheres in free-fall [5]. Their imaging system resolved the trajectory throughout the collision process and the rotational motion generated by the tangent interaction force. However, for micro-scale objects, the diffraction of visible light demands the use of very accurate focusing, requiring short focal lenses. These lenses greatly reduce the depth of field to a sub-millimeter region, which is then too small to allow accurate control of the sphere's collision within the imaging field.

Here, we present a new experimental system in which the spheres are placed within a v-shaped groove. The confinement of the collision process in the groove is advantageous as it allows the positions of the particles to be precisely set. This feature

W.-H. Lin (✉)
Department of Physics, California Institute of Technology, Pasadena, CA 91125, USA
e-mail: wlin@caltech.edu

C. Daraio
Graduate Aerospace Laboratories, California Institute of Technology, Pasadena, CA 91125, USA

Applied Physics, California Institute of Technology, Pasadena, CA 91125, USA

V. Chalivendra et al. (eds.), *Dynamic Behavior of Materials, Volume 1: Proceedings of the 2012 Annual Conference on Experimental and Applied Mechanics*, Conference Proceedings of the Society for Experimental Mechanics Series, DOI 10.1007/978-1-4614-4238-7_61, 

enables the application of non-contact measurement techniques such as laser vibrometry for accurate velocity measurement. However, this technique also presents a drawback as it introduces frictional coupling between the particles and the groove, which might alter the dynamics of the collision process. In the following sections we will discuss the experimental implementation of such a system, we will present preliminary results characterizing the effects of the groove support, and we will show measurements of the coefficient of restitution of an impact between two particles at different impact velocities.

## 61.2 Experiment

We studied the excitation of a particle in a groove and the collision between two particles. Stainless steel particles (grade 440 C with diameter 229 μm and mass 47.9 μg, and grade 316 with diameter 300 μm and mass 113 μg) were aligned in v-shaped grooves fabricated on silicon wafers (Fig. 61.1). The grooves were fabricated by an anisotropic wet chemical etching process using a solution of potassium hydroxide (KOH). A 1 μm thick layer of silicon nitride ($Si_3N_4$) was deposited on the surface of [100] silicon wafers via chemical vapor deposition (CVD), which was then etched to create patterns of desired dimensions via photolithography and reactive-ion etching (RIE) processes. The wafer was subsequently treated with a 50% KOH solution at 85°C. Because of the anisotropic etching rate of different crystal plane in silicon by KOH, the etching process removed the unmasked silicon and revealed the [111] planes. The grooves had a width of 240 μm and the angle between the two inclined planes was 70.6°, with an estimated surface roughness of ~0.1 μm [7].

Our dynamic measurement system consists of a laser excitation source to trigger particles into motion and it is equipped with two independent methods to measure particle displacements. In the first method we use a high-speed camera (Phantom v12.1) to image the motion of the particles. Magnification is provided by a Leica S6D microscope with additional achromatic lens sets between the camera and the sample stage, which provides a spatial resolution of 4.3 μm per pixel and the maximum field of view is 5.5 × 3.4 mm$^2$. We operate the high-speed camera at 25,000fps with a 39 μs exposure time and a total pixel resolution of 960 × 120. The maximum available frame rate is limited by the available illumination and a 150 W quartz halogen, fiber optic illuminator (Fiber-Lite PL800 150 W) is used to provide illumination for the sample stage.

The second displacement measurement method employs a laser vibrometer system (Polytec OFV-534, OFV-5000) focused on the surface of a particle. The system can achieve a resolution of <1 pm and has a maximum sampling frequency of 2.5 MHz. An objective lens is installed to reduce the laser spot size to 3 μm, and is capable of measuring the displacements of particles of similar size. Due to the physical geometry of the system, the laser vibrometer has to be mounted at an inclined angle of 7° between the laser beam of the vibrometer and the horizontal plane, which introduces a scaling factor (cos7° = 0.99) between the measured and the real displacement of the spheres. A second correction factor must also be included to account for the curvature of the surface on which the laser is focused. When the displacement of the sphere is comparable to (or larger than) its radius, the laser vibrometer measures a nonlinear displacement that is non-proportional to the real motion of the sphere. This limits the range of the displacement acquisition only to the first 5 μm of particle's motion. In our experiments, we obtain an effective scaling factor of ~0.59 when comparing the displacement measured by the high-speed camera and the one obtained with the vibrometer in the first 5 μm of the particle's motion.

The dynamic interactions between two particles are excited using a Q-switched Nd: YAG nanosecond laser (Quantel Brilliant, maximum power 3.2 W, 532 nm wavelength, pulse duration 4 ns) operated at varying intensities. The beam is split

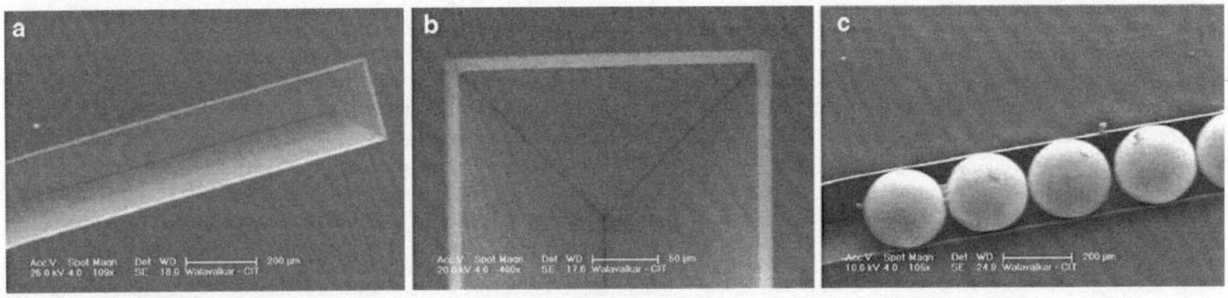


**Fig. 61.1** SEM images of a v-shaped groove and stainless steel spheres on the silicon wafer. The scale bars in these images are 200, 50, 200 μm, respectively. **(a)** The width of the groove is 240 μm. **(b)** The surfaces of the groove are the silicon crystal planes [111]. The inclined angle between each crystal plane is 70.6°. The white margin visible around the groove shows the silicon nitride mask, which was undercut by the KOH solution. **(c)** Stainless steel particles (440 C) aligned within the groove. The particles are positioned on the wafer with antimagnetic tweezers, and are then pushed into the groove with a small diameter stainless steel wire, which was clamped on a translation stage

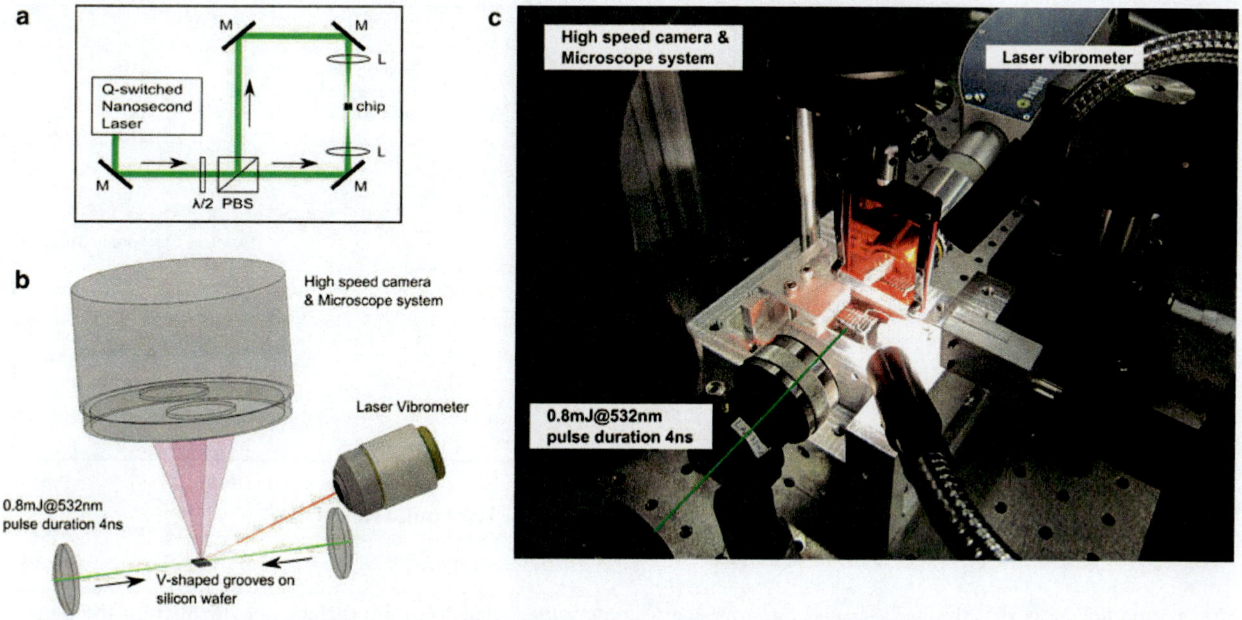


**Fig. 61.2** Schematics of the experimental setup. (**a**) The laser beam output from a Nd-YAG Q-switched nanosecond laser is split into two by passing it through a half-wave plate and a polarized beam splitter. The two beams are focused and aligned with the groove. The beam waist at the focal point is 40 μm. (**b**) High-speed microscopic photography and a laser vibrometer based on the Doppler effect are employed to measure the displacements and velocities of the particles in the groove. (**c**) A picture of the sample stage. A long-pass filter is installed to protect the laser vibrometer (633 nm) from the Nd-YAG laser (532 nm)

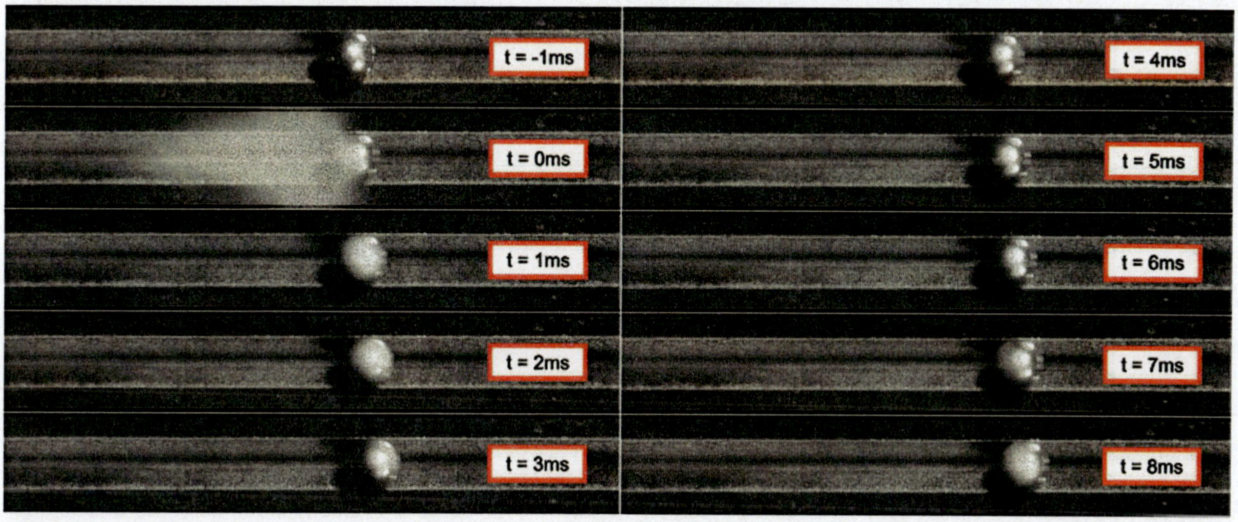


**Fig. 61.3** Images of a single particle driven by the laser taken by the high speed camera at 1,000 frames per second, exposure time 990 μs. Within 1 μs before the second image (t = 0 ms), the 440 C stainless steel particle was shot by the laser with a pulse energy 0.1 mJ. After ablation, the particle has an initial velocity of 0.03 m/s

into two, aligned with the groove at both ends, and focused at the centers of the test particles, with a spot size of 40 μm. By passing the beam through a half-wave plate and a polarized beam-splitter, the related laser intensity in the two arms could be controlled. The half-wave plate rotates the polarization of the beam and therefore controls the related intensity of the two outputs of the polarized beam splitter. When the laser beam excites a particle, a thin layer of material is ablated from the particle's surface, resulting in reaction forces that push the particle forward in the groove [8] (Fig. 61.2).

In Fig. 61.3 we show typical experimental images obtained with the high-speed camera (fps = 1,000, exposure time = 990 μs) when a single sphere is positioned in the groove. The second image (t = 0 ms) was taken less than a microsecond

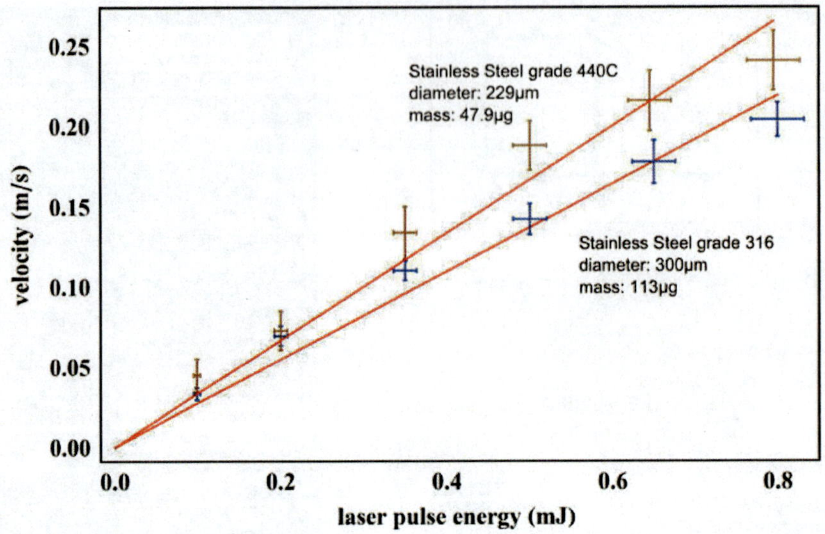


**Fig. 61.4** Relation between the initial translational velocity of the excited particle and the laser pulse energy. Two different stainless steel particles were tested and the results show a linear dependency between the velocity and the pulse energy

after the particle was shot by the laser and the presence of materials ejected from the surface is evident from the plume trailing the particle. Immediately after ablation, the particle has a velocity of 0.03 m/s. The location and velocity of the particle on the groove are retrieved via image deconvolution. From the evolution of the particle in the images it is evident that the particles are rotating in the groove after being excited. The motion of the particles in a groove is classical mechanics problem [9–11]. Here we discuss a minimal model of the translational and rotational motion of a sphere in a groove.

Let the inclined angle between the two surfaces of the groove be $\theta$, the moment of inertia be $I$, and the radius of the ball $r$. Let us assume that the interaction between the spheres and the surface of the grooves is described by a Coulomb frictional force, with a kinetic friction coefficient $\mu_{bg}$. Let the velocity of the sphere be $v$ and the effective velocity of the angular motion of the sphere be $v_a = r\omega$, with initial values $v^{(0)}$ and ${v_a}^{(0)}$. The equations of motion of a particle moving on the groove are

$$\frac{d}{dt}v = -sign(v - sv_a)g\mu_{bg}/s$$
$$\frac{d}{dt}v_a = sign(v - sv_a)kg\mu_{bg} \tag{61.1}$$

where $sign(x) = x/|x|$, $g = 9.8m/s^2$, $s \equiv \sin(\theta/2)$ and $k \equiv mr^2/I$ (with $s = 0.578$, $k = 5/2$ for our system). After a time $\Delta t = \dfrac{v^{(0)} - s{v_a}^{(0)}}{g\mu_{bg}(1/s - sk)}$, the particle reaches its steady state velocity:

$$v^{(1)} = \frac{ks^2}{ks^2+1}v^{(0)} + \frac{1}{ks^2+1}s{v_a}^{(0)} \tag{61.2}$$

This model predicts a linear decay of the particle velocity. However, the particle's velocity is not observed to decrease linearly in the data corresponding to Fig. 61.3, and the particle is observed to start rotating immediately after being excited by the laser. This indicates that the kinetic frictional force between the particles and the grooves could be substantially greater than what the Coulomb frictional force predicts.

To measure the relationship between laser energy and particle velocity, we varied the pulse energy from 0 to 0.8 mJ and measured the motion of the particle using the laser vibrometer. The resulting velocity is shown to be proportional to the input energy (Fig. 61.4). This behavior can be explained by the mechanism of laser ablation on the surface of metals. The ablation depth was estimated to be ~250 nm at 0.8 mJ from the data in Ref [12]. Considering the resulting particle's velocity and the total ablated volume from the stainless steel 316 particle, the eject velocity is calculated to be ~$10^4$ m/s, which agrees with the typical ablation velocity on metal (~$10^4$ to $10^5$ m/s) [8]. The total kinetic energy of the ejected mass and the particle is estimated to be 0.07 mJ, which is reasonable compared with the energy of the input laser pulse (0.8 mJ). The momentum of the photon plays no role in this scenario, and in the experiment we find that the initial velocity of the particles is solely

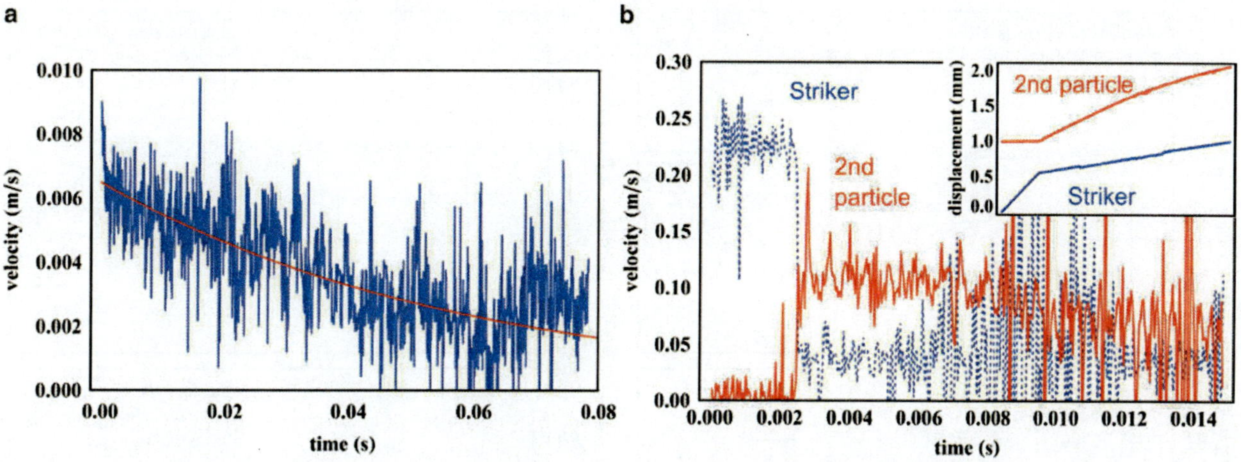


**Fig. 61.5** (**a**) Typical data of the velocity of a particle after being excited by the laser as determined by high speed camera. (**b**) Typical data of the velocity of the particle collision experiments. Both of the tests were run with stainless steel 440 C particles

determined by the location on which laser is focused, and independent of the laser direction. In Fig. 61.4 we see that despite a factor of 2 difference in the volume of the two types of particles tested, the resulting velocity is similar, indicating that the stainless steel 316 particles can be excited more effectively by laser ablation.

In Fig. 61.5a we plot a typical velocity profile of a particle after being excited by the laser. The data is calculated from the images taken by the high-speed camera. The particle is excited at $t = 0$ and moves 0.5 mm in the groove in 0.08 s. At first glance, the velocity decays linearly; however, further investigations show that the velocity decay has a fixed time constant and is independent from the initial velocity. This indicates that the Coulomb frictional force between the surfaces of the particle and the groove is not sufficient to fully capture the dynamics of the particles. We describe this decay behavior with an empirical frictional force of the form $f = -mv/T$, where $T \approx 0.06 \pm 0.01s$.

The images taken by the high speed camera can also be processed to track multiple particles in the same instance. We performed experiments on the collisions between two micro-particles. Two particles are placed in the groove, with a center to center distance of 1 mm. The first particle (striker particle) is excited by the laser and then collides with the second particle with a varying velocity controlled by laser pulse energy. The displacements and velocities are measured using the high-speed camera – a typical run is shown in Fig. 61.5b. We can see that the total momentum right after the collision is roughly 3/4 of the initial momentum. The translational kinetic energy is ~1/3 of the input translational kinetic energy; hence the collision process is lossy. One possible reason for the losses is the presence of frictional forces between the groove and the particles. Another interesting observation is that the velocity of the second particle decayed after collision, but the striker velocity remains nearly constant after the collision. A more in-depth understanding of this dynamic interaction will be pursued in further investigations.

The coefficient of restitution for the particles' collision (the ratio between the difference in velocities before and after the collision) and the ratio of final to initial translational kinetic energy at different striker velocity right before the collisions are plotted in Fig. 61.6. We tested two sets of particles (stainless steel 440 C and stainless steel 316). We find that the coefficient of restitution varies with the striker velocity and both types of particles have a maximum restitution value around $v = 0.14$ m/s in the velocity range tested. For 440 C particles the coefficient of restitution approaches to a constant value of 0.3 at higher velocity. For the 316 particles at low velocity, dissipation becomes more relevant. In this experiment the collisions occur when the striker is rotating, so the influence of the rotational coupling need to be considered.

In conclusion, we used a new experimental setup to study the dynamic collision between micro-particles. The system uses laser ablation as a method to excite mechanical motion with high repeatability and uses non-contact methods to detect particles velocities and displacements. The smallest possible working size of particles is defined by the optical limits of the excitation and detection systems, which is ~10 μm. The laser ablation excitation technique is applicable to non-transparent particles and the system can be adapted to simultaneously excite multiple particles by simply splitting the laser beams.

We presented preliminary experimental results of micro-particles collisions on a v-shape groove. High-speed microscopic photography and laser vibrometry provided noncontact measurements on the particle's velocity and displacement. Our preliminary data showed that modeling the dynamics of the system including only simple Coulomb frictional forces between the particles and grooves fails to quantitatively predict the behavior of the particles' motion. The coefficient of

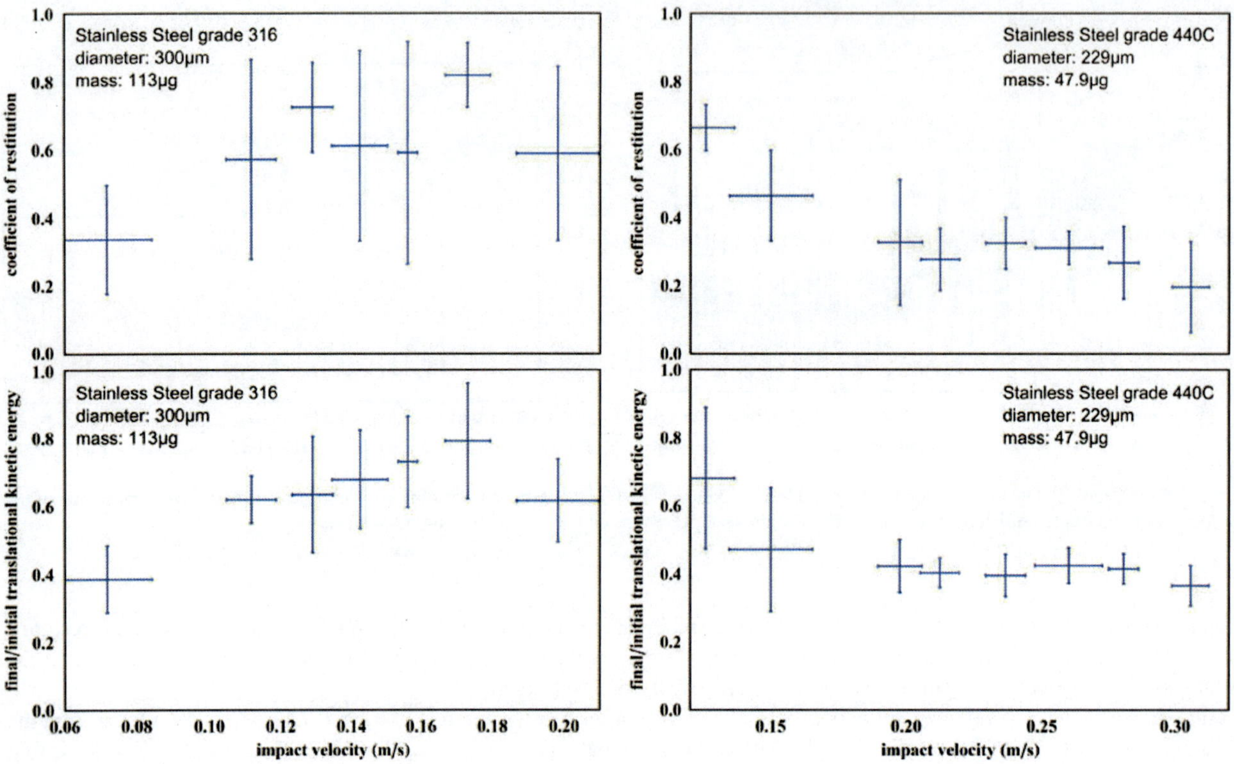


**Fig. 61.6** Coefficient of restitution (*top plots*) and ratio of output to input translational kinetic energy at different impact velocity (*bottom plots*)

restitution of the collision between two particles was measured and showed a functional dependence on the striker velocity. More detailed studies of the frictional and adhesive forces between the particles and the groove are necessary for revealing the dynamics of the collisions in this system.

**Acknowledgement** The financial support provided by the Office of Naval Research (YIP) and the National Science Foundation (Career) is gratefully acknowledged. We thank Melissa Melendes from the Kavli nanoscience institute at Caltech for her advice on the anisotropic etching process, and Petros Arakelian for his technical assistance in the experiments.

# References

1. Campbell CS (1990) Rapid granular flows. Ann Rev Fluid Mech 22:57
2. Torquato S et al (2010) Jammed hard-particle packings: from Kepler to Bernal and beyond. Rev Mod Phys 82:2633
3. Jaeger HM et al (1996) Granular solids, liquids, and gases. Rev Mod Phys 68:1259
4. Harada S et al (2003) Wave propagation in a dynamic system of soft granular materials. Phys Rev E 67:061305
5. Foerster SF et al (1994) Measurements of the collision properties of small spheres. Phys Fluids 6:1106
6. Rioual F et al (2000) Experimental study of the collision process of a grain on a two-dimensional granular bed. Phys Rev E 62:2450
7. Satoa K et al (1999) Roughening of single-crystal silicon surface etched by KOH water solution. Sens Actuators A 73:122
8. Bauerle D (2011) Laser processing and chemistry. Springer, Berlin/New York
9. Halsey TC et al (1999) A ball in a groove. Phys Rev Lett 83:5007
10. Song QG (1988) The requirement of a sphere rolling without slipping down a grooved track for the coefficient of static friction. Am J Phys 56:1145
11. Domenech A et al (1991) Frontal impact of rolling spheres. Phys Educ 26:186
12. Dupont A et al (1995) Enhancement of material ablation using 248, 308, 532, 1064 nm laser pulse with a water film on the treated surface. J Appl Phys 78:2022

# Chapter 62
# Low Velocity Impact Behaviour of the Epoxy Matrix Composites Reinforced with Metallic and Oxide Particles

**E. Bayraktar, D. Katundi, A. Guinault, A. Kosecki, and A. Larbi**

**Abstract** Like parts or structures assembled from the common isotropic engineering materials, very fine particles reinforced epoxy resin matrix composite structures can be subjected to impact loading during their designed life. These composites have different interaction with the externally applied load. In fact, several damage mechanisms can be operative such as matrix cracking, reinforcement-matrix interface rupture etc.

In order to vary the toughness of the composites while maintaining a constant volume fraction of different reinforcement, three different epoxy matrix composites were manufactured using SiC, $Fe_2O_3$, $Al_2O_3$, $SiO_2$, $TiO_2$ as reinforcements, respectively. Three different compositions were prepared for the test specimen in this study that manufactured with vacuum environment in order to eliminate air bulb during the solidification. A special series was prepared only Epoxy matrix without reinforcement.

An instrumented drop-weight test apparatus was developed for the impact tests, and a special software program was also developed to record the load-time curves. It consists of a typical bloc weighting 900 g containing a load cell inside of hemispherical punch with a diameter of 10 mm. From these data, the total energy vs. time curves and load vs. displacement curves could be obtained, enabling a complete analysis of the behaviour of each composite.

Detail damage analyses are carried out on the scanning electron microscope (SEM) after impact tests of each composition.

**Keywords** Epoxy composites • Drop weight test • Damage analysis • SEM

## 62.1 Introduction

Composite materials can be effectively used as a suitable alternative of structural material, replacing with advantage not only steel but even light alloys in the construction of some parts, such as vehicular or aircraft body. Structural design of composite parts has already passed the so-called “replacement phase” and in a number of applications moves toward a “design phase” with a complete redesign of the part in order to take advantage of the customs of this class of materials. However, impact design of composite material structures is still in the early stage of the “demonstrative phase”: the designer is still trying to confirm whether some parts could be constructed with a composite material as energy absorber or not [1–4].

In other way, upon impact, a series of physical phenomena takes place: elastic, shock, and plastic wave propagation, fracture and fragmentation, perforation, etc. In the literature, very often papers published on the subjects of stress waves and impact mechanics essentially after the Second World War (from 1950 up to now). Today, many scientific papers concentrate on the present dynamic behaviors of materials focusing on the analysis and design of energy absorbing materials and structures capable of resisting impact and mitigating blast [2–7]. Catastrophic failures of aerospace structures due to impact damage, such as that of the Columbia space shuttle, demonstrated the pressing need for continued efforts from the research community on impact resistant materials. Civil and military applications require designing high performance and high

E. Bayraktar (✉) • D. Katundi • A. Kosecki • A. Larbi
Mechanical and Manufacturing Engineering School, Supmeca-LISMMA-Paris, Paris, France
e-mail: bayraktar@supmeca.fr

A. Guinault
Chair of Materials Science & Polymer, CNAM-Paris, Arts et Metiers (PIMM), Paris, France

V. Chalivendra et al. (eds.), *Dynamic Behavior of Materials, Volume 1: Proceedings of the 2012 Annual Conference on Experimental and Applied Mechanics*, Conference Proceedings of the Society for Experimental Mechanics Series,
DOI 10.1007/978-1-4614-4238-7_62, © The Society for Experimental Mechanics, Inc. 2013

energy absorbing materials and structures. For example, in the aerospace industry, new composite wing structures resisting hail and bird impacts are of relevance. Designing energy absorbing materials is challenging since various mechanisms – wave propagation, dynamic cracks, and delamination, thermal effects, dislocation generation, growth, and motion, etc. – are acting concurrently, at different material scales, and intertwined during impact.

New research is needed especially in modeling and simulation of impact to be able to reach a point where models are predictive of the experimental results. Armed with such models, one can expect that optimal design of novel heterogeneous and anisotropic material systems with high energy absorption capabilities can become reality. Heterogeneous and anisotropic materials have useful features for justifying damage when impacted because of many mechanisms of transforming the kinetic energy of the penetrator into: opening extensive crack surfaces and spreading (de-localizing) the damage zone; plastic deformation; buckling and other dynamic instabilities etc. that can change the material response [1–8].

In the same way, thermosetting resins, such as epoxy and phenol resins, are widely used as structural materials and adhesives in the aerospace and electronics industries for their high strength, high elastic modulus, and good heat and solvent resistance. The good heat and solvent resistance have been reasoned to originate from their cross-linked chemical structure [8]. One simple way to increase the toughenability of thermosetting resins is to reduce the cross-link density. In many cases, however, reducing the cross-link density causes a decrease in the glass transition temperature, of the cured resin. To maintain a high with reduced cross-link density, uncured resin monomers with a more rigid chemical backbone structure need to be used [7, 8]. However, epoxy monomers which have rigid chemical structures tend to be in a solid state at room temperature. This reduces their abilities of the process and limits their applicability. More research is therefore needed in order to overcome the trade-o between toughness, heat resistance and the ability of the processes.

The objective of our current research was to determine whether the ductility and toughenability of a highly cross-linked epoxy matrix composites containing SiC, $Fe_2O_3$, $Al_2O_3$, $SiO_2$, $TiO_2$ as reinforcements. This paper contains partially preliminary experimental results of the research project for French aeronautic industry going on the LISMMA-Paris.

## 62.2 Experimental Conditions

Three different epoxy matrix composites were manufactured using SiC, $Fe_2O_3$, $Al_2O_3$, $SiO_2$, $TiO_2$ as reinforcements, respectively. Three different compositions were prepared for the test specimen in this study that manufactured with vacuum environment in order to eliminate air bulb during the solidification. A special series was prepared only Epoxy matrix without reinforcement (Table 62.1). Differential Scanning Calorimetry test analyses were made for their glass transition temperature $T_g$ and also the absolute density for each composition has been carried out by Archimed's method.

An instrumented drop-weight test apparatus was developed for the impact tests in the LISMMA materials laboratory and a special software program was also developed to record the load-time curves. It consists of a typical bloc weighting 900 g containing a load cell inside of hemispherical punch with a diameter of 10 mm. All of the impact tests have been carried out from 1 m. From these data, the maximum force at the failure vs. time curves could be obtained, enabling a complete analysis of the behaviour of each composite (Fig. 62.1).

An image analysis system (software attached to the microscope) were used to detect the outlined regions in order determine the projected damaged areas in both of the faces of the test specimens; front and rear side.

### *62.2.1 Drop Weight Test Setup Description*

Measuring an impact force is never an easy task. For this experiment, we have chosen to use a piezoelectric force sensor. The advantages of such sensors are (1) low price (2) high natural frequency that enables measuring of short pulses and (3)

**Table 62.1** Three compositions with matrix prepared for epoxy composites with different reinforcements

| | Reinforcement as % | | | |
|---|---|---|---|---|
| Composition with hardener | SiC | $Al_2O_3$ | $Fe_2O_3$ | $TiO_2$ |
| Matrix neat epoxy | – | – | – | – |
| C1 | 10 | 20 | | |
| C2 | 10 | 20 | 10 | – |
| C3 | – | 20 | 10 | 10 |

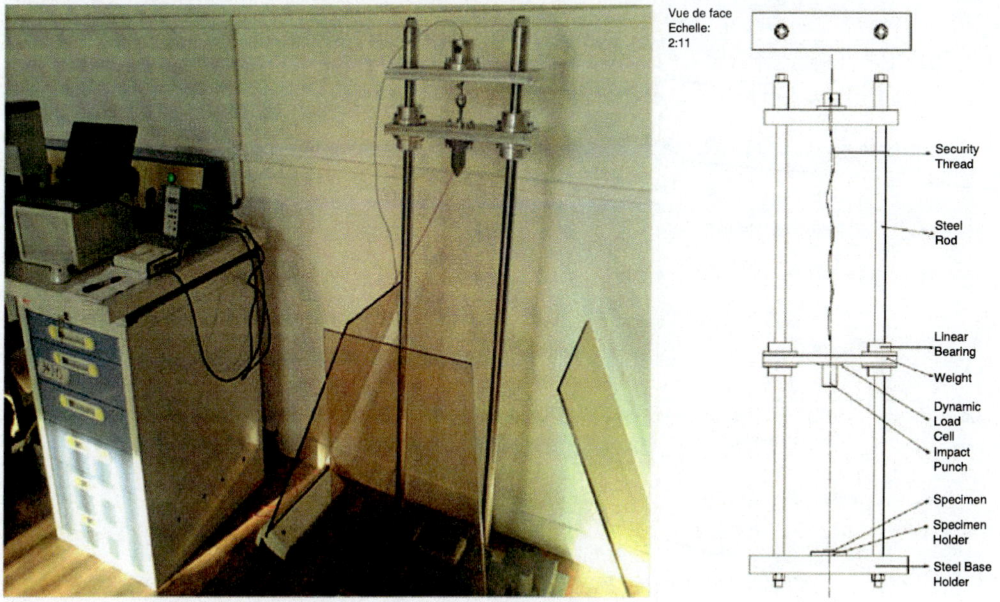


**Fig. 62.1** Drop weight test apparatus developed in the LISMMA materials laboratory (*left*) and schematic view (*right*) [11]

high stiffness that allows direct integration of the sensor directly into impact head. The last property allows placing the sensor as close as possible to the impact point [11].

In the present study, a Dytran - 1200C model force sensor was used for all of the tests. This is a low-profile, ring-style, piezoelectric sensor intended for measuring forces up to 89 kN. The ring is inserted into impact head in such way that its active surface is about 0.2 mm above the upper surface of the ring [9]. The impact head is then screwed to the retaining plate with two bolts that create a rigid connection and preload the force sensor. Any force that appears on the peak of the impact head compresses the sensor allowing us to measure this force. As the sensor is charge-output type, we have used a Kistler 5,011 precision charge amplifier in order to obtain a voltage output. The signal acquisition was carried out using a National Instruments USB 4431dynamic signal acquisition board that allowed us to sample the signal at 102.4 kHz which corresponds to a effective measured bandwidth of about 40 kHz (this means that the shortest measurable pulse is of 25 μs).

While the preliminary results have shown that the acquisition setup is correct with a high margin of both amplitude and frequency ranges, the mechanical setup apparently suffers from some flaws. The major problem identified is that the sensor seems to be too far from the impact point and the sensor preloading is not perfect. This can result in very scattered first pulse, suffering from spurious oscillations as will be seen in the next section (see Fig. 62.3a–d). This problem is likely not very dependent on the nature of the shock – the same problems were observed in calibration "steel on steel" impacts. Despite on these errors, the overall maximum force seems to be reliable, but has to be taken as relative and not absolute measurement i. e. it can be compared for the examined samples but should not be compared to maximum force obtained with other experimental devices. Some improvements to the experimental setup are planned for further experiments. We will try to place the force sensor closer to the impact point and we will redesign the preloading method.

## 62.3 Results and Discussion

The entire test specimens were measured (60 mm diameter 10 mm thickness) before the impact test. First of all the absolute densities were measured by Archimed's method. The densities measured for all of the compositions are given as: Matrix 1,141 kg/m$^3$, C1 1,286 kg/m$^3$, C2 1,544 kg/m$^3$, C3 1,472 kg/m$^3$ respectively.

Figure 62.2a–b show the Differential Scanning Calorimetry measurements to determine the glass transition temperature, Tg for each composition. The calculated results from the curves indicate that DSC has great potential utility for the characterization of epoxy composites. The results obtained from each composition are given as follows;

Matrix-Tg: 67.62°C, C1-Tg: 46.95°C, C2-Tg: 59.98°C and C3-Tg: 36.28°C

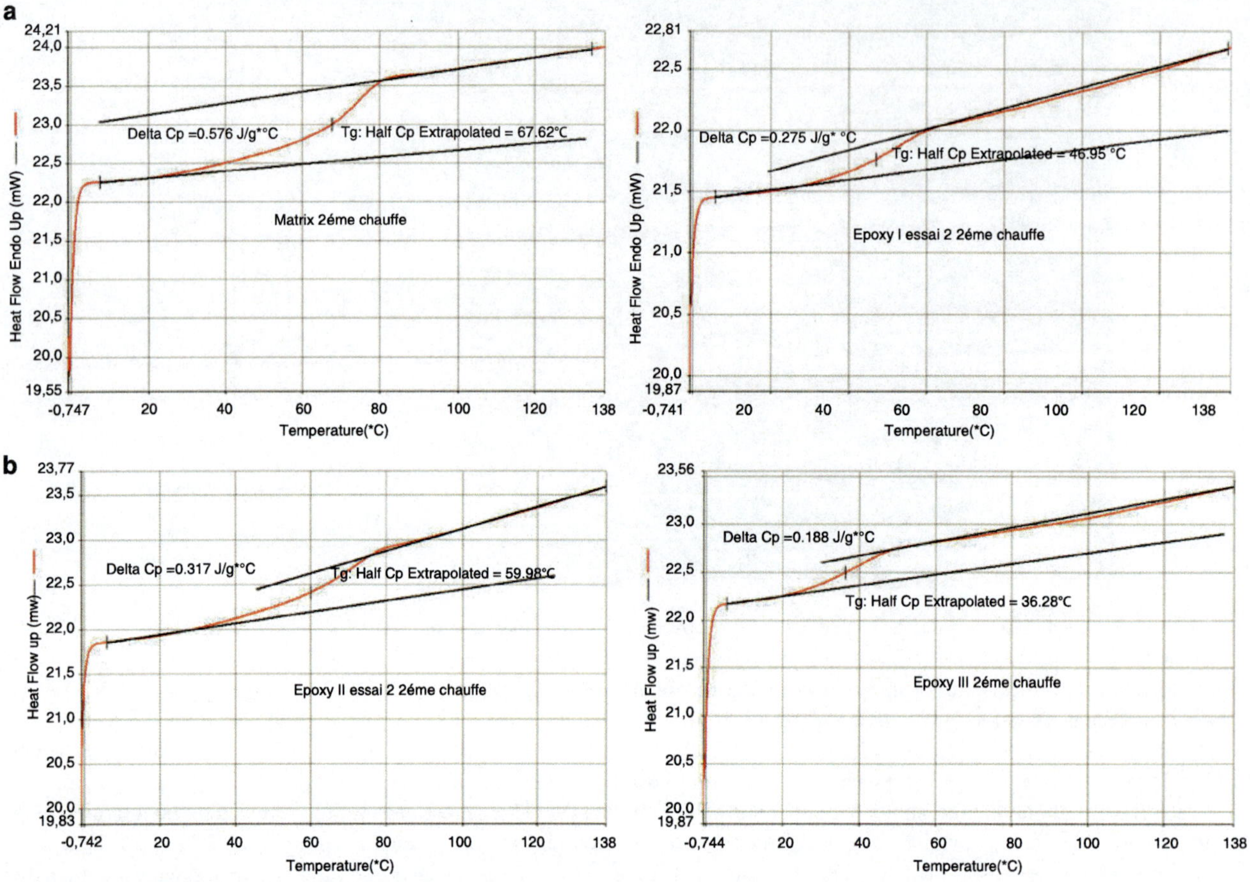


**Fig. 62.2** (**a**) Differential Scanning Calorimetry measurements to determine the glass transition temperature, Tg, for the matrix (*left*) and for the composition C1 (*right* graph). (**b**) Dynamic Scanning Calorimetry measurements to determine the glass transition temperature, Tg, for the composition C2 (*left*) and for the composition C3 (*right* graph)

Typical load vs. time curves obtained for the epoxy composites after the drop weight test from 1 m only one impact event are shown in Fig. 62.3a–d. It can be seen from these curves that, the maximum loads for the composition with different reinforcement regarding to matrix. Additionally, the time duration of the impact event changes for each composition. Here the maximum loads are given for the matrix and for the each composition:

Matrix: 2,020 N, C1: 2,602 N, C2: 2,655 N, and C3: 2,814 N.

It can also be seen that the amplitude of the oscillations in each curve is variable depending on the composition. These aspects were common to all composites tested and reflect the interaction of the composites´ microstructure, and the geometry of the test specimen and clamping device, with the incident indenter.

Fractographic aspects of the epoxy composites as projected damaged area front and rear side are shown in Fig. 62.4 and also measurements of the projected damaged area front and rear side for each epoxy composition and % of the surface damaged zone regarding to the total area of the specimens are indicated in Table 62.2.

Fracture surfaces are shown in Fig. 62.5 as low and high magnification. In reality, epoxy composites do not exhibit the same brittle fracture as metallic materials. At high magnifications, it is always possible to see plastic deformations in films and deformed fibrils, occasionally one may see on each specimen. Elliptical cap, steps and striped patterns can also be observed. These microstructures that are located in narrow stressed layers of the fracture surface give obvious information about the initiation peak and propagation and fracture directions. The formation of these brittle failure bands is due to the crack crossing over heterogeneities. This behaviour can be observed very often in epoxy based composites. Similar results carried out on the different types of impact test can be found in the literature [8, 10].

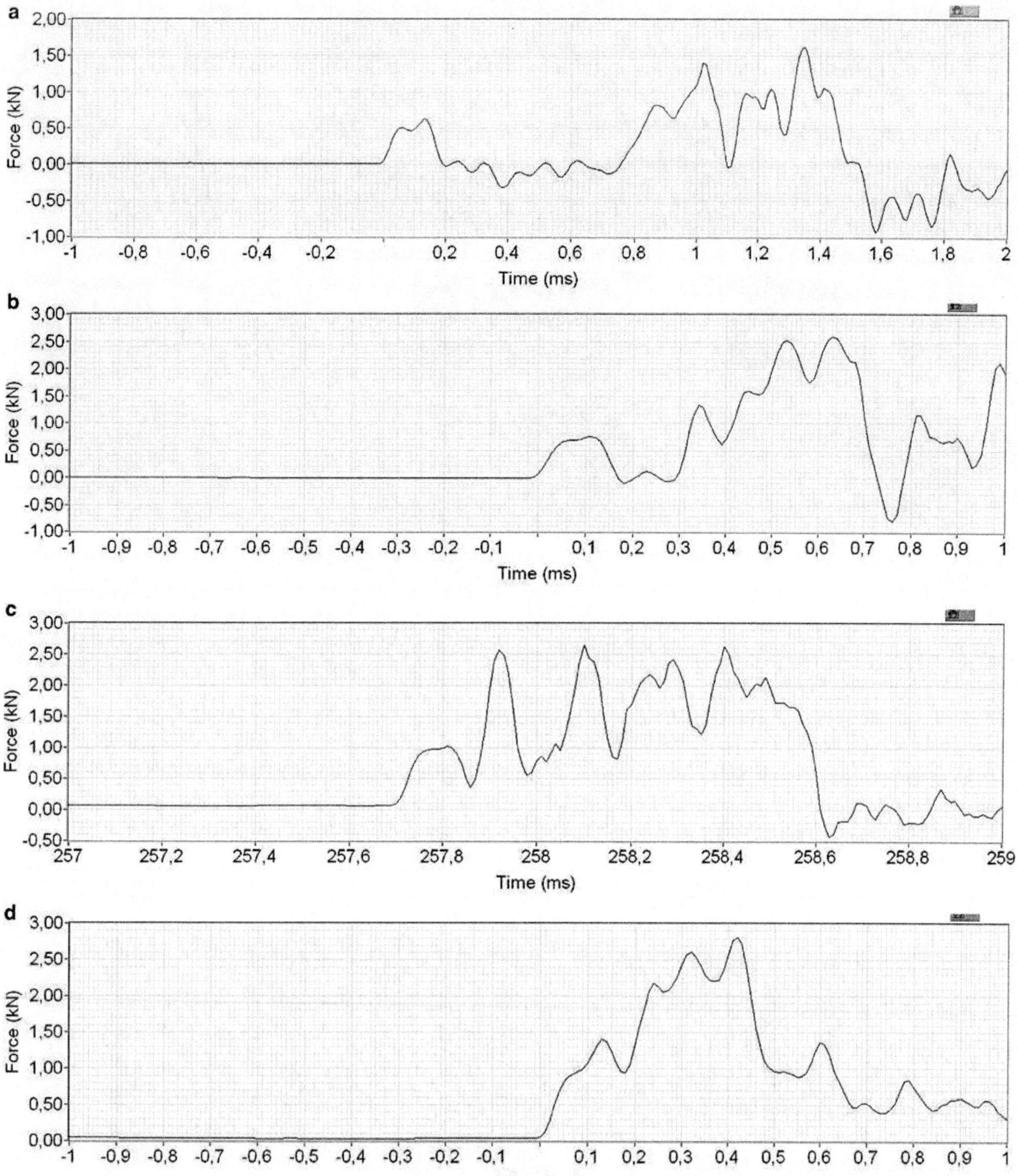


**Fig. 62.3** (**a**) Typical load versus time curve obtained for the matrix (max force 2,020 N) tested. Observation of the attenuation of the oscillations on the load signal with the one number of impact events. (**b**) Typical load versus time curve obtained for the composition C1 (max force 2,602 N) tested. Observation of the attenuation of the oscillations on the load signal with the one number of impact events. (**c**) Typical load versus time curve obtained for the composition C2 (max force 2,655 N) tested. Observation of the attenuation of the oscillations on the load signal with the one number of impact events. (**d**) Typical load versus time curve obtained for the composition C3 (max force 2,814 N) tested. Observation of the attenuation of the oscillations on the load signal with the one number of impact events

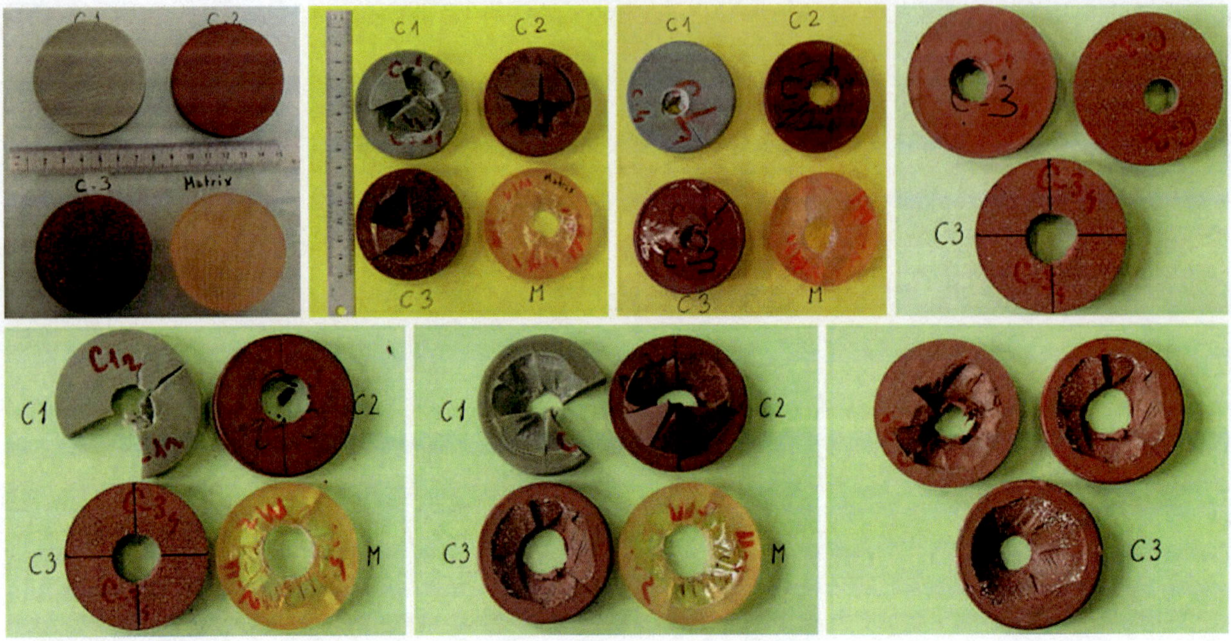


**Fig. 62.4** Fractographic aspects of the epoxy based composites before and after the drop weight test (composites n° C1, C2 and C3 and also matrix (*front* and *rear* side))

**Table 62.2** Measurement of the projected damaged area front and rear side for each epoxy composition and % of the surface damaged zone regarding to the total area of the specimens

| Damage zone "A" projected damaged area, front side | | |
|---|---|---|
| | Surface of damage zone ($mm^2$) | % of surface damaged zone/specimen (disc) |
| C1-A | 1307 | 46.26 |
| C2-A | 1296 | 45.86 |
| C3-A | 1336 | 47.29 |
| M-A | 1374 | 48.63 |
| Damage zone "B" rear side of projected damaged area | | |
| | Surface of damage zone ($mm^2$) | % of surface damaged zone/specimen (disc) |
| C1-B | 192 | 6.80 |
| C2-B | 230 | 8.13 |
| C3-B | 201 | 7.12 |
| M1-B | 241 | 8.54 |

## 62.4 Conclusion Remarks

The behaviour of epoxy composite plates subjected to low velocity impact loading has been examined. The presentation against low energy impact events of the epoxy composites tested were dominated by the intrinsic characteristics of the metallic and oxide powder reinforcements.

In fact, this type of the experimental setup allows acquiring the impacting running force versus time signals; the relevant kinematic quantities can then been obtained by simple numerical integration of the acceleration history.

The histories of the kinematic quantities and of the contact force at the moment of impact drop weight test have been examined in all the three compositions above mentioned situation.

Composition C1 left and right

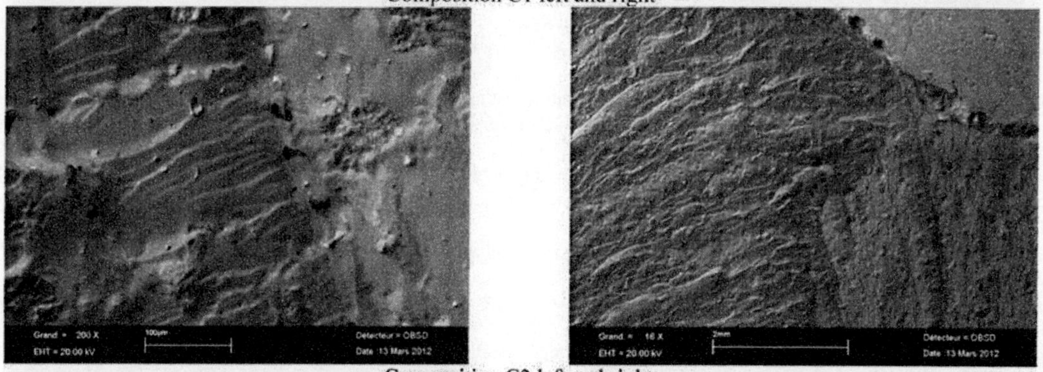

Composition C2 left and right

Composition C3 left and right

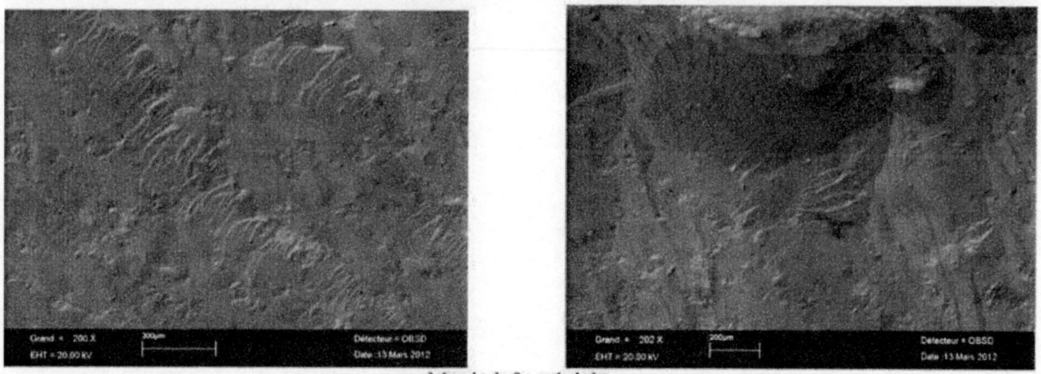

Matrix left and right

**Fig. 62.5** Typical Scanning Electron Microscopy (SEM) fracture surface images, at low and high magnifications

SEM images of the fracture surfaces allow the reinforcement and complement of the data obtained with the instrumented drop weight tests. This paper contains partially preliminary experimental results of the research project going on. In the frame of the content of the project, experimental study and modelling will continue for two aspects; materials processing for optimization of the composition and also modelling and numerical analysis for final design of the epoxy based composites for aeronautical applications.

**Acknowledgment** Authors thank to LISMMA and CNAM materials - Polymer research foundation for the financial support for the research project going on.

## References

1. Abrate S (1998) Impact on composite structures. Cambridge University Press, Cambridge
2. Choi HI, Chang FK (1992) A model for predicting damage in graphite/epoxy laminated composites resulting from low velocity point impact. J Compos Mater 26(14):2134–2169
3. Sjoblom P, Hartness JT, Cordell TM (1988) On low-velocity impact testing of composite materials. J Compos Mater 22:30–52
4. Belingardi G, Vadori R (2002) Low velocity impact tests of laminate glass-fiber-epoxy matrix composite material plates International. J Impact Eng 27:213–229
5. Pizhong Q, Mijia Y, Florin B (2008) Impact mechanics and high-energy absorbing materials: review. J Aerosp Eng 21(4):235–248
6. Robinson P, Davies GAO (1992) Impactor mass and specimen geometry effects in low-velocity impact of laminated composites. Int J Impact Eng 12(2):189–207
7. Lagace PA, Williamson JE, Tsang PH, Wolf E, Thomas S (1993) A preliminary proposition for a test method to measure (impact) damage resistance. J Reinf Plast Compos 12(5):584–601
8. Vasconcelos PV, Lino FJ, Magalhaes A, Neto RJL (2005) Impact fracture study of epoxy-based composites with aluminum particles and milled fibers. J Mater Process Technol 170:277–283
9. Kishi H, Shi Y-B, Huang J, Yee AF (1997) Shear ductility and toughenability study of highly cross-linked epoxy/polyether sulphone. J Mater Sci 32:761–771
10. Morais WA, d'Almeida JRM, Godefroid LB (2003) Effect of the fiber reinforcement on the low energy impact behaviour of fabric reinforced resin matrix composite materials. J Braz Soc Mech Sci Eng 25(4):325–328
11. ASTM D3029 (1982) Standard test method for impact resistance (falling weight). American Society for Testing and Materials, Philadelphia

# Chapter 63
# Experimental Study on Dynamic Mechanical Properties of a Mock PBX

**J.G. Qin, Y.L. Lin, F.Y. Lu, Zh.B. Zhou, and R. Chen**

**Abstract** The mechanical properties and fracture behavior of a mock polymer bonded explosives (PBX) material were studied using the split Hopkinson pressure bar (SHPB) apparatus. The compressive response of the material is measured from strain rate 110–500 $s^{-1}$. A dynamic Brazilian test, in which a disc-shaped specimen is loaded diametrically, was chosen to investigate the tensile properties of the material. The fracture behavior of the material was performed on a notched semi-circular bending (SCB) specimen. With a high speed camera and the digital image correlation (DIC) technique, the displacement and strain field history of specimen during loading were observed. The results exhibit that the compressive, tensile and fracture behaviors of the material are all dependency on loading rate, and the DIC technique provides a quantified and non-contact measurement of displacement field and strain field of the specimen under high strain rate loading.

**Keywords** Mock polymer bonded explosives • Hopkinson bar • Digital image correlation

## 63.1 Introduction

Polymer bonded explosives are highly particle filled composite materials comprised of 90–95% by weight of powerful secondary explosive crystals held together by a polymeric binder (5–10% by weight). They are used in a wide variety of applications, ranging from rocket propellants to the main explosive charge in conventional munitions. Of great concern to explosive formulations is the possibility of accidental explosion during manufacture, transport or handling. The mechanical properties of explosive materials subjected to a range of conditions are important criteria to determine a safe working life. To realize the missions of detonation, the related components of PBX may be mechanically manufactured from the granular-based composite to sustain external loads and to install the structures for usage. The failure prediction of the material and structures with fracture, therefore, is very important for the durability and safety of the solid explosives under environmental conditions. For most usages of munitions, PBXs are under impulse loading and high strain rate deformation, and the behaviors of PBXs under these cases are notable different with being quasi-static loaded, so it is significant to investigate the mechanical properties and fracture behavior of PBXs under high strain rate loadings. The split Hopkinson pressure bar (SHPB) is a standard testing apparatus for strain rates between 100 and 10,000 $s^{-1}$[1]. These rates are equivalent to those that may occur in automobile, or aviation crashes, and an understanding of the properties of explosives at these rates is important for safe transportation.

Failure in particulate composites is heavily dependent upon the interaction between the particles and the matrix, specifically on the separation of particles and binder. In addition, failure also depends on the volume ratio of particle to matrix. With a much higher concentration of particles than conventional structural polymers (usually in the range of 15–40%), PBXs may exhibit different failure mechanisms to conventional particulate composites. The study of mechanical properties and the failure mechanisms of explosives have drawn tremendous attention in recent years [2–5].In view of the

J.G. Qin (✉) • Y.L. Lin • F.Y. Lu • R. Chen
College of Science, National University of Defense Technology, 410073 Changsha, Hunan, People's Republic of China
e-mail: jg_qin@yahoo.cn

Zh.B. Zhou
National Key Laboratory of Explosion Science and Technology, Beijing Institute of Technology,
100081 Beijing, People's Republic of China

V. Chalivendra et al. (eds.), *Dynamic Behavior of Materials, Volume 1: Proceedings of the 2012 Annual Conference on Experimental and Applied Mechanics*, Conference Proceedings of the Society for Experimental Mechanics Series,
DOI 10.1007/978-1-4614-4238-7_63, © The Society for Experimental Mechanics, Inc. 2013

hazardous nature of the explosive materials and the lack of facilities for handling large quantities, tests sometimes have to be performed on small samples. Low strengths and safety concerns bring additional difficulties in conducting such tests for explosive materials, sometimes even making conventional tests, e.g. uniaxial tests impossible. One of the most suitable tests for this is a Brazilian test, known as the diametric compression test or the indirect tensile test. Diametric compression tests, in which a disc of material is loaded diametrically in compression, can be used to generate tensile stress and estimate the tensile failure stress of a material without the inconvenience of uniaxial dumbbell testing [6]. In the investigation of the fracture behaviours of PBXs, one of the important parameters is dynamic fracture toughness. A notched core-based semi-circular bend (SCB) specimen has been successful loaded dynamically with SHPB to test the dynamic fracture toughness of rock [7].

When performing dynamic Brazilian test and fracture test using Hopkinson bar, the transmitted signal is too small due the mismatch between the acoustic impedance and the contact area of the bar with materials. The X-cut quartz has been shown to be much more sensitive in detecting stresses in its X-direction than the strain gauge in Hopkinson bar tests, with three orders of magnitude increase in the sensitivity [8]. The traditional measuring techniques are only capable of provided a single point strain by means of a strain gauge or an average strain in a specific direction using a displacement sensor [9, 10]. Recently, Gonzalez and Knauss [11] applied the digital image correlation DIC technique to solve the strain field surrounding the tip of a tensile crack in a solid propellant rocket fuel, which showed that the method could be used to analyze the highly inhomogeneous deformations of the particulate composite involved in the globally homogenous deformation of the specimen. DIC method is an efficient optical method for measuring inhomogeneous deformation by providing whole field contour maps of displacements with high sensitivity.

In the present work, the mechanical properties and fracture behavior of a mock PBX under high strain rate was experimentally studied by using the SHPB apparatus under the uniaxial compression test, the Brazilian test and the fracture test. Tensile strength and fracture toughness of specimens are obtained by using the diagnostics of a quartz crystal transducer. DIC technique is used to observe whole field deformations of specimen in real time to study the tensile and fracture behaviour of specimen.

## 63.2 Experimental Method and Procedure

### *63.2.1 Materials*

A mock PBX consists of 89% KCl and 11% polymeric binder by weight, was investigated in the experiments. The specimen was pressed in a steel die under the temperatures of 100 °C and pressures of 100 MPa and duration of pressing was 1.5 h to produce specimen. Three shapes specimens were used in three different set tests. The block specimen with the geometry of 13 mm × 13 mm × 13 mm was used in the uniaxial compression test, the disc specimen with 20 mm in diameter and 10 mm in thickness was tested in the dynamic Brazilian test. The SCB specimen also with 20 mm in diameter and 10 mm, and a notch of 4 mm length and 1 mm width was used in the fracture test. The tip of the notch was sharpened with a diamond wire saw to achieve the diameter of 0.5 mm which was on the order of grain size of the specimen. This will ensure accurate measurements of fracture toughness.

### *63.2.2 Experimental Device*

The SHPB used for this study was equipped with 20 mm diameter 7075A All-alloy. In the uniaxial compression test, the stress waves are measured in the bars by means of strain gauges, mounted halfway down the bar. The transmitted part of the wave gives a measure of the load on the specimen and the reflected part gives a measure of the specimen deformation. In order to achieve the stress equilibrium of two sides of this low sound speed polymer specimen, a rubber as a pulse shaper was used to reduce the slope of the loading pulse and thus allowed more time for the specimen to achieve an almost stress equilibrium state during the loading. Figure 63.1 shows the schematic diagram of the three types tests, Fig. 63.1a for uniaxial compression test, Fig. 63.1b for dynamic Brazilian test, Fig. 63.1c for dynamic fracture test.

In the dynamic Brazilian test and fracture test, a quartz crystal transducer is mounted at the head of the transmitted bar to measured the very mild signal of loading in specimen. The quartz crystal is calibrated by the strain gauge signal. And the whole field deformations of specimens are measured by DIC. A random speckle pattern was applied to the disc specimen and

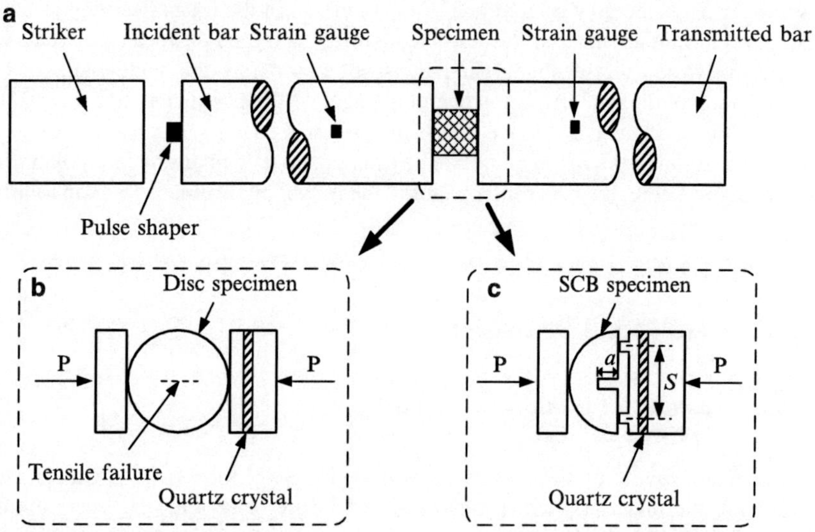


**Fig. 63.1** A schematic diagram of the SHPB; (**a**) uniaxial compression test; (**b**) dynamic Brazilian test; (**c**) dynamic fracture test

SCB specimen using an artist's airbrush and black spray paint. Photographs were taken using a FASTCAMSA-1 high-speed camera.

In Fig. 63.1b, the imposition of a compressive load, *P*, from the two bars on a disc of material causes a tensile failure on the loading axis. According to simple elasticity theory, the stress distribution along the loading diameter is described by the following formula:

$$\sigma_t(t) = \frac{P(t)}{\pi RB} \tag{63.1}$$

where *R* and *B* are the specimen radius and the thickness, respectively.

The notched SCB specimen sandwiched in the SHPB system is schematically shown in Fig. 63.1c, based on the ASTM standard E399-06e2 for rectangular three-point bending sample, we propose a similar equation for calculating the stress intensity factor for mode-I fracture in current SCB specimen:

$$K_I(t) = \frac{P(t)S}{BR^{3/2}} \cdot Y(a/R) \tag{63.2}$$

Where *R* is the radius of the disc, *S* is the span of two supporting points, *a* is the length of the notched, and *B* is the thickness of the rock disc. *Y*(*a*/*R*), *a* dimensionless geometry factor, is a function of the crack geometry, which can be calculated with a standard finite element software package (e.g., ANSYS).

### *63.2.3 Deformation Measurement by DIC Methods*

As an image correlation technique, the DIC method uses surface images where speckle points of light may change their positions during any movement or deformation, and speckle image correlation processing searches for displaced speckle points in a pair of images. To obtain the in-plane displacement components, *u* and *v*, in the *x* and *y* directions for a pair of such images, a pixel subset with $m \times n$ pixels is used as a searching sub-region. To eliminate the influence of image background, a normalized cross-correlation coefficient, C*(*u*,*v*) of the grey intensity is commonly used to define the correlation degree as [12]:

$$C^*(u^*, v^*) = \frac{\sum_{i=1}^{m}\sum_{j=1}^{n}\left[f(x_i, y_i - \bar{f})\right]\left[g(x_i', y_j') - \bar{g}\right]}{\sqrt{\sum_{i=1}^{m}\sum_{j=1}^{n}\left[f(x_i, y_i - \bar{f})\right]^2}\sqrt{\sum_{i=1}^{m}\sum_{j=1}^{n}\left[g(x_i', y_j') - \bar{g}\right]^2}} \tag{63.3}$$

where $f(x_i,y_i)$ is the grey level at point $F(x_i,y_i)$ in one of the images serving as the reference frame; $g(x'_i, y'_i)$ is that at point $G(x'_i, y'_i)$ in another image as the deformed frame; $\bar{f}$ and $\bar{g}$ are the mean grey scale values of the pixels involved in the subsets, respectively; and the searching parameters, $u^*$ and $v^*$, are the components of the vector linking the subset centers. The correlation coefficient, $C^*(u,v)$ is computed against the parameters $u^*$, $v^*$ and the matching spots are obtained when the correlation coefficient, $C^*(u,v)$ reaches a maximum, then $u^* = u \equiv x'_i - x_i$, and $v^* = v \equiv y'_i - y_i$, respectively. The displacement resolution of the DIC method can be extended to 0.05 pixels by interpolating the correlation map to fractional locations of the pixels. Here, the linear interpolation, the polynomial spline approximation, and other interpolations may be utilized.

## 63.3 Results and Discussion

### *63.3.1 Uniaxial Compression Test*

Figure 63.2 shows the compressive stress strain curves of the mock PBX under strain rate from 110 to 500 $s^{-1}$. It is seen in Fig. 63.2, the material presents typical elastic-brittle behaviour, the stress climbs on the nearly constant slope with the increasing of strain, as the strain comes to the small failure strain, the stress reaches maximum, after that, the stress decreases quickly as the strain increasing. This behaviour is mainly due to the contribution of the high weigh fraction of the grain, which is tough and brittle. Figure 63.3 presents the compressive strength and failure strain varies with the different strain rate. The result shows that, the compressive strengths of material increase from 68.2 to 74.5 MPa, and the failure strain decrease from 1.67% to 1.37% with an increase of strain rate from 110 to 500 $s^{-1}$. The strain rate dependent behavior of material is mainly due to the contribution of the binder, demonstrating that material still exhibits visco-elastic properties despite a low concentration of binder.

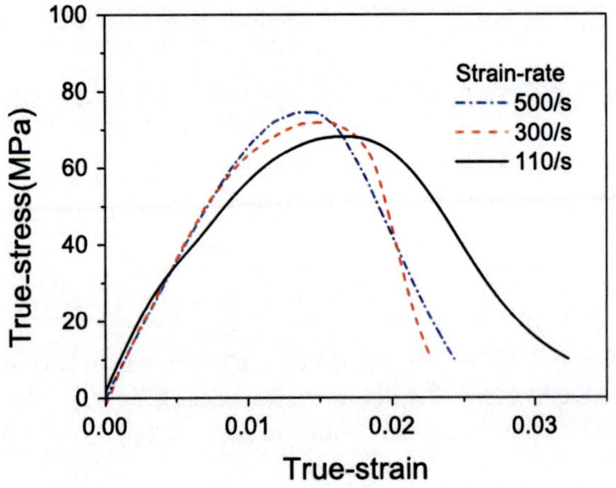


**Fig. 63.2** Compressive stress/strain curves for specimen under strain rate 110 ~ 500 $s^{-1}$

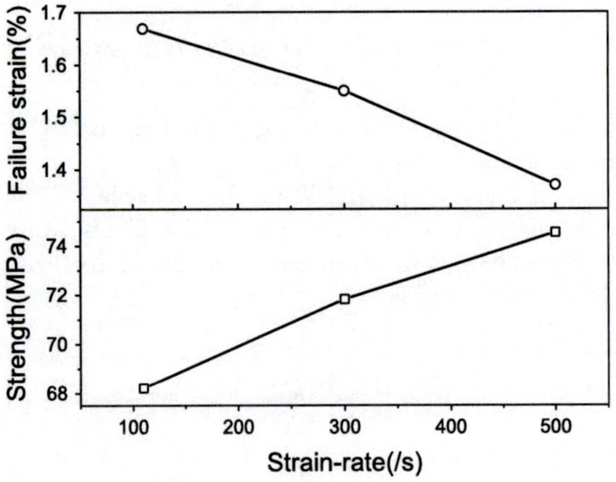


**Fig. 63.3** Strain rate dependency of compressive strength and failure strain for specimen

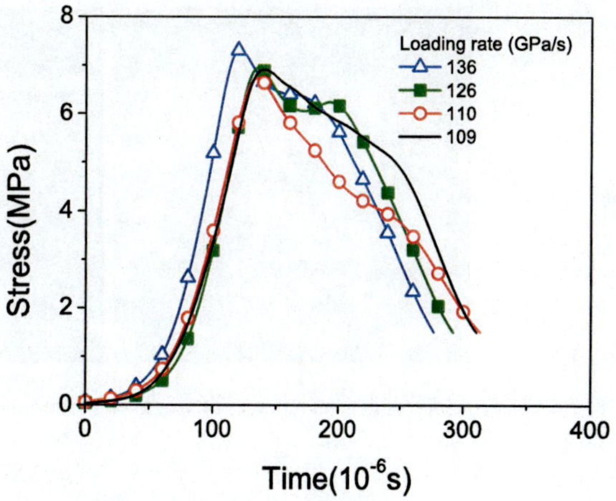


Fig. 63.4 Tensile stress/time curves for specimen under dynamic Brazilian tests

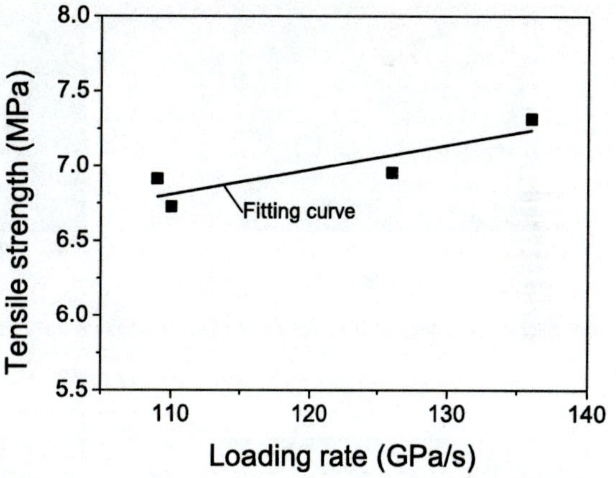


Fig. 63.5 Loading rate dependency of tensile strength for specimen

## *63.3.2 Dynamic Brazilian Test*

A plot of the tensile stress of the specimen as a function of time during the dynamic Brazilian test for different loading rates is shown in Fig. 63.4. There is an approximately linear region in tensile stress vs. time curve, and its slope is taken as the tensile loading rate. Figure 63.4 also shows elastic-brittle behaviour of the material, once the stress reaches the tensile strength, it decreases quickly as the continued deformation. Figure 63.5 shows tensile strength as a function of loading rate for the material. The result exhibits that the tensile behaviour is also clear rate dependent, tensile strength increase with an increase of loading rate. And the maximum of tensile strength of the material under the range of loading rates considered here is 7.3 MPa. The value of tensile strength of the material is about one tenth of the compressive strength.

Figure 63.6 shows the high-speed sequence obtained for experiments on the material under loading rate of 126 GPa/s. The inter-frame and exposure time were both 28 $\mu s$. Figures 63.7 and 63.8 show the displacement and strain of direction $y$ distribution of the box region (see in Fig. 63.6a) calculated by using DIC, respectively. The result shows that at the early stage of loading ($t = 28\ \mu s$), there was little displacement and strain of the specimen, the maximum displacement is only 0.06 mm. At the middle stage of loading ($t = 56\ \mu s$), there was clear deformation in the specimen, the horizontal line divides the specimen into upper part and lower part, the displacement of upper part is positive whereas lower part is negative. The maximum displacement in this stage reaches about 0.17 mm. The strain of the center of specimen reaches approximately 1.1%, nearly the failure strain of the material. At the third stage of loading ($t = 84\ \mu s$), there was a clear crack in the center of specimen in Fig. 63.6c, the maximum displacement in this stage comes to about 2.7 mm. Because of the fracture of specimen, the value of strain obtained in this stage has no significant to consider. The contours of the displacement and strain in Figs. 63.7 and 63.8 show that the displacements and strain through the specimen are consistent and symmetrical.

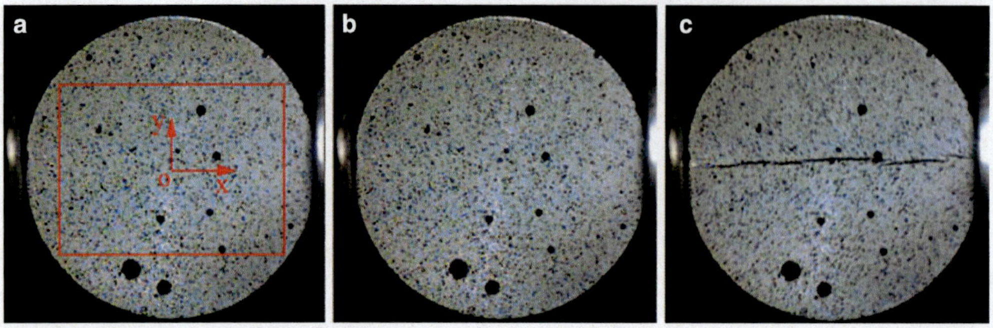


**Fig. 63.6** Sequence of dynamic images of specimen deformation under dynamic Brazilian test (**a**) $t = 28\ \mu s$ (**b**) $t = 56\ \mu s$ (**c**) $t = 84\ \mu s$

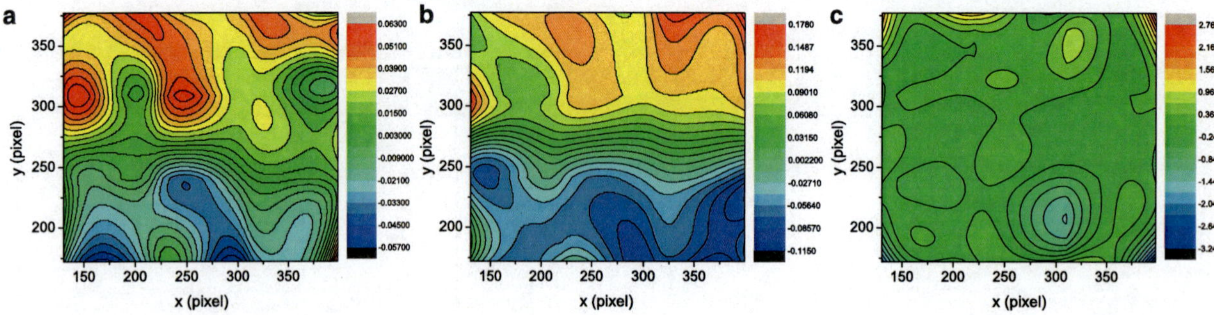


**Fig. 63.7** Displacement distribution in the box Area for specimen under dynamic Brazilian test (**a**) $t = 28\ \mu s$ (**b**) $t = 56\ \mu s$ (**c**) $t = 84\ \mu s$

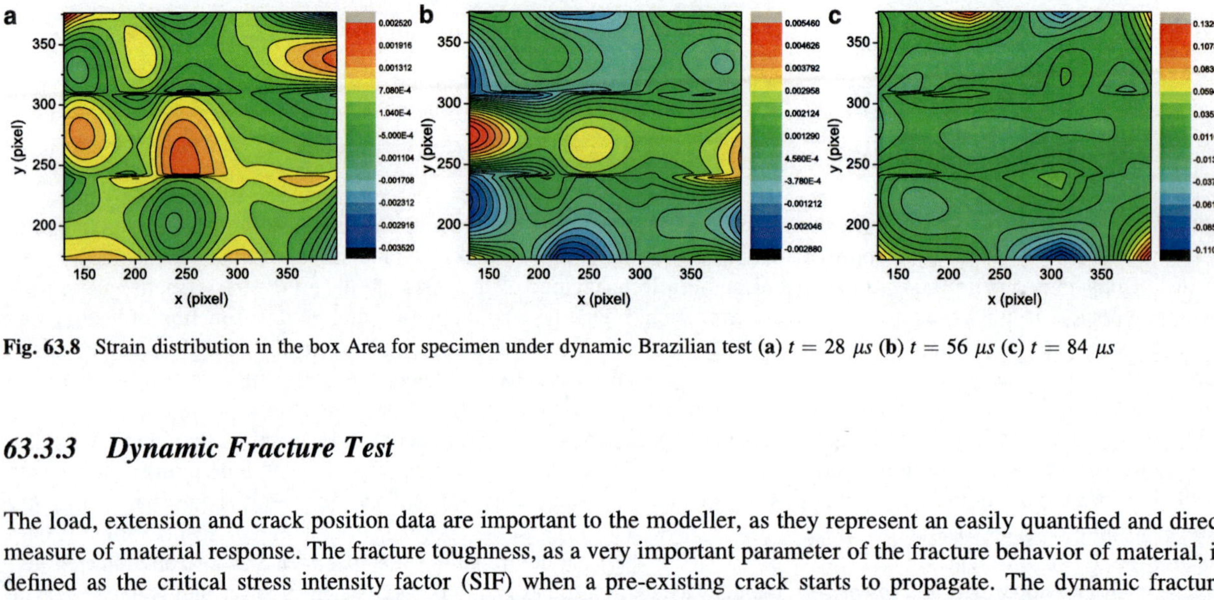


**Fig. 63.8** Strain distribution in the box Area for specimen under dynamic Brazilian test (**a**) $t = 28\ \mu s$ (**b**) $t = 56\ \mu s$ (**c**) $t = 84\ \mu s$

### *63.3.3 Dynamic Fracture Test*

The load, extension and crack position data are important to the modeller, as they represent an easily quantified and direct measure of material response. The fracture toughness, as a very important parameter of the fracture behavior of material, is defined as the critical stress intensity factor (SIF) when a pre-existing crack starts to propagate. The dynamic fracture toughness is then determined as the SIF at the crack initiation time, equals the maximum value of the evolution of SIF. Figure 63.9 shows the evolution of SIF of the SCB specimen during dynamic fracture test. Loading rate is defined as the slope of approximately linear region of the evolution curve of SIF. Figure 63.10 shows the dynamic fracture toughness ($K_{IC}$) as a function of loading rate for the material, it exhibits the dynamic fracture toughness increases with an increase of loading rate. In another word, when under higher loading rate, the material is more difficult to crack. Under the range of loading rates considered here, the maximum of dynamic fracture toughness of the material is 1.0 MPa $m^{1/2}$.

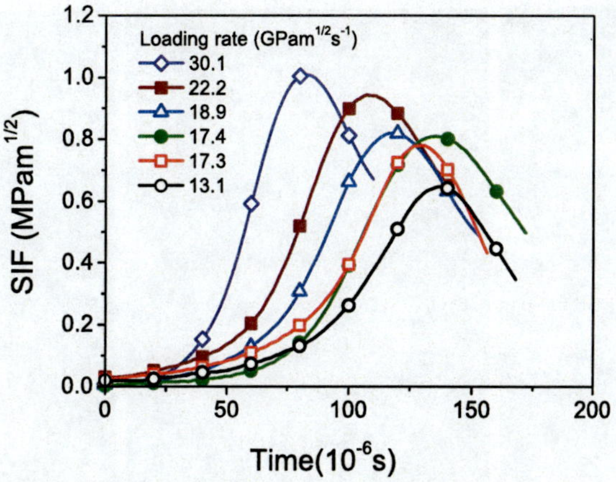


**Fig. 63.9** Stress intensity factor (SIF)/time curves for specimen under dynamic fracture tests

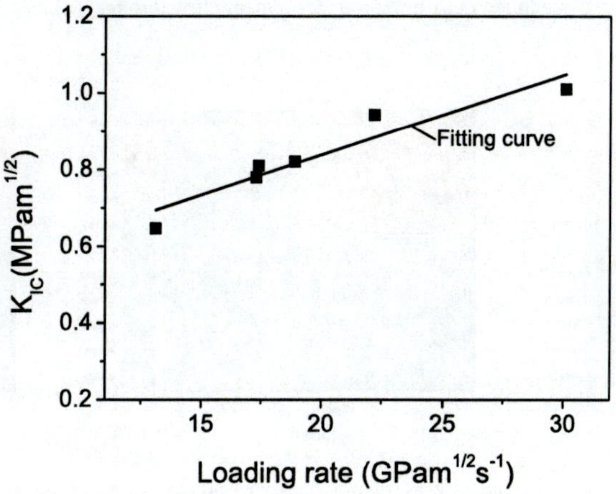


**Fig. 63.10** Loading rate dependency of dynamic fracture toughness ($K_{IC}$) for specimen

Figure 63.11 shows the high-speed sequence obtained for experiments on the material under loading rate of 17.3 GPam$^{1/2}$ s$^{-1}$. The inter-frame and exposure time were both 22 $\mu s$. Figures 63.12 and 63.13 show the displacement and strain of direction $y$ distribution of the box region (see in Fig. 63.11a) calculated by using DIC, respectively. At the early stage of loading ($t = 22$ $\mu s$), there was little displacement and strain of the specimen, the maximum displacement is also only 0.06 mm. At the middle stage of loading ($t = 44$ $\mu s$), there was clear deformation in the specimen, the displacement of left upper part is positive and left lower part is negative, the value of the right part is very small compare to the left part. The maximum displacement in this stage reaches about 0.15 mm. At the third stage of loading ($t = 66$ $\mu s$), there was a clear crack at the tip of notch of specimen in Fig. 63.11c, that means the SIF has reached the peak point and begin to decrease. The contours of the displacement and strain also show that the displacements and strain through the specimen are consistent and symmetrical.

## 63.4 Conclusion

Three type tests of uniaxial compression, dynamic Brazilian and fracture test have been performed on a mock PBX by using the SHPB apparatus in order to investigate the mechanical and fracture behavior of the material. High-speed photography was used, in conjunction with the DIC technique to measure the specimen displacements and strains. The result shows that the material presents brittle and visco-elastic properties in the experiments. The compressive strength, tensile strength and

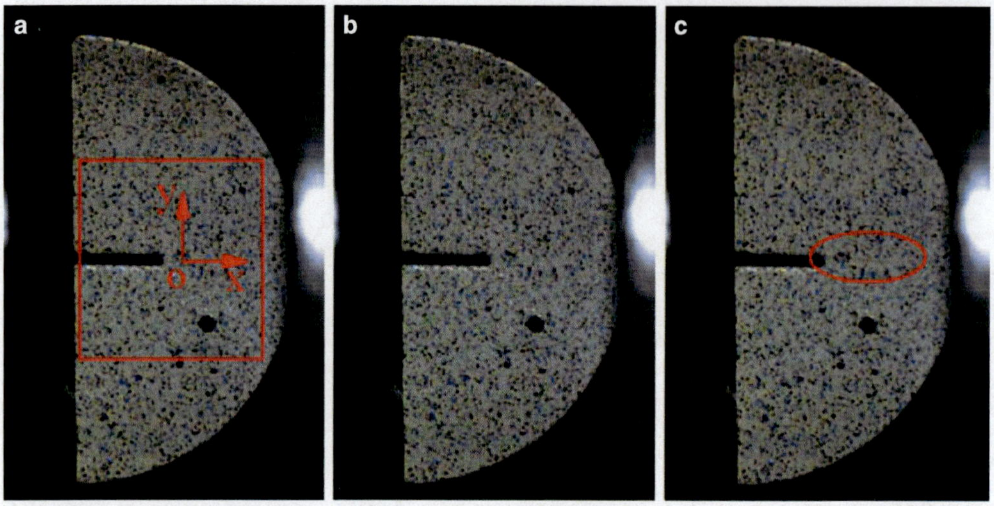


**Fig. 63.11** Sequence of dynamic images of specimen deformation under dynamic fracture test **(a)** $t = 22\ \mu s$ **(b)** $t = 44\ \mu s$ **(c)** $t = 66\ \mu s$

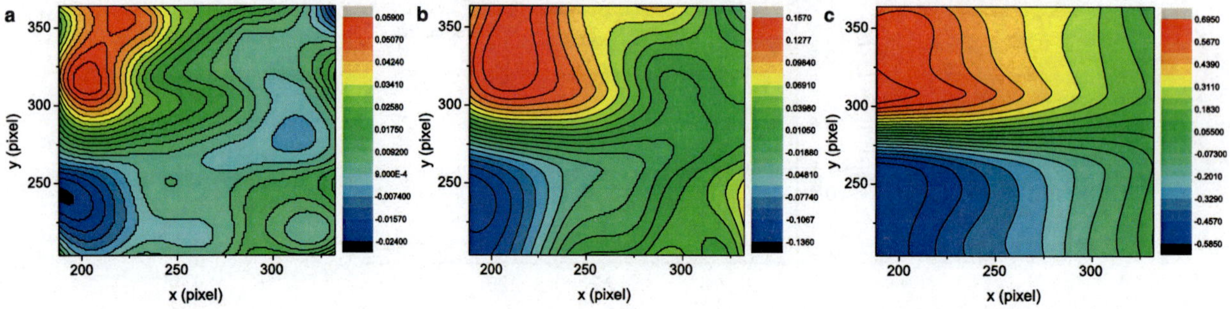


**Fig. 63.12** Displacement distribution in the box Area for specimen under dynamic fracture test **(a)** $t = 22\ \mu s$ **(b)** $t = 44\ \mu s$ **(c)** $t = 66\ \mu s$

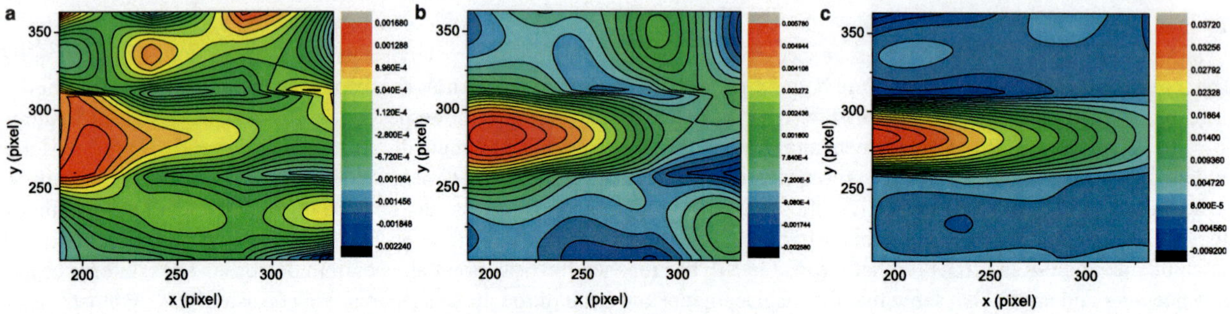


**Fig. 63.13** Strain distribution in the box Area for specimen under dynamic fracture test **(a)** $t = 22\ \mu s$ **(b)** $t = 44\ \mu s$ **(c)** $t = 66\ \mu s$

fracture toughness of the material are all clear strain rate dependent due to the contribution of the binder. The compressive strength of the specimen was observed to be ten times the tensile strength. The DIC technique provides a quantified and non-contact measurement of displacement field and strain field of specimen under high strain rate loading. With the result from the DIC technique, it can be seen that the specimen deformed consistent and symmetrical once the stress equilibrium in the specimen is achieved.

**Acknowledgment** The authors gratefully acknowledge the financial support of the Natural Science Foundation of China (NSFC) through Grant No.11172328&11132012, and the foundation of National University of Defense Technology through Grant No. JC-11-02-17.

## References

1. Gray GT III (2000) Classic split-Hopkinson pressure bar testing ASM handbook: mechanical testing and evaluation, vol 8. ASM International, Materials Park, pp 462–476
2. Wiegand DA (1998) Mechanical properties and mechanical failure of composite plastic bonded explosives and other energetic materials. In: Proceedings of the eleventh international detonation symposium. Boston, pp 744–750
3. Gray III GT, Idar DJ, Blumenthal WR, Cady CM, Peterson PD (1998) High- and low-strain rate compression properties of several energetic material composites as a function of strain rate and temperature. In: Proceedings of the eleventh international detonation symposium. Boston, pp 76–84
4. Kauly T, Kimmel T (1998) Failure mechanism in PBX. In: Proceedings of the energetic materials, 29th international annual conference of ICT. Karlsruhe, p 19:1–11
5. Peterson PD, Mortensen KS, Idar DJ, Asay BW, Funk DJ (2001) Strain field formation in plastic bonded explosives under compressional punch loading. J Mater Sci 36:1395–1400
6. Awaji H, Sato S (1979) Diametral compressive testing method. J Eng Mater Technol 101:139–147
7. Dai F, Chen R, Xia K (2009) A semi-circular bend technique for determining dynamic fracture toughness. Experimental Mechanics 50:783–791
8. Chen R, Huang S, Xia K, Lu F (2009) A modified Kolsky bar system for testing ultra-soft materials under intermediate strain rates. Rev Sci Instrum 80:076108
9. Johnson HD (1978) Mechanical properties of LX-10-1 evaluated with the diametric disc test. MHSMP-78-08
10. Johnson HD (1981) Diametric disc and standard tensile test correlation study. MHSMP-81-22
11. Gonzalez J, Knauss WG (1998) Strain inhomogeneity and discontinuous crack growth in particulate composite. J Mech Phys Solids 46(10):1981–1995
12. Sutton MA, Wolters WJ, Peters WH, Ranson WF, McNeill SR (1983) Determination of displacements using an improved digital correlation method. Image Vis Comput 1:33–39

# Chapter 64
# Dynamic Fracture Toughness Test Using Hopkinson Bar

**Wei-Yang Lu, Bo Song, and Kenneth Gwinn**

**Abstract** Various loading and measuring configurations have been developed in Hopkinson bar fracture toughness experimental techniques. It is well known that several fundamental issues, such as force equilibrium, pulse shaping, stress-wave propagation, etc., must be evaluated in order to obtain a reliable measurement. In our previous work of characterizing Mode II dynamic fracture toughness of a woven composite, highly sensitive polyvinylidene fluoride (PVDF) force transducers were employed to check the forces on the front wedge and back spans in a SHPB ENF experiment. The results show that proper pulse shaping is necessary so the specimen can achieve stress equilibrium before the crack starts to propagate. This study addresses the issue that stress wave propagates through the non-uniform section, which is between the incident and transmission bars including the specimen, loading wedge, and supporting fixture. The transmitted signals are compared with PVDF measurements, and also with numerical simulations of stress waves propagate through supporting fixture and down to the transmission bar.

**Keywords** Mode II fracture • Dynamic fracture toughness • PVDF • Woven glass composite • Hopkinson Bar

Various loading and measuring configurations have been developed in Hopkinson bar fracture toughness experimental techniques. It is well known that several fundamental issues, such as force equilibrium, pulse shaping, stress-wave propagation, etc., must be evaluated in order to obtain a reliable measurement. A typical experimental setup for Mode II dynamic fracture toughness is shown in Fig. 64.1. In our previous work of characterizing a woven composite [1, 2], highly sensitive polyvinylidene fluoride (PVDF) force transducers were employed to check the forces on the front wedge and back spans in a high rate end notch flexure (ENF) fracture experiment, shown in Fig. 64.2. The results show that proper pulse shaping is necessary to achieve stress equilibrium in the specimen before the crack starts to propagate. This study addresses the issue that stress wave propagates through the non-uniform section, Fig. 64.3, which is between the incident and transmission bars including the specimen, loading wedge, and supporting fixture. Except the composite specimen, all are made of steel.

Note that force equilibrium of the specimen, $F_1 = 2\,F_2 = 2\,F_3$, is not equivalent to the equilibrium of the test section, $F_4 = F_5$, which is usually the requirement of Hopkinson bar force measurement under dynamic loading. PVDF transducers measure the forces $F_1$, $F_2$ and $F_3$ at contact interfaces; Hopkinson bars evaluate $F_4$ and $F_5$ at a distance away from the contact points. In an experiment, where a pulse shaper was used, the force histories measured from PVDF transducers are shown in Fig. 64.4. It takes the transverse wave approximately 40 μs to propagate from the central wedge to the end rail supports. Both forces $F_1$ and $F_2 + F_3$ increased slowly in the beginning but at a different rate, where the specimen was not at equilibrium. When $t > 150$ μs, $F_1$ showed a steady increase, the oscillation of $F_2 + F_3$ became small, and two curves converged indicating the forces were equilibrated. Examining the forces $F_2 + F_3$ and $F_5$, shown in Fig. 64.5, both wave shape and timing are quite different. In this case, the stress wave propagating in the rail support complicates the measurement. Force measurement from the transmission bar does not represent the force acting on the specimen. Not shown in the figures, the measured $F_1$ and $F_4$ are basically the same. The loading wedge has little effect on transmitted and reflected stress waves.

W.-Y. Lu (✉) • B. Song • K. Gwinn
Sandia National Laboratories, Livermore, CA 94551-0969, USA
e-mail: wlu@sandia.gov; bsong@sandia.gov

V. Chalivendra et al. (eds.), *Dynamic Behavior of Materials, Volume 1: Proceedings of the 2012 Annual Conference on Experimental and Applied Mechanics*, Conference Proceedings of the Society for Experimental Mechanics Series,
DOI 10.1007/978-1-4614-4238-7_64, 

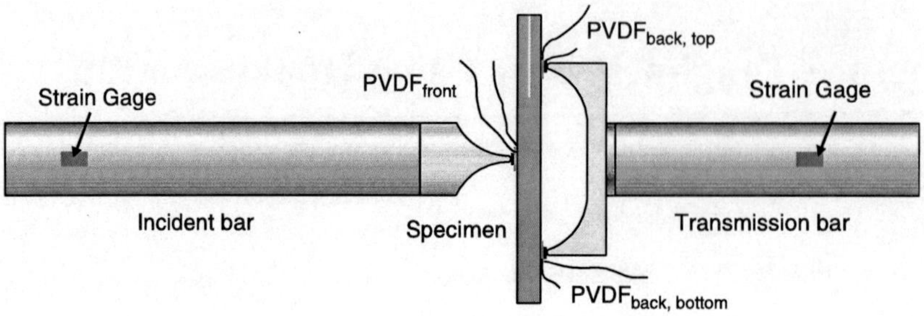


**Fig. 64.1** Schematic of ENF experiment on Hopkinson bar

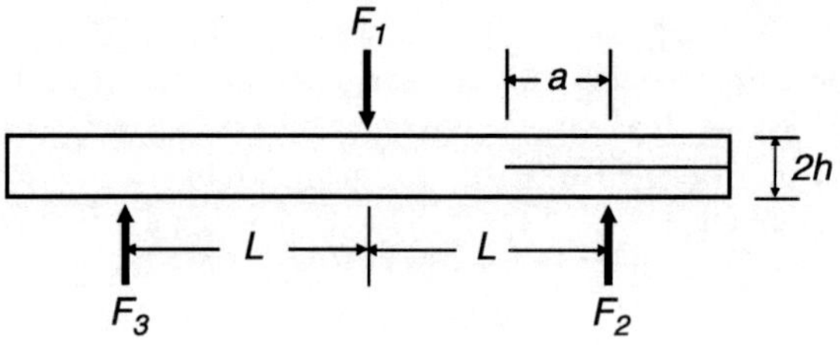


**Fig. 64.2** ENF specimen

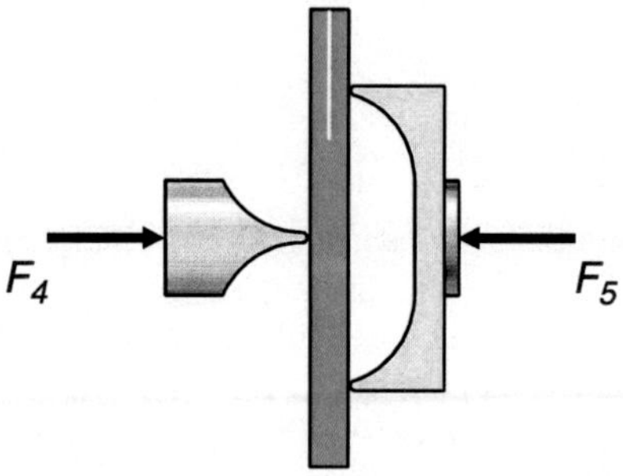


**Fig. 64.3** The non-uniform section

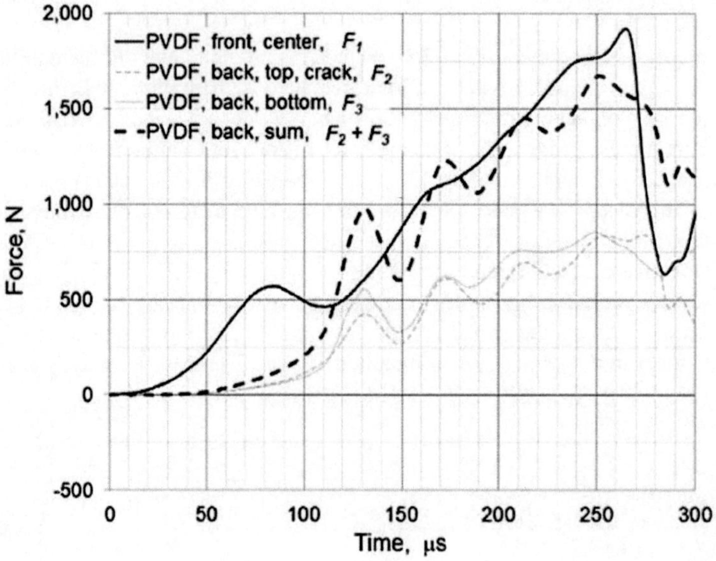


**Fig. 64.4** Force histories at contact points, case 1

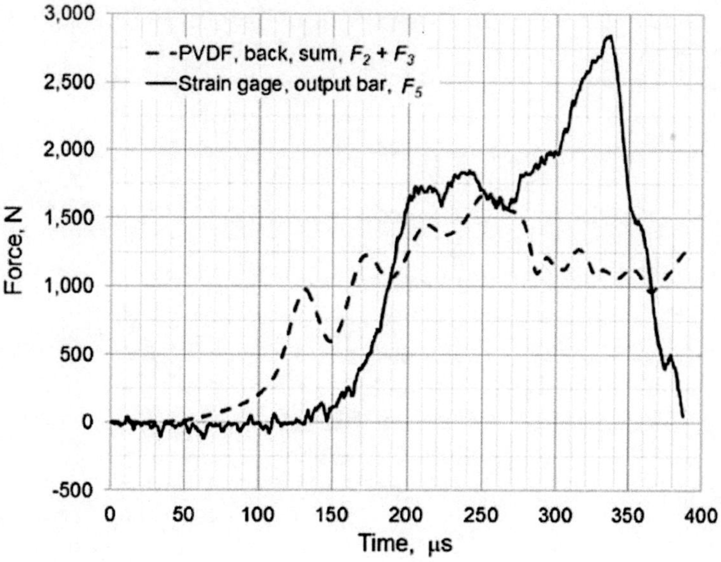


**Fig. 64.5** Force calculated from strain gage, case 1

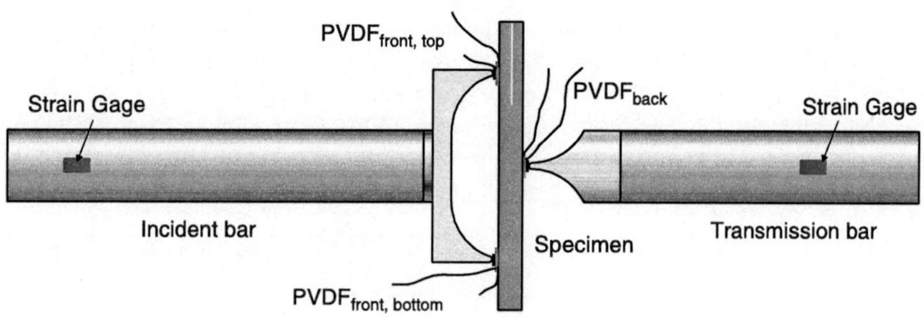


**Fig. 64.6** The second configuration, where the non-uniform section is reversed

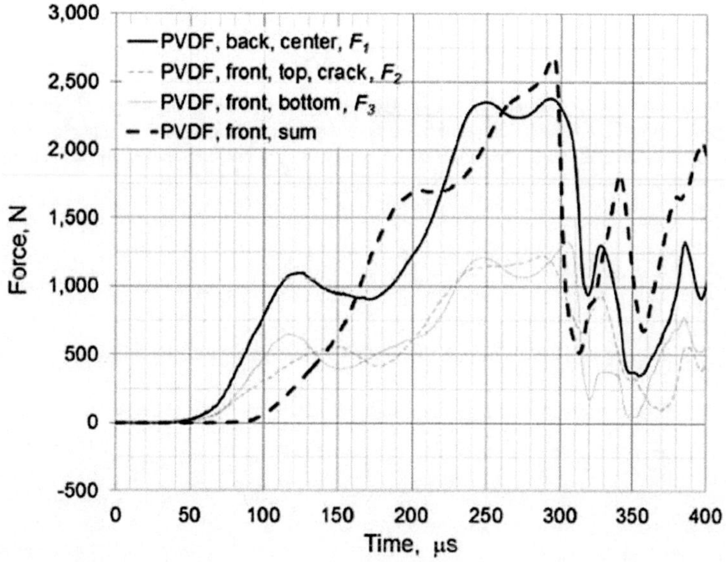


**Fig. 64.7** Force histories at contact points, case 2

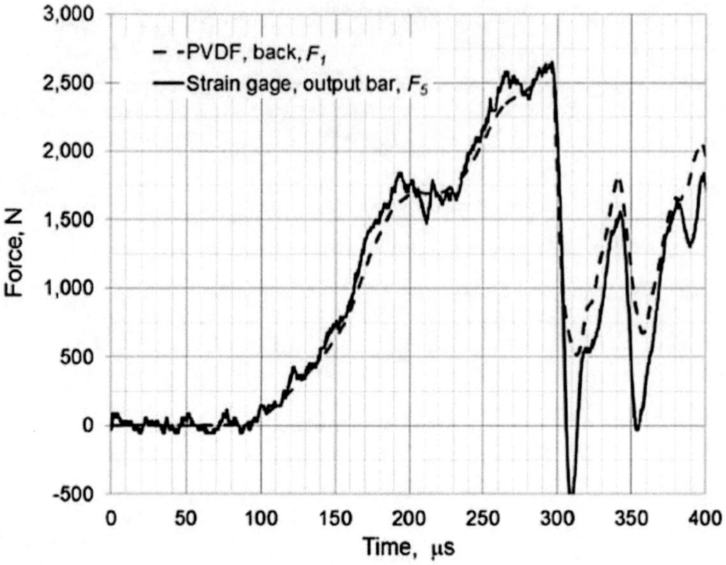


**Fig. 64.8** Force calculated from strain gage, case 2

Since the wedge does not disturb Hopkinson bar measurement and the equilibrium condition can be achieved by pulse shaping, will the test with the reversed non-uniform section, Fig. 64.6, produce useable data? A test of the new configuration was performed with the exactly same conditions (pulse shaping, striker speed, etc.) as the first test. The force histories measured at the specimen and from the transmission bar are shown in Figs. 64.7 and 64.8, respectively. The transmission bar measurement is consistent with the force at the specimen, which again shows that the wedge does not affect 1-D wave propagation in the bar. Unfortunately, the condition of specimen equilibrium was not reached as shown in Fig. 64.7. $F_1$ and $F_2 + F_3$ curves do not agree.

Numerical simulations of stress waves propagate through the non-uniform section and down to the transmission bar are currently under investigation.

**Acknowledgements** Sandia National Laboratories is a multi-program laboratory managed and operated by Sandia Corporation, a wholly owned subsidiary of Lockheed Martin Corporation, for the U.S. Department of Energy's National Nuclear Security Administration under contract DE-AC04-94AL85000.

## References

1. Lu W-Y, Song B, Jin H (2010) A revisit to high-rate mode-II fracture characterization of composites with Kolsky bar techniques. In: Proceedings of the 2010 SEM annual conference and exposition on experimental and applied mechanics, Indianapolis, 7–10 June 2010
2. Lu W-Y, Song B, Jin H (2011) Dynamic mode-II characterization of a woven glass composite. In: Proceedings of the 2010 SEM annual conference and exposition on experimental and applied mechanics, Uncasville, 13–16 June 2011

# Chapter 65
# Damage Assessments of Ballistic Impact Behaviors of GLARE 5 (3/2) Beams with Various Stacking Sequences

**A. Seyed Yaghoubi and B. Liaw**

**Abstract** Fiber-metal laminated beams of GLARE 5 (3/2) lay-up configuration of dimensions: 254 × 25.4 mm (10″ × 1″) with various stacking sequences were impacted by a 0.22 caliber bullet shaped projectile using a high-speed gas gun. The effect of stacking sequence was considered on FMLs with unidirectional [$0_4$°], unidirectional [$90_4$°], cross-ply [0°/90°]$_s$ and quasi-isotropic [0°/±45°/90°] lay-up orientations. Optical imaging and mechanical sectioning techniques were used to assess the impact damages. The results showed that the failure pattern was changed by altering the specimen stacking sequences. A visible longitudinal crack along the 0° fiber direction was observed for the unidirectional [$0_4$°] specimen. Subjected to the ballistic impact, damage development in the unidirectional [$90_4$°] FMLs caused the specimens to split into two halves. A Phantom V710 high-speed camera was used to monitor the crack propagation and damage growth on the backside of the specimens with unidirectional [$0_4$°] and cross-ply [0°/90°]$_s$ lay-up sequences. In particular, it was illustrated experimentally that a small circular region of damage was created on the non-impacted side of the specimens. In addition, for the unidirectional specimen, the damage progressed as a lip-shaped zone; whereas, it advanced as a diamond-shaped contour for the cross-ply specimen. The 3D dynamic nonlinear finite element (FE) code, LS-DYNA, was used to validate the experimental results. The finite element results were compared to those obtained experimentally. Good agreement between the experimental and FE result was obtained.

**Keywords** Ballistic impact • GLARE 5 (3/2) FMLs • Stacking sequence • LS-DYNA

## 65.1 Introduction

GLARE 5 belongs to the family of fiber–metal laminates (FMLs) made of alternating layers of 2024-T3 aluminum alloy sheets and unidirectional S2-glass/epoxy laminates. By combining the beneficial properties of monolithic metals and fiber-reinforced composites, FMLs provide drastically improved mechanical properties compared to conventional polymer matrix composites or aluminum alloys [1, 2]. Seyed Yaghoubi et al. [3] considered velocity measuring approaches for the determination of ballistic limits of GLARE 5 FMLs plates. They experimentally showed that the projectile was further accelerated along the ballistic trajectory after emerging from the gun barrel. Hoo Fatt et al. [4] studied the ballistic impact on GLARE FMLs. Their result showed the energy dissipated in delamination represented 2–9% of the total absorbed energy. Gama et al. [5] investigated impact, damage evolution and penetration of thick-section composites using explicit finite element analyses. They reported that 75% of the kinetic energy of the projectile was transferred to the composite laminate in the first 25–30 μs of time. At around $V_{50}$, about 80% of the impact energy was converted into the strain energy or internal energy of the composite laminate and the rest of the energy was converted into sliding energy and kinetic energy of the projectile-composite system. The incident versus residual velocity of GLARE 5 FML beams with various stacking sequences was studied by Seyed Yaghoubi et al. [6]. It was shown that unidirectional [$90°_4$] specimen offered the least resistance to the projectile perforation. In addition, except for the unidirectional [$90°_4$] specimen, ballistic limit was almost invariant with respect to change in stacking sequence. They also demonstrated that the results obtained from their finite element simulation were in good agreement with the experimental counterparts. Sevkat et al. [7] estimated ballistic limit

A. Seyed Yaghoubi (✉) • B. Liaw
Department of Mechanical Engineering, The City College of New York, Convent Avenue and 138th Street, New York, NY 10031, USA
e-mail: ASeyed_Yaghoubi@gc.cuny.edu; a_syaghoubi@hotmail.com

V. Chalivendra et al. (eds.), *Dynamic Behavior of Materials, Volume 1: Proceedings of the 2012 Annual Conference on Experimental and Applied Mechanics*, Conference Proceedings of the Society for Experimental Mechanics Series, DOI 10.1007/978-1-4614-4238-7_65, © The Society for Experimental Mechanics, Inc. 2013

velocity with sufficient accuracy by combining results obtained from ballistic tests and finite element predictions for S2-glass fiber/toughened epoxy composite beams.

The main objective of this study was to investigate damage evaluation in the GLARE 5 (3/2) FML beams with various stacking sequences subjected to ballistic impact at room temperature. The 3D dynamic nonlinear FE code, LS-DYNA, was used to validate the experimental results such as incident versus residual velocity data, post-impact damage patterns and bullet residual lengths.

## 65.2 Experimental Procedures

All GLARE panels considered in this study consist of 2024-T3 aluminum alloy with a thickness of 0.305 mm (0.012″) per layer and unidirectional S2-glass/epoxy laminated layers, each with a thickness of 0.508 mm (0.020″). Each S2-glass/epoxy layer has a lay-up orientation according to Table 65.1. GLARE 5 panels with various stacking sequences were cut into rectangular beam specimens with dimensions of 254 × 25.4 mm (10″ × 1″). The configuration of GLARE 5 panels with different stacking sequences, MVF and typical mechanical properties of constituents are described in Tables 65.1 and 65.2, respectively [8–10]. In Table 65.1, the term MVF represents metal volume fraction and is defined as the ratio of the sum of the thicknesses of all aluminum layers over the total thickness of the fiber-metal laminate [11]:

$$\mathrm{MVF} = \frac{\sum t_{\mathrm{aluminum}}}{t_{\mathrm{FML}}} \tag{65.1}$$

The ballistic impact tests were performed using an in-house-designed gas gun. The gun consists of a pressurized tank, a gas storage vessel, a solenoid valve and a stainless-steel barrel. The pictures of the gas gun setup and the specimen container are shown in Fig. 65.1. Compressed helium/air was used as the propellant. A 0.22 caliber copper bullet was chosen as the projectile. The bullet mounted inside a pair of plastic sabots and then inserted into the gun barrel. Figure 65.2 illustrates the 0.22 caliber copper bullet and the sabot. For the same type of sabots, the projectile initial velocity can be controlled by changing its position inside the barrel and/or by varying the input pressure of the gun. The beam specimen was clamped over

**Table 65.1** GLARE 5 specimens tested in this study with different stacking sequences

| Lay-up configuration | Prepreg plies and orientation | Total thickness | MVF |
|---|---|---|---|
| 3/2 | $[0°_4]$ unidirectional | 1.930 mm (0.076″) | 0.474 |
| | $[90°_4]$ unidirectional | | |
| | $[0°/90°]_s$ cross-ply | | |
| | [0°/±45°/90°] quasi-isotropic | | |

The configuration notation *3/2* means the panel is composed of three aluminum-alloy layers interlaced with two fiber-reinforced epoxy layers

**Table 65.2** Typical values of mechanical properties [8–10]

| Mechanical property | | Aluminum 2024-T3 | UD S2 glass/FM 94 epoxy prepreg |
|---|---|---|---|
| Tensile ultimate strength (MPa) | L | 455 | 1,900 |
| | T | 448 | 57 |
| Tensile yield strength (MPa) | L | 359 | – |
| | T | 324 | – |
| Tensile modulus (GPa) | L | 72 | 54 |
| | T | – | 9.4 |
| Ultimate strain (%) | L | 19 | 3.5 |
| | T | – | 0.6 |
| Compressive yield strength (MPa) | L | 303 | – |
| | T | 345 | – |
| Compressive modulus (GPa) | L | 74 | 54.4 |
| | T | – | 11 |
| Density (g/cm$^3$) | | 2.79 | 2.00 |

The symbols, *L* and *T*, stand for longitudinal (the rolling direction for the metal) and transverse directions, respectively

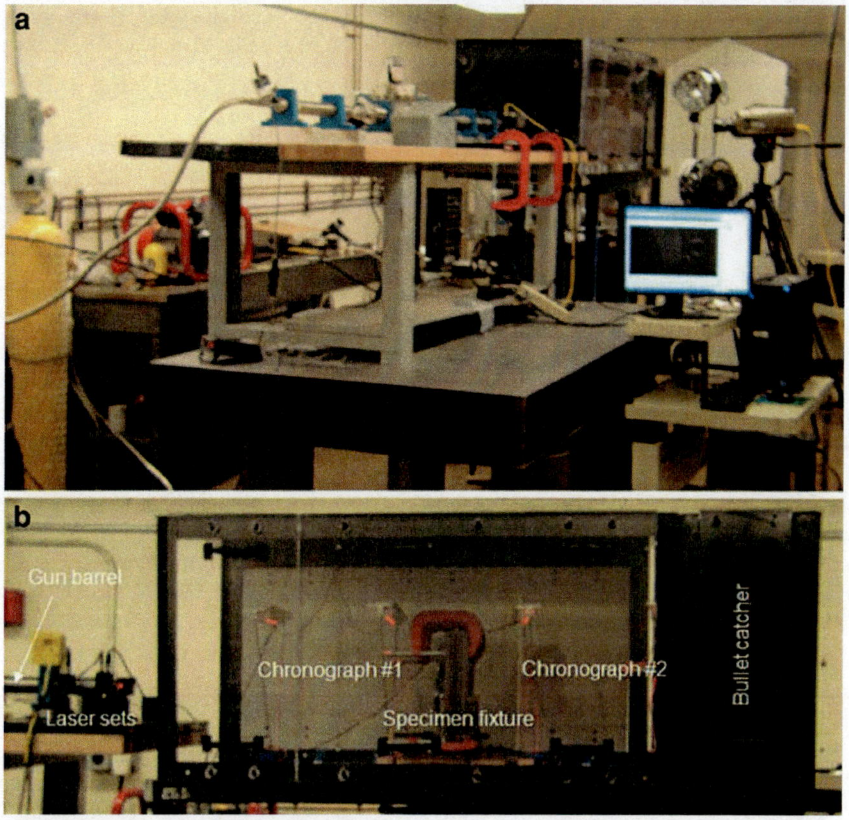


**Fig. 65.1** (**a**) The high-speed gas gun setup for ballistic impact tests and (**b**) specimen container

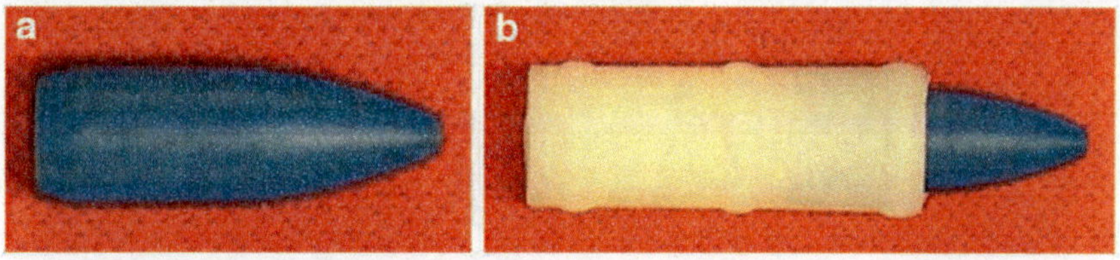


**Fig. 65.2** (**a**) 0.22 caliber copper bullet and (**b**) bullet mounted in a pair of plastic sabots

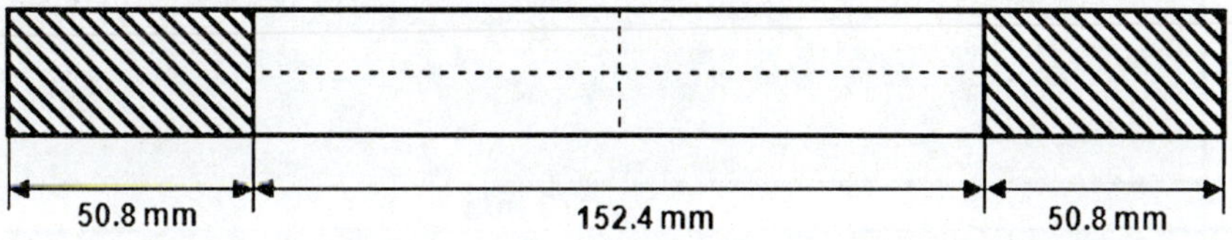


**Fig. 65.3** A composite beam specimen clamped at the two ends

a length of 50.8 mm from each side (Fig. 65.3). The specimen fixture was extra supported by four C-clamps inside the specimen container (Fig. 65.1b). The composite beam specimen was then impacted by the projectile at the center.

The speed of the projectile was measured in three distinct ways. A pair of diode-lasers/amplified-photodiodes, separated by 101.6 mm (4″), was located near the exit of the gun barrel, forming two optical gates to measure the speed of the projectile at the gun muzzle. For the second measurement, two pairs of chronographs were used to measure the projectile speed before and after impact; one pair was located in front of the target, while the second was located behind the target. Thirdly, a high-speed camera was used to monitor the bullet motion during the test. In this study, the high-speed camera was set perpendicular to the projectile ballistic trajectory. Using the captured high-speed video, the bullet speed was then determined before and after the impact.

## 65.3 Results and Discussion

Figure 65.4 represents impacted and non-impacted side views of the GLARE 5 (3/2) FML beams with unidirectional [0°$_4$], unidirectional [90°$_4$], cross-ply [0°/90°]$_s$ and quasi-isotropic [0°/±45°/90°] stacking sequences. The corresponding cross-sectional views are depicted in Fig. 65.5. As is apparent from these figures, aluminum layers failed through petaling, bulging and plugging. In addition, the fiber breakage/splitting/pullout, matrix cracking, debonding/delamination and plastic deformations in the specimens were observed (Figs. 65.4 and 65.5). For both of the unidirectional specimens the crack on the non-impacted side of the specimens propagated along the 0° fiber direction. It should be noted that for the unidirectional [90°$_4$] specimen the crack reached the specimen edges, causing it to split into two halves. Moreover, different damage zones on the non-impacted side of the specimens were noticeable (Fig. 65.4). For both of the unidirectional specimens, i.e. [0°$_4$] and [90°$_4$], a lip-shaped damage contour was induced; whereas it had a diamond shape for the cross-ply specimen. For the quasi-isotropic specimen, the damage zone on the non-impacted side of the specimen was a mixture of the damages introduced in the unidirectional and cross-ply specimen.

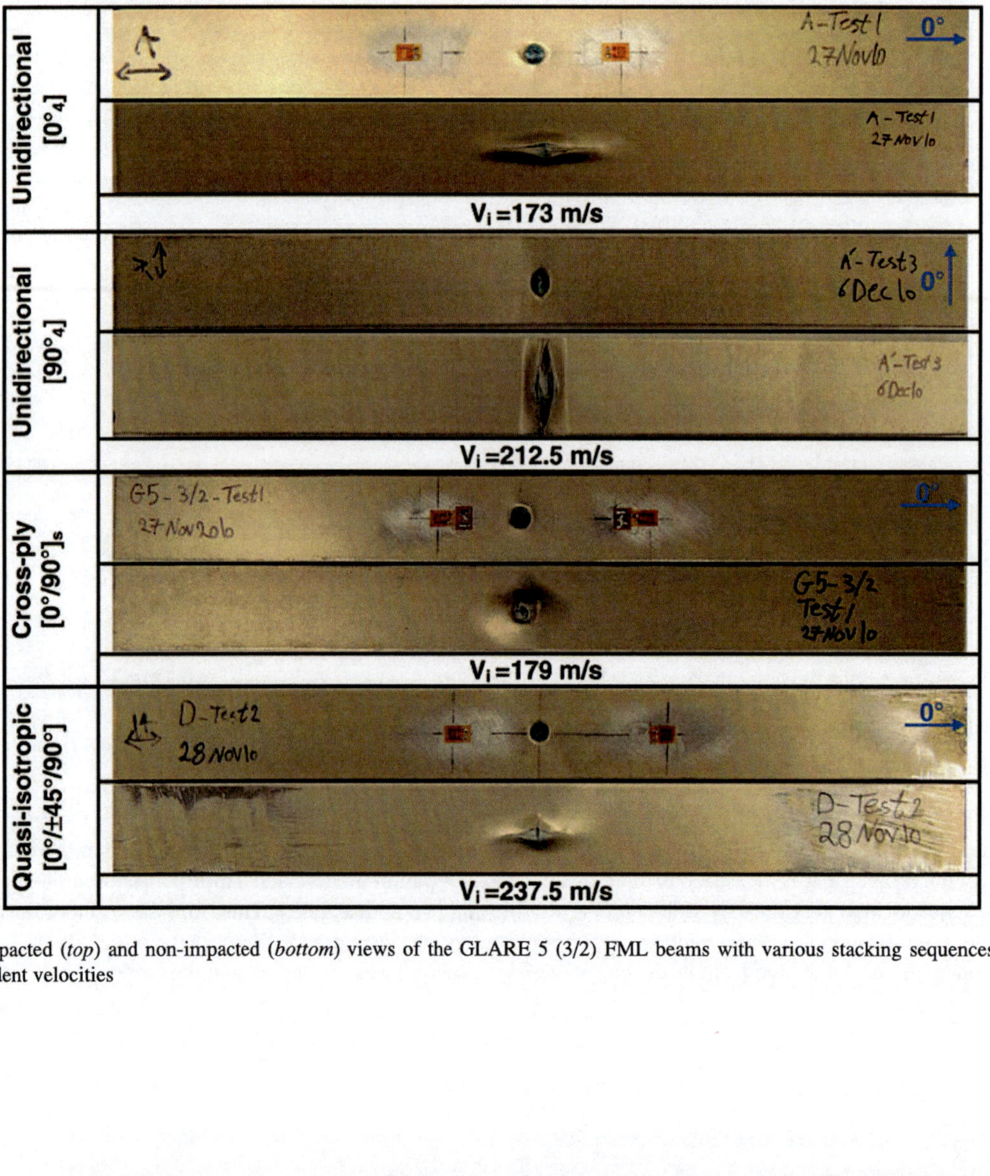


**Fig. 65.4** Impacted (*top*) and non-impacted (*bottom*) views of the GLARE 5 (3/2) FML beams with various stacking sequences impacted at different incident velocities

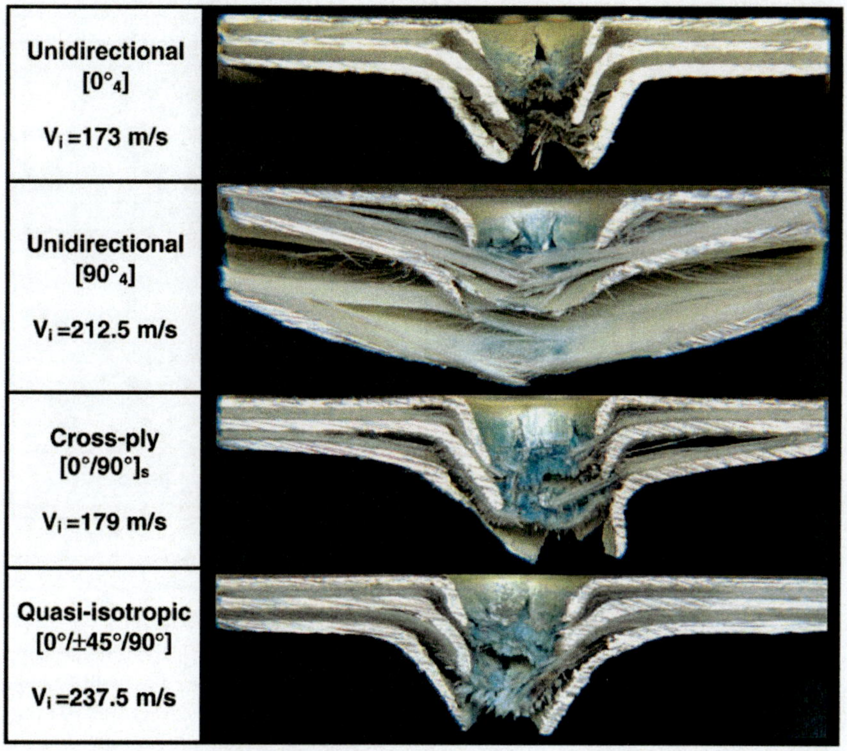


**Fig. 65.5** Cross-sectional views of the GLARE 5 (3/2) FML beams with various stacking sequences impacted at different incident velocities

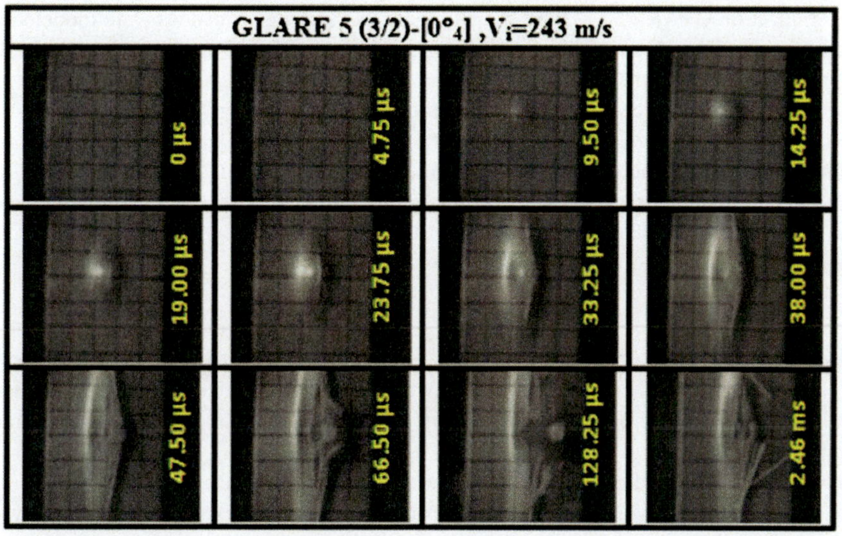


**Fig. 65.6** High-speed camera footage of the GLARE 5 (3/2)-[0°$_4$] FML beam impacted at 243 m/s

Figures 65.6 and 65.7 illustrate damage progressions on the non-impacted side of the GLARE 5 (3/2) FML beams with unidirectional [0°$_4$] and cross-ply [0°/90°]$_s$ stacking sequences impacted at 243 m/s and 240 m/s, respectively. These pictures were obtained using a high-speed camera. The bulging initiated during the first 4.75 µs of contact time for the unidirectional [0°$_4$] specimen; where, it occurred during the first 4.63 µs for the cross-ply specimen. As apparent from these figures, the

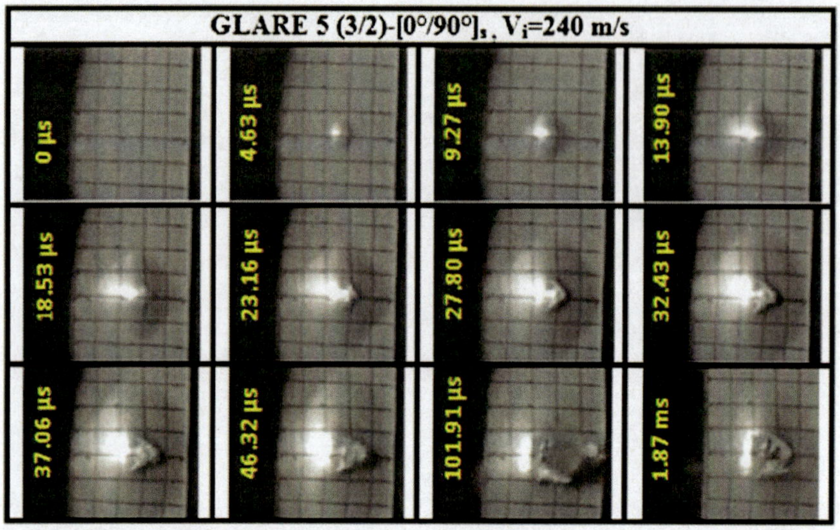


**Fig. 65.7** High-speed camera footage of the GLARE 5 (3/2)-[0°/90°]$_s$ FML beam impacted at 240 m/s

induced damage on the non-impacted side of unidirectional [0°$_4$] and cross-ply [0°/90°]$_s$ specimens propagated in the form of a lip – and diamond-shaped damage contours, respectively. Furthermore, a crack along the 0° fiber direction initiated at the non-impacted side of the unidirectional [0°$_4$] specimen during the time interval of 14.25 μs and 19.00 μs (Fig. 65.6).

## 65.4 Finite Element Simulation

In this study, ballistic impact on GLARE 5 FML beams with various stacking sequences were modeled and analyzed using the finite element software, LS-DYNA. Figure 65.8 exhibits perspective views of the composite beam. Please refer to Ref. [6] for more details on the FE simulation.

The FE prediction was validated based on the following factors: (1) incident vs. residual velocity data, (2) post-impact cross-sectional views and (3) bullet residual lengths. Figures 65.9 and 65.10 compare the FE prediction with experimental residual velocity versus incident velocity data for the GLARE 5 (3/2) FML beams with unidirectional [0°$_4$] and quasi-isotropic [0°/±45°/90°] stacking sequences, respectively. The experimental and FE result are shown by a diamond and circular symbol, respectively. The experimental data was fitted by least-square regression according to the classical Lambert-Jonas equation [12] for the positive residual velocity values:

$$V_R^P = A\left(V_I^P - V_{50}^P\right) = A \cdot V_I^P - B \tag{65.2}$$

where A and B are two regression coefficients and P is a power. $V_R$ and $V_I$ are the residual and incident velocities of the projectile, respectively, while $V_{50}$ is the ballistic limit velocity, which is defined as the velocity required for a projectle to perforate the target 50% of the time.

The Lambert–Jonas curves with different P values are represented by dashed lines [5]. The intersection of the Lambert-Jonas curve with the incident-velocity axis was considered as the $V_{50}$ velocity for a given P value. As is apparent from the figures, the FE predictions are in excellent agreement with the experimental results.

Figures 65.11 and 65.12 illustrate the comparison of the post-impact damage patterns between experimental and FE results for the GLARE 5 (3/2) FML beams with unidirectional [0°$_4$] and quasi-isotropic [0°/±45°/90°] stacking sequences impacted at 173 m/s and 237.5 m/s, respectively. The FE predicted bullet residual length was also compared to the experimental result in the figures. Good agreement between the experimental and FE result was obtained. Furthermore, it should be noted that the

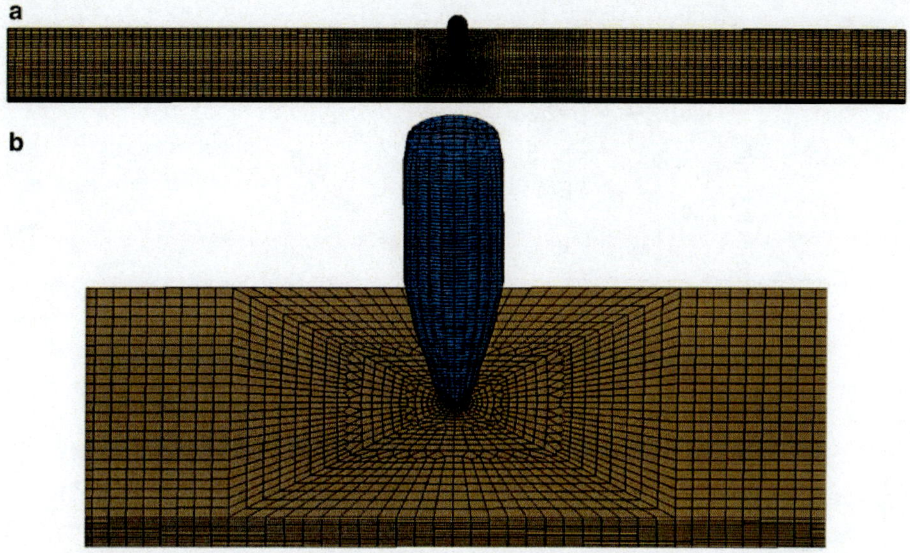


**Fig. 65.8** Perspective views of the finite element model: (**a**) far-off view and (**b**) close-up view

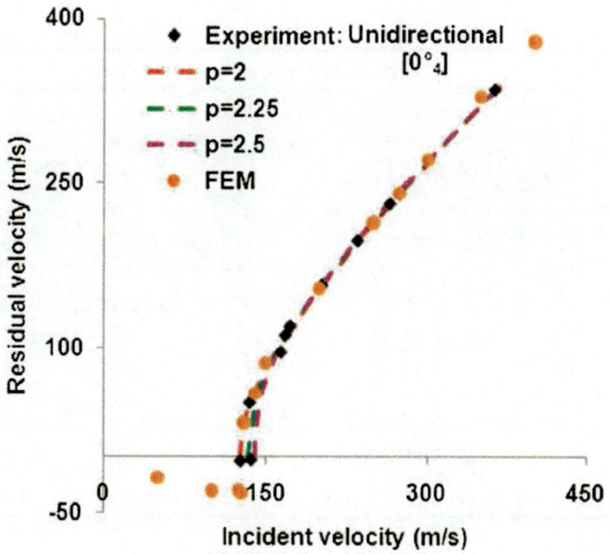


**Fig. 65.9** Comparison between FE prediction and experimental residual-incident velocity data for the GLARE 5 (3/2)-[0°$_4$] FML beam

broken/split fibers, which appeared around the impact area of the specimens, were not seen that much in the FE predictions (Figs. 65.11 and 65.12). This was due to the strain-based-element-erosion criterion used in the FE models. The failed elements were eroded and removed from the FE calculations. Therefore, they are not presented in the comparisons.

After the FE model was validated based on the abovementioned criteria, i.e. incident versus residual velocity data, post-impact damage patterns and bullet residual lengths, it was then used to obtain some insights to the ballistic impacts. As an example, Fig. 65.13 depicts the FE predicted contact force histories for the GLARE 5 (3/2) FML beams with unidirectional [0°$_4$] and quasi-isotropic [0°/±45°/90°] stacking sequences impacted at 350 m/s. As the figure suggest, there was a relatively negligible difference between the force history curves.

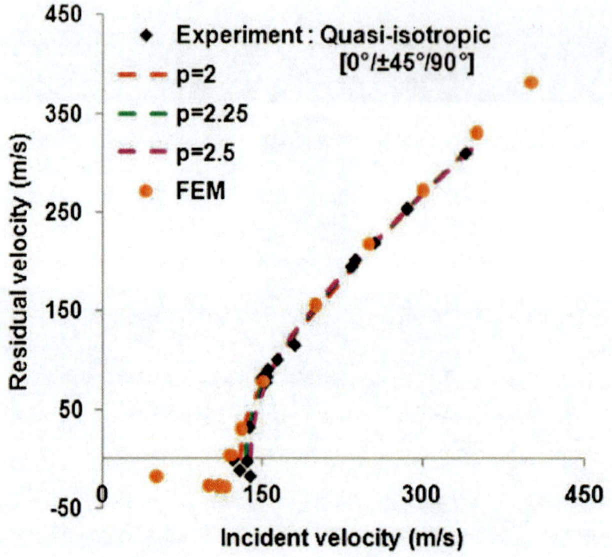


**Fig. 65.10** Comparison between FE prediction and experimental residual-incident velocity data for the GLARE 5 (3/2)-[0°/±45°/90°] FML beam

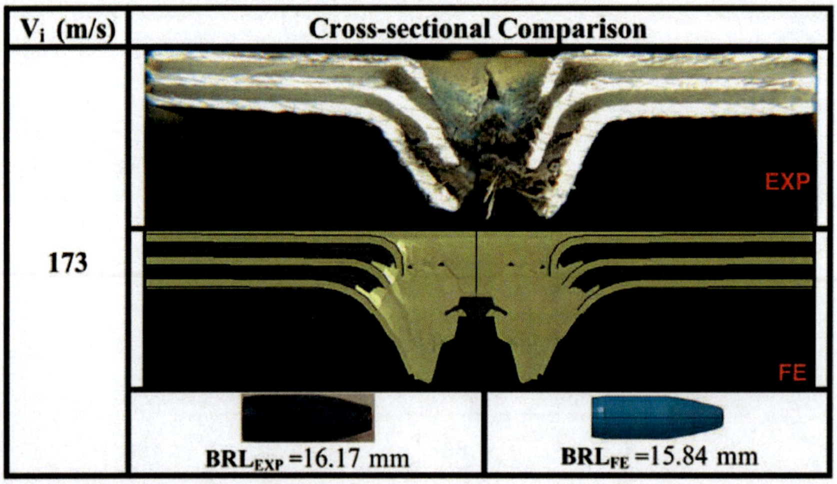


**Fig. 65.11** Comparison of the post-impact damage pattern for the GLARE 5 (3/2)-[0°$_4$] FML beam impacted at 173 m/s. In the figure BRL stands for the bullet residual length

## 65.5 Conclusions

This study presented experimental and numerical investigations on damage evaluation in the GLARE 5 (3/2) FML beams with various stacking sequences subjected to ballistic impact at room temperature. The following remarks could be concluded from this study.

- The aluminum layers failed through petaling, bulging and plugging. In addition, the fiber breakage/splitting/pullout, matrix cracking, debonding/delamination and plastic deformation in the specimens were observed.
- For both of the unidirectional specimens, i.e. [0°$_4$] and [90°$_4$], a lip-shaped damage contour was induced; whereas it had a diamond shape for the cross-ply specimen. For the quasi-isotropic specimen, the damage zone on the non-impacted side of the specimen was a mixture of the damages introduced in the unidirectional and cross-ply specimen.

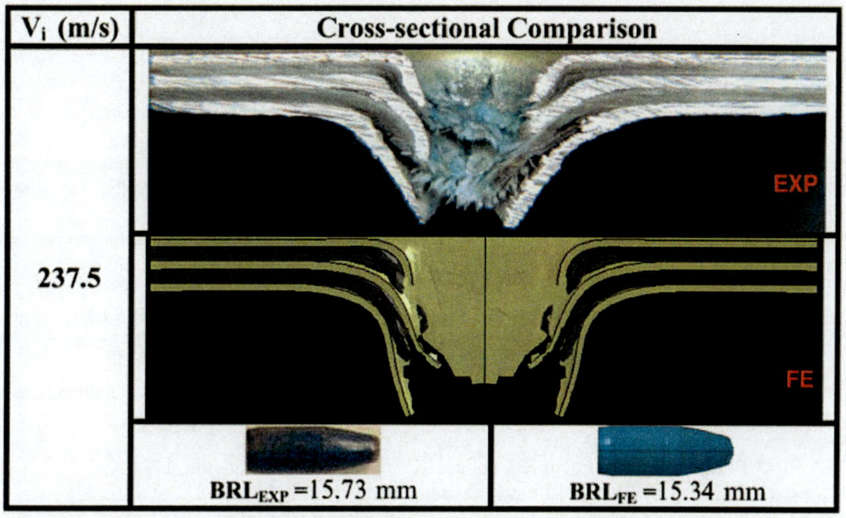


**Fig. 65.12** Comparison of the post-impact damage pattern for the GLARE 5 (3/2)-[0°/±45°/90°] FML beam impacted at 237.5 m/s. In the figure BRL stands for the bullet residual length

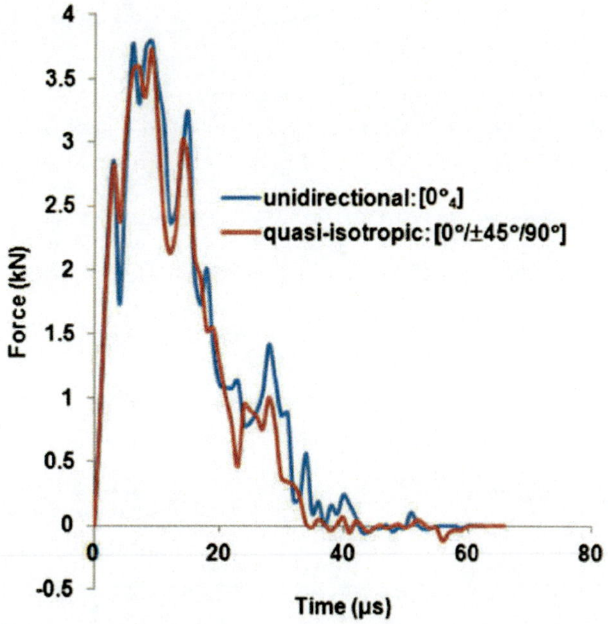


**Fig. 65.13** FE predicted contact force histories for the GLARE 5 (3/2) FML beams with unidirectional [0°$_4$] and quasi-isotropic [0°/±45°/90°] stacking sequences impacted at 350 m/s

- Good agreement between the FE predictions and experimental results was achieved in terms of the projectile residual velocity versus incident velocity data, damage patterns and bullet residual lengths.

**Acknowledgments** The authors would like to thank the supports from NASA Faculty Award for Research (FAR) under Grant No. NAG3-2259 and from PSC-CUNY under Grants 63168-00 41 and 64543-00 42. Dr. Kenneth J. Bowles and Dr. John P. Gyekenyesi were the Technical Monitors of the NASA grant. Part of the equipment was procured through ARO Grant No. DAAD19-03-1-0086, of which Dr. Bruce LaMattina was the Program Manager.

## References

1. Vlot A (2001) Glare-history of the development of a new aircraft material. Kluwer, Dordrecht
2. Vermeeren CAJR, Beumler T, De Kanter JLCG, Van Der Jagt OC, Out BCL (2003) Glare design aspects and philosophies. Appl Compos Mater 10:257–276
3. Seyed Yaghoubi A, Chow MF, Liaw BM (2011) Velocity measuring approaches for the determination of ballistic limits of GLARE 5 fiber-metal laminate plates. Experimental mechanics in the age of modeling and simulation. In: SEM annual conference and exposition on experimental and applied mechanics, Mohegan Sun, Uncasville, 13–16 June 2011
4. Hoo Fatt MS, Lin C, Revilock DM Jr, Hopkins DA (2003) Ballistic impact of GLARETM fiber-metal laminates. Compos Struct 61:73–88
5. Gama BA, Gillespire JW Jr (2011) Finite element modeling of impact, damage evolution and penetration of thick-section composites. Int J Impact Eng 38:181–197
6. Seyed Yaghoubi A, Liaw B (2012) Experimental and numerical approaches on behavior of GLARE 5 beams: influences of thickness and stacking sequence. In: IMAC XXX A conference and exposition on structural dynamics. Hyatt Regency Jacksonville Riverfront Jacksonville, 30 Jan – 2 Feb 2012
7. Sevkat E, Liaw B, Delale F, Raju BB (2009) A combined experimental and numerical approach to study ballistic impact response of S2-glass fiber/toughened epoxy composite beams. Compos Sci Technol 69(7–8):965–982
8. Data sheets (1998) Aviation equipment structures, Inc., Costa Mesa
9. QA reports B0319B-2, B1008B-1, B0904A-3. Structural Laminates Company, New Kensington, 1994
10. Alloy 7475 plate and sheet, ACRP-053-B. Alcoa Mill Products, Bettendorf
11. Vlot A, Gunnink JW (eds) (2001) Fiber metal laminates, an introduction. Kluwer, Dordrecht
12. Lambert JP, Jonas GH (1976) Towards standardization of in terminal ballistics testing: velocity representation. Ballistic Research Laboratories, Aberdeen Proving Ground, Report no. BRL-R-1852

# Chapter 66
# Effect of Filler Particle Shape on Dynamic Fracture Behavior of Glass-Filled Epoxy

**Vinod Kushvaha and Hareesh Tippur**

**Abstract** The effect of filler shape and filler volume fraction (micron sized rods, flakes and spheres) on dynamic fracture behavior of particulate polymer composites (PPC) is studied. The mode-I dynamic fracture experiments are carried out on pre-notched glass-filled epoxy specimens. An experimental set-up comprised of a long-bar apparatus to deliver one-point impact loading to an unconstrained specimen, is used in conjunction with a gas-gun. A controlled stress pulse is delivered to the specimen by impacting the long-bar by a striker launched using a gas-gun. A crack propagates into the specimen dynamically and is captured using high-speed photography (~300,000 frames per second). Using the Digital Image Correlation (DIC) method, in-plane displacement fields around the crack tip are determined from the speckle images recorded during the fracture event. With these, dynamic fracture toughness histories are evaluated to examine the filler shape effects. The results show pronounced improvement fracture toughness for all filler types with rod-shaped fillers producing ~145% increase in crack initiation toughness over unfilled epoxy at 15% $V_f$ with flakes and spheres showing ~97% and ~67% improvement, respectively.

**Keywords** Dynamic fracture • Particulate composites • Filler shape effects • DIC • High-speed photography

## 66.1 Introduction

The particle-filled polymer composites have been widely used in various engineering fields due to their excellent mechanical characteristics, chemical resistance, electric insulation, and adhesion properties. They are also relatively easy to process at a relatively low cost. Their structural simplicity in terms of macroscopic isotropy, unlike fiber reinforced composite, is another aspect which often makes it quite desirable for mechanical design. In particulate composites, the fillers modify their mechanical performance by reinforcing the polymer matrix. Therefore, the filler concentration, filler size and shape, and filler interfacial strength with the polymer matrix have a significant impact on the mechanical properties of PPC. Among the various modifications of mechanical properties, the stiffening, strengthening and toughening of the polymer matrix are the most useful for engineering applications.

Polymers are normally modified by the addition of inorganic-particulate fillers, such as alumina, mica or silica etc. Song et al. studied the particle shape effects on the fracture and ductility of a spherical and an irregular particulate-reinforced 6061-Al composite containing $Al_2O_3$ 20% by volume under quasi-static tensile loading [1]. The spherical particles produced a slightly lower yield strength and work hardening rate but a considerably higher ductility than the irregular particle counterpart. The finite element analysis results indicate that the distinction between the failure modes for these two composites can be attributed to the differences in the development of internal stresses and strains within the composites due to particle shape. Nakamura et al. examined the effect of particle size on the static fracture behavior of epoxy filled with different size (ranging from 6 to 42 μm) spherical silica particles [2]. They observed increase in both energy release rate and fracture toughness with particle size. Fractography showed a relatively smooth fracture surface with small particles (6 μm) and a rough surface with large particles (caused by crack deflections around large particles 42 μm). Wu analytically studied the effect of inclusion shape on elastic modulus of two phase solids [3]. The disk-shaped inclusions showed the maximum

V. Kushvaha • H. Tippur (✉)
Department of Mechanical Engineering, Auburn University, Auburn, AL 36849, USA
e-mail: htippur@eng.auburn.edu

V. Chalivendra et al. (eds.), *Dynamic Behavior of Materials, Volume 1: Proceedings of the 2012 Annual Conference on Experimental and Applied Mechanics*, Conference Proceedings of the Society for Experimental Mechanics Series,
DOI 10.1007/978-1-4614-4238-7_66, 

enhancement in elastic modulus compared to needles and spheres. The effect of particle size (4.5–62 μm) on the elastic modulus of epoxy/spherical glass particle composites were examined by Spanoudakis et al. [4]. At lower $V_f$ (10–18%), the modulus was nearly independent of particle size. For higher $V_f$ (30–46%), there was a slight decrease in modulus with increasing particle size. In a similar study, the effect of particle size on the modulus of epoxy/spherical and irregularly-shaped silica composites have been explored [5] for a size range of 2–30 μm and the modulus was observed to remain constant with particle size. Kitey et al. studied the role of particle size in the dynamic fracture behavior of glass-filled epoxy using optical interferrometry and high-speed photography [6]. Spherical particles (ranging from 7 to 200 μm) were used to reinforce the epoxy matrix at a constant 10% $V_f$. Elastic characteristics were unaffected by the particle size, whereas fracture toughness increased with size from 7 to 35 μm and then decreased from 35 to 200 μm.

The prior works in this area suggest that the particle shape effect on fracture toughness for particulate composites under dynamic loading remain largely unexplored. Therefore, the focus of the present study is to understand the effect of two basic parameters, filler shape and their volume fraction on fracture behavior of glass filled epoxy composite under dynamic conditions.

## 66.2 Specimen Preparation

Flakes (average size 50 μm and average thickness 5 μm; aspect ratio 10), rods (average length 800 μm and average diameter 10 μm; aspect ratio 80) and spheres (average diameter 35 μm; aspect ratio 1) glass fillers (Fig. 66.1) were chosen to study their relative shape effects on dynamic fracture mechanics of PPC. The glass fillers were dispersed into a low-viscosity epoxy (Bisphenol-A resin and Amine based hardener; densities 1,130 and 961 kg/m$^3$, respectively). To carry out the comparative dynamic fracture study, glass-filled epoxy (containing 0% (neat epoxy), 5%, 10% and 15% glass filler by volume, respectively) sheets were cast. The sheets were machined into specimens having dimensions of 60 × 30 × 9 mm (Fig. 66.2). An edge notch of 6 mm length was introduced at the mid-span of each specimen and the notch tip sharpened using a razor blade.

**Fig. 66.1** SEM micrographs of *sphere*, *flake* and *rod* shaped glass fillers used in present study (left to right respectively; scale bar: 50 μm)

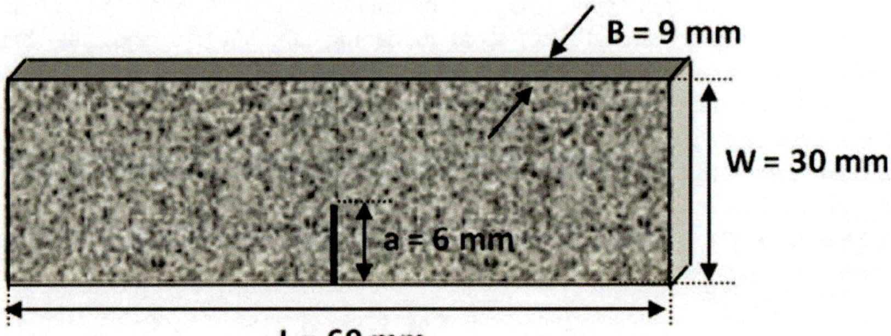


**Fig. 66.2** Schematic of test specimen

## 66.3 Material Property Measurement

Physical and elastic properties were measured for all PPC and are tabulated in Table 66.1. The stress wave transducers were used to perform pulse-echo measurements to determine longitudinal ($C_l$) and shear wave ($C_s$) speeds at discrete locations. Dynamic elastic modulus and Poisson's ratio were calculated using,

$$C_l = \sqrt{\frac{E_d(1-\nu_d)}{\rho(1+\nu_d)(1-2\nu_d)}},\ C_s = \sqrt{\frac{E_d}{2\rho(1+\nu_d)}}$$

## 66.4 Experimental Setup and Procedure

A schematic of the experimental setup used for dynamic fracture tests is shown in Fig. 66.3. The setup included a 6 ft long, 1 in. diameter long-bar with 1/4 in. diameter bull-nose tip against the unconstrained specimen and a 12 in. long, 1 in. diameter striker held inside the barrel of a gas gun. Both the long-bar and the striker were of the same diameter and made

**Table 66.1** Material properties of glass-filled epoxy composite

| Particle type | Density $\rho$ (kg/m$^3$) | Longitudinal wave speed $C_l$ (m/s) | Shear wave speed $C_s$ (m/s) | Elastic modulus $E_d$ (GPa) | Poisson's ratio $\upsilon_d$ |
|---|---|---|---|---|---|
| (a) $V_f = 5$ % | | | | | |
| Epoxy | 1,146 | 2,481 | 1,128 | 3.99 | 0.369 |
| Sphere | 1,222 | 2,534 | 1,177 | 4.62 | 0.362 |
| Flake | 1,225 | 2,526 | 1,187 | 4.69 | 0.358 |
| Rod | 1,226 | 2,534 | 1,188 | 4.68 | 0.359 |
| (b) $V_f = 10$ % | | | | | |
| Sphere | 1,286 | 2,553 | 1,207 | 5.08 | 0.356 |
| Flake | 1,288 | 2,571 | 1,248 | 5.40 | 0.346 |
| Rod | 1,285 | 2,534 | 1,243 | 5.33 | 0.341 |
| (c) $V_f = 15$ % | | | | | |
| Sphere | 1,358 | 2,600 | 1,243 | 5.67 | 0.352 |
| Flake | 1,361 | 2,630 | 1,300 | 6.16 | 0.338 |
| Rod | 1,375 | 2,598 | 1,286 | 6.08 | 0.337 |

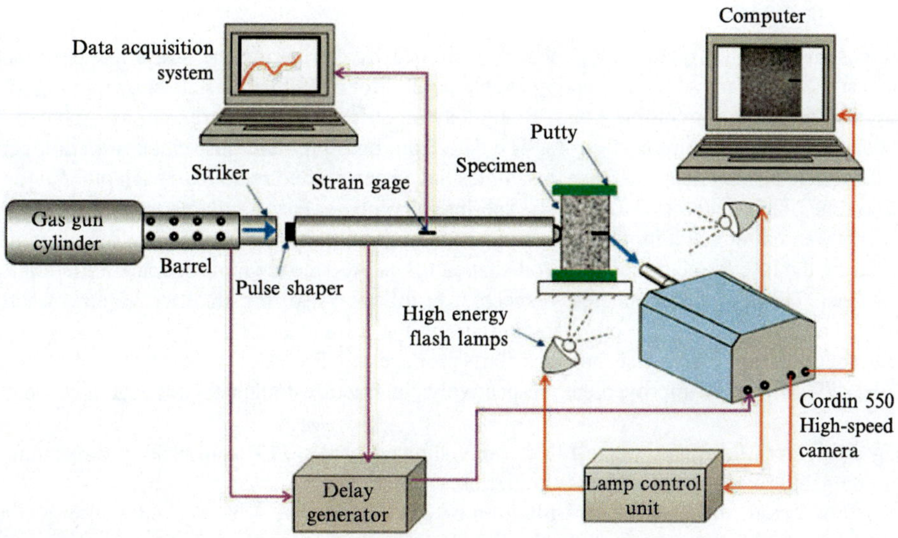


**Fig. 66.3** Schematic of experimental setup

of aluminum 7,075 to eliminate the impedance mismatch between them. A 1 mm thick and 9/32 in. diameter pulse shaper made of soft aluminum 1,100 was used at the front end of the long-bar. The role of the pulse shaper is to ramp up the stress wave in a nearly linear fashion in the long-bar during impact. The striker is launched towards the long-bar using the gas-gun at a velocity of 16 m/s. When the striker contacts the long-bar, a compressive stress wave propagates through the bar and gets transmitted into the specimen.

A stochastic black and white speckle pattern was sprayed on to the specimen for DIC measurements. The pattern is photographed using a Cordin-550 ultrahigh-speed digital camera. It has 32 independent CCD image sensors positioned radially around a rotating mirror which sweeps light over these sensors. Prior to loading, a set of 32 undeformed images were recorded at a rate of ~300,000 frames per second at a resolution of 1,000 × 1,000 pixels per image. When the striker contacts the long-bar and completes the circuit, a signal is sent to the delay generator which in turn triggers the camera. The camera triggers a high energy flash lamp control unit and initiates image capture. Thus, a second set of 32 deformed images were captured at the same framing rate. Therefore each image in the undeformed set had a corresponding image in the deformed set. The recorded sets of deformed and undeformed images are then correlated to obtain in-plane displacements.

## 66.5 Optical Data Analysis

The recorded undeformed set and deformed set of images were correlated using a digital image correlation tool ARAMIS™ and the in-plane crack-opening ($u_y$) and crack-sliding ($u_x$) displacement fields were estimated near the crack tip. In order to extract SIF history, ($u_y$, $u_x$) displacement fields were digitized by identifying the current crack-tip location and subsequently establishing the Cartesian and polar coordinates. A number of data points were collected in the vicinity of the current crack-tip $0.5 < r/B < 1.5$ (where $B$ is sample thickness) and $[(-150° < \theta < -90°)$ and $(90° < \theta < 150°)]$ as it has been shown that 3D deformations and far field effects are minimum in this region [7]. At each data point, ($u_y$, $u_x$) displacement values as well as the location of these points were stored. The asymptotic expressions for a dynamically loaded stationary crack given in [8] were used to extract mode-I and –II stress intensity factors using an over-deterministic least-squares analysis of the optical data. In doing so, the rigid body translation and rotation terms were taken into account. Once the crack starts propagating, the asymptotic expressions for sliding and opening displacements for a steadily growing crack (assuming transient effects are negligible) are evaluated using expressions in Ref. [9].

For mode I problem, the crack-opening displacements ($u_y$) can be viewed as with mode I-rich information whereas crack sliding displacements ($u_x$) with mode II-rich information. Thus, $u_y$ and $u_x$ were used to extract mode-I and –II stress intensity factors $K_I$ and $K_{II}$, respectively (Fig. 66.4).

## 66.6 Effect of Filler Shape

The filler shape effect on dynamic fracture toughness was studied by comparing the stress intensity factor histories for different glass fillers at 5%, 10% and 15% $V_f$ in epoxy (Table 66.2). Plot of $K_I$ histories are shown in Fig. 66.5. In each plot, histories for $K_{II}$ are also provided for completeness. Being a symmetric loading experiment, as expected, $K_{II}$ histories are essentially negligible within the measurement errors. The data from both pre- and post-crack initiation periods are shown. It should be noted that the measured histories of each samples have been shifted relative to each other along the time axis to make crack initiation time (identified as $t = 0$) of each experiment coincide. For all volume fractions, the $K_I$ histories show a monotonic increase until and after crack initiation. In some cases, a modest kink in the history is noticeable based on when the crack initiated during the interframe period. The rod-shaped fillers produce the highest crack initiation toughness as well post-initiation $K_I$ values. The flakes and spheres, respectively, follow. Also, for all three volume fractions, the rate of increase of $K_I$ followed a similar trend up to pre-crack initiation time instant. The rate of increase of $K_I$ was approximately constant for all different shapes of fillers, once the crack starts propagating. For 5% $V_f$ (Fig. 66.5); rod, flake and sphere-filled epoxy showed 82%, 47% and 18% improvement, respectively, in fracture toughness compared to neat epoxy at crack initiation.

For 10% $V_f$ (Fig. 66.6); rod, flake and sphere-filled epoxy showed 118%, 71% and 39% improvement, respectively, in fracture toughness compared to neat epoxy at crack initiation.

For 15% $V_f$ (Fig. 66.7); rod, flake and sphere-filled epoxy showed 145%, 97% and 68% improvement respectively in fracture toughness compared to neat epoxy at crack initiation.

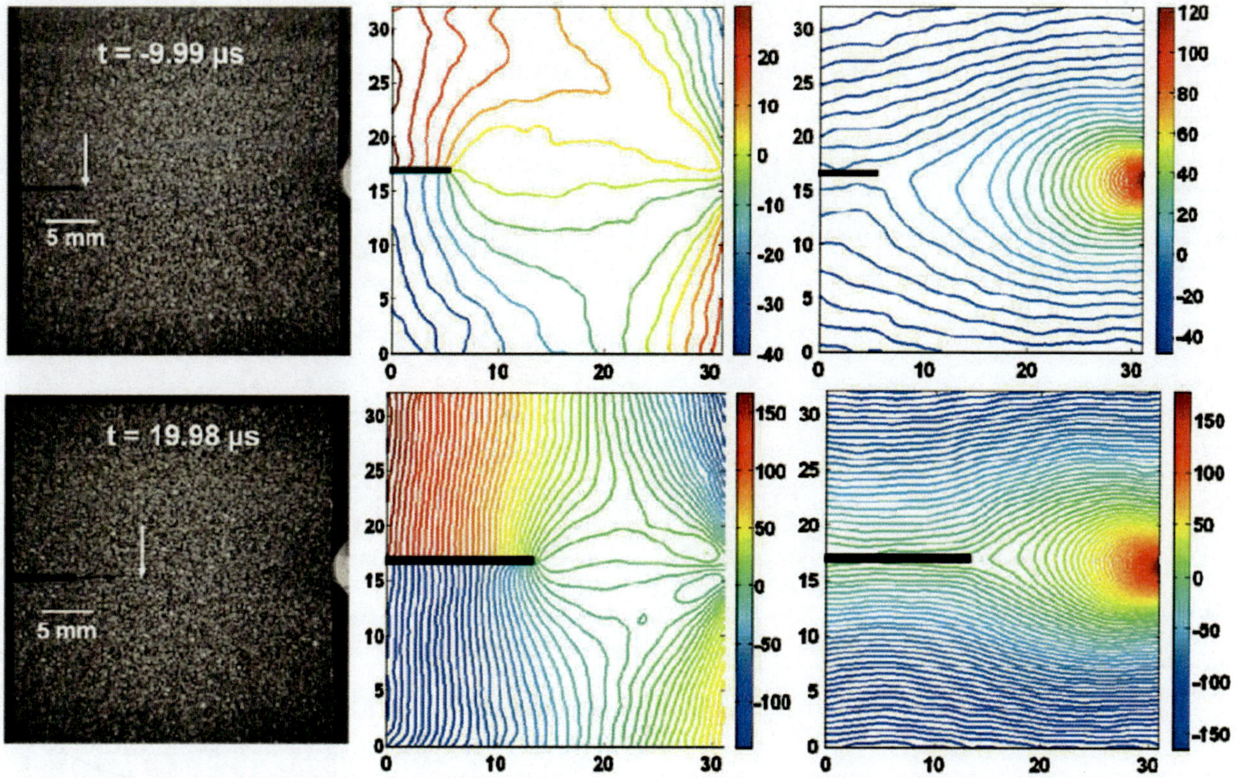


**Fig. 66.4** Acquired speckle images for pre ($t = -9.99$ μs) and post-crack initiation ($t = 19.98$ μs) time instants with crack opening ($u_y$) and crack sliding ($u_x$) displacement contours (contour interval 5 μm)

**Table 66.2** Fracture toughness (in MPa√m) comparison at crack initiation (% improvement is relative to neat epoxy)

| Volume fraction | | Sphere | Flakes | Rods |
|---|---|---|---|---|
| 0% (neat epoxy) | Fracture toughness ($K_I$) | 1.42 | 1.42 | 1.42 |
| 5% | Fracture toughness ($K_I$) | 1.68 | 2.09 | 2.59 |
| | % improvement | 18 | 47 | 82 |
| 10% | Fracture toughness ($K_I$) | 1.97 | 2.43 | 3.10 |
| | % improvement | 39 | 71 | 118 |
| 15% | Fracture toughness ($K_I$) | 2.38 | 2.80 | 3.48 |
| | % improvement | 68 | 97 | 145 |

## 66.7 Effect of Filler Volume Fraction

For each filler shape, dynamic fracture toughness was compared for 0%, 5%, 10% and 15% $V_f$ respectively (Table 66.2). In each case, $K_I$ showed increasing trend with increase in volume fraction of the filler during the whole event. The rate of increase of $K_I$ followed the above volume fractions trend up to pre-crack initiation. The rate of increase of $K_I$ was approximately constant for all different volume fractions, once the crack starts propagating. For spherical-fillers (Fig. 66.8), 5%, 10% and 15% $V_f$ showed 18%, 39% and 68% improvement, respectively, in crack initiation toughness compared to neat epoxy.

For flakes (Fig. 66.9), 5%, 10% and 15% $V_f$ showed 47%, 71% and 97% improvement, respectively, in crack initiation toughness compared to neat epoxy.

For rods (Fig. 66.10), 5%, 10% and 15% $V_f$ showed 82%, 118% and 145% improvement, respectively, in crack initiation toughness compared to neat epoxy.

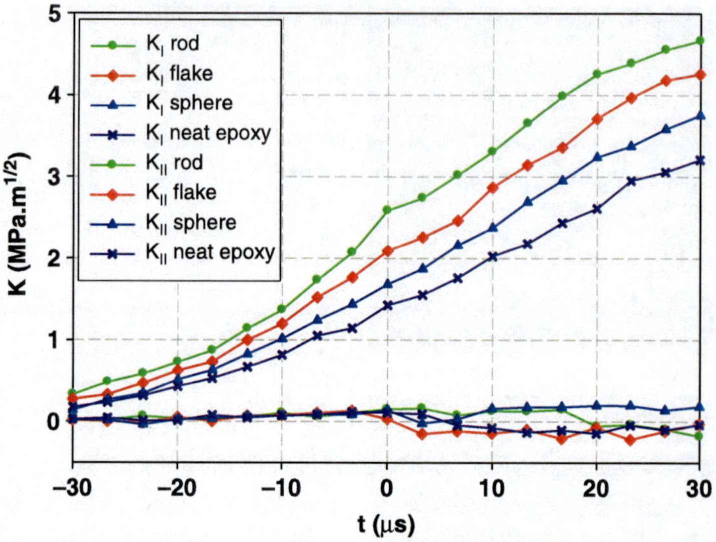


**Fig. 66.5** $K_I$ and $K_{II}$ histories for 5% $V_f$ glass-filled epoxy with different filler particle shape. (Histories for $K_{II}$ are also provided for completeness. Being a symmetric loading experiment, as expected, $K_{II}$ histories are essentially negligible within the measurement errors)

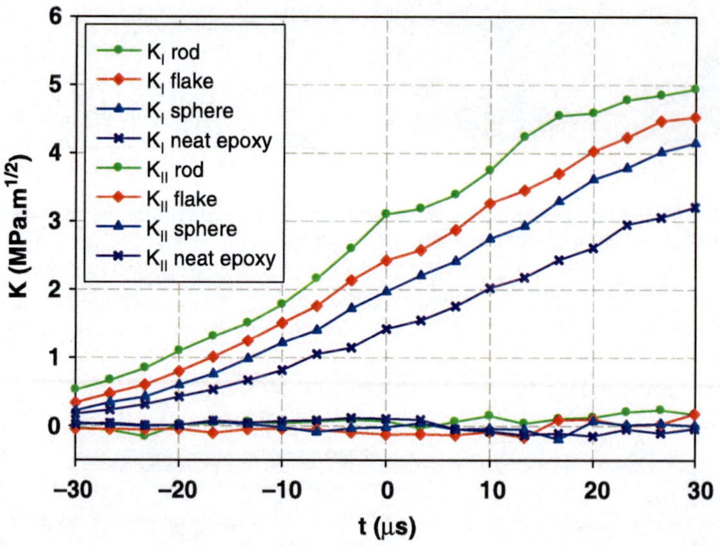


**Fig. 66.6** $K_I$ and $K_{II}$ histories for 10% $V_f$ glass-filled epoxy with different filler particle shape

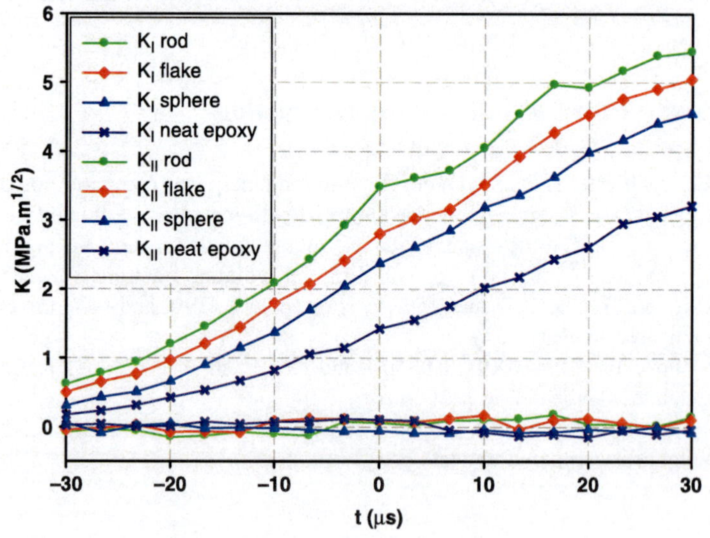


**Fig. 66.7** $K_I$ and $K_{II}$ histories for 15% $V_f$ glass-filled epoxy with different filler particle shape

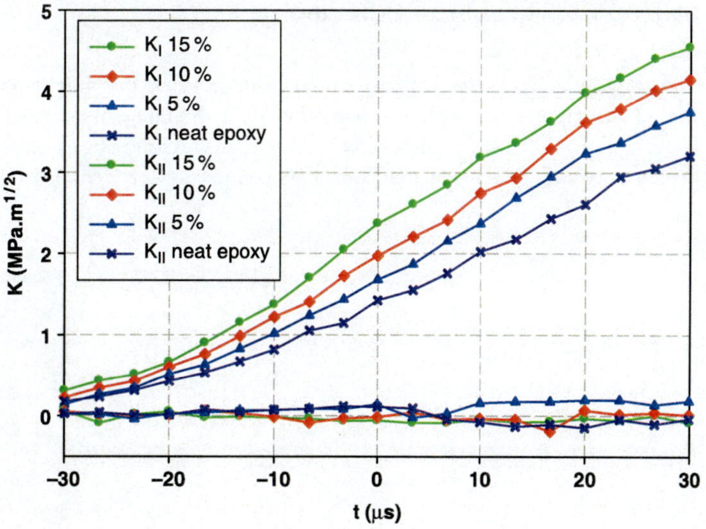


**Fig. 66.8** $K_{\mathrm{I}}$ and $K_{\mathrm{II}}$ histories for spheres-filled epoxy with different volume fractions

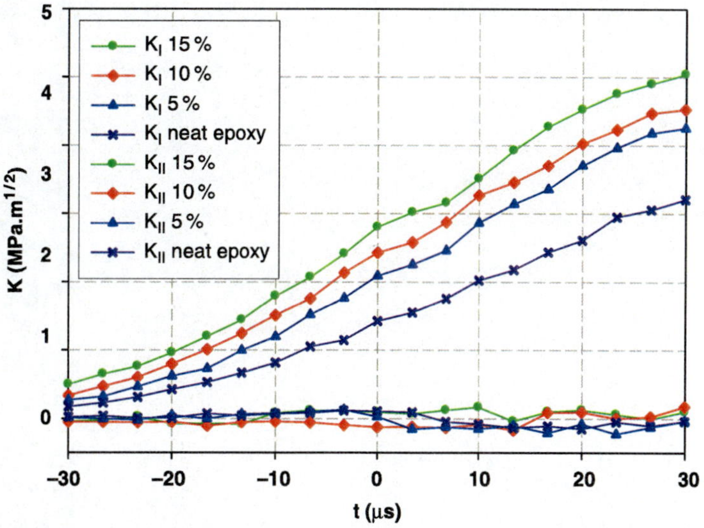


**Fig. 66.9** $K_{\mathrm{I}}$ and $K_{\mathrm{II}}$ histories for flakes-filled epoxy with different volume fractions

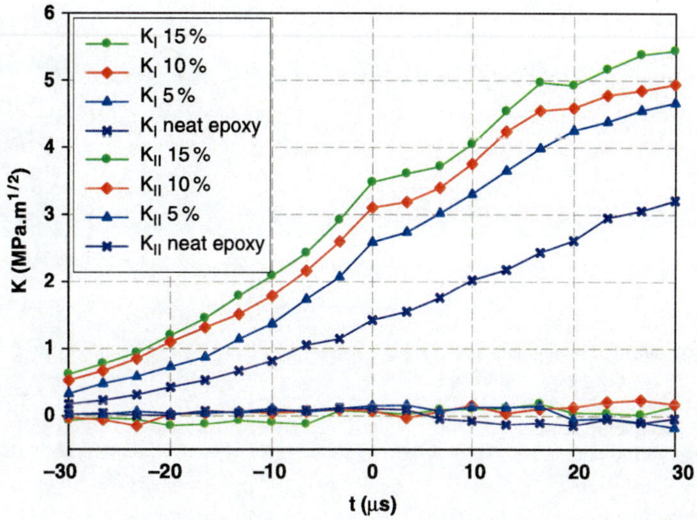


**Fig. 66.10** $K_{\mathrm{I}}$ and $K_{\mathrm{II}}$ histories for rods-filled epoxy with different volume fractions

## 66.8 Fractographic Evaluation

The fractographic study was carried out to understand the toughening mechanisms behind the dramatic improvement in fracture toughness due to filler's shape change. It should be first noted that fracture surface micrographs (Fig. 66.11) show no agglomeration of filler particles in all three cases. As the stress wave propagates through the specimen, the crack front is driven forward. When the crack front encounters filler particles during propagation, it interacts with the filler phase.

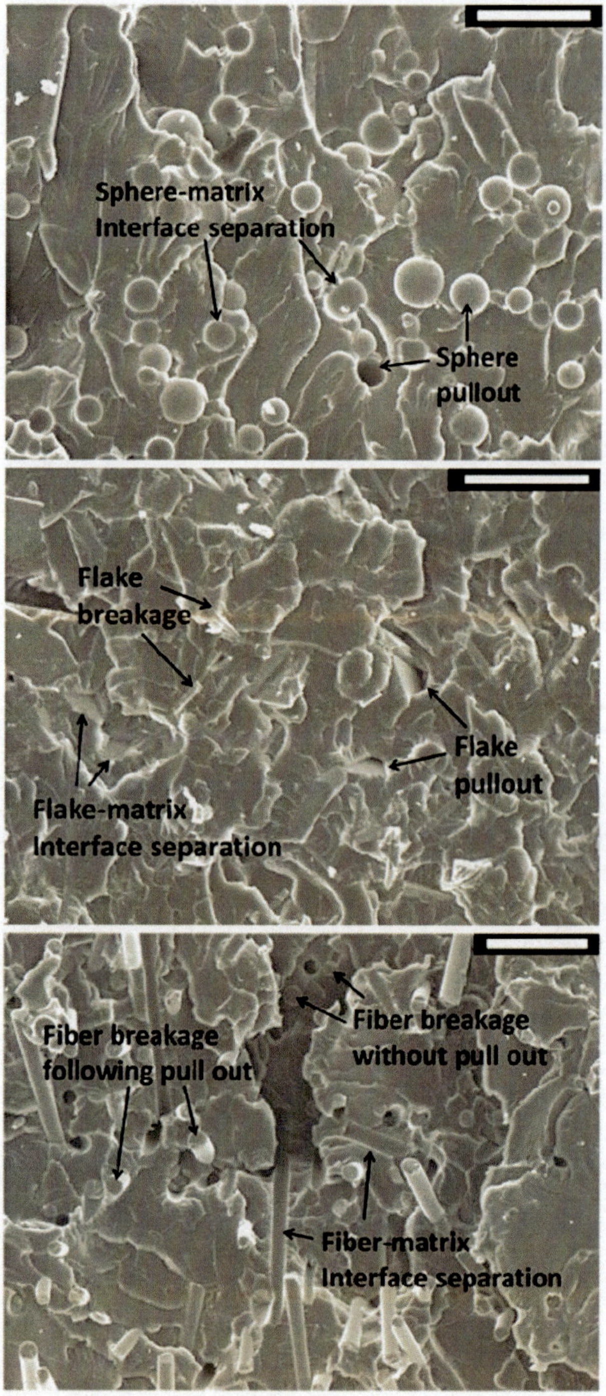


**Fig. 66.11** Fractographs of 10% *sphere*, *flakes*, *rod*-filled epoxy (top to bottom respectively); scale bar = 100 μm

A careful observation of fractographs suggests that the interaction of crack front with filler phase leads to following primary failure modes:

- Matrix cracking
- Filler-matrix interface separation
- Filler pullout
- Filler breakage

Hence, during crack growth each existing failure mode contributes to energy dissipation which adds to the overall increase in fracture toughness. In case of spherical filler, matrix cracking and sphere-matrix interface separation were found to be the dominant modes. A few spherical fillers were pulled out but none of them were found fractured. Therefore due to the filler's low aspect ratio (aspect ratio 1), it did not bridge the crack well and crack progressed by circumventing the inclusion and/or debonding the interface causing momentary crack front arrest and reinitiation. In case of flakes, fracture surface is dominated by matrix cracking, flake-matrix interface separation and flake breakage. The flake breakage contributes to extra energy dissipation and also leads to higher surface roughness. All these contribute to an increase in fracture toughness when compared to spherical filler counterparts. However, the filler breakage without pullout is less preferred because it does not offer crack-face bridging for a longer duration which would contribute further to energy dissipation.

For rod shaped filler, the failure modes were matrix cracking, fiber-matrix separation and fiber breakage with evidence of fiber pullout. On the fracture surface, several broken short fibers were identified. Since the average fiber length was ~800 μm, it is likely that the fibers running across the crack front resisted crack growth, but eventually were pulled out and fractured as the crack opened. The tensile strength of the fiber's being about 3.45 GPa, it seems to have contributed to a rather high apparent fracture toughness of the composite.

## 66.9 Conclusion

In present study, the effect of filler shape (flakes, rods and spheres) and filler volume fraction on dynamic fracture behavior of glass-filled epoxy composites was investigated using Digital Image Correlation in conjunction with high speed photography. The dynamic fracture experiments were carried out on pre-notched glass-filled epoxy specimens using a set-up comprised of a long-bar apparatus which delivers one-point impact loading to unconstrained specimen, is used in conjunction with a gas-gun.

For all three different volume fractions used, the rod-shaped fillers produce the highest crack initiation toughness as well post-initiation $K_I$ values followed by flakes and spheres, respectively. For 5% $V_f$; rod, flake and sphere-filled epoxy showed 82%, 47% and 18% improvement, respectively, for 10% $V_f$; rod, flake and sphere-filled epoxy showed 118%, 71% and 39% improvement, respectively and for 15% $V_f$; rod, flake and sphere-filled epoxy showed 145%, 97% and 68% improvement, respectively, in crack initiation fracture toughness compared to neat epoxy. For all shapes, there was increase in fracture toughness with increase in filler volume fraction. For sphere-filled epoxy, 5%, 10% and 15% $V_f$ showed 18%, 39% and 68% improvement, respectively; for flake-filled epoxy, 5%, 10% and 15% $V_f$ showed 47%, 71% and 97% improvement, respectively and for rod-filled epoxy; 5%, 10% and 15% $V_f$ showed 82%, 118% and 145% improvement, respectively, in crack initiation fracture toughness compared to neat epoxy. The fractographic evaluation explained the mechanisms associated with different shape fillers during dynamic fracture and their contribution in enhancing fracture toughness.

**Acknowledgements** Partial support of this research through grant NSF-CMMI - 1100700 is gratefully acknowledged.

## References

1. Song SG et al (1996) Reinforcement shape effects on the fracture behavior and ductility of particulate-reinforced 6061-Al matrix composites. Metall Mater Trans Phys Metall Mater Sci 27(11):3739–3746
2. Nakamura Y et al (1992) Effect of particle-size on the fracture toughness of epoxy-resin filled with spherical silica. Polymer 33(16):3415–3426
3. Wu TT (1966) The effect of inclusion shape on the elastic moduli of a two-phase material. Int J Solids Struct 2:1–8
4. Spanoudakis J, Young RJ (1984) Crack-propagation in a glass particle-filled epoxy-resin. 1. Effect of particle-volume fraction and size. J Mater Sci 19(2):473–486

5. Nakamura Y et al (1992) Effect of particle-size on mechanical properties of epoxy-resin filled with angular-shaped silica. J Appl Polym Sci 44(1):151–158
6. Kitey R, Tippur HV (2005) Role of particle size and filler-matrix adhesion on dynamic fracture of glass-filled epoxy. I. Macromeasurements. Acta Mater 53(4):1153–1165
7. Tippur HV, Krishnaswamy S, Rosakis AJ (1991) Optical mapping of crack tip deformations using the methods of transmission and reflection coherent gradient sensing-A study of crack-tip K-dominance. Int J Fract 52(2):91–117
8. Westergaard HM (1939) Bearing pressure and cracks. ASME Appl Mech 6:49–53
9. Nishioka T, Atluri SN (1983) Path-independent integrals, energy-release rates, and general-solutions of near-tip fields in mixed-mode dynamic fracture-mechanics. Eng Fract Mech 18(1):1–22

# Chapter 67
# Effect of Particle Size on the Compressive Behavior of Dry Sand under Confinement at High Strain Rates

**Hongbing Lu, Huiyang Luo, William L. Cooper, and Ranga Komanduri**

**Abstract** Dynamic compressive behavior of dry sand (Quikrete #1961 sand quarried in Pensacola, FL) under confinement was characterized using a split Hopkinson pressure bar (SHPB). The as-received dry Eglin sand was sieved into different grain sizes, following the ASTM D2487 standard. The sand were sorted into 0.60, 0.50, 0.42, 0.30, 0.212, 0.15, and 0.106 mm grains through a set of ASTM E11 sieves of #30, #35, #40, #50, #70, #100 and #140, respectively. Sand grains were confined inside a hollow cylinder of hardened steel and capped by cemented tungsten carbide cylindrical rods. This assembly was subjected to repeated manual shaking to consolidate sand to attain the desired bulk mass densities. It is then sandwiched between incident and transmission bars on SHPB for dynamic compression under high strain rate. Sand specimens of seven particle sizes (shaking into maximum mass density) were characterized to determine the volumetric and deviatoric behaviors at high strain rates, and the particle size effect was discussed. The stress-density relationship and the specific energy absorption were determined as a function of sand grain sizes. The void ratio-axial pressure relationships (*e*-log *p* curves) were obtained and the compressibility of sand was determined as a function of sand particle sizes.

**Keywords** Long split hopkinson pressure bar (SHPB) • Granular materials, dynamic compaction • Confinement • Hydrostatic pressure

## 67.1 Introduction

Dense compacted sand has high compressive strength and high energy absorption capacity. Sand is often used to provide bulk ballistic protection for military structures, and as construction materials widely used in civil engineering structures and highways. Numerous researches on the behavior of sand for geological and civil engineering applications focus primarily on the mechanical behavior at low stress levels and under quasi-static loading conditions [1–4]. Its mechanical behavior under relatively high stresses and high strain rates has not been well understood, due that the experimental technique at that time encountered more difficulties in dynamic characterization of sand than in quasi-static tests. Recently, Luo and Lu et al., successfully resolved this difficulty on the preparation of sand samples with consistent mass densities for dynamic compression on a split Hopkinson pressure bar (SHPB) [5].

The mechanical behaviors of sand and soil have been characterized at high strain rates for moisted water [6], soft soil [7], clayey/silty soil [8], and dry sand [9], etc. However, the pioneer data on sand/soil at high strain rate may be suspectible due to the shortcoming of dynamic experimental technique then. Pulse shaping techniques, developed in recent years for SHPB, can facilitate in establishing dynamic equilibrium and constant strain rate conditions necessary in a valid dynamic

H. Lu (✉) • H. Luo
Department of Mechanical Engineering, The University of Texas at Dallas, Richardson, TX 75080, USA
e-mail: hongbing.lu@utdallas.edu

W.L. Cooper
Air Force Research Laboratory, Eglin Air Force Base, FL 32542, USA

R. Komanduri
School of Mechanical and Aerospace Engineering, Oklahoma State University, Stillwater, OK 74078, USA

V. Chalivendra et al. (eds.), *Dynamic Behavior of Materials, Volume 1: Proceedings of the 2012 Annual Conference on Experimental and Applied Mechanics*, Conference Proceedings of the Society for Experimental Mechanics Series, DOI 10.1007/978-1-4614-4238-7_67, © The Society for Experimental Mechanics, Inc. 2013

compression test [10]. For the dynamic behavior of sand, a modified SHPB using pulse shaping technique to facilitate establishing the dynamic stress equilibrium and constant strain rate conditions, has been to characterize the behavior of sand confined in a steel sleeve to determine the effects of moisture content [11], strain rate and confinement [12], and initial mass density effect [5]. In our work, we employed our technique to prepare sorted dry sand specimens with dense mass densities for experiments on a modified long SHPB, combined with pulse shaper and strain gage attached on the outer surface of confining cylinder tube. It allows the determination of the particle size effects of dense sand on the compressive behavior under high strain rates. We report the results on the compressive behavior of sorted dry sand with dense initial mass densities under stiff confinement provided by a hollow steel cylinder (12.7 mm inner diameter and 19.3 mm outer diameter) on a modified SHPB, to determine the dependence of dynamic mechanical behavior on the sand particle sizes.

## 67.2 Experimental

In this investigation, a 11 m long modified SHPB was used with a pulse shaping technique to apply dynamic compression over relatively long loading times (up to ~1.5 ms) and to reach relatively high compressive strains [13, 14]. The SHPB, as shown schematically in Fig. 67.1a, consists of a solid stainless steel bars (7.5 m long for the incident bar and 3.66 m long for the transmission bar) with an outer diameter of 19 mm and a strain data acquisition system. To allow preparation of sand samples with consistent mass densities, we employed an assembly to prepare a sand specimen, and then directly used it in the SHPB apparatus for dynamic compression. The assembly consists of two cemented WC rods with a diameter of 12.7 mm and a hollow high carbon steel cylinder. The WC rods have a mechanical impedance that matches well with both the incident and transmission stainless steel bars with a diameter of 19 mm. Sand grains were first poured into the hollow steel cylinder with both ends closed by the WC rods. A schematic diagram of sand assembly is shown in Fig. 67.1b. The total mass of the assembly, including sand, is less than 200 g, so that it can be easily shaken manually to consolidate the sand sample to attain different initial mass densities. The net mass of the sand of 2 g was accurately added into the hollow steel cylinder capped with one WC rods, monitored the weighting by digital balance (Denver Instrument APX-200, 0.1 mg resolution). The mass and the length were then converted to bulk mass density of the sand sample. By shaking the assembly sufficiently (more than 30 min), sand samples with maximum initial mass densities can be prepared, through monitoring the length of the enclosed sand sample.

A strain gage was attached on the external surface of the confining steel sleeve [5, 8]. The lateral deformation of the sand sample was measured by the strain gage, and the radial/circumferential stress was determined. This arrangement allows characterization of dynamic response for the volumetric and deviatoric behaviors of sand. Further analysis yields the energy absorption diagram and the void ratio – pressure relationship at high strain rates. In each experiment, the sand assembly was sandwiched between incident and transmission bars on SHPB. Compressive loading was applied on the sand specimen through WC rods in direct contact with the sand sample. The as-received unsorted sand filled in the hollow cylinder assembly had a dense mass density of 1.75 g/cm$^3$ with length/diameter ratio of 0.715. For the seven sorted sands, namely, 0.6, 0.5, 0.425, 0.3, 0.212, 0.15 and 0.105 mm grains, the ratios of length to diameter are varied in the range of 0.735 ~ 0.690. When the length of sand specimen reached a stable value during shaking, the mass density reached the maximum value for a given size sand.

A copper disks was used as pulse shaper in the SHPB experiments by placing it on the impact surface of the incident bar. A 1.22 m long steel striker, with the same diameter as that of both the incident and transmission bars, launched by

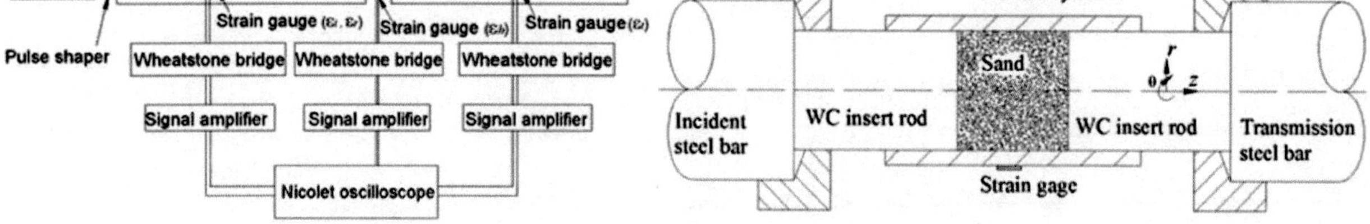


**Fig. 67.1** Schematic SHPB setup for testing confined sand. (**a**) SHPB diagram; (**b**) test section for sand assembly

compressed air in a gas gun, impacts the pulse shaper and generates a desired compressive wave traveling in the incident bar to induce a nearly constant strain rate in the sand sample. A Nicolet Sigma-30 digital oscilloscope (12-bit resolution, 10 Ms/s sampling rate) was used to acquire strain signals through Wheatstone bridges and a Vishay 2310A signal conditioning amplifier (125 kHz frequency response) connected to the strain gages. In a valid SHPB experiment in which a specimen is loaded by equal forces from both ends, the compressive stress in the specimen, the axial stress was determined from transmission signals, and the axial strain rate and strain were calculated from reflected signals [15].

For a cylindrical specimen confined by a hollow cylinder within the elastic range [16, 17], the confining radial and circumferential pressures, as well as circumferential strain on a sample are determined. The hardened steel hollow cylinder (4340 hardened steel) was used in this work, to prevent sand particle from indenting into the inner surface of steel cylinder. In this study, we used the confinement method developed by Ravi-Chandar *et al.* under quasi-static compression of polymer for dynamic compression of sand on SHPB to characterize the hydrostatic and deviatoric behaviors of sand at high strain rates. With the known axial, confining stress and strain histories, the hydrostatic pressure (isotropic stress) component and the volumetric dilatation are determined. The maximum shear stress is applied on any plane inclined at 45° angle with respect to the longitudinal direction, and the maximum shear strains were well documented [5]. With the known values for hydrostatic and shear stresses, the hydrostatic and deviatoric behaviors of sand under multi-axial stress state can be determined at high strain rates. Thus, the known dynamic 3-D stresses can be used as input for the future modeling works.

In this investigation all tests were conducted at room temperature (~23°C) and near 26% relative humidity environment. We first conducted dynamic compaction of the as-received, assorted dry Eglin sand on SHPB. Following the ASTM D2487 standard [18], the sand grains were sorted into 0.60, 0.50, 0.42, 0.30, 0.212, 0.15, and 0.106 mm through shaking with a set of stacked ASTM E11 sieves of #30, #35, #40, #50, #70, #100 and #140, respectively. Then we conducted the tests of the seven particle sizes of sands as its maximum mass density on SHPB. Under each testing condition, eight samples were tested to evaluate reproducibility of experimental results, and the average values are reported.

## 67.3 Results and Discussion

All as-received dry sand used in this work was silica based fine grain Quikrete #1961 sand quarried in Pensacola, FL, washed and kiln-dried by Eglin Air Force Base. It may be noted that Eglin sand has been characterized by the Air Force Research Laboratory at Eglin Air Force Base extensively and is currently widely used by researchers in the U.S. and Europe [19]. Nanoindentation on single sand grain shows the mechanical properties in a significant Weibull statistical distribution [20]. The sand grains are irregular and identified to be comprised primarily of alpha-quartz crystals. The particle size distributions of the as-received sand has the $d_{60}$ value as 0.41 mm, and the effective grain size $d_{10}$ as 0.18 mm; and the uniformity $U$ (defined as $d_{60}/d_{10}$) of 2.33. After SHPB experiments on 1.75 g/cm$^3$ unsorting sand, the crushed sand samples were sorted and the particle size distributions were determined [5]. The average values for $d_{60}$, $d_{10}$ and $U$ of the compacted sand are determined as 0.225, 0.033, and 6.82, respectively, indicating a large size variation for the crushed sand [5]. As the sorted seven sand was shaking inside the sand specimen assembly, the initial dense mass densities were achieved as 1.69, 1.71, 1.74, 1.76, 1.78, 1.79 and 1.80 g/cm$^3$, for the corresponding assorted sand in sizes of 0.6, 0.5, 0.425, 0.3, 0.212, 0.15 and 0.105 mm, respectively; while the as-received dry sand reached 1.75 g/cm$^3$ mass density.

Typical recorded input and output signals from strain gages attached to the bars, and hoop signal from strain gages mounted on the confinement sleeve in circumferential surface in a SHPB test, are similar to our previous results on assorted sand [5]. In the dynamic compaction testing of sand using SHPB apparatus, dynamic equilibrium state is established when the stresses applied on both ends of the specimen are equal. To examine the dynamic equilibrium condition, the front stress and the back stress on the specimen are calculated following the 1-wave, and 2-wave methods [15]. The dynamic equilibrium and strain rate history is also similar to the former one [5]. In a valid SHPB experiment, the incident, transmitted, and reflected signals are processed further to determine the stress and strain histories at high strain rates.

An Eglin sand specimen is comprised of sand grains primarily in the range of 0.1–1 mm with irregular shape. Different sand grains are randomly stacked together, giving large variation in sand packing, including variation in end contact conditions with WC rods, from one sample to another, potentially leading to large variation in data. When assorted sand was consolidated through shaking to reach a maximum density of 1.75 g/cm$^3$, all the fine sand grains were allowed to move and rotate to fill the cavities between large sand grains, giving much lower porosity with less variation from sample to sample in packing configuration than the case with the loose sand. Therefore, we choose the dense sand density for all sizes of sands. So for

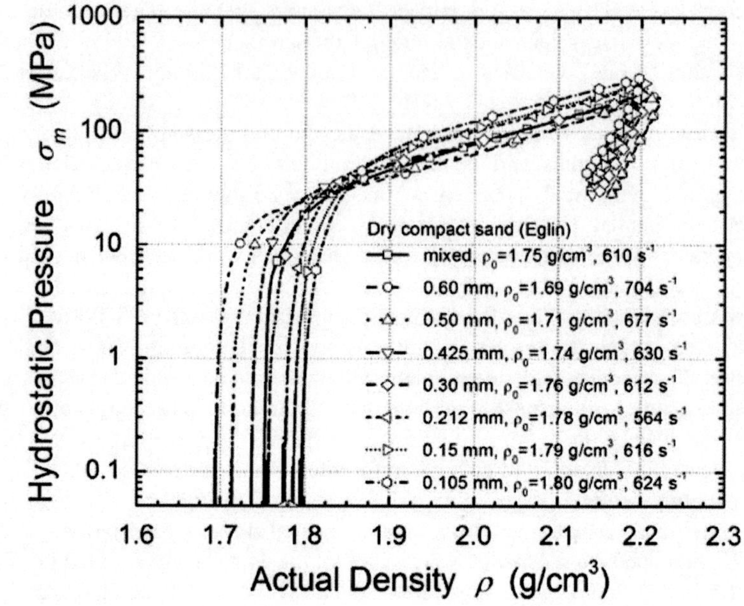


**Fig. 67.2** Results of hydrostatic pressure versus actual mass density of dry sorted sands

the smaller size of sand grain, its maximum mass density is larger than that of larger size sand grains. Since the initial mass densities are different, the initial void ratios are also different. The mass density history during the dynamic compaction will be calculated through three measured principal strains on a sand specimen.

### *67.3.1 Volumetric and Deviatoric Behaviors*

Since the initial mass density has significant effects on the mechanical behavior of assorted sand in form of exponent of 8.25, the stress-strain curves were plotted as stress-actual mass density curves, in order to compare obviously the particle size effects. To evaluate the compressibility of sand we adapt the approach often used in soil mechanics, in which the void ratio is plotted as a function of axial pressure, to give *e*-log *p* curve (under quasi-static loading conditions). For dry sand under confinement, the hydrostatic pressure – actual mass density, and shear stress – actual mass density curves under confinement are plotted at high strain rates ~600 s$^{-1}$, as shown in Figs. 67.2 and 67.3, respectively. In Figs. 67.2 and 67.3, the stress components are plotted in logarithmic scale while the mass density is plotted in linear scale. In these semi-logarithmic plots, the volumetric and deviatoric stresses follow approximately a two-segment linear relationship with the mass density. The first linear segment occurs within the initial 4 MPa stress, most likely due to elastic deformation of sand grains. It can be referred to as sand grain-level elastic behavior, similar to the situation under quasi-static compression [21]. They all behave similarly at all sizes in this segment except the different initial mass densities. The second linear segment on a semi-logarithmic scale occurs at compressive stress larger than focus stress (40 MPa for hydrostatic pressure, and 13 MPa for shear stress). It is perhaps associated with sand grain compaction through grain crushing, rearrangement of crushed fragments and intact grains under higher pressure. This stage can be referred to as grain-level plastic behavior. The smaller fragments fill the gaps between large sand grains to provide resistance for compression. The behavior in this region depends highly on the mass density and initial particle size. At higher initial mass density, the sand grains are closely packed. Hence, they provide higher resistance for gain rearrangement, and consequently crushing becomes dominant. At compressive stress between 4 MPa to focus stress, there is a gradual transition region in Figs. 67.2 and 67.3 between the two linear segments in the semi-logarithmic scale for hydrostatic pressure and deviatoric stress. The behavior shows mass density compaction at the transition of the two linear segments. The shear behavior shows a yield-like step (Fig. 67.3), oscillating more at sand size of 0.6 mm, than 0.5 mm and other sizes of sands.

In Fig. 67.2, at compacted mass density lower 1.85 g/cm$^3$, the curves intersect at about 40 MPa. The curves shift to the right as the sand particle size decreases. Beyond the intersecting point (1.85 g/cm$^3$, 40 MPa), smaller size sands shift upwards and bear more compressive stress than the larger size sand with the increase of mass density; this trends is different

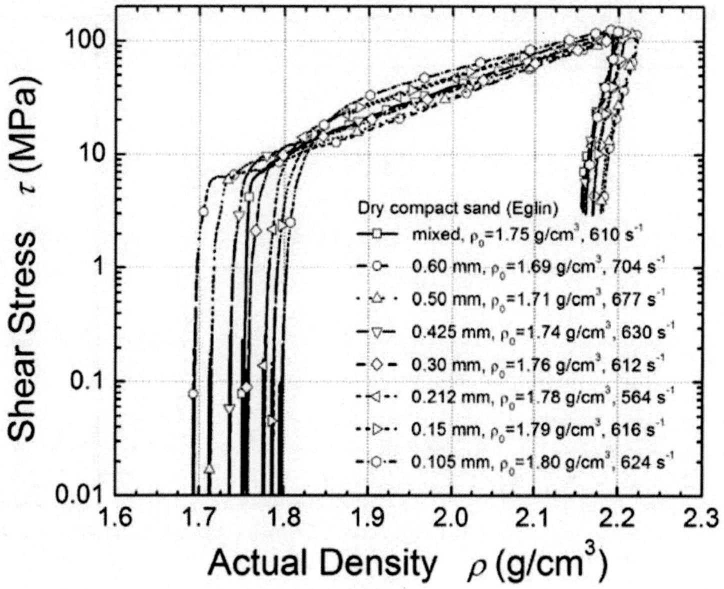


**Fig. 67.3** Results of shear stress versus actual density of dry sorted sands

from the trend in the stress – strain curves which shift upwards with the increase of initial mass density below actual mass density of 1.85 g/cm$^3$. The unloading curve shows a steep reduction in stress. Similar behaviors were found for deviatoric response (Fig. 67.3) with the intersecting point of (1.83 g/cm$^3$, 13 MPa). The stress-mass density curve of assorted sand was in between the curve of the 0.425 mm size sand and 0.3 mm size sand. It is noted that at grain-level plastic deformations shown in Figs. 67.2 and 67.3, curves of two groups of sand sizes (0. 212 and 0.15 mm) and (0.6 and 0.5 mm) were overlapped, respectively. And the curve of the assorted sand was overlapped with that of 0.425 and 0.3 mm size sands.

### *67.3.2 Energy Absorption of Sorted Sand*

The energy absorption capability of sand under dynamic loading conditions was rarely reported in literature. In this work, the energy absorption capability is estimated in confined compression. The energy absorption density is calculated in term of the area enclosed by the axial stress-strain curve. The energy absorption, defined as the energy density $W$ (area enclosed by the stress-strain curve up to a peak stress), is plotted against a function of the peak stress at a given initial density [22]. The specific energy absorption is then calculated in terms of the energy absorption density divided by the initial mass density [22]. The specific energy absorptions under axial deformations under these situations are calculated under the maximum axial stress and strain attained in experiments. For full characterization of the energy absorption capability of sand at different axial peak stress levels, we have replaced the constructed energy absorption diagrams into energy absorption vs actual mass density, due to different initial mass density for all sizes of dense sands, as shown in Fig. 67.4. For all sizes of sands, when the sand was compacted to higher stress level or higher compacted density (2.17 ~ 2.23 g/cm$^3$), the smaller size sands show more energy absorption than the larger size sands. This behavior is consistent with size dependence in energy absorption for nanofoam [13, 14] and engineering foam [22, 23].

The specific energy absorption values of sand due to axial deformations are 14.2 J/g, under a maximum axial compressive stress of 350 MPa for assorted sand with 1.75 g/cm$^3$ initial density. When the sand gap/void completely vanished and all sand fractured fragments were compacted. The energy absorption values of sorted sand are in the range of 14.8 ~ 20.2 J/g, under a maximum axial compressive stress of 360 ~ 470 MPa for sorted sands with initial density of 1.69 ~ 1.80 g/cm$^3$. For the sorted sands, the energy absorption diagram has a similar focus point at (1.93 g/cm$^3$, 5 MPa.m/m). Below the intersecting point, larger size sands provide more energy absorption than smaller size sands; beyond the intersecting point, the larger size sands show lower energy absorption capability than the smaller size sands; while the assorted (mixed) sand shows an overlap between sand sizes of 0. 25 and 0.3 mm, located at the center (Fig. 67.4).

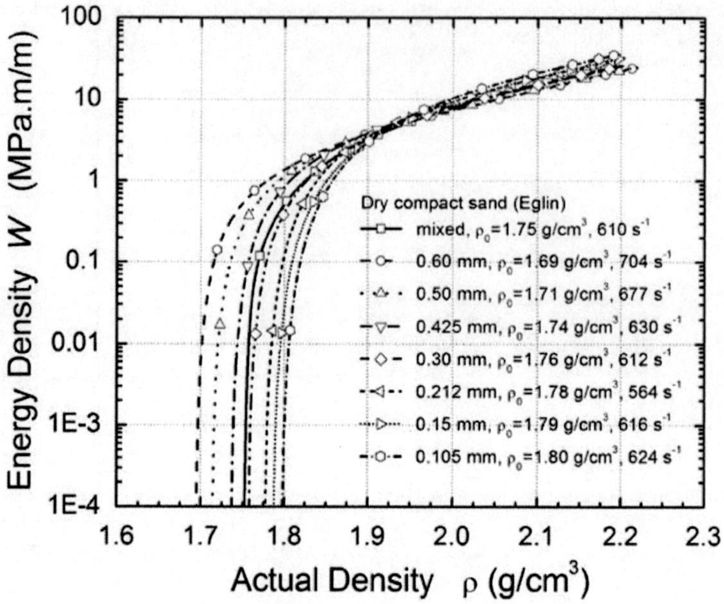


**Fig. 67.4** Energy absorption diagrams for dry sand with effects of sand sizing

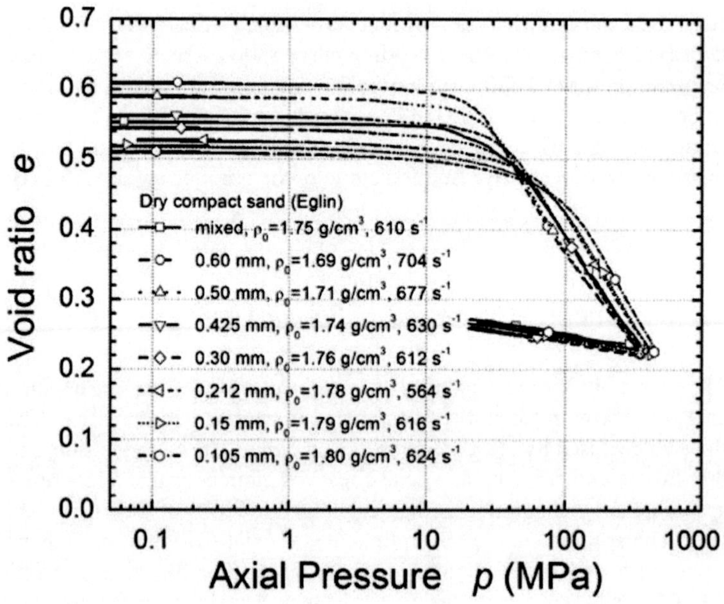


**Fig. 67.5** *e*-log *p* curves of dry sands in seven different sizes under confinement

### *67.3.3 Compressibility of Sorted Sands*

We next investigate the compressibility of sand as a function of axial pressure. The relationship is given in terms of the void ratio, *e*, and the applied axial stress, often termed as *p*, in order to make it consistent with the term used in soil mechanics [1, 2]. It may be noted that results on the *e*-log *p* relationship of sand at high strain rates was sparkly reported in the literature. The measurement of both radial and circumferential stresses and strains in the sand sample in this work made it possible to report the relationship.

In Fig. 67.5, two linear segments appear in the *e*-log *p* curve, representing a power law relationship between *e* and *p*, similar to what was observed under quasi-static loading conditions [1–4]. Two important parameters of soil mechanics [24, 25], namely, the pre-consolidation pressure $p_c$, representing the effective vertical stress to cause the destructuration of natural soil and the transitional pressure between the two linear segments, and the compression index $C_c$, representing the magnitude of the slope in the linear segments of the *e*-log *p* curve after pre-consolidation, can be determined. The *e*-log *p* curve during unloading also follows a linear relationship, indicating that *e* follows a power law relationship with *p* during unloading. At these high strain rates, with the increase of the initial void ratio (decrease of mass density), the pre-consolidation pressure decreases and the compression index increases. It is noted that there is also an intersecting point at (50 MPa, 0.48) after pre-consolidation, where the smaller size sand is harder to compress than the larger size sand. The unloading shows nearly identical behavior.

## 67.4 Conclusion

The dynamic compressive behavior of dry sand (from Eglin AFB) under stiff confinement was investigated using a modified long split Hopkinson pressure bar (SHPB) at high strain rates (~600 $s^{-1}$) up to 470 MPa compressive stress, using a sand specimen assembly with WC inserts inside a steel tube. The assembly allows shaking to consolidate sand to prepare sand samples with maximum dense initial mass densities. Sand samples of unsorted sand and seven sizes of sand in dense initial density were compressed on SHPB at high strain rates. A strain gage was attached on the external surface of the confining steel sleeve to measure the circumferential strain, to determine both the bulk and the deviatoric responses. The curves of hydrostatic pressure-mass density, shear stress-mass density of sand under the 3D stress state are determined up to 24% compressive strain and up to 470 MPa axial compressive stress at high strain rates. The specific energy absorption under axial deformations reach 20.2 J/g and 14.8 J/g for 0.105 mm sand and 0.6 mm sand, respectively. The dense smaller size sand shows more energy absorption than the dense larger size sand. The *e*-log *p* curves representing the sand compressibility as a function of sand grain particle sizes, were compared with the assorted sand. Significant particle size effects have been identified. All curves show an intersecting point; where the mechanical behavior was gradually inversed with increasing sand particle sizes. These results can be analyzed further for constitutive modeling and for mesoscale simulations to understand the dynamics of sand under high pressure and high rate deformation.

**Acknowledgements** We acknowledge the support of ONR Multidisciplinary University Research Initiative program (MURI) BAA 10-026.

## References

1. Yamamuro JA, Bopp PA, Lade PV (1996) One-dimensional compression of sands at high pressure. J Geotech Eng 122:147–154
2. Russell AR, Khalili N (2004) A bounding surface plasticity model for sands exhibiting particle crushing. Can Geotech J 41:1179–1192
3. Vallejos J (2008) Hydrostatic compression model for sandy soils. Can Geotech J 45:1169–1179
4. Graham J, Alfaro M, Ferris G (2004) Compression and strength of dense sand at high pressure and elevated temperatures. Can Geotech J 41:1206–1212
5. Luo H, Lu H, Cooper WL, Komanduri R (2011) Effect of mass density on the compressive behavior of dry sand under confinement at high strain rates. Exp Mech 51:1499–1510
6. Charlie WA, Ross CA, Pierce SJ (1990) Split-Hopkinson pressure bar testing of unsaturated sand. Geotech Test J 13:291–300
7. Bragov AM, Grushevsky GM, Lomunov AK (1996) Use of the Kolsky method for confined tests of soft soils. Exp Mech 36:237–242
8. Felice CW, Gaffney ES, Brown JA, Olsen JM (1987) Dynamic high stress experiments on soil. Geotech Test J 10:192–202
9. Bragov AM, Lomunov AK, Sergeichev IV, Tsembelis K, Proud WG (2008) Determination of physicomechanical properties of soft soils from medium to high strain rates. Int J Impact Eng 35:967–976
10. Frew DJ, Forrestal MJ, Chen W (2002) Pulse shaping techniques for testing brittle materials with a split Hopkinson pressure bar. Exp Mech 42:93–106
11. Martin BE, Chen W, Song B, Akers SA (2009) Moisture effects on the high strain-rate behavior of sand. Mech Mater 41:786–798
12. Song B, Chen W, Luk V (2009) Impact compressive response of dry sand. Mech Mater 41:777–785
13. Luo H, Lu H, Leventis N (2006) The compressive behavior of isocyanate-crosslinked silica aerogel at high strain rates. Mech Time Depend Mater 10:83–111
14. Luo H, Churu G, Fabrizio EF, Schnobrich J, Hobbs A, Dass A, Mulik S, Zhang Y, Grady BP, Capecelatro A, Sotiriou-Leventis S, Lu H, Leventis N (2008) Synthesis and characterization of the physical, chemical and mechanical properties of isocyanate-crosslinked vanadia aerogels. J Sol Gel Sci Tech 48:113–134
15. Gray GT III (2000) Classic split-Hopkinson pressure bar testing. In: Mechanical testing and evaluation, vol 8, Metals handbook. American Society for Metals, Materials Park, pp 462–476
16. Ravi-Chandar K, Ma Z (2000) Inelastic deformation in polymers under multiaxial compression. Mech Time Depend Mater 4:333–357
17. Knauss WG, Emri I, Lu H (2008) Mechanics of polymers: viscoelasticity. In: Sharpe WN (ed) Handbook of experimental solid mechanics. Springer, New York, pp 49–95

18. Cooper WL, Breaux BA (2010) Grain fracture in rapid particulate media deformation and a particulate media research roadmap from the PMEE workshops. Int J Fract 162:137–150
19. Daphalapurkar NP, Wang F, Fu B, Lu H, Komanduri R (2010) Determination of mechanical properties of sand grains by nanoindentation. Exp Mech 51:719–728
20. ASTM D2487-06 (2006) Standard practice for classification of soils for engineering purposes (Unified Soil Classification System). Annual Book of ASTM Standards–Soil and Rock 04.08, Conshohocken
21. Hagerty MM, Hite DR, Ullrich CR, Hagerty DJ (1993) One-dimensional high-pressure compression of granular material. J Geotech Eng Trans ASCE 119:1–18
22. Gibson LJ, Ashby MF (1997) Celluar solids: structure and properties, 2nd edn. Cambridge University Press, Cambridge
23. Daphalapurkar NP, Hanan JC, Phelps NB, Bale H, Lu H (2008) Tomography and simulation of microstructure evolution of a closed-cell polymer foam in compression. Mech Adv Mater Struct 15:594–611
24. Lambe TW, Whitman RV (1969) Soil mechanics, 3rd edn. Wiley, New York
25. Terzaghi K, Peck RB, Mesri G (1996) Soil mechanics in engineering practice. Wiley-IEEE, New York

# Chapter 68
# A Testing Technique to Characterise the Shear Behaviour of Concrete at High Strain-Rates

**P. Forquin and L. Sallier**

**Abstract** Shear fracture mechanisms (in mode II) are usually observed in plain-concrete structures when subjected to impact loading of a rigid projectile. In the present work, an experimental method is proposed to characterise the confined dynamic shear strength and softening behaviour of concrete. A cylindrical specimen in which to circular notches are preformed is subjected to a dynamic loading through a Split Hopkinson Pressure Bar (SHPB) device. A cylindrical Hopkinson bar is used as input bar whereas a tube is used as output bar. The particle velocities measured by laser interferometers on both input and output bars allow the deduction of the shear stress in the ligament (fractured zone) and the axial displacement in the central zone beside the peripheral zones. Moreover an instrumented confining ring is used to induce a confining pressure on the fractured zone. This experimental method has been employed to characterize the dynamic shear strength of a common concrete at strain-rates as high as 100/s.

**Keywords** Shear failure • Dynamic testing • Concrete • Confined loading • Free water

## 68.1 Introduction

Concrete structures as nuclear power plants, dams, bridges or bunkers can be exposed to intensive loadings such as earthquake, blasting or projectile-impact. During such dynamic loading, tensile and shear damage modes can be observed as cratering, radial and spalling fracturing of shear failure [1, 2]. Shear fracturing in mode II has been also observed in high strength concrete targets during edge-on impact tests and normal impact tests in confined zones [3, 4]. In a case of penetration of a rigid projectile within massive target, triaxial compression and shear stresses are generated whereas the inertia of the surrounding material creates also a passive confinement near the front face of the target. To improve the understanding and modelling of concrete behaviour under such confined loading, quasi-static and dynamic shear tests have been developed in LEM3 laboratory.

In the last decades, an experimental technique, the Punch Through Shear (PTS) test, was proposed by Watkins [5]. A cylindrical specimen is preformed by two cylindrical notches on the top and bottom faces. The specimen is first subjected to a pure hydrostatic pressure whereas an axial load is added in a second step to create a shear fracturing of the ligament. This technique was more recently used by Montenegro et al. [6] to evaluate the fracture energy of concrete as function of the hydrostatic pressure. In the present work, a passive confining cell has been applied to PTS specimens instead of a hydraulic pressure in a similar way than in quasi-oedometric compression tests [7, 8]. The experimental method used for the dynamic experiments is presented below as well as some experimental results. Influence of strain-rate and free water content is briefly set-out.

P. Forquin (✉) • L. Sallier
LEM3 laboratory, Lorraine University, Ile du Saulcy, Metz 57045, France
e-mail: pascal.forquin@univ-metz.fr

V. Chalivendra et al. (eds.), *Dynamic Behavior of Materials, Volume 1: Proceedings of the 2012 Annual Conference on Experimental and Applied Mechanics*, Conference Proceedings of the Society for Experimental Mechanics Series,
DOI 10.1007/978-1-4614-4238-7_68, © The Society for Experimental Mechanics, Inc. 2013

## 68.2 Experimental Method

### 68.2.1 Concrete Sample

The specimen geometry is shown in Fig. 68.1a. Inner diameter of the lower notch and outer diameter of upper notch coincide to obtain a straight cylindrical fracture surface. The length of the ligament (10 mm) allows ensuring an almost homogeneous shear stress field in the ligament. Moreover the confining ring is not centred toward the symmetry plane of the specimen to get a uniform pressure between the confining ring and the concrete sample (Fig. 68.1b). The aluminium confining rings considered for quasi-static and dynamic experiments are 10 mm thick and 15 mm in height. The gap between the sample and the confining ring is filled by a bi-component resin (Chrysor®). Furthermore, four radial notches at 90° through the outer part of the specimen are preformed in the sample to avoid a self confinement of the sample.

The tested "R30A7" concrete is a common concrete that contains hard siliceous aggregates from 2 to 8 mm, sand, cement and water. The water to cement ratio is 0.64. All the specimens have been cored out from large blocks and stored in water saturated by lime in order to avoid the dissolution of portlandite in water. Two sets of specimens have been considered for shear testing: saturated specimens have been kept in water until few minutes before testing. "Dry" specimens were oven-dried at 60 °C during several weeks until a constant weight has been achieved.

### 68.2.2 Experimental Procedure

The experimental devices used in dynamic shear testing is presented on Fig. 68.2. The specimen is subjected to a dynamic compression loading through a Split Hopkinson Pressure Bar (SHPB) equipment. A cylindrical Hopkinson bar 1.5 m in length, 45 mm in diameter is used as input bar whereas a tube 1 m in length, 50 mm as inner diameter is used as output bar (Fig. 68.2a, b). The cylindrical Hopkinson bar and the tube are both instrumented by laser interferometers instead of strain-gauges (Fig. 68.2c). Moreover a specific pulse-shaper technique was developed to decrease the loading rate in the sample and to improve the mechanical balance of the sample. The method used to process the experimental data is the following. First, the mean radial stress in the ligament (sheared zone) is deduced from the contact pressure between the sample and the confining vessel knowing the contact surface between the specimen and the ring ($S_{contact}$) and the area of the fracture surface ($S_{ligament}$):

$$\sigma_{radial} = \frac{S_{contact}}{S_{ligament}} P_{contact} \tag{68.1}$$

Second, the particle velocity measured on the input bar provides the incident and reflected waves so the input force between the specimen and the Hopkinson bar is obtained:

$$F_{input} = S_{input} \rho_{bar} C_{bar} \left( V_{incident} - V_{reflected} \right) \tag{68.2}$$

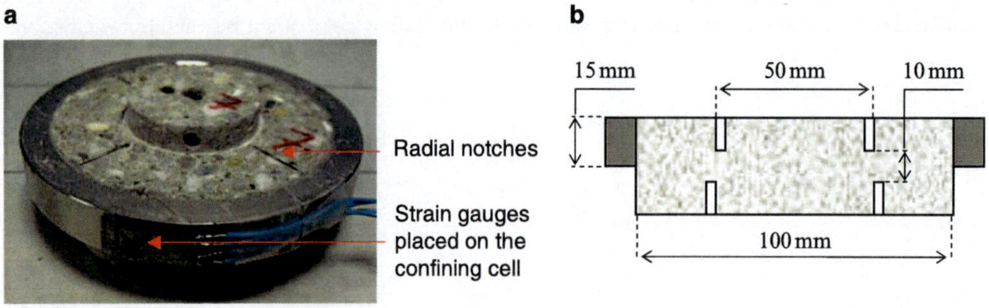


**Fig. 68.1** (**a**) Geometry of the sample used for the QS and dynamic experiments. (**b**) Concrete sample after shear testing

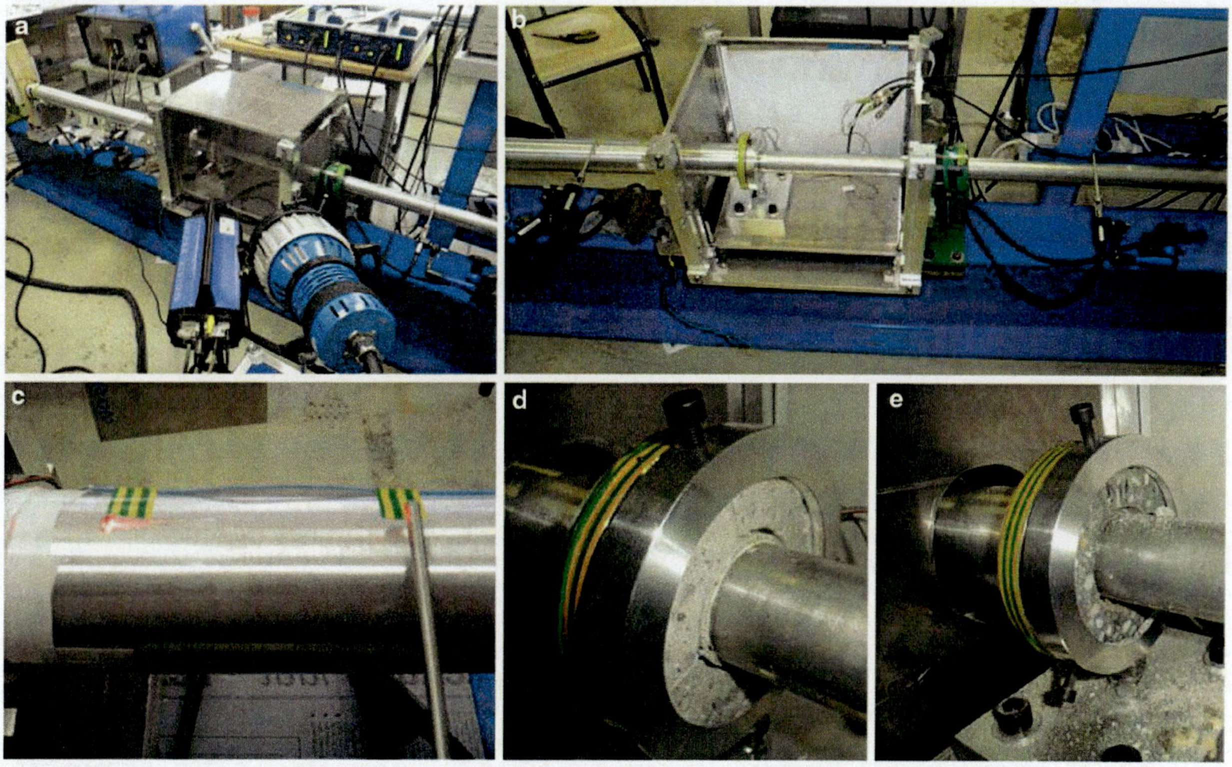


**Fig. 68.2** Experimental device used in LEM3 to perform dynamic shear tests. **(a, b)** Cylindrical Hopkinson bar and tube used as input and output bars. **(c)** Laser interferometer applied to the input bar. **(d)** Concrete sample before testing, **(e)** after dynamic shear testing

In a similar way, the laser interferometer pointed out toward the tube and strain-gauges glued on the tube both provide the transmitted pulse and the output force:

$$\begin{cases} F_{output} = S_{output}\rho_{bar}C_{bar}(V_{transmitted}) \\ \quad or \\ F_{output} = S_{output}E_{bar}(\varepsilon_{transmitted}) \end{cases} \tag{68.3}$$

Finally the shear stress in the ligament is deduced by averaging the input and output forces:

$$\sigma_{shear}^{input} = \frac{F_{input} + F_{output}}{2S_{ligament}} \tag{68.4}$$

The displacement is given by integrating the particle velocity provided by laser interferometers directed toward the input and output bars.

$$V_{axial} = V_{reflected} + V_{incident} - V_{transmitted} \tag{68.5}$$

This experimental method was applied to a series of dynamic experiments performed with dry and wet specimens. Some experimental results are presented in the next part.

## 68.3 Measurement Results

### *68.3.1 Quasi-Static Tests*

Two experimental results obtained with dry and wet specimens in quasi-static loading with an aluminium alloy confining vessel are reported on Fig. 68.3. In wet specimen the maximum shear stress reached 41 MPa whereas a maximum value of 54 MPa was obtained in dry sample. A strong decrease of shear stress is observed after the stress peak. The change of lateral pressure between the confining cell and the specimen is reported on the same plot as well as the resulting radial stress in the shear zone. A similar radial stress about 30 MPa (in absolute value) is observed in both tests (dry and wet specimens) in the shear zone when the peak shear stress is reached.

### *68.3.2 Dynamic Tests*

Two experimental results obtained with dry and wet specimens in dynamic loading are reported on Fig. 68.4. As in quasi-static test the shear strength of the wet sample is markedly lower than that of the dry specimen. As in quasi-static conditions a strong decrease of shear stress is observed after the stress peak. Again the change of lateral pressure between the confining cell and the specimen is reported on the same plot as well as the resulting radial stress in the shear zone. A radial stress about 40–50 MPa (in absolute value) is observed in both tests (dry and wet specimens) in the shear zone when the peak shear stress is reached. Finally, a very limited increase of strength with strain-rate is pointed out for both sets of specimens (dry and wet sets). This result significantly differs to the conclusion obtained for concrete under tensile loading. Indeed, as shown by several authors [9–11] the increase of tensile strength of wet concrete is much more pronounced than in the case of dry specimens in a similar strain-rate range (1e-6/s–100/s). Thus, the so-called Stephan effect supposed to explain the strain-rate sensitivity of wet concrete in tension below 1/s is supposedly inoperative under shear loading.

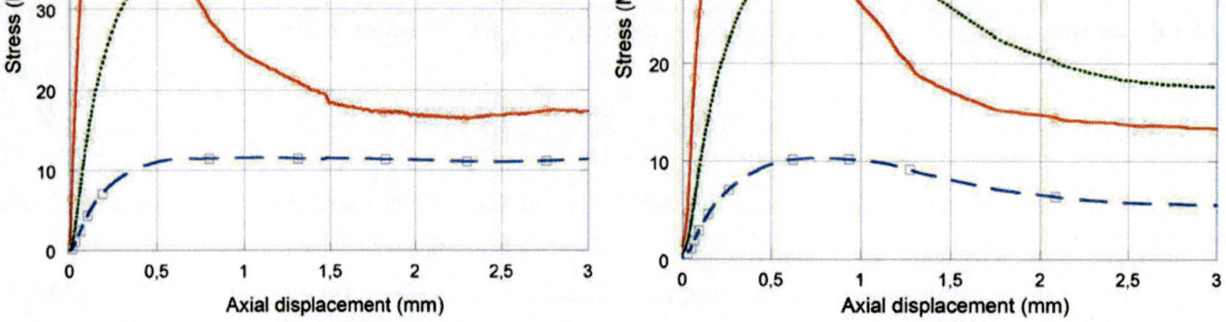


**Fig. 68.3** Results of quasi-static shear tests performed on dry (**a**) and wet (**b**) specimens. Strain-rate: $2^{e}$-5/s, aluminium confining cell, concrete samples with radial notches

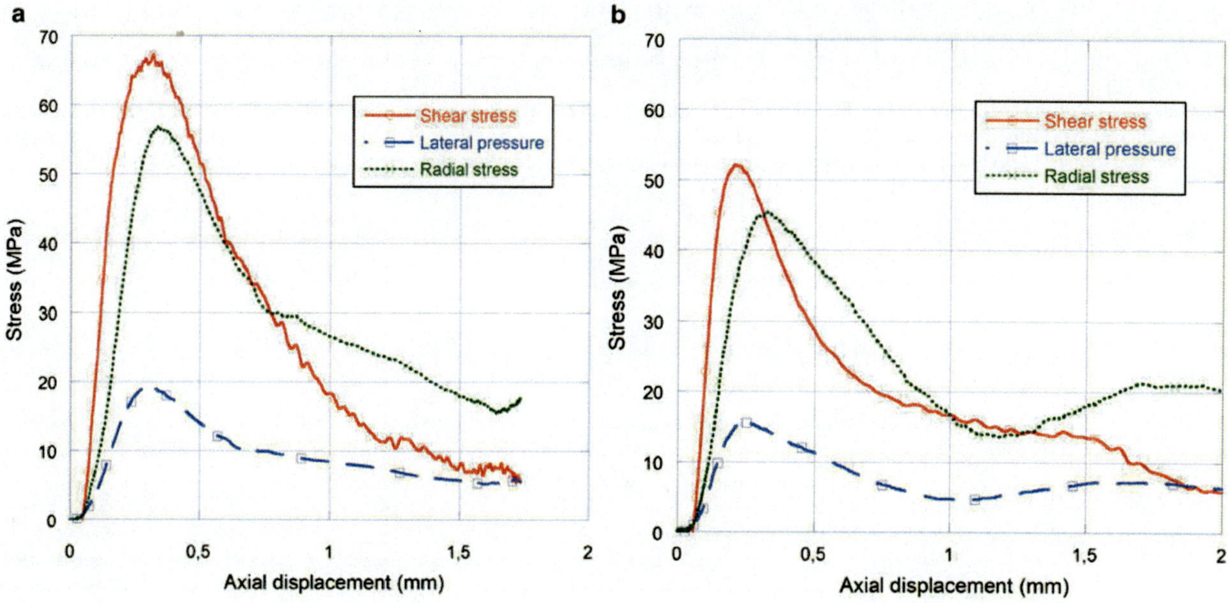


**Fig. 68.4** Results of quasi-static shear tests performed on dry (**a**) and wet (**b**) specimens. Strain-rate: 60/s, aluminium confining cell, concrete samples with radial notches

## 68.4 Summary

An experimental method is proposed to investigate the dynamic shear behaviour of concrete. The set-up based on PST (Punch Through Shear) testing technique relies on the use of a passive confining ring. The specimen is subjected to a dynamic compression loading through a Split Hopkinson Pressure Bar (SHPB) equipment. A cylindrical Hopkinson bar is used as input bar whereas a tube is used as output bar. The cylindrical Hopkinson bar and the tube are both instrumented by laser interferometers instead of strain-gauges. A pulse-shaper technique is employed to improve the mechanical balance of the PTS sample.

Experiments have been conducted on dry and water saturated concrete samples in quasi-static and dynamic loading. The obtained results show a higher strength with dry samples than in wet ones. Furthermore, oppositely to previous results obtained in the literature in tension, both sets of concrete (dry and wet sets) show a very small strain-rate sensitivity in the considered range of strain-rate. The so-called Stephan effect is though to be inoperative under shear loading.

**Acknowledgements** This work was performed under a program entitled “Experimental approach and modelling of the dynamic behavior of brittle materials (Exam-Brittles)” sponsored by the French National Agency for Research (ANR) and supported by the CEA, DAM, GRAMAT. These supports are gratefully acknowledge.

## References

1. Li QM, Reid SR, Wen HM, Telford AR (2005) Local impact effects of hard missiles on concrete targets. Int J Impact Eng 32:224–284
2. Forquin P, Arias A, Zaera R (2008) Role of porosity in controlling the mechanical and impact behaviours of cement-based materials. Int J Impact Eng 35(3):133–146
3. Forquin P (2003) Endommagement et fissuration de matériaux fragiles sous impact balistique, rôle de la microstructure. Ph.D. dissertation, Ecole Normale Supérieure de Cachan, France (in French)
4. Forquin P, Hild F (2008) Dynamic fragmentation of an ultra-high strength concrete during edge-on impact tests. ASCE J Eng Mech 134 (4):302–315
5. Watkins J (1983) Fracture toughness test for soil-cement samples in mode II. Int J Fract 23:135–138
6. Montenegro O, Sfer Carol I (2007) Characterization of concrete in mixed mode fracture under confined conditions. In: Proceedings of the ICEM13 conference, Alexandroupolis
7. Forquin P, Arias A, Zaera R (2007) An experimental method of measuring the confined compression strength of geomaterials. Int J Solids Struct 44(13):4291–4317

8. Forquin P, Gary G, Gatuingt F (2008) A testing technique for concrete under confinement at high rates of strain. Int J Impact Eng 35 (6):425–446
9. Rossi P (1991) A physical phenomenon which can explain the mechanical behaviour of concrete under high strain rates. Mater Struct 24:422–424
10. Cadoni E, Labibes K, Albertini C, Berr M, Giangrasso M (2001) Strain-rate effect on the tensile behaviour of concrete at different relative humidity levels. Mater Struct 34:21–26
11. Erzar B, Forquin P (2011) Experiments and mesoscopic modelling of dynamic testing of concrete. Mech Mater 43:505–527

# Chapter 69
# Is Nano Always Better? Static and Dynamic Fracture Study of Particulate Polymer Composites

**K.C. Jajam and H.V. Tippur**

**Abstract** In this work, the role of nano- versus micro-filler particle size-scale on static and dynamic fracture behaviors of silica-filled epoxy is examined. Particulate composites of epoxy matrix are studied under quasi-static and stress-wave loading conditions. Mode-I crack initiation and crack growth behaviors are examined using 2D digital image correlation method and high-speed photography in symmetrically impacted specimens. The measured displacement fields are analyzed using 2D crack-tip fields for dynamically propagating cracks in brittle solids to extract stress intensity factor histories ($K_I{}^d$) and crack velocity histories ($V$). The quasi-static fracture tests show enhanced fracture toughness for nanocomposites relative to micro-particle filled ones. On the other hand, the dynamic crack-initiation toughness is consistently higher for micro-particle filled composites relative to the nano-filler counterparts. The crack velocities in nanocomposites are significantly higher than that observed in micro-filler cases. Also, the post-mortem analyses of fracture surfaces reveal greater surface ruggedness for nanocomposites under quasi-static conditions. However, the opposite is evident under dynamic loading conditions. The qualitative and quantitative fractographic measurements correlate well with the measured fracture parameters for both quasi-static and dynamic fracture tests.

**Keywords** Particulate composites • Size effects • Loading-rate effects • Nanocomposites • Fracture • Fractography • Optical measurements • High-speed photography

## 69.1 Introduction

Particulate polymer composites are used in a variety of engineering applications. These are generally multiphase materials with polymer matrix phase reinforced with fillers of different sizes and shapes. The addition of rigid fillers play vital role in optimizing overall stiffness, impact energy absorption and thermo-mechanical characteristics of the resulting composite system. However, the effect of particle size-scale (nano- vs. micro-) could vary the fracture response of these materials depending upon the nature of loading, quasi-static or dynamic.

A review [1–4] on fracture behavior of micron-size rigid filler composites shows improved fracture toughness relative to unfilled matrix. Also, a modest improvement in quasi-static fracture toughness $K_{Ic}$ and stiffness have been reported by several researchers [5–10] using nano-fillers in the last few years. It should be noted that most studies to date on fracture behavior of particle-filled composites have been performed quasi-statically and very limited data exists from a dynamic failure perspective. Among the few dynamic fracture investigations reported on particulate composites include the study of filler particle size on the mode-I fracture behavior of glass-filled epoxy conducted by Kitey and Tippur [11]. Their work shows that micron-size spherical glass beads of different mean diameter (11–200 μm) improved the fracture toughness of epoxy and noted a specific particle size (35 μm) for optimum fracture toughness at a constant volume fraction (10%). In regards to nanocomposites, dynamic fracture work has been reported by Shukla et al. [12] using $TiO_2$ (35 nm) and $Al_2O_3$ (14 nm) nano-particles. They observed enhanced fracture toughness and higher crack velocity in nano-filler cases than neat resin. Note that most of the works reported on fracture behavior of particulate (nano- or micro-) composites do not deal with dynamic fracture caused by impact loading and this gap needs to be bridge if nanocomposites are to find engineering

K.C. Jajam • H.V. Tippur (✉)
Department of Mechanical Engineering, Auburn University, Auburn, AL 36849, USA
e-mail: jajamkc@auburn.edu; tippuhv@auburn.edu; htippur@eng.auburn.edu

V. Chalivendra et al. (eds.), *Dynamic Behavior of Materials, Volume 1: Proceedings of the 2012 Annual Conference on Experimental and Applied Mechanics*, Conference Proceedings of the Society for Experimental Mechanics Series, DOI 10.1007/978-1-4614-4238-7_69, © The Society for Experimental Mechanics, Inc. 2013

applications where stress-wave loading as in aerospace and automotive structures, sporting equipments, and high-speed trains is expected. In view of this, the authors [13] have recently conducted an experimental study of nano- versus micro-size fillers and loading-rate effects on static and dynamic fracture behavior of particulate polymer composites. This paper presents some of the major observations of their investigation.

## 69.2 Experimental Approach

In this work, the dynamic fracture behavior was studied using the 2D digital image correlation (DIC) method. In this method, decorated random speckle patterns on a specimen surface were monitored during a fracture event. The gray scales were recorded at two different states, one before and another during deformation. The corresponding images from deformed and undeformed sets were paired and analyzed separately. That is, a sub-image in an undeformed image was chosen and its location in the corresponding deformed image was identified. Once the location of this sub-image in the deformed image was found, the local displacements of this sub-image could be readily quantified. Further details of this method can be found in [14, 15].

## 69.3 Material Preparation and Specimen Geometry

Two types of spherical rigid fillers were used in this study, one is nano- size silica particles of average diameter 20 nm (Nanopox F400 from Nanoresins Germany) and the other is micron- size soda-lime A-glass particles of average diameter 35 μm (Spheriglass®, Potters Industries Inc.). Each type of fillers was dispersed in a low viscosity two-part epoxy (Epo-Thin™ from Beuhler, Inc. USA) as a matrix material. Enough care was exercised to ensure uniform distribution of filler throughout the matrix. It should be noted that neat epoxy and two types of particle filled epoxies were cast for quasi-static and dynamic tests. The nanocomposites and the micron-size particle filled composites were loaded with 3%, 5%, 7%, and 10% volume fraction ($V_f$) of fillers. The material was cured at room temperature for about one week. The cured material was machined into rectangular coupons of dimensions 106 × 20 × 8 mm for quasi-static fracture tests (span 90 mm) and 212 × 50 × 8 mm dynamic fracture experiments (span 196 mm). An edge notch of 4 and 10 mm in length was cut into the samples for quasi-static and dynamic fracture tests, respectively. The notch tip was then sharpened using a razor blade in order to have a naturally sharp crack-tip to achieve a consistent crack initiation followed by a steady growth [16]. As noted earlier, the dynamic fracture experiments were performed using the method of DIC to quantify crack-tip deformations and crack growth parameters. Therefore, a random speckle pattern was created on the surface of the sample by spraying a fine mist of black and white paints alternatively. Figure 69.1 depicts the specimen geometry, dimensions and loading configuration with crack-tip coordinate system with an illustration of random speckle pattern on it. The region in the dotted box represents 30 × 30 mm$^2$ region-of-interest.

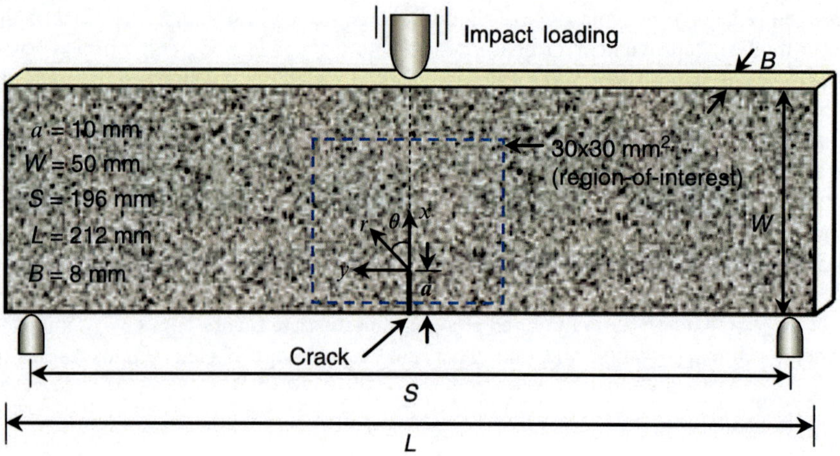


**Fig. 69.1** Dynamic fracture specimen geometry and impact loading configuration

## 69.4 Experimental Setup and Testing Procedure

For quasi-static fracture tests, the single edge notched bend (SENB) specimens were loaded in displacement control mode with testing speed of 0.01 in./min in a symmetric three point bending configuration using Instron 4465 testing machine. The load *vs.* deflection data was recorded up to complete fracture of the specimens and the quasi-static crack initiation toughness, $K_{Ic}$ was calculated using the load at fracture.

The schematic of the experimental setup used for dynamic fracture tests is shown in Fig. 69.2. It included a drop tower for impacting the specimen and a Cordin 550 ultra high-speed digital framing camera for capturing speckle images during the fracture event. Two high-energy flash lamps, triggered by the camera, were employed to illuminate the sample surface. The digital high-speed camera is capable of recoding images at rates of up to two million frames per second on 32 individual 1,000 × 1,000 pixels CCD image sensors positioned circumferentially around a rotating mirror which reflects and sweeps light over these sensors. The camera was focused on a 30 × 30 mm$^2$ region-of-interest (ROI) in the vicinity of crack-tip. Prior to impacting the specimen, a set of 32 images of the ROI were recorded at a chosen framing rate. While maintaining all camera settings same, another set of 32 images was captured when the specimen was impacted at a velocity of 4.5 m/s. In order to capture the entire fracture event, the images in the case of micro-particle filled composites were recorded at a framing rate of 250,000 frames per second, whereas framing rates ranging from 250,000–300,000 frames per second were used for nanocomposite specimens. A total of 32 images were recorded for undeformed and deformed sets. The corresponding two images for each sensor were paired from undeformed and deformed sets and each of these 32 matched pairs was analyzed individually. Further details regarding testing procedure and image analysis can be found in [13].

## 69.5 Results and Discussion

In order to ensure uniform distribution of filler particles, microscopy was performed using transmission electron microscope (TEM: Zeiss EM10) for nanocomposites and optical microscope (Nikon: Eclipse L150) for micro-filler specimens. The images in Fig. 69.3a and b show well-dispersed nano- and micro-fillers, respectively. It can be seen that *particle agglomeration is absent in both micrographs at 10% volume fraction.*

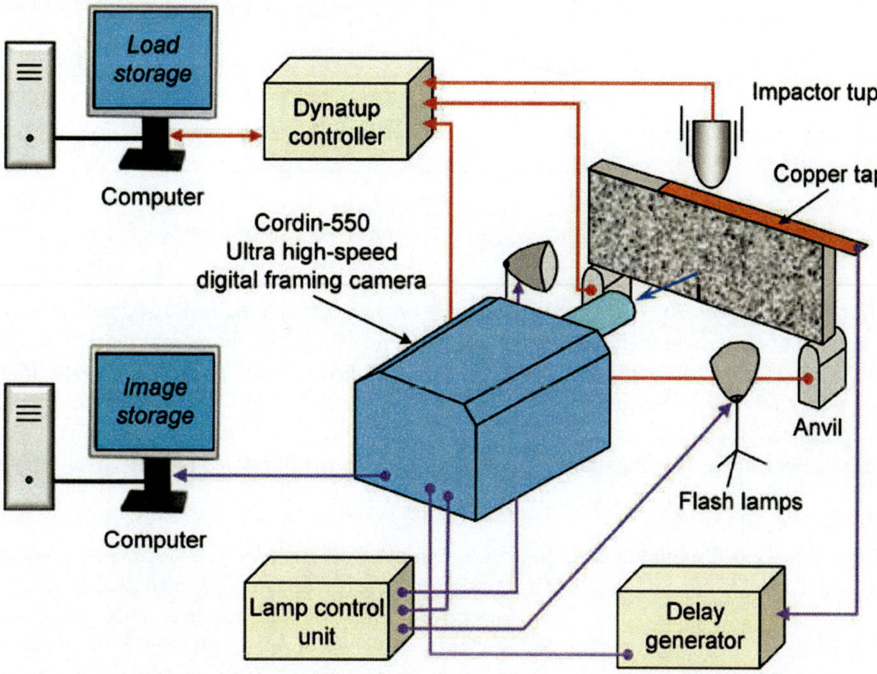


**Fig. 69.2** Schematic of the dynamic fracture experimental setup

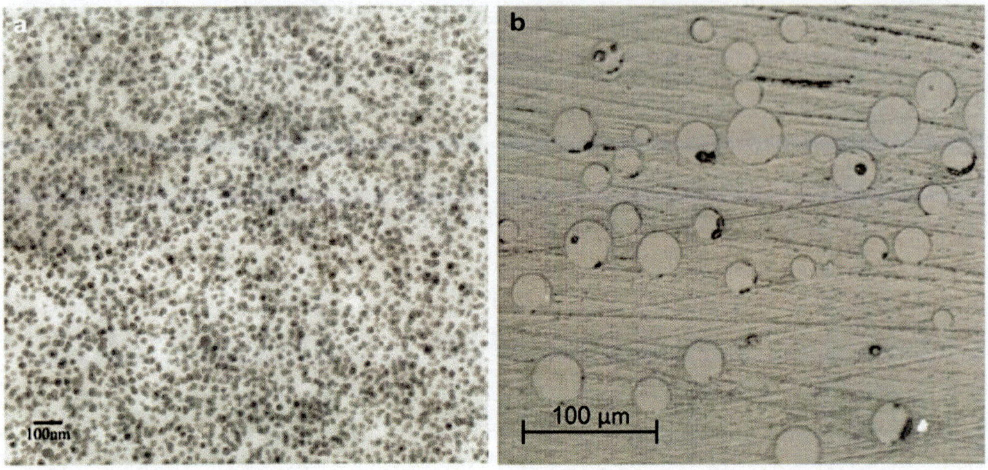


**Fig. 69.3** Micrographs showing particle dispersion: (**a**) TEM image of 10% $V_f$ nano-filler. (**b**) Optical micrograph of 10% $V_f$ micro-filler

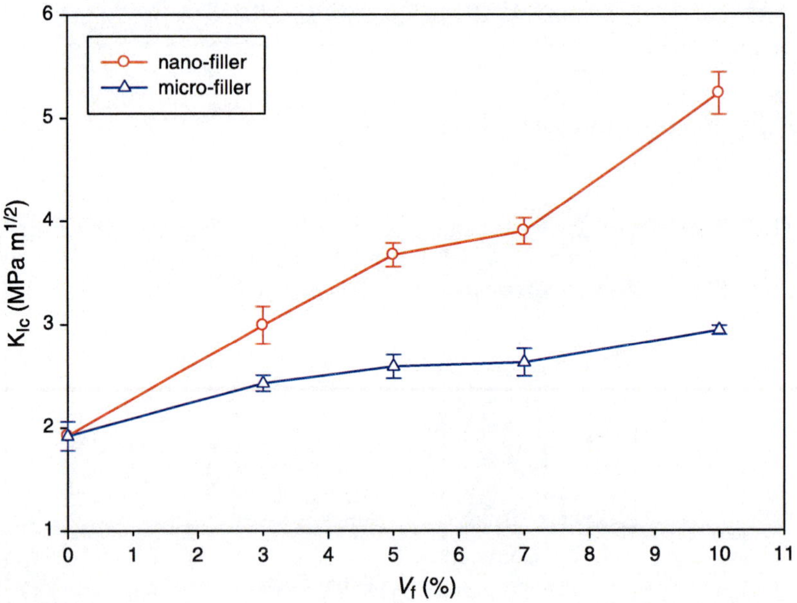


**Fig. 69.4** Variation of quasi-static fracture toughness ($K_{Ic}$) with filler volume fraction ($V_f$) for nano- and micro-filler modified epoxies

The quasi-static fracture toughness, $K_{Ic}$, measured at the onset of crack growth in each case, and the variation of $K_{Ic}$ as a function of volume fraction of nano- and micro-fillers is plotted in Fig. 69.4. Each data point represents an average of four measured values of $K_{Ic}$ and the error bars indicate their standard deviations. For both types of filler loadings, the fracture toughness increases relative to neat epoxy as $V_f$ increases. Figure 69.4 also shows that the fracture toughness of nano-particle filled specimens increases dramatically with volume fraction relative to the micro-filler counterparts. Furthermore, the enhancement in $K_{Ic}$ is significantly higher than the micro-filler cases for any given volume fraction. For example, a ~42% and ~78% enhancement in $K_{Ic}$ values at 5% and 10% volume fraction, respectively, for nano-sized filler relative to micro-filler case is evident.

As mentioned earlier, the 2D DIC method was used to carry out dynamic fracture tests. From each experiment 64 images were available, 32 from the undeformed set, and 32 from the deformed set, each having a resolution of 1,000 × 1,000 pixels. In this work, a subimage size of 26 × 26 pixels was chosen for correlation and displacements fields were obtained as a 37 × 37 matrix of data points for each image pair. Subsequently, full-field in-plane displacement contours with 5 µm per contour interval were generated. A few selected speckle images with corresponding full-field crack-opening displacement, $v$ (displacement along $y$-axis) and sliding displacement, $u$ (displacement along $x$-axis) for nano-filler (10% $V_f$) and micro-filler

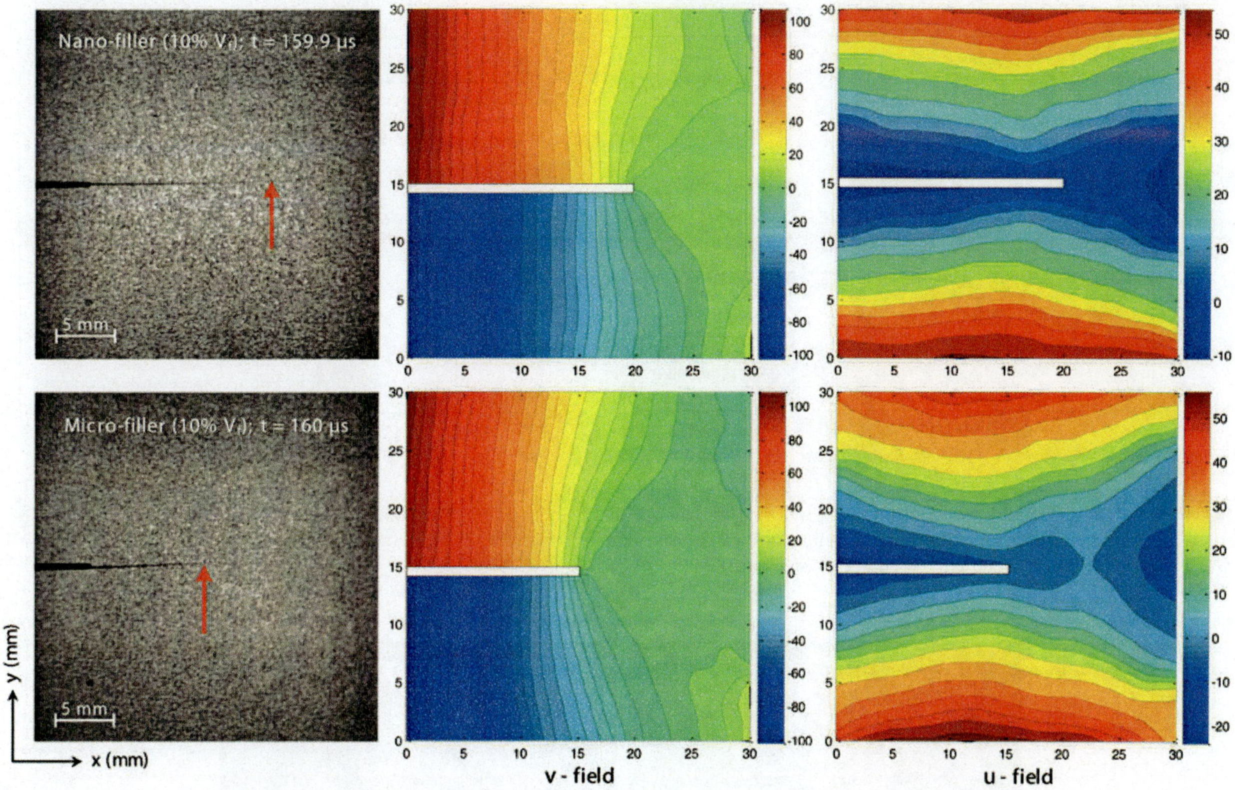


**Fig. 69.5** Acquired speckle images of 30 × 30 mm$^2$ region-of-interest at time instant, t ~160 µs with corresponding crack-opening ($v$-field) and crack-sliding ($u$-field) displacement contours. Contour interval is 5 µm. The *arrow* in speckle images indicates current crack-tip position. *Gray-scale bars* represent displacement in µm

(10% $V_f$) specimens can be seen in Fig. 69.5. It should be noted that, at a particular time instant, $t$ ~160 µs, crack extension in nano-filler specimen is larger than the micro-filler specimen. The $v$- and $u$-fields show that contour lines and magnitude of displacement (in µm shown by color-bars) are nearly symmetric relative to the crack path suggesting dominant mode-I fracture conditions.

The instantaneous values of crack extension were measured by locating the crack-tip from the digitized images and crack extension histories were used to estimate crack-tip velocity. The crack velocity histories for each type of composite at 10% $V_f$ are shown in Fig. 69.6a. It can be seen that following initiation, the crack accelerated to a maximum value and subsequently propagates at an average velocity of ~570 and ~230 m/s for nano- and micro-filler specimens, respectively. A modest oscillatory crack growth behavior due to the transient nature of impact induced fracture can be noticed. It is also worth noting that crack growth behavior (length, velocity) is consistent with speckle images shown in Fig. 69.5.

The mode-I dynamic stress intensity factors ($K_I^d$) were evaluated using an over-deterministic least-squares analysis of crack-opening displacement fields near the tip of a dynamically loaded stationary crack and for a steadily growing crack according to the methodology described in [13]. Figure 69.6b shows the mode-I SIF ($K_I^d$) histories for nano- and micro-filler composites at 10% $V_f$. The vertical broken line indicates crack initiation time as well as the corresponding $K_I^d$ values at initiation. In each case, the monotonic increase in mode-I SIF can be seen until crack initiation. At initiation the $K_I^d$ values are ~1.85 and, ~2.13 MPa m$^{1/2}$ for nano- and micro-filler specimens, respectively. After initiation, $K_I^d$ values are increasing relatively slowly in the nano-filler case than the micro-filler counterpart. It should be noted that, the micro-filler specimen shows higher $K_I^d$ values throughout the fracture event when compared to nano-filler case. This $K_I^d$ trend is consistent with crack velocity histories shown in Fig. 69.6a, as it can be seen that crack growth is the slowest in micron-size filler case suggesting higher crack growth resistance, resulting in higher apparent $K_I^d$ values and vice-versa in nano-filler counterpart.

Microscopy was performed for qualitative examination of fracture surfaces. Note that the surface features of quasi-statically fractured specimens were evident even to the naked-eye; hence, optical microscope was found suitable for imaging surface features. Figure 69.7 shows optical micrographs of fracture surfaces of quasi-statically failed nano- and micro-filler composites at 10% $V_f$. The arrow in each image indicates the initial crack front as well as the direction

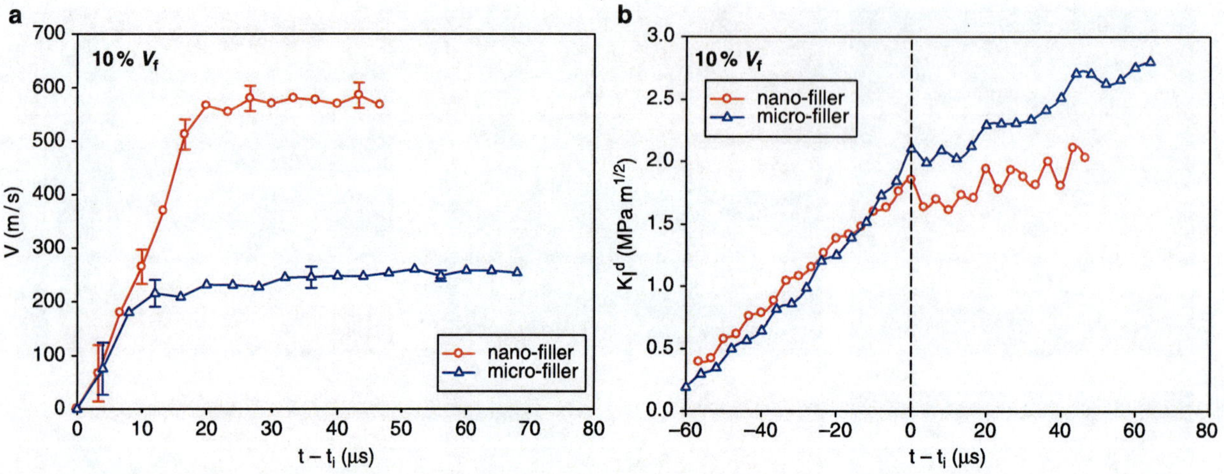


**Fig. 69.6** Dynamic fracture response of nano- and micro-filler modified epoxies at 10 % volume fraction: (**a**) crack-tip velocity histories. (**b**) Mode-I dynamic stress intensity factor ($K_I^d$) histories. (Time base is shifted such that $t - t_i = 0$ corresponds to crack initiation. The *vertical broken line* indicates crack initiation time in each specimen)

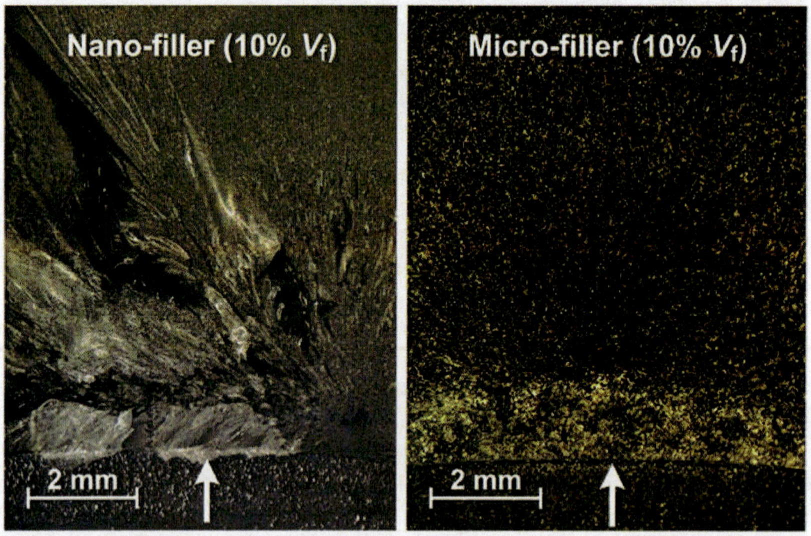


**Fig. 69.7** Optical micrographs of quasi-static fracture surfaces of nano- and micro-filler modified epoxies at 10 % $V_f$

of crack growth. These micrographs can be viewed as a global representation of surface features as the images were captured at low magnification in order to cover maximum region of fractured surfaces. The effect of particle size-scale can be clearly seen in nano- and micro-filler modified epoxies. The nanocomposites show very rough, highly textured/torn surfaces containing a large number of macro as well as micro cracks with deep valleys and furrows with steep ridges. On the other hand, these features are absent in the case of micro-filler counterpart indicating lower quasi-static fracture toughness relative to nanocomposites.

In order to examine dynamically fractured surfaces, scanning electron microscope (SEM) was used because visual inspection was insufficient as the fracture surface features were mainly confined to a very small spatial domain. Figure. 69.8a, b show SEM micrographs of dynamically fracturedsurfaces of nano- and micro-filler epoxies at 10% $V_f$, respectively. The arrow in each micrograph indicates crack propagation direction. The differences between the surface roughness features are rather striking. The fracture surface roughness and ruggedness in Fig. 69.8a for nano-particle filled epoxy is lower than the micro-filler one. Some features such as crack front bowing and pinning can be seen but are less prominent than the micro-filler epoxy. The micrograph in Fig. 69.8b reveals particle-matrix debonding and particle pullout.

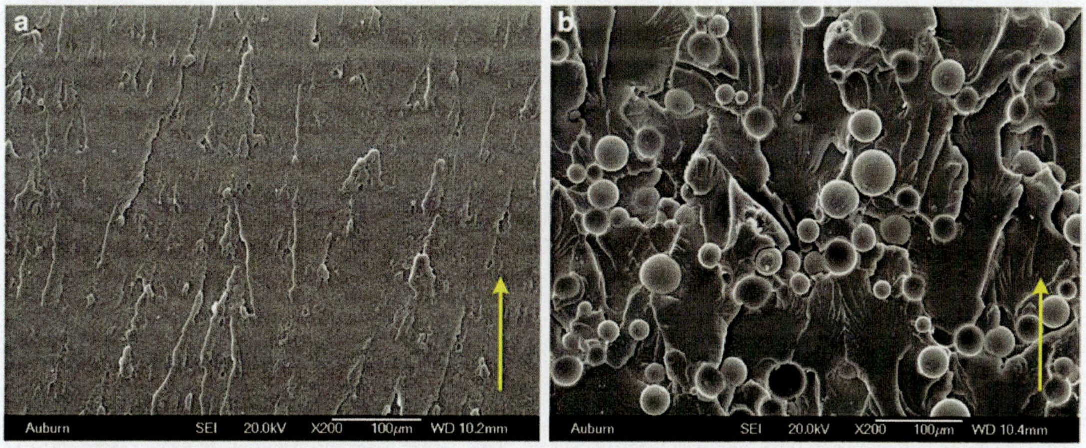


**Fig. 69.8** SEM micrographs of dynamically fractured surfaces: (**a**) nano-particle filled epoxy (10% $V_f$). (**b**) Micro-particle filled epoxy (10% $V_f$)

Thus, a propagating crack front is attracted and trapped by the debonded particle-matrix interfaces resulting in momentary deceleration and arrest [17]. It can also be noticed that the crack bows between micron-size particles indicating crack pinning. As micro-cracks leave the pinned positions, tail lines as well as step patterns in the direction of crack propagation appear.

## 69.6 Conclusions

The role of particle size-scale (nano- vs. micro-) on quasi-static and dynamic fracture behavior of particulate composites with uniformly dispersed particles in a low-viscosity, room temperature curing epoxy matrix was studied. Both nano- and micro-fillers improved fracture toughness of epoxy under quasi-static loading. The quasi-static fracture toughness of nanocomposites was significantly higher than the micro-particle filled epoxies for any given volume fraction. Nanocomposite specimens containing 10% $V_f$ filler yielded the greatest enhancement of fracture toughness (by about 78%) compared to the micro-filler counterparts at the same volume percentage. In dynamic fracture experiments, the crack propagated at significantly higher velocity in nanocomposite (~570 m/s) than micron-size filler composite (~230 m/s) case at 10% $V_f$. Contrary to the quasi-static fracture tests, the mode-I dynamic crack-initiation toughness is higher for micro-particle filled composites relative to the nano-filler counterpart at 10% $V_f$. The post-mortem analyses of fracture surfaces reveal greater surface ruggedness for nanocomposites under quasi-static conditions. However, the opposite is evident under dynamic loading conditions. The qualitative fractographic observations correlate well with the measured fracture parameters for both quasi-static and dynamic fracture tests.

**Acknowledgements** The authors gratefully acknowledge the support for this research through a grant NSF-CMMI-0653816 from the National Science Foundation, USA.

## References

1. Spanoudakis J, Young RJ (1984) Crack propagation in a glass particle-filled epoxy resin. J Mater Sci 19(2):473–486
2. Moloney AC, Kausch HH, Stieger HR (1983) The fracture of particulate-filled epoxide resins. J Mater Sci 18(1):208–216
3. Nakamura Y, Yamaguchi M, Okubo M, Matsumoto T (1992) Effects of particle size on mechanical and impact properties of epoxy resin filled with spherical silica. J Appl Poly Sci 45(7):1281–1289
4. Nakamura Y, Okabe S, Iida T (1999) Effects of particle shape, size and interfacial adhesion on the fracture strength of silica-filled epoxy resin. Anglais 7(3):177–186
5. Hussain M, Nakahira A, Nishijima S, Niihara K (1996) Fracture behavior and fracture toughness of particulate filled epoxy composites. Mater Lett 27(1–2):21–25
6. Kinloch A, Mohammed R, Taylor A, Eger C, Sprenger S, Egan D (2005) The effect of silica nano particles and rubber particles on the toughness of multiphase thermosetting epoxy polymers. J Mater Sci 40(18):5083–5086

7. Johnsen BB, Kinloch AJ, Mohammed RD, Taylor AC, Sprenger S (2007) Toughening mechanisms of nanoparticle-modified epoxy polymers. Polymer 48(2):530–541
8. Liang YL, Pearson RA (2009) Toughening mechanisms in epoxy–silica nanocomposites (ESNs). Polymer 50(20):4895–4905
9. Reynaud E, Jouen T, Gauthier C, Vigier G, Varlet J (2001) Nanofillers in polymeric matrix: a study on silica reinforced PA6. Polymer 42(21):8759–8768
10. Wetzel B, Rosso P, Haupert F, Friedrich K (2006) Epoxy nanocomposites – fracture and toughening mechanisms. Eng Fract Mech 73(16):2375–2398
11. Kitey R, Tippur HV (2005) Role of particle size and filler–matrix adhesion on dynamic fracture of glass-filled epoxyl. I. Macromeasurements. Acta Mater 53(4):1153–1165
12. Shukla A, Parameswaran V, Du Y, Evora V (2006) Dynamic crack initiation and propagation in nanocomposite materials. Rev Adv Mater Sci13(1):47–58
13. Jajam KC, Tippur HV (2012) Quasi-static and dynamic fracture behavior of particulate polymer composites: a study of nano- vs. micro-size filler and loading-rate effects. Compos Part B Eng. doi:10.1016/j.compositesb.2012.01.042, (Available online)
14. Kirugulige MS, Tippur HV, Denney TS (2007) Measurement of transient deformations using digital image correlation method and high-speed photography: application to dynamic fracture. Appl Optics 46(22):5083–5096
15. Lee D, Tippur H, Kirugulige M, Bogert P (2009) Experimental study of dynamic crack growth in unidirectional graphite/epoxy composites using digital image correlation method and high-speed photography. J Compos Mater 43(19):2081–2108
16. Jajam KC, Tippur HV (2011) An experimental investigation of dynamic crack growth past a stiff inclusion. Eng Fract Mech 78(6):1289–1305
17. Jajam KC, Tippur HV (2012) Role of inclusion stiffness and interfacial strength on dynamic matrix crack growth: an experimental study. Int J Solids Struct 49(9):1127–1146

## Chapter 70
# A New Approach to Determine the Quasi-Static and Dynamic Fracture Toughness of Engineering Materials

**Jy-An John Wang and Addis Kidane**

**Abstract** There are well developed standard techniques that can be used to determine the fracture initiation toughness of materials under quasi-static loading. One of the drawbacks on the conventional methods to determine the fracture toughness of materials, even under quasi-static loading, is the sample size requirement to get an acceptable accuracy. The ASTM standard test methods used to determine the fracture toughness of material required a certain sample thickness and volume to ensure the plan-strain condition ahead of the crack tips. This limits the use of the conventional fracture toughness test methods to a certain material that their properties are size independent. On the other hand there are no standard testing methods to determine the dynamic fracture properties of materials. There is a need to design a new technique, which avoid the sample size requirement, to determine the fracture toughness of material under quasi-static and dynamic loading. In this paper a new approach, a spiral notch sample loaded under pure torsion, is proposed to determine the fracture toughness of different engineering materials. The fracture toughness of different engineering materials has been investigated and demonstrated by the proposed approach. The proposed experimental approach is extended to the dynamic fracture toughness investigation of engineering materials.

**Keywords** Fracture intensity factor • Dynamic fracture • Spiral notched-grove sample • Torsional loading

## 70.1 Introduction

The objective of this paper is to propose a novel experimental technique for the study of the dynamic fracture properties of engineering materials for different applications. Usually, due to manufacturing imperfection and other related problem, there will be a flaw in materials that raises the local stress level and produces fracture when the stress level reaches a critical value. This critical fracture parameter can be represented in terms of stress-intensity factor (K) in different modes (mode I, mode II and mode III) depending on the loading conditions. Fracture toughness ($K_{IC}$) is a material property and determines the maximum mode-I stress intensity factor under plane-strain conditions that can be applied in the material before fracture growth initiation occur. There are different experimental techniques and specimen geometry for determining the fracture initiation toughness of materials under quasi-static loading. Although the ASTM standard techniques used to determine the quasi-static fracture toughness is well established, there is no standard experimental technique to determine the dynamic fracture toughness of materials. Usually an extended quasi-static method is used to determine the dynamic fracture of materials. For example, a tree point bend specimen along a modified split Hopkinson pressure bar apparatus is widely used to determine the dynamic fracture intensity factor of materials. For the dynamic three-point bending test, a quasi-static relation is used to estimate the dynamic fracture parameters by assuming the inertia effect is small and negligible. However, the technique only works up to a certain low strain rate level and fails to satisfy the main assumption at higher strain rate, where the inertia effect is large.

J.-A.J. Wang
Mechanical Properties and Mechanics Group, Oak Ridge National Laboratory, Oak Ridge, TN 37831-6069, USA
e-mail: wangja@ornl.gov

A. Kidane (✉)
Department of Mechanical Engineering, University of South Carolina, Columbia, SC 29208, USA
e-mail: kidanea@cec.sc.edu

V. Chalivendra et al. (eds.), *Dynamic Behavior of Materials, Volume 1: Proceedings of the 2012 Annual Conference on Experimental and Applied Mechanics*, Conference Proceedings of the Society for Experimental Mechanics Series, DOI 10.1007/978-1-4614-4238-7_70, 

On the other hand, the ASTM standard test methods used to determine the quasi-static fracture toughness of materials required a certain sample thickness and volume to ensure the plane-strain condition ahead of the crack tip. This limits the use of the conventional fracture toughness test methods only to a certain size independent materials. These techniques cannot be used in the case of materials that are geometrically unsuitable and/or have insufficient volume for fabricating the standard ASTM specimens [1–3]. Therefore, there is a need to design a new technique that can be used to determine the fracture toughness of engineering materials and eliminates the sample size requirement under quasi-static loading and avoids the inertial effect during dynamic loading.

Recently a new experimental approaches is proposed by Wang [1], using a spiral notch sample (SNTT) loaded under pure torsion [1–3], to determine the quasi-static fracture toughness of structural materials. The technique has been implemented to investigate the quasi-static fracture toughness of different materials, such as concrete, metals, polymer composites and ceramics [1–3]. In this paper, the fracture toughness obtained from the new technique is presented along with results from conventional methods. The advantage of the proposed technique over the conventional method is also been discussed. The experimental approach is extended to determine the dynamic fracture properties of engineering materials. The exceptional advantage using the spiral notch sample loaded under pure torsion for dynamic fracture analysis is, the nonexistence of inertia effect during the entire loading event, which is a major problem in the existing technique. Hence, by combining the recently developed spiral notch sample [1–3] and torsional Hopkinson bar (THBar) apparatus [4], the proposed approach is extended to determine the dynamic fracture properties of engineering materials.

## 70.2 Experimental

Unlike the conventional mode-I fracture experiment methods, where a direct opening-mode load is applied, in the proposed approach the samples are loaded under pure torsion and mode-I and Mode-II fracture toughness of the materials is investigated. The load is applied in to a cylindrical specimen that has spiral notch lines around the specimen at a 45° pitch angle. The pure torsion load creates a uniform equi-biaxial stress fields (compression and tension) on the concentric cylinder of the sample. This result in a tension load perpendicular to the grooved line which creates a mode I crack opening fields. The mixed mode fracture toughness can be investigated by the proposed approach by simply adjusting the pitch angle. The experimental procedure and the sample geometry are discussed in the next sections.

## 70.3 Material and Specimen Geometry

A spiral Notch Torsion fracture Test (SNTT) method uses a cylindrical sample having a v-notched groove line spiraling around the surface at 45° pitch angel as shown in Fig. 70.1. In this technique the size of the sample doesn't matter and only a spiral grove in ductile materials and a spiral line in brittle materials are needed. The main advantage of

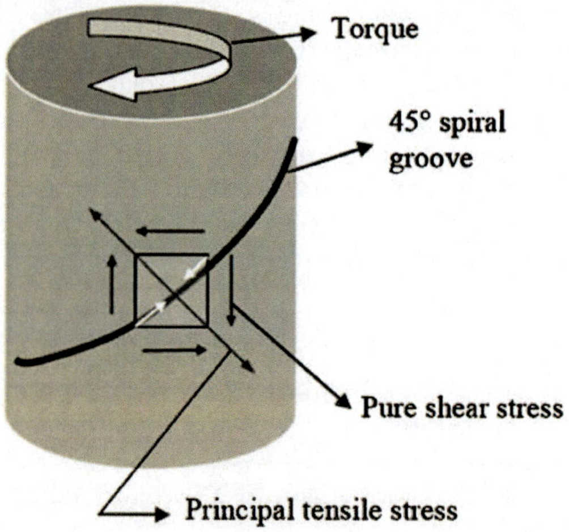


**Fig. 70.1** Typical SNTT specimen geometry and configuration

spiral sample compared to a convention sample is a plane-strain condition is achieved on every plane normal to the spiral groove. Furthermore, the crack always propagates along the spiral direction, ensures the crack only advance in the opening mode.

## 70.4 Experimental Setup

### *70.4.1 Quasi-Static*

The cylindrical specimen shown in Fig. 70.1 is attached on a closed-loop controlled, electro-hydraulic, biaxial testing system shown in Fig. 70.2. The specimen was secured between two grips with a biaxial load cell on one side of the grip and LVDT and RVDT on the other end of the grip. A biaxial extensometer is placed on the specimen where each prop of the extensometer are aligned on the other end of the groove. A rosette strain gage was also mounted on the specimen to measure the shear strain.

The torsion load was applied slowly, while the shear load and the shear strain are measured. A pure torsion was achieved with a zero axial force in control. In some material, a pre-crack was generated by cyclic torsion loading, using Haversine wave form. The maximum torque required to generate the pre-cracking depends on the material and must be determined experimentally. The size of the sample along with the torque loading determines the number of cycle required to generate the pre-crack. Different samples made of ceramic, metal and polymer was tested. A detail experimental procedure that are followed for testing SNTT samples made of different materials can be found in reference [1–3].

### *70.4.2 Dynamic Experiment*

A similar approach as discussed above is followed in the case of dynamic loading, except the loading mechanism. A detail procedure and theory of THBar can be found in the literature [4], for example [3], and here briefly discussed for brevity. The THBar consists of two elastic bars, a clamp with a sudden releasing mechanism, and a torque mechanism. The specimen is sandwiched between the bars as shown in Fig. 70.3. The torque mechanism is attached in to the end of the input bar, and the clamping mechanism is attached on the middle of the same bar. A twist is applied by the torque mechanism at the end of

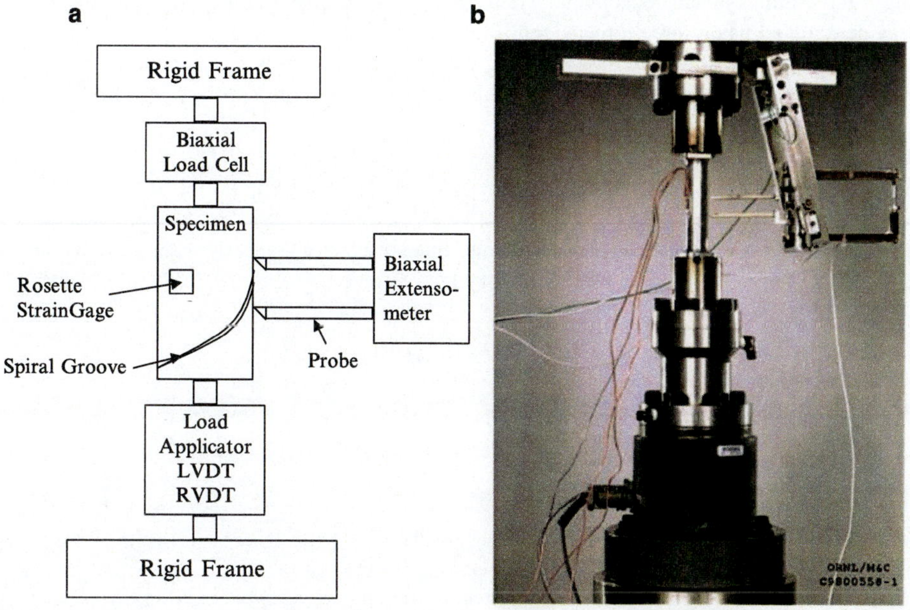


**Fig. 70.2** Spiral notch torsion test rig with specimen installed. (**a**) Schematic drawing. (**b**) Actual picture

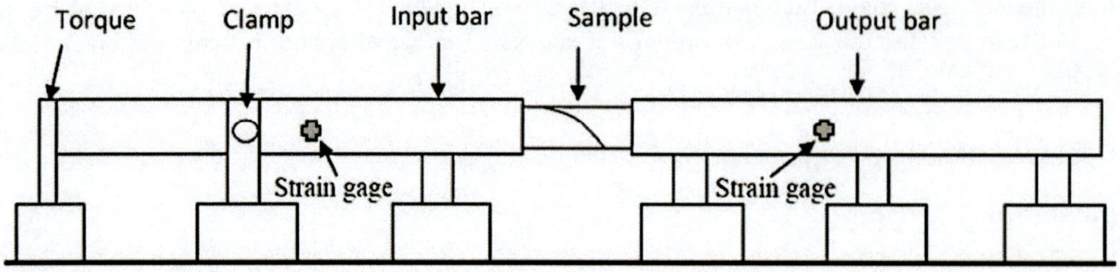


**Fig. 70.3** Schematic drawing of THBs with spiral specimen

the input bar, while the clamp in middle of the bar is used to hold the bar from rotating. This creates a stored torque on the portion of the bar, between the clamp and the torque mechanism, while the other end of the bar is unaffected. By a sudden release of the clamp, a shear stress wave is propagated towards the specimen through the unloaded part of the bar. When the wave reaches the specimen, some of the wave gets reflected from the specimen and input bar interface and the rest of the wave propagates to the output bar loading the spiral specimen in pure torsion. The strain generated in the bar due to these waves can be measured using strain gages bonded in the bars. Using one dimensional wave theory, these strains can be used to calculate the torsion force and strain on the specimen.

## 70.5 Results and Discussion

The fracture surface of SNTT samples made of different materials is shown in Fig. 70.4. As shown in the figure, the fracture originates from the pre-crack surface and propagates directly to the center and result in a pure mode I fracture.

## 70.6 Basic Theory of SNTT Configuration

During loading, a tri-axial tensile stress fields will be evolved in the neighborhood of the notch root area as shown in Fig. 70.4, creating a three-dimensional noncoplanar crack front. Due to unavailability of a closed form solution, $K_{IC}$ of SNTT method was determined using three-dimensional finite element analysis and derived from minimum strain energy density criterion. According to the strain energy density criterion, the crack growth occurs when the strain energy density reach a critical value and propagates in the direction of minimum strain energy density [5]. The three-dimensional strain energy density (S) can be written as Eq. 70.1.

$$S = a_{11}K_I^2 + 2a_{12}K_IK_{II} + a_{22}K_{II}^2 + a_{33}K_{III}^2 \tag{70.1}$$

Where $K_I$, $K_{II}$ and $K_{III}$ are mode-I, mode-II and mode-III stress intensity factor respectively and $a_{11}$, $a_{22}$ and $a_{33}$ are constants that can be determined from material property and geometry factors given by Eqs. 70.2, 70.3, 70.4 and 70.5.

$$a_{11} = \frac{1}{16\mu}[(3 - 4v - \cos\theta)(1 + \cos\theta)] \tag{70.2}$$

$$a_{22} = \frac{1}{16\mu}[4(1 - v)(1 - \cos\theta) + (3\cos\theta - 1)(1 + \cos\theta)] \tag{70.3}$$

$$a_{12} = \frac{1}{8\mu}[\sin\theta(\cos\theta - 1 + 2v)] \tag{70.4}$$

$$a_{33} = \frac{1}{4\mu} \tag{70.5}$$

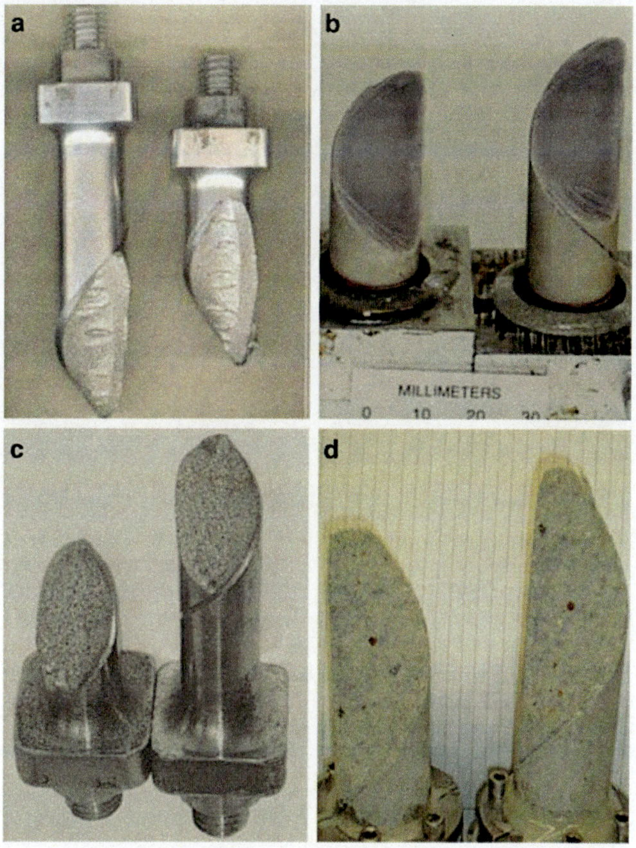


**Fig. 70.4** Typical fracture surface (**a**) 7475-T7351 aluminum. (**b**) Mullite ceramic. (**c**) A302B steel. (**d**) Mortar

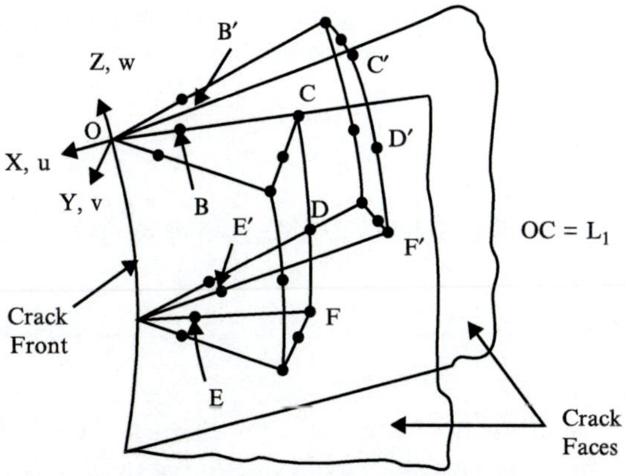


**Fig. 70.5** Singular prismatic elements [1]

In the proposed approach, the stress intensity factor $K_I$, $K_{II}$ and $K_{III}$ are evaluated from the crack tip opening displacements (CTODs) of a singular prismatic element around the crack front shown in Fig. 70.5.

The CTODs are determined from a three-dimensional finite element analysis and the stress intensity factor obtained from Eqs. 70.6, 70.7 and 70.8. The complete formation of the 3D finite element analysis and the detail fracture toughness evaluation procedure can be found from references [1–3].

**Table 70.1** $K_{IC}$ of different materials obtained from SNTT and compared with convention method

| Materials | SNTT (MPa √m) | Conventional method |
|---|---|---|
| A302B steel | 55.8 | 55.0 compact tension |
| 7475-T7351 Al | 51.3 | 51.0 compact tension |
| Mullite ceramic | 2.21 | 2.20 3-point bending test |
| Mortar | 0.360 ± 0.017 | – |

$$K_I = \frac{E}{4(1-\upsilon^2)}\left(\sqrt{\frac{\pi}{2L_1}}\right)\Big[\big(2v_B - v_C + 2v_E - v_F + v_D - 2v_{B'} + v_{C'} - 2v_{E'} + v_{F'} - v_{D'}\big). \\ + \frac{\eta}{2}(-4v_B + v_C + 4v_E - v_F + 4v_{B'} - v_{C'} - 4v_{E'} + v_{F'}) \\ + \frac{\eta^2}{2}(v_C + v_F - 2v_D - v_{C'} - v_{F'} + 2v_{D'})\Big] \tag{70.6}$$

$$K_{II} = \frac{E}{4(1-\upsilon^2)}\left(\sqrt{\frac{\pi}{2L_1}}\right)\Big[\big(2u_B - u_C + 2u_E - u_F + u_D - 2u_{B'} + u_{C'} - 2u_{E'} + u_{F'} - u_{D'}\big). \\ + \frac{\eta}{2}(-4u_B + u_C + 4u_E - u_F + 4u_{B'} - u_{C'} - 4u_{E'} + u_{F'}) \\ + \frac{\eta^2}{2}(u_C + u_F - 2u_D - u_{C'} - u_{F'} + 2u_{D'})\Big] \tag{70.7}$$

$$K_{III} = \frac{E}{4(1-\upsilon^2)}\left(\sqrt{\frac{\pi}{2L_1}}\right)\Big[\big(2w_B - w_C + 2w_E - w_F + w_D - 2w_{B'} + w_{C'} - 2w_{E'} + w_{F'} - w_{D'}\big). \\ + \frac{\eta}{2}(-4w_B + w_C + 4w_E - w_F + 4w_{B'} - w_{C'} - 4w_{E'} + w_{F'}) \\ + \frac{\eta^2}{2}(w_C + w_F - 2w_D - w_{C'} - w_{F'} + 2w_{D'})\Big] \tag{70.8}$$

Where variable *u, v* and *w* are the displacements in x, y and z axis, respectively and the subscript represent the element nodes on the crack face. $\eta = 1$ at corner node along crack front, $\mu$ is shear modulus, $\upsilon$ is Poisson's ratio and $L_1$ is singular element length.

According to the minimum strain energy density criterion, the crack growth occur when it reach a critical value. This critical value can be related to the fracture toughness as

$$S = S_{cr} = \frac{1-2v}{4\mu}K_{IC}^2 \tag{70.9}$$

From Eqs. 70.1 and 70.9, $K_{IC}$ can be evaluated as

$$K_{IC} = \sqrt{\left(\frac{4\mu}{1-2v}\right)(a_{11}K_I^2 + 2a_{12}K_IK_{II} + a_{22}K_{II}^2 + a_{33}K_{III}^2} \tag{70.10}$$

A summary of KIC value for different material found from SNTT method is shown on Table. 70.1 along with the conventional method. From Table 70.1, it can be seen that, the results from proposed technique is well agreed with the results from conventional methods.

## 70.7 Summary

A new approach to study the quasi-static and dynamic fracture toughness of material is demonstrated. The approach is based on applying a pure torsion load on a cylindrical sample with 45° spiral notch grove. In the proposed approach, a pure mode–I condition can be maintained and the plain-strain condition can be satisfied automatically, indicating the sample size required by conventional method can be avoided by using SNTT. The approach is tested and demonstrated using different materials such as, mullite ceramic, mortar, 7475-T7351 aluminum and A302B steel, and the agreement between the SNTT data and reported data from conventional method is remarkable. The SNTT system overcomes many of the limitations inherent in conventional techniques and introduces new possibilities for fracture toughness testing of unconventional materials. The approach is been extended to the dynamic fracture analysis of material, and currently the dynamic work is under progress.

## References

1. Wang J-AJ, Wright IG, Lance MJ, Liu KC (2006) A new approach for evaluating thin film interface fracture toughness. Mater Sci Eng A 426:332–345
2. Wang JA, Liu KC (2004) A new approach to evaluate fracture toughness of structural materials. J Press Ves Technol Trans ASME 126:534–540
3. Wang J-A, Liu K, Naus D (2010) A new test method for determining the fracture toughness of concrete materials. Cement Concrete Res 40:497–499
4. Naik NK, Kidane AA, Kavala VR, Veerraju Ch (2007) Interlaminar shear properties of polymer matrix composites: strain rate effect. Mech Mater 39:1043–1052
5. Sih GC (1974) Strain energy density factor applied to mixed mode crack problem. Int J Fract 10:305–321

# Chapter 71
# Experimental Investigation on Dynamic Behavior of Ultra High Performance Composite Concrete with Kolsky Bar

**Hwun Park, Weinong Chen, Bradley Martin, and William Heard**

**Abstract** Cor-Tuf is ultra high performance composite concrete consisting of pozzolanic powder, fine aggregates and steel fibers. Its compressive strength is more than 200 MPa, five times larger than conventional concrete. We conducted compressive Kolsky bar experiments on Cor-Tuf to investigate the behavior of Cor-Tuf under dynamic loading. Considering the inhomogeneity of concrete, we used two types of samples with 19 and large 76 mm diameters. To see the effect of steel fibers to bulk properties of concrete, both samples with and without steel fibers were used. We found that the strength was affected by strain rate.

**Keywords** Cor-Tuf • Kolsky bar • UHPC • Dynamic behavior

## 71.1 Extended Abstract

A Kolsky bar made with Maraging steel was employed to characterize the mechanical properties of Cor-Tuf under dynamic loadings as shown in Fig. 71.1 [1]. Uni-axial compressive stress is generated by the striker, and travels through the incident bar, the Cor-Tuf sample and the transmitted bar in order. The stress applied onto the sample is obtained from the strain gages attached on the bars.

The stress($\sigma$) and strain($\varepsilon$) of the sample are obtained from the measured strain from the gages ($\varepsilon_T$ and $\varepsilon_R$) on the bar with following equations [2].

$$\sigma = -E_o \frac{A_o}{A_s} \varepsilon_T \quad \text{and} \; \varepsilon = -\frac{2C_o}{L_s} \int \varepsilon_R dt$$

where $C_o$, $A_o$, $E_o$, $L_s$ and $A_s$ are longitudinal wave speed of the bar, cross-sectional area of the bar, elastic modulus of the bar, length of the sample and cross-sectional area of the sample respectively.

Figure 71.2a shows the comparison between transmitted stress and the sum of incident and reflected stress. They show good agreement, which means that dynamic equilibrium is achieved. Figure 71.2b represents the stress versus strain of Cor-Tuf without steel fiber. In the experiment, the strain rate is nearly constant as 140 $s^{-1}$, and the corresponding strength is 297 MPa.

Figure 71.3 summaries the strength of both types of Cor-Tuf with and without steel fiber versus strain rate. The strength decreases with strain rate regardless existence of steel fiber. Cor-Tuf with fiber shows slightly high strength.

H. Park (✉) • W. Chen
School of Aeronautics and Astronautics, Purdue University, 701 West Stadium Avenue West, Lafayette, IN 47907, USA
e-mail: hwun.park@gmail.com

B. Martin
Air Force Research Laboratory, Eglin AFB, FL 32542, USA

W. Heard
US Army Engineer Research and Development Center, 3909 Hall Ferry Road, Vicksburg, MS 39180, USA
e-mail: wchen@purdue.edu

V. Chalivendra et al. (eds.), *Dynamic Behavior of Materials, Volume 1: Proceedings of the 2012 Annual Conference on Experimental and Applied Mechanics*, Conference Proceedings of the Society for Experimental Mechanics Series, DOI 10.1007/978-1-4614-4238-7_71, © The Society for Experimental Mechanics, Inc. 2013

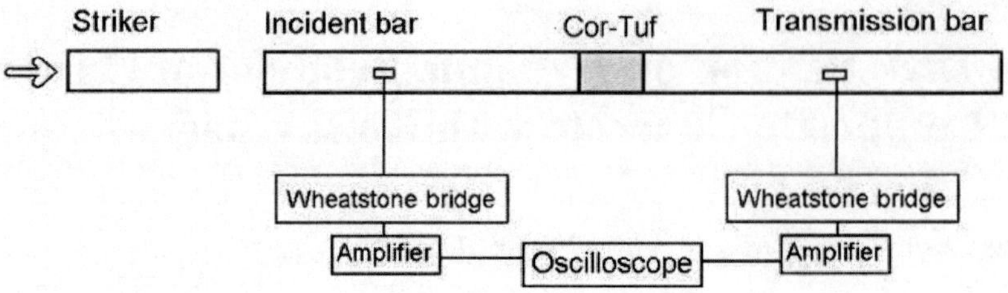


**Fig. 71.1** Configuration of Kolsky bar with Cor-Tuf sample

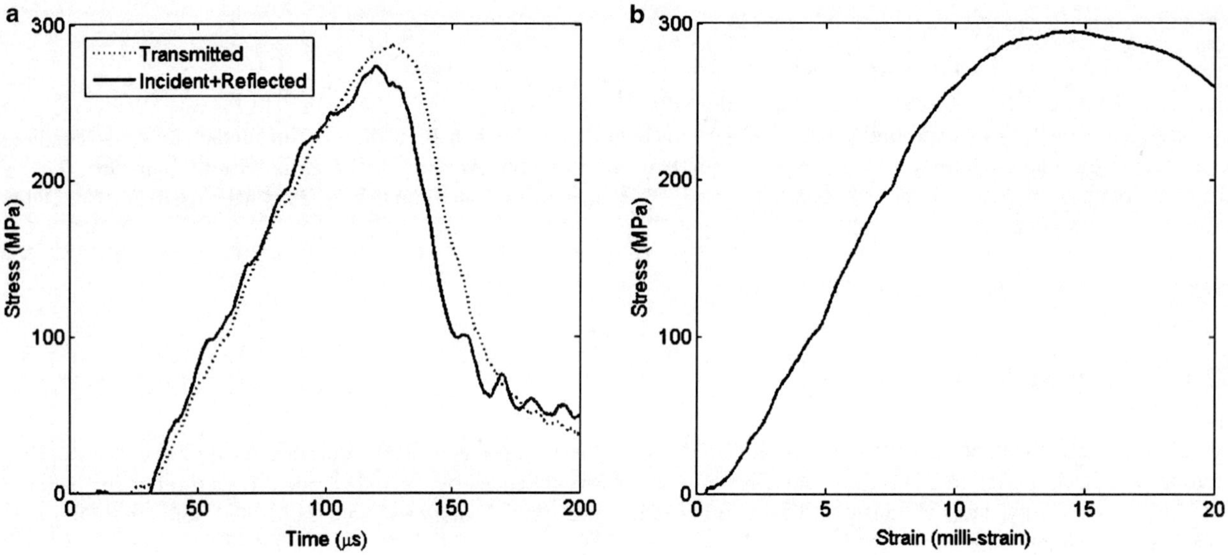


**Fig. 71.2** Dynamic equilibrium and stress versus strain. (**a**) Verification of dynamic equilibrium. (**b**) Stress versus (strain rate: 140/1/s)

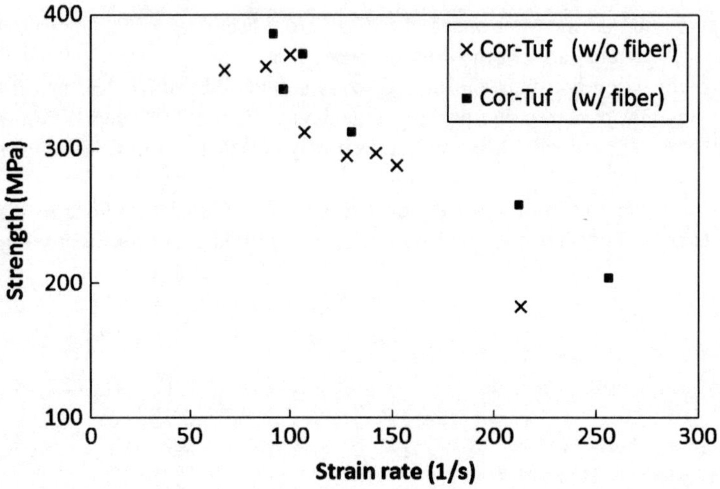


**Fig. 71.3** Strength versus strain-rate

**Acknowledgements** This research was supported by Air Force Research Laboratory at Eglin AFL, FL and Army Engineer Research and Development Center at Vicksburg, MS.

## References

1. Williams EM, Graham SS, Akers SA, Reed PA, Rushing TS (2009) Mechanical properties of a baseline UHPC with and without steel fibers. Comput Method Exp Mater Charact IV 93:104
2. Chen W, Song B (2011) Split Hopkinson (Kolsky) bar design, testing and applications. Springer, NY, USA

# Chapter 72
# Numerical Simulations of the Mechanical Behavior of Adobe

**Christopher S. Meyer**

**Abstract** Numerical simulation of geomaterials in the literature is sparse, and little data is available on modeling and simulation of high-velocity penetration of geomaterials using constitutive models built into hydrocodes. The brittle fracture kinetics (BFK) concrete material model in continuum mechanics codes has been used to simulate adobe targets; however the material parameters utilized for adobe were those of concrete with the unconfined compressive strength adjusted to match the adobe, which does not capture the pressure-dependent strength behavior of adobe.

In this work, material parameters for adobe are derived from available mechanical test data. The resulting parameters are for use in the Holmquist-Johnson-Cook (HJC) constitutive model for concrete, and this work includes an exploration of the material's pressure dependent strength behavior. The HJC model captures the pressure and strain-rate-dependent strength behavior of geomaterials reasonably well, is readily available in many continuum mechanics codes, and is commonly used to simulate high-velocity penetration. This work also presents numerical simulations of high-velocity penetration of adobe targets and compares the results to available experimental penetration data.

**Keywords** Numerical simulations • Adobe brick • Penetration

## 72.1 Introduction

The U.S. Army's interest in urban operations has led to efforts at the U.S. Army Research Laboratory to develop an initial set of material model parameters to enable physics-based penetration simulations of high-fidelity adobe brick wall models. These material model parameters take advantage of the Holmquist-Johnson-Cook (HJC) model for concrete [1].

A search of the open literature turned up very little involving experiments or modeling and simulation of weapon effects against adobe targets. While dynamic material properties for concrete are well characterized, the dynamic material properties of adobe are in the early stages of investigation. Research involving penetration and perforation experiments of adobe bricks has only recently begun to be published, for example Heine, et al. [2]. To the author's knowledge, research involving numerical simulation of penetration and perforation of adobe remains unavailable in the open literature. In early numerical simulations, adobe has been simulated using the Brittle Failure Kinetics model [3] with concrete material properties adjusted by the input of the adobe material's unconfined compressive strength (Volkmann and Moser (2011) Alliant techsystems, private communication); while an acceptable option for initial investigations, this approach does not capture the behavior of adobe in penetration and perforation simulations with an adequate degree of fidelity. The subject work seeks to develop a means of numerically simulating adobe penetration and perforation: this aim requires a constitutive material model that captures the geomaterial family of behavior, material parameters for adobe derived from experimental data, and experimental penetration data for validation of the approach. Development of a constitutive material model for adobe is beyond the scope of this work, which instead developed material parameters for the existing HJC constitutive model for concrete. Considering the pressure dependent strength behavior of geomaterials, it is assumed that the HJC concrete model, which is widely available in Lagrangian and Eulerian simulation codes and has been shown to produce reasonable residual velocity results in high-velocity concrete penetration simulations when compared with experiments [4], captures

C.S. Meyer (✉)
U.S. Army Research Laboratory, ATTN: RDRL-WML-H, Aberdeen Proving Ground, MD, USA
e-mail: christopher.s.meyer12.civ@mail.mil

V. Chalivendra et al. (eds.), *Dynamic Behavior of Materials, Volume 1: Proceedings of the 2012 Annual Conference on Experimental and Applied Mechanics*, Conference Proceedings of the Society for Experimental Mechanics Series,
DOI 10.1007/978-1-4614-4238-7_72, 

the pressure dependent behavior of geomaterials sufficiently well to simulate physics-based penetration of adobe. This work developed a complete set of HJC constitutive model material parameters for adobe from mechanical characterization data provided by the U.S. Army Corps of Engineers Engineer Research and Development Center (ERDC) [5]. Numerical simulations of select experiments carried out by Heine, et al. were performed using CTH [6] to validate the approach. To perform these simulations, the adobe material parameters developed from ERDC test data were modified with unconfined compressive strength, Young's modulus, density, and sound speed reported by Heine, et al. for sphere and rod penetration and perforation experiments.

## 72.2 Development of Material Parameters

As described in Holmquist, et al., the constitutive Eq. 72.1 describes the pressure- and strain-rate-dependent strength of the subject material; Eq. 72.1 states that the material's yield strength, when the material is confined, will increase with increasing pressure and with increasing strain rate. The constants $A$, $B$, $C$, and $N$ are material parameters determined by fitting the model to test data, $D$ is scalar damage, $P^*$ is normalized pressure and $\sigma^*$ is the normalized compressive yield strength of the material. Strength and pressure are normalized by dividing by the unconfined compressive strength, $f'_c$, of the material, which was determined from unconfined, axial compression mechanical test data provided by ERDC. Strain rate is made dimensionless by dividing the actual strain rate, $\dot{\varepsilon}$, by a reference strain rate, $\dot{\varepsilon}_o$, typically 1.0 $s^{-1}$. Damage is a value from 0 to 1 that describes the accumulation of damage as a percentage of full cohesive strength that the material possesses such that at $D = 0$ the material is undamaged and exhibits its full strength, but at $D = 1$ the material is fully damaged and retains no cohesive strength but will continue to exhibit pressure-dependent and strain-rate-dependent strength. The first term of the constitutive Eq. 72.1 describes the pressure-dependent strength behavior of the subject materials, the second term describes the strain rate effects, and a third term, not shown here, describes the temperature-dependent strength of the material; temperature effects are not explored in the present work.

$$\sigma^* = \left(A \cdot (1 - D) + B \cdot P^{*N}\right)\left(1 + C \cdot \ln\left(\frac{\dot{\varepsilon}}{\dot{\varepsilon}_o}\right)\right) \tag{72.1}$$

The pressure-dependent strength behavior of the material was fit to uniaxial strain (UX) test data and triaxial compression (TXC) test data provided by ERDC. UX data was found by loading confined cylindrical samples of material in the axial direction. Hydrostatic compression (HC) data was found by loading confined cylindrical samples of material equally in all directions to failure. TXC data was found by loading confined cylindrical samples of material hydrostatically to a preset level and then holding the hydrostatic pressure while loading the sample axially to failure. The strength fits are shown in Fig. 72.1 for the original adobe (Model 1) as well as the strengths of the parameters for rod (Model 2) and sphere (Model 3) penetrations, which were adapted from the original fit according to mechanical properties reported by Heine, et al. To determine material strength constants, first the cohesive strength constant, $A$, was determined as the ratio of the failure strength to the undamaged strength at a particular strain rate, in this case $10^{-5}$ $s^{-1}$, which was then normalized to the reference strain rate, 1.0 $s^{-1}$, by scaling up using the dynamic increase factors presented for mortar by Hao and Tarasov [7]. Mortar dynamic data was used because dynamic behavior of adobe is to date unavailable, adobe and mortar have similar compositions (sand and a binder, clay and cement respectively, are the major constituents), and it is assumed the dynamic behavior of adobe is similar to that of mortar since the static mechanical behavior of adobe and mortar are similar as seen in Fig. 72.2 [5, 8–12]. The strain rate coefficient, $C$, was also determined from strain-rate-dependent strength data for mortar published by Hao and Tarasov. To determine $C$, the material failure strength for mortar as reported in Hao and Tarasov was normalized to the average unconfined compressive strength of the tested materials and plotted versus strain rate; the slope of a linear fit through the data was used as the strain rate coefficient, $C$, for adobe. Once $A$ and $C$ were determined, the pressure hardening constant, $B$, and the pressure hardening exponent, $N$, were fit to the uniaxial test data using Eq. 72.1.

Damage is defined as the accumulation of equivalent plastic strain, $\Delta\varepsilon_p$, and plastic volumetric strain, $\Delta\mu_p$, over each time step; damage accumulation is expressed as Eq. 72.2 [1]. The model calculates damage by summing the equivalent plastic strain and plastic volumetric strain and dividing the sum by the plastic strain to fracture under a constant pressure, determined by Eq. 72.3. The plastic strain to fracture under constant pressure may be found from whichever is greater, EFMIN (minimum failure strain) or plastic strain to fracture found from Eq. 72.3. EFMIN provides the minimum plastic strain value required to cause the material to fracture; since cyclic unconfined compressive failure data is not available for these materials, the default value of 0.01 is used in this work. Plastic strain to fracture is determined by the model from two

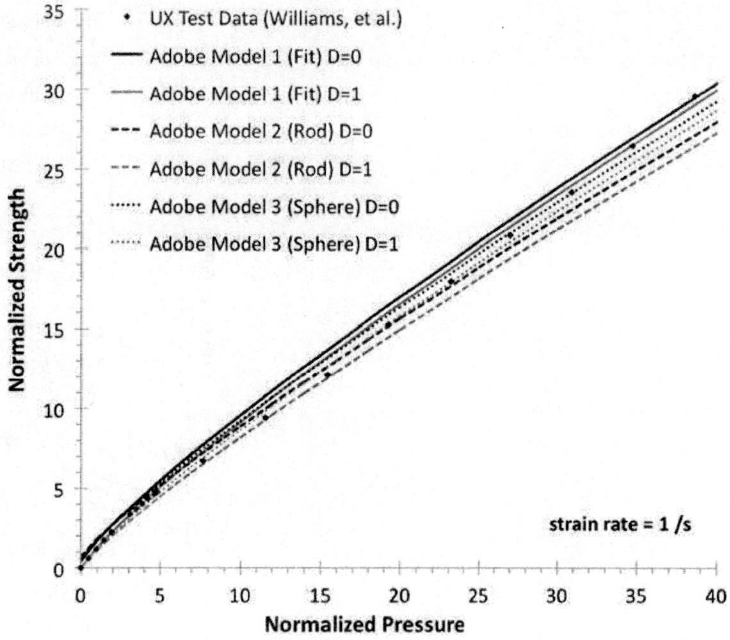


**Fig. 72.1** HJC material model strength fit for adobe

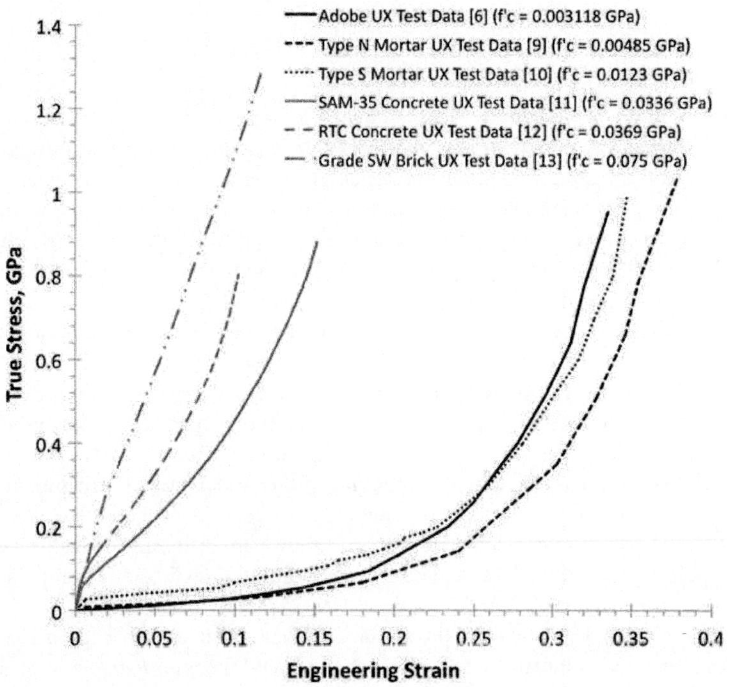


**Fig. 72.2** Stress–strain curves for various geomaterials

damage constants, *D1* and *D2*, and from the normalized pressure, *P**, and normalized tensile strength, *T**. The tensile strength is found for adobe from direct pull test data. *D1* was found from unconfined compression test data, and *D2* was assumed to be 1.0. Further study of the fracture behavior of these materials is necessary to fully understand the effect damage has on strength, particularly the assumption of linearly increasing failure strain with increasing pressure, but fracture data is presently unavailable for adobe.

$$D = \sum \frac{\Delta \varepsilon_p + \Delta \mu_p}{\varepsilon_f} \tag{72.2}$$

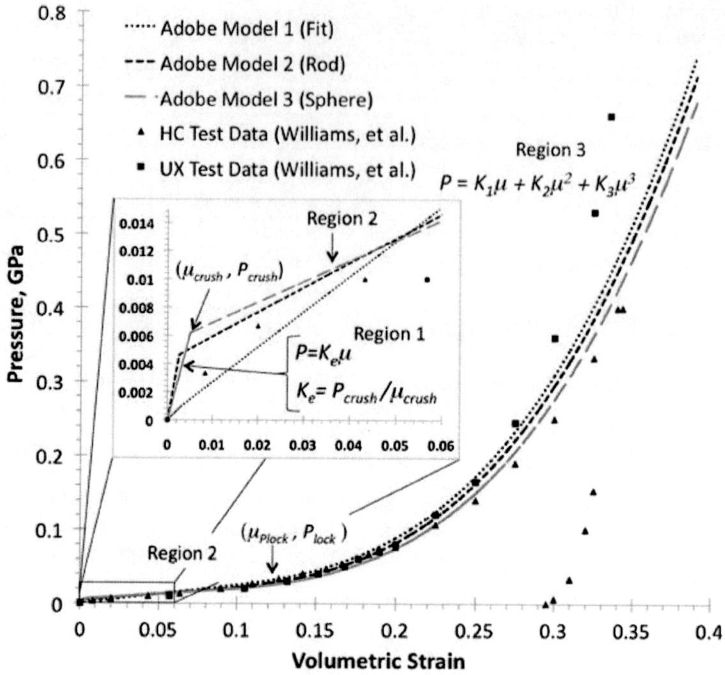


**Fig. 72.3** HJC material model pressure-volume response for adobe

$$\varepsilon_f = \mathrm{D1}(P^* + T^*)^{\mathrm{D2}} \tag{72.3}$$

Damage accumulates in geomaterials under load as the material fractures (the accumulation of equivalent plastic strain) and as the voids in the porous material compress and collapse (the accumulation of plastic volumetric strain). The majority of damage is accumulated through fracturing of the material with crushing of the material accounting for a small amount of the total damage. As damage accumulates, the scalar damage parameter from Eqs. 72.1 and 72.2, $D$, will approach a value of 1 (or 100%); as damage approaches 100%, the normalized cohesive strength term, $A$, from Eq. 72.1 approaches zero: a complete loss of cohesive strength of the material.

Holmquist et al. describes the materials' hydrostatic pressure response to volumetric strain, $\mu$, in three distinct regions: (1) elastic, (2) plastic, (3) fully compressed. The material behaves differently as pressure increases. At low pressure, less than crush pressure, $P_{crush}$, the material undergoes reversible, elastic deformation. The model predicts this elastic behavior with a simple linear relationship controlled by the user's input of parameters, $P_{crush}$ and $\mu_{crush}$. The model calculates elastic bulk modulus, $K_e$, as the crush pressure, $P_{crush}$, divided by the crush volumetric strain, $\mu_{crush}$. This region of elastic material behavior is known as Region 1; in Region 1 the model predicts pressure according to a simple linear relationship in which the pressure is equal to the applied volumetric strain multiplied by the elastic bulk modulus. Figure 72.3 illustrates the pressure-volume response of adobe; Region 1 is shown in an inset. For brittle geomaterials, the elastic region of behavior is very small. $P_{crush}$ and $\mu_{crush}$ were determined by fitting to hydrostatic compression (HC) and uniaxial strain (UX) mechanical test data shown in Fig. 72.3.

Region 2 begins and the linear elastic behavior governing Region 1 ends with parameters, $P_{crush}$ and $\mu_{crush}$, which is the pressure and volumetric strain at which the material begins to undergo plastic deformation. Permanent, plastic deformation in brittle geomaterials, which are riddled with tiny pores and air voids, may be described as micro-cracking and air void crushing as pressure and volumetric strain increase beyond $P_{crush}$ and $\mu_{crush}$. Plastic deformation (cracking) and void collapse occur in Region 2, shown in Fig. 72.3. Holmquist, et al. define Region 2 as a transition region; the model interpolates the material behavior in this region between Regions 1 and 3. The user of the model cannot directly control the predicted material behavior in this region other than by controlling the model parameters that govern behavior for Regions 1 and 3.

The onset of Region 3, shown in Fig. 72.3, is controlled by the parameters, $P_{lock}$ and $\mu_{Plock}$, which are the pressure and volumetric strain at which all air voids have been crushed out of the material. $P_{lock}$ and $\mu_{Plock}$ were determined by fitting to hydrostatic compression and uniaxial strain mechanical test data. In the HJC model, the volumetric strain, $\mu_{lock}$, defines the residual volumetric strain that the fully dense material will reduce to if unloaded. In this work, $\mu_{Plock}$ and $\mu_{lock}$, were assumed

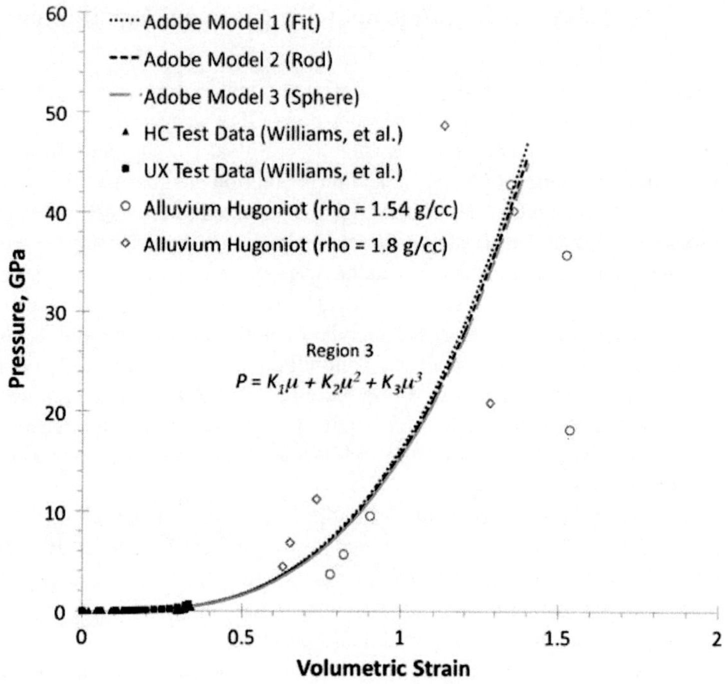


**Fig. 72.4** HJC material model high-pressure-volume response for adobe

equal as this assumption produced a significantly better low-pressure fit to the data. This affects the low-pressure unloading behavior of the material in the model such that volumetric strain returns to near zero rather than a larger residual volumetric strain that would be expected for a fully compacted material. This assumption is acceptable because the focus of this work is the simulation of high-velocity adobe penetration and, under the velocities and high-pressures produced in penetration events, the material will unload after the penetrator has passed through but the unloading will not affect a singular penetration event. However, the residual strength and pressure-volume state of the adobe will be somewhat incorrect in the area of undamaged material away from the penetration channel where there should be some retained volumetric strain. Region 3 is described in Holmquist et al. as the behavior for fully dense material, where all air voids have been crushed out of the material. In this region the material is locked and cannot compress any further in either plastic deformation or void collapse, and since it is fully densified, the material is able to undergo shock in Region 3. Since the material is under hydrostatic compression and the material is fully compressed and has nowhere to go, the pressure begins to increase dramatically for small changes in volumetric strain. The material behavior in this region is governed by Eq. 72.4, which is a fit between high-pressure hydrostatic compression data and shock hugoniot data [13]. In Eq. 72.4, pressure is a function of volumetric strain, $\mu$, and three coefficients used to fit a cubic equation to the data seen in Figs. 72.3 and 72.4. The use of modified volumetric strain described by Holmquist et al. shifts the high-pressure region back to the origin so that there is no apparent softening due to void collapse under very high pressure stimuli.

$$P = K_1\mu + K_2\mu^2 + K_3\mu^3 \tag{72.4}$$

Shock hugoniot data for alluvium from Marsh, et al. was used for the adobe high-pressure fit; alluvium exhibits an average initial density of 1.54 g/cm$^3$ (similar to the adobe density of 1.599 g/cm$^3$) and alluvium may also possess average initial density of 1.8 g/cm$^3$ (the same as the adobe density of 1.8 g/cm$^3$ that was reported by Heine, et al.). Alluvium was chosen for the similarity in densities to the adobes of interest, because like adobe it is a geological material (as opposed to other materials in Marsh, et al. that have similar densities to adobe), and because alluvium is composed of clay and sand (similar to adobe), silt and gravel [14]. At the high pressures of the shock hugoniot data, density driven inertial effects, rather than strength effects, dominate the material behavior; thus, at these pressures, the density is more important than the composition or strength of the material. Since shock hugoniot data was not available specifically for adobe, using alluvium data is assumed to be acceptable since the densities are similar; however, while these data only affect the constants for Eq. 72.4, which governs the high-pressure behavior of the adobe.

## 72.3 Numerical Simulations with Material Parameters

Numerical simulations using the material parameters for adobe were compared with true stress and engineering strain data provided by ERDC. The shock physics code CTH was used to stress a 1-cm$^3$ cube with ten uniformly distributed cells per side. Use of a flow code such as CTH is not ideal for simple deformation induced stress and strain, but time constraints necessitated the use of CTH as a near term solution as it has the HJC model readily available and the model had been in use for geomaterial simulations. Future work is expected to use EPIC [15] to test material parameters. However, CTH provided a rapid first look at the efficacy of the material parameters.

The material parameters for adobe were added to the cube and strain was applied to the cube using the prescribed deformation (PRDEF) utility in CTH. Three different mechanical tests were simulated by applying strain to the cube to produce compression against set boundary conditions. In these simulations, strain was applied in different configurations of the three directions to produce the following loading scenarios: (1) uniaxial compression (UX) – strain was applied to the cube axially while the two sides were confined, (2) hydrostatic compression (HC) – strain was applied equally in all three directions, and (3) triaxial compression (TXC) – strain was applied in hydrostatic compression to a set level, in this case 200 MPa, and then held constant while the axial strain was increased to failure. Failure strains were determined from mechanical testing.

Figure 72.5 shows the results of low-pressure uniaxial simulations compared with ERDC test data for adobe. Figure 72.6 shows the results of hydrostatic simulations compared with ERDC test data for adobe. Figure 72.7 shows the results of triaxial compression simulations compared with ERDC test data for adobe. Unfortunately, high-pressure mechanical test data was not available for adobe, however the high-pressure performance of the model parameters was evaluated by numerical simulation of penetration and perforation.

In accordance with the work by Heine, et al., numerical simulation of penetration and perforation of a simple tungsten rod penetrator and of a steel sphere were performed in two-dimensional (2D) simulations using the CTH code, and the results of these simulations are shown in Tables 72.1 and 72.2. As may be seen from these results, this method is able to predict residual velocity and depth of penetration reasonably well in 2D penetration simulations.

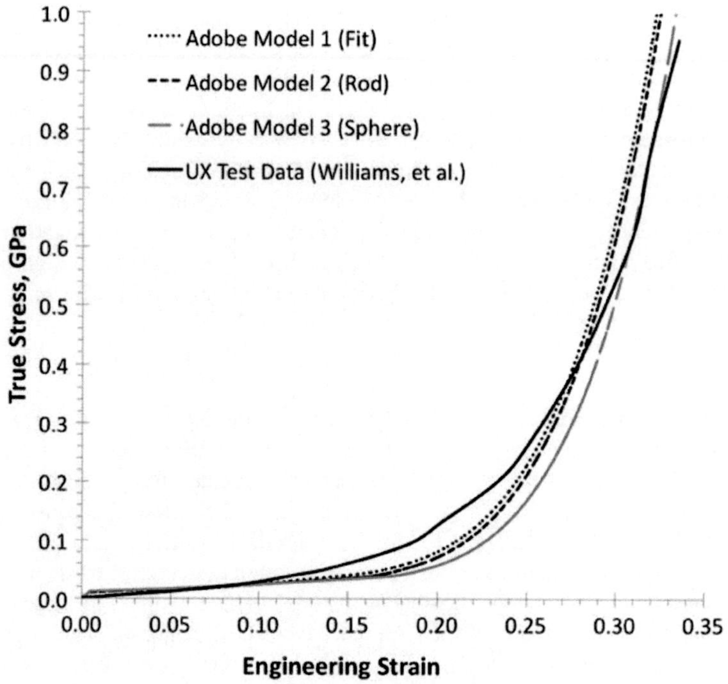


**Fig. 72.5** Uniaxial compression (UX) true stress and engineering strain test data compared with simulation results for adobe

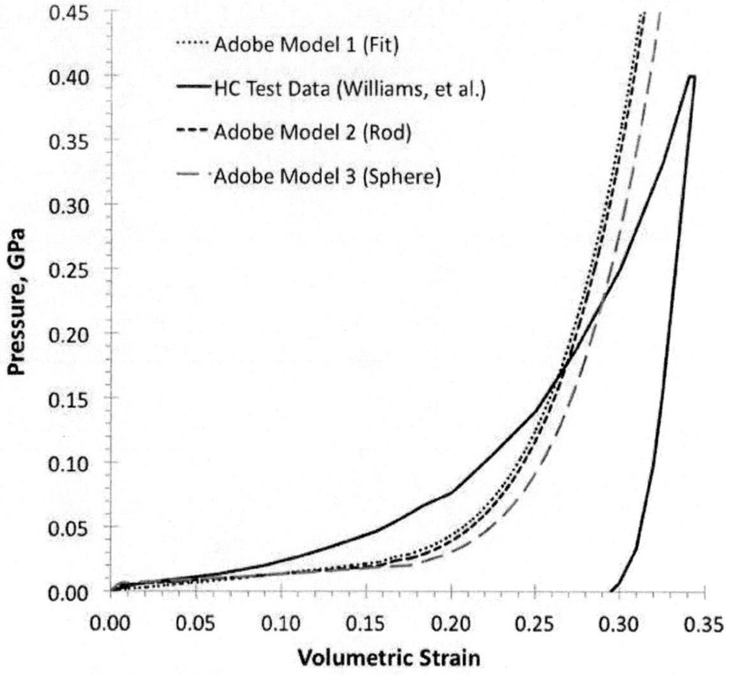


**Fig. 72.6** Hydrostatic compression (HC) pressure and volumetric strain test data compared with simulation results for adobe

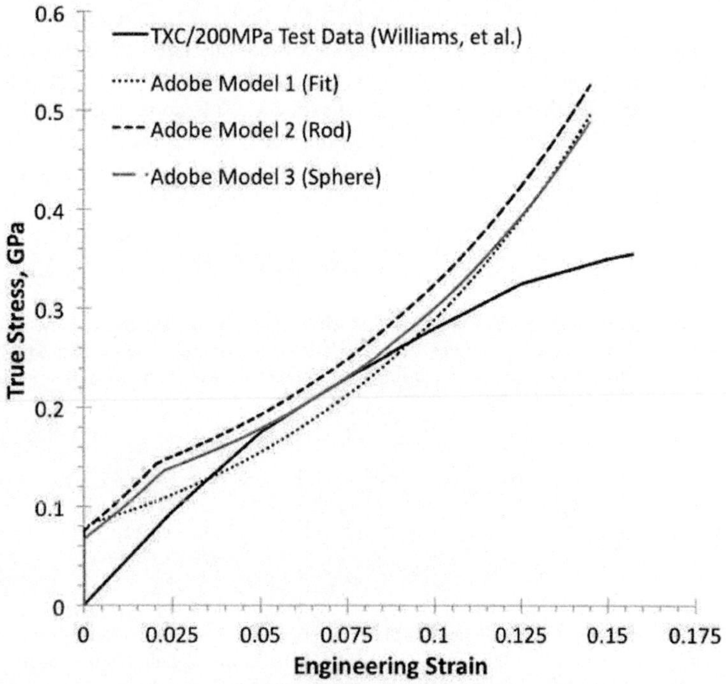


**Fig. 72.7** Triaxial compression (TXC) true stress and engineering strain test data compared with simulation results for adobe

**Table 72.1** L/D 15 chamfered tungsten rod perforation simulation results

| Target thickness (cm) | Initial velocity (m/s) | Experiment residual velocity (m/s) | Simulation residual velocity (m/s) |
|---|---|---|---|
| 7.1 | 810 | 790 | 795 |
| 14.2 | 820 | 780 | 785 |
| 35.5 | 800 | 680 | 709 |

**Table 72.2** 13.5 mm steel sphere penetration simulation results

| Initial velocity (m/s) | Experiment penetration (cm) | Simulation penetration (cm) |
|---|---|---|
| 610 | 6.1 | 8.8 |
| 900 | 9.0 | 11.7 |
| 1,200 | 12.9 | 13.6 |

## 72.4 Summary and Conclusions

Sample numerical simulations were performed to compare the model parameters to test data. Simulations included uniaxial compression, hydrostatic compression, and triaxial compression loading scenarios. Simulations also included simple penetration and perforation calculations using parameters adapted from the original fit by shifting the fit to match unconfined compressive strength and Young's modulus reported by Heine, et al. Although the HJC model has been reported to overpredict the penetration rate for high velocity impacts into concrete as well as to miss the wide variation in entry and exit hole size, it was also reported to predict residual velocity fairly well; the HJC model remains one of a limited number of tools available for use in hydrocode simulations of geomaterials, but it provides an expedient solution with minimal computational cost [16]. Since the use of a Eulerian code such as CTH is not ideal for simulations of simple strain induced deformations, additional arbitrary Lagrangian–Eulerian or additional Lagrangian simulations are recommended for further examination of these parameters. However the CTH code does provide the PRDEF subroutine for use in examining the behavior of material models in CTH simulations, and the results of the sample numerical simulations conducted here predict stress–strain relationships and penetration and perforation reasonably well when compared with experimental data. Sample numerical simulations were limited in scope due to time constraints and did not explore damage; further study of the damage behavior of these materials is recommended. This method provides a solution for modeling adobe in penetration simulations with a higher degree of fidelity than previously possible.

**Acknowledgements** Special thanks to Ms. Erin M. Williams, Dr. Stephen A. Akers, and Mr. Paul A. Reed, all of ERDC, for performing mechanical characterization of the subject materials, and for providing the data to ARL, without which this work would not have been possible. Thanks to the ARL DSRC for providing the resources on which simulations were performed. Finally, thank you Nathan Reichenbach for the technical review of this paper.

## References

1. Holmquist TJ, Johnson GR, Cook WH (1993) A computational constitutive model for concrete subjected to large strains, high strain rates, and high pressures. In: Proceedings of the 14th international symposium on ballistics, Quebec, pp 591–600
2. Heine A, Weber KE, Wickert M (2011) Experimental investigation of the penetration and perforation of building materials by projectiles. In: Proceedings of the 26th international symposium on ballistics, Miami
3. Silling SA (1997) Brittle failure kinetics model for concrete, SAND97–0439°C. Sandia National Laboratories, Albuquerque
4. Dawson A, Bless S, Levinson S, Pedersen B, Satapathy S (2008) Hypervelocity penetration of concrete. Int J Impact Eng 35:1484–1489
5. Williams EM, Akers SA, Reed PA (2008) Laboratory characterization of Adobe. U. S. Army Corps of Engineers Engineer Research and Development Center, Geotechnical and Structures Laboratory, ERDC/GSL TR-08-11
6. McGlaun JM, Thompson SL, Elrick MG (1990) CTH: a three-dimensional shock wave physics code. Int J Impact Eng 10:351–360

7. Hao H, Tarasov BG (2008) Experimental study of dynamic material properties of clay brick and mortar at different strain rates. Aust J Struct Eng 8(2):117–131
8. Williams EM, Akers SA, Reed PA (2009) Laboratory characterization of type N Mortar. U. S. Army Corps of Engineers Engineer Research and Development Center, Geotechnical and Structures Laboratory, ERDC/GSL TR-09-5
9. Williams EM, Akers SA, Reed PA (2008) Laboratory characterization of type S Mortar. U. S. Army Corps of Engineers Engineer Research and Development Center, Geotechnical and Structures Laboratory, ERDC/GSL TR-08-10
10. Williams EM, Akers SA, Reed PA (2006) Laboratory characterization of SAM-35 concrete. U. S. Army Corps of Engineers Engineer Research and Development Center, Geotechnical and Structures Laboratory, ERDC/GSL TR-06-15
11. RTC concrete data provided by Akers SA, U. S. Army Corps of Engineers Engineer Research and Development Center, Geotechnical and Structures Laboratory, 2011
12. Williams EM, Akers SA, Reed PA (2007) Laboratory characterization of solid grade SW Brick. U. S. Army Corps of Engineers Engineer Research and Development Center, Geotechnical and Structures Laboratory, ERDC/GSL TR-07-24
13. Marsh SP (ed) (1980) LASL shock Hugoniot data. University of California Press, Berkeley
14. The American Heritage Science Dictionary (2005) Alluvium. Houghton Mifflin, Boston
15. Johnson GR, Beissel SR, Stryk RA, Gerlach CA, Holmquist TJ (2006) User instructions for the 2006 version of the EPIC code, U. S. Army Research Laboratory
16. Frank AO, Adley MD (2007) On the importance of a three-invariant model for simulating the perforation of concrete slabs. In: Proceedings of the 78th shock and vibration symposium, Philadelphia

# Chapter 73
# Influence of the Confined Behaviour and the Tensile Strength of Concrete Slabs Under Projectile-Impact

**L. Sallier and P. Forquin**

**Abstract** Severe damage modes are observed in concrete structures when subjected to impulse loading as projectile-impact or blast loading. For instance, the impact of a kinetic penetrator generates scabbing on the front face, radial cracking in the whole target and, for high impact velocities or a thin target, spalling on the rear face. These damage mechanisms may strongly decrease the penetration resistance of concrete structures. On the other hand, high confining pressures in front of the projectile may lead to a micro-cracking and collapse of pores in the vicinity of the tunnel. Moreover both confined and tensile behaviours are influenced by humidity level of the concrete. Therefore the numerical modelling of concrete slabs under projectile-impact relies on the use of coupled modelling in which the hydrostatic and deviatoric behaviours of the concrete under high pressure is modelled in addition to the anisotropic damage induced in dynamic tensile loading. In the present work, a series of numerical simulations of impact tests have been conducted considering different humidity levels and target thicknesses. The confined behaviour of the concrete material is modelled through the KST (Krieg-Sweenson-Taylor) plasticity model. The so-called DFH (Denoual-Forquin-Hild) damage model combined with a cohesion model is used to describe the tensile strength and softening behaviour of concrete in tension. A series of computations have been performed to assess the influence of the confined behaviour and tensile strength on the ballistic performance of concrete slabs.

**Keywords** Impact • Concrete • Damage Model • Modelling

## 73.1 Introduction

In the last decades several studies have been dedicated to the ballistic performances of concrete targets subjected to the impact of a kinetic penetrator [1]. Rapidly, closed-form solutions have shown their limit because of the complexity of damage modes involved during such impact loading. Indeed, scabbing on the front face, radial cracking around the penetration tunnel and spalling fracture in the vicinity of the rear face may have a huge influence on the ballistic performances of the target [2]. These damage modes are mainly due to tensile loadings. Moreover, in front of the projectile a confined triaxial stress state develops [2]. This is the reason why a coupled modelling that takes into account the hydrostatic and deviatoric behaviours of concrete under confining pressure and the anisotropic damage induced by tensile loading is needed to simulate numerically such kinds of dynamic loading. In the present work, numerical simulations of impact tests against concrete slabs have been conducted with a coupled plasticity-anisotropic damage model. The KST (Krieg-Swenson-Taylor) [3, 4] plasticity model is used to account for the confined behaviour of concrete. On the other hand, the DFH (Denoual-Forquin-Hild) [5, 6] damage model combined with a cohesion law allows a description of the dynamic strength and softening behaviour of concrete in tension. First, experimental data provided by quasi-oedometric compression tests and spalling tests performed on the so-called "R30A7" common concrete have been used to identify the model parameters. Next, two impact configurations involving different impact velocities and target thicknesses have been simulated with the Abaqus-explicit commercial code. These computations allow assessing the role of free water on the confined behaviour and tensile strength that influence the ballistic performance of concrete slabs.

L. Sallier • P. Forquin (✉)
LEM3 laboratory, Lorraine University, Ile du Saulcy, Metz 57045, France
e-mail: pascal.forquin@univ-metz.fr

V. Chalivendra et al. (eds.), *Dynamic Behavior of Materials, Volume 1: Proceedings of the 2012 Annual Conference on Experimental and Applied Mechanics*, Conference Proceedings of the Society for Experimental Mechanics Series, DOI 10.1007/978-1-4614-4238-7_73, © The Society for Experimental Mechanics, Inc. 2013

## 73.2 The Coupled DFHcoh-KST Model

The coupled DFHcoh-KST model has been used to simulate the behaviour of R30A7 concrete under impact. On the one hand, the KST plasticity model provides a description of the hydrostatic and deviatoric behaviour under high confining pressure [7]. It includes an equation of state linking the volumetric strain to the hydrostatic pressure. This equation of state is a piecewise linear curve entered point by point. Moreover, the model accounts for the increase of deviatoric strength ($\sigma_{eq}$) of concrete with confining pressure (P).

$$\sigma_{eq} = \begin{cases} min\left(\sigma_{eq}^{max}, \left|a_0 + a_1 P + a_2 P^2\right|\right) \\ max\left(\sigma_{eq}^{min}, \left|a_0 + a_1 P + a_2 P^2\right|\right) \end{cases} \tag{73.1}$$

where $a_0, a_1, a_2, \sigma_{eq}^{max}$ are parameters identified from quasi-oedometric compression tests performed with dry and wet R30A7 samples [8]. In particular $\sigma_{eq}^{max}$ allows to reproduce the limited strength observed in water saturated concrete under high hydrostatic pressure [9]. The KST model parameters are gathered in Table 73.1. On the other hand, the DFHcoh model is used to describe the dynamic tensile behaviour of R30A7 concrete. In this model the macroscopic strength is expressed as:

$$\Sigma = (1 - P_O)\sigma + (P_O)^{\alpha_D}\sigma^{coh}(\varepsilon) \tag{73.2}$$

where $P_O$ denotes the obscuration probability. In non-obscured zones the microscopic stress is used whereas a cohesion strength is considered in obscured zones:

$$\sigma^{coh} = \sigma_O^d Exp\left(-\left(\frac{\varepsilon}{\varepsilon_0^d}\right)^{n_d}\right) + \sigma_O^w Exp\left(-\left(\frac{\varepsilon}{\varepsilon_0^w}\right)^{n_w}\right) \tag{73.3}$$

First part of Eq. 73.3 provides the cohesion strength of the dry concrete whereas the second part describes the additional strength provided by the free water. The parameters of this cohesion law $\sigma_O{}^d, \sigma_O{}^w, n_d, n_w, \varepsilon_0{}^d, \varepsilon_0{}^w$ were identified by comparing the experimental data of spalling tests to the results given by numerical simulation. These parameters are gathered in Table 73.1.

**Table 73.1** DFHcoh-KST parameters for R30A7 concrete

| | Wet concrete | Dry concrete |
|---|---|---|
| Density (kg/m$^3$) | 2,390 | 2,290 |
| Poisson's ratio | 0.2 | 0.16 |
| Young modulus (GPa) | 42 | 32 |
| $K_{final}$ (Pa) | $2.3 \cdot 10^{10}$ | $2.3 \cdot 10^{10}$ |
| {$\varepsilon_i$; $P_i$ (MPa)} | (0; 0) | (0; 0) |
| | (−0.0017; 30) | (−0.0025; 30) |
| | (−0.025; 100) | (−0.028; 150) |
| | (−0.045; 195) | (−0.050; 230) |
| | (−0.055; 300) | (−0.072; 350) |
| | (−0.065; 500) | (−0.086; 450) |
| | (−0.073; 700) | (−0.107; 700) |
| | (−0.080; 900) | (−0.120; 950) |
| $a_0$ (Pa$^2$) | $1.61 \cdot 10^{15}$ | $1.61 \cdot 10^{15}$ |
| $a_1$ (Pa) | $9.8 \cdot 10^7$ | $9.8 \cdot 10^7$ |
| $a_2$ | 0.54 | 0.54 |
| $\sigma_{max}^{eq}$ (MPa) | 210 | – |
| $\varepsilon_0$ (s$^{-1}$) | $10^{-5}$ | $10^{-5}$ |
| $n_{QS}$ | 0.05 | 0.01 |
| $\sigma_0^d$ (MPa) | 11 | 8 |
| $\varepsilon_0^d$ | $4 \cdot 10^{-3}$ | $2 \cdot 10^{-3}$ |
| $n^d$ | 1 | 1 |
| $\sigma_0^w$ (MPa) | 4 | 0 |
| $\varepsilon_0^w$ | $1 \cdot 10^{-3}$ | – |
| $n^w$ | 1 | – |

## 73.3 Numerical Simulation of Two Impact Configurations

Two impact configurations have been considered in the present study. In the first one, a kinetic penetrator hits a concrete slab 300 mm thick at 333 m/s. In the second configuration a concrete slab 800 m thick is impacted at 227 m/s. The projectile, 2.44 Kg, ogival-nose (3CRH), is 52 mm in diameter and 300 mm length.

Each configuration has been simulated considering the parameters of dry and wet R30A7 concrete. Projectile deceleration profiles are represented in Fig. 73.1 for the thinnest target (300 mm). On the one hand, a higher deceleration level is noted with the dry target in the first 400 μs. Conversely, after 500 μs, a greater deceleration is observed in the case of a wet target. The confined behaviour of the target is supposed to drive the deceleration of the projectile at the beginning of impact. Nevertheless, according to damage fields, a fracturing develops toward the rear face after 500 μs leading to a decrease of the deceleration level specially in the case of dry concrete. Finally, the dry target exhibits a more intense damage than the wet target (Fig. 73.2).

The deceleration curves of the second configuration (thickness: 800 mm, impact velocity: 227 m/s) are presented in Fig. 73.3 considering again the parameters of a wet and dry target. On the one hand, a higher deceleration level is noted with the dry target in the first 300 μs. On the second one a higher deceleration is noted in the case of a wet target as it was observed with the target 300 mm thick. The Fig. 73.4 shows the damage fields at the end of impact for the second configuration. Finally, as for thinner target (300 mm) dry concrete exhibits a more intense damage and more radial cracks than the wet target (Fig. 73.4).

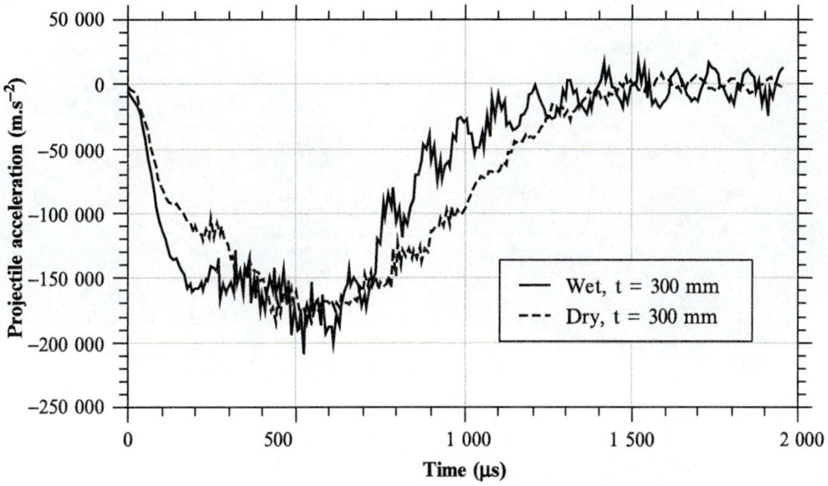


**Fig. 73.1** Projectile acceleration profiles versus time for a target 300mm thick. $V_{impact} = 333$ m/s

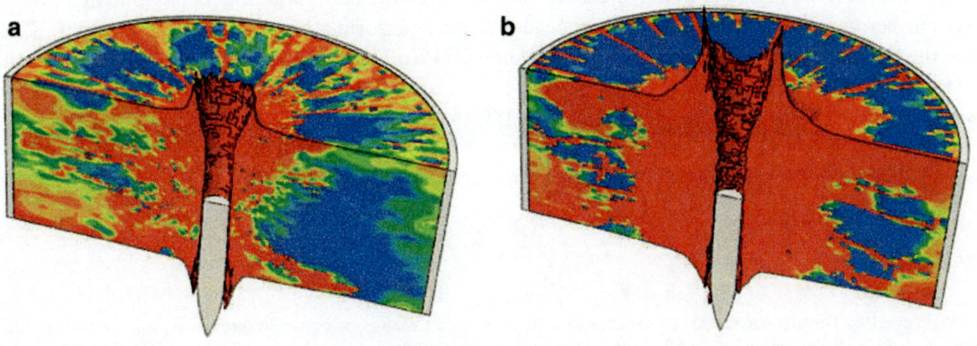


**Fig. 73.2** Damage field for a target 300 mm thick: (**a**) wet target, (**b**) dry target. $V_{impact} = 333$ m/s

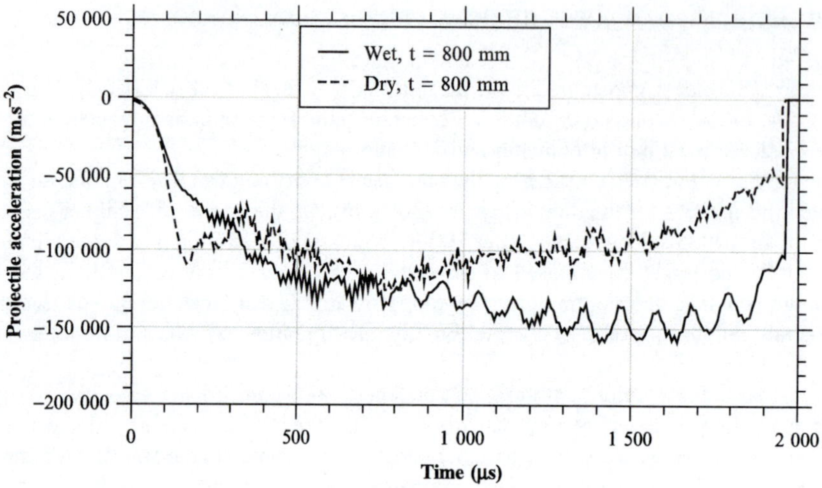


**Fig. 73.3** Projectile acceleration profiles versus time for a target 800 mm thick. $V_{impact} = 227$ m/s

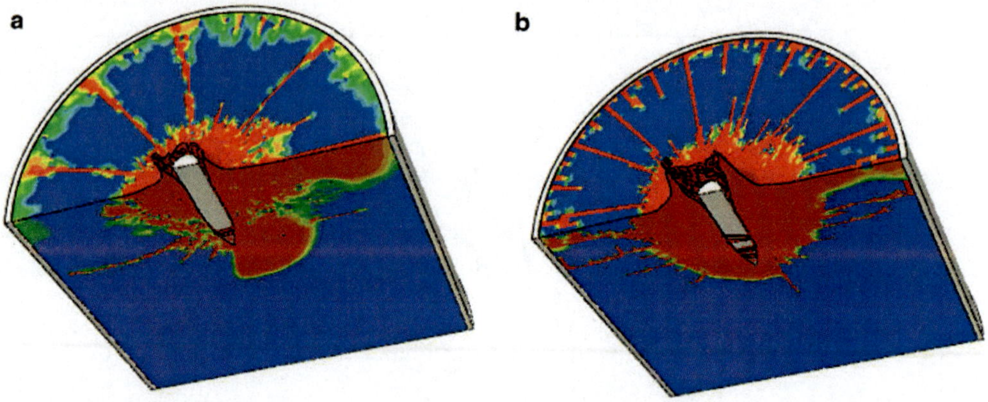


**Fig. 73.4** Damage field for a target 800 mm thick: (**a**) wet target, (**b**) dry target. $V_{impact} = 227$ m/s

## 73.4 Conclusion

To conclude the behaviour in high confining pressure, tensile strength and softening behaviour are influenced by the humidity level. Furthermore both behaviours influence the impact performance of the concrete target. This is why the consideration of the humidity level is needed to complete numerical simulations of projectile-impact against concrete slabs.

**Acknowledgements** This work was supported by the CEA, DAM, GRAMAT. This support is gratefully acknowledged.

## References

1. Li QM, Reid SR, Wen HM, Telford AR (2005) Local impact effects of hard missiles on concrete targets. Int J Impact Eng 32:224–284
2. Beppu M, Miwa K, Itoh M, Katayama M, Ohno T (2008) Damage evaluation of concrete plates by high-velocity impact. Int J Impact Eng 35 (12):1419–1426

3. Swenson DV, Taylor LM (1983) A finite element model for the analysis of tailored pulse stimulation of boreholes. Int J Numer Anal Meth Geomech 7:469–48
4. Krieg RD (1978) A simple constitutive description for soils and crushable foams. Report, SC-DR-7260883, Sandia National Laboratory
5. Denoual C, Hild F (2000) A damage model for the dynamic fragmentation of brittle solids. Comp Meth Appl Mech Eng 183:247–258
6. Forquin P, Hild F (2010) A probabilistic damage model of the dynamic fragmentation process in brittle materials. Adv Appl Mech 44:1–72. Academic Press, San Diego
7. Forquin P, Arias A, Zaera R (2008) Role of porosity in controlling the mechanical and impact behaviours of cement-based materials. Int J Impact Eng 35(3):133–146
8. Erzar B, Forquin P (2011) Experiments and mesoscopic modelling of dynamic testing of concrete. Mech Mater 43(9):505–527
9. Forquin P, Safa K, Gary G (2010) Influence of free water on the quasi-static and dynamic strength of concrete in confined compression tests. Cement Concrete Res 40(2):321–333

# Chapter 74
# Temperature- and Rate-Dependent Failure in Dynamically Loaded Viscoplastic Materials

**Patrice Longère, Anne-Gaëlle Geffroy, Bruno Leblé, and André Dragon**

**Abstract** This work aims at studying experimentally and reproducing numerically the failure mechanisms of a ship structure constitutive material when submitted to airblast loading. With this aim in view, a physically motivated approach has been developed and applied in order to describe the transition of behaviour between dense metal plasticity and micro-porous metal plasticity in the context of dynamic plasticity and failure. The viscoplastic material is supposed to be initially exempt of micro-voids. Subjected to a monotonic loading involving a positive or null stress triaxiality, it behaves elastic-(visco) plastically. As soon as the conditions of plastic strain, plastic strain rate, temperature and stress triaxiality for the germination of a given volume fraction of voids are satisfied, the material behaviour becomes pressure dependent. Its damage-plastic yielding is consequently described using a micro-porous metal plasticity potential. The latter is proposed in the form of a modified version of the Gurson-Tvergaard-Needleman model allowing for describing cavity growth and further fracture under shear loading. The 3D constitutive equations are implemented as user material in the engineering finite element computation code Abaqus®. Numerical simulations are conducted and compared with experiments considering airblast loaded ship structures. The numerical results show clearly the influence of the hole nucleation criterion related constants and of the loading conditions (temperature, strain rate) on the damage and further fracture of the material.

**Keywords** Dynamics • Air-blast • Viscoplasticity • Ductile fracture • Numerical simulation

## 74.1 Introduction

Several authors have attempted to describe the consequences of ductile damage on the bulk material behaviour. These consequences are double: a progressive loss of the overall properties of the bulk material and the appearance, in addition to the isochoric plastic deformation due to dislocation glide in the matrix material, of an inelastic dilatancy due to void growth. Based on a micromechanical analysis, Gurson [1] developed a plastic potential for slightly porous and perfectly plastic metals accounting explicitly for the concentration of voids and hydrostatic stress. Tvergaard and Needleman [2] modified Gurson's model in order to take notably into account isotropic strain hardening and strain rate effects, see also [3]. The so-called GTN (for Gurson-Tvergaard-Needleman) model is widely used by the community of researchers dealing with ductile damage. Using the rate type formulation coupled with physical concepts, Perzyna [4] developed an elliptic plastic potential taking into consideration cooperative effects of void growth, strain rate sensitivity and heating. Finally, relying on Rice and

P. Longère (✉)
Université de Toulouse, Institut Supérieur de l'Aéronautique et de l'Espace – Institut Clément Ader (EA 814), 10 avenue E. Belin, BP 54032, Toulouse cedex 4 31055, France
e-mail: patrice.longere@isae.fr

A.-G. Geffroy • B. Leblé
DCNS NA Ingénierie, Rue de Choiseul, Lorient 56311, France
e-mail: anne-gaelle.geffroy@dcnsgroup.com; bruno.leble@dcnsgroup.com

A. Dragon
CNRS-Institut Pprime (UPR 3346), Ecole Nationale Supérieure de Mécanique et d'Aérotechnique – Université de Poitiers, 1 avenue C. Ader, BP 40109, Futuroscope – Chasseneuil du Poitou 86961, France
e-mail: andre.dragon@ensma.fr

V. Chalivendra et al. (eds.), *Dynamic Behavior of Materials, Volume 1: Proceedings of the 2012 Annual Conference on Experimental and Applied Mechanics*, Conference Proceedings of the Society for Experimental Mechanics Series, DOI 10.1007/978-1-4614-4238-7_74, © The Society for Experimental Mechanics, Inc. 2013

Tracey [5] results, Rousselier [6] established a plastic potential from the internal variable procedure in the context of thermodynamics of irreversible processes. These approaches all apply to micro-porous metals, i.e. all suppose initially the presence of micro voids, or equivalently assume mostly that void expansion incipience and plastic deformation occurrence are concomitant. Independently of the practical character of this assumption from the modelling viewpoint, it is clear that such a hypothesis is not supported physically, because cavity nucleation requires a certain amount of plastic deformation under moderate to low stress triaxiality. This is particularly true for materials which may be considered as pure, i.e. containing a very low concentration of inclusions, and/or for which void nucleation results from cavitation at grain boundary. It must be noted that, whereas the attempts of modelling the void growth and coalescence are numerous in literature, see e.g. [7] for the first item and [8] for the second one, a satisfying modelling of the transition from dense metal plasticity to micro-porous metal plasticity deserves still deeper investigation. To face up to deficiencies concerning notably the lack of a physically satisfying description of the transition from dense metal plasticity to micro-porous metal plasticity, we are considering a solid material which does not contain initially any voids, i.e. which is not initially micro-porous. Hole nucleation under a positive or null stress triaxiality is supposed to require a certain amount of plastic deformation, as suggested in [9]. In a way consistent with the present modelling approach, the magnitude of the critical plastic strain at hole nucleation depends on strain rate, temperature and stress triaxiality.

The approach is here directly applied to a mild steel for which the micro-voids have been shown to appear at an advanced stage of plastic deformation, i.e. in the necking zone of a cylindrical sample or in front of the macro-crack tip of a plate sample, both submitted to a remote uniaxial tensile loading. This work indeed aims at studying experimentally and reproducing numerically the failure mechanisms of a ship structure constitutive material when submitted to airblast loading. The high-purity, ferritic-pearlitic mild steel considered here is employed as structural material in panels of naval structures. To ensure integrity of panels regarding extreme loading conditions, the strain-rate as well as the temperature effects are accounted for. This application context is thus expected to involve a large strain-rate spectrum, from moderate to high strain-rate levels. For the latter case, the adiabatic conditions are assumed further. The 3D constitutive equations are implemented as user material in the engineering finite element computation code Abaqus®. Numerical simulations are conducted and compared with experiments considering airblast loaded ship structures.

## 74.2 Experimental Results

The material of the present work is a ferritic-pearlitic mild steel (DH36) widely used in the automotive and railway industries as well as in shipbuilding. An important experimental campaign, including laboratory thermomechanical tests as well as airblast experiments allowed for characterizing the behaviour of a ship structure constitutive material under high strain rate conditions. The mechanisms of damage and cracking which occur during the deformation processes until the ultimate failure were identified from micrographic analyses of partially and totally fractured specimens.

### *74.2.1 Thermomechanical Characterisation*

Quasi-static experiments were performed on laboratory specimens, at strain rates ranging from $10^{-4}$ s$^{-1}$ to $10^{-2}$ s$^{-1}$ for the tension tests, and from $10^{-3}$ s$^{-1}$ to $10^{-1}$ s$^{-1}$ for the compression tests, see Fig. 74.1a. In order to better quantify the strain rate effects on the material behaviour, some strain rate jump tests were also carried out at room temperature. The same procedure was followed to quantify the influence of the initial temperature, where tests with temperature jumps were performed at low equivalent plastic strain rate $\dot{\kappa}$, see Fig. 74.1b. Finally, dynamic compression and tension tests have been realised at room temperature using a split Hopkinson pressure bar (SHPB) device, see Fig. 74.1a. The former were carried out at 450 s$^{-1}$ and 3,500 s$^{-1}$ and the latter at 1,500 s$^{-1}$, leading to the determination of the effects of high strain rate loading on the material behaviour. Only the most significant experimental results are shown and commented in the following.

### *74.2.2 Airblast Experiments*

To qualify the material under consideration for explosion loading, airblast experiments were carried out. The samples were machined in the form of square thin plates. The steel plate was held down by two frames fixed to the underlying structure.

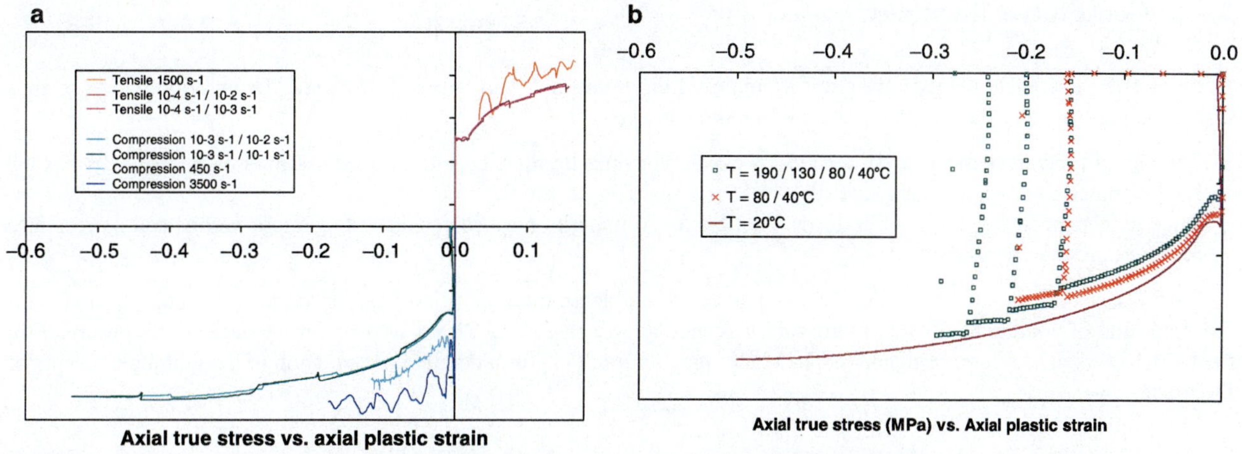


**Fig. 74.1** Experimental results on smooth specimens. (**a**) Strain rate effects in tension and compression-$T_0 = 20°C$. (**b**) Temperature effects in compression. Monotonic and temperature jump tests $-\dot{\kappa} = 10^{-3} s^{-1}$

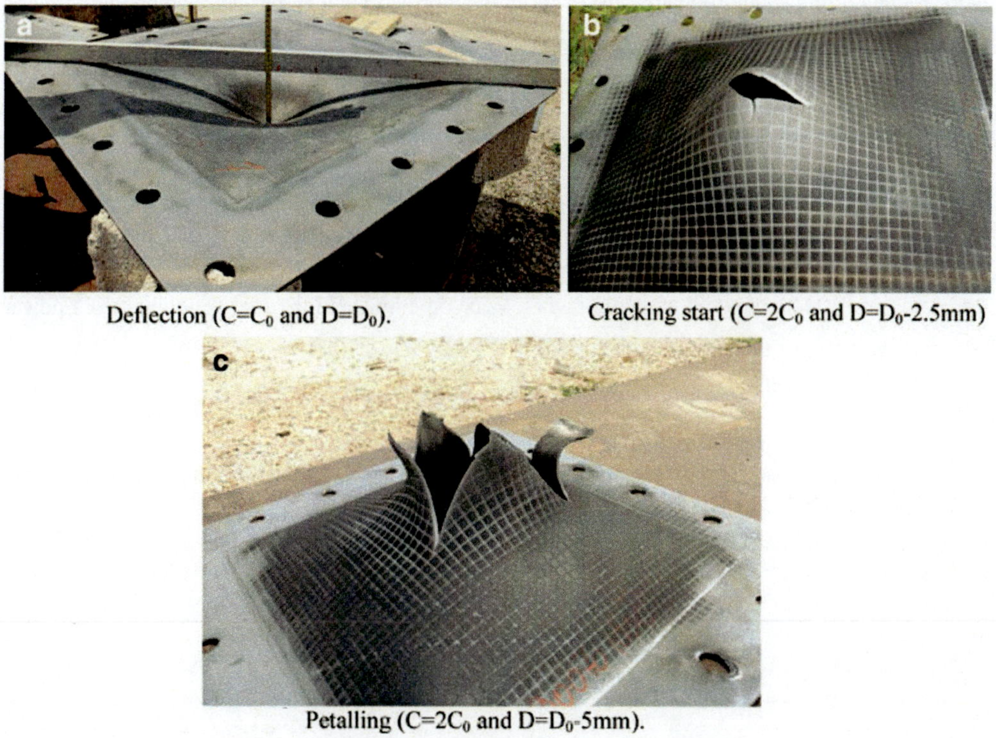


**Fig. 74.2** Experimental aspects of the plate after airblast for various loading configurations (DCNS-DGA). (**a**) Deflection ($C = C_0$ and $D = D_0$). (**b**) Cracking start ($C = 2C_0$ and $D = D_0 - 2.5$ mm). (**c**) Petalling ($C = 2C_0$ and $D = D_0 - 5$ mm)

The explosive was spherical, hung on a post and braces. The mass of the explosive and the distance between the explosive and the plate are controlled parameters. For further details on the experimental set-up, the reader may refer to [10].

Depending on the mass of the explosive C, and on the distance between the plate and the explosive D, one can typically observe three states resulting from the explosion loading: the deflection of the plate (Fig. 74.2a), the macrocracks incipience and growth (Fig. 74.2b), and the phenomenon of petalling (Fig. 74.2c).

## 74.3 Constitutive Relations

In the present approach, defect nucleation is supposed to require a certain amount of plastic deformation. One can thus distinguish at least two steps in the process of ductile damage:

- The sound (undamaged) material behaves in elastic-viscoplastic manner and its response may be described using the J2-flow theory, i.e. via a HUBER-MISES type yield criterion and potential;
- Once defects nucleate and begin to grow, the (damaged) material may be considered as micro-porous and its response becomes pressure dependent.

Such a scenario implies a transition between conventional dense metal plasticity and micro-porous metal plasticity and the definition of conditions for the micro-void induced damage initiation. The case of the mild steel briefly described in Sect. 74.1 is taken as reference material. The reader may refer to [11] for a detailed presentation of the guiding concepts of the model.

### *74.3.1 Constitutive Equations for the Sound Material*

The viscoplastic yielding of the sound (undamaged) material is described via a Huber-Mises type yield function:

$$\Phi_0 = \tilde{\sigma}_{eq}^2 - 1 = 0;\ \tilde{\sigma}_{eq} = \sigma_{eq}/\bar{\sigma}_y \tag{74.1}$$

The yield stress $\bar{\sigma}_y$ in (74.1) includes a rate independent contribution $\sigma_y$ and a strain rate induced overstress $\sigma_{vp}$:

$$\bar{\sigma}_y = \sigma_y + \sigma_{vp} \tag{74.2}$$

Based on the experimental campaign summarized in Sect. 74.1.1, the rate independent contribution $\sigma_y$ in (74.2) incorporates the combined effects of strain hardening, via a Voce type law, and thermal softening, via a power law:

$$\sigma_y = [R_0 + h'(\kappa)]g(T);\ h'(\kappa) = R_\infty[1 - \exp(-k\kappa)]^\beta;\ g(T) = 1 - \left(\frac{T}{T_{melt}}\right)^m \tag{74.3}$$

where $(R_0, R_\infty, k, \beta)$ represent isotropic hardening related constants and $(T_{melt}, m)$ thermal softening related constants. With (74.3), the rate independent contribution $\sigma_y$ in (74.2) takes thus the form

$$\sigma_y = \left\{R_0 + R_\infty[1 - \exp(-k\kappa)]^\beta\right\}\left[1 - \left(\frac{T}{T_{melt}}\right)^m\right] \tag{74.4}$$

The isotropic hardening force $r$ is expressed as follows

$$r = h'(\kappa)g(T) = R_\infty[1 - \exp(-k\kappa)]^\beta\left[1 - \left(\frac{T}{T_{melt}}\right)^m\right] \tag{74.5}$$

The tension/compression asymmetry is assumed to be controlled by a thermally activated mechanism involving the mean stress, as it is postulated in [12]. The strain rate induced overstress $\sigma_{vp}$ in (74.2) is accordingly expressed in the form

$$\sigma_{vp} = Y\left[\dot{\kappa}\exp\left(\frac{V_a p_m}{k_B T}\right)\right]^{1/n} \tag{74.6}$$

where $(Y, n)$ are viscosity related constants and $(V_a, k_B)$ behaviour asymmetry related constants.

Assuming the normality rule, the evolution laws are given by:

$$\begin{cases} \dot{\varepsilon}_0^{pM} = -\Lambda \dfrac{\partial \Phi_0}{\partial p_m} = 0 \\ \dot{\varepsilon}_0^{pD} = \Lambda \dfrac{\partial \Phi_0}{\partial \sigma_{eq}} = 2\Lambda \dfrac{\tilde{\sigma}_{\acute{e}q}}{\bar{\sigma}_y} \end{cases}; \quad \begin{cases} \dot{\kappa} = \dot{\varepsilon}_0^{pD} \\ \dot{T} = \dfrac{(\sigma_{eq} - r)}{\rho C} \dot{\kappa} \end{cases} \tag{74.7}$$

### 74.3.2 Constitutive Equations for the Damaged Material

The behaviour of the micro porous material is described via a Gurson-Tvergaard-Needleman (GTN) like model. The approach adopted here aims at palliating the GTN model incapacity to describe void growth related damage under shear and under low and small negative triaxiality in a way preserving the classical meaning of $f$ (i.e. the microporosity related damage) and allowing for void growth under shear-dominated deformation. The proposed modification is based on the hypothesis of a kinematic mean stress mechanism which governs a shift of the yield locus centre towards negative stress triaxiality values along the increasing damage. The modified GTN yield locus is given, according to [11], in the form

$$\Phi_G = \tilde{\sigma}_{\acute{e}q}{}^2 + 2q_1 f \cosh[-(3/2)q_2(\tilde{p}_m + \tilde{p}_r)] - (1 + q_3 f^2) = 0; \ \tilde{p}_m = p_m/\bar{\sigma}_y; \ \tilde{p}_r = p_r/\bar{\sigma}_y \tag{74.8}$$

Assuming the kinematic pressure $p_r$ as vanishing when the porosity $f$ tends to the unity, the former is given by

$$p_r = b \ln(q_1 f) \tag{74.9}$$

where $b$ is a non-negative constant.

For the damaged material, also assumed as being standard in the thermodynamic sense (obeying the normality rule), the evolution laws are given by:

$$\begin{cases} \dot{\varepsilon}_G^{pM} = -\Lambda \dfrac{\partial \Phi_G}{\partial p_m} = 3q_1 q_2 f \Lambda \dfrac{\sinh\left[-\frac{3}{2} q_2(\tilde{p}_m + \tilde{p}_r)\right]}{\bar{\sigma}_y} \\ \dot{\varepsilon}_G^{pD} = \Lambda \dfrac{\partial \Phi_G}{\partial \sigma_{eq}} = 2\Lambda \dfrac{\tilde{\sigma}_{\acute{e}q}}{\bar{\sigma}_y} \end{cases}; \quad \begin{cases} \dot{\kappa} = \dfrac{\sigma_{eq}\dot{\varepsilon}_G^{pD} - p_m \dot{\varepsilon}_G^{pM}}{(1-f)\bar{\sigma}_y} \\ \dot{T} = \dfrac{\sigma_{eq}\dot{\varepsilon}_G^{pD} - p_m \dot{\varepsilon}_G^{pM} - r\dot{\kappa}}{\rho C} \\ \dot{f} = \dot{f}_n + \dot{f}_g \end{cases} \tag{74.10}$$

where $\dot{f}_g$ and $\dot{f}_n$ represent the growing rate of existing (primary) voids and the nucleation rate of secondary voids, respectively. The expression for the void growth rate is deduced from the matrix material incompressibility whereas the kinetic law for secondary nucleation is assumed to be well described by a Weibull type distribution function, see [13].

$$\begin{cases} \dot{f}_n = B\langle \dot{\sigma}_y \rangle \\ \dot{f}_g = (1-f) Tr \underline{\underline{\dot{\varepsilon}}}^p = (1-f)\dot{\varepsilon}_G^{pM} \end{cases}; \quad \begin{cases} f_n(0) = 0 \\ f_g(0) = f_0 \end{cases} \tag{74.11}$$

where the quantity $B$ involves the hole nucleation function $\Phi_{I_0}$ and the critical stress $\sigma_c$ both detailed in the following sub section:

$$B = f_{\sup} \frac{p}{\sigma_c} \langle \Phi_{I_0} \rangle^{p-1} \exp\left(-\Phi_{I_0}^p\right) \tag{74.12}$$

### 74.3.3 Complementary Damage Initiation Criterion

The hole nucleation criterion describes the conditions for which a specific volume fraction of 'primary' voids $f_0$ instantaneously germinates. In order to ensure the instantaneous transition between dense metal plasticity and microporous metal plasticity, the hole nucleation criterion and the microporous metal potential have the similar aspect and the corresponding expressions:

$$\Phi_{I_0} = \hat{\sigma}^2 + 2q_1 f_0 \cosh(-(3/2)q_2(\hat{p}_m + \hat{p}_r)) - (1 + q_3 f_0^2) = 0;\ \hat{\sigma} = \sigma_y/\sigma_c;\ \hat{p}_m = p_m/\sigma_c;\ \hat{p}_r = p_r/\sigma_c \qquad (74.13)$$

where $\sigma_c$ is a critical stress. To describe the delaying effects of temperature and strain rate the latter is assumed in the form:

$$\sigma_c = \sigma_I + Y[\dot{\kappa}\exp(V_a p_m/(k_B T))]^{1/n};\ \sigma_I = \alpha(R_0 + R_\infty) \qquad (74.14)$$

The expression of the equivalent plastic strain at 'primary' hole nucleation $\kappa_0 = \kappa(f_0)$ may be explicitly deduced from (74.13) as:

$$\kappa_0 = -\frac{1}{k}\ln\left(1 - \left\{\frac{\sigma_c}{R_\infty}\left[1 - \left(\frac{T}{T_{fusion}}\right)^m\right]^{-1}\left[(1 + q_3 f_0^2) - 2q_1 f_0 \cosh\left(-\frac{3}{2}q_2\left(\frac{p_m + p_r}{\sigma_c}\right)\right)\right]^{1/2} - \frac{R_0}{R_\infty}\right\}^{1/\beta}\right) \qquad (74.15)$$

The plastic strain at 'primary' hole nucleation $\kappa_0$ in (74.15) is drawn in Fig. 74.3 as a function of the stress triaxiality for various values of temperature, strain rate and coefficient $\alpha$, see Eq. (74.14). The graphs in Fig. 3 show clearly that the hole nucleation criterion given in (74.13) is able to reproduce, at least qualitatively, the accelerating effects of stress triaxiality and the delaying effects of temperature, see Fig. 74.3a, and strain rate, see Fig. 74.3b. The influence of the ratio $\alpha$ in Eq. (74.14) is shown in Fig. 74.3c, where it can be clearly seen that a large value of $\alpha$ delays 'primary' holes germination.

## 74.4 Numerical Simulations

The model detailed in Sect. 74.2 was implemented as user material (Vumat) in the engineering finite element computation code Abaqus®.

### 74.4.1 Tension Tests on Notched Samples

This subsection deals with the tension of notched cylindrical specimens. The configurations with a notch radius value of 1.2 mm (stress triaxiality close to 1.15) and 6.2 mm (stress triaxiality close to 0.55), are considered here. Quasi-static tests were performed at room temperature and at 5 mm/min displacement rate on these samples.

Concerning numerical simulations, the spatial discretisation consisted in meshing one eighth of the samples using solid finite elements with reduced integration C3D8R. The vertical translation of the lower face nodes is constrained while a vertical velocity is imposed to the upper face nodes. The sample material behaviour is described via the model detailed in Sect. 74.2 and implemented as user material (Vumat) in Abaqus®. The material constant values are reported in Table 74.1. The time integration scheme is explicit.

Figure 74.4 clearly shows the effect of the critical stress $\sigma_c$ via the ratio $\alpha$, see Eq. (74.14), on the overall response of the notched round sample. Acting on the 'primary' void germination occurrence and on the 'secondary' void nucleation rate, see (74.11) and (74.12), the ratio $\alpha$ is consequently a key parameter in the model detailed in Sect. 74.4. For early damage conditions, the ratio $\alpha$ value must be low, and for late damage conditions, the ratio $\alpha$ value must be large.

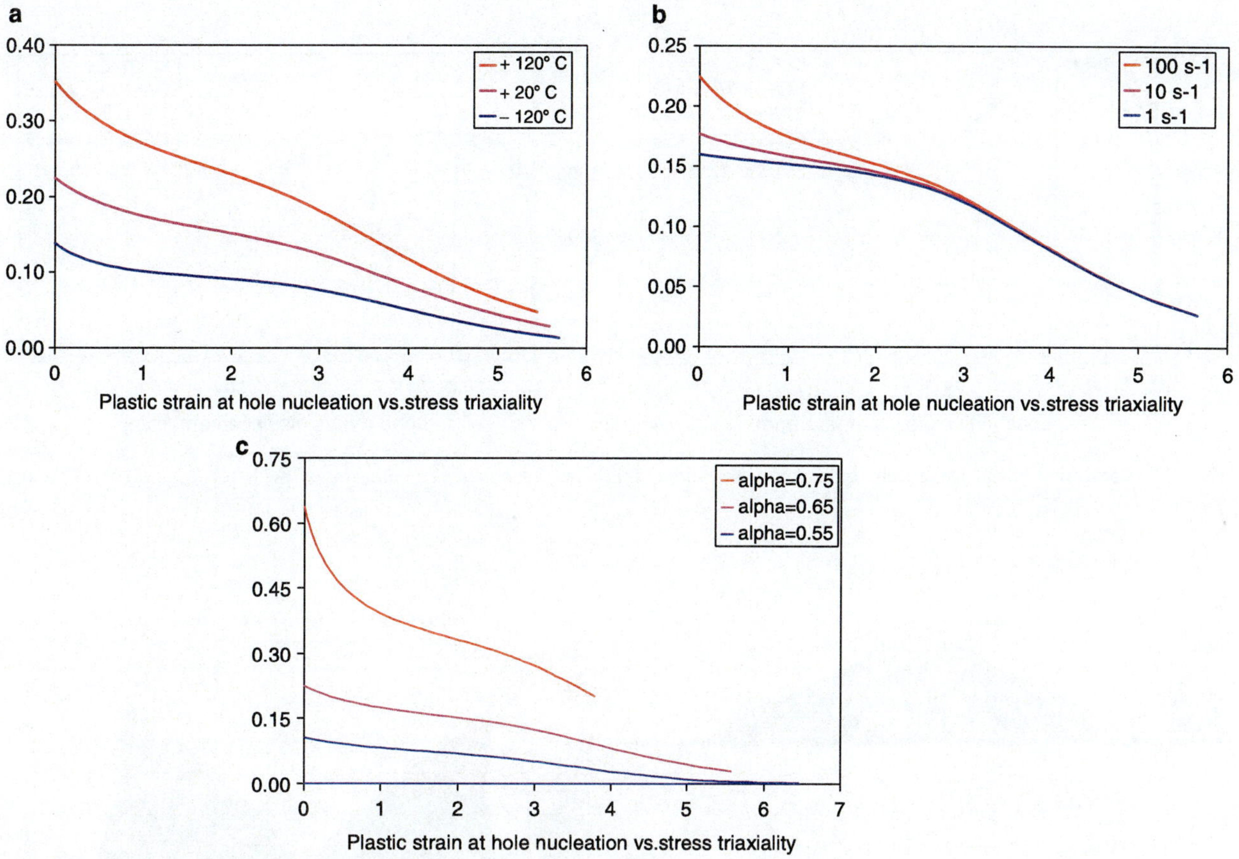


**Fig. 74.3** Influence of temperature (**a**), strain rate (**b**) and α (**c**) on the plastic strain at hole nucleation. In this figure, the stress triaxiality is defined by $-p_m/\bar{\sigma}_y$ (**a**) $\alpha = 0.65$; $b = 115$ *MPa*; $f_0 = 10^{-4}$; $\dot{\kappa} = 10^2 s^{-1}$. (**b**) $\alpha = 0.65$; $b = 115$ *MPa*; $f_0 = 10^{-4}$; $T_0 = 20°C$. (**c**) $b = 115$ *MPa*; $f_0 = 10^{-4}$; $\dot{\kappa} = 10^2 s^{-1}$; $T_0 = 20°C$

**Table 74.1** Microporous model related constants for the numerical simulation of notched samples under tension

| $q_1$ | $q_2$ | $q_3$ | $f_0$ | α | *b (MPa)* | $f_{sup}$ | $f_r$ |
|---|---|---|---|---|---|---|---|
| 1 | 1 | 1 | $10^{-3}$ | / | 115 | 0.03 | 0.2 |

## *74.4.2 Airblast Experiments*

The numerical results for various loading conditions, in terms of explosive mass and plate/explosive distance, corresponding to those used in the experiments (see Sect. 74.1.2) are presented in Fig. 74.5. According to Fig. 74.5 one can note a relatively satisfying (conservative) correlation between experimental and numerical results. It must be noted that the release of burnt gases is not possible when the Abaqus® Conwep card, reproducing numerically approximately the fluid/structure interaction, is used. This implies that the explosion induced pressure keeps on being applied on the plate surface all along the loading duration, provoking a petal reversal more important than in reality. Consequently, for a better description of the airblast induced failure of the plate, a more accurate reproduction of the fluid–structure interaction is needed.

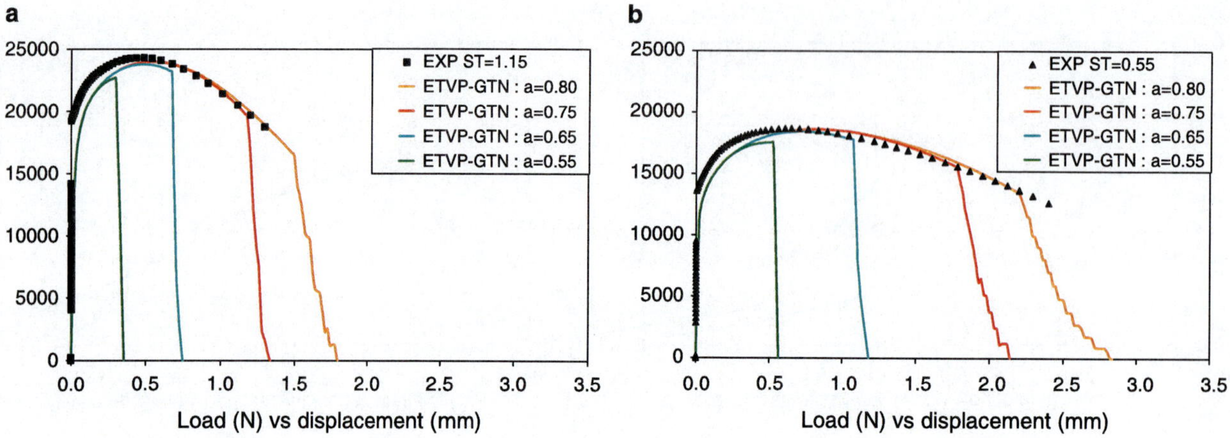


**Fig. 74.4** Comparison model-experiment. Influence of the ratio $\alpha$. Tensile test on notched cylindrical samples; $\alpha$ =//- $f_{sup} = 0.03 - f_r = 0.2$. (**a**) ST = 1.15. (**b**) ST = 0.55

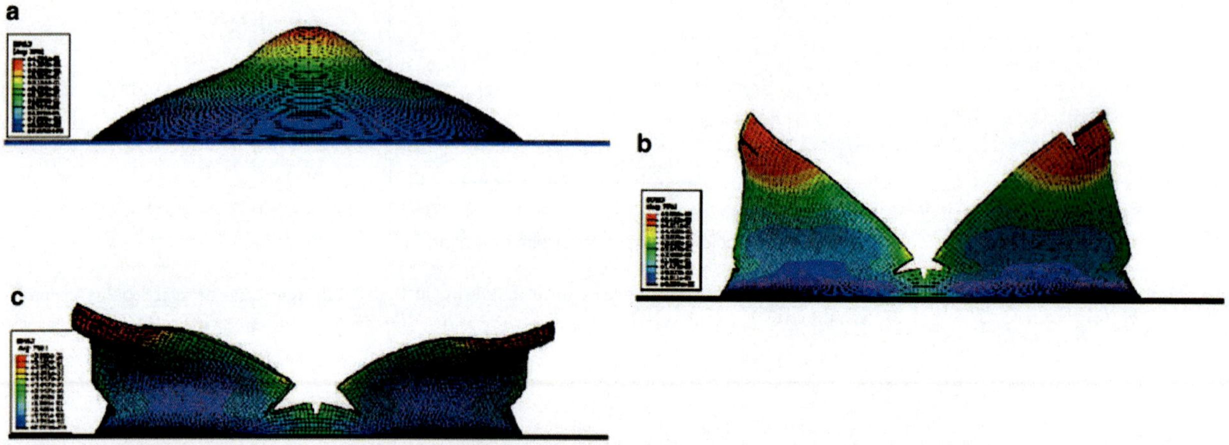


**Fig. 74.5** Experimental aspects of the plate after airblast for various loading configurations (DCNS-DGA). (**a**) Deflection ($C = C_0$ and $D = D_0$). (**b**) Cracking start ($C = 2C_0$ et $D = D_0{-}3$ mm). (**c**) Petalling ($C = 2C_0$ et $D = D_0{-}5$ mm)

## 74.5 Concluding Remarks

An important experimental campaign, including laboratory thermomechanical tests as well as airblast experiments allowed for characterizing the behaviour of a ship structure constitutive material. The mechanisms of damage and cracking which occur during the deformation processes until the ultimate failure were identified from micrographic analyses of partially and totally fractured specimens.

Some experimental results were used to build a multi-surface based constitutive model able to describe the transition between dense metal plasticity and microporous metal plasticity. The criterion of damage initiation accounts for the accelerating effects of stress triaxiality and for the delaying effects of temperature and strain rate. A modification of the GTN model has also been proposed permitting to describe the cavity growth under shear loading.

After identifying a set of material constants from standard experiments, the model was validated considering numerical experiments of explosion loaded plates. The corresponding results are encouraging when compared qualitatively with experimental data. A better description of the fluid/structure interaction, numerically reproduced in the present work via the Abaqus® Conwep card, could allow for improving the predictive capacity of the numerical model.

## References


1. Gurson AL (1977) Continuum theory of ductile rupture by void nucleation and growth: part I – yield criteria and flow rules for porous ductile media. J Eng Mat Tech ASME 99:2–15
2. Tvergaard V, Needleman A (1984) Analysis of the cup-cone fracture in a round tensile bar. Acta Metall 32(1):157–169
3. Becker R, Needleman A, Richmond O, Tvergaard V (1988) Void growth and failure in notched bars. J Mech Phys Solids 36:317–351
4. Perzyna P (1984) Stability of flow processes for dissipative solids with internal imperfections. J Appl Math Phys 35:848–867
5. Rice JR, Tracey DM (1969) On the ductile enlargement of voids in triaxial stress fields. J Mech Phys Solids 17:201–217
6. Rousselier G (1987) Ductile fracture models and their potential in local approach of fracture. Nucl Eng Des 105:97–111
7. Gologanu M, Leblond JB, Perrin G, Devaux J (1997) Recent extension of Gurson's model for porous ductile metals. In: Suquet P (ed) Continuum micromechanics. CISM courses and lectures no. 377, Springer, New York, pp. 61–130
8. Pardoen T, Hutchinson JW (2000) An extended model for void growth and coalescence. J Mech Phys Solids 48:2467–2512
9. Goods SH, Brown LM (1979) The nucleation of cavities by plastic deformation. Acta Metall 27:1–15
10. Geffroy A-G, Longère P, Leblé B (2011) Fracture analysis and constitutive modelling of ship structure steel behaviour regarding underwater explosion. Eng Fail Anal 18:670–681
11. Longère P, Geffroy A-G, Leblé B, Dragon A (2011) Modelling the transition between dense metal and damaged (micro-porous) metal viscoplasticity. Int J Dam Mech. doi: 10.1177/1056789511427472
12. Graff S, Forest S, Strudel S, Prioul JL, Pilvin P, Béchade JL (2004) Strain localization phenomena associated with static and dynamic strain ageing in notched specimens: experiments and finite element simulations. Mat Sci Eng A 387–389:181–185
13. Molinari A, Wright TW (2005) A physical model for nucleation and early growth of voids in ductile materials under dynamic loading. J Mech Phys Solids 53:1476–1504

# Chapter 75
# Deformation Mechanisms in $M_{n+1}AX_n$ Phase Ternary Ceramics at High Strain Rates

**Riddhiman Bhattacharya and N.C. Goulbourne**

**Abstract** The present study is concerned with understanding the high strain rate deformation mechanisms and the role of microstructural characteristics in $M_{n+1}AX_n$ phase ternary ceramics. These materials crystallize in a Hexagonal Close Packed (HCP) structure with a c/a ratio greater than 1.67 which results in kink band formations when subjected to loading. In this work, we report the high strain rate deformation and fracture mechanisms observed in $Ti_2AlC$, a ternary ceramic, when subjected to dynamic compressive loading (~1,200 $s^{-1}$) using a Split Hopkinson Pressure Bar (SHPB) technique and 2-D high speed Digital Image Correlation (DIC) to measure in-plane displacements. Specimens with different geometries are tested due to the tendency of these materials to fracture prematurely and also due to lack of established experimental protocols for testing ternary ceramics using SHPB (in comparison to metals, ceramics, polymers), which is another objective of the study. The fractography analysis, done using a Scanning Electron Microscope (SEM), reveals the dominant fracture mechanisms to consist of kink band formations, grain folding and delamination, cavity formation, delamination of the nano-laminates, grain pullout, and tearing etc. These features, though observed under static loading conditions, initiate and propagate in the material by a completely different mechanism. Qualitative and quantitative comparison show dynamic loading leads to more *advanced* deformation modes.

**Keywords** MAX phase ternary ceramics • Split Hopkinson Pressure Bar (SHPB) • Digital Image Correlation (DIC) • High strain rate

## 75.1 Introduction

$M_{n+1}AX_n$ phase materials represent layered ternary ceramics where M is an early transition metal, A is an A-group element and X is C or N. The value of n typically varies from 1 to 3 [1–3]. These materials, like many carbides, show high temperature stability of up to 2,000°C [1, 3] but with increased machinability and ductility. This unique combination of properties makes them attractive materials for structural applications at room and elevated temperatures. These materials are also classified as Kinking Nonlinear Elastic (KNE) solids, which have a hexagonal crystal structure [4]. Such behavior can be attributed to the hexagonal structure with a high c/a ratio, greater than 1.67, and hence these materials deform by extensive kink band formations [2–6]. $Ti_2AlC$ is one such material which belongs to this class of layered ternary KNEs. While the behavior and response of these materials under quasi-static loading conditions ($10^{-4}$ $s^{-1}$) has been investigated [3, 7, 8], mechanical dynamic characterization has not been reported. In this paper we report on the high strain rate (up to ~2,000 $s^{-1}$) response using a Split Hopkinson Pressure Bar (SHPB) technique.

SHPB testing is typically used to characterize the constitutive stress–strain response in a material at high strain rates ($10^2$ to $10^4$ $s^{-1}$) [9]. The data reduction technique is based on assumptions that deformation in the specimen is: (1) uniform

R. Bhattacharya (✉)
Department of Materials Science and Engineering, The University of Michigan, 2006 FXB Building, 1320 Beal Avenue, Ann Arbor, MI 48109, USA
e-mail: ribh@umich.edu

N.C. Goulbourne
Department of Aerospace Engineering, The University of Michigan, 2006 FXB Building, 1320 Beal Avenue, Ann Arbor, MI 48109, USA

V. Chalivendra et al. (eds.), *Dynamic Behavior of Materials, Volume 1: Proceedings of the 2012 Annual Conference on Experimental and Applied Mechanics*, Conference Proceedings of the Society for Experimental Mechanics Series, DOI 10.1007/978-1-4614-4238-7_75, 

and evenly distributed, (2) the mechanical impedance of the material tested is similar to the incident/transmission bar, (3) the bars remain elastic, a uniaxial stress state exists in the specimen and the material is isotropic, homogenous, and without inherent flaws. Since the test is material dependent, proper specimen design is important [10], especially for hard and brittle ceramics. Any flaw in these materials would lead to premature failure and hence lead to an erroneous measurement of the stress–strain response. Therefore, a tolerance to high degree of flatness and parallelism of the specimens is recommended. The typical values of length-diameter ratio have been recommended to be between 0.5 and 1 for metals, and between 1 and 2 for ceramics [11]. The specimens can be tested under different loading modes such as compression, tension, torsion at both room temperature as well as elevated temperatures [12]. All the tests reported in this paper are conducted under compressive loading. Further details regarding this technique, specimen preparation and experimental set-up are presented in a subsequent section. Apart from measuring and calculating the strains using strain-gauge signals, a full field method, 2D Digital Image Correlation (DIC) was used (the high speed camera capability is up to a frame rate of one million frames per second).

The principle of DIC is based on the identification of an object point on the surface of the material being tested using an indistinguishable intensity pattern, which is applied to the material before testing (speckle patterning). A correlation step then divides the images into several sub-images known as facets. At every loading step, the facet deformation is tracked. In 2D DIC, only one camera is used whereas 3D DIC involves two cameras, in which the correlation algorithm first performs the task of finding a homologous point on the surface of the specimen, from images obtained from the two cameras [13–18]. In this paper we report 2D DIC at high speeds ($10^5$ frames per second) and hence neglect any out-of-plane motions. The specimen is therefore assumed to be nominally planar and deforms only in the original object plane. Though this method is highly advanced and uses a non-contact algorithm to calculate the strain fields, errors due to image distortion and correlation are possible. Distortion is the deviations in the image position with respect to a specific model. The sources of distortion could be from the aberrations of the lenses and misalignment of optical elements. Parametric models have been developed to correct these distortions and these methods are dealt with and described in detail in [16]. Correlation errors are typically classified into statistical and systematic errors. Statistical errors are due to camera image noise and also due to a limited number of pixels caused by camera digitization. Systematic errors are correlation errors caused by several reasons when the facet transformation model does not consider the real situation. This error creeps into measurements when a curved specimen surface is analyzed using DIC, assuming planar surface. This situation is particularly relevant to the tests reported here since cylindrical specimens, having a curved cross-section are analyzed using 2D DIC. A more detailed discussion of this error and ways to tackle them is presented elsewhere [17] and not considered here further.

This paper is organized into three broad sections, with sub-sections. Section 2 outlines the experimental steps with Sect. 2.1 describing the sample preparation, with Sects. 2.2 and 2.3 outlining a detailed description of the SHPB set up used for the tests, coupled with the high speed camera, data processing and analysis, etc. The results and discussions from the tests are reported in Sect. 3, with Sect. 3.1 reporting the strain calculations and the constitutive stress–strain behavior of $Ti_2AlC$ under high strain rate loading. The microstructural analysis of the specimens before and after deformation (with and without fracture) is reported in Sect. 3.2 and a discussion on the results is presented in Sect. 3.3. Section 4 outlines the conclusions from the results obtained with possible scope and direction of future work.

## 75.2 Experimental

### *75.2.1 Sample Preparation*

The material used in this work is polycrystalline $Ti_2AlC$, which was received from Texas A&M University (TAMU-MURI 18 team lead and collaborator). The material has been synthesized using a pressureless sintering process. The details of the synthesis process can be found in [1–7, 19]. Cylindrical specimens were received with a diameter of 12.7 mm (0.5 in.), which was prepared for SHPB testing, using Electro Discharge Machining (EDM) to ensure perfectly flat and smooth surfaces with minimal flaws. Specimens with Length (L)-Diameter (D) ratios of 0.2, 0.4 and 0.8 were tested to study the effect of specimen size on the high strain rate response of the material. The ratios were decided based on reported literature [13, 14, 20–22] on SHPB testing in ceramics and ductile metals, since MAX phase materials exhibit a unique combination of both properties. The microstructure of the specimens before and after deformation and fracture was examined and the results are presented in a subsequent section. The specimens were speckled with black paint on a white coating to provide the contrast for subsequent analysis using DIC.

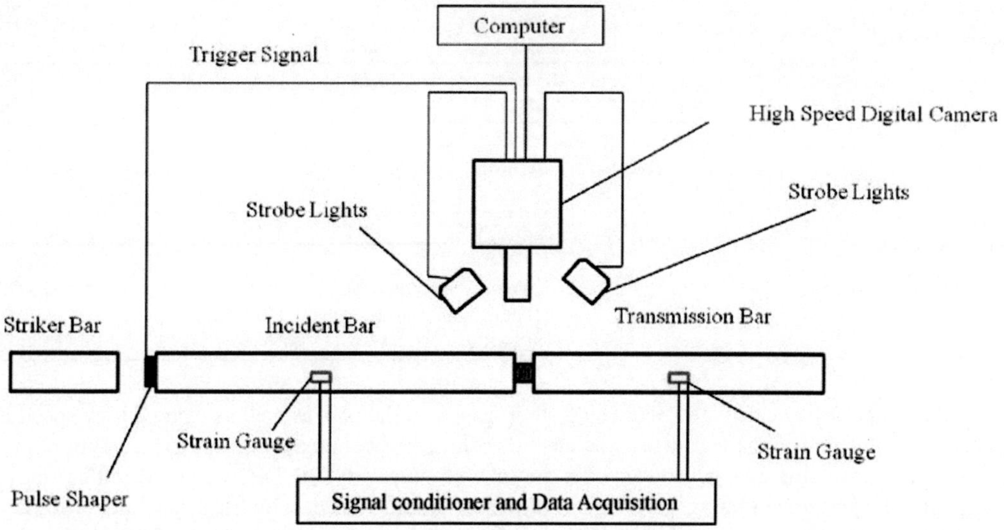


**Fig. 75.1** Schematic representation of the Split Hopkinson Pressure Bar (SHPB) experimental set-up

### *75.2.2 SHPB Experimental Set-up*

The Split Hopkinson Pressure Bar (SHPB) set up consists of two bars, an incident bar and a transmission bar, on which strain gauges are placed. The specimen is "sandwiched" between the two bars. A third bar, much shorter in length compared to both the incident and transmission bars is used to impact the incident bar. This leads to the creation of a strain pulse which travels through the incident bar and reaches the specimen. A portion of the pulse reaching the specimen gets transmitted into the transmission bar while the remaining portion is reflected back into the incident bar. The reflected pulse is due to the mechanical impedance mismatch of the specimen and the bar. The incident bar has a greater length than the transmission bar. The strain pulses obtained from the test is analyzed and processed assuming 1D wave propagation theory. It is assumed that the bars are infinitely long (compared to size of the specimen), the system is in mechanical equilibrium and is free from vibrations. A detailed derivation of the equations is presented in the [12, 23]. A 2-wave analysis was been used to process the strain pulses and obtain dynamic stress–strain curves, which are presented here (without derivation), as follows:

$$\dot{\varepsilon}(t) = 2\frac{C_0}{L}\varepsilon_r(t), \tag{75.1}$$

$$\varepsilon(t) = \int_0^t \dot{\varepsilon}(\tau)d\tau, \tag{75.2}$$

$$\sigma(t) = \frac{A_b E_b}{A_s}\varepsilon_t(\mathrm{t}), \tag{75.3}$$

where $A_b$ is the cross sectional area of the bar, $E_b$ is the elastic modulus of the bar material, L is the length of the specimen, $A_s$ is the area of the specimen, $\varepsilon_r$ (t) and $\varepsilon_t$ (t) are reflected and transmitted strain pulses respectively in the axial directions. Additionally $\tau$ is the "dummy" variable used for integration, $C_0$ is the wave speed of the bar given by

$$C_0 = \sqrt{\frac{E_b}{\rho}}, \tag{75.4}$$

where ρ is the density of the bar. Figure 75.1 below shows a schematic of the SHPB set-up used for the tests reported here. The high speed camera is triggered using a trigger signal, as shown and the strobe lights are used for resolving the speckle patterns effectively at high frame rates.

**Table 75.1** Dimensions of the bars in the experimental set-ups

| | Incident bar length (m) | Transmission bar length (m) | Striker bar length (mm) | Diameter of the bars (mm) |
|---|---|---|---|---|
| Set-up I | 1.83 | 1.22 | 305 | 12.7 |
| Set-up II | 2.43 | 1.77 | 700 | 38.1 |

**Table 75.2** Physical and mechanical properties of the bar material used for data analysis

| Property | Value |
|---|---|
| Material | 440 C stainless steel |
| Young's modulus | 220 GPa |
| Ultimate tensile strength | 1,965 MPa |
| Density | 7,620 kg/m$^3$ |

The strain in the material was calculated both using strain gages on the bars as well as using a high speed DIC method (Fig. 75.1). This method of analyzing the data to calculate the strain is free from the errors due to assumptions included in the wave analysis of the strain pulse signals. However for the sake of simplicity, 2D DIC was used in this study, which can measure only in-plane displacements thus neglecting out-of-plane motion. Since a cylindrical specimen with curve surface is used, this method is not fully accurate and can be improved by (a) using two cameras and 3D DIC, (b) 3D DIC using one camera [16] or (c) using a specimen with planar (rectangle/square) cross section [24].

### 75.2.3 *Experimental Procedure*

For the tests reported here, the striker bar was fired with pressure in the range 4–6 MPa which resulted in striker velocities of about 11–16.5 m/s and strain rates in the range 500–2500 s$^{-1}$. Two different set-ups were used having different bar diameter and lengths (Table 75.1) and two specimens were used for each testing condition. In both set-ups the bar material was case hardened stainless steel. However for setup II (See Table 75.2), premature failure of specimens was observed and hence stress-strain data was not obtained for these tests. In the experiments reported here, no modifications were made to prevent such premature failure as has been reported earlier for high strength ceramics [26].

The strain pulse signals obtained from the test were analyzed using a 2-wave analysis. The strain and displacement fields were also calculated using 2D DIC using images obtained at a rate of 100,000 frames per second. A Photron® SA5 camera with a lens of focal length 105 mm was used to capture the images of the speckled specimens. The results of the strains obtained from these two analyses are compared and presented in the subsequent section.

## 75.3 Results and Discussion

### 75.3.1 *Strain and Displacement Fields Using DIC*

Cylindrical specimens of three different L-D ratios were machined from a bulk cylindrical specimen. Tests were conducted using the two different experimental set-ups on these specimens to determine the effect of specimen geometry on the response of these materials. Figure 75.2 below shows the strain-time relationship in the different materials using different set-ups. Figure 75.2a shows the strain-time history for specimen of L-D ratio 0.2 using set-up I. Testing of this specimen on set-up II resulted in a premature failure and pulverization which led to lack of appreciable strain in the material. Figures 75.2b, c show the strain-time history for specimen with L-D ratio 0.4 tested using set-up I at strain rates of 500 and 700 s$^{-1}$ respectively. In all the three cases a comparison between the strain calculation from strain gauges and DIC (using ARAMIS ®) is shown. Comparisons between Fig. 75.2a and b show that the thicker specimens achieve the stress equilibrium at relatively lower strains whereas Fig. 75.2b and c show the effect of strain rate on a specimen with the same L-D ratio. A comparison between the strains calculated using DIC and that calculated from analysis of gauge data shows some mismatch (error ~5%). The trend, however, is similar i.e. the strain gauge signals overlap well with the DIC calculations during the rise time but shows some deviation and goes to relatively higher strains. It must be noted that the scales used in the plots are magnified and thus the mismatch is exaggerated. The cause for this slight mismatch is attributed

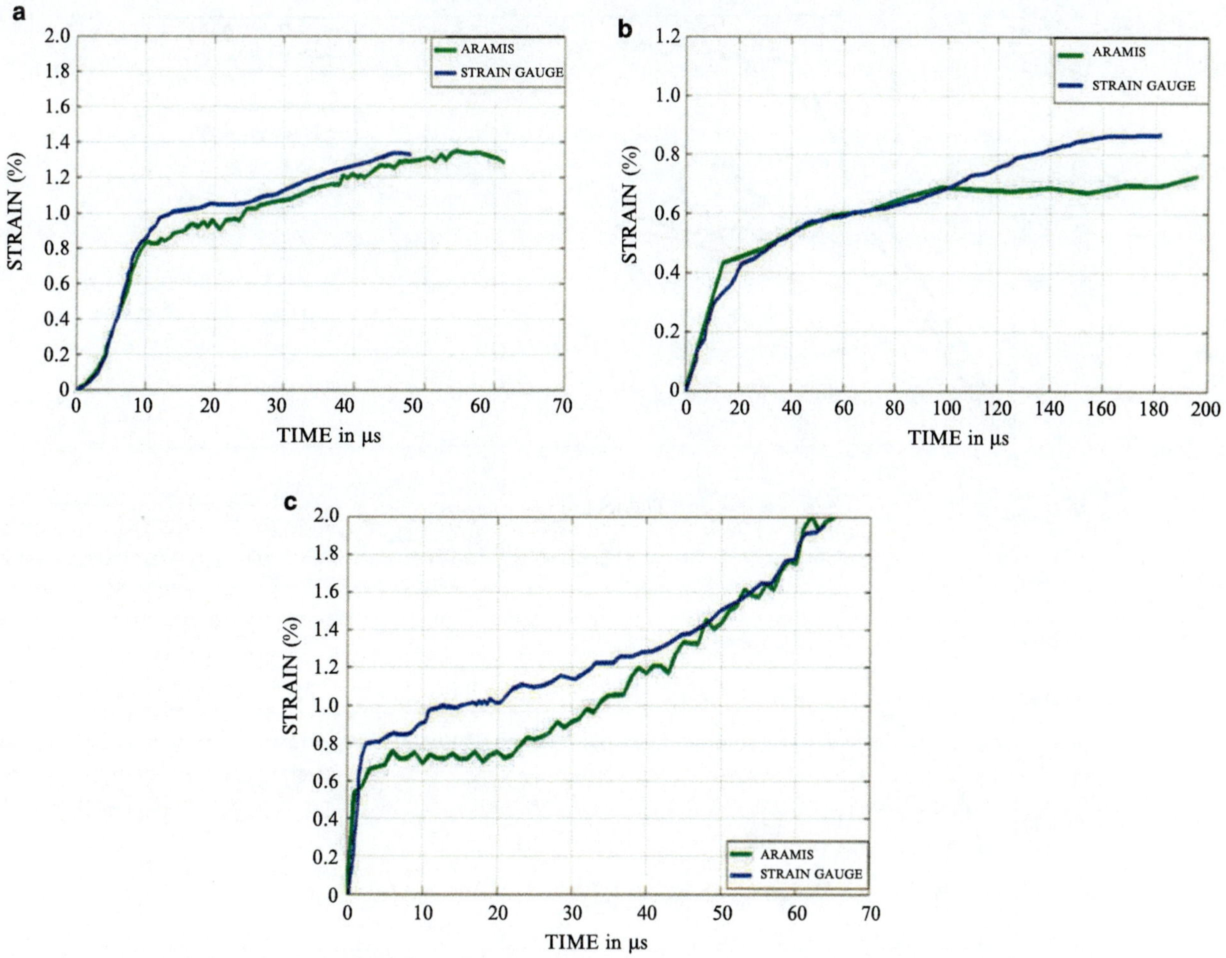


**Fig. 75.2** Strain-time curve calculated using strain gauge signals (*blue*) and DIC (*green*) using ARAMIS ® for (**a**) L-D = 0.2 at strain rate 500 1/s; (**b**) L-D = 0.4 at strain rate 500 1/s; (**c**) L: D = 0.4 at strain rate 700 1/s (color online)

to the 2-wave analysis, which has been used to analyze the strain gauge data. A more accurate method would be using a 3-wave analysis and also using more sophisticated methods for determining the rise time of the pulses and dispersion corrections (using an FFT algorithm). The details of these methods are outlined in [25]. In this paper the method described in [23] is used to analyze the data. Figure 75.2c shows higher amount of strain in the material and slightly more mismatch compared to the previous data. The error in the plots can also be attributed to the experimental challenges in using 2D DIC which considers only in-plane motion, as pointed out in an earlier section. Therefore accuracy of the results can be improved by using 3D DIC, using two cameras or an arrangement described in [16].

Figure 75.3a below shows the stress–strain behavior of tests done at L-D ratios of 0.2 and 0.4. The stress–strain behavior at an L-D ratio 0.8 cannot be used to calculate the modulus of the material due to the premature propagation of cracks and subsequent fracture, which will be treated separately in a subsequent section. The stress–strain curves reveal that the material goes to higher strain value at lower L-D ratios. It also clearly shows that specimens having identical L-D ratios, when tested at different strain rates, reach higher stress values at higher strain rates. The dynamic modulus of the material is calculated for each case and summarized in Table 75.3 below. Figure 75.3b shows a comparison between an aluminum alloy (Al-2024 alloy) and Ti2AlC, having L-D values of 0.55 and 0.4 respectively tested at the same strain rate.

For a particular L-D ratio, there was a difference in the modulus value for different strain rates at which the test was done, as is clearly shown. Hence the response is strain rate dependent. The modulus also varies for different L-D ratios at a particular strain rate value in the range 9–20 GPa. This variation is likely due to experimental error (and data analysis) as a dependence on geometry of the specimen is not expected. Although a theoretical value of ~300 GPa is reported for elastic modulus in pure $Ti_2AlC$ [2], modulus measurements from quasi static loading ($10^{-4}$ $s^{-1}$) [7] results in a value of ~13 GPa.

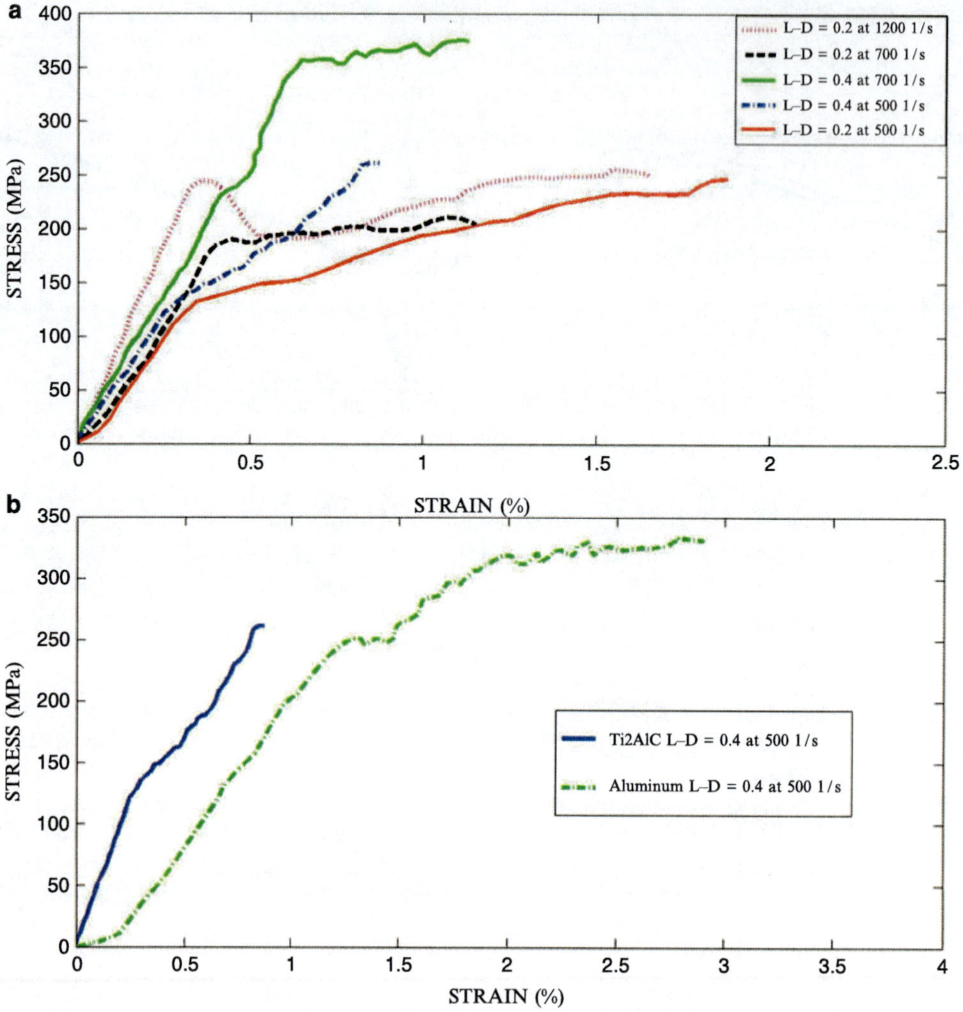


**Fig. 75.3** (**a**) Stress–strain behavior at high strain rates showing effect of varying length (L)-diameter (D) ratio and strain rate (color online). (**b**) Stress–strain behavior at high strain rates of Aluminum (*green*) and $Ti_2AlC$ (*blue*) at 500 $s^{-1}$ (color online)

The modulus values presented here are higher compared to quasi-static conditions under similar loading conditions. The stress–strain plots shown in Fig. 75.3 also shows a strain rate dependence of the behavior. There is some amount of ductility in the specimens due to which it yields and deforms plastically. The yield strength values observed from the stress–strain behavior in Fig. 75.3 are summarized in Table 75.3 above. No definite yield point was observed in the specimen with L-D ratio 0.4 tested at 500 $s^{-1}$. For a given L-D ratio, it is observed that the yield strength increases with increasing strain rate. Yielding occurs in these materials due to formation of fully formed kink bands which are permanent microstructural rearrangements and cause plastic deformation behavior. The curve termination points do not depict fracture of the material but a loss of contact of the specimen with the bars. The maximum elongation in the material does not have a systematic variation with strain rate as observed in the values in Table 75.3. However it is observed that thicker specimens elongate less compared to a specimen with lower L-D ratio. The behavior reveals strain hardening, with the rate of hardening being higher compared to that of a metal, as can be inferred from the plots shown in Fig. 75.3b.

Figures 75.4 and 75.5 show the displacement and strain fields in the x direction for a specimen tested at L-D ratio of 0.4, at different time intervals of 10 μs. The scale (shown on the last image of each row) is uniform for all the images in the series and has a range of 0–0.1 mm (in the negative sense indicating compression in the x direction) for the displacement. The displacement fields are not completely homogenous along the length of the specimen. The strain field heterogeneity between the "hot spots" and "cold spots" are within a range of 0.3–0.4%. Based on microstructural characterization of the deformed specimens (without obvious macroscale fracture), which show numerous rough area formations throughout the surface, we hypothesize that these rough areas act as sources of stress concentrations and are "hot-spots". (These results are shown in detail in Sect. 3.2).

**Table 75.3** Dynamic modulus at different L-D ratios and strain rates

| L-D ratio | Strain rate ($s^{-1}$) | Modulus (GPa) | Yield strength (MPa) | Max. elongation (%) |
|---|---|---|---|---|
| 0.2 | 500 | 35.7 | 130 | 1.8 |
| 0.2 | 700 | 39.8 | 200 | 1.2 |
| 0.2 | 1,200 | 75.6 | 240 | 1.6 |
| 0.4 | 500 | 46.6 | – | 0.8 |
| 0.4 | 700 | 59.4 | 350 | 1.2 |

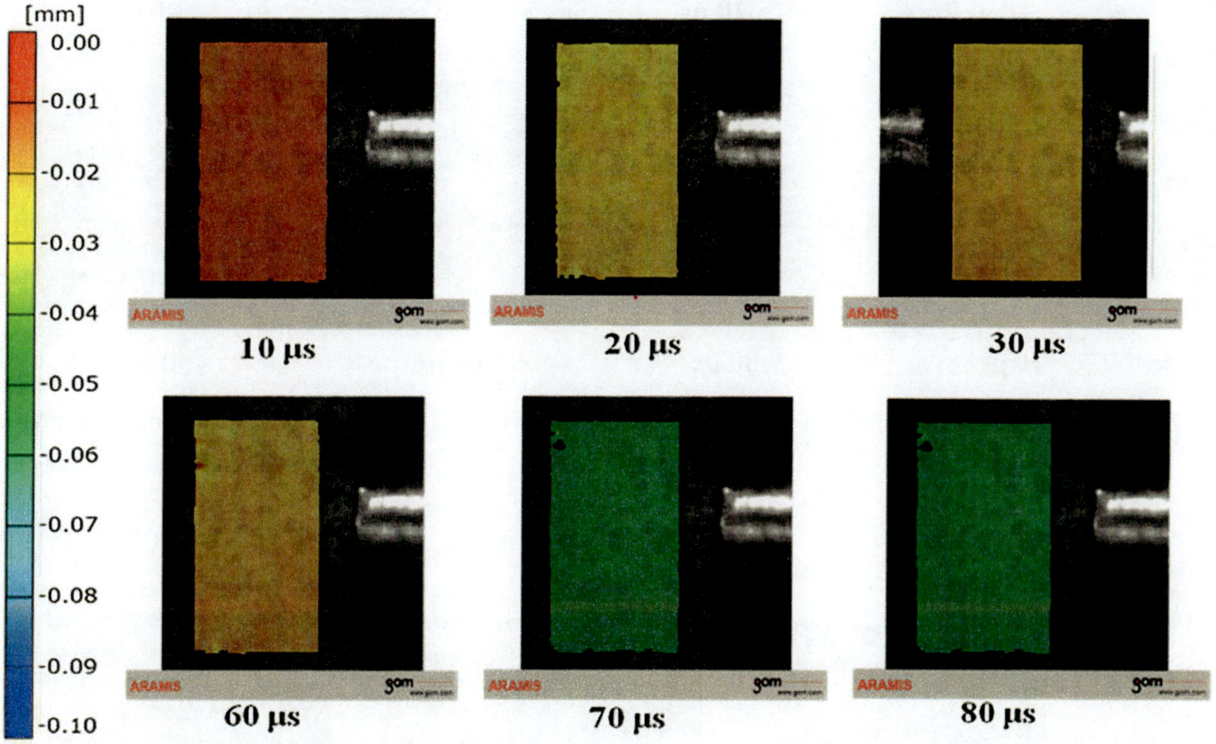


**Fig. 75.4** Displacement field in the x direction (L-D ratio 0.4), at time intervals of 10 μs, starting with 10 μs from the start of the test. Scale: 0 (*Red*) to 0.1 (*Blue*) mm (negative x direction, color online)

Figures 75.6 and 75.7 below shows the displacement and strain fields for the same specimen in the perpendicular direction. The displacement, as is expected, is tensile in the y direction, with values smaller than the displacement in the x direction. Heterogeneities in the range 0.3–0.4% are observed throughout the surface of the specimen.

Figures 75.8 and 75.9 show the displacement fields for the specimen with an L-D ratio of 0.8 tested at 2,500 $s^{-1}$. The displacement fields shown in these figures show heterogeneities to a lesser degree and smooth transitions between areas having different displacements, as compared to the case of elasto-plastic deformation (shown in Figs. 75.4 and 75.6 above). The images shown below are taken at intervals of 5 μs as the specimen starts developing cracks within 30 μs from the start of the test. Even though this test is also done under compressive loading, the range of the scales have been modified (with respect to Figs. 75.4 and 75.6) and in the following figures from 0 to 0.07 mm for displacements in the negative-x direction and from 0 to 0.006 mm for displacements in positive-y direction. These modifications have been made to facilitate clarity in observing the displacement fields in the vicinity of the crack nucleation and propagation. A sudden increase in strain (compressive) in the x direction and subsequent loss of correlation indicates the formation and propagation of a large crack throughout the specimen, which leads to its subsequent failure. The displacement in the positive y direction clearly shows the direction and propagation of the cracks, with two cracks nucleating, one at the left edge of the specimen and the other at the upper right corner. Subsequently these cracks join and propagate, which then branches out to pulverize the material into small pieces. In the images below, the displacement fields are overlaid on the specimen and are also shown separately to clearly demonstrate the gradual loss of correlation during crack formations and also to indicate the strain fields in the region surrounding the cracks.

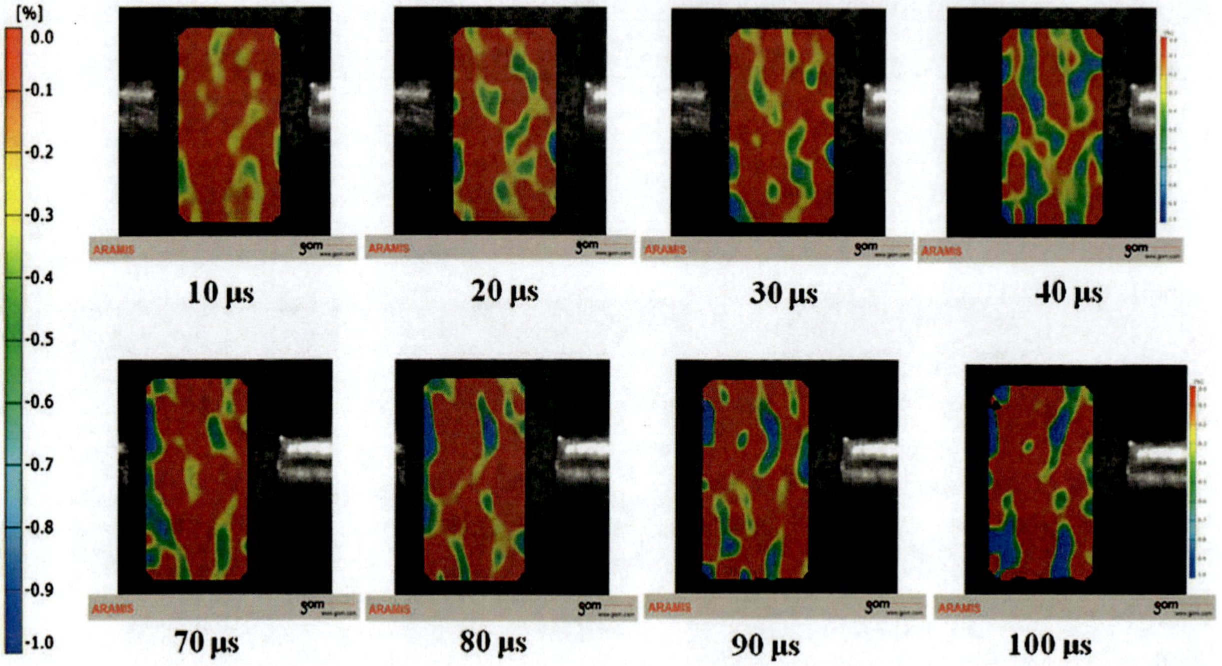


**Fig. 75.5** Strain field in x direction for specimen of L-D ratio 0.4, at time intervals of 10 μs, starting with 10 μs from the start of the test. Scale: 0 (*Red*) to 1 (*Blue*) % (negative x direction, color online)

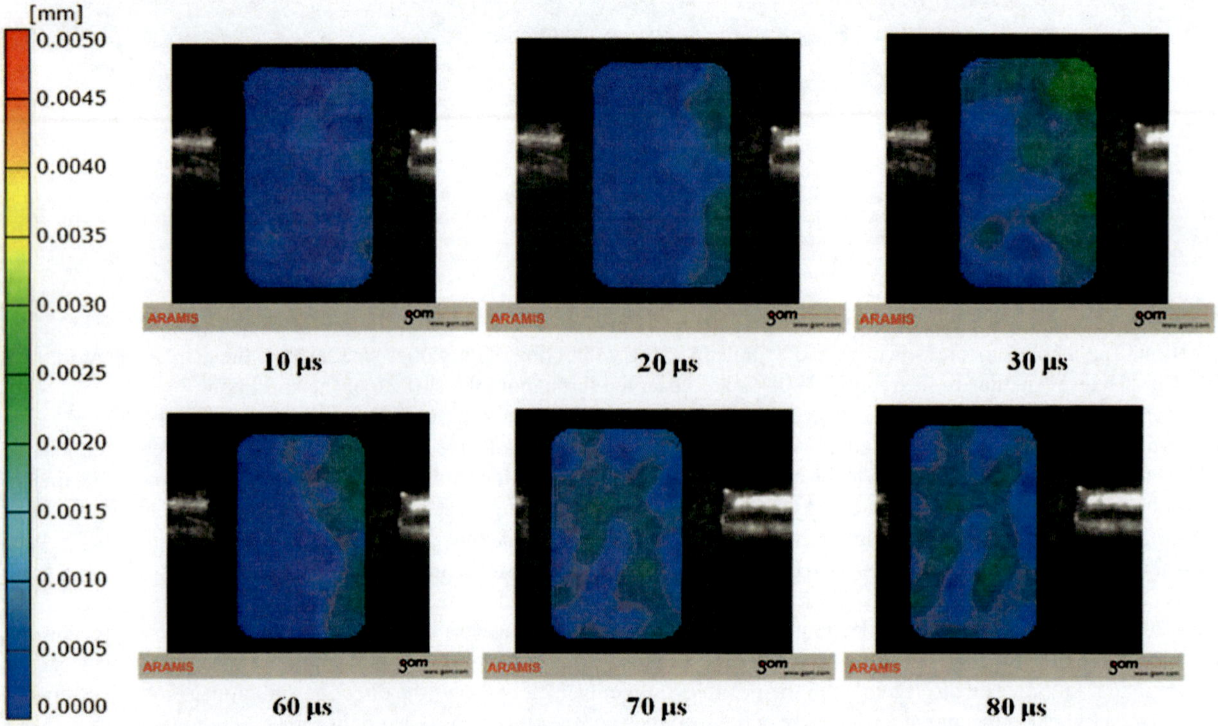


**Fig. 75.6** Displacement field in y direction for specimen of L-D ratio 0.4, at time intervals of 10 μs, starting with 10 μs from the start of the test. Scale: 0 (*Blue*) to 0.05 (*Red*) mm (Positive y direction, color online)

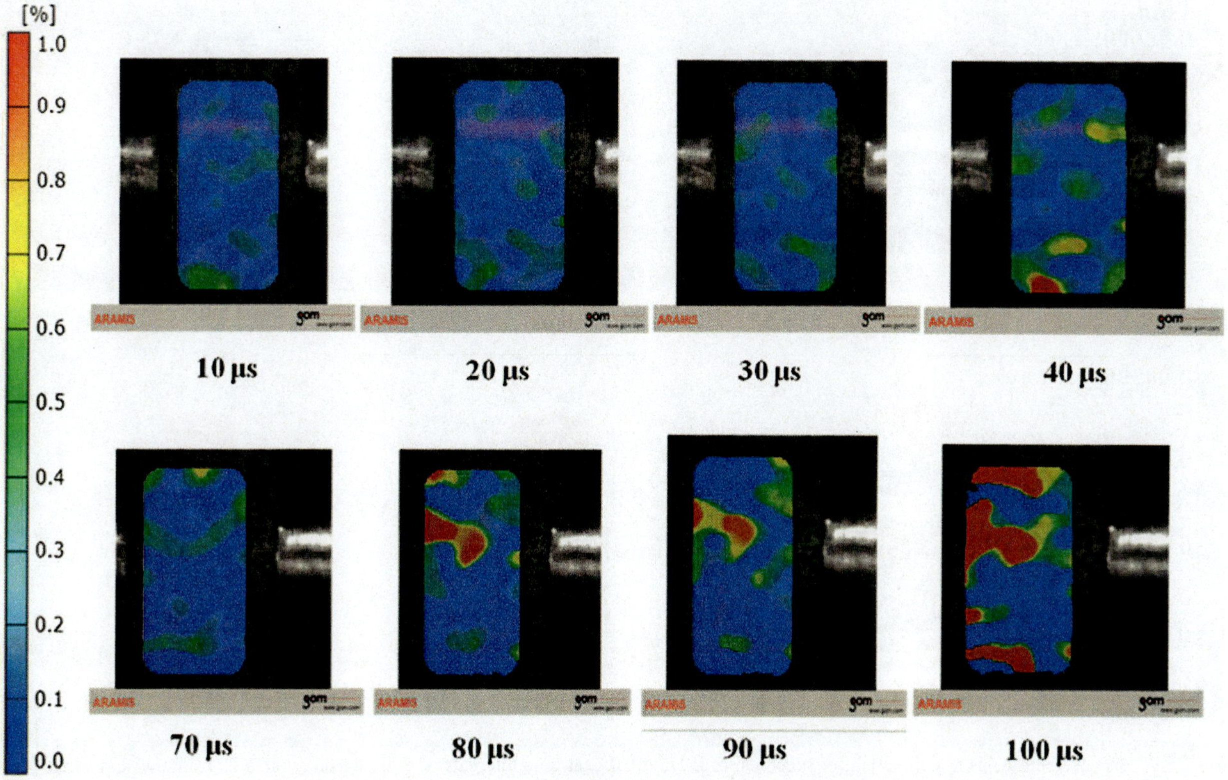


**Fig. 75.7** Strain fields in the y direction for a specimen of L-D ratio 0.4, at time intervals of 10 μs, starting with 10 μs from the start of the test, shows heterogeneity in the strains. Scale: 0 (*Blue*) to 1 (*Red*) % (Positive y direction, color online)

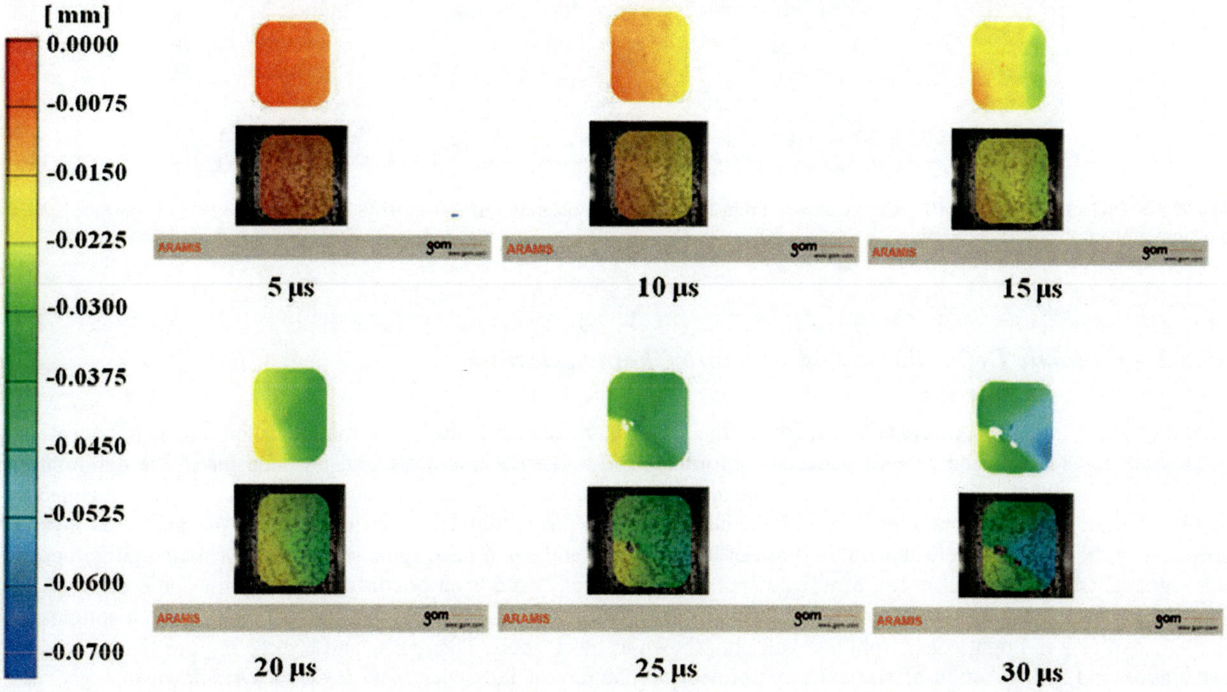


**Fig. 75.8** Displacement fields in x direction for specimen of L-D ratio 0.8, at time intervals of 5 μs, starting with 5 μs from the start of the test, showing strain fields leading to formation of crack. Scale: −0.07 (*Blue*) to 0 mm (*Red*) (color online)

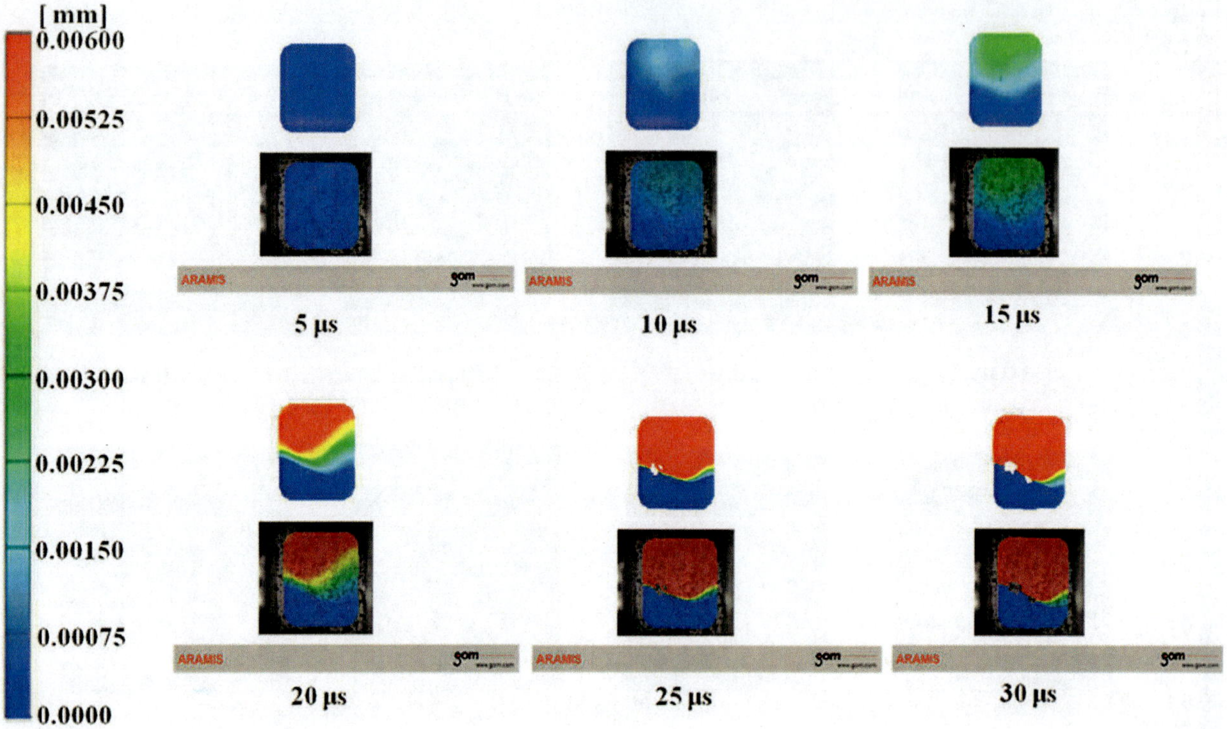


**Fig. 75.9** Displacement fields in y direction for specimen of L-D ratio 0.8, at time intervals of 5 μs, starting with 5 μs from the start of the test, showing displacement fields leading to formation of crack. Scale: 0 (*Blue*) to 0.006 mm (*Red*) (color online)

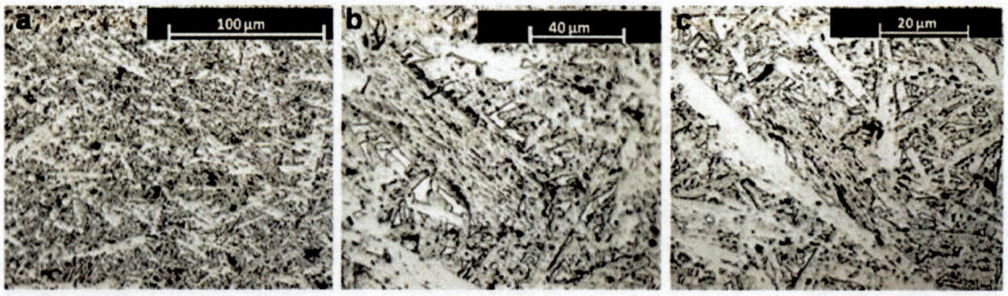


**Fig. 75.10** Optical micrographs of $Ti_2AlC$as received (From TAMU) at magnifications (**a**) 200×, (**b**) 400× and (**c**) 600×, highlighting the layered structure of the material before loading

## *75.3.2 Dynamic Deformation and Fracture Characterization*

In this section, microscopic characterization of the material before and after deformation is presented to reveal the microstructural features and provide necessary information to relate the microstuctural mechanisms to the deformation observed at macroscale. The microstructure before deformation, shown in Figs. 75.10 and 75.11 (Sect. 3.2.1) are obtained by etching it in a solution containing HCl, $HNO_3$ and $H_2O$ in the proportion 1:1:1, which revealed the grain and layered structure in the material. Inclusions in the form of TiC particles and small dark spots were seen distributed in the optical micrographs (shown in black). The nanoscale spots (~1 nm) could have been possibly formed during processing of the specimens. The inclusions were distinguished from pits using SEM. An SEM micrograph, shown in Fig. 75.11, distinguishes the TiC particles as bright zones from the dark pits, shown as black spots. The grains and layered regions have a random orientation and a distribution of sizes. The deformed morphology of the specimens (L−D = 0.4) shown in Fig. 75.12 (Sect. 3.2.2) revealed numerous rough patches (much larger than pits observed in Fig. 75.11) throughout the surface of the

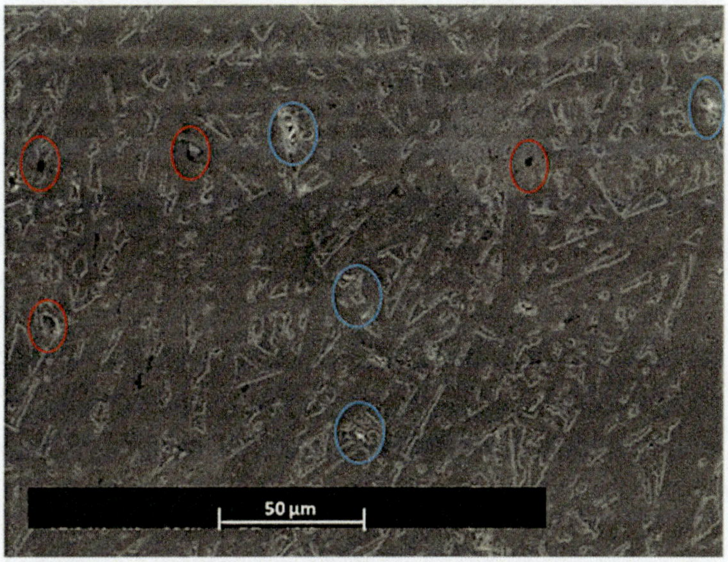


**Fig. 75.11** SEM micrograph showing TiC particles as bright spots (*blue circles*) and pits as dark spots (*red circles*) distributed throughout the specimen surface

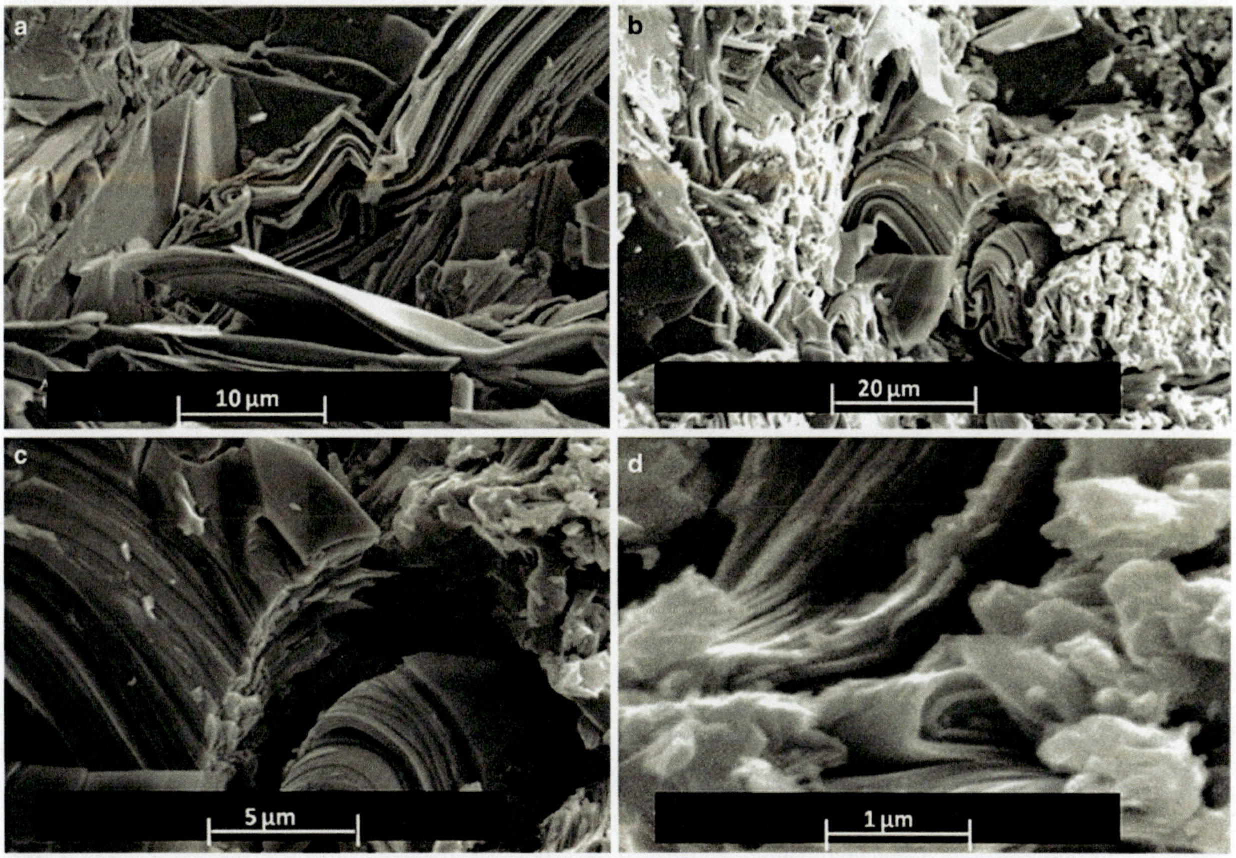


**Fig. 75.12** SEM micrographs of $Ti_2AlC$ after fracture using a specimen of L-D ratio 0.2 at magnifications (**a**) 2000× showing kink band formations with well defined boundaries, (**b**) 1500× showing grain folding and void formations, (**c**) 5000× showing cracking through the lamellae at the kink boundary and (**d**) 25000× showing complete folding of the layers making an angle of almost 0°

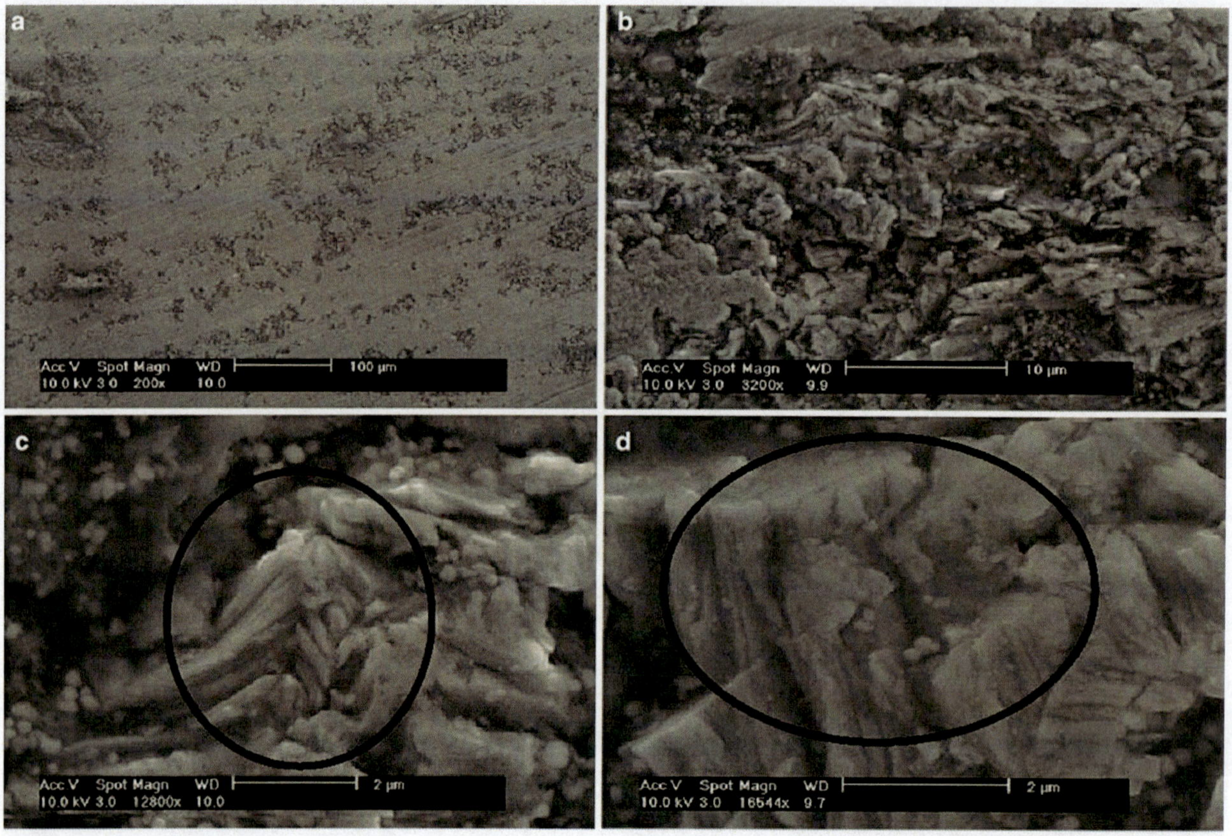


**Fig. 75.13** SEM micrographs of $Ti_2AlC$ after deformation without fracture showing (a) distribution of rough patches throughout the surface, (b) a zoomed in portion of a pit showing the deformed surface, (c) kink band formation and (d) sliding of the lamellae

specimen, which are possible areas of stress concentration. These patches are areas on the surface of the specimen from where surface smoothness achieved after machining was lost. Fractography analysis for the remaining specimens (L–D = 0.2) and (L–D = 0.8), shown in Figs. 75.13 and 75.14 (Sect. 3.2.2), reveal mechanisms of initiation and propagation of fracture throughout the material, some of which are different from the mechanisms observed under quasi-static loading. These will be discussed in detail in Sect. 3.3.

#### 75.3.2.1 Microstructural Characterization

This sub-section demonstrates the surface condition (etched conditions) to reveal the layered structure of the material, before subjecting it to compressive loading at high strain rates. Figure 75.10 shows optical micrographs at magnifications of up to 600× and Fig. 75.11 shows the SEM (Philips XL 30 SEM) micrograph at a higher magnification. The black spots in Fig. 75.10 are due to inclusions and pits, which are clearly distinguished in the SEM as the inclusions (TiC) appear as bright spots compared to the pits (induced during processing) that appear dark.

#### 75.3.2.2 Deformed and Fractured Surface Analysis (Fractography)

In this sub-section, SEM microstructural characterization of deformed and failed specimens is reported, which reveal the microscopic features associated with deformation. Figure 75.12a–d shows the fractured surface of specimens with L-D 0.2 which shows kink band formation with a delaminating layer in the foreground (Fig. 75.12a), grain and nano-laminate *folding* (Fig. 75.12b), with a fibrous appearance, which clearly indicates the ductile nature of this material. Figure 75.12c shows the mechanism of grain tearing. A portion of a grain is separated into two, in a direction perpendicular to the layered structure.

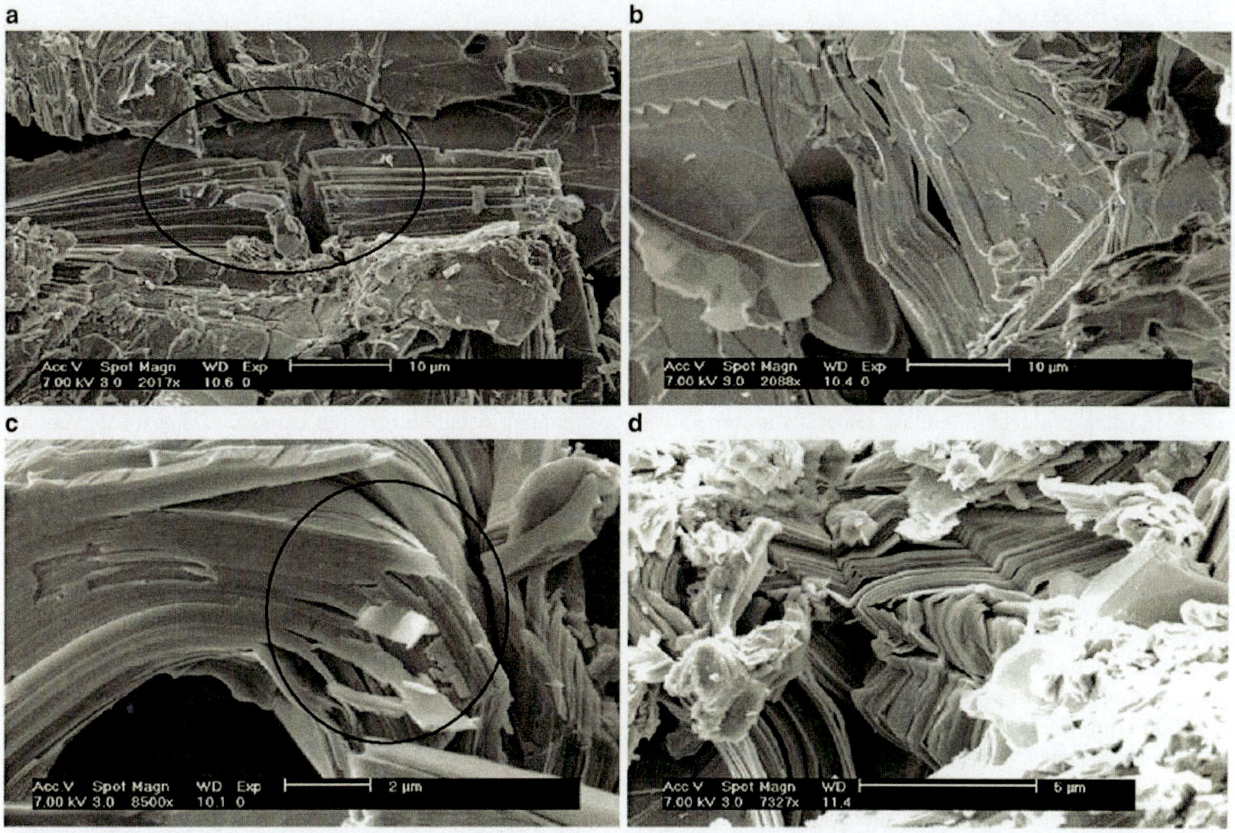


**Fig. 75.14** SEM micrographs of fractured surface of specimen tested at L-D ratio of 0.8 showing (**a**) cleavage like fracture through the lamellae, (**b**) kink band formations with void and well defined kink boundaries (**c**) grain folding with void formation, delamination, surface peeling, kink banding and (**d**) extensive kink band formations with interlamellar cracking and grain buckling

This crack formation is in the vicinity of a kink band, as shown, and possibly initiated at the kink boundary, which are areas of high energy since it contains an array of dislocations. Figure 75.12d shows a complete folding of nano-laminates and a portion of it lies completely on top of the other portion. Figure 75.13a–d shows the microstructural features of the deformed specimen without fracture for specimens tested at L-D 0.4, which show the distribution of rough patches on the surface (Fig. 75.13a), with magnified views of patches (Fig. 75.13b–d). These images clearly show features of deformation similar to kink bands (Fig. 75.13c) but these are much less extensive compared to fractured specimens. Figure 75.13d shows relative sliding of layers on top of each other, which is a process initiated by dislocation movement leading to the creation of kink bands. Figure 75.14a–d show the fractured surface analysis of specimen with L-D 0.8. The modes of fracture are similar to the specimen with L-D 0.2 (shown in Fig. 75.12) with some differences. Figure 75.14a shows the splitting of the grain into two halves similar in nature to grain tearing observed in Fig. 75.12c. Grain tearing, as described earlier, is observed in the vicinity of kink bands and hence has a ductile appearance whereas the splitting here is a cleavage type fracture and no kink bands or buckled grains are observed in the vicinity. Figure 75.14b shows the process of grain buckling and kink band formations, with a large gap formation between two grains – this process is initiated by relative sliding of nano-layers associated with dislocation movements during plastic deformation. Figure 75.14c shows a *surface peeling* mechanism observed in a kinked set of nano-laminates. The exact mechanism and cause of this feature is not known currently. Figure 75.14d shows a fibrous nature of the failure, with kink band formations along the nano-laminates, with formation of a few cracks at high areas of dislocation densities (kink boundaries), grain buckling and delaminations. The presence of such a region showing cleavage type fracture (Fig. 75.14a) clearly indicates that both ductile and brittle fracture mechanisms were operating during the failure of this material.

### *75.3.3 Discussion*

The high strain rate response of $Ti_2AlC$ was studied at L-D ratios of 0.2, 0.4 and 0.8 (Sect. 2) at strain rates of 500–2,500 $s^{-1}$. The test at L-D ratios of 0.2 and 0.4 on set-up I resulted in elasto-plastic deformation without fracture and the strain signals from the strain gauges were used to obtain strain-time and dynamic stress–strain constitutive behavior, from which material properties were calculated. The strain-time behavior was compared to the data obtained using a full-field 2D DIC, which showed a good match (mean error <5%). The constitutive stress–strain behavior reveals a variation of the modulus in the range 4–13 GPa for the same L-D ratio at different strain rate values and a variation of 9–20 GPa for different L-D ratios at same strain rate. The calculation of modulus from the dynamic stress–strain curves, therefore, is not completely accurate, even though the value is higher compared to the quasi-static value (~13 GPa) and is consistent with expectations. SHPB test results are highly dependent on the mechanical properties of materials and proper test design and protocols are necessary to accurately quantify the material behavior. Since $M_{n+1}AX_n$ phase ceramics exhibit properties of both ceramics and metals, and test protocols for ceramics and metals have been reported separately in literature [11, 13, 14, 20–22, 24], development of suitable testing protocols for these materials is imperative for obtaining accurate material behavior. This would require extensive study of different L-D ratios across a wide range of strain rates and development of correct experimental methods and this work is currently ongoing. The displacement and strain fields obtained using DIC for specimens with L-D 0.4 showed certain amount of heterogeneity (~0.5%), which can be explained based on the microstructural observations.

A comparison of the micrographs before deformation (Figs. 75.10 and 75.11) to the micrographs after deformation without fracture (Fig. 75.12) show heterogeneity in the microstructure with rough patch formations of different sizes scattered throughout the specimen surface. The patches observed in the deformed specimen varied in size from 0.1 to 1 mm and the heterogeneous strain field regions (Figs. 75.5 and 75.7) had variation regimes from ~0.2–1.4 mm. From this analysis, it was inferred that these rough patches were areas of stress concentrations which led to relatively higher strains compared to the surrounding areas and hence resulted in a heterogeneous strain field. A closer analysis (at higher magnification of 12000×-Fig. 75.12c) of these rough patches reveal ridge like formations, which are similar in structure to kink bands. A possible mechanism of these patch formation is from the growth and subsequent coalescence of minuscule pits, initially present in the microstructure. Tests were also performed on specimens with L-D ratios of 0.2 and 0.8 using set-up II, which resulted in premature failure with almost no transmitted strain pulse. Because of this, the constitutive stress–strain behavior could not be determined from these tests. However, the fractured surfaces was characterized optically using a Philips XL30FEG SEM, which revealed the microstructural features associated with fracture at strain rates of 1,500 $s^{-1}$ and 2,500 $s^{-1}$. The fracture surface (L-D = 0.2) shows kink band formations and delamination resulting from sliding of basal planes (Fig. 75.12a), grain and lamellar *folding* which are similar to kink bands but have only one boundary. Two types of folding are shown in Fig. 75.12b, d. In the former, the angle of the fold is about 45° whereas the latter shows complete folding and one portion of the lamellar stack rests completely on the other portion resulting with an angle of almost 0°. At the same time, there were regions where the lamellar stacks formed kink bands, with two distinct boundaries and then cracks developed separating the kinked lamellar stack into two halves, as shown in Fig. 75.12c. The fibrous appearance of the cracked surface (cross-sectional) suggests that it failed in a ductile manner. Hence, this is termed *grain-tearing*. The fractography analysis of the 0.8 L-D specimens showed similar mechanisms with some additional features (Fig. 75.14). A closer analysis into the micrographs shown in Fig. 75.14 reveals cleavage type brittle fracture (Fig. 75.14a) in the lamellae, formation of a more well-defined kink band (Fig. 75.14b) in comparison to the previous case. Figure 75.14c revealed a new feature of *peeling* at the surface of a kinked lamellae. Under quasi-static compressive loading reported earlier in [7, 8], $Ti_2AlC$ specimens fractured due to a single crack, which initiated at one corner of the specimen and propagated through the material at 45° splitting the material into two equal halves. Zhou [7] reported that after fracture the material didn't the fractured pieces remained together and didn't pulverize into smaller fragments. It was also reported that the microstructural features responsible for fracture (kink bands, grain pullout, etc) were localized in the vicinity of the crack. The high strain rate fracture, on the other hand, showed a completely different mechanism of crack initiation and propagation and resulted in a catastrophic failure with pulverized and fragmented fracture pieces, extremely small in size. The first crack initiation step was similar to the quasi-static case in which a crack started propagating almost at 45° but this crack started propagating from approximately the centre of the left edge (Figs. 75.8 and 75.9) of the specimen as opposed to the quasi-static case. As this crack was propagating across the width, another crack nucleated at the opposite corner, joined this crack which led to widening of the crack and subsequent branch formations to pulverize the specimen into smaller fragments (for L-D = 0.2, a big central piece having a radius of approximately 10 mm with smaller fragments of size 1–5 mm, for L-D = 0.8 all pieces were in the range 5–8 mm). In this case, the kink bands, delaminations and other microstructural features associated with fracture were distributed throughout the specimen, instead of localized failure observed in the quasi static case. To generalize this statement for the entire class of ternary ceramics, experiments at different L-D ratios and for other $M_{n+1}AX_n$ phase materials such as $Ti_3SiC_2$, should be investigated.

## 75.4 Conclusion

$M_{n+1}AX_n$ phase ternary ceramics are nano-laminated materials which exhibit unique combination of properties of both metals and ceramics and hence are attractive for structural applications. In this paper, we reported SHPB testing of $Ti_2AlC$, a $M_{n+1}AX_n$ phase ternary ceramic at strain rates of 500–2,500 $s^{-1}$ using specimens of different L-D ratios (0.2, 0.4 and 0.8) to characterize its behavior under dynamic loading conditions. At L-D ratio of 0.2 and 0.8, the specimens fractured in a brittle manner at a strain rate of 1,500 $s^{-1}$ and 2,500 $s^{-1}$ respectively, using the same experimental set-up (Set-up II). For a different experimental set-up (Set-up I), however, at L-D of 0.2, up to 1,200 $s^{-1}$ the material showed a ductile behavior with clearly defined yield points at some strain rates. The dynamic stress–strain constitutive behavior was analyzed for these specimens. The modulus calculated from this analysis had values (35–75 GPa) higher than the modulus obtained from quasi-static loading (13 GPa). The mode of deformation during this combined ductile-brittle behavior is due to the formation of kink bands. In order to investigate the nature, size, distribution of kink bands and associated advanced modes of deformation and fracture, microstructural characterization of specimens was done before and after subjecting them to loading. The features revealed advanced deformation modes and a slightly different mechanism of initiation and propagation of fracture processes compared to loading under quasi-static conditions.

**Acknowledgements** This work was supported by AFOSR under MURI (FA 9550-09-1-0686) and the authors would like to thank the program manager, Dr. David Stargel. The authors would like to thank Dr. M Radovic and Dr. I Karaman (TAMU) for providing the $Ti_2AlC$ samples which have been used in this work. The authors also acknowledge Prof. Anthony M Waas and Dr. Mark Pankow for use of the SHPB lab set up and fruitful discussions on the experimental techniques.

## References

1. Barsoum MW, El-Raghy T, Radovic M (2000) Ti3SiC2: a layered machinable ductile carbide. Interceram 49(4):226–233
2. Barsoum MW, Radovic M (2011) Elastic and mechanical properties of the MAX Phases. Annu Rev Mater Res 41:195–227
3. Zhou AG et al (2006) Incipient and regular kink bands in fully dense and 10 vol.% porous Ti2AlC. Acta Mater 54:1631–1639
4. Zhou AG, Basu S, Barsoum MW (2008) Kinking nonlinear elasticity, damping and microyielding of hexagonal close-packed materials. Acta Mater 56:60–67
5. Barsoum MW, El-Raghy T (1999) Room temperature ductile carbides. Metall Mater Trans A 30A:363–369
6. Barsoum MW et al (2010) Kinking non-linear elastic solids. In: Encyclopedia of materials: science and technology, Elsevier, Oxford, pp 1–23
7. Zhou YC, Wang XH (2001) Deformation of polycrystalline Ti2AlC under compression. Mater Res Innovat 5:87–93
8. Poon et al (2011) Damage accumulation and hysteretic behavior of MAX phase materials. J Mech Phys Solid 59(10):2238–2257
9. Kolsky H (1949) An investigation of the mechanical properties of materials at very high rates of loading. Proc Phys Soc B 62:676–700
10. Gama BA et al (2004) Hopkinson bar experimental technique: a critical review. Appl Mech Rev 57(4):223–250
11. Subhash G, Ravichandran G (2000) Split-Hopkinson pressure bar testing of ceramics. In: ASM handbook. Mechanical testing and evaluation, ASM International, Materials Park, OH, 8:488–496
12. Al-Mousawi MM, Reid SR, Deand WF (1997) Proceedings of institution of mechanical engineers, 211(Part C), pp 273–292
13. Ravichandran G, Subhash G (1994) Critical appraisal of limiting strain rates for compression testing of ceramics in a split Hopkinson pressure bar. J Am Ceram Soc 77(1):263–267
14. Lambros J (2010) Dynamic response of materials, Encyclopedia of aerospace engineering. Wiley, New York. ISBN 978-0-470-68665-2
15. Sutton MA, Orteu JJ, Schreier HW (2009) Image correlation for shape, motion and deformation measurements: basic concepts, theory and applications. Springer, New York. ISBN 978-0-387-78746-6
16. Pankow M, Justusson B, Waas AM (2010) Three-dimensional digital image correlation technique using single high speed camera for measuring large out-of-plane displacements at high framing rates. J Appl Optic 49(17)
17. Tiwari V, Sutton MA, McNeill SR (2007) Assessment of high speed imaging systems for 2D and 3D deformation measurements: methodology development and validation. Exp Mech 47:561–579
18. Siebert T et al (2007) High speed digital image correlation: error estimations and applications. Opt Eng 46(5):051004-1–051004-7
19. Basu S et al (2011) Processing and structural characterization of MAX phase/metal composites. In: ASME international mechanical engineering congress and exposition, Denver
20. Gupta N, Luong DD, Rohtagi PK (2011) A method for intermediate strain rate compression testing and study of compressive failure mechanism of Mg-Al-Zn alloy. J Appl Phys 109:103512-1–103512-7
21. Armstrong RW, Walley SM (2008) High strain rate properties of metals and alloys. Int Mater Rev 53(3):105–128
22. Song B, Chen W, Antoun BR, Frew DJ (2007) Determination of early flow stress for ductile specimens at high strain rates by using a SHPB. Exp Mech 47:671–679
23. Yuan J, Takeda N, Waas A (1998) A note on data processing in the split pressure bar tests. Exp Tech 22:21–24
24. Pankow M, Attard C, Waas AM (2004) Specimen size and shape effect in split Hopkinson pressure bar testing. J Strain Anal Eng Des 4:689–698
25. Lifshitz JM, Leber H (1998) Data processing in the split Hopkinson pressure bar tests. Int J Impact Eng 15(6):723–733
26. Lankford J et al (1998) The role of plasticity as a limiting factor in the compressive failure of high strength ceramics. Mech Mater 29:205–218